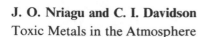

FOOD CONTAMINATION FROM ENVIRONMENTAL SOURCES

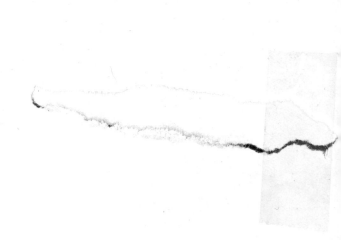

Volume

23

in the Wiley Series in

Advances in Environmental
Science and Technology

JEROME O. NRIAGU, Series Editor

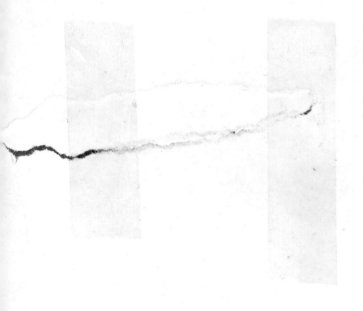

FOOD CONTAMINATION FROM ENVIRONMENTAL SOURCES

Edited by

Jerome O. Nriagu

National Water Research Institute
Burlington, Ontario, Canada

and

Milagros S. Simmons

University of Michigan
Ann Arbor, Michigan

A WILEY-INTERSCIENCE PUBLICATION

John Wiley & Sons, Inc.

NEW YORK / CHICHESTER / BRISBANE / TORONTO / SINGAPORE

Library of Congress Cataloging-in-Publication Data:
    Food contamination from environmental sources / edited by Jerome O.
    Nriagu, Milagros S. Simmons.
        p. cm. — (Advances in environmental science and technology,
    ISSN 0194-0287; v. 23)
        "A Wiley-interscience publication."
    Bibliography: p.
    ISBN 0-471-50891-8
    1. Food contamination.   2. Pollutants.   I. Nriagu, Jerome O.
    II. Simmons, Mila S.   III. Series.
    TD180.A38 vol. 23   47970
    [TX531]
    628 s—dc19   89-5671
    [363.1'92]   CIP

Printed in the United States of America
10 9 8 7 6 5 4 3 2 1

CONTRIBUTORS

ALBERTS, L. A., Instituto Nacional de Investigaciones Sobre Recursos Bioticos, Xalapa, Ver., Mexico

ASHOKE, M., Department of Environmental and Occupational Health, University of South Florida, Tampa, Florida

AZCUE, J. M., National Water Research Institute, Burlington, Ontario, Canada

CAPPON, C. J., Environmental Health Sciences Center, University of Rochester Medical Center, Rochester, New York

CASALE, W. L., Department of Food Science and Human Nutrition, Michigan State University, East Lansing, Michigan

CONACHER, H. B. S., Health and Welfare Canada, Banting Research Centre, Ottawa, Ontario, Canada

DAVIES, K., Department of Public Health, City Hall, Toronto, Ontario

DE LEON, R., Department of Microbiology and Immunology, University of Arizona, Tucson, Arizona

DHALIWAL, G. S., College of Agriculture, Punjab Agricultural University, Ludhiana, India

ELIAS, R. W., United States Environmental Protection Agency, Research Triangle Park, North Carolina

FLEGAL, A. R., Institute of Marine Sciences, University of California, Santa Cruz, California

FRANK, R., Ontario Ministry of Agriculture and Food, University of Guelph, Guelph, Ontario

GERBA, C. P., Department of Microbiology and Immunology, University of Arizona, Tucson, Arizona

GRIFFITHS, M. W., The Hannah Research Institute, Ayr, Scotland

HARGITAI, F., Department of Plant Protection and Agrochemistry, Ministry of Agriculture and Food, Budapest, Hungary

HEIKES, D. L., Department of Health and Human Services, Food and Drug Administration, Kansas City, Missouri

KHAYAT, A., Scientific and Analytical Services, Beatrice/Hunt–Wesson Inc., Fullerton, California

KULKARNI, A. P., Department of Environmental and Occupational Health, University of South Florida, Tampa, Florida

LAWRENCE, J. F., Bureau of Chemical Safety, Health and Welfare Canada, Banting Research Centre, Ottawa, Ontario

LUPIEN, J. R., Food Quality and Standards Service, FAO – United Nations, Rome, Italy

MATTHEES, D. P., Department of Chemistry, South Dakota State University, Brookings, South Dakota

NRIAGU, J. O., National Water Research Institute, Burlington, Ontario

PAGE, B. D., Health and Welfare Canada, Banting Research Centre, Ottawa, Ontario, Canada

PESTKA, J. J., Department of Food Science and Human Nutrition, Michigan State University, East Lansing, Michigan

PHILLIPS, J. D., The Hannah Research Institute, Ayr, Scotland

RIPLEY, B. D., Ontario Ministry of Agriculture and Food, Guelph, Ontario, Canada

SINGH, B., Department of Entomology, Punjab Agricultural University, Ludhiana, India

SINGH, P. P., Department of Entomology, Punjab Agricultural University, Ludhiana, India

SMART, N. A., Ministry of Agriculture, Fisheries and Food, Harpenden Laboratory, Hatching Green, Harpenden, United Kingdom

SMITH, D. R., Department of Biology, University of California, Santa Cruz, California

STACEY, C. I., School of Applied Chemistry, Curtin University of Technology, Bentley, Western Australia

TODD, E. C. D., Health Protection Branch, Health and Welfare Canada, Ottawa, Ontario

INTRODUCTION
TO THE SERIES

The deterioration of environmental quality, which began when mankind first congregated into villages, has existed as a serious problem since the industrial revolution. In the second half of the twentieth century, under the ever increasing impacts of exponentially growing population and of industrializing society, environmental contamination of the air, water, soil, and food has become a threat to the continued existence of many plant and animal communities of various ecosystems and may ultimately threaten the very survival of the human race. Understandably, many scientific, industrial, and governmental communities have recently committed large resources of money and human power to the problems of environmental pollution and pollution abatement by effective control measures.

Advances in Environmental Sciences and Technology deals with creative reviews and critical assessments of all studies pertaining to the quality of the environment and to the technology of its conservation. The volumes published in the series are expected to service several objectives: (1) stimulate interdisciplinary cooperation and understanding among the environmental scientists; (2) provide the scientists with a periodic overview of environmental developments that are of general concern or that are of relevance to their own work or interests; (3) provide the graduate student with a critical assessment of past accomplishment, which may help stimulate him or her toward the career opportunities in this vital area; and (4) provide the research manager and the legislative or administrative official with an assured awareness of newly developing research work on the critical pollutants and with the background information important to their responsibility.

As the skills and techniques of many scientific disciplines are brought to bear on the fundamental and applied aspects of the environmental issues, there is a heightened need to draw together the numerous threads and to present a coherent picture of the various research endeavors. This need and the recent tremendous growth in the field of environmental studies have clearly made some editorial adjustments necessary. Apart from the changes in style and format, each future volume in the series will focus on one particular theme or timely topic, starting with Volume 12. The author(s) of each pertinent section will be expected to critically review the literature and the most important recent developments in the particular field; to critically evaluate new concepts, methods, and data; and to focus attention on important unresolved or controversial questions and on probable future trends. Monographs embodying the results of unusually extensive and well-rounded investigations will also be published in the series. The net result of the new editorial policy should be more integrative and comprehensive volumes on key environmental issues and pollutants. Indeed, the development of realistic standards of environmental quality for many pollutants often entails such a holistic treatment.

JEROME O. NRIAGU, Series Editor

PREFACE

Every chemical released into the ecosystem has the potential of getting into the human food chain. It is therefore not surprising that our foods now contain a large number of chemical and biological contaminants derived either directly or indirectly from the environment. For the general population, the diet has become the major exposure route for most of the known toxic contaminants in the environment. What we now eat probably represents a good index of the persistent contaminants being discharged into our environment. Although food contamination has been of public health concern for many years, most of the publications on this topic deal with food additives used as preservatives, colorants, and flavoring, or with contamination problems associated with food production, processing, packaging, or storage. Little attention has been paid to the accumulation and transfer of contaminants through the food chain to the human consumer.

This volume, aimed at filling this critical gap in knowledge, deals with the accumulation and persistence of toxic contaminants in the human food chain; it addresses the violation of the long-held belief that "if it's natural, it must be good." It provides critical and timely reviews on the detection, sources, distribution, speciation and bioavailability of environmental contaminants and their metabolites in various food products. The toxicological and health significance of the observed contaminant levels in foods has also been discussed. Thus, the volume covers a subject of great interest in food science and public health. It should also be fundamental reading for anyone concerned about the quality of our foods and how it is being affected by industrial discharges into the environment.

Any success of this volume belongs to our distinguished authors. We thank the staff of John Wiley & Sons, Inc., New York for invaluable editorial assistance.

Burlington, Ontario

Ann Arbor, Michigan

Jerome O. Nriagu

Milagros S. Simmons

CONTENTS

1

AGRICULTURAL AND RELATED CHEMICAL CONTAMINATION IN FOODS: A HISTORICAL PERSPECTIVE

Duane P. Matthees

Department of Chemistry, South Dakota State University, Brookings, South Dakota 57007–1217

---

1. CONCEPT OF ENVIRONMENTAL CONTAMINANTS IN FOODS

Environmental contaminants in foods are considered to be compounds that are not normal food constituents and that inadvertantly find their way into food products and cannot readily be removed by the usual manufacturing practices. These compounds are sometimes used intentionally, such as pesticides, and, in this case, the amounts remaining in food products are usually referred to as residues (which does not evoke the same negative connotations as the term "contaminant"). Some people would prefer to consider contaminants as any synthetic compound in foods, although others would reserve this term for harmful or illegal amounts of various chemicals. In any case, most would agree that large amounts of residues are considered to be contaminants. Many individuals are uncomfortable with the idea that they are ingesting varying amounts of pesticides and industrial chemicals, and the controversy over residues and the concept of legal tolerances has been a long one.

Many people think of environmental contaminants as extremely durable, persistent, bioaccumulating compounds that are excreted only very slowly. Although some compounds, such as the polychlorinated biphenyls and hexachlorobenzene, fit this description, there are other compounds that are excreted more rapidly, but are still toxic when ingested in small amounts; these compounds can be potential threats to the food supply. Although compounds used today are less persistent than some of the compounds used in the past, their biological activity, although of shorter duration, may be much greater.

2. CHEMICAL ANALYSIS OF FOOD CONTAMINANTS

2.1. Introduction

The history of environmental contamination of foods might to a large extent be termed the history of the *detection* of environmental contaminants, since contamination is otherwise not detected unless illness results. Moreover, unless enough individuals are affected so that an incident is studied from an epidemiological viewpoint, an isolated instance of toxicity may well be overlooked, in the absence of chemical evidence.

The pesticide residue analyst of the early twentieth century needed methods for the analysis of only arsenic and possibly of copper in Bordeaux mixture; the emphasis in food analysis at that time was the detection of adulterants more than the detection of spray residues. Although pesticides were undoubtedly toxic, it was apparently felt that the residues were low anyway and thus possibly of less concern than the preservatives and adulterants that might be present in fairly large amounts. Moreover, while spraying of crops for insect control was not particularly controversial for many people, adulteration involved dishonesty and possible health effects at the same time. As will be seen later, there was some discussion on the effects of arsenic residues, but the proposed tolerances for arsenic were the issue rather than the use of insecticides per se.

For these and other reasons, food analysis earlier in this century was not unduly preoccupied with trace analysis at all, let alone pesticide residue analysis.

Organic chemical analysis was slow to develop. Until recently, many introductory quantitative analysis textbooks scarcely acknowledged the existence of organic compounds, let alone any need for their analysis. Organic analysis, especially trace analysis, presents some unique problems. There are so many organic compounds, relative to inorganic, that it is more difficult to use a standard sample treatment. In contrast to inorganic analysis, in which organic matter can be decomposed to facilitate the analytical process, organic analysis requires that the compounds be extracted in unaltered form from a variety of sample matrices. Thus, environmental and food samples are difficult to extract quantitatively. In addition, there were a limited number of analytical methods that would identify or quantify organic compounds in a mixture. Barger (1914) noted that chemical methods were of little value in the determination of organic poisons in foods; he preferred physiological methods of detection. Higgins (1914) assayed for food toxins by subcutaneous injection of extracts into guinea pigs. The need for better methods of organic chemical analysis became more obvious in the 1930s and later as the synthetic chemical industry and petrochemical industry grew.

2.2. Colorimetric Methods

The use of spectroscopic methods of analysis in connection with color-producing reactions is one of the first methods used that had the sensitivity needed for analysis at the part per million (ppm) level. Although developments in colorimetry were gradual, a large number of methods were eventually developed and became the starting point for much of the trace analysis in the first part of the twentieth century. For example, the Gutzeit method was based on the color reaction of generated arsine gas from a sample as it reacted with a strip of treated paper, and it eventually became official for arsenic determination. Schechter et al. (1945) carried out the first practical residue analysis for DDT using colorimetric methods, and these methods of analysis persisted until most were displaced by chromatographic methods.

2.3. Development of Chromatography

Environmental and residue analysis was extremely difficult after organic compounds replaced the arsenicals as pesticides, as the extracted sample would now contain a complex mixture of organic compounds. The primary approach in such cases would normally be to try to carry out some spot tests for functional groups and hope that there were no serious interferences. The development of chemical separation methods based on chromatography dates at least to the turn of the century and probably before that; no single individual was responsible for its discovery. Adsorption and paper chromatography were in use in the first part of the twentieth century, but although they were quite useful in the purification of compounds on a preparative scale, separation of compounds without the ability to detect what has just been separated does not in itself allow trace analysis. This would later be accomplished by instrumental separation methods in the form of gas and liquid chromatography.

2.4. Instrumental Methods in Organic Trace Analysis

At about the 1920s and 1930s it was widely recognized that the usual gravimetric and volumetric methods in classical quantitative analysis were not sufficiently sensitive as commonly carried out for certain purposes, such as when sample quantity was limited. As a result, the art of ultramicroquantitative analysis gradually developed. This was a scaling down of conventional methods to the point at which quantities well under a microgram could be analyzed. However, the science was not developed to the point of extensive use in organic trace analysis in part because of the long-standing emphasis on the analysis of inorganic substances that dominated quantitative analytical chemistry at the time. Moreover, although very good results can be obtained by ultramicroanalysis, relatively few individuals had the skill and knowledge required (to say nothing of the patience!). It is not easy to advance the practice of a branch of chemistry when the standard deviation is so strongly dependent on the individual making the measurement. For typical applications, refer to a text such as that of Kirk (1950).

One of the earlier applications of an instrumental method to food analysis is the organic halide gas detector of Martinek and Marti (1931). Concerned about the possible presence of methyl chloride (a refrigerant) in air and foods, they developed a flame test based on the change in color of a gas flame in contact with copper (a type of modified Beilstein test).

Mass spectrometry had been developed in the first decade of the twentieth century by J. J. Thomson, but his instrument was not an analytical prototype but rather a device to show that elements could exist as isotopes. (Remember that although chemists in particular often used the atomic theory, there was little direct evidence of the existence of atoms. Much of the evidence fell into place about this time, with Thomson's mass spectrometer, Millikin's oil drop experiment, etc.). An instrument that is more nearly contemporary in application would not come until the 1930s and later.

Polarography was invented by Heyrovsky in the 1920s and later developed into a widely used analytical method for inorganic compounds, but, as noted by Gunther and Blinn (1955), it was not commonly used in residue analysis, perhaps because of excessive background interference.

The development of gas chromatography with sensitive and selective detectors revolutionized organic trace analysis. Martin and James (1956) equipped their gas–liquid chromatography apparatus with a gas density meter rather than the automatic titrator first used. This extended the applicability of gas chromatography to a much wider range of compounds, and, within several years, commercial units were available. Coulson et al. (1959) showed that the separation of pesticides was possible using a gas chromatograph with a thermal conductivity detector, although they had to work with milligram amounts. The usefullness was recognized, and within a year, Lovelock and Lipsky (1960) and Lovelock (1961) introduced the argon activation and electron capture detectors, which allowed analysts to go from the milligram range to picograms. Coulson et al. (1960) interfaced coulometry with gas chromatography, resulting in the microcoulometric detector for selective detection of the halides and sulfur compounds, and Giuffrida (1964) found that a selective response to phosphorus compounds could be obtained by modifying the flame ionization detector with alkali metal salts. (This was later modified into our nitrogen–phosphorus thermionic detector.) As most of the compounds screened for contained halogen, nitrogen, or phosphorus or sulfur, the selective detectors allowed their analysis without excessive interference by other compounds extracted from the sample.

However, extreme sensitivity can also lead to errors in interpretation, as chromatographic methods are not completely specific and do not directly determine identity of an unknown. The need for still better specificity and for the identification of compounds in which standards might not be available led to the development of gas chromatography interfaced with a mass spectrometer (GC–MS). Ryhage (1964) and Watson and Biemann (1964) led in the development of GC–MS techniques, but it remained to the microelectronics and computer revolution to develop the data systems needed to prevent the analyst from drowning in data that could not be readily interpreted.

Liquid chromatography gradually developed into the analytical method known as high-performance liquid chromatography over a period of time, again with no one individual being primarily responsible. Detection systems are still in a state of development, with electrochemical, optical, and mass spectrometric interfacing being promising approaches to organic trace analysis, especially for nonvolatile compounds and metabolites.

3. AGRICULTURAL AND INDUSTRIAL CHEMICALS IN FOODS

Environmental contamination of foods has paralleled the growth of the chemical industry, but only in the sense that we now have synthetic chemical contaminants. Ancient man, as well as his later descendants, had to deal with toxic plants and fungi, bacterial toxins due to poor storage, filth, rat feces, insect parts, and, most seriously, lack of food due to environmental causes. Only in comparatively recent generations have humans progressed from helplessness or resignation in the face of famine, pests, and disease, to the realization that something could be done about it, to actual attempts to control their environment. It was perhaps inevitable that there would be casualties in the effort. As a society we have learned the hard way that it is very difficult to anticipate all possible problems associated with a new technology. For this reason, extensive testing of agricultural chemicals is carried out before registration as well as afterward. This does not guarantee that there will be no unpleasant surprises, but it will certainly reduce the number of problems arising in use.

In this section of the chapter, a number of contamination incidents will be noted, particularly as they relate historically to the growth of the chemical industry. There are certainly tragic and pathetic incidents involving pesticide and industrial chemical toxicity, and reciting these would be a sad and depressing narrative indeed. This is not at all the intent of the chapter. Rather, we have been taught some very expensive environmental lessons, but even these are not always clear without additional study and research. Moreover, there is an element of human suffering that has arisen in our efforts to control natural adversaries, but it is not entirely in vain if we are able to recognize the problems and correct them.

3.1. Arsenical Agrochemicals

Although arsenic compounds were used as early as 1681 as ant poisons, phytotoxicity prevented use on crops until the less soluble metal arsenates were formulated. Paris green was in use in the United States by the late 1860s, and lead arsenate eventually became the most widely used arsenical insecticide (Frear, 1948a). Arsenic, of course, had a reputation for toxicity since the Middle Ages or before, but it also had some medicinal uses. This duality of purposes tended to give people strong opinions concerning the danger or harmlessness of arsenical pesticides on crops, as one might regard these crops as being "sprayed with poison" or else "having no more arsenic than a fraction of a medicinal dose" and thus harmless.

Whorton (1974) has written an excellent book detailing the pre-DDT era, which featured arsenical pesticides. The controversy that would later arise over the use of DDT was paralleled several decades earlier by a similar controversy over arsenical pesticide use, particularly with regard to arsenic residues on fruit. As Whorton explained, entomologists in the United States tended to promote arsenicals rather strongly, as they were effective insecticides, and residues were regarded as being far below the amounts required to cause toxic symptoms. On the other hand, Europeans, particularly the British and French, were highly suspicious of the use of arsenic and, in Europe, the views of the medical profession tended toward concern over possible chronic toxicity of arsenic. A world tolerance of arsenic residues was apparently in the process of being accepted in Europe, and as these levels were lower than the American tolerances, there was a considerable amount of disagreement and potential trade restrictions on arsenic-sprayed crops.

The British experience with arsenic undoubtedly had much to do with the cautious attitude toward residues. Although food adulteration was widespread in many areas in the nineteenth century, the British public was first sensitized to the arsenic issue in 1855, when peppermint lozenges were treated accidentally with white arsenic instead of plaster, and a number of deaths occurred. Then, in 1900, a number of cases of arsenic poisoning, some fatal, occurred among certain beer drinkers in northwestern England. Invert sugar used in some batches of beer contained large amounts of arsenic, apparently from contaminated sulfuric acid used in the inversion. In another location, malt was contaminated from drying with fuels containing appreciable amounts of arsenic. After the investigation, chaired by none other than Lord Kelvin, standards for arsenic were set, not primarily from toxicological data, which were sparse, but from arsenic levels consistent with good manufacturing practices (Martin, 1963).

The arsenic residue controversy was debated for some time and was never resolved to everyone's satisfaction. The 1.43ppm residue limit set by the English Royal Commission was not always enforced, and after reported cases of illness following the eating of American apples, shipments in which excessive residues were found were rejected. Since arsenic residues were considered to be primarily on the fruit surface, a number of residue removal techniques were used to ensure that fruit would meet tolerance requirements (Frear, 1948b).

The issue was not yet resolved; concern over the lead in lead arsenate sprays soon became the dominant issue. A study of orchardists and their families was published in 1941. Although these individuals had lived in an apple-growing region, and were exposed to large amounts of arsenic relative to the population as a whole, no illness or adverse health effects were noted; the only difference appeared to be in the increased excretion of lead and arsenic (National Academy of Sciences, 1972). This seemed to calm some fears for a time, and the entire arsenical issue became an academic question somewhat later when the organic insecticides came into wide use.

3.2. Other Agricultural Chemicals in the Pre-DDT Era

Partly because of the controversy over arsenical pesticides, the expanded use of fluorine-containing compounds was explored. Marcovitch (1926), for example, proposed the use of various fluosilicates as agricultural insecticides. It was soon recognized that fluorine insecticides presented residue problems of their own, and they never seemed to compete seriously with the arsenicals. De Eds (1933) commented on the need to obtain baseline data on natural fluoride levels in foods. He suggested that fluorinated insecticides be used only where residues could be removed and also proposed that a tolerance be set on the amount of added fluoride in foodstuffs.

In addition to the inorganic pesticides, there was some use of the botanicals such as rotenone and pyrethrum. These had been known earlier, and some had hoped that they would eventually replace the arsenicals as major insecticides, but problems with steady supplies, consistency of quality, and potency, etc. would have limited their role as major pesticides in any case. Ironically, although rotenone is promoted in some circles as a safe, natural pesticide today, Lightbody and Mathews (1936) were concerned over possible cumulative toxicity associated with its use. At this time, however, residue analysis of organic compounds was not well developed, so there was no evidence for actual amounts in foods.

Roark (1932) mentions some pesticides used in food production and includes ethylene oxide, dichloropropane, rotenone, and pyrethrum. At about this time, fumigation of foods for insect infestation control was becoming more prevalent, and there was evidently some public concern over the practice. Williams (1933) defended the use of fumigants and stated that it was not a public health menace.

3.3.   Tricresyl Phosphate

In spite of the economic depression of the 1930s, this was a time of growth and expansion of the chemical industry, particularly in the area of synthetic organic compounds and petrochemicals. One compound that was widely used because of desirable high temperature stability was tricresyl (or tritolyl) phosphate, which was an ingredient in lubricants, some plasticizers, hydraulic fluids, and other products.

During the early 1930s prohibition was a fact of life in the United States, and people had to come up with some creative substitutes for the usual alcoholic beverages. Ginger extract was one of the common substitutes in some areas. Starting around 1930, there was an epidemic of polyneuropathy, a delayed paralysis usually affecting the extremities. Some hand function often returned after a while, but many victims never walked unassisted again.

Smith et al. (1930) demonstrated that this outbreak was due to the contamination or adulteration of lots of ginger extract with tricresyl phosphate, with the ortho isomer showing the greatest toxicity. They had few methods of analysis available for organic compounds, as noted earlier, but they were able to demonstrate a typical phenolic color test after hydrolysis of distilled fractions from the ginger extracts.

The so-called "ginger jake" paralysis is considered one of the classic examples of delayed neurotoxicity characteristic of some of the organic phosphate esters. This incident is often cited as the typical example of organophosphate neuropathy, but what is perhaps not so well known is the rash of poisonings that followed. Heavily adulterated apiole (evidently in demand as an abortifacient) led to a number of cases of paralysis in the Netherlands (van Itallie, 1932; van Itallie et al., 1932). The adulteration of cooking oils led to additional outbreaks in Mauritius (Kirk et al., 1939), in South Africa (Sampson, 1942), on a ship (Debre and Bloc, 1938), and in Morocco (Smith and Spalding, 1959), among other places. The Morocco epidemic is interesting in that the tricresyl phosphate contaminant in the oil was depleted of the ortho isomer. (Manufacturers, aware of the toxicity of the ortho isomer, had usually tried to minimize its content.) Neuropathy occurred in spite of the newer formulation, although medical authorities thought that most of the victims would recover in this case. It did not take long after the introduction of the tricresyl phosphates into the chemical industry before the compound was grossly abused; its solubility in and association with oils and lubricants evidently suggested use as an edible oil contaminant. However, "ginger paralysis" appears to have had some publicity, and it would seem that the individuals who adulterated cooking oils must have had some idea of the problems they were causing.

Although the triaryl phosphates, including tricresyl phosphate, are not agricultural chemicals in the usual sense (on the farm, they would most likely be encountered in hydraulic fluids for some machinery), the tricresyl phosphate edisode was very important in later concerns and studies relating to the organophosphate insecticides and their neuropathic potential. The triaryl phosphates, incidentally, are somewhat persistent in the environment and are sometimes encountered in sediments and aquatic organisms (Boethling and Cooper, 1985).

3.4. Polychlorinated Biphenyls (PCBs) and Naphthalenes (PCNs)

The development of plastics, electrical components, and high-temperature fluids required materials that were stable to heat and oxidation, preferably with high boiling points and low tendency toward flammability. The halogenated hydrocarbons were well suited for these purposes, and two of the best candidates developed during the 1930s were the polychlorinated biphenyls (PCBs) and the polychlorinated naphthalenes (PCNs). Penning (1930), for example, wrote an article noting the physical and chemical properties of the PCBs and suggested that this substance would find many uses in industry and commerce. One disadvantage, he noted, was that the vapors tended to be irritating and caused headaches in some workers. Shortly thereafter, some other problems were noticed. Workers engaged in the manufacture of the PCBs and PCNs sometimes developed a dermatitis resembling severe acne (Wedroff and Dolgoff, 1935; Jones and Alden, 1936). This condition, known as "oil acne" or "chloracne," had been observed as early as the 1890s in workers at a chlorine manufacturing plant, and it was though at first that chlorine itself was involved in the cause of the condition (Baughman, 1979). It was in the 1930s, however, that the appearance of chloracne was found to be associated more closely with industrial production of the PCBs, PCNs, polychlorinated diphenyl oxides, crude chlorinated phenols, and some petroleum oils (Adams et al., 1941). In another incident, PCBs and PCNs were blamed in the deaths of some workers, and it was suggested that some individuals, especially those with preexisting health problems, should avoid these compounds (Greenburg et al., 1939). There was apparently some uncertainty about whether the skin condition was caused by a contaminant or the compounds themselves. In any case, there was increased awareness of hazardous compounds in an industrial setting, and more references on industrial hygiene and toxicology appear in the literature by the 1940s.

In about 1941 in the United States, possibly even earlier, a serious livestock disease of unknown etiology appeared. It was characterized grossly by the appearance of skin lesions with extreme thickening and wrinkling. The ill effects of this hyperkeratosis were not confined to the skin; there was also loss of appetite, depression, and liver damage. From 1941 to 1953 over $20,000 000 in losses was sustained in the United States. The cause was finally traced to the use of polychlorinated naphthalenes in lubricants used on machinery; these were ingested or otherwise absorbed by the cattle. Although environmental contamination and food residues were never measured, the PCNs would be presumed to have been persistent in the fatty tissues of the livestock. It was also noted that the PCNs are excreted into milk and are toxic to humans as well (Olafson, 1947; Clarke and Clarke, 1967). Although uses of PCNs seemed to decline somewhat after this time, the use of PCBs continued.

The PCB's came to the world's attention in 1966 and the years following. In a short note, Sören Jensen voiced his concern over the possible toxicity of the PCBs, which he was finding in a number of locations (Anon., 1966). These samples included a dead eagle, fish, and his own and his family's hair. An analysis of museum specimens indicated contamination starting around 1944. Reynolds (1969) suggested that the use of the PCBs in pesticide formulations might have been a factor in contamination, although there were also many other uses. He commented on the possibility of PCB interference in pesticide analysis, since both tend to behave alike in sample processing. This would cast doubt on the accuracy of some pesticide analyses by gas–liquid chromatography in the 1960s, depending on the sample history, the extraction, and chromatographic conditions.

In 1968, rice oil contamination with PCBs from a leaking heat exchanger was eaten by a number of people in Japan. This poisoning epidemic, commonly referred to as Yusho disease, is unique as a food-borne human exposure that produced obvious health consequences. Fries (1972) noted that about 0.5 g was required to produce symptoms, although the average intake in Yusho disease was closer to 2 g. Food contamination incidents also occurred in the United States, either from animal feed contamination due to leaks of PCB-containing oils, or from the use of PCB-containing silo sealers, which resulted in milk residues (Fries, 1972). An interesting note on the Yusho victims is that the symptoms showed considerable variability. Some noticed fatigue, nausea, vomiting, abdominal pain, and jaundice, with or without chloracne. Others had lung involvement, with coughing, bronchitis, and asthma or pneumonia, whereas still others showed no overt symptoms, but had residues in their fatty tissues. In some, effects were delayed as long as 3 years. However, the official diagnosis depended on the presence of abnormal skin pigmentation or chloracne (Di Nardi and Desmarais, 1976). This illustrates the problem with correlating disease conditions with chemical toxicity, even with a known exposure. In a study including farmers who had used PCB silo coatings, reported health problems included a greater frequency of numbness and joint problems, as well as a higher serum PCB level, but it was also noted that the subjectivity of the symptoms makes it difficult to draw firm conclusions (Humphrey, 1983).

In any case, the extreme persistence of the PCBs in the environment and in organisms, including man, has resulted in a halt to manufacturing of PCBs in a number of countries. In some areas, freshwater fish still contain appreciable residues, and they are expected to retain these for some time.

3.5. Chlorinated Hydrocarbon Pesticides

Even prior to the development of DDT as an insecticide in the mid-1940s, there had been efforts at insect control by the use of synthetic organic compounds as an alternative to the inorganic insecticides that had been used for several decades. Although DDT was not the only insecticidal compound tested, it was certainly the greatest success story in insecticides. The need for an effective insecticide that could be used on human parasites and pests was acute during wartime, and DDT not only was effective, but it was safe in spite of great variability of dose and application methods. Lindane, the gamma isomer of hexachlorocyclohexane (often called benzene hexachloride), was developed at about the same time. Following the close of World War II, DDT and lindane were available to the civilian sector on a broad scale, and other chlorinated hydrocarbon insecticides, such as chlordane and toxaphene, followed shortly thereafter (Frear, 1948c). The controversy over environmental persistence and toxicity to wildlife, Rachel Carson's *Silent Spring*, and the often-cited bioaccumulation and biomagnification through the food chain eventually resulted in the restriction and banning of many of the organochlorine pesticides in several Western nations by the 1970s.

The literature on pesticide residues is now so extensive that it can hardly be reviewed any longer; animal studies and discussions over the significance of food residues and the regulatory measures that resulted are equally extensive and still continue. However, from a historical perspective it is necessary to point out that contrary to what is sometimes heard, the chlorinated hydrocarbon insecticides were not foisted on the world by people lacking any concern for health issues. We may not yet know everything about these compounds, but we are hardly human guinea pigs in a great uncontrolled study.

Fennah (1945) simulated a DDT food contamination situation to show the safety of residues that might occur in foods. Telford and Guthrie (1945) demonstrated that enough DDT could be excreted in milk to be toxic to test animals, and the fat storage of DDT was shown at the same time (Ofner and Calvery, 1945). Calvery (1945) warned that DDT was toxic and cumulative, and that tolerances were needed for this reason. As in the earlier experience with the arsenicals, it was assumed from the beginning that residues on crops or in animal products were inevitable if the insecticides were used; the only real question at the time was what tolerance should be in effect. The mammalian toxicity was unquestionably quite low, safety in use was better than anything in the past, and pesticide use therefore was not particularly controversial at the time. However, the relative safety of DDT led to many creative efforts in its use, including insecticidal fabrics, paper, floor waxes and polishes, paints, and other uses in addition to the usual dust and emulsion formulations (Frear, 1948d). Perhaps the abuse of DDT was inevitable under these conditions, but they also reflect a lot of creativity in potential uses for what seemed at the time to be a safe, persistent, inexpensive pesticide. Perhaps if DDT were not used so widely and in such quantity that it distributed itself into the environment to such an extent, it might still be used in countries that now prohibit its use. In retrospect, DDT, the predecessor to most of our modern insecticides, was tested prior to widespread use. As noted earlier, all risks cannot be eliminated by testing, but there are likely to be fewer surprises if testing is done. The science of toxicology was also changing, however, and new developments in that area may lead to reevaluation of compounds once considered perfectly safe. Starting with acute toxicity, toxicologists evaluated materials for chronic toxicity; we saw how this difference in approach led to differences of opinion on the safety of the arsenical pesticides (Section 3.1). Later, the science of environmental toxicology was developed, and reproductive toxicity, mutagenesis, and behavioral toxicology are now considerations in evaluation of chemical toxicity.

Human poisoning has occurred as a result of contamination by the chlorinated hydrocarbons, but the worst outbreaks have been associated with endrin contamination of flour, presumably in shipping, not from agricultural use (Hayes, 1975a). Another major outbreak involved the fungicide hexachlorobenzene, which was used as a seed treatment. The endrin poisoning epidemic of 1967 in Qatar and Saudi Arabia resulted in 874 cases with 26 deaths from contaminated flour and bread; analysis of bread showed from 48 to 1807 ppm endrin (Curley et al., 1970). Gross misuse of toxaphene application to vegetables resulted in illness, with over 3000 ppm present as a residue (McGee et al., 1952).

Hexachlorobenzene toxicity is of particular interest as a chronic disease caused by a substance that is normally considered to be relatively nontoxic in single doses. From 1955 to 1959 in southeastern Turkey, people developed a mysterious disease characterized by blisters with later scarring, photosensitivity, and lesions of the hand joints. Liver damage and porphyria also developed, along with skin pigmentation and abnormal hair growth (sometimes referred to by the peasants as "monkey disease"). The mortality rate was about 10% generally, but about 95% in infants. The cause proved to be ingestion of hexachlorobenzene-treated seed grain by the victims (Cam and Nigogosyan, 1963; Hayes, 1982a).

The Kepone incident in 1973–1975 illustrates widespread contamination from a point source. Kepone (chlorodecone) was manufactured for a time in a totally inadequate plant, with much waste material flushed into the municipal sewer system. Eventually, the poor hygiene led to worker illness, and the plant was closed, but not before contaminating the James River in Virginia and closing down the associated fisheries (Raloff, 1976).

Finally, a residue problem that does not, fortunately, involve illness is the flavor or quality change associated with some pesticides. Not long after the introduction of lindane and the other benzene hexachloride (BHC) products, it was noted that use of BHC as a soil insecticide caused some crops to retain an off-flavor. This was particularly the case for the technical BHC. A number of studies have been carried out on various pesticides; some affected the flavor of crops grown on treated plots, some had no effect, and still others had effects that depended on the use rate and the particular crop (Mahoney, 1962). It was also noted that poor quality and off-flavors can occur if excessive insect infestation is not controlled.

The most significant contribution of the chlorinated hydrocarbon insecticides to the history of pest control was twofold: (1) the chlorinated hydrocarbon insecticides demonstrated the utility and flexibility in use of synthetic organic pesticides, and (2) their persistence in the environment virtually guaranteed that future agricultural chemicals and many industrial chemicals would be designed with capability for biodegradation.

3.6. Organophosphate and Carbamate Insecticides

During World War II, Gerhard Schrader developed an organic phosphorus compound that was later widely known as parathion. This compound was the success story for the organophosphates that DDT was for the organochlorine insecticides. It was also extremely toxic to mammals and humans as well as to insects, and this led to a number of serious poisonings. Eventually, parathion was restricted in use, but it and other organophosphate insecticides introduced from the early 1950s were so successful that they are still used, and new organophosphates are still being introduced today. Although the organophosphates cover a wide spectrum of toxicity, species selectivity, and uses, they all are capable of degradation by chemical or enzymatic hydrolysis. Thus, they are less persistent in the environment than the chlorinated hydrocarbons were, and, for that reason, the organophosphates are the leading insecticides in use today.

The rapid decomposition of the organophosphates in plants and animals tends to prevent accumulation of large residues in most cases. An additional factor in the safety of the organophosphates is that these compounds are cholinesterase inhibitors, but since cholinesterase is regenerated in the blood, it should be possible to metabolize a certain amount of these pesticides without any significant effect, i.e., chronic toxicity should not be much of a problem at reasonable use rates.

Unfortunately, acute toxicity can be a serious problem, and Hayes (1975a) has listed several food contamination incidents with parathion. Parathion is not the only toxic member of this group, but its long use history in many countries in addition to its high toxicity have involved it in some serious poisonings.

One of these cases, an epidemic poisoning in Singapore, is noteworthy because of some observations on the susceptibility of some age groups to poisoning. In this case, involving contaminated barley, there was no indication that anything was wrong until comatose patients, primarily children, began arriving at the hospital. In fact, coma was sometimes the only symptom noticed, whereas in typical organophosphate poisoning other signs such as twitching, salivation, or nausea usually precede loss of consciousness. When the cause of the problem was finally recognized as organophosphate poisoning, fast action by the authorities in seizing suspect grain and warning the public undoubtedly prevented an even worse tragedy. Chemical analysis indicated that the barley was contaminated to the extent of 0.044–0.18% parathion. Analysis of an actual portion of cooked barley provided an estimated lethal dose of about 2 mg (about 0.1 mg/kg) for a 5- to 6-year-old child. A two-year-old child died after a very long (56 hour) onset. Other members in the family eating the same food were not even affected, demonstrating the extreme danger to young children (Kanagaratnam et al., 1960).

This case, and others referenced by Hayes were contamination incidents, not residue poisonings, but the Singapore incident shows that for some age groups, the extent of contamination does not have to be very great to result in serious poisoning.

The delayed neurotoxicity associated with ingestion of tricresyl phosphates was noted earlier (Section 3.3). When the organophosphate insecticides were developed, the question concerning their potential neurotoxicity arose, particularly since these cholinesterase inhibitors act on the nervous system. Later research would indicate that the delayed neurotoxicity and cholinesterase inhibition are not necessarily related phenomena. Some incidents of delayed neurotoxicity occurred in workers at a plant making mipafox (a fluorophosphate insecticide), indicating that the delayed neurotoxicity syndrome was not confined to tricresyl phosphate (Bidstrup et al., 1953). Abou-Donia and Graham (1978) demonstrated some signs of neurotoxicity in hens with as little as 0.1 mg/kg/day of EPN, and the concern that continuous low-level ingestion of organophosphate residues might lead to neurological problems was undoubtedly in some peoples' minds. Fortunately, the vast majority of organophosphates in use as insecticides have not proven to be neurotoxic in this sense, and use experience has tended to reinforce the lack of neurotoxicity in contrast to a compound such as tricresyl phosphate.

The carbamate insecticides, which are N-methyl carbamate esters of phenols or N-methyl carbamoyl oximes, are also cholinesterase inhibitors, as are the organophosphates. Most of these were developed in the late 1950s and the 1960s and were modeled after the naturally occurring methyl carbamate, physostigmine. Although the carbamates also display a range of toxicities, they are generally considered to be readily biodegradable and also hydrolyze to give biologically inactive compounds.

The soil insecticide aldicarb is perhaps unique in this group of insecticides in the fact that it has been associated with residue poisoning in humans, although from unauthorized uses rather than labeled applications. Hayes (1982b) cited a case from 1966 during the experimental trials of aldicarb. The wife of a scientist had access to aldicarb and applied some to a rose bush. Some time later she ate some leaves from a mint plant growing nearby and became ill with typical anticholinesterase symptoms. The mint plant's root zone had entered treated soil and accumulated aldicarb to the extent of 186 to 318 ppm; leaves from the mint plant killed rabbits, whereas mint plants farther from the treated rose were nontoxic, as were tomatoes grown in treated soil. Aaronson et al. (1979) cited some instances of poisoning by aldicarb that was accumulated in hydroponically grown cucumbers, which contained about 6–10 ppm. (The hydroponic solution contained only 1.8 ppm.)

More recently, a larger poisoning epidemic occurred in the western United States and Canada from aldicarb illegally used on watermelons (Anon., 1985). Over 300 people became ill from aldicarb residues of up to 3 ppm; 0.2 ppm is reportedly sufficient to cause illness.

Although aldicarb toxicity symptoms are unpleasant, there were no fatalities in these incidents. These occurred only because of the extremely high toxicity of aldicarb and because of its systemic action. In addition, mint, watermelons, and cucumbers are vegetables that are harvested earlier in the season, perhaps before aldicarb could degrade. Again, it should be noted that none of these incidents involved approved uses of the pesticide. The aldicarb incidents illustrate another important issue is pesticide residues: the use of systemic pesticides in general. A number of organophosphates and some carbamates are used as systemic insecticides, i.e., they are translocated in the plant and either the parent compound or toxic metabolites are active in the plant for a period of time. There is current and past research into the use of systemic fungicides to protect plants against pathogens. In these cases, the pesticide is not removed by washing, and the ingestion of residues could occur if proper preharvest application restrictions are not followed, or, as in the aldicarb cases, if the plants are able to retain toxic levels of the pesticide in question.

3.7.   Mercury and Thallium Pesticides

Inorganic mercury salts have had limited use in agriculture due to their toxicity to plants, but are organic mercury compounds, notably alkyl mercury salts, proved to be excellent fungicides. They are also extremely toxic and, unlike inorganic mercury salts, the organomercury compounds have a peculiar affinity for the brain, which can be severely injured by exposure.

The world was alerted to the environmental contamination hazards of organomercury compounds by the notorious outbreak of "Minamata disease" in the 1950s in Japan (Kurland et al., 1960). This was caused by discharge of industrial mercury wastes into the bay, followed by uptake into fish, which were eaten by the victims or their mothers (mental retardation and motor disturbances can be caused by intrauterine exposure).

Although the Minamata disease epidemic was caused by industrial effluent rather than agricultural chemical use, another incident illustrated the danger involved in feeding of grain treated with organomercury fungicides to livestock. In New Mexico in 1969, children in a family became comatose. The mother, though not ill, gave birth to an infant with severe central nervous system defects. After some investigation, it was found that the father had obtained seed grain screenings and had fed it to his hogs. Although the animals became blind, the family ate some of the pork. The secondary poisoning by mercury-contaminated meat was sufficient to cause illness (Curley et al., 1971). In addition to the New Mexico case, Hayes (1975a) listed five other organomercury poisoning epidemics: in Iraq, Pakistan, Guatemala, and the USSR from 1960 to the early 1970s. Hayes commented on the use of seed grain as food. The people who ate the seed grain knew that it was treated and that it was intended for planting; they ate it because they were hungry, sometimes washing it in an attempt to remove the fungicidal treatment. In another case, seed grain was issued to the farmers too late for planting; they had already planted their own grain and had nothing left for food but the seed grain they had been given.

Thallium salts have been used as rodenticides since the early 1930s; in a short time, there were fatalities from its use, including from the ingestion of thallium-treated bait (Munch et al., 1933). In another case somewhat similar to the New Mexico mercury poisoning case, a Bedouin family ate a goat that had eaten thallium-treated poison bait; those who ate more than one meal became ill. Fortunately, the poisoning was relatively mild (Ben-Assa, 1962). Thallium has proven to be so dangerous that its use has essentially been discontinued in a number of countries.

3.8.   Phenols and Their Derivatives; the Chlorinated Dioxins

Most phenolic materials used as agricultural chemicals are the chlorinated phenols, which are most commonly used as wood preservatives or industrial biocides, as opposed to use on food crops or animals. Large quantities of phenols are also starting materials in the herbicides 2, 4-D, MCPA, 2, 4, 5-T, and related compounds.

In spite of the fact that the chlorophenols are not in general use on food crops or livestock, they have from time to time posed a residue problem. Engel et al. (1966) found that a musty taste in eggs and poultry was due to tetrachloroanisole, the methyl ether of tetrachlorophenol. Apparently, when wood chips or shavings are used as bedding for livestock or pountry, any chlorophenols used as a wood treatment may be methylated by bacteria to the corresponding chloroanisoles. These do not apparently pose a health problem, but they produce disagreeable flavors in meat and egg products, impairing or ruining the market value (National Research Council of Canada, 1982).

Since their synthesis in 1941 and commercial registration several years later, the phenoxy herbicides, particularly 2, 4-D and 2, 4, 5-T, have been extensively used in agriculture. In spite of the low environmental persistence of the compounds and their infrequent occurrence in foods, there has been a considerable amount of controversy over these herbicides, in particular, 2, 4, 5-T, which has been the object of many use restrictions (Bovey and Young, 1980). Although the phenoxy herbicides themselves are not highly toxic, a number of compounds derived from 2, 4, 5-trichlorophenol came under suspicion due to the presence of the highly toxic condensation product, 2, 3, 7, 8-tetrachlorodibenzo-p-dioxin. The 2, 4, 5-T restrictions exemplify the unusual case in which a pesticide is restricted because of potential or actual contaminants rather than because of the toxicity of the active ingredients.

The term "dioxin" is virtually a household word in many industrialized countries because of some notorious contamination incidents. It should be noted that the "dioxin" of news releases is the 2, 3, 7, 8-tetrachloro isomer, which is by far the most toxic. Unfortunately, this has tended to draw attention away from the fact that there are a large number of chlorinated dibenzo-p-dioxins and the closely associated chlorinated dibenzofurans. Most of these are much less toxic than the 2, 3, 7, 8-tetrachloro isomer, which presents a problem if all chlorinated dioxins are to be regulated. Some of the other members of the family are widely distributed and are much more likely to occur in foods than the tetrachloro isomer, and it will be necessary to determine their significance and what, if anything, will be done about them.

The story of the chlorinated dibenzodioxins is interesting in that a number of isolated incidents had a common denominator of this group of compounds. The problems that are posed by the chlorinated dioxins and furans are both old and very new, and it is only in recent years that the many diverse leads and pieces of information are beginning to come together to give a more nearly complete picture of the environmental and toxicological aspects of these compounds.

As was noted earlier, chloracne was a problem in workers engaged in the manufacture of some of the chlorinated hydrocarbons and phenols (Section 3.4). Baughman (1978) reviewed some of the industrial aspects of chloracne, noting its appearance since the 1890s. As early as 1937, it had been suggested that the chlorinated phenols not be used as biocides until the chloracne problem was understood. Although one compound known to cause problems, chlorinated naphthalene, was declining in use by the 1950s, the production of phenoxy herbicides required large amounts of chlorinated phenols, and the chloracne continued. Baughman noted that 2, 4, 5-T manufacturing plants had particular problems with chloracne in their workers. In addition, some chemists working on the halogenation reactions of dibenzodioxin became extremely ill, although the connection was not immediately made.

(Chloracne should not be confused with typical acne, although the skin lesions are superficially similar. Chloracne is of prolonged duration and is often accompanied by liver damage and porphyria in poisoning by these compounds. Chloracne lessions are a sensitive indicator of exposure to some of the chlorinated hydrocarbons of a particular geometry and halogen substitution, hence the emphasis on it.)

Kimming and Schulz (1957), in their investigation of 2, 4, 5-T plant incidents, arrived at the conclusion that the chlorinated dibenzodioxins and chlorinated dibenzofurans were the active toxic materials in the skin lesions. Attempts were then made to reduce the levels of the chlorinated dioxins in 2, 4, 5-T, but it was nearly inevitable that some would be formed in the process, whether any ended up in the product or not.

Meanwhile, in 1957 there was an outbreak of a disease in chickens in the United States. There was an accumulation of fluid in the pericardium, hence the name "chicken edema disease." Brew et al. (1959) traced the problem to certain lots of feed grade fats, whereas Friedman et al. (1959) noted that the toxic principle was found in the unsaponifiable lipid fraction. They also noted that the toxic fat fraction tended to deposit in the chicken flesh, although it was not known whether or not the toxic material itself did. Brew et al. (1959) and Wootton and Alexander (1959) tended to believe that the chicken edema factor was a cholesterol alteration product, but later work by Wootton et al. (1982) showed the presence of an aromatic nucleus and six chlorine atoms from the mass spectral evidence and from chemical analysis of a carefully fractionated sample of the chicken edema factor. Because of the chlorinated nature of the toxic material,

Firestone et al. (1963) were able to develop a microcoulometric gas chromatographic method for the detection of the factor, even though the structure was still uncertain. Finally, X-ray crystallographic data showed that the chicken edema factor was 1, 2, 3, 7, 8, 9-hexachlorodibenzo-*p*-dioxin (Cantrell et al., 1969). The original source of the contamination was then traced to the use of pentachlorophenol in preserving hides for leather manufacture. In addition to the 1957 epidemic, additional outbreaks occurred in the United States in 1960, again in 1969 from leakage of phenols into chicken feed fats, and in Japan in 1968 from contamination with PCBs—from the same factory that contaminated food grade oil in the Yusho disease epidemic (see Section 3.4). It was observed that a variety of dioxins were found in fats, some being more active in producing edema and others being more acutely toxic (Hayes, 1982c).

Although the chicken edema losses and worker health problems certainly were felt within the industries affected and within families, the dioxin issue really became public when the 1971 incident of industrial waste dumping occurred in Missouri. The hexachlorophene wastes were sprayed onto a horse arena, and the horses and other animals died. Further investigation disclosed additional improperly stored waste in various locations, along with the finding that the dioxins from the earlier incident had not degraded as expected, and that a persistent contamination situation existed. These events eventually led to the evacuation of Times Beach, Missouri (Long and Hanson, 1983).

In 1976, the industrial accident at Seveso, Italy was not confined, and the 2, 3, 7, 8-tetrachlorodibenzodioxin contaminant was released over residential areas, resulting in evacuation of areas near the plant accident. Thus, what might otherwise have been just another chemical plant incident became a news item around the world and a public issue.

The events thus far have been relatively serious accidents in which economic loss and health problems occurred. In the late 1970s, another controversy over the chlorinated dioxins received fairly wide publicity, this time at Midland, Michigan at the Dow Chemical plant. In routine environmental monitoring, low level tetrachlorodibenzodioxin residues (in the picogram per gram range) were found in fish from nearby waters. In an effort to find the source, sampling was carried out on many samples, and the evidence seemed to point to the incinerator. This touched off widespread monitoring in many combustion sources; various chlorinated dioxins were found in many of them, even where no prior use history would have suggested their presence (Long and Hanson, 1983). Thus, the "trace chemistry of fire" theory on the environmental occurrence of the chlorinated dioxins originated.

There is evidence that the chlorinated dioxins and dibenzofurans have been present for many years, but sediment samples from Lake Huron indicated that the levels increased significantly after 1940 (Anon., 1983). It was believed that this corresponded to the use of chlorinated precursors that then pyrolyzed to form the dioxins and furans.

The foregoing evidence confuses the issue of whether or not the phenoxy herbicides are sources of chlorinated dioxins in the food chain, but it appears that they may only be a minor source in comparison with a number of possibilities. Ryan et al. (1985) found that in their survey of Canadian chicken and pork a large fraction contained chlorinated dibenzodioxins, some also contained chlorinated dibenzofurans but many also contained pentachlorophenol, which seemed to be the source of the dioxins and furans. They commented that the wide use of pentachlorophenol on wood, along with the wide use of such treated wood products in agricultural production, makes pentachlorophenol nearly ubiquitous in the United States, and it was certainly common in the eastern Canadian provinces from which the samples were taken. The fact that the dioxins and furans found contained from six to eight chlorines, and none were tetrachlorinated species, suggests their origin in pentachlorophenol.

The analysis of picogram per gram levels of the chlorinated dioxins and furans is difficult and expensive, and this tends to limit the amount of environmental and food monitoring that can be done. It has also been only since about 1980 that isomers specific analyses were available (this is needed if they are to be toxicologically evaluated) (National Research Council of Canada, 1981). However, it appears that much more monitoring of the chlorinated dibenzodioxins and related compounds will need to be done to establish the significance and the sources of such contamination.

4. MYCOTOXINS

4.1. Mycotoxins in Human Food Supplies

In contrast to the previously mentioned compounds, the mycotoxins occur naturally and thus, it may be assumed, have been in the food chain for a long time. A *mycotoxicosis* should be distinguished from a *mycosis*; the former is an intoxication by a toxic substance produced by fungi with or without the presence of viable fungi, whereas the latter is an actual invasion of the tissues by living parasitic fungi. Although fungal infections may be associated with toxin production in the tissues, there is little evidence for this in the case of animal infections, and the presence of a toxin is not necessary to explain the harmful effects of the mycosis. A mycotoxicosis should also not be confused with other allergic reactions to fungi or their spores, such as may occur in pulmonary conditions such as farmer's lung, which is often attributed to the inhalation of dust and molds.

There are a number of theories concerning the reason for the production of mycotoxins, which are secondary metabolites of fungi in most cases. These metabolites are not always produced, as they are strain characteristics rather than species characteristics; not all members of a particular species produce toxins. Thus, they are not essential to the growth or welfare of the species. For this reason also, identification of a mold that is cultured from foods or feeds is not indicative of its potential toxicity. In other words, one cannot say that there is such a thing as "toxic molds" in a generic sense. Current views seem to be arriving at a consensus that mycotoxicoses are specific interactions between a particular strain of a fungus grown in a particular medium under specific conditions and affecting a species of animal. In addition, it is thought that in some cases of fungal toxicity, there is a degree of synergism and that the crude mold culture is more toxic than individual compounds. Of course, this greatly complicates interpretation of mycotoxin analyses and experimental studies of moldy feeds versus administration of purified toxins.

Most mycotoxicoses are studied in the context of veterinary toxicology. Human cases are apt to be much less common for a number of reasons, including the fact that people frequently obtain food from a variety of sources, reducing any risk from excessive intake of a single contaminated food. Human cases of mycotoxin poisoning have occurred, however, and some of these will be discussed.

4.2.   Ergotism

Intoxication by ergot, generally from ingestion of ergot infested cereal grains, is old, but the major epidemics for which there is some evidence occurred during the Middle Ages, although the cause was not recognized. The use of ergot as a drug apparently dates to 1582, but the fungal nature of ergot was not recognized until 1764 (Gröger, 1972). Human ergotism has practically disappeared; the last recorded epidemic allegedly occurred in France in 1951, but there is much doubt concerning the actual toxic agent in that case (Wilson and Hayes, 1973). In man, ergotism was known to produce central nervous system disturbances, but circulatory disturbances and gangrene were conspicuous features due to the vasoconstrictive properties of the ergot alkaloids. Although ergotism is now quite rare, the pharmacological properties of the alkaloids have made them useful in medicine.

4.3.   *Fusarium* Mycotoxicoses in the USSR

The *Fusarium* toxins have been the most prominent in human mycotoxicoses and have been the best documented. Although illness has occurred due to ingestion of grain infested with *Fusarium* species, the USSR has been unique in suffering endemic mycotoxicoses from this agent and in experiencing significant mortality. The following description is from Joffe (1986), who is perhaps the world's leading authority on the subject, and whose efforts have brought the matter to the attention of scientists in the West.

In parts of eastern Siberia and other locations, people eating bread made from the locally grown cereal grains during certain years sometimes experienced dizziness, nausea, vomiting, and shivering. This condition, known as the "inebriated bread" syndrome, was first investigated by scientists in the 1890s and was associated with certain fungi, notably members of the *Fusaria*. Investigations continued sporadically into the twentieth century; the problems were apparently more common in climatic situations with high moisture and low temperatures.

It was not long before more problems appeared. In the 1930s, a disease outbreak characterized by the appearance of necrotic lesions on the skin and mucous membranes and hemorrhagic manifestations appeared in parts of Siberia. At this time, the cause was not certain and was attributed to other diseases such as diphtheria.

This "septic angina" of the 1930s reappeared with a vengeance during the World War II years, notably in 1943–1944. The farm labor shortage was acute because of the war situation, and food was scarce, so that many people had to subsist on overwintered grain that had been under snow cover and was heavily molded. Climatic conditions were ideal for *Fusarium* toxin production, and a terrible outbreak of the disease now known as alimentary toxic aleukia occurred. A collaborative effort by physicians, mycologists (A.Z. Joffe was actively involved in this and later documented the history of the epidemic), chemists, and others resulted in a good understanding of the mycotoxicological nature of the disease, its course, and its treatment.

Even after the incidence of alimentary toxic aleukia decreased to the vanishing point, research on the chemical nature of the toxins involved continued through the 1950s and 1960s. The Soviet scientists believed that the toxins were steroidal glycosides, whereas investigators in the United States and Japan, who now were active in study of the *Fusarium* toxins as well, tended to favor the trichothecene structure. There was legitimate reason for difference in opinion: the steroidal glycosides are known natural products, and color reactions of the toxic grain extracts seemed to be consistent with such a structure. Bamburg and Strong (1971) summarized their own and other work supporting evidence for a number of *Fusarium* toxins based on the 12, 13-epoxytrichothecene nucleus. There were still some doubts about the identity of the toxins, as the strains from the USSR seemed extraordinarily potent in comparison with many of the strains from other sources. Finally, Mirocha and Pathre (1973) determined that the major constituent of an authentic sample of the toxin from USSR strains of fungi was identical with the toxin known as T-2 toxin, which had been characterized previously as one of the trichothecene toxins. Since then, the trichothecenes have been generally accepted as the major toxic principles in the *Fusaria*, which give rise to the irritant and hemorrhagic symptoms. There are currently about 50 of these related toxins that can occur naturally, although fungi other than the *Fusaria* produce some of them (Joffe, 1986).

Although there was much work on the fungal metabolites following the discovery of penicillin, this was directed toward antibiotic research, and the mycotoxicoses tended to be neglected, even in the field of veterinary medicine, in most countries.

4.4. Rice Toxins—Japan

Fusarium toxins had caused some sporadic problems in the Japanese grain crops, and some illness had resulted, but in Japan and other Asian countries there was a disease referred to as Shoshin-kakke, which was largely urban and appeared to be associated with changes in diet. It was characterized by severe pain near the heart, followed by difficulty in breathing and an ascending paralysis. It was considered to be a fulminating manifestation of beriberi, but the high mortality rate did not seem consistent with that disease. Moreover, the incidence of the disease decreased sharply in 1910, even though beriberi continued. In that year, systematic rice inspection was carried out to prevent moldy rice from reaching the market. Uraguchi (1971), after 30 years of research on rice toxins, suggested that this was actually a mycotoxicosis, and he isolated a toxin, citreoviridin, that reproduced a comparable condition in animal models. This was a difficult retrospective study of the epidemiology and mycology of the food supply associated with the disease, and Uraguchi has some substantial data supporting his theory. Like some other human mycotoxicoses, however, evidence is circumstantial and epidemiological, and it is difficult to prove that mycotoxins are acting in the absence of other factors.

4.5. Aflatoxins

An outbreak of a severe epidemic of unknown etiology in poultry occurred in England in 1960, leading to a concerted effort to determine the cause of the problem. The occurrence of the disease seemed to correlate with certain lots of feed, and the presence of fungal hyphae in a sample of the toxic feed eventually led to the identification of the fungus *Aspergillus flavus* as the factor responsible. The chemical structure and properties of the fungal toxins, called aflatoxins, were determined shortly thereafter. However, the 1960 outbreak was not the first incident of aflatoxicosis; the cause was simply not recognized at the time until the "turkey X disease" led to studies that associated certain symptoms with the toxin (Allcroft, 1969).

This might have been left to the veterinary journals if it had not been for the fact that the aflatoxins were shown to be potent liver carcinogens, and that the aflatoxins were present in foods consumed by humans. Shank (1978) collected evidence suggesting possible links between dietary aflatoxins and the occurrence of liver tumors (particularly in sub-Saharan Africa), the condition known as Indian childhood cirrhosis, and possibly Reye's syndrome in children. The aflatoxins are considered enough of a health hazard that there are restrictions in a number of countries with regard to the tolerable amounts in foods, notably those made from peanuts or corn. In contrast to the *Fusarium* toxins, which predominate in cooler climates, the aflatoxins are produced in warm, humid situations; they are particularly a hazard in tropical and subtropical climates around the world.

Although limits on the permissible levels of aflatoxins in foods might protect consumers in some countries, locally produced foods are not inspected, and many of the nations that might be expected to experience the worst aflatoxin problems include those that have marginal food situations anyway. It makes sense to have tolerances for aflatoxins only if one is in the position to be able to discard contaminated foods, and many individuals cannot do this. Although there have been many research projects engaged in the decontamination of aflatoxin-contaminated feeds, these have been designed for livestock production use rather than the human diet, for the most part.

4.6.   Other Mycotoxins in the Human Food Chain

Fortunately, there have been relatively few epidemics of mycotoxicoses in human populations. However, sporadic incidents involving illness after contamination of foods by fungi have occurred at times. Although concern over aflatoxins related to liver tumor induction has prompted food inspection and analysis for aflatoxins, Shank (1978) also noted fatalities in children in Taiwan, Thailand, and Uganda, following the consumption of moldy foods. Although the evidence was not complete, these were attributed to aflatoxicosis, in which levels may have been high enough to cause acute toxicity.

Another fatality in Thailand following ingestion of moldy food led to the isolation of cytochalasin E from *Aspergillus clavatus* cultured from the moldy food in question (Glinsukon et al., 1974).

Lest one be tempted to assume that mycotoxicoses occur only in restricted geographical areas, an individual consuming a small quantity of commercial canned beer developed nausea, vomiting, weakness, and a tremor. A fungal mass, identified as *Penicillium crustosum*, was found in the can. The cultured fungus produced alkaloids of the clavine group and also some roquefortine, which has also been detected in blue cheese (Cole et al., 1983). Whether or not these alkaloids produced the illness could not be proven, since samples of the contaminated beer were not available, but the fungus isolated certainly had the ability to produce the alkaloids.

Deoxynivalenol, also known as vomitoxin, is a mycotoxin of the trichothecene group that is associated with scab on cereal grains and with some corn diseases. Hart and Braselton (1983) showed that naturally infected wheat samples containing vomitoxin yielded fractions after milling that all contained some of the toxin. In addition, it was noted that the vomitoxin is relatively stable to heat, and that it conceivably could persist to a degree even after processing.

5.   CURRENT AND FUTURE CONSIDERATIONS

In view of the fact that most of the accidents and illnesses with pesticide poisoning have involved gross contamination, application exposure, or children, Hayes (1975b) suggested that concern for human health might be more effectively expressed in taking measures to avoid poisoning by accidental ingestion and industrial exposure rather than spending excessive amounts of limited resources on monitoring of residues. Parke and Truhaut (1985) have suggested the following as persistent and current problems in pesticide safety: (1) delayed neurotoxicity concerns, (2) interpretation of animal carcinogenicity data, which may be specific for the mice often used in such studies, (3) possible toxicity of pesticide impurities and decomposition products, (3) evaluation of mixtures of pesticides, (5) investigation of properties of the more complex pesticides having stereoisomers of greatly different biological activities, (6) possible interactions of pesticides with normal food constituents before and after cooking, (7) pesticides and fumigants on stored grain, and (8) ensuring reliability of pesticide testing data. Some of these topics are timely issues. The chlorinated dibenzodioxins are contaminants of some pesticides, or at least have been in the past, whereas fumigant residues are of some concern. In the United States in 1982–1984, the focus shifted to ethylene dibromide in grain and in processed food products made from the milled flour. Parke and Truhaut (1985) also noted the formation of chlorohydrins from ethylene oxide fumigation of foods. At the same time, measures are needed to prevent insect infestation of stored food products; these losses can be as severe as losses in the field.

Another current area of concern that will have to be addressed in the near future involves the growing international agricultural chemical market. We may expect a growing problem with substandard and contaminated pesticides that are sold at discounts to individuals who cannot afford quantitites of standard grade pesticides, but who can ill afford a crop loss due to diluted or incorrect materials. Moreover, some contaminants pose the possibility of health and environmental problems due to the contaminants they contain. Although both developed and developing nations are victims and perpetrators of this fraud, the poorer nations obviously have even more to lose and are more likely to be driven to the use of cheap imitations because of the lack of finances (Freistadt, 1987). The lack of laboratories in many of the developing countries makes this type of activity more difficult to detect and stop.

Most mycotoxicoses are sporadic problems, but we can unfortunately count on aflatoxins to be an endemic problem in the warmer regions of the world. Here, the entire food production and processing chain would have to be modified to decrease the risk of aflatoxicosis; toxin production may begin in the field, continue after harvest, and can occur even in cooked foods stored under conditions favorable to mold growth. Unlike the pesticides and industrial chemicals, mold spores are truly ubiquitous and cannot reasonably be excluded from the environment. It is necessary to maintain conditions under which those spores cannot maintain active growth and toxin production.

One trend noticeable in the history of food contamination is that some of the environmental contaminants caused worker health problems before they were released into the environment and became a more widespread problem. This suggests that there are prior warnings of potential problems, and it is urgent that we not ignore them when they appear.

Just as the history of nations is not merely a record of warfare and conflict, the story of environmental contamination is not merely a bleak recitation of man's failures in the struggle against pests, hunger, and disease. Egginton (1980) has related the now well-known story of the contamination of much of Michigan's food supply by the polybrominated biphenyls, a fire retardant that was accidentally mixed into animal feed. The persistent and courageous efforts of a number of people prevented the disaster from becoming even worse. In a note in the *Federal Register* (Taylor, 1987), the action levels for the polybrominated biphenyls were dropped. This event had much less publicity than the original contamination, but it is a very significant event. The action levels were dropped because the extent of contamination had declined to a point at which the action levels were no longer necessary; these contaminants have apparently disappeared in spite of high persistence in fat. Along similar lines, most of the persistent bioaccumulating compounds are no longer made in many of the countries formerly using them, and levels in food and in human tissue can be expected to decline steadily into the future.

If nothing else, we perhaps communicate better than some of our predecessors. We noted earlier how the conflict between the economic entomologists and the medical community raged over arsenic residues. Each side apparently was not very interested in acknowledging the other's point of view. Today, we are more likely to see cooperation and advice solicited from experts in diverse fields, so that chemists, entomologists, weed scientists, physicians, and environmental scientists all work together on problems of mutual interest. We are not too provincial to read the international literature and to seek out other points of view. Partly as a result of this increased awareness of research by many different individuals, we are now in a much better position to understand what went wrong in previous years, and what can be done to use chemicals effectively and safely.

REFERENCES

Aaronson, M., Ford, S. A., Goes, E. A., Savage, E. P., Wheeler, H. W., Gibbons, G., and Stoesz, P. A. (1979) "Suspected carbamate intoxications-Nebraska." *Morb. Mort. Week. Rep.* **28**, 133–134.

Abou-Donia, M. B., and Graham, D. G. (1978). "Delayed neurotoxicity of O-ethyl-O-4-nitrophenyl phenylphosphonothioate: Subchronic (90 day) oral administration in hens." *Toxicol. App. Pharmacol.* **45**, 685–700.

Adams, E. M., Irish, D. D., Spencer, H. C., and Rowe, V. K. (1941). "The response of rabbit skin to compounds reported to have caused acneform dermatitis." *Ind. Med.* **10**, Ind. Hyg. Sect. 2, 1–4.

Allcroft, R. (1969). "Aflatoxicosis in farm animals." In L. A. Goldblatt, ed., *Aflatoxin.* Academic Press, New York, pp. 237–261.

Anon. (1966). "Report of a new chemical hazard." *New Sci.* **32** (Dec. 15), 612.

Anon. (1983). "Symposium updates health effects of dioxins, benzofurans." *Chem. Eng. News* Sept. 12, 26–30.

Anon. (1985). Melon contamination: Toxic effects raise pesticide use issue." *Chem. Eng. News* July 15, 3–4.

Bamburg, J. R., and Strong, F. M. (1971). "12, 13-Epoxytrichothecenes." In S. Kadis, A. Ciegler, and S. J. Ajl, eds., *Microbial Toxins.* Academic Press, New York, Vol. 8, pp. 207–292.

Barger, G. (1914). "The detection of organic poisons (toxins and the like) in foods." *Pharm. J.* **91**, 572. (*CA* **8**, 968).

Baughman, R. W. (1978). "TCDD and industrial accidents." In T. Whiteside ed., *The Pendulum and the Toxic Cloud.* Yale University Press, New Haven, Appendix, pp. 145–158.

Ben-Assa, B. (1962). "Indirect thallium poisoning in a Bedouin family." *Harefuah* **62**, 378–380. Cited in Hayes, W. J. Jr. (1982). *Pesticides Studied in Man.* Williams & Wilkins, Baltimore, pp. 27, 63.

Bidstrup, P. L., Bonnell, J. A., and Beckett, A. G. (1953). Paralysis following poisoning by a new organic phosphorus insecticide (mipafox)." *Br. Med. J.* **1**, 1068–1072.

Boethling, R. S., and Cooper, J. C. (1985). "Environmental fate and effects of triaryl and tri-alkyl/aryl phosphate esters." In F. A. Gunther and J. D. Gunther, eds., *Residue Reviews*. Springer-Verlag, New York, Vol. 94, pp. 49–99.

Bovey, R. W., and Young, A. L. (1980). *The Science of 2, 4, 5-T and Associated Phenoxy Herbicides*. Wiley, New York, pp. 1–26, 335–345.

Brew, W. B., Dore, J. B., Benedict, J. A., Potter, G. C., and Sipos, E. (1959). "Characterization of a type of unidentified compound producing edema in chicks." *J. Assoc. Off. Agr. Chem.* **42**, 120–128.

Calvery, H. O. (1945). "DDT is poisonous." *Food Packer* **26**, 61–62.

Cam, C., and Nigogosyan, G. (1963). "Acquired toxic porphyria cutanea tarda due to hexachlorobenzene." *J. Am. Med. Assoc.* **183**, 88–91.

Clarke, E. G. C., and Clarke, M. L. (1967). *Garner's Veterinary Toxicology*. Williams & Wilkins, Baltimore, 3rd ed., pp. 287–291.

Cole, R. J., Dorner, J. W., Cox, R. H., and Raymond, L. W. (1983). "Two classes of alkaloid mycotoxins produced by *Penicillium crustosum* Thom isolated from contaminated beer." *J. Agr. Food Chem.* **31**, 655–657.

Coulson, D. M., Cavanagh, L. A., and Stuart, J. (1959). "Gas chromatography of pesticides." *J. Agr. Food Chem.* **7**, 250–251.

Coulson, D. M., Cavanagh, L. A., de Vries, J. E., and Walther, B. (1960). "Microcoulometric gas chromatography of pesticides." *J. Agr. Food Chem.* **8**, 399–402.

Curley, A., Jennings, R. W., Mann, H. T., and Sedlak, V. (1970). "Measurement of endrin in flour following epidemics of poisoning." *Bull. Environ. Contam. Toxicol.* **5**, 24–29.

Curley, A., Sedlak, V. A., Girling, E. F., Hawk, R. E., Barthel, W. F., Pierce, P. E., and Likosky, W. H. (1971). "Organic mercury identified as the cause of poisoning in humans and hogs." *Science* **172**, 65–67.

Debre, R., and Bloc, H. (1938). "Mass poisoning in the form of polyneuritis on shipboard caused by the ingestion of oil containing tritolyl phosphate." *Bull. Mem. Soc. Med. Hop. Paris* **54**, 1726–1733. (*CA* **33**, 3877).

De Eds, F. (1933). "Chronic fluorine intoxication. A review." *Medicine* **12**, 1–60. (*CA* **27**, 4590).

Di Nardi, S. R., and Desmarais, A. M. (1976). "Polychlorinated biphenyls in the environment." *Chemistry* **49**, 14–17.

Egginton, J. (1980). *The Poisoning of Michigan*. Norton, New York, pp. 13–344.

Engel, C., de Groot, A. P., and Weurman, C. (1966). "Tetrachloroanisole: A source of musty taste in eggs and broilers." *Science* **154**, 270–271.

Fennah, R. G. (1945). "Preliminary tests with DDT against insect pests of foodcrops in the Lesser Antilles." *Trop. Agric.* **22**, 222–226.

Firestone, D., Ibrahim, W., and Horwitz, W. (1963). "Chick edema factor. III. Application of microcoulometric gas chromatography to detection of chick edema factor." *J. Assoc. Off. Agr. Chem.* **46**, 384–395.

Frear, D. E. H. (1948a). *Chemistry of Insecticides, Fungicides, and Herbicides*. D. Van Nostrand, New York, 2nd ed., pp. 11–30.

Frear, D. E. H. (1948b). *Chemistry of Insecticides, Fungicides, and Herbicides*. D. Van Nostrand, New York, 2nd ed., pp. 293–300.

Frear, D. E. H. (1948c). *Chemistry of Insecticides, Fungicides, and Herbicides*. D. Van Nostrand, New York, 2nd ed., pp. 57–84.

Frear, D. E. H. (1948d). *Chemistry of Insecticides, Fungicides, and Herbicides*. D. Van Nostrand, New York, 2nd ed., pp. 65–70.

Freistadt, M. (1987). "Bootleg chemicals threaten farmers—and the industry itself." *Agrichem. Age* **31**(8), 8–9.

Friedman, L., Firestone, D., Horwitz, W., Banes, D., Anstead, M., and Shue, G. (1959). "Studies of the chicken edema factor." *J. Assoc. Off. Agr. Chem.* **42**, 129–140.

Fries, G. F. (1972). "PCB residues: Their significance to animal agriculture." *Agr. Sci. Rev.*, Third Quarter (U.S. Dept. Agric.), pp. 19–24.

Giuffrida, L. (1964). "A flame ionization detector highly selective and sensitive to phosphorus—a sodium thermionic detector." *J. Assoc. Off. Agr. Chem.* **47**, 293–300.

Glinsukon, T., Yuan, S. S., Wightman, R., Kitaura, Y., Büchi, G., Shank, R. C., Wogan, G. N., and Christensen, C. M. (1974). "Isolation and purification of cytochalasin E and two tremorgens from Aspergillus clavatus." *Plant Foods Man* **1**, 113–119.

Greenburg, L., Mayers, M. R., and Smith, A. R. (1939). "The systemic effects resulting from exposure to certain chlorinated hydrocarbons." *J. Ind. Hyg. Toxicol.* **21**, 29–38. (*CA* **33**, 2608).

Gröger, D. (1972). "Ergot." In S. Kadis, A. Ciegler, and S. J. Ajl, eds., *Microbial Toxins*. Academic Press, New York, Vol. 8, pp. 321–323.

Gunther, F. A., and Blinn, R. C. (1955). *Analysis of Insecticides and Acaricides*. Interscience Publishers, New York, pp. 101–103.

Hart, L. P., and Braselton, W. E., Jr. (1983). "Distribution of vomitoxin in dry milled fractions of wheat infected with *Gibberella zeae*." *J. Agr. Food Chem.* **31**, 657–659.

Hayes, W. J., Jr. (1975a). *Toxicology of Pesticides*. Williams & Wilkins, Baltimore, pp. 322–326.

Hayes, W. J., Jr. (1975b). *Toxicology of Pesticides*. Williams & Wilkins, Baltimore, p. 371.

Hayes, W. J., Jr. (1982a). *Pesticides Studied in Man*. Williams & Wilkins, Baltimore, pp. 593–594.

Hayes, W. J., Jr. (1982b). *Pesticides Studied in Man*. Williams & Wilkins, Baltimore, pp. 447–448.

Hayes, W. J., Jr. (1982c). *Pesticides Studied in Man*. Williams & Wilkins, Baltimore, p. 479.

Higgins, C. H. (1914). "Toxic products in food and their detection." *Centr. Bakt. Parasitenk.* Abt. I, **74**, 193–197. (*CA* **8**, 968).

Humphrey, H. E. B. (1983). "Population studies of PCB's in Michigan residents." In F. M. D'Itri and M. A. Kamrin, eds. *PCB's: Human and Environmental Hazards*. Butterworth, Boston, pp. 299–309.

Jones, J. W., and Alden, H. S. (1936). "An acneform dermatergosis." *Arch. Dermat. Syphilol.* **33**, 1022–1034. (*CA* **30**, 7719).

Joffe, A. Z. (1986). *Fusarium Species: Their Biology and Toxicology*. Wiley, New York, pp. 1–8, 225–298.

Kanagaratnam, K., Boon, W. H., and Hoh, T. K. (1960). "Parathion poisoning from contaminated barley." *Lancet* **1**, 538–542.

Kimmig, J., and Schulz, K. H. (1957). "Chloracne from chlorinated aromatic cyclic ether." *Dermatologica* **115**, 540–546. (*CA* **52**, 4026).

Kirk, P. L. (1950). *Quantitative Ultramicroanalysis*. Wiley, New York, pp. 1–305.

Kirk, J. B., and Lavoipierre, R. (1939). "Epidemic of ascending peripheral polyneuritis at Mauritius in March and April, 1938." *Bull. Office Intern. Hyg. Publ.* **30**, 1480–1484. (*CA* **34**, 2949).

Kurland, L. T., Faro, S. N., and Siedler, H. (1960). "Minamata disease." *World Neurol.* **1**, 370–394.

Lightbody, H. D., and Mathews, J. A. (1936). "Toxicology of rotenone." *Ind. Eng. Chem.* **28**, 809–811.

Long, J. R., and Hanson, D. J. (1983). "Dioxin issue focuses on three major controversies in U.S." *Chem. Eng. News* June 6, pp. 23–36.

Lovelock, J. E. (1961). "Ionization methods for the analysis of gases and vapors." *Anal. Chem.* **33**, 162–177.

Lovelock, J. E., and Lipsky, S. R. (1960). "Electron affinity spectroscopy—a new method for the identification of functional groups in chemical compounds separated by gas chromatography." *J. Am. Chem. Soc.* **82**, 431–433.

Mahoney, C. H. (1962). "Flavor and quality changes in fruits and vegetables in the United States caused by application of pesticide chemicals." In F. A. Gunther, ed., *Residue Reviews*. Springer-Verlag, New York, Vol. 1, pp. 11–20.

Marcovitch, S. (1926). "The fluosilicates as insecticides." *Ind. Eng. Chem.* **18**, 572–573.

Martin, H. (1963). "Present safeguards in Great Britain against pesticide residues and hazards." In F. A. Gunther, ed., *Residue Reviews.* Springer-Verlag, New York, Vol. 4, pp. 18–26.

Martin, A. J. P., and James, A. T. (1956). "Gas-liquid chromatography: The gas density meter, a new apparatus for the detection of vapours in flowing gas streams." *Biochem. J.* **63**, 138–143.

Martinek, M. J., and Marti, W. C. (1931). "Practical methods of detecting and estimating methyl chloride in air and foods." *Ind. Eng. Chem., Anal. Ed.* **3**, 408–410.

McGee, L. C., Reed, H. L., and Fleming, J. P. (1952). "Accidental poisoning by toxaphene. Review of toxicology and case reports." *J. Am. Med. Assoc.* **149**, 1124–1126.

Mirocha, C. J., and Pathre, S. (1973). "Identification of the toxic principle in a sample of poaefusarin." *Appl. Microbiol.* **26**, 719–724.

Munch, J. C., Ginsburg, H. M., and Nixon, C. E. (1933). "The 1932 thallotoxicosis outbreak in California." *J. Am. Med. Assoc.* **100**, 1315–1319.

National Academy of Sciences, Committee on Biologic Effects of Atmospheric Pollutants, Division of Medical Science, National Research Council. (1972). *Lead*, National Academy of Sciences, Washington, D.C., pp. 145–146.

National Research Council of Canada, NRC Associate Committee on Scientific Criteria for Environmental Quality. (1982). *Chlorinated Phenols: Criteria for Environmental Quality.* NRCC No. 18578, NRCC, Ottawa, Canada, pp. 93–99.

National Research Council of Canada, Associate Committee on Scientific Criteria for Environmental Quality. (1981). *Polychlorinated Dibenzo-p-Dioxins: Limitations to the Current Analytical Techniques.* NRCC No. 18576, NRCC, Ottawa, Canada, pp. 1–7.

Ofner, R. R., and Calvery, H. O. (1945). "Determination of DDT (2, 2-bis-(p-chlorophenyl)-1, 1, 1-trichloroethane) and its metabolite in biological materials by use of the Schechter-Haller method." *J. Pharmacol. Exp. Ther.* **85**, 363–370.

Olafson, P. (1947). "Hyperkeratosis (X-disease) of cattle." *Cornell Vet.* **37**, 279–291.

Parke, D. V., and Truhaut, R. (1985). "Evaluation of risks from pesticide residues in food." In G. G. Gibson and R. Walker, eds., *Food Toxicology—Real or Imaginary Problems?* Taylor & Francis, London, pp. 259–270.

Penning, C. H. (1930). "Physical characteristics and commercial possibilities of chlorinated diphenyl." *Ind. Eng. Chem.* **22**, 1180–1182.

Raloff, J. (1976). "The Kepone episode." *Chemistry* **49**, 20–21.

Reynolds, L. M. (1969). "Polychlorinated biphenyls and their interference with pesticide residue analysis." *Bull. Environ. Contam. Toxicol.* **4**, 128–143.

Roark, R. C. (1932). "Chemically combating insect pests of foodstuffs." *Ind. Eng. Chem.* **24**, 646–648.

Ryan, J. J., Lizotte, R., Sakuma, T., and Mori, B. (1985). "Chlorinated dibenzo-p-dioxins, chlorinated dibenzofurans, and pentachlorophenol in Canadian chicken and pork samples." *J. Agr. Food. Chem.* **33**, 1021–1026.

Ryhage, R. (1964). "Use of a mass spectrometer as a detector and analyzer for effluents emerging from a high temperature gas liquid chromatography column." *Anal. Chem.* **36**, 759–764.

Sampson, B. F. (1942). "The strange Durban epidemic of 1937." *S. African Med. J.* **16**, 1–9. (*CA* **36**, 5556).

Schechter, M. S., Soloway, S. B., Hayes, R. A., and Haller, H. L. (1945). "Colorimetric determination of DDT." *Ind. Eng. Chem., Anal. Ed.* **17**, 704–709.

Shank, R. C. (1978). "Mycotoxicoses of man: Dietary and epidemiological conditions." In T.D. Wyllie and L.G. Morehouse, eds., *Mycotoxic Fungi, Mycotoxins, Mycotoxicoses, an Encyclopedic Handbook.* Dekker, New York, pp. 1–19.

Smith, H. V., and Spalding, J. M. K. (1959). "Outbreak of paralysis in Morocco due to *ortho*-cresyl phosphate poisoning." *Lancet* **2**, 1019–1021.

Smith, M. I., Elvove, E., Uglaer, P. J., Frazier, W. H., and Mallory, G. E. (1930). "Pharmacological and chemical studies on the cause of so-called 'ginger paralysis.'" *U.S. Pub. Health Rept.* **45**, 1703–1716.

Taylor, J. M. (1987). "Revocation of action levels for polybrominated biphenyls." Department of Health and Human Services, Food and Drug Administration. In *Fed. Reg.* **52**(13), Wed., Jan. 21, 1987, pp. 2296–2297.

Telford, H. S., and Guthrie, J. E. (1945). "Transmission of the toxicity of DDT through milk of white rats and goats." *Science* **102**, 647.

Uraguchi, K. (1971). "Yellowed rice toxins." In A. Ciegler, S. Kadis, and S. J. Ajl, eds., *Microbial Toxins*. Academic Press, New York, pp. 367–380.

van Itallie, L. (1932). "Paralysis caused by tri-O-tolyl phosphate." *Bull. Acad. Med.* **107**, 278–280. (*CA* **27**, 4590).

van Itallie, L., Harnsma, A., and van Esveld, L. W. (1932). "Abortifacients, particularly apiole." *Arch. Exp. Pathol. Pharmakol.* **165**, 84–100. (*CA* **26**, 4380).

Watson, J. T., and Biemann, K. (1964). "High resolution mass spectra of compounds emerging from a gas chromatograph." *Anal. Chem.* **36**, 1135–1137.

Wedroff, N. S., and Dolgoff, A. P. (1935). "The pathology of oil acne." *Arch. Gewerbepath. Gewerbehyg.* **6**, 428–436. (*CA* **30**, 7719).

Whorton, J. (1974). *Before Silent Spring*. Princeton University Press, Princeton, pp. 3–175.

Williams, C. L. (1933). "Fumigation of foodstuffs. Public health aspects of an increasing commercial practice." *Am. Pub. Health* **23**, 561–566. (*CA* **27**, 3756).

Wilson, B. J., and Hayes, A. W. (1973). "Microbial Toxins." In Committee on Food Protection, Food and Nutrition Board, National Research Council, *Toxicants Occurring Naturally in Foods*. National Academy of Sciences, Washington, D.C., 2nd ed., p. 382.

Wootton, J. C., and Alexander, J. C. (1959). "Some chemical characteristics of the chicken edema disease factor." *J. Assoc. Off. Agr. Chem.* **42**, 141–148.

Wootton, J. C., Artman, N. R., and Alexander, J. C. (1962). "Isolation of three hydropericardium-producing factors from a toxic fat." *J. Assoc. Off. Agr. Chem.* **45**, 739–746.

2

ENVIRONMENTAL CONTAMINANTS IN TABLE-READY FOODS FROM THE TOTAL DIET PROGRAM OF THE FOOD AND DRUG ADMINISTRATION

David L. Heikes

Total Diet Research Center Food and Drug Administration Kansas City, Missouri

---

1. INTRODUCTION

1.1. History of Total Diet Program

Through the implementation of several diversified programs the United States Food and Drug Administration (FDA) surveys residue levels of environmental contaminants in our diets. Of these studies the Total Diet Program represents one of the oldest surveillance efforts. In its modest beginning in May of 1961, the program focused on radioactive contamination from atmospheric testing and pesticide residues in foods. Over the years it has evolved to a study that monitors the foods of this nation for pesticides, industrial chemicals, metals, and nutrients (Johnson et al., 1981). This program is unique in that all foods analyzed are "table-ready." In a manner of speaking, the samples are taken from the consumer's dinner plate. Essentially every grocery item found in a supermarket, including alcoholic beverages, is included in the "market basket" samples. Each food is prepared for consumption (table-ready), just as it might be in the average home. For example, oranges are peeled; meats are roasted, baked, or fried; and potatoes are separated into three weighings for baking, boiling, and frying. Foods requiring processing are prepared by dieticians in institutionalized kitchens. Certain recipe items such as meatloaf, lasagna, and soup are also prepared.

Initially, market basket samples consisted of food items purchased from retail stores in several regions of the country. The shopping list was prepared in cooperation with the Household Ecomonics Research Division of the Department of Agriculture. Totalling 117 items, each market basket represented a 14-day food supply for a 16- to 19-year-old male (the nation's largest consumer). Samples were similar but reflective of regional dietary patterns. After preparation, the foods were separated into 12 categories (e.g., dairy products, meats, oils and fats, potatoes, root vegetables, and beverages). A mixture of specified amounts of each food was blended or chopped to form 12 homogeneous composites. Aliquots were taken directly from these composites for analysis.

Beginning in 1974, the diets of infants (6 months of age) and toddlers (2 years old) were also examined. As with the adult diet study, samples were collected from retail markets throughout the continental United States. Approximately 160 food items, consisting mainly of commercially available infant and junior foods, comprised the market basket samples. As with the adult samples, food items were arrayed in categories and formed into composits prior to analysis.

1.2. Current Total Diet Program

A revised diet program was implemented in April 1982 and is currently in use. A market basket sample presently consists of 232 retail grocery items from each of three cities of one of four geographic regions of the country. Identical food items from the three cities are combined and represent this food for one region of the country. Sampling is based on data obtained from the Nationwide Food Consumption Survey (United States Department of Agriculture) and the Second National Health and Nutrition Examination Survey (National Center for Health Statistics). Each sample reflects the dietary preferences for eight different age–sex groups represented in this country—infants, young children, male and female teenagers, male and female adults, and male and female older persons. Average daily intakes of the selected foods were determined for each age–sex group. The Total Diet Study allows the FDA to assess the actual dietary exposure of selected age–sex groups to environmental contaminants and to observe trends in consumption of these substances over time. The study also possesses the potential to identify health-related problems (Pennington, 1983).

These foods are prepared to table-ready status as before, but now are examined individually or as recipe items rather than composites. Additionally, some fast-food items (milkshakes and 1/4 pound hamburger) and "off-the-shelf" infant and junior foods are included. Thus, 234 table-ready items (including 13 recipe items) are surveyed for pesticides, herbicides, industrial chemicals, toxic metals, and selected nutrients. A more extensive description of the Total Diet Program has been presented by Pennington and Gunderson (1987).

2. ANALYTICAL PROCEDURES

Several extraction and cleanup procedures are used for the isolation of the approximately 500 pesticides, herbicides, and industrial chemicals that are surveyed in each market basket sample. With the exception of N-methyl carbamates, which are determined through use of a liquid chromatographic method (Krause, 1980), organic contaminants are determined by gas chromatography (GC).

In one procedure, all 234 food items are individually subjected to a general extraction and cleanup procedure that retains most of the surveyed organochlorinated and organophosphorus pesticides and industrial chemicals (Williams, 1984). Aliquots of chopped and blended foods are extracted with acetonitrile. Extracts are diluted with water and the analytes partitioned into petroleum ether. Further cleanup is subsequently accomplished with Florisil column chromatography. Food items high in fat (e.g., dairy products and meats) are treated somewhat differently. Samples are extracted with ether to isolate the analyte-containing fat that in turn is subjected to gel permeation chromatography (Hopper, 1982). The resultant eluates, which are nearly fat free, are partitioned with petroleum ether and subjected to Florisil column cleanup as above. Pesticide and industrial chemical analytes are determined with gas–liquid chromatography (GC) with electron capture (EC), Hall electrolytic conductivity (HECD), microcoulometric (MC), and flame photometric (FPD) detection. Organochlorinated pesticides (e.g., aldrin, DDT, and lindane), organophosphorus pesticides (e.g., parathion and diazinon), and industrial chemicals (e.g., polychlorinated biphenyls and industrial phosphates) are determined through the use of this procedure.

Many grain–based foods as well as most fruits and vegetables (a total of 128 items) examined for organophosphorus pesticides and industrial phosphates using two additional procedures. Generally, those items that are low in fat are analyzed using methods developed by Luke et al. (1975, 1981). Chopped or blended samples are extracted with acetone and partitioned into petroleum ether–methylene chloride prior to determination by GC. Because no cleanup procedure is employed a wide range of analytes (more polar) is determined (e.g., acephate, omethoate, and oxygen analogs of several pesticides). Those items with a higher fat level and low in moisture are examined through the use of a procedure developed by Storherr et al. (1971) and modified by Carson (1981). Samples are extracted with acetonitrile or water–acetonitrile, partitioned into methylene chloride, and passed through a cleanup column consisting of an adsorbant mixture of charcoal, magnesium oxide, and Celite.

Analytes from both procedures are determined by GC with FPD (phosphorus mode) detection. Many of organophosphorus pesticides analyzed by the above general procedure are thus confirmed by these two methods. However, several important industrial chemicals, pesticides, and metabolites are uniquely determined through these procedures.

A class of herbicides, chlorophenoxy acids, and pentachlorophenol are also determined in all 234 food items using a fourth procedure. Blended samples are acidified and extracted using various techniques (Williams, 1984) depending on the nature of the food item (i.e., fat and/or moisture content). Eluants are cleaned up using a gel permeation chromatographic system described by Hopper (1982). Concentrated extracts are methylated using an ion-pair alkylation procedure (Hopper, 1987) and further cleanup accomplished with Florisil column chromatography (Williams, 1984). Determination is achieved through the use of GC with EC detection.

Volatile chemicals occur as residues in our food supply and represent a class of compounds with characteristics and properties far different than those of other pesticides and industrial chemicals. A method that involves purging and trapping of these compounds has been developed (Heikes and Hopper, 1986). Samples are stirred with water and purged with nitrogen in a boiling water bath. The volatile chemicals are collected on a duplex trap consisting of two adsorbants. After elution from the trap, the compounds are determined by GC using wide-bore columns and ECD and/or HECD detection.

3. ANALYTICAL RESULTS

Two recently analyzed market basket samples contained a total of 896 residues in the 468 food items represented (234 food items per sample). Of these residues, there were 68 different chemicals represented, 42 of which were halogenated compounds, 23 chemicals classed as organophosphates, and 3 different carbamates. As Table 1 indicates residues with the highest frequency of occurance are malathion (111 times), p, p'-DDE (107), diazinon (77), dieldren (53), chlorpyrifos (51) and hexachlorobenzene (46). These two samples are representative of all market basket samples.

The vast majority of residues were at levels below 1 ppm. However, several food items contained residues with levels at several parts per million (see Table 2). Some food items rarely contain a pesticide or industrial chemical residue, and several others have never been contained a single residue. Table 3 lists food items that have contained a residue on only one occasion or have never contained a residue in the last 18 market basket samples.

Findings of this study are published periodically (Gartrell et al., 1986) and have been used to compose historical baseline data that highlight current trends in pesticide and industrial chemical uses and consumptions. Additionally, regulatory considerations and evaluations of both FDA and other government agencies have been influenced by the data generated by the Total Diet Program.

4.   EMERGING AND NOVEL CONTAMINANTS

The number of contaminant types and incident of residue detection have increased significantly over the tenure of the Total Diet Program. This trend has closely paralleled advancement of analytical instrumentation in both sensitivity and selectivity.

Retention data, element-specific detector responses, cleanup column elution patterns, and expert interpretation by experienced residue chemists combine to identify all but the most unique chemical contaminants of our food supply. Those residues that defy confirmation by gas chromatographic (GC) analysis are classified as unidentified analytical responses (UARs) and are referred to the Total Diet Research Center for possible mass spectrometric (MS) analysis. UARs become prime candidates for MS characterization if they yield an element-specific GC response, are present at moderate to high levels (i.e., > 0.01 ppm, estimated), and are a recurring contaminant.

Several emerging and novel pesticides and industrial chemical contaminants have been identified in market basket samples over the past 10 years. Mass spectrometry has been instrumental in the characterization of these compounds. Some of these measurements have been reported previously (Heikes, 1986) and will not be recounted in detail here.

The remainder of this chapter will relate the accounts of the investigative process used to characterize and determine industrial processing contaminants, metabolites of registered pesticides, volatile chemicals, and several unusual residues.

4.1.   Industrial Processing Contaminants

Additives are used to preserve foods, enhance their flavor, appearance, and so on. These compounds are deliberately added for an expressed purpose and appear as

TABLE 1   Residues Detected and Frequency of
Detection in Two Recent Market Basket Samples
(468 Total Food Items)

Residue	Frequency
Malathion	111
DDE, *p, p'*-	107
Diazinon	77
Dieldren	53
Chlorpyrifos	51
Hexachlorobenzene	46
BHC, α	27
Lindane	27
Heptachlor epoxide	25
Endosulfan II	23
Octachlor epoxide	22
Endosulfan I	21
Methamidophos	19
Carbaryl	17
2-Chloroethyl linoleate	14
Dicloran	13
Dicofol, *p, p'*-	13
Chlorpropham	12
Endosulfan sulfate	12
Phosalone	12
Acephate	11
Dimethoate	10
Parathion	10
Pentachloroanisole	10
2-Chloroethyl palmitate	9
DCPA	9
Ethion	9
Chlorpyrifos methyl	8
Diphenyl 2-ethylhexyl phosphate	8
Toxaphene	8
Pentachloroaniline	7
TDE, *p, p'*-	7
2-Chloroethyl myristate	6
Omethoate	6
Permethrin, *cis*	6
Permethrin, *trans*	6
DDT, *p, p'*-	5
Chlorobenzilate	4
Pentachlorobenzene	4
Pentachlorothioanisole	4
Quintozene	4
Tributyl phosphate	4
Tri-(2-ethylhexyl) phosphate	4

TABLE 1 (*Continued*)

Residue	Frequency
Fenitrothion	3
Phosmet	3
Pirimiphos methyl	3
Chlordane	2
2,4-Dichloro-6-nitroaniline	2
Pentachloroaniline	2
Polychlorinated biphenyls	2
Captan	1
2-Chloroethyl laurate	1
Chlorpropham, *p*-methoxy	1
DDT, *o,p*-	1
Dicofol, *o,p*-	1
Endrin	1
Ethion oxygen analog	1
Fonophos	1
Gardona	1
Methiocarb	1
Methomyl	1
Methoxychlor, *p,p'*-	1
Nonachlor, *trans*	1
Parathion methyl	1
Perthane	1
Tri-(2-butoxyethyl) phosphate	1
Triphenyl phosphate	1
Vinclozolin	1

Totals: 896 residues of 68 different chemicals

TABLE 2 Highest Levels of Individual Residues in Two Recent Market Basket Samples

Food Item	Contaminant	Level (ppm)
Caramel candy	Diphenyl 2-ethylhexyl phosphate	8.5
Chili soup	2-Chloroethyl linoleate	3.5
Canned spinach	*cis*- and *trans*-Permethrin	2.4
Raw peaches	Dicloran	2.3
Potato chips	Chlorpropham	1.1
Raw peaches	Endosulfan II	0.90
Raw cherries	Perthane	0.78
Chili soup	2-Chloroethyl palmitate	0.71
Green peppers	Acephate	0.37
Raw pears	Ethion	0.27

TABLE 3   Food Items with Lowest Number of Residues in Last 18 Market Basket Samples

Food	Number
Canned peas	1
Canned pears	1
Canned pineapple	1
Pineapple juice	1
Coffee	1
Water	1
Infant formula with iron[a]	1
Banana and pineapple tapioca[a]	1
Raw onions	0
Bananas	0
Vegetable beef soup	0
Cola soda	0
Cherry soda	0
Lemon lime soda	0
Decaffeinated coffee	0
Canned beer	0
Whiskey	0

[a]Infant and junior food items.

part of the list of ingredients. There are, however, a number of industrial chemicals that occur in foods and are incidental to manufacturing and processing. Many of these compounds are of unknown toxicity, and many remain unidentified as to source or character.

This type of compound is occasionally detected by the Total Diet Study. They are of a recurring nature being detected in the same food item of each market basket sample and usually at consistent residue levels.

4.1.1. Chloroethyl Esters

Under the old diet survey, a particularly difficult composite to extract and clean up was the mixture with the inclusive title of "fats and oils." Several complex food items such as salad dressings (including mayonnaise), peanut butter, margarine, and shortening were ingredients of this composite. Although gel permeation chromatography (GPC) was used to remove most of the fatty material, eluates from this composite contained relatively high levels of lipids. Gas chromatography with microcoulometric detection (GC/MC), which has an inherent tolerance for fatty eluates, was used for quantitation.

The fats and oils composite of several market basket samples showed evidence of two UARs (Heikes and Griffitt, 1979a). These UARs had been much more responsive (50 times) to GC/MC than electron capture detection (^{63}Ni). This phenomenon tended to class these compounds as alkyl halides rather than aryl halides.

Examination of the individual items of this composite isolated the source of the UARs to French dressing. Because of the complexity of this product, individual ingredients were obtained from the manufacturer and analyzed separately. It was determined that the compounds in question were found only in the paprika oleoresin.

Electron impact mass spectra of these UARs were generated. The larger and later eluting peak showed little evidence of the halogenation the GC/MC response had indicated. In fact, the spectrum resembled that of the ethyl ester of linoleic acid, an 18-carbon fatty acid with two points of unsaturation. In contrast, the earlier eluting GC peak exhibited an abundance of chlorinated ions. Elucidation of this mass spectrum indicated this compound to be the chloroethyl ester of palmitic acid (saturated, 16 carbons).

With the earlier eluting GC peak tentatively identified as chloroethyl palmitate, the other UAR was suspected of being chloroethyl linoleate. (Palmitic and linoleic acids are common to plants and plant products.) The formation of chloroesters of these acids, however, remained a mystery. A logical reaction for the formation of these compounds is the simple acid-catalyzed transesterification of these acids with a chlorinated alcohol.

A literature search on this topic revealed that 2-chloroethanol (ethylene chlorohydrin) has been found in whole and ground spices following fumigation with ethylene oxide. Ethylene oxide is used to destroy insects and microorganisms in a variety of foods including spices. It is effective and leaves little or no residue. However, the fumigant does combine with natural moisture and inorganic chloride to form the 2-chloroethanol (Wesley et al., 1965). More recent studies confirmed the presence of 2-chloroethanol in foods fumigated with ethylene oxide (Richardson and Smee, 1975; Pfeilsticker and Siddiqui, 1975). This evidence lended support to the tentative identifications of 2-chloroethyl palmitate and 2-chloroethyl linoleate.

Positive identification was acheived with the mass spectral comparison of the sample and the synthesized esters of these two acids. Synthesis consisted of the mineral acid-catalyzed reaction of the fatty acids with an excess of 2-chloroethanol. Purification of the esters was accomplished through the use of a Florisil adsorbant chromatographic column. Chloroethyl esters of several other acids, which are indigenous to plants, were also synthesized.

Since literature sources had indicated the presence of 2-chloroethanol in spices, a limited survey of spices was initiated (Heikes and Griffitt, 1979b). Seven different spices from commercial retail sources were analyzed by procedures similar to those used to examine market basket samples. Samples from each of three producers—a total of 21 samples—were examined for the presence of 2-chloroethyl esters of natural fatty acids. Individual values ranged as high as 1400 ppm of 2-chloroethyl linoleate in a paprika sample. Chloroethyl esters of myristic acid (saturated, 14 carbons), lauric acid (saturated, 12 carbons), and capric acid (saturated, 10 carbons) were also found.

It was postulated that these esters were being formed from the natural fatty acids of the spices. If this were true, the concentration and types of esters found in the spices fumigated with ethylene oxide should be dependent on the concentrations and types of natural fatty acids indigenous to the spices. The correlation of natural fatty acid content of these spices to the levels of 2-chloroethyl esters found previously was examined. Using the same esterification reaction, 2-chloroethanol was added to each spice with a drop of mineral acid. The resultant esters were extracted and determined by GC/MC as before. Comparison of acid and ester levels for each spice sample yielded a convincing correlation, > 0.99, for each of the spice types (see Table 4).

With only one exception, the amount of fatty acid in each spice increased with the length of the carbon chain of the acid. Nutmeg, which finds use as an ingredient of pumpkin pie, contains an abundance of myristic acid. Consequently, 2-chloroethyl myristate is frequently reported in this food item.

Although evidence has been presented to trace the origin of 2-chloroethyl esters in foods to fumigation with ethylene oxide, absolute proof was lacking. With the intent of conclusively establishing this association, two samples of spice that have been found to be free of these esters were fumigated with ethylene oxide in the laboratory. The samples (5 g) were placed in chromatographic tubes and the ethylene oxide (1 L) was allowed to vaporize up through the spice. The fumigated samples were analyzed for the presence of 2-chloroethyl esters at elapsed times of 16 and 64 hr. Allspice, after 64 hr elapsed time, was found to contain 77 ppm 2-chloroethyl linoleate. This same sample contained only 15 ppm of this ester after 16 hr (Table 5). Although it is not clear whether the formation of 2-chloroethanol or the actual transesterification is the rate-determining step, it is evident the formation of these esters is the result of ethylene oxide fumigation.

Paprika oleoresin is frequently used as a "natural" food coloring agent because it imparts an eye-pleasing orange color to food products. In addition to French dressing, a number of snack food items such as cheese swirls, barbecue potato chips, and cheese crackers declare paprika oleoresin as an ingredient. Any brightly colored orange food items (except oranges) might contain paprika oleoresin and, therefore, the associated 2-chloroethyl esters. In reality, any food item with a high spice content is suspect. Additionally, dried or dehydrated products such as peppers or onions are also frequently fumigated with ethylene oxide and have been shown to contain 2-chloroethyl esters of their naturally occurring fatty acids. The occurrence and residue levels of these esters in two recent market basket samples are displayed in Table 6.

The presence of 2-chloroethyl esters has been consistently reported in a variety of foods from the Total Diet Study since they were first identified in 1978. The toxicity of these compounds has not been established.

4.2. Metabolites of Registered Pesticides

The metabolism of pesticides and other agricultural chemicals in the environment is well documented (Menzie, 1969, 1978, 1980; Aizawa, 1982). Metabolism and degradation are often means of detoxifying or limiting the action of agricultural chemicals. However, metabolism may also intensify the toxicity and

TABLE 4 Results of Analysis of 21 Samples of Spices for Fatty Acids and 2-Chloroethyl Esters of Fatty Acids (ppm Acid per ppm Ester)

Acid (Ester)	Producer		
	A	B	C
Allispice			
Palmitic	1000/0.95	2000/0.12	890/0.0
Linoleic	6600/8.5	12000/0.73	3900/0.0
Cinnamon			
Myristic	50/0.0	0.0/0.0	0.0/0.0
Palmitic	240/0.25	310/0.0	170/0.0
Linoleic	540/0.43	680/0.0	550/0.0
Ginger			
Palmitic	490/0.82	280/2.8	160/0.51
Linoleic	1700/3.2	980/13	430/2.3
Nutmeg			
Capric	20/0.0	20/0.0	20/0.0
Lauric	2000/0.56	2600/0.66	1800/0.03
Myristic	30000/6.6	40000/11	33000/0.51
Palmitic	5400/1.5	5400/1.3	4400/0.09
Linoleic	13000/3.5	14000/4.0	14000/0.27
Paprike			
Capric	20/0.0	180/0.0	100/0.04
Lauric	190/0.0	200/0.0	190/1.1
Myristic	500/0.0	420/0.15	190/5.0
Palmitic	3400/0.81	5800/3.2	2600/190
Linoleic	36000/8.6	40000/48	19000/1400
Pepper			
Palmitic	260/0.26	340/0.24	330/0.09
Linoleic	1300/0.98	960/0.74	1100/0.25
Sage			
Palmitic	830/0.17	800/0.40	280/0.0
Linoleic	1600/0.33	1300/0.54	1400/0.0

TABLE 5   Results of Analysis of Two Spices for 2-Chloroethyl Esters after Fumigation with Ethylene Oxide (in ppm)

Ester	After 16 hr	After 64 hr
Cinnamon		
Myristate	0.044	0.12
Palmitate	0.68	3.1
Linoleate	1.1	6.8
Allspice		
Palmitate	3.5	11
Linoleate	15	77

TABLE 6   Results of Analysis of Two Recent Market Basket Samples for 2-Chloroethyl Esters (in ppm)

	Ester				
Food	Caprate	Laurate	Myristate	Palmitate	Linoleate
Pork sausage	0.0	0.0	0.0	0.0090	0.045
Pork sausage	0.0	0.0	0.0080	0.18	1.1
Frankfurters	0.0	0.0	0.0050	0.0	0.026
Bologna	0.0	0.0	0.0	0.0	0.034
Salami	0.0	0.0	0.026	0.051	0.59
Peanuts	0.0	0.0	0.0	0.061	0.27
Rye bread	0.0	0.0	0.0	0.0080	0.11
Coleslaw	0.0	0.0	0.0	0.0050	0.017
Tomato sauce	0.0	0.0	0.0	0.0060	0.018
Tomato sauce	0.0	0.0	0.0	0.015	0.11
Pizza	0.0060	0.0040	0.0060	0.032	0.11
Chili soup	0.0030	0.0030	0.033	0.71	3.5
Chili soup	0.0020	0.0030	0.019	0.38	1.7
Meat loaf	0.0	0.0	0.0	0.0	0.0060
Salad dressing	0.0	0.0	0.0	0.018	0.080
Salad dressing	0.0	0.0	0.0	0.019	0.041
Pumpkin pie	0.0	0.0	0.0090	0.0	0.0

thereby sustain the action of the pesticide. Because it cannot be safely assumed that metabolites are relatively nontoxic or innocuous, identification of these compounds in our food supply is of primary importance. Although not universally true, metabolites are often at relatively low residue levels and accompanied by relatively high residue levels of the parent compound.

4.2.1. Sprout Suppressants

The chemical 2, 3, 5, 6-tetrachloronitrobenzene (TCNB) has found use as a sprout suppressant on potatoes in storage. The recommended postharvest applications rate is approximately 100 ppm actual (Thomson, 1986). It is not suprising, then, that TCNB is a common residue on potatoes. The potato composites of the old diet program consisted of baked, boiled, and fried potatoes, as well as french fried potatoes, potato chips, and yams. Five small UARs seemed to be associated with the residue levels of TCNB found in these composites. These compounds were suspected to being metabolites of TCNB as residue levels of these compounds were found to be statistically correlated with residue levels of TCNB. Tentative identification through the use of GC retention indices accounted for three of these compounds: 2, 3, 5, 6-tetrachloroanisole (tetrachloromethoxybenzene), 2, 3, 5, 6-tetrachloroaniline, and 2, 3, 5, 6-tetrachlorothioanisole. Identification was confirmed with GC/MS. Through the use of mass spectrometry the remaining two UARs were identified as tetrachloronitroanisole (tetrachloro-p-methoxynitrobenzene) and tetrachloroanisidine (tetrachloro-p-methoxy aniline). None of these five compounds had been previously reported as metabolites of TCNB in potatoes.

Although these compounds were assumed to be the para isomers, only by comparison to the positional isomer standards could unambiguous identification be achieved using the techniques available. Consequently, the several positional isomers were synthesized using classical procedures (Peters et al., 1943).

A sample of a commercial preparation of TCNB (Fusarex[R]) was examined by GC/MS. In addition to TCNB, the sample contained trace amounts of 1, 2, 4-trichloro-5-nitrobenzene, 1, 2, 4, 5-tetrachlorobenzene, pentachloronitrobenzene, and pentachlorobenzene. No evidence of the five suspected metabolites was seen.

These metabolites were initially identified and reported in the era of the old Total Diet Program (Heikes et al., 1979). The values presented as Table 7 depict the distribution of TCNB and its metabolites in baked, boiled, and fried potatoes. These values are also typical of those reported from the current market basket samples. Nearly one-third of all the potato items contain at least a trace of TCNB with associated metabolites. Only baked potatoes, which have the highest levels of TCNB, are analyzed unpeeled. Although no study has been conducted it seems

TABLE 7   Results of Analysis of Boiled, Fried, and Baked Potatoes from One Market Basket Sample (in ppb)

Compound	Boiled	Fried	Baked
Tetrachloronitrobenzene	280	210	1800
Tetrachloroanisole	2.0	1.0	3.0
Tetrachloroaniline	49	62	140
Tetrachlorothioanisole	9.0	9.0	36
Tetrachloronitroanisole	0.0	0.0	Trace
Tetrachloroanisidine	5.0	6.0	8.8

reasonable to assume that a higher level of this sprout suppressant would be found in the peel.

A second sprout suppressant frequently applied to potatoes in storage is chlorphopham, isopropyl *N*-(3-chlorophenyl) carbamate. This compound also finds use as a selective herbicide. Nearly all potato products examined in the Total Diet Study contain residues of either TCNB, chlorpropham, or both. Table 8 describes the incident and levels of sprout suppressants and their metabolites encountered in potato items from four recent market basket samples.

A small UAR was found to be present occasionally when residues of chlorpropham were found in potatoes. Although this compound was suspected of

TABLE 8   Incident and Levels of Sprout Suppressants and Their Metabolites Encountered in Potato Items from Four Recent Market Basket Samples (in ppb)

Item/Compound	Market Basket Sample			
	1	2	3	4
Mashed potatoes				
Chloropropham	87	53	55	1.0
Methoxychlorpropham	3.0	0.0	0.0	0.0
French fried potatoes				
Chlorpropham	11	3.0	93	150
Methoxychlorpropham	0.0	0.0	0.0	9.0
Tetrachloronitrobenzene	0.9	0.0	0.0	0.0
Tetrachloroaniline	0.2	0.0	0.0	0.0
Tetrachloroanisidine	0.4	0.0	0.0	0.0
Potato chips				
Chlorpropham	0.0	0.0	410	2100
Methoxychlorpropham	0.0	0.0	0.0	29
Boiled potatoes				
Chlorpropham	15	240	350	100
Tetrachloronitrobenzene	20	2.0	0.0	0.0
Tetrachloroaniline	1.0	8.0	0.0	0.0
Tetrachloroanisole	0.1	0.0	0.0	0.0
Tetrachlorothioanisole	0.5	0.0	0.0	0.0
Tetrachloroanisidine	2.0	0.0	0.0	0.0
Baked potatoes				
Chlorpropham	120	0.0	130	1400
Methoxychlorpropham	0.0	0.0	0.0	2.0
Tetrachloronirobenzene	130	10	0.0	0.0
Tetrachloroaniline	37	4.0	0.0	0.0
Tetrachlorothioanisole	3.0	0.0	0.0	0.0
Tetrachloroanisidine	2.0	0.0	0.0	0.0
Scalloped potatoes				
Chlorpropham	9.0	17	16	150
Tetrachloronitrobenzene	8.0	0.0	0.0	0.0

being a metabolite, the erratic frequency of occurrence did not support a statistical correlation. However, this UAR was found only in potatoes and only when residue levels of chlorpropham were also present.

Because of the elusive nature of this UAR, an experiment was conducted to substantiate its identity as a metabolite of chlorpropham. Red Pontiac potatoes from a home garden were washed at harvest and stored for 2 weeks. Approximating commercial application, 12 tubers of nearly equal weight were dipped in a freshly prepared solution of chlorpropham. The samples were allowed to dry and were stored in the dark at room temperature. Potatoes were analyzed in duplicate at 1-week intervals. Two undipped tubers comprised a control. No detectable amounts of the potential metabolite appeared until trace levels (1.5 ppb) appeared at 21 days. Subsequent analyses showed increasing concentrations of the UAR to a level of 170 ppb for the final determination at the end of 6 weeks. The character of the UAR was thus established to be that of a metabolite of chlorpropham.

GC/MS analysis of the metabolite from this test indicated the compound was the methoxy (aryl) derivative of chlorpropham. Extensive mass spectral characterization is described elsewhere (Heikes, 1985a). The positional isomers of o-methoxychlorpropham and p-methoxychlorpropham were synthesized by adaptation of a procedure by Strain (1956) for preparation of N-(chlorophenyl) carbamates. Although the synthesized standards produced similar mass spectra, only the GC retention data of the para derivative are identical with those of the metabolite found in potatoes.

Potato items have one of the highest number of pesticide residues of the market basket items. In addition to sprout suppressants and their metabolites, low levels of diazinon, dieldren, malathion, and p,p'-DDE are frequently encountered.

4.2.2. Pentachloronitrobenzene

The single food item from the Total Diet Program that consistently contains the highest number of pesticides and industrial chemicals is peanut butter. As with potatoes, a number of these residues are metabolites of applied pesticides.

Pentachloronitrobenzene (PCNB) is used as a soil fungicide and seed disinfectant on peanuts. The recommended application rate is 5 to 200 pounds (actual) per acre or 1/2 to 3/4 pounds (actual) per bushel of seed (Thompson, 1985).

Residues of PCNB have been reported from the analysis of several peanut butter samples. The gas–liquid chromatograms of these samples often contain several other peaks thought to be related to PCNB. The identity of these compounds has been determined by GC/MS (Heikes, 1980). Table 9 lists seven compounds found in peanut butter that appear to be associated with PCNB.

A sample of a commercial preparation of PCNB (Terraclor[R]) was examined by GC/MS. In addition to PCNB, the sample was found to contained low levels of hexachlorobenzene (HCB) (0.46%), pentachlorobenzene (QCB) (0.17%), and

TABLE 9   Compounds Associated with
Pentachloronitrobenzene Residues in Peanut Butter

Pentachlorobenzene (QCB)
Tetrachloronitrobenzene (TCNB)
Hexachlorobenzene (HCB)
Pentachloroanisole (PCAS)
Pentachloroaniline (PCA)
Pentachlorothioanisole (PCTA)
Pentachlorophenol (PCP)

tetrachloronitrobenzene (TCNB) (0.06%). No evidence of the four remaining compounds from Table 9 was observed.

PCNB is fairly rapidly converted to pentachloroaniline (PCA) by various soil microorganisms. Although PCA is less inhibitory to soil fungi, it is much more stable than the parent compound (Ko and Farley, 1969). The metabolism of PCNB to pentachlorothioanisole (PCTA) in soil has been established by Dejonckheere et al. (1975). However, although the metabolism of PCNB to PCA is common to various microorganisms, the metabolism of PCNB to PCTA is specific for only a few filimentous fungi (Nakanishi and Oku, 1969).

Hexachlorobenzene (HCB) and pentachlorobenzene (QCB) have been reported at significant levels in both soil treated with PCNB and in crops grown on these soils. These two persistent compounds are impurities in PCNB formulations and accumulate in treated soils (Smelt and Leistra, 1974; Haefner, 1975). The level of QCB has been found to be higher than that of HCB in peanut butter samples examined in the Total Diet Survey. This was not expected as HCB was found at higher concentrations in the Terraclor[R] sample. It has, however, been shown that PCNB may be reduced to QCB by UV light (Crosby and Hamadmad, 1971). This photochemical reaction may account for the higher than anticipated levels of QCB. The fairly wide use of tetrachloronitrobenzene (TCNB) as a fungicide and growth regulator may account for its significant residue levels in peanut butter samples.

Pentachlorophenol (PCP) and pentachloroanisole (pentachloromethoxybenzene)(PCAS) had not been previously reported as impurities or metabolites of PCNB. However, their consistent presence in these samples led to speculation that these chemicals may be related to PCNB. A statistical correlation using environmental levels of PCNB is probably futile as it is quite rapidly metabolized. However, environmental levels of HCB should reflect initial levels of PCNB as HCB is a stable impurity. This theory was tested by statistical comparison of residue levels of HCB with those of two known metabolites of PCNB—PCA and PCTA. A correlation coefficient of 0.92 was found by comparison of HCB residue levels from an extensive survey of commercially available peanut butter with those of PCA. Similarly, a correlation coefficient of 0.81 was achieved by comparison of HCB and PCTA residue levels. Thus, it was established that the measurements of HCB are reasonably representative of the initial levels of

TABLE 10   Incident and Levels of Pentachloronitrobenzene and Its
Metabolites and Other Residues Encountered in Peanut Butter from Four
Recent Market Basket Samples (in ppb)

Compound	Market Basket Sample			
	1	2	3	4
Pentachloronitrobenzene	4.0	6.0	0.7	4.0
Pentachlorobenzene	4.0	8.0	4.0	9.0
Tetrachloronitrobenzene	0.0	1.0	0.0	0.0
Hexachlorobenzene	4.0	7.0	2.0	5.0
Pentachloroanisole	2.0	8.0	1.0	6.0
Pentachloroaniline	8.0	35	3.0	26
Pentachlorothioanisole	5.0	13	2.0	1.0
Pentachlorophenol	8.0	17	0.0	12
Dieldren	1.0	2.0	3.0	0.0
Malathion	89	52	55	6.0
Toxaphene	54	78	130	130
Heptachlor epoxide	0.0	0.5	0.0	0.0
Dicloran	2.0	0.0	0.0	39
Chlorpyrifos	0.0	2.0	2.0	1.0
Fonofos	0.0	0.0	1.0	0.0
p,p'-DDE	0.0	0.0	3.0	0.0

PCNB. Consequently, the correlation coefficient of 0.95 for the comparison of
residue levels of PCAS with those of HCB strongly suggests that PCAS is a
metabolite of PCNB.

Essentially no correlation was found between the levels of HCB and
pentachlorophenol (PCP). PCP is nearly ubiquitous in the environment, having
several sources. However, the fact that PCP is metabolized to PCAS in soil
(Kuwatsuka and Igarashi, 1975) may suggest that at least a portion of the applied
PCNB is first metabolized to PCP, which, in turn, is metabolized to PCAS.

The results of analysis of peanut butter from four recent market basket
samples is presented as Table 10. Other common residues, as well as PCNB and
its associated compounds, are all reported to illustrate the plethora of residues
found in peanut butter.

4.3.   Volatile Chemicals

Although analysis for volatile chemicals is not a standard part of the Total Diet
Program, the recent focus on ethylene dibromide (EDB) has prompted a study of
these compounds.

4.3.1.   Volatile Halocarbons

EDB has recently been established as a potent carcinogen and its wide use on a
variety of foods has caused natural concern. EDB in its use as grain fumigants has

been shown to carry over to intermediate grain-based products (Heikes, 1985b) and even finished, table-ready, cooked, and uncooked foods (Rains and Holder, 1981; Heikes, 1985c). Use of EDB has been abolished, and although several other fumigants that are carcinogens or potential carcinogens continue to be used, modifications or elimination of their application are being legislated.

A recent survey of whole grains, milled grain products, and intermediate grain-based foods (Heikes and Hopper, 1986) revealed that chloroform and carbon tetrachloride had replaced EDB as grain fumigants to a large degree. These VHCs appear to be carried into intermediate grain-based foods (e.g., flour and cake mixes) although at a much lower level.

The source of VHCs in table-ready foods is probably not exclusively from their use as fumigants. VHCs are also used in large quantities as chemical intermediates and solvents. Regulations allow various VHCs as indirect food additives from sources such as adhesives and components of coatings, plastics (polymers), and paper and paperboard used in food packaging (Byrne, 1987).

Although VHCs had been found in grain-based foods, their occurrence in other food items has not been established. A limited survey was recently conducted to determine both level and frequency of VHCs in a variety of foods from the Total Diet Study (Heikes, 1987). Initially, 19 food items (including drinking water), which are representative of the 234 table-ready food items examined in the Total Diet Study, were analyzed for methylene chloride ($MeCl_2$), chloroform ($CHCl_3$), 1,2-dichloroethane (EDC), methyl chloroform (MC), carbon tetrachloride (CCl_4), trichloroethylene (TCE), EDB, and tetrachloroethylene (PCE). The results are listed as Table 11. Those food types shown to contain higher levels of VHCs were designated for further study. Limited surveys of these foods were conducted on food items from current market basket samples, representing two or three regions of the country. Butter, margarine, ready-to-eat (RTE) cereal products, cheese, peanut butter, highly processed foods, and drinking water were examined. The average values of VHCs in these products are displayed in Table 12. The highest levels of VHCs were found in butter, margarine, and cheese. No evidence of EDB was found in any of the food samples and EDC occurred only in ready-to-eat cereal products.

Drinking water was analyzed as one of the original representative food items. Two GC peaks that matched none of the VHC standards were found in this sample and identified by GC/MS as bromodichloromethane (BDCM) and chlorodibromomethane (CDBM), respectively.

It is well established that chlorination of water containing organic compounds results in the formation of trihalomethanes (THMs) such as chloroform, BDCM, CDBM, and bromoform (Rook, 1974; Symons et al., 1975). In this study, drinking water from nine cities across the country was analyzed for VHCs including THMs. Levels as high as 58 ppb ($CHCl_3$) were found (Table 13). However, none of the samples analyzed exceeded the maximum contaminant level (0.10 mg/L) for total THMs established by the Environmental Protection Agency (EPA). Although process water has been suspected as a possible source of VHCs in

TABLE 11 **Results of Analysis of Representative Table-Ready Food Items for Volatile Halocarbons (in ppb)**

Item	MeCl$_2$	CHCl$_3$	EDC	MC	CCl$_4$	TCD	EDB	PCE
				Analyte				
Chocolate chip cookies	1.6	22	0.0	4.7	1.3	2.9	0.0	3.8
Plain granola	53	57	0.31	8.6	3.4	8.0	0.0	3.3
Cheddar cheese	71	80	0.0	13	1.1	3.1	0.0	3.2
Soft boiled eggs	0.0	0.0	0.0	0.0	0.0	0.0	0.0	0.0
Peanut butter	26	29	0.0	7.0	0.44	1.7	0.0	2.2
Butter	67	670	0.0	45	6.0	12	0.0	12
Evaporated milk	0.0	0.0	0.0	0.0	0.10	1.7	0.0	0.0
Boiled green peas	0.0	0.0	0.0	1.2	0.18	0.0	0.0	0.72
Fried breaded shrimp	5.0	24	0.0	0.0	0.88	0.0	0.0	0.86
Pork sausage	0.0	0.0	0.0	3.5	0.44	5.2	0.0	2.0
Scalloped potatoes	0.0	7.1	0.0	0.0	0.0	0.0	0.0	0.0
Cream style corn	0.0	6.1	0.0	0.0	0.0	0.0	0.0	0.0
Brewed coffee	0.0	0.0	0.0	0.0	0.0	0.0	0.0	0.0
Frozen dinner	8.2	29	0.0	5.2	0.76	1.7	0.0	9.4
Toddler high meat Dinner	0.0	17	0.0	0.0	0.0	0.0	0.0	0.0
Raw cantaloupe	0.0	0.0	0.0	0.0	0.0	0.0	0.0	0.0
Canned spinach	0.0	0.0	0.0	0.0	0.0	0.0	0.0	0.0
Orange juice	0.0	0.0	0.0	0.0	0.0	0.0	0.0	0.0
Drinking water	1.4	8.1	0.0	0.0	0.0	0.0	0.0	0.074

TABLE 12   Average Values of Volative Halocarbons in Selected Categories of Foods (in ppb)

Item	Analyte							
	MeCl$_2$	CHCl$_3$	EDC	MC	CCl$_4$	TCE	EDB	PCE
Butter	84	420	0.0	55	2.1	79	0.0	15
Margarine	27	310	0.0	320	0.81	150	0.0	17
RTE cereals	95	60	1.8	30	0.67	2.9	0.0	5.2
Cheese	45	180	0.0	490	0.65	4.0	0.0	4.2
Peanut butter	19	51	0.0	13	0.087	0.48	0.0	2.0
Processed foods	34	120	0.0	6.0	1.0	1.3	0.0	3.6
Drinking water	0.91	17	0.0	0.024	0.0	0 0.0	0.0	0.024

TABLE 13   Results of Analysis of Drinking Water for Trihalomethanes (THMs) (in ppb)

City	Analyte			
	CHCl$_3$	BDCM	CDBM	CHBr$_3$
1	8.1	0.48	0.021	0.0
2	58	9.8	3.0	0.0
3	3.6	1.5	0.58	0.0
4	7.6	0.76	0.096	0.0
5	17	1.2	0.39	0.0
6	0.81	1.0	0.0	0.0
7	29	1.1	1.2	0.069
8	0.0	0.18	0.95	3.0
9	29	16	9.9	1.1

food items, none of the brominated THMs found in drinking water was present in the food items that were analyzed.

4.3.2. *Residual Chlorinated Solvents*

Although residual chlorinated solvents are also volatile chemicals, their origin in our food supply is probably quite different form those of VHCs. Chlorinated solvents have long been used in the production of decaffeinated instant and ground roasted coffees (Sivetz, 1963; Valle-Riestra, 1974). Trichloroethylene, once the chief chlorinated solvent used in the decaffeination process, has been indicted as toxic and has fallen from use. Currently, methylene chloride is the only chlorinated solvent used in the United States and Canada. However, recent information has indicated that methylene chloride may present a risk of human

TABLE 14   Average Levels of Residual Chlorinated Solvents in Decaffeinated Coffees (in ppb)

Sample	Analyte			
	MeCl$_2$	CHCl$_3$	MC	TCE
Instant coffees				
Brand A	20	5.2	0.0	0.0
Brand B	16	11	0.0	0.0
Brand C	17	24	0.0	0.0
Brand D	32	10	0.0	0.0
Brand E	9.0	0.0	0.0	0.0
Brand F	34	6.5	0.0	0.0
Brand G	44	0.0	0.0	0.0
Ground coffees				
Brand A	26	0.0	0.0	0.0
Brand B	28	0.0	0.0	0.0
Brand C	167	0.0	9.8	14
Brand D	11	0.0	0.0	0.0
Brand H	175	9.2	25	0.0
Brand I	375	19	0.0	0.0

cancer from certain exposures (Cohen et al., 1980). Consequently, the U.S. Food and Drug Administration has established an allowable level of 10 ppm methylene chloride in decaffeinated ground roasted and decaffeinated instant coffees.

A study, undertaken in an effort to determine methylene chloride and other chlorinated solvents in instant decaffeinated and ground, roasted, decaffeinated, coffee, used a modified purge and trap procedure developed by Heikes and Hopper (1986). Coffee samples used for this survey were obtained from local commercial sources. Seven brands of decaffeinated instant coffee and six brands of decaffeinated, ground, roasted coffee were sampled. A total of 25 lots representing nine different brands were examined. Sample weights were taken from freshly opened glass or metal containers and most determinations were performed in duplicate. Table 14 contains the results of these analysis. The highest level of chlorinated residual solvents in instant coffees was 49 ppb methylene chloride. Ground, roasted, coffees contained up to 640 ppb methylene chloride. These values are substantially lower than those reported by Page and Charbonneau (1984). In their study, conducted in 1982, six coffee samples with North American origin labels were examined by a headspace GC technique; four samples were found to contain levels greater than 800 ppb. The highest level reported was 2130 ppb methylene chloride. It is suggested that perhaps recent, more efficient manufacturing techniques for the removal of chlorinated decaffeination solvents may be responsible for the variations between the two studies.

In the second phase of this study, representative instant and ground coffees were brewed according to the recipe suggested by the manufacturers. Although two ground coffees had shown significant levels of methyl chloroform (MC) and

**TABLE 15   Results of Analysis of Ground
Coffees, Brews and Grounds for Residual
Chlorinated Solvents (in ppb)**

Sample	Analyte			
	MeCl$_2$	CHCl$_3$	MC	TCE
Brand C				
Coffee	42	5.7	0.0	42
Brew	0.65	0.0	0.0	0.0
Grounds	7.4	3.4	0.0	13
Brand H				
Coffee	230	28	65	0.0
Brew	2.6	4.3	0.0	0.0
Grounds	8.6	8.6	6.5	0.0
Brand I				
Coffee	630	7.1	0.0	0.0
Brew	4.6	0.0	0.0	0.0
Grounds	13	7.0	0.0	0.0

trichloroethylene (TCE), only traces of these residues were present in the brews. In an effort to account for this loss, the grounds resulting from brewing were analyzed. The results of analysis of the brews and grounds from three ground coffees are listed in Table 15 and indicate that these two solvents tend to remain in the grounds. It is suggested that MC and TCE tend to remain in the coffee grounds because of their relative insolubility in water.

4.4.   Unusual Residues

A number of residues encountered in the Total Diet Program are perplexing as to their origin and/or identity. Two compounds that represent unusual residues are pentachlorophenol, which has been found in Mason jar lids, and cyclooctasulfur, which has been found as a residue in raw cabbage.

4.4.1.   *Pentachlorophenol*

As part of the Total Diet Program, drinking water from various regions of the nation is examined. Several years ago a sample of water was found to have a relatively high level of pentachlorophenol (PCP). The source of this contaminant was traced to the Mason jar lid of the collection jar. Customarily, an aluminum foil liner is used to avoid contamination of the food item by the lid. The liner had not been used with this particular sample. As a result of this error, it was discovered that PCP is present in at least some Mason jar lids and, consequently, an informal survey was conducted (Heikes and Griffitt, 1980).

A rapid method for determination of PCP in the seals and enamels of Mason lids was developed. The samples were digested with acidic methylene chloride, methylated with diazomethane reagent, and determined by GC with electron

TABLE 16   Average Levels PCP in Seals and Enamel of Nine Brands of Mason Lids (in μg/Lid)

Brand	Seal	Enamel	Total
A	130	2.2	132
B	198	0.15	198
C	0.16	0.024	0.18
Six other brands	<0.10	<0.010	<0.11

TABLE 17   Levels of PCP in Six Fruits and Vegetables Canned with Lids containing PCP (107 μg/Lid)

Food Item	ppb	μg/quart
Pears	1.0	0.82
Tomatoes	1.1	0.99
Peaches	1.8	1.5
Plums	4.3	2.7
Green beans	17	14
Carrots	21	16

capture detection. The seals and enamel from nine brands of Mason lids were examined. The results are presented in Table 16. Only two brands of lids contained significant levels of PCP.

Sodium and potassium pentachlorophenates are allowed in closures with sealing gaskets at levels not to exceed 0.05% (500 ppm) (Byrne, 1987). These salts have been used in the production of certain types of sealing material. Fortunately, this type of sealing material is no longer used on Mason lids.

Six fruits and vegetables, which were analyzed and found to be free of PCP, were canned using home canning techniques. The lids used were those of brand A from Table 16. After standing for 4 days the canned foods were examined for the presence of PCP. The results are listed in Table 17.

4.4.2. Cyclooctasulfur

Under the old diet program, what appeared to be a series of unidentified compounds was discovered in the leafy vegetable composite. These UARs appeared as three GC peaks, two of which were similar to α-hexa-chlorocyclohexane (BHC) and aldrin. Analysis of the individual food items isolated the UARs to raw cabbage. The mass spectrum of the latest eluting peak exhibited a probable molecular ion at 256 mass units with a series of losses of 32 amu. Because of the pattern of losses and relatively large $A+2$ ions, the spectrum was readily interpreted as that of molecular sulfur. Using reagent grade precipitated sulfur as a standard, the level of the residue in raw cabbage was found to be 0.71 ppm.

Molecular sulfur has long been recognized as an interfering coextractive in pesticide analysis. Pearson et al. (1967) showed sulfur to exhibit three GC peaks using an electron capture detector (ECD). The peak that elutes second has a retention time similar to α-hexachlorocyclohexane and the latter (and largest) peak mimics aldrin. A total of 10 sulfur peaks on optimized GC with ECD were reported by Chen and Gupta (1973). The identity of the sulfur homologs was determined by plotting the log of retention time versus the number of sulfur atoms in the molecules. The major peaks represent S_4, S_6, and S_8, with minute quantities of S_2, S_5, S_7, S_{11}, S_{14}, and S_{16}. The number and intensity of the peaks changed with a change in GC parameters.

Cyclooctasulfur or common sulfur occurs in three allotropic forms. Orthorohombic (sulfur) is the only therodynamically stable form. The remaining two allotropes are β and γ sulfur, both monoclinic forms that slowly transform to α sulfur (Tuinstra, 1967; Rahman et al., 1970).

A mass spectrum of all three peaks of the compounds found in raw cabbage showed each to have a spectrum essentially identical to each other and identical to cyclooctasulfur. This phenomenon suggests the three peaks may represent the allotropic forms of cyclooctasulfur.

5.   CONCLUSION

Improved instrumentation resulting in lower levels of detection and more specific determinations has resulted in an increase in the incidence and types of residues reported. This trend has given the appearence that our food supply is losing quality due to environmental contamination with pesticides and industrial chemicals. This may not be the case. On the contrary, there is substantial evidence that the foods of this nation are sound. Even the highest levels of residues found in our foods generally fall far short of acceptable daily intakes (ADI) established by the Food and Agriculture Organization of the United Nations and the World Health Organization (FAO/WHO). These ADIs are based on the toxicity of the contaminant and are generally at least two orders of magnitude higher than the levels found in this nation's food supply. The ADI for each chemical is the maximum daily intake which, during a lifetime, appears to be without appreciable risk. The Total Diet Study, from its inception to the present day, has pronounced the recurring conclusion that the food supply of this nation is relatively free from serious environmental contamination.

REFERENCES

Aizawa, H. (1982). *Metabolic Maps of Pesticides*. Academic Press, New York, 243 pp.

Byrne, J. E. (1987). *Code of Federal Regulations*. National Archives and Records Administration, Washington, D.C., Title 21, Parts 175–177.

Carson, L. J. (1981). "Modified Storherr method for determination of organophosphorus pesticides in nonfatty food total diet composites." *J. Assoc. Off. Anal. Chem.* **64**, 714–719.

Chen, K. Y., and Gupta, S. K. (1973). "Formation of polysulfides in aqueous solutions." *Environ. Lett.* **4**, 187–200.

Cohen, J. M., Dawson, R., and Koketsu, M. (1980). "Extent-of-exposure survey of methylene chloride." DHHS (NIOSH) Publ. (U.S.), pp. 80–131.

Crosby, D. G., and Hamadmad N. (1971). "The photoreduction of pentachlorobenzene." *J. Agr. Food Chem.* **19**, 1171–1174.

Dejonckheere, W., Willcox, J., Strirbaut, W., Kips, R. H., Voets, J. P., and Virstraite, W. (1975). "Changes in the rate of metabolism of quintozene in the soil under varying microbial conditions." *Meded. Fac. Landbouwet.* **40**, 1187–1197.

Gartrell, M. J., Craun, J. C., Podrebarac, D. S., and Gunderson, E. L. (1986). " Pesticides, selected elements, and other chemicals in infant and toddler total diet samples, October 1980–March 1982," and "Pesticides, selected elements, and other chemicals in adult total diet samples, October 1980–March 1982." *J. Assoc. Off. Anal. Chem.* **69**, 123–161.

Haefner, M. (1975). "Contamination of garden soil and agriculturally useful soil with hexachlorobenzene and pentachloronitrobenzene." *Gesunde Pflanz* **27**, 88–97.

Heikes, D. L. (1980). "Residues of pentachloronitrobenzene and related compounds in peanut butter." *Bull. Environ. Contam. Toxicol.* **24**, 338–343.

Heikes, D. L. (1985a). "Mass spectral identification of a metabolite of chlorpropham in potatoes." *J. Agr. Food Chem.* **33**, 246–249.

Heikes, D. L. (1985b). "Purge and trap method for determination of ethylene dibromide in whole grains, milled grain products, intermediate grain-based foods, and animal feeds." *J. Assoc. Off. Anal. Chem.* **68**, 1108–1111.

Heikes, D. L. (1985c). "Purge and trap method for determination of ethylene dibromide in table-ready foods." *J. Assoc. Off. Anal. Chem.* **68**, 431–436.

Heikes, D. L. (1986). "Mass spectral identification of residues in the Total Diet Program." *Appl. Spectr. Rev.* **22**, 111–136.

Heikes, D. L. (1987). "Purge and trap method for determination of volatile halocarbons and carbon disulfide in table-ready foods." *J. Assoc. Off. Anal. Chem.* **70**, 215–226.

Heikes, D. L., and Griffitt, K. R. (1979a). "Identification of 2-chloroethyl palmitate and 2-chloroethyl linoleate in French dressing." *Bull. Environ. Contam. Toxicol.* **21**, 98–101.

Heikes, D. L., and Griffitt, K. R. (1979b). "Mass spectrometric identification and gas-liquid chromatographic determination of 2-chloroethyl esters of fatty acids in spices and foods." *J. Assoc. Off. Anal. Chem.* **62**, 786–791.

Heikes, D. L., and Griffitt, K. R. (1980). "Gas-liquid chromatographic determination of pentachlorophenol in Mason jar lids and home canned foods." *J. Assoc. Off. Anal. Chem.* **63**, 1125–1127.

Heikes, D. L., and Hopper, M. L. (1986). "Purge and trap method for determination of fumigants in whole grains, milled grain products, and intermediate grain-based foods." *J. Assoc. Off. Anal. Chem.* **69**, 990–998.

Heikes, D. L., Griffitt, K. R. and Craun, J. C. (1979). "Residues of tetrachloronitrobenzene and related compounds in potatoes." *Bull. Environ. Contam. Toxicol.* **21**, 536–568.

Hopper, M. L. (1982). "Automated gel permeation system for rapid separation of industrial chemicals and organophosphate and chlorinated pesticides from fats." *J. Agr. Food Chem.* **30**, 1038–1041.

Hopper, M. L. (1987). "Methylation of chlorophenoxy acid herbicides and pentachlorophenol residues in foods using ion-pair alkylation." *J. Agr. Food Chem.* **35**, 267–269.

Johnson, R. D., Manske, D. D., and Podrebarac, D. S. (1981). "Pesticide, metal, and other chemical residues in adult total diet samples—(XII)—August 1975–July 1976." *Pestic. Monit. J.* **15**, 54–69.

Ko, W., and Farley, J. D. (1969). "Conversion of pentachloronitrobenzene to pentachloroaniline in soil and the effect of these compounds on soil microorganisms." *Phytopathology* **59**, 64–67.

Krause, R. T. (1980). "Multiresidue method for determining *N*-methylcarbamate insecticides in crops, using high performance liquid chromatography." *J. Assoc. Off. Chem.* **63**, 1114–1124.

Kuwatsuka, S., and Igarashi,M. (1975). "Degradation of PCP in soils." *Soil Sci. Plant Nutr.* **21**, 405–414.

Luke, M. A., Froberg, J. E., and Masumoto, H. T. (1975). "Extraction and cleanup of organochlorine, organophosphate, organonitrogen, and hydrocarbon pesticides in produce for determination by gas-liquid chromatography." *J. Assoc. Off. Chem.* **58**, 1020–1026.

Luke, M. A., Froberg. J. E., Doose, G. M., and Masumoto, H. T. (1981). "Improved multiresidue gas chromatographic determination of organophosphorus, organonitrogen, and organohalogen pesticides in produce, using flame photometric and electrolytic conductivity detectors." *J. Assoc. Off. Chem.* **64**, 1187–1195.

Menzie, C. M. (1969). *Metabolism of Pesticides.* Bureau of Sport Fisheries and Wildlife, Washington D.C., 187 pp.

Menzie, C. M. (1978). *Metabolism of Pesticides Update II.* Bureau of Sport Fisheries and Wildlife, Washington D.C., 381 pp.

Menzie, C. M. (1980). *Metabolism of Pesticides Update III.* Bureau of Sport Fisheries and Wildlife, Washington D.C., 709 pp.

Nakanishi, T., and Oku, H. (1969). "Mechanism of selective toxicity of fungicides. Metabolism of pentachloronitrobenzene by phytopathogenic fungi." *Nippon Skokubutsu Byor Gakkaiho* **35**, 339–346.

Page, B. D., and Charbonneau, C. F. (1984). "Headspace gas chromatographic determination of methylene chloride in decaffeinated tea and coffee, with electrolytic conductivity detection." *J. Assoc. Off. Anal. Chem.* **67**, 757–761.

Pearson, J. R., Aldrich, F. D., and Stone, A. W. (1967). "Identification of the aldrin artifact." *J. Agr. Food Chem.* **15**, 938–939.

Pennington, J. A.T. (1983). "Revision of the Total Diet Study food list and diets." *J. Am. Diet. Assoc.* **82**, 166–173.

Pennington, J. A. T., and Gunderson, E. L. (1987). "History of the Food and Drug Administration's Total Diet Study—1961 to 1987." *J. Assoc. Off. Chem.* **70**, 772–782.

Peters, A. T., Rowe, F. M., and Stead D. M. (1943). "Derivatives of 1:2:4:5-tetrachlorobenzenes. Part I. Nitro- and amino-compounds." *J. Chem. Soc.* **23**, 233–235.

Pfeilsticker K., and Siddiqui, I. R. (1975). "Retention of ethylene oxide in the fumigation of food." *Z. Lebensm.-Unters. Forsch.* **158**, 157–167.

Rahman, R., Safe, S., and Taylor, A. (1970). "Stereochemistry of polysulfides." *Q. Rev.* **24**, 208–237.

Rains, D. M., and Holder, J. W. (1981). "Ethylene dibromide residues in biscuits and commercial flour." *J. Assoc. Off. Anal. Chem.* **64**, 1252–1254.

Richardson, H., and Smee, L. (1975). "Freight container fumigation and wood penetration by ethylene oxide mixture against quarantinable termites and other insects." *Pest Control* **43**, 22–31.

Rook, J. J. (1974). "Formation and occurrence of haloforms in drinking water." *J. Water Treatment Exam.* **23**, 234–243.

Sivetz, M. (1963). *Coffee Processing Technology*, Vol. II. AVI Publishing Westport, pp. 207–215.

Smelt, J. S., and Leistra, M. (1974). "Hexachlorobenzene in soils and crops after soil treatment with pentachloronitrobenzene." *Agr. Environ.* **1**, 65–71.

Storherr, R. W., Ott, P., and Watts R. (1971). "A general method for organophosphorus pesticide residues in nonfatty foods." *J. Assoc. Off. Chem.* **54**, 513–516.

Strain, F. (1956). U. S. Patent 2 734 911, Feb. 14, 1956 assigned to Columbia-Southern Chemical Corp., Allegheny County, PA.

Symons, J. M., Bellar, T. A., Carswell, J. K., DeMarco, J. Kropp, K. L., Robeck, G. G., Seeger, D. R., Slocum, C. J., Smith, B. L., and Stevens, A. A. (1975). "National organics reconnaissance survey for halogenated organics." *J. Am. Water Works Assoc.* **67**, 634–647.

Thomson, W. T. (1985). *Agricultural Chemicals Book IV — Fungicides.* Thomson, Fresno, CA, pp. 61–62.

Thomson, W. T. (1986). *Agricultural Chemicals Book III — Fumigants, Growth Regulators, Repellents, and Rodenticides.* Thomson, Fresno, CA, pp. 60–62.

Tuinstra, F. (1967). *Structural Aspects of the Allotropy of Sulfur and the Other Divalent Elements.* Waltman, Delft, Netherlands, p. 112.

Valle-Riestra, J. F. (1974). "Food processing with chlorinated solvents." *Food Technol.* **28** (2), 25–32.

Wesley, F., Rourke, B., and Darbishire, O. (1965). "The formation of persistent toxic chlorohydrins in foodstuffs by fumigation with ethylene oxide and propylene oxide." *J. Food Sci.* **30**, 1037–1042.

Williams, S. (1984). *Official Methods of Analysis.* Association of Official Analytical Chemists, Arlington, VA, pp. 533–562.

3

FOOD CONTAMINATION WITH CADMIUM IN THE ENVIRONMENT

Jerome O. Nriagu

National Water Research Institute, Burlington, Ontario, Canada

1. INTRODUCTION

Cadmium is a very rare element, its average concentration of $0.1-0.2\ \mu g/g$ making it the sixty-seventh element in the earth's crust in terms of abundance (Mason and Moore, 1982). The baseline concentrations of cadmium in most environmental media are also notoriously low, some of the most common values being $2-20\ ng/L$ dissolved in lakes (Lum, 1987), $5-15\ ng/L$ dissolved in rivers (Shiller and Boyle, 1987), $< 5\ ng/m^3$ in the atmosphere (Nriagu, 1980), and $0.35-0.62\ \mu g/g$ in soils

(Bowen, 1979; Ure and Berrow, 1982). Therefore, the background reservoirs of cadmium in the different segments of any given ecosystem should be small, suggesting that the biogeochemical cycling of cadmium can be altered significantly by small contributions from anthropogenic sources.

The recent epidemic of cadmium poisoning (Itai-Itai disease) was a clear manifestation of the effects of wanton discharge of cadmium into the local environment and its subsequent transfer to the local food chain (for details of this incidence, see Kobayashi, 1978; Nogawa, 1981; Tsuchiya, 1984). There is now growing evidence to suggest that the levels of cadium in air, water, and soils in many parts of the world have increased several-fold as a result of emissions from industrial activities and that the natural biogeochemical cycle of cadmium has been overwhelmed. This chapter explores the possibility that the cadmium pollution is being transferred to the human food chain, resulting in elevated cadmium levels in our diets. Long-term exposure to the elevated levels of cadium in the environment has apparently increased the accumulation of cadmium in certain body organs (notably the kidney and liver). Today, there is some concern that the safety margin for cadium intake is small or nonexistent for special population groups (such as smokers and occupationally exposed persons; see Friberg et al., 1985) and that further increase in the environmental exposure level poses a real threat to the health of such a population.

2. SOURCE OF POLLUTANT CADMIUM IN FOODS

Environmental contamination with cadmium began with the systematic exploitation of the zinc and base metal ores in prehistoric times. The all-time world production of cadmium has been estimated to be about 320,000 tonnes (Nriagu, 1979). In recent years (since about 1970), worldwide consumption of cadmium has remained fairly constant at 16,000–19,000 tonnes/year (Nriagu, 1980, 1988). Most of the cadmium produced is used in electroplating or rustproofing (about 35%), in pigments, mainly as the orange coloring agent in enamels and paints (about 25%), as stabilizers in plastic (about 15%), and in batteries and cells (about 15%). Less than 10% of the cadmium used commercially is recycled, implying that of most of the cadmium mined is dissipated in the environment (Nriagu, 1980).

Primary cadmium is recovered exclusively as a by-product from the zinc (primarily) and other base metal ores. Before the World War II, when there was little demand, most of the cadmium associated with the zinc ores was allowed to go to waste, a practice that no doubt resulted in extensive cadmium contamination of the area surrounding a zinc smelter (Franzin, 1984). As an integral part of the zinc industry, cadmium recovery now accounts for 4–10% of the income from zinc sales (Nriagu, 1980). Nevertheless, zinc (and to a much lesser extent, other base metal) smelters still constitute an important source of cadmium released into the environment (see following).

TABLE 1   Global Emissions of Cadmium to the Atmosphere in 1983[a]

Source Category	Cadmium Emission (tonnes/yr[a])[b]
Coal combustion	
Electric utility	232 (77–387)
Industry and domestic	297 (99–495)
Oil combustion	143 (41–246)
Lead production	117 (39–195)
Copper/nickel production	2550 (1700–3400)
Zinc/cadmium production	2760 (920–4600)
Secondary nonferrous metals	3.0 (2.3–3.6)
Iron and steel industry	156 (28–284)
Refuse incineration	
Municipal	728 (56–1400)
Sewage sludge	20 (3.0–36)
Phosphate fertilizer industry	171 (68–274)
Cement production	272 (8.9–534)
Wood combustion	120 (60–180)
Total emissions	7570 (3100–12,040)

[a] Nriagu and Pacyna (1988).
[b] Median values, with ranges in parentheses.

2.1. Atmospheric Emissions

Worldwide emissions of cadmium to the atmosphere in 1983 have been estimated to total about 7570 tonnes (Table 1). About 70% of all the cadmium emitted can be attributed to the smelting of base metal ores. Large quantities of cadmium are also released by refuse incineration (about 750 tonnes/year), coal combustion (about 530 tonnes), cement production (about 270 tonnes), and fertilizer production (about 170 tonnes/year). On the other hand, worldwide emission of cadmium from *natural sources* has been estimated to be 960 tonnes/year (Nriagu,1979; Pacyna, 1986). Thus, anthropogenic sources now account for nearly 90% of the total annual flux of cadmium to the atmosphere.

Studies in many parts of the world have clearly documented the existence of haloes of elevated levels of cadmium in ecosystems near many point sources, especially around base metal smelting and sintering plants (see Table 2). Within a 1.0 km radius of a cadmium-emitting smelter, the daily cadmium fallout rate can exceed 1.0 mg/m² yr; the deposition rate rapidly decreases with distance away from the plant. Table 2 compares the cadmium fallout rates near point sources with the rates at urban, rural, and remote locations. The heavy contamination of ecosystems in urban areas and near point sources with cadmium is obvious. On the average, the ratio of cadmium deposition in urban areas exceed those of rural and remote locations by about 6-fold and over 700-fold, respectively. The ratio of rural to remote concentrations is about 116.

TABLE 2 Atmospheric (Bulk) Deposition of Cadmium

Location	Fallout Rate (mg/m² yr)	Reference
Point Sources		
Amax smelter, Missouri		
250 m	>200	Gale and Wixson (1979)
600 m	12	Gale and Wixson (1979)
13 km	1.6	Gale and Wixson (1979)
Flin Flon, smelter, Manitoba		Franzin et al. (1979)
4 km	4.5	
9 km	2.1	
29 km	0.6	
Amaka smelter, Japan		Asami (1984)
500 m	6.6	
1000 m	4.0	
2000 m	2.2	
Wollongong, NSW, Australia		Beavington (1977)
100 m	19	
600 m	7.0	
1.9 km	3.0	
Urban/Industrial Areas		
New York City	2.7	Kleinman et al. (1977)
Tulsa, Oklahoma	0.95	Tate and Bates (1984)
Gary, Indiana (industrial	1.5	Peyton et al. (1976)
London, U.K.	2.9	Harrison et al. (1975)
Ardenne, Belgium (industrial	1.4	Hallet et al. (1982)
Several cities, Holland	0.7	Vermeulen (1977)
Hamilton, Ontario	3.0	Jeffries and Snyder (1981)
Manitoba–Sask patchewan (industrial)	1.1	Franzin et al. (1979)
Rural and Semirural		
Southern Ontario	0.14	Chan et al. (1986)
Central Ontario	0.13	Chan et al. (1986)
Northern Ontario	0.11	Chan et al. (1986)
Northeast Minnesota	0.4	Ritchie and Thingvold (1985)
Rural Indiana	0.3	Peyton et al. (1976)
Chedron, Nebraska	0.12	Struempler (1976)
Skinface Pond, South Carolina	0.68	Wiener (1979)
Northeast United States	0.19	Groets (1976)
New York Bight	0.3	Duce et al. (1976)
Greifensee, Switzerland	0.6	Imboden et al. (1975)
Rural sites, Holland	0.4	Vermeulen (1977)
North Sea	0.43	Buat-Menard (1986)
Western Mediterranean	0.13	Buat-Menard (1986)
South Atlantic Bight	0.09	Buat-Menard (1986)

TABLE 2 (*Continued*)

Location	Fallout Rate (mg/m² yr)	Reference
	Remote Locations (μg/m²/yr)	
Tropical North Pacific	3.5	Buat-Menard (1986)
Dome C, Antarctica	0.23	Boutron (1979)
Centrale, Greenland	3.7	Boutron (1979)

The ratios point to the excessive contamination of urban areas with cadmium and one may, in fact, regard some urban areas as *hot spots* of cadmium.

The actual concentration of cadmium in the atmosphere is often below $10\,ng/m^3$ even in urban areas (see Nriagu, 1980). Nevertheless, the atmosphere serves as a key medium in the transmission of cadmium from anthropogenic sources to human food chains at considerable distances away. For example, the ambient levels of cadmium in the Arctic haze can reach levels normally found in rural and urban areas of North American (Nriagu and Davidson, 1986). Most of the cadmium in the haze is associated with the long-range transported pollutants from the Eurasian sources (Barrie, 1986).

2.1.1.   Water Pollution with Cadmium

The principal industrial operations that release cadmium into natural waters are listed in Table 3. The estimated total input of 9400 tonnes/year is somewhat higher than the atmospheric flux. In fact, the atmospheric deposition accounts for about 23% of the pollutant cadmium released into the water. On a global basis, the major sources of pollution are the effluents from industries that use cadmium, such as textile mills, pulp and paper mills, porcelain enameling plants,

TABLE 3   Anthropogenic Releases of Cadmium into Aquatic Ecosystems[a]

Source Category	Cadmium input (× 1000 tonnes/y[a])[b]
Domestic wastewater	1.8 (0.48–3.0)
Steam electric	0.12 (0.01–0.24)
Base metal mining and dressing	0.15 (0–0.3)
Smelting and refining, nonferrous metals	1.8 (0.01–3.6)
Manufacturing processes	
Metals	1.2 (0.5–1.8)
Chemicals	1.3 (0.1–2.5)
Atmospheric fallout	2.2 (0.9–3.6)
Sewage sludge disposal	0.7 (0.08–1.3)
Total input, water	9.4 (2.1–17)

[a] Nriagu and Pacyna (1988).
[b] Median value, with ranges in parentheses.

TABLE 4   Authropogenic Inputs of Cadmium into the Great Lakes[a]

Source	L. Ontario	L. Erie	L. Huron	L. Michigan	L. Superior
Atmospheric deposition	28	75	60	58	82
Discharge of industrial effluents	3.0	3.8	0.3	19	0.6
Input of municipal effluents	12	19	1.5	8.4	0.6
Miscellaneous discharges	4.3	9.8	0.6	8.1	0.8
Total, input	47	108	62	94	84

[a]Nriagu (1986).

rubber processing, and paint and ink plants (Table 3). On a local scale, however, the most concentrated sources of cadmium pollution are likely to include domestic wastewaters and urban runoff. The importance of domestic chores in generating cadmium-contaminated waters has been well illustrated in a mass balance study in Japan that found that about $128\,\mu g$ Cd/day person was discharged. Of this amount, about 18, 42, 18, 46, and $4\,\mu g$/day person were derived from the kitchen, laundry, bath, lavatory, and miscellaneous household wastewaters, respectively (Chino and Mori, 1982). Drainage from fertilized agricultural soils and from waste disposal sites can also contain elevated levels of cadmium.

The relative strengths of the sources of cadmium pollution often depend on the proximity of a surface water to an urban center or a point source. For example, about 98% of the cadmium flux into Lake Superior comes via the atmosphere compared to about 62% for Lake Michigan, which receives massive quantities of effluents from the huge industrial and urban centers in its southern borders. The atmosphere however remains the primary route of cadmium input into all of the Great Lakes (Table 4).

2.1.2.   Contamination of Soils with Cadmium

Soils generally function as a major sink for most of the cadmium released into the terrestrial environment (Adriano, 1986). The principal sources of pollutant cadmium in soils include atmospheric fallout (about 23% of total), urban refuse disposal, dumping of fly ash, use of sewage sludge as source of nutrients or organic ammendments for improving soil physical properties, disposal of agricultural and animal wastes, and the application of fertilizers (Table 5). The relative contributions from these sources will vary in different countries. Of the 42 tonnes or so of cadmium discharged annually into arable lands in Britain, about 42% is derived from phosphate fertilizers, about 36% is deposited from the atmosphere, and only about 12% comes from sewage sludge (Hutton and Symon, 1986). It should be pointed out that Britain traditionally discharges a large portion of her waste effluents into the ocean.

If it is assumed that the 22,000 tonnes of cadmium is spread uniformly over the earth's land surface area of $140 \times 10^{12}\,m^2$ (Atjay et al., 1979), the average rate of pollutant cadmium input into soils is estimated to be $147\,\mu g/m^2$ yr. By

TABLE 5 Worldwide Inputs of Cadmium Pollution into Soils (× 1000 tonnes/yr)[a]

Source Category	Cadmium Emission[b]
Agricultural and food wastes	1.5 (0–3.0)
Animal wastes, manure	0.7 (0.2–1.2)
Logging and other wood wastes	1.1 (0–2.2)
Urban refuse	4.2 (0.88–7.5)
Municipal sewage sludge	0.18 (0.02–0.34)
Coal fly ash and bottom ash	7.2 (1.5–13)
Wastage of commercial products	1.2 (0.78–1.6)
Atmospheric fallout	5.3 (2.2–8.4)
Fertilizer production	0.14 (0.03–0.25)
Total input, soils	22 (5.6–38)

[a] Nriagu and Pacyna (1988).
[b] Renges in parentheses.

comparison, the input of cadmium into agricultural lands in Europe varies from about 16,000 (Italy) to 94,000 (Belgium) and average about 80,000 $\mu g/m^2$ yr (Hutton, 1983). The average cadmium concentration in soils is 0.3–0.6 $\mu g/g$ (Bowen, 1979; Ure and Berrow, 1982). If it is assumed that the average density of the soil is 2.5 g/cm^3, the cadmium content of the soils may be expected to double every 50–80 years at the current rate of anthropogenic input. As previously noted, the contamination of soils in Europe (and other developed countries) with cadmium is very intensive and the current rate of pollutant cadmium addition into soils in Europe of 16–94 mg/m^2 yr easily dwarfs the background concentration in most soils. Analysis of archived soil samples from the Rothamsted Experimental Station in Britain indeed revealed an increase in cadmium concentration of 19,000–54,000 $\mu g/m^2$ yr since the 1850s (Jones et al., 1987).

Contamination of soils has justifiably been studied extensively (see Purves, 1985; Adriano, 1986; Page et al., 1987). In the top layers of many urban soils, cadmium concentrations below 1.0 $\mu g/g$ have become rare (see Purves, 1985). A recent survey shows that about 9.5% of the paddy soils, 3.3% of upland soils, and 7.5% of orchard soils in Japan have become severely contaminated with cadmium and are unusable for rice cultivation (Asami, 1984). Cadmium pollution in Japan comes primarily from the huge base metal industry that has been in operation for centuries (Yagamata, 1979; Tsuchiya, 1984). Arable soils in many parts of central Europe have also become contaminated with cadmium derived from the application of sewage sludge and fertilizers (Kloke et al., 1984). In addition to arable fields, soils in urban areas, near point sources (such as smelters, incinerators, and power plants) and near major roadways, generally receive large inputs of pollutant cadmium. On a global basis, one can surmise that the soil acreage already contaminated with cadmium is considerable. The technology for decontaminating such soil has yet to be developed.

3. THE TRANSFER OF CADMIUM POLLUTION TO HUMAN FOODS

Of all the toxic metals released in large quantities to the environment, cadmium is generally regarded as the one most likely to accumulate in the human food chain. The pollutant cadmium is selectively concentrated by certain food crops, notably the root crops, leafy vegetables, and tobacco plants. Increased concentrations of cadmium in other foodstuff including seafood, which can be attributed to environmental pollution with cadmium have been reported in many parts of the world (Friberg et al., 1985). The most notorious example has to be the contamination of rice grown on the 41 cadmium-polluted areas covering over 3000 ha of Japan (Asami, 1984).

3.1. Air → Food Transfer

The most obvious route of transfer of atmospheric cadmium to the human food chain is direct deposition on fruits and vegetables. This transfer route can be particularly important near point sources. The foliar retention of cadmium depends on the particulate deposition velocity and the interception velocity; these parameters are determined by particle size, climatic factors, and leaf characteristics (Koranda and Robinson, 1978). Washing often reduces the cadmium content of some vegetables suggesting the cadmium has been deposited on the surface of the leaves.

The multimedia exposure to the airborne cadmium needs to be emphasized (Figure 1). Highly elevated concentrations of cadmium have been observed in household, road, and playground dusts (Table 6). These cadmium-laden dusts can be ingested through hand-to-mouth transfer or transferred to the food and water by any number of processes. Urban storm waters contain up to $20\,\mu g/L$

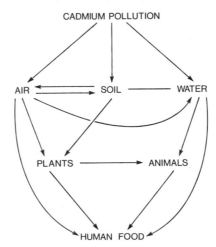

Figure 1. Routes of pollutant cadmium flow into human foods.

TABLE 6   Cadmium Contents of Household Dusts[a]

Site	Number of Samples	Cadmium (ppm)	Cadmium ($\mu g/m^2$)
Residential sites			
A. Rugs	7	24	43
Nonrugs	13	29	17
B. Rugs	5	17	20
Nonrugs	10	25	6
F. Rugs	6	66	71
Nonrugs	1	105	22
G. Rugs	11	14	7
Nonrugs	3	71	8
Average, 12 sites		18	25
Nonresidential sites			
Offices, hallways	36	9	8
Offices, nonrugs	11	13	7
Classrooms, nonrugs	8	1	0
Offices, rugs	6	1033	2443
Offices, nonrugs	1	1060	200
Public school, rugs/mats	10	29	122
Public school nonrugs	17	7	4
Public school, rugs/mats	12	30	115
Public school, nonrugs	26	19	5
Hospitals, entry areas, rugs	11	20	390
Hospitals, corridors, rooms	11	22	4
Supermarkets	34	9	2
Chemical laboratories	35	185	57
All floors	254	44	110

[a]Solomon and Hartford (1976).

cadmium (Harrison and Wilson, 1985) and therefore represent one of the major sources of cadmium pollution in surface waters. Because inhalation of airborne cadmium is regarded as a minor exposure route, the ingestion of cadmium deposited on leaves may be responsible for the elevated cadmium levels observed in the visceral organs of the deer and moose in many parts of the world susceptible to acid rains (see following). Thus, the atmosphere represents a critical pathway in the contamination of foods with cadmium.

3.2.   Water → Food Transfer

The cadmium concentration in drinking water is generally below 2.0 $\mu g/L$ (Sharrett et al., 1982; Ryan et al., 1982), implying that this is not an important source of cadmium exposure for most people (Friberg et al., 1985). The recent claim by Mance and O'Donnell (1983) that about 15% of the surface and brackish waters of the European Community Countries exceed 5.0 $\mu g/L$ may

TABLE 7   Bioaccumulation Factors for the Food
Chain in Contaminated Waters[a]

Organism	Bioaccumulation Factor[b]
Aquatic macrophytes	
Leaves	12,000–15,000 (14,000)
Roots	10,000—28,000 (19,000)
Bacteria	500–7,000 (4,000)
Phytoplankton	1,500–20,000 (11,000)
Zooplankton	600–12,000 (6,000)
Aquatic insects	800–9,000 (5,000)
Mollusks	500–9,000 (5,000)
Crustaceans	300–7,000 (4,000)
Fish	
Whole body	200–700 (500)
Gill	500–3,000 (2,000)
Liver	1,000–5,000 (3,000)
Kidney	1,000–8,000 (4,500)

[a]Compiled from the literature.
[b]Estimated average in parentheses.

be unfounded considering that most of the cadmium in the drinking water is acquired from the plumbing system (Sharrett et al., 1982).

Pollutant cadmium in the surface water can be transferred to the aquatic food resources and subsequently to the human diet. Aquatic fauna and flora, with very few exceptions, accumulate cadmium to levels well above the concentrations in the water. Table 7 summarizes the accumulation factors (defined as the ratio of concentration in organism or its parts to the concentration in water) for cadmium in the aquatic food chain. In general, the first trophic level (acquatic plants) possesses the highest accumulation potential for cadmium. For example, water hyacinth (*Eichonia crassipes*) showed a linear uptake that was directly related to the cadmium concentraions in water (McCracken, 1987) whereas cadmium concentrations in seaweeds in excess of $10 \mu g/g$ are common (Phillips, 1980; Bernhard and Andreae, 1984). Seaweeds are consumed in some parts of Europe (the so-called layered bread from Irish Sea or agar consists of seaweed) and especially in the Far East (Berhard and Andreae, 1984). Such eating habits can increase the dietary cadmium intake.

Mollusks and crustaceans also have a propinquity for bioaccumulating cadmium. In fact, these two classes of organisms are now widely used as the test organisms for biomonitoring the levels of cadmium and other trace metal pollution in aquatic ecosystems (MARC, 1986). When exposed to 3.0, 8.3, and $28 \mu g/L$ Cd for 28 days, the snail (*Physa integra*) accumulated 30, 45, and $225 \mu g/g$ Cd, respectively (Spehar et al., 1978). Cadmium concentrations of up to $60 \mu g/g$ have been reported in *Mytilus edulis* from the British Channel, whereas concentrations over $120 \mu g/g$ have been found in a Pacific oyster from the contaminated Tama River, Tasmania (Engel and Fowler, 1979). The concentration of cadmium in shrimps from the Severn Estuary, Britain, varies from 5 to 124 μg/g (Hardisty et al., 1974). Cadmium concentrations of up to 30 μg/g have also been reported in crab meat and some mushrooms (see Friberg et al., 1985).

The published literature generally shows that, in contaminated waters, the cadmium concentrations in the whole body of crustaceans and mollusks fall in the range of 2–10 μg/g, with the levels in the kidney, gills, and liver being much higher (Phillips, 1980; McCracken, 1987). In view of this, consumption of large quantities of shellfish can result in undue exposure to cadmium. The cadmium concentration in edible fish parts is often less than 0.2 μg/g although levels above 1.0 μg/g have been reported in fish samples from contaminated waters (McCracken, 1987). The pattern of cadmium accumulation in the different trophic levels (Table 7) thus suggests that there is a biopurification of this element in the higher trophic levels of the aquatic food chain. The levels of cadmium in the visceral organs however can exceed 10 μg/g even in fish samples from uncontaminated waters.

Marine foods for human consumption are harvested from many trophic levels. It has been estimated that aquatic plants and seaweeds make up about 4% of the world's harvest of aquatic food resources (Bernhard and Andreae, 1984). Of the 73 million tonnes of aquatic animals harvested annually, about 10% are caught in freshwater and the 90% from the ocean consists of 75% fish, 7% mollusks, and 4% crustaceans. Pelagic plankton feeders such as sardines, herrings, and anchovy make up about 28% of the fish landings followed by benthopelagic fishes (haddock, cod, and hake) that feed on fish larvae, invertebrates, clupoids, and anchovy (Bernhard and Andreae, 1984). The cadmium concentrations in the plankton feeders are likely to be higher than those of the benthopelagic fishes.

Small segments of the human population (especially in fishing villages) in many countries consume inordinate quantities of seafoods. For example, the Canadian Indians are known to consume up to 1300 g/day of seafoods during the fishing season whereas up to 800 g/day of fish can be consumed on fishing boats in southern Italy (Bernhard and Andreae, 1984). These excessive consumption rates are rather unusual, however. A comparison of the average seafood consumption in several nations (Table 8) shows that the populations of two nations, Faeroe Islands and Maldives, consume more than one fish meal a day, equivalent to 150 g of edible tissue or 30 g of protein (Bernhard and Andreae, 1984). Seafood consumption averages about three-quarters of a meal a day in Japan but forms only a small part of the total diet in the United States, Soviet Union, and Australia. Assuming that the average cadmium concentrations in the edible parts of fish, crustaceans, mollusks, and freshwater biota are 0.05, 0.3, 0.3, and 0.1 μg/g respectively, the dietary intake of cadmium with seafoods can be estimated to range from 2.0 μg/day in Australia and the United States to about 18 μg/ day in the Faeroe Islands (Table 8). In the Maldives, Faeroe Islands, and Iceland, seafoods probably account for well over 50% of the daily dietary cadmium intake. The total dietary intake of cadmium by urban residents of the United States has been estimated to be 14 μg/day (Piscator, 1985), and the intake by residents of unpolluted areas of Japan is about 42–69 μg/day (Nogawa, 1984). Thus, seafoods can contribute about 15% of dietary cadmium intake in urban areas of the United States and 14–24% of the total intake by the population that resides in the

TABLE 8   Intake of Cadmium Associated with the Consumption of Fish and Seafood by Populations of Selected Countries[a]

Country	Population (in Millions)	Edible portion Consumed[b]	Freshwater (Biota)[c]	Seawater Fish[c]	Seawater Crustaceans[c]	Seawater Mollusks[c]	Estimated Cd Ingested (μg/day)[d]
Faeroe Islands	0.04	193(38.6)	0.5	157(82)	30(16)	5.5(3)	18
Maldives	0.14	186(37.1)	—	186(100)	—	—	9.3
Vanuata	0.1	116(23.2)	—	115(99)	1.0(2)	—	6.1
Japan	114	113(22.5)	7.5(7)	87(77)	5.0(4)	2(10)	10
Iceland	0.1	96(19.2)	3.0(3)	80(83)	12(13)	—	8.1
Bermuda	0.06	79(15.8)	—	69(87)	10(13)	—	6.4
Yemen Republic	1.9	65(13.0)	—	65(100)	—	—	3.2
Portugal	8.8	51(10.2)	—	69(87)	1.0(2)	1.5(3)	4.2
USSR	260	47(9.4)	8.0(17)	38(81)	—	0.5(1)	2.8
Australia	13.8	17(3.4)	1.5(9)	12(68)	4.0(24)	—	1.9
United States	220	16(3.2)	1.0(6)	10(63)	3.5(22)	1.0(6)	2.0

[a]Seafood and fish consumption data from Bernhard and Andreae (1984).
[b]In grams living weight per day; equivalent grams of protein per day given in parentheses (see Bernhard and Andreae, 1984).
[c]Percentage in relation to total consumption is given in parentheses.
[d]Estimated assuming that the Cd concentrations in edible parts of fish, crustaceans, mollusks, and freshwater biota are 0.05, 0.3, 0.3 and 0.1 μg/g, respectively.

unpolluted regions of Japan. It should be noted that the intake of seafood-associated cadmium by the inhabitants of the Faeroe Islands is comparable to the total dietary intake by the urban population of the United States. These data clearly suggest that the ingestion of seafoods represents an important route by which many people become exposed to cadmium.

3.3. Soil → Plant → Food Transfer

> What is done with this golden dung? It is swept into the abyss.... All the human and animal manure which the world wastes, if returned to the land, instead of being thrown into the sea, would suffice to nourish the world.

> Those piles of ordure collected at street corners, those carts of mud jolted at night through the streets, those frightful barrels of the night-man, those fatid streams—do you know what they are? They are the flowering field, the green grass, the mint and thyme and sage, the game, the cattle, the satisfied lowing of the heavy oxen at night, the perfumed hay, the golden wheat, the bread on your table, the warm blood in your veins, health, joy, life. (Victor Hugo, undated; cited by Hamlin, 1980).

The song above in praise of sewage sludge was written over a century ago by Victor Hugo, a French author. His optimism has subsequently been tempered by studies on cadmium (and other trace metals) that found concentrations in excess of 200 μg/g, with levels of 10–20 μg/g being regarded as normal (Torrey, 1979; Mumma et al., 1984; Mininni and Santori, 1987). When taken up in a large

TABLE 9   Cadmium Concentrations in Selected Vegetables Grown on Composted Sludge-Treated Soils[a]

Treatment	Cadmium Content (μg/g)				
	Carrot	Lettuce	Radish	Swiss Chard	Turnip
Control					
1976	0.44	0.60	0.13	0.24	0.11
1977	0.10	0.65	0.17	0.20	0.16
1978	0.60	0.40	0.12	0.23	0.10
1979	0.60	0.61	0.25	0.27	0.20
1.2 kg Cd/ha yr[b]					
1976	0.91	1.25	0.16	0.56	0.17
1977	0.91	1.27	0.32	0.55	0.27
1978	0.60	0.90	0.38	0.55	0.20
1979	1.40	1.00	0.68	0.90	0.30
2.4 kg Cd/ha yr[b]					
1976	1.51	2.39	0.27	0.66	0.12
1977	1.64	3.05	0.46	1.35	0.27
1978	0.80	1.25	0.56	1.65	0.30
1979	1.10	2.80	0.78	1.91	0.60

[a] Page et al. (1987).
[b] The cadmium comes from the application of the municipal sewage sludge.

quantity, the metals in the sewage can turn the vegetables into a health hazard.

Numerous studies have dealt with the uptake and accumulation of cadmium by plants, both in greenhouses and under natural conditions (Haghiri, 1974; Chaney and Hornick, 1978; Chang et al., 1982; Lund et al., 1981; Munshower, 1977; Street et al., 1978; Valdares et al., 1983; Mortvedt, 1987; Shariatpanahi, 1986; and the recent excellent reviews by Purves, 1985; Adriano, 1986; Page et al., 1987). In general, it has been found that for many soils, the concentration of cadmium in plants shows a highly significant positive correlation with the level of cadmium present in or added to the soil. The relationship holds even at very low rates of cadmium addition to the soils (Table 9). In terms of cadmium uptake, it thus appears that there is no known threshold for safe concentration of cadmium in the soil.

Soil pH is particularly important in determining the amount of cadmium

TABLE 10   Effects of Soil pH and Cadmium Concentration on Uptake of Cadmium by Swiss Chard and Oat Grain[a]

Treatment	Soil Cd Concentration (mg/kg)	Soil pH	Cd Concentration[b]	
			Swiss Chard	Oat Grain
Control				
Farm 1	0.18	4.9	3.34	0.104
Farm 2	0.10	5.3	2.65	0.076
Farm 3	0.22	5.4	1.48	0.034
Farm 4	0.07	5.9	0.44	0.051
Mean			1.98	0.066
Control plus limestone				
Farm 1	0.15	6.4	0.94	0.052
Farm 2	0.10	6.1	0.37	0.064
Farm 3	0.16	6.4	0.63	0.025
Farm 4	0.07	6.3	0.33	0.044
Mean			0.57	0.046
Sludge-treated				
Farm 1	1.66	4.9	94.80	1.960
Farm 2	9.10	5.5	54.30	2.240
Farm 3	0.98	4.9	3.24	0.209
Farm 4	3.26	5.5	9.51	0.299
Mean			40.50	1.180
Sludge-treated plus limestone				
Farm 1	2.10	6.3	2.20	0.259
Farm 2	7.02	6.2	5.34	0.277
Farm 3	0.94	6.0	0.82	0.065
Farm 4	4.50	6.2	3.68	0.193
Mean			3.01	0.198

[a] Page et al. (1987).
[b] Concentration in $\mu g/g$ dry wt.

taken up from the soils (Jastrow and Koeppe, 1980). A reduction in pH (soil acidification) often enhances the cadmium uptake by plants. For example, a reduction in soil pH from 5.9 to 4.9 in control plots has been shown to increase the cadmium content of Swiss chard by about 6-fold, and that of the oat grain by 100% (Table 10); apparently the native cadmium in the soil was readily available to plants. Addition of sludge to the sample soils increased the cadmium content of the chard by about 22-fold at pH of 5.5–5.9, and by about 28-fold at pH of 4.9 (Table 10). Such observations serve to heighten the growing concern that the availability of cadmium in poorly buffered soils can be increased by acid rain, resulting in further accumulation of the element in agricultural crops. Evidence that this may be happening comes from the elevated concentrations of cadmium in the liver and kidney of herbivorous animals (especially deer and moose) that forage in acid sensitive areas of Ontario (Glooschenko et al., 1988), Quebec (Crete et al., 1986), northeastern United States (Scanlon et al., 1986), and Scandinavia (Frank, 1986; Froslie et al., 1986). Large cervides are particularly suitable for biomonitoring the bioaccumulative pollutants (like cadmium) because they are widely distributed, can be sampled easily through hunting harvest, and they are long-lived. Table 11 compares the cadmium concentrations in the kidney and liver of deer and moose in areas of Ontario that contain buffered and poorly buffered soils. The effect of acid rain in the soil → plant → deer/moose transfer of cadmium is clearly demonstrated. In view of such findings, advisories have been issued against human consumption of deer and moose kidneys or livers from acid-sensitive areas of Ontario (Glooschenko et al., 1988) and Quebec (Crete et al., 1986). Most of the cadmium pollution in the areas studied presumably can be attributed to the deposition of long-range transported cadmium aerosols as there are no known local sources for the cadmium. It would be interesting to compare the cadmium contents of agricultural crops from soils with different sensitivities to acid rain.

TABLE 11   Cadmium Concentrations (μg/g) in Tissues of Moose and Deer in Buffered (B) and Acid-Sensitive (AS) Areas of Ontario[a]

Tissue	Age	Moose		Deer	
		Algonquin (AS)[b]	St. Joseph (B)[b]	Loring (AS)[b]	Huronia (B)[b]
Kidney	Calf	3.0(17)	4.4(5)	2.0(22)	0.8(23)
	Yearling	13(16)	4.4(2)	9.2(20)	1.4(24)
	Young adult	20(34)	4.4(3)	15(32)	4.4(23)
	Adult	51(14)	25(1)	34(6)	4.6(5)
Liver	Calf	1.4(6)	1.2(5)	0.3(23)	0.1(9)
	Yearling	3.4(7)	0.4(2)	1.5(27)	0.2(31)
	Young adult	4.1(20)	1.1(2)	1.0(43)	0.3(34)
	Adult	5.7(7)	3.5(1)	1.0(8)	0.5(4)

[a]Glooschenko et al. (1988).
[b]Sample size given in parentheses.

Other physicochemical properties of soils that affect the uptake of cadmium by crops include the (1) oxidation–reduction potential of soils, (2) soil temperature, (3) clay content and cation exchange capacity, (4) concentrations of iron, manganese, and aluminum oxides and hydroxides, (5) presence of anions such phosphate, silicate, and reduced sulfur species that can immobilize the cadmium, (6) organic matter concentration, and (7) interactive effects of zinc, copper, nickel, manganese, and selenium. Details about the roles of these factors in cadmium uptake can be found in a large number of references such as Page et al. (1987), Adriano, (1986), Haghiri (1974), Asami (1984), Chang et al. (1982), Valdares et al. (1983), and Jastrow and Koeppe (1980).

The average concentrations of cadmium in crops grown in a wide variety of soils and climatic conditions in the United States are shown in Table 12. Samples of the cultivars were generally obtained away from any major point sources of contamination. From a critical assessment of the available data bases, Page et al. (1987) derived the following average cadmium concentrations in different classes of crop in the United States (in μg/g): fruits = 0.005, seed crops = 0.028, grains = 0.047, root/bulb (excluding peanuts) = 0.208, vegetable fruits (tomatoes, cucumbers etc.) and leafy vegetables = 0.560. On the other hand, the 1977 market basket survey of the adult diets in the United States found the following cadmium concentrations (in μg/g, with ranges in parentheses): grains and cereal products, 0.021 (0.007–0.030); potatoes, 0.044 (0.012–0.111); leaf vegetables, 0.046 (0.010–0.107); legume vegetables, 0.005 (0.001–0.050); root vegetables, 0.027 (0.002–0.126); garden vegetables, 0.014 (0.001–0.033); and fruits, 0.006 (0.001–0.031) (Johnson et al., 1983). It would appear that the grooming (washing) of agricultural produce sold in the grocery store helps in reducing the cadmium burden in the human diet.

The United States data (Table 12) may be compared with the following average cadmium concentrations (in μg/g) in crops grown in uncontaminated areas of Japan: potato = 0.18, onion = 0.11, carrot = 0.42, cabbage = 0.12, cucumber = 0.19, tomato = 0.63, beans = 0.029, apple = 0.03, and mandarin orange = 0.03 (Kobayashi, 1978). The higher concentrations in Japanese produce reflect the more extensive contamination of the Japanese environment with cadmium (Yamagata, 1979), and the fact that the baseline concentrations of cadmium in Japanese soils are much higher than the levels found in soils in the United States (see Kobayashi, 1978; Asami, 1984).

Other data published in the literature also show that food crops grown in contaminated soils can acquire cadmium concentrations that are 2- over 20-fold higher than those from unpolluted soils (for example, see Purves, 1985; Adriano, 1986). The most famous example is the accumulation of cadmium in rice grown in the contaminated paddy soils of Japan (Kobayashi, 1978; Asami, 1984). This point, however, can best be illustrated using data from Shipham (Dorset, England), a village contaminated by tailings from an old zinc mining operation. In this village, the most contaminated vegetables (notably cabbage, kale, spinach, spring greens, lettuce, and rhubab) showed cadmium concentrations that were about 15 times above the normal values (Table 13). Even the Shipham potatoes

TABLE 12 Representative Concentrations of Cadmium in Food Crops Grown in Different Soils of the United States

Crop	Number of Samples	Average Concentration ($\mu g/g$)	Reference
Leafy vegetables			
Cabbage	24	0.093	Shacklette (1980)
Lettuce	40	0.420	Shacklette (1980)
Spinach	104	0.800	Wolnick et al. (1983)
Swiss chard	16	0.470	Page et al. (1987)
Seed vegetables			
Sweet corn	268	0.016	Wolnick et al. (1983)
Dry beans	35	0.110	Shacklette (1980)
Snap beans	42	0.024	Shacklette (1980)
Vegetable fruits			
Cucumber	22	0.093	Shacklette (1980)
Eggplant	2	0.380	Shacklette (1980)
Peppers	2	0.021	Shacklette (1980)
Tomatoes	231	0.270	Wolnick et al. (1985)
Fruits			
Apples	36	0.034	Shacklette (1980)
Grapefruit	23	0.006	Shacklette (1980)
Grapes	21	0.011	Shacklette (1980)
Oranges	20	0.005	Shacklette (1980)
Peaches	24	0.011	Shacklette (1980)
Pears	38	0.006	Shacklette (1980)
Plums	16	0.006	Shacklette (1980)
Field crops			
Wheat			
United States	288	0.048	Wolnick et al. (1983)
Canada	—	0.045	Kobayashi (1978)
Argentina	—	0.013	Kobayashi (1978)
Australia	—	0.015	Kobayashi (1978)
Japan	—	0.085	Kobayashi (1978)
Barley	29	0.027	Chang et al. (1979)
Rice	166	0.013	Wolnick et al. (1985)
Field corn	277	0.014	Wolnick et al. (1985)
Sorghum	36	0.033	Chang et al. (1979)
Soy beans	322	0.064	Wolnick et al. (1983)
Root and bulb crops			
Asparagus	10	0.032	Shacklette (1980)
Carrots	207	0.250	Wolnick et al. (1983)
Onion	10	0.050	Shacklette (1980)
Potatoes	297	0.170	Wolnick et al. (1983)
Peanuts	320	0.090	Wolnick et al. (1983)
Raddish	17	0.310	Page et al. (1987)
Turnips	5	0.170	Page et al. (1987)

TABLE 13   Cadmium in Vegetables from Shipham (Polluted) and from Noncontaminated Soils[a]

Vegetable/fruit	From Noncontaminated Soils		From Shipham	
	Number of Samples	Mean Cadmium Concentration (μg/g) (range)	Number of Samples	Mean Cadmium Concentration (μg/g) (range)
Beans broad	11	< 0.02 (< 0.01–0.04)	25	0.09 (0.07–0.10)
Beans, french	11	< 0.01 (< 0.01–0.03)	12	0.02 (0.01–0.04)
Beans, runner	9	< 0.01 (< 0.01–0.01)	60	0.03 (0.002–0.30)
Beetroot	15	< 0.03 (< 0.01–0.06)	30	0.36 (0.04–1.14)
Brussels sprouts	9	< 0.01 (< 0.01–0.02)	46	0.17 (0.02–0.79)
Cabbage	22	< 0.01 (< 0.01–0.03)	68	0.61 (0.02–8.24)
Carrots	13	< 0.05 (< 0.01–0.11)	75	0.34 (0.03–1.58)
Kale	12	< 0.04 (< 0.01–0.10)	19	0.75 (0.08–1.23)
Leeks	14	< 0.04 (< 0.01–0.16)	52	0.39 (0.06–1.77)
Lettuce	17	< 0.06 (< 0.01–0.39)	79	0.68 (0.03–2.90)
Onions	9	< 0.02 (< 0.01–0.04)	33	0.22 (0.03–0.44)
Parsley	6	0.03 (0.02–0.06)	15	0.41 (0.15–1.02)
Parsnip	10	< 0.08 (< 0.01–0.37)	41	0.30 (0.09–0.68)
Peas	23	< 0.03 (< 0.01–0.06)	25	0.06 (0.01–0.18)
Potato	20	< 0.03 (< 0.01–0.06)	62	0.13 (0.03–0.03)
Tomato	22	< 0.01 (< 0.01–0.04)	11	0.06 (0.02–0.12)

[a]Sherlock et al. (1983).

contained about 4 times more cadmium than potatoes grown in unpolluted soils (Sherlock et al., 1983). Many other instances have been reported linking environmental cadmium pollution with elevated levels of this element in the food crops; it would be otiose to cover the case histories in this report.

Tobacco is a special "food" crop with an unusual propinquity to bioaccumulate cadmium from soils. A recent extensive study of cigarettes produced in various parts of the world found a range in cadmium concentration of 0.29–3.4 μg/g, the mean value being 1.45 μg/g or about 1.1 μg cadmium per cigarette (Watanbe et al., 1987). There were no discernible differences in the cadmium content of cigarettes produced in the various countries (Table 14). The high cadmium content of cigarettes points to smoking as an important source of cadmium exposure, and a number of studies have indeed found increased cadmium levels in the blood and target organs of smokers compared to the levels in the general population (Cherry, 1981; Moreau et al., 1983; Friberg et al., 1985). It is generally believed that about 10% of the cadmium in a cigarette is inhaled and that the inhaled cadmium is more readily absorbed into the body than the ingested cadmium (Friberg et al., 1985).

Soil → Plant → Animal → Food Transfer

Meat and animal products form a vital part of the human diet. Even bones and other inedible organs are often added to animal feeds to provide essential

TABLE 14 Concentrations of Cadmium in Cigarettes and Beede Made in Different Parts of the World[a]

Source	Number of Samples	Cadmium Content (μg)[b]	
		Per Cigarette or Beedi	Per Gram Weight
Cigarettes			
Canada	10	1.57 (1.39–1.66)	2.01 (1.74–2.18)
China	56	1.22 (0.51–2.79)	1.27 (0.59–2.96)
Finland	6	1.10 (0.80–1.30)	1.54 (1.22–1.90)
France	8	1.22 (0.68–1.90)	1.66 (0.76–2.83)
West Germany	15	1.06 (0.79–1.25)	1.36 (1.13–1.65)
India	18	0.35 (0.21–0.55)	0.43 (0.29–0.60)
Indonesia	8	0.92 (0.72–1.21)	0.79 (0.63–1.12)
Ireland	8	0.70 (0.56–0.79)	0.86 (0.77–1.10)
Italy	14	1.01 (0.42–1.27)	1.17 (0.42–1.53)
Japan	60	1.25 (0.64–1.74)	1.71 (0.86–2.50)
Korea	7	1.03 (0.90–1.29)	1.41 (1.27–1.59)
Mexico	19	2.03 (1.69–2.63)	2.70 (1.87–3.38)
Pakistan	8	0.48 (0.36–0.64)	0.51 (0.40–0.70)
Philipines	6	1.17 (0.96–1.44)	1.36 (1.24–1.44)
Singapore	12	1.04 (0.78–1.41)	1.33 (1.03–1.82)
Spain	10	1.28 (0.64–1.78)	1.59 (0.86–2.23)
Thailand	9	1.22 (1.09–1.84)	1.39 (1.19–1.98)
U.K.	6	0.68 (0.51–0.87)	0.86 (0.65–1.04)
United States	38	1.07 (0.84–1.30)	1.48 (1.23–1.76)
USSR	13	1.37 (0.78–1.86)	1.88 (1.02–2.77)
Total	331	1.15 (0.21–2.79)	1.45 (0.29–3.38)
Beedin (India)	9	0.09 (0.05–0.11)	0.25 (0.16–0.36)

[a] From Watanabe et al. (1987).
[b] Arithmetic mean with ranges in parentheses.

nutrients. The feeds as well as the general habitat of the diary animals are now also being contaminated with cadmium. Highly significant correlations have been found, for example, among the cadmium concentrations in soils, plants, and animals at 18 study sites in the United States (Sherma, 1980). Although the cadmium content of a typical cattle feedlot diet has been determined to be only 0.05 μg/g (Capar et al., 1978), the concentration in the cattle manure has been reported to be 0.24–0.80 μg/g, and the concentration in poultry waste to be 0.42–0.58 μg/g (Furr et al., 1976; Capar et al., 1978). The high cadmium content of the wastes certainly suggests that the animals are being exposed to elevated levels of cadmium. Whether the animals accumulate enough cadmium so that their meats pose a health hazard to the consumer remains an open-ended question.

The contribution of animal products to the total dietary intake of cadmium tends to be small. Eggs, milk, muscle, and red meat are known to be particularly low (< 0.05 μg/g) in cadmium (Sherma, 1980; Baker et al., 1979; Sherma et al., 1979). The mean cadmium concentration in fresh, whole milk sold in the grocery stores of the United States in 1978 was less than 2 μg/L (Podrebarac, 1984). The average concentrations (μg/g, with ranges parentheses) of cadmium found in the 1978 market basket survey of adult diets in the United States were 0.002 (0.001–0.019) in diary products, 0.007 (0.002–0.042) in meat, fish, and poultry, and 0.016 (0.006–0.034) in oils, fats, and shortening (Podrebarac, 1984). These three food classes together accounted for less than 10% of the estimated total dietary intake of cadmium in the general adult population of the United States (Johnson et al., 1983).

Some disconcerting dose-related increases in the cadmium contents of the liver and kidney of chicken (Baker et al., 1979; Vogt et al., 1977; Sherma et al., 1979), meadow vole (Williams et al., 1976), lambs (Doyle et al., 1974), mouse (Exon, 1977), and cattle and swine (Sherma, 1980) have been reported when the animals are given cadmium-supplemented diets. Because of the highly selected accumulation of cadmium in the visceral organs, the animal body, in general, tends to serve as a filter for cadmium transfer to human foods. Visceral organs represent the meat product whose consumption may result in undue cadmium exposure. The high concentrations of cadmium reported in sliced bacon, frankfurters, ground beef, and canned meat products (see Sherma et al., 1979) probably stem from contamination during food processing, a topic that is outside the scope of this chapter.

4.   EVIDENCE FOR HISTORICAL CHANGES IN CADMIUM CONTENT OF FOODS

Retrospective monitoring using archieved food samples has yielded equivocal results (MARC, 1985). Analysis of late nineteenth- and early twentieth-century vegetable, mushroom, berry, fruit, meat, and fish samples collected from various sources in Sweden showed a lot of noise in the data although there was a general tendency for the cadmium concentration to decrease with time from the present. The observed temporal trend was statistically significant for only the fall wheat, with the average concentration doubling between the 1920s and 1970s (Kjellstrom, 1979).

There is, however, little doubt that environmental pollution is responsible for the well-documented, time-dependent elevations in the concentration of cadmium in human organs and in its overall body burden. The half-lives of cadmium in the critical organs are usually long (about 20 years in the kidney, for example) so that changes in cadmium intake with time would be reflected by age-related and cohort-related variations in the cadmium concentrations in human tissues. Analysis of autopsy specimens shows that the concentration of cadmium in the renal cortex increases from extremely low values at birth to about 30–70 μg/g at the age of 50 years (Kjellstrom, 1979; Friberg et al., 1985). Age-dependent increases in cadmium have also been noted in the liver, muscle, pancreas, lung, and prostate (Friberg et al., 1985). These data show that cadmium is highly accumulative in the human body and that the tissue concentrations may be used in retrospective assessment of dietary cadmium intake.

Drasch (1983) found that the average concentration of cadmium in old kidney specimens from 1897 to 1939 was 0.86 μg/g (for the renal cortex) compared to 41 μg/g for the 1980 autopsy material, a 47-fold increase. No corresponding increase in the cadmium content of the liver was observed during the same period, however. Elinder and Kjellstrom (1977) showed that the average cadmium concentration in the nineteenth-century kidney specimens from Sweden of 15 μg/g was about 4-fold lower than the average level in nonsmoking adults who died in 1974. Analysis of bones collected from different parts of France showed that the cadmium concentration was constant from the Neolithic period (about 5600 B.P.) but has increased drastically (by over an order of magnitude) since the seventeenth century (Jaworowski et al., 1985). It is believed that the historical changes in levels of cadmium in human tissues reflect the increasing intensity of environmental cadmium pollution that is being transferred to the human food chain.

REFERENCES

Adriano, D. C. (1986). *Trace Elements in the Terrestrial Environment.* Springer-Verlag, New York.

Asami, T. (1984). "Pollution of soils by cadmium." In J. O. Nriagu, ed., *Changing Metal Cycles and Human Health.* Springer-Verlag, Berlin, pp. 95–111.

Atjay, G. L., Ketner, P., and Duvigneaud, P. (1979). "Terrestrial primary production and phytomass." In B. Bolin, E. T. Degens, S. Kempe, and P. Ketner, eds. *The Global Carbon Cycle.* Wiley, Chichester, pp. 129–189.

Barrie, L. A. (1986). "Arctic air pollution: An overview of current knowledge." *Atmospheric Environ.* **20**, 643–663.

Baker, D. E., Amacher, M. C., and Leach, R. M. (1979). "Sewage sludge as a source of cadmium in soil-plant-animal systems." *Environ. Health Perspect.* **20**, 45–49.

Beavington, F. (1977). "Trace elements in rainwater and dry deposition around a smelting complex." *Environ. Pollut.* **13**, 127–131.

Bernhard, M., and Andreae, M. O. (1984). "Transport of trace metals in marine food chains. In J. O. Nriagu, ed., *Changing Metal Cycles and Human Health.* Springer-Verlag, Berlin, pp. 143–167.

Boutron, C. (1979). "Past and present day tropospheric fluxes of Pb, Cd, Cu, Zn and Ag in Antarctica and Greenland. *Geophys. Res. Lett.* **6**, 159–164.

Bowen, H. J. M. (1979). *Environmental Chemistry of the Elements.* Academic Press, New York.

Buat-Menard, P. (1986). "Air to sea transfer of anthropogenic trace metals." In P. Buat-Menard, ed., *The Role of Air-Sea Exchange in Geochemical Cycling.* D. Reidel, Dordrecht, pp. 477–496.

Capar, S. G., Tanner, J. T., Friedman, M. H., and Boyer, K. W. (1978). "Multielement analysis of animal feed, animal wastes and sewage sludge." *Environ. Sci. Technol.* **12**, 785–790.

Chan, W. H., Tang, A. J. S., Chung, D. H. S., and Lusis, M. A. (1986). "Concentration and deposition of trace metals in Ontario—1982. *Air, Water Soil Pollut.* **29**, 374–389.

Chaney, R. L., and Hornick, S. B. (1978). "Accumulation and effects of cadmium on crops." In *Cadmium 77: Proceedings of the First International Cadmium Conference.* Metal Bulletin Ltd., London, pp. 125–138.

Chang, A. C., Page, A. L., Lund, L. J., Pratt, P. F., and Bradford, G. R. (1979). "Land application of sewage sludge—a field demonstration." *Final Report, Regional Wastewater and Solids Management Program.* Los Angeles/Orange County Metropolitan Area (LA/OMA Project), 311 pp.

Chang, A. C., Page, A. L., Warneke, J. E., and Johanson, J. B. (1982). "Effects of sewage sludge application on the Cd, Pb and Zn levels in selected vegetable plants." *Hilgardia* **50**, 1–14.

Cherry, W. H. (1981). "Distribution of cadmium in human tissues." In J. O. Nriagu, ed., *Cadmium in the Environment.* Wiley, New York, Vol. 2, pp. 69–536.

Chino, M., and Mori, T. (1982). Cited in Adriano, 1986, p. 141.

Crete, M., Potvin, F., Walsh, P., Benedett, J., Lefebvre, M. A., Weber, J., Paillard, G., and Gagnon, J. (1986). "Pattern of cadmium contamination in the liver and kidneys of moose and white-tailed deer in Quebec." *Sci. Total Environ.* **66**, 45–53.

Doyle, J. J., Pfander, W. H., Grebing, S. E., and Dorn, J. O. (1974). "Effects of dietary cadmium on growth, cadmium absorption, and cadmium tissue levels in growing lamb." *J. Nutr.* **104**, 160–166.

Drasch, G. A. (1983). "An increase of cadmium body burden for this century—An investigation on human tissues." *Sci. Total Environ.* **26**, 111–119.

Duce, R. A., Wallace, G. T., and Ray, B. G. (1976). *Atmospheric Trace Metals over the New York Bight.* NOAA Technical Report No. ERL-361-MESA4.

Elinder, C. G., and Kjellstrom, T. (1977). "Cadmium concentration in samples of human kidney cortex from the 19th century." *Ambio* **6**, 270–272.

Engel, D. W., and Fowler, B. A. (1979). "Factors influencing cadmium accumulation and its toxicity to marine organisms." *Environ. Health Perspect.* **28**, 81–88.

Exon, J. H., Lamberton, J. G., and Koller, L. D. (1977). "Effects of chronic oral cadmium residues in organs of mice." *Bull. Environ. Contam. Toxicol.* **18**, 74–76.

Frank, A. (1986). "In search of biomonitors for cadmium: Cadmium content of wild Swedish fauna during 1973–1976." *Sci. Total Environ.* **57**, 57–65.

Franzin, W. G. (1984). "Aquatic contamination in the vicinity of the base metal smelter at Flin Flon, Manitoba, Canada—A case history. *Adv. Environ. Sci. Technol.* **15**, 523–550.

Franzin, W. G., McFarlane, G. A., and Lutz, A. (1979). "Atmospheric fallout in the vicinity of a base metal smelter at Flin Flon, Manitoba, Canada." *Environ. Sci. Technol.* **13**, 1514–1521.

Friberg, L., Elinder, C. G., Kjellstrom, T., and Nordberg, G. F. (1985). *Cadmium and Health: A Toxicological and Epidemiological Appraisal.* CRC Press, Boca Raton, F.

Froslie, A., Haugen, A., Holt, G., and Norheim, G. (1986). "Levels of cadmium in liver and kidneys from Norwegian cervides." *Bull. Environ. Contam. Toxicol.* **37**, 453–460.

Furr, A. K., Lawrence, A. W., Tong, S. S. C., Grandolfo, M. C., Hofstader, R. A., Bache, C. A., Gutenmann, W. H., and Lisk, D. J. (1976). Multielement and chlorinated hydrocarbon analysis of municipal sludges of American cities. *Environ. Sci. Technol.* **10**, 683–691.

Gale, N. L., and Wixson, B. G. (1979). "Cadmium in forest ecosystems around lead smelters in Missouri." *Environ. Health Perspect.* **28**, 23–37.

Glooschenko, V., Downes, C., Frank, R., Braun. H. E., Addison, E. M., and Hickle, J. (1988). "Cadmium levels in Ontario moose and deer in relation to soil sensitivity to acid precipitation." *Sci. Total Environ.,* **71**, 173–186.

Groets, S. S. (1976). "Regional and local variations in heavy metal concentrations of bryophytes in the northeastern United States." *OIKOS* **27**, 445–456.

Haghiri, F. (1974). "Plant uptake of cadmium as influenced by cation exchange capacity, organic matter, zinc and soil temperature." *J. Environ. Qual.* **3**, 180–183.

Hallet, J. P., Lardinois, P., Ronneau, C., and Cara, J. (1982). "Elemental deposition as a function of distance from an industrial zone." *Sci Total Environ.* **25**, 99–109.

Hamlin, C. (1980). "Sewage: waste or resource?" *Environment* **22**, 16–42.

Hardisty, M. W., Huggins, R. J., Kartar, S., and Sainsbury, M. (1974). "Ecological implications of heavy metal in fish from the Severn Estuary." *Mar. Pollut. Bull.* **5**, 12–15.

Harrison, R. M., Perry, R., and Wellings, R. A. (1975). "Lead and cadmium in precipitation: Their contribution to pollution." *J. Air Pollut. Control Assoc.* **25**, 627–630.

Harrison, R. M., and Wilson, S. J. (1985). "The chemical composition of highway drainage waters. I. Major ions and selected trace metals." *Sci Total Environ.* **43**, 63–77.

Hutton, M. (1983). "Sources of cadmium in Europe, inputs into agricultural land and implications for crop uptake and dietary exposure." In *Proceedings, 4th International Conference on Heavy Metals in the Environment, Heidelberg.* CEP Consultants, Edinburgh, Vol. 1, pp. 430–433.

Hutton, M., and Symon, S. (1986). "The quantities of cadmium, lead, mercury and arsenic entering the U.K. environment from human activities." *Sci. Total Environ.* **57**, 129–150.

Imboden, D. M., Hegi, H. R., and Zobrist, J. (1975). "Atmospheric loading of metals in Switzerland." *EAWAG News*, Swiss Federal Inst. of Technology, EAWAG, CH-8600 Dubendorf, Switzerland, Sept. Issue, pp. 5–7.

Jastrow, J. D., and Koeppe, D. E. (1980). "Uptake and effects of cadmium in higher plants." In J. O. Nriagu, ed., *Cadmium in the Environment.* Wiley, New York, Vol. I, pp. 607–638.

Jaworowski, Z., Barbalat, F., Blain, C., and Peyre, E. (1985). "Heavy metals in human and animal bones from ancient and contemporary France." *Sci. Total Environ.* **43**, 103–126.

Jeffries, D. S., and Snyder, W. R. (1981). "Atmospheric deposition of heavy metals in central Ontario." *Water, Air Soil Pollut.* **15**, 127–152.

Johnson, R. D., Manske, D. D., New, D. H., and Podrebarac, D. S. (1983). "Pesticide, metal, and other chemical residues in adult total diet samples. (XII), August 1976–September 1977." *J. Assoc. Off. Anal. Chem.* **67**, 154–166.

Jones, K. C., Symon, C. J., and Johnson, A. E. (1987). "Retrospective analysis of an archived soil collection, II, cadmium." *Sci. Total Environ.* **67**, 75–89.

Kjellstrom, T. (1979). "Exposure and accumulation of cadmium in populations from Japan, the United States, and Sweden." *Environ. Health Perspect.* **28**, 169–197.

Kleinman, M. T., Kneip, T. J., Bernstein, D. M., and Eisenbud, M. (1977). "Fallout of toxic metals in New York City." In *Biological Implications of Metals in the Environment.* National Technical Information Service, Conf-750929/RAS, Springfield, VA, pp. 144–152.

Kloke, A., Sauerbeck, D. R., and Vetter, H. (1984). "The contamination of plants and soils with heavy metals and the transport of metals in terrestrial food chains." In J. O. Nriagu, ed., *Changing Metal Cycles and Human Health.* Springer-Verlag, Berlin, pp. 113–141.

Kobayashi, J. (1978). "Pollution by cadmium and the Itai–Itai disease in Japan." In F. W. Oehme, ed., *Toxicity of Heavy Metals in the Environment.* Dekker, New York, pp. 199–260.

Koranda, J. J., and Robinson, W. L. (1978). "Accumulation of radionuclides by plants as a monitor system." *Environ. Health Perspect.* **27**, 165–179.

Lum, K. R. (1987). "Cadmium in freshwaters: The Great Lakes and St. Lawrence River." *Adv. Environ. Sci. Technol.* **19**, 35–50.

Lund, L. J., Betty, E. E., Page, A. I., and Elliot, R. A. (1981). "Occurrence of high Cd levels in soil and its accumulation by vegetation." *J. Environ. Qual.* **10**, 551–556.

Mance, G., and O'Donnell, A. R. (1983). "Contamination of the aquatic environment by heavy metals in some EC member states—A comparison." In *Proceedings, 4th International Conference on Heavy Metals in the Environment, Heidelberg.* CEP Consultants Ltd., Edinburgh, Vol. 2, pp. 1281–1284.

MARC (1985). *Historical Monitoring.* Monitoring and Assessment Research Center, Chelsea College, University of London Report No. 31.

MARC (1986). *Biological Monitoring.* Monitoring and Assessment Research Center, Chelsea College, University of London, Report No. 32.

Mason, B. and Moore, C. B. (1982). *Principles of Geochemistry*. Wiley, New York.

McCracken, I. R. (1987). "Biological cycling of cadmium in freshwater." *Adv. Environ. Sci. Technol.* **19**, 89–116.

Mininni, G., and Santori, M. (1987). "Problems and perspectives of sludge utilization in agriculture." *Agr. Ecosyst. Environ.* **18**, 291–311.

Moreau, T., Orssaud, G., Lellouch, J., Claude, J. R., Juguet, N., and Festy, B. (1983). "Blood cadmium levels in a general male population with special reference to smoking." *Arch. Environ. Health* **38**, 163–167.

Mortvedt, J. J. (1987). "Cadmium levels in soils and plants from some long-term soil fertility experiments in the United States of America." *J. Environ. Qual.* **16**, 137–142.

Mumma, R. O., Raupach, D. C., et al. (1984). "National survey of elements and constituents in municipal sewage sludges." *Arch. Environ. Contam. Toxicol.* **13**, 75–83.

Munshower. F. F. (1977). "Cadmium accumulation in plants and animals of polluted and nonpolluted grasslands." *J. Environ. Qual.* **6**, 411–413.

Nogawa, K. (1981). "Itai–Itai disease and follow-up studies." In J. O. Nriagu, ed., *Cadmium in the Environment*. Wiley, New York, Vol. 2, pp. 1–37.

Nogawa, K. (1984). "Cadmium." In J. O. Nriagu, ed., *Changing Metal Cycles and Human Health*. Springer Verlag, Berlin, pp. 275–284.

Nriagu, J. O. (1979). "Global inventory of natural and anthropogenic emissions of trace metals to the atmosphere." *Nature (London)* **279**, 409–411.

Nriagu, J. O. (1980). "Cadmium in the atmosphere and in precipitation." In J. O. Nriagu, ed., *Cadmium in the Environment*. Wiley, New York, Vol. 1, pp. 71–114.

Nriagu, J. O. (1986). "Metal pollution in the Great Lakes in relation to their carrying capacity." In G. Kullenberg, ed., *The Role of the Oceans as a Waste Disposal Option*. D. Reidel, New York, pp. 441–468.

Nriagu, J. O. (1988). "A silent epidemic of environmental metal poisoning?" *Environ. Pollut.* **50**, 139–161.

Nriagu, J. O., and Davidson, C. I., eds. (1986). *Toxic Metals in the Atmosphere*. Wiley, New York.

Nriagu, J. O., and Pacyna, J. M. (1988). "Quantitative assessment of worldwide contamination of the air, water and soil with trace metals." *Nature (London)* **333**, 134–139.

Pacyna, J. M. (1986). "Atmospheric trace elements from natural and anthropogenic sources." *Adv. Environ. Sci. Technol.* **17**, 33–52.

Page, A. L., Chang, A. C., and El-Amamy, M. (1987). "Cadmium levels in soils and crops in the United States." In T. C. Hutchinson and K. M. Meema, eds., *Lead, Mercury, Cadmium and Arsenic in the Environment*. Wiley, Chichester, pp. 119–145.

Peyton, T., McIntosh, A., Anderson, V., and Yost, K. (1976). "Aerial input of heavy metals into an aquatic ecosystem." *Water, Air Soil Pollut.* **5**, 39–47.

Phillips, D. J. H. (1980). "Toxicity and accumulation of cadmium in marine and estuarine biota." In J. O. Nriagu, ed., *Cadmium in the Environment*. Wiley, New York, Vol. 1, pp. 425–569.

Piscator, M. (1985). "Dietary exposure to cadmium and health effects: Impact of environmental change." *Environ. Health Perspect.* **63**, 127–132.

Podrebarac, D. S. (1984). "Pesticide, metal, and other chemical residues in adult total diet samples. (XIV), October 1977–September 1988." *J. Assoc. Off. Anal. Chem.* **67**, 176–185.

Purves, D. (1985). *Trace-Element Contamination of the Environment*. Elsevier, Amsterdam.

Ritchie, I. M., and Thingvold, D. A. (1985). "Assessment of atmospheric impacts of large-scale copper-nickel development in northeastern Minnesota." *Water, Air Soil Pollut.* **25**, 145–160.

Ryan, J. A., Pahren, H. R., and Lucas, J. B. (1982). "Controlling cadmium in the human food chain: A review and rationale based on health effects." *Environ. Res.* **28**, 251–302.

Scanlon, P. F., Morris, K. I., Clark, A. G., Fimreite, N., and Lierhagen, S. (1986). "Cadmium in moose tissue: Comparison of data from Maine, USA and from Telemark, Norway." Presented at the North American Moose Conference, New Brunswick, Canada.

Shacklette, H. T. (1980). *Elements in Fruits and Vegetables from Areas of Commercial Production in the Conterminous United States.* U.S. Geological Survey Professional Paper No. 1178, U.S. Government Printing Office, Washington, D.C., 149 pp.

Shariatpanahi, M. (1986). "Accumulation of cadmium, mercury and lead by vegetation following long-term land application of wastewater." *Sci. Total Environ.* **52**, 41–47.

Sharrett, A. R., Carter, A. P., Orheim, R. M., and Feinleib, M. (1982). "Daily intake of lead, cadmium, copper, and zinc from drinking water: The Seattle study of trace metal exposure." *Environ. Res.* **28**, 456–475.

Sherlock, J. C., Smart, G. A., Walters, B., Evans, W. H., McWeeny, D. J., and Cassidy, W. (1983). "Dietary surveys on a population at Shipham, Dorset, United Kingdom." *Sci. Total Environ.* **29**, 121–142.

Sherma, R. P. (1980). "Soil-plant-animal distribution of cadmium in the environment." In J. O. Nriagu, ed., *Cadmium in the Environment.* Wiley, New York, Vol. 1, pp. 587–605.

Sherma, R. P., Street, J. P., Verma, M. P., and Shupe, J. L. (1979). "Cadmium uptake from feed and its distribution to food products of livestock." *Environ. Health Perspect.* **28**, 59–66.

Shiller, A. M., and Boyle, E. A. (1987). "Variability of dissolved trace metals in the Mississippi River." *Geochim. Cosmochim. Acta* **51**, 3273–3277.

Solomon, R. L., and Hartford, J. W. (1976). "Lead and cadmium in dusts and soils in a small urban community." *Environ. Sci. Technol.* **10**, 773–777.

Spehar, R. L., Anderson, R. L., and Fiandt, J. T. (1978). "Toxicity and bioaccumulation of cadmium and lead in aquatic envertebrates." *Environ. Pollut.* **15**, 195–208.

Street, J. J., Sabey, B. R., and Lindsay, W. L. (1978). "Influence of pH, phosphorus, cadmium, sewage sludge, and incubation time on the solubility and plant uptake of cadmium." *J. Environ. Qual.* **7**, 286–290.

Struempler, A. W. (1976). "Trace metals in rain and snow during 1973 at Chadron, Nebraska." *Atmos. Environ.* **10**, 33–37.

Tate, M. B., and Bates, M. H. (1984). "Bulk deposition of metals in Tulsa, Oklahoma." *Water, Air Soil Pollut.* **33**, 15–26.

Torrey, S. (1979). *Sludge Disposal by Landspreading Techniques.* Noyes Data Corporation, Park Ridge, NJ.

Tsuchiya, K. (1984). *Toxicology of Cadmium.* Personal Publication, University of Occupational and Environmental Health, Kitakyushu, Japan.

Ure, A. M., and Berrow, M. L. (1982). "The elemental constituents of soils." In H. J. M. Browen, ed., *Environmental Chemistry.* Royal Society of Chemistry, Burlington House, London, Vol. 2, pp. 94–204.

Valdares, J. M. A. S., Gal, M., Mingelgrin, U., and Page, A. L. (1983). "Some heavy metals in soils treated with sewage sludge, their effects on yield and their uptake by plants." *J. Environ. Qual.* **12**, 49–57.

Vermeulen, A. J. (1977). "Immieeieonderzoek met behulp van regenvangers, opzet, ervaringen en resultaten. In *Dienst voor de Milieuhydiene.* Prov. Waterstaat van Noord-Holland, Haarlem.

Vogt, H., Nezel, K., and Matthes, S. (1977). "Effects of various lead and cadmium levels in broiler and laying rations on the performance of the birds and on the residues in tissues and eggs." *Natr. Metab.* **21** (Suppl. 1), 203–204.

Watanabe, T., Kasahara, M., Nakatsuka, H., and Ikeda, M. (1987). "Cadmium and lead contents of cigarettes produced in various areas of the world." *Sci. Total Environ.* **66**, 29–37.

Wienner, J. G. (1979). "Aerial inputs of cadmium, copper, lead, and manganese into a freshwater pond in the vicinity of a coal-fired power pland." *Water, Air Soil Pollut.* **12**, 343–353.

Williams, P. H., Shenk, J. S., and Baker, D. E. (1976). *Application of Sewage Sludge to Cropland:*

Appraisal of Potential Hazards of the Heavy Metals to Plants and Animals. U.S. Environmental Protection Agency, Report No. MCD-33, p. 58.

Wolnick, K. A., Frickle, F. L., Capar, S. G., Braude, G. L., Meyer, M. W., Satzger, R. D., and Bonnin, E. (1983). "Elements in major raw agricultural crops in the United States. I. Cadmium, and lead in lettuce, peanuts, potatoes, soybeans, sweet corn and wheat." *J. Agr. Food Chem.* **31**, 1240–1244.

Wolnick, K. A., Frickle, F. L., Capar, S. G., Braude, G. L., Meyer, M. W., Satzger, R. D., and Bonnin, E. (1985). "Elements in major raw agricultural crops in the United States. III. Cadmium, lead and other elements in carrots, field corn, onion, rice, spinach and tomatoes." *J. Agr. Food Chem.* **33**, 807–811.

Yamagata, N. (1979). "Industrial emission of cadmium in Japan." *Environ. Health Perspect.* **28**, 17–22.

4

LEAD CONTAMINATION IN FOOD

A. Russell Flegal

Institute of Marine Sciences, University of California, Santa Cruz, Santa Cruz, California

Donald R. Smith

Department of Biology, University of California, Santa Cruz, Santa Cruz, California

Robert W. Elias*

Environmental Criteria and Assessment Office, United States Environmental Protection Agency, Research Triangle Park, North Carolina

*The views expressed here are those of the authors and do not necessarily reflect the views or policies of the U.S. Environmental Protection Agency.

1.   INTRODUCTION

People in North America consume an average of 50 µg of lead per day in food, beverages, and dust. Thirty to fifty percent of this can be attributed to lead in food and beverages, and most of the rest comes from dust of various types. This chapter discusses the source of lead in food and the historical significance of anthropogenic lead in food, in the context of total lead exposure in the human environment.

The simplest and most direct approach to evaluating lead contamination in food is to estimate the amount of lead consumed in food and examine the sources of this lead. Because of a history of analytical problems that produced inaccurate data, it is important to screen all data to eliminate inaccurate estimates of the amount of lead consumed. Using analytically acceptable data, this chapter presents three models that relate to total human exposure to lead. These are the Biogeochemical Model of Natural Lead, the Multiple Source Food Model, and the Total Human Exposure Model for lead. These models are sequential, in that the second relies on information from the first, and the third from the second.

The two major sources of lead in food are atmospheric aerosols and lead-based solder. The methodology for distinguishing between these sources was first described in 1986, using data from 1982 to 1983 (U.S. Environmental Protection Agency, 1986a). Many assumptions were made concerning the contribution of atmospheric lead at each stage of the food processing sequence. These assumptions appear to be validated by new data for 1984 to 1985 that show food categories with the greatest atmospheric exposure have decreased lead concentrations corresponding to the decreases in atmospheric lead concentration. This review makes a further evaluation of the Multiple Source Food Model in an attempt to provide some guidelines on the expected trends of lead in food.

From a historical perspective, humans have been exposed to anthropogenic lead since before 3000 B.C. The full extent of this exposure was confirmed during the 1970s, with the development of a Biogeochemical Model of Natural Lead concentrations by Patterson and his colleagues, who showed that modern man is contaminated with anthropogenic lead by a factor between 300 and 2000. Our increasing awareness of the impact of lead exposure on past civilizations, the relatively high exposure of our present civilization, and the potential for subtle toxic effects at very low lead exposure have driven concerned scientists to urge their governments to control the release of lead into the human environment.

Data are presented in this chapter that suggest a modest downward trend in human lead exposure in the United States since 1978–1980. This trend can be directly related to the requirement to provide unleaded gasoline, for the removal of lead from leaded gasoline, and the voluntary removal of lead-based solder from food cans. It is important to know whether these recent changes have truly reversed the historical upward trend in the total human exposure to lead, because there are still many other sources of lead exposure that must be considered. This question cannot be answered on the basis of contamination of food alone, without consideration of other sources of lead, especially dust, in the human environment. For this purpose, a simple model of total human exposure to lead will be discussed. This Total Human Exposure Model incorporates the Multiple Source Food Model.

Finally, the significance of lead contamination in food becomes more apparent as the level at which toxic effects are first observed continues to drop. This contrasts with the previous, singular focus on sources of high lead concentrations in food that actuated acute lead toxicity (Nriagu, 1985b; Lin-Fu, 1985). Both primary and secondary sources are addressed in recent reports on lead contamination in food (U.S. Environmental Protection Agency, 1986a, b; U.S. Department of Health and Human Services, 1988). Much of that new data, as well as biogeochemical calculations of the natural concentrations of lead in the biosphere, have been incorporated in this review of lead contamination in food.

2. INACCURATE MEASUREMENTS OF LEAD CONCENTRATIONS IN FOOD

There is ample evidence that many reported measurements of low concentrations of environmental lead were inaccurate prior to the mid-1970s. Successive improvements in the reduction and control of the level of contamination introduced during sampling, storage, and analysis (Patterson and Settle, 1976) revealed errors of one to three orders of magnitude in seawater (Bruland, 1983), prehistoric human bones (Ericson et al., 1979), and food (Settle and Patterson, 1980). These studies set ultraclean analytical standards that serve as a basis for accepting or rejecting data on environmental lead. Other examples of the problems of inaccurate measurements of lead are described in recent reviews on lead concentrations in the environment (Patterson, 1982; Rutter, 1983; U.S. Environmental Protection Agency, 1986a, b).

The report on lead contamination in tuna (Settle and Patterson, 1980) demonstrated that accurate measurements of the lead concentration of fresh tuna (0.3 ng/g wet weight) were three orders of magnitude lower than the concentration (400 ng/g wet weight) measured with previously accepted standard techniques. Their report prompted the U.S. Food and Drug Administration to revise its procedures for food analysis, and to produce, since 1982, acceptable data on lead in food.

3.   THE BIOGEOCHEMICAL MODEL FOR NATURAL LEAD

The significance of such inaccurate data was also addressed by Settle and Patterson (1980). They compared the apparent and true magnitude of lead contamination in canned tuna relative to the lead concentration in fresh tuna. Concentrations in tuna from lead-soldered cans (1400 ng/g wet weight) were somewhat similar to the erroneously high concentration reported for fresh tuna (400 ng/g wet weight), thereby apparently indicating a relatively modest increase from processing and canning of less than 4-fold. This contrasted with the more than 4000-fold increase in the lead content of canned tuna shown by accurate

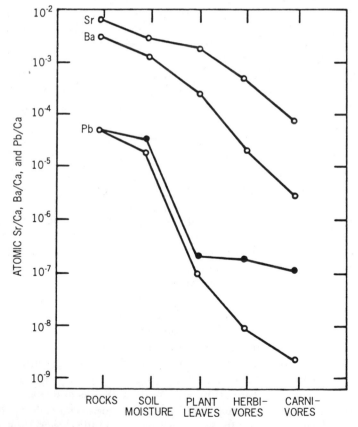

Figure 1. The natural lead biogeochemical model, applied to a terrestrial grazing food chain, predicts nearly two orders of magnitude difference between the observed (solid dots) and natural (open dots) Pb/Ca ratio (Elias et al., 1982).

measurements of the lead content of fresh tuna. The biogeochemical model further indicated that the natural lead concentration of the fish was 0.03 ng/g wet weight and that lead contamination had increased that concentration more than 40,000-fold.

The natural lead biogeochemical model uses the ratios of alkaline earth metals, Sr/Ca and Ba/Ca, to determine the expected natural Pb/Ca ratio, based on the assumption that there is no anthropogenic strontium, barium, or calcium in the sample. The natural Pb/Ca ratio should be lower than the Ba/Ca ratio. The model works best when a sequence of samples is taken that represents a typical food chain. This sequence permits the quantitative estimate of the difference between the Ba/Ca ratio and the natural Pb/Ca as it relates to adjacent food chain components. An example of the model for a typical terrestrial grazing food chain is illustrated in Figure 1. From this diagram, the degree of contamination can be determined as the difference between the observed Pb/Ca ratio with anthropogenic lead, and the predicted natural Pb/Ca ratio. In this study, the natural Pb/Ca ratios were confirmed by independent chemical measurements, including precise lead isotopic composition measurements that can clearly distinguish between natural and anthropogenic lead.

4. HISTORICAL RECORD OF ATMOSPHERIC LEAD CONTAMINATION

In contrast to the relatively recent release of organic contaminants into the environment, the history of elemental lead contamination began with the development of metallurgical science over 5000 years ago (Patterson, 1982; Nriagu, 1985a). The discovery of cupellation, a process for separating silver from lead ores, began a series of economic and industrial changes that resulted in an exponential increase in the release of anthropogenic lead into the environment. There is some evidence of even earlier uses of lead in anthropological artifacts, such as lead beads in the Hittite ruins of Catal Huyuck from 6500 B.C. and a lead statuette from the temple of Osiris in Abydos from 3800 B.C. that may indicate endemic lead contamination was present before 3000 B.C.

Environmental lead contamination began with the first mining and smelting of lead ores. Nriagu (1983a) quipped that the first person to develop the cupellation process was probably the first individual to suffer from lead toxicity. His review of the history of lead contamination in the environment has been substantiated by reports of acute lead toxicity in lead miners and refiners that date back to the Roman Empire (Eisinger, 1977; Winder, 1984; Nriagu, 1985b). The historical record of lead production over the past five millennia was deciphered by Settle and Patterson (1980), and is plotted in Figure 2. They illustrated the exponential increase in lead production over much of that period.

By the twentieth century, these anthropogenic emissions to the atmosphere exceeded natural emissions by two orders of magnitude, and may have finally peaked in the last decade (Nriagu, 1978, 1979; Settle and Patterson, 1980; Boyle

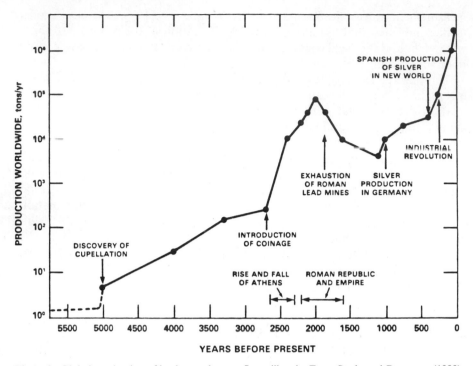

Figure 2. Global production of lead over the past five millennia. From Settle and Patterson (1980).

et al., 1986). Current production rates are approximately 3.2×10^9 kg/yr (U.S. Bureau of Mines, 1987). About half is recycled and half is permanently released into the human environment (National Academy of Sciences, 1980). Much of this released lead (0.33×10^9 kg/yr) is discharged directly to the atmosphere (Nriagu and Pacyna, 1988). Those anthropogenic emissions are approximately 700-fold greater than the natural emissions of lead (0.00045×10^9 kg/yr) to the atmosphere (Nriagu, 1979; Patterson and Settle, 1987).

Anthropogenic emissions have, in turn, elevated lead concentrations within the biosphere by at least one order of magnitude above natural concentrations (Burnett et al., 1980; Settle and Patterson, 1980; Elias et al., 1982; Patterson, 1982; Flegal, 1985; Shen and Boyle, 1987; Flegal and Patterson, 1989). Consequently, that part of lead in food that is considered natural must be deduced from analyses of prehistoric samples and the natural lead biogeochemical model (Patterson, 1982; U.S. Environmental Protection Agency, 1986a).

The historic increase in the global dispersion of anthropogenic lead aerosols was initially documented by the 230-fold increase in lead deposition rates in Greenland ice cores (Figure 3), from prehistoric (800 B.C.) levels of 0.03 ng Pb/cm^2/yr to present levels of approximately 7 ng Pb/cm^2/yr (Murozumi et al., 1969). That increase has been confirmed by subsequent arctic ice core data (Ng and Patterson, 1981; Wolff and Peel, 1985). Comparable increases have been

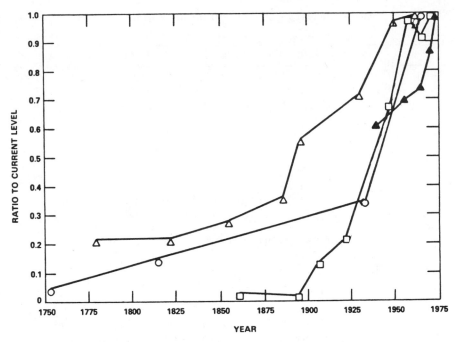

Figure 3. Historical record of lead contamination from industrial aerosols in recorded chronological strata. Open circles are Greenland snow from Murozumi et al. (1969), open squares are a dated pond sediment from the remote Sierras reported by Shirahata et al. (1980), open triangles are lake sediments from Edgington and Robbins (1976), and closed triangles are marine sediments from Ng and Patterson (1982).

detected in pond and lake sediments (Lee and Tallis, 1973; Edgington and Robbins, 1976; Robbins, 1978; Livett et al., 1979; Davis et al., 1982; Shirahata et al., 1980; Schell, 1986), seawater (Schaule and Patterson, 1981, 1983; Flegal and Patterson, 1983; Boyle et al., 1986), oceanic sediments (Hamelin et al., 1987; Veron et al., 1987), and marine corals (Shen and Boyle, 1987) in the northern hemisphere. Smaller increases in Antarctic ice cores (Boutron and Patterson, 1983, 1986, 1987) and South Pacific seawater (Flegal and Patterson, 1983) reflect the fact that 90% of the global atmospheric lead emissions are in the northern hemisphere, with a residence time of about 10 days, compared to the inter-hemispheric mixing rate of 1–2 years (Turekian et al., 1977). Yet, even in Antarctica, there is at least a 2-fold increase over natural concentrations in environmental lead from anthropogenic lead aerosols.

Lead contamination in urban areas is often orders of magnitude greater than in remote areas (Table 1). This is due to the extensive use of lead in industry, and its relatively limited mobility in the environment. Since the industrial revolution, lead has been used in solders, wall and window construction, paint pigments, cosmetics, sheathing of ships, roofs, gutters, cisterns, water pipes, glazes, tableware, containers, potions, additives in wine, sealants, protective coatings,

TABLE 1   Environmental Lead Concentrations from Remote/Rural and Urban Areas[a]

	Remote/Rural	Reference[b]	Urban[c]	Reference[b]
Air	0.05	1	2.3	2, 3
Freshwater	1.7×10^{-5}	4	0.005–0.030	5
Soil	10–30	5	150–300	5
Plants	0.18[d]	4	950[e]	6
Herbivores	2.0[e]	4	38[e]	7 (bone)
Omnivores	1.3[e]	4	67[e]	7 (bone)
Carnivores	1.4[e]	4	193[e]	7 (bone)

[a] Units in μg Pb/g, except for air, which are μg Pb/m^3.
[b] (1) Lindberg and Harriss (1981); (2) Facchetti et al. (1982); (3) Galloway et al. (1982); (4) Elias et al. (1982); (5) U.S. Environmental Protection Agency (1986a); (6) Graham and Kalman (1974); (7) Chmiel and Harrison (1981).
[c] Values can be highly variable, depending on organism and habitat location.
[d] Fresh weight basis.
[e] Dry weight basis.

type, insecticides, plastics, lubricants, ceramics, machine alloys, and gasoline additives (National Academy of Sciences, 1980). Aeolian and fluvial concentration gradients demonstrate that most of that lead has remained as a contaminant in urban areas (Huntzinger et al., 1975; Roberts, 1975; Biggins and Harrison, 1979; Ragaini et al., 1977; Palmer and Kucera, 1980; Harrison and

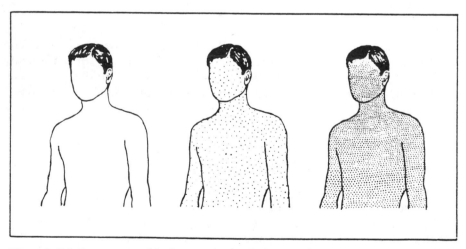

Figure 4. Relative amounts of lead contamination in humans (from Patterson, 1982). Each dot represents 300 μg Pb/70 kg person. The figure on the left with one dot represents the natural concentration of lead in humans (300 μg/70 kg). The figure in the center with 500 dots represents the current average concentration of adults in urban environments (150,000 μg/70 kg). The figure on the right with 2000 dots represents the minimum concentration that will actuate classical lead poisoning in a significant fraction of the population (600,000 μg/70 kg).

Williams, 1982; Ng and Patterson, 1982; Elbaz-Poulichet et al., 1984; Flegal and Smith, 1988).

The resultant lead contamination of humans is illustrated in Figure 4. It shows the skeletal lead/calcium atomic ratios of prehistoric humans (3×10^{-8}) is 500-fold less than the current average of adults from industrialized regions in the northern hemisphere (1500×10^{-8}). Those ratios were derived from ultraclean measurements of lead concentrations in prehistoric South American Indians (Ericson et al., 1979), and were substantiated by comparable measurements of lead concentrations in prehistoric North American Indians (Flegal et al., 1989). The prehistoric ratios are 2000-fold less than the minimum amount of lead required to exhibit classic lead poisoning in humans (6000×10^{-8}). Higher natural lead concentrations reported in previous studies are considered invalid because of contamination during collection, storage, and analysis (Patterson et al., 1987).

5. MODERN PARALLELS WITH ANCIENT ROME

Reviews of lead contamination in food invariably recall those of the Roman Empire, where plumbism was pandemic. The significance of lead poisoning in ancient Rome has been vigorously debated since it was hypothesized that chronic lead toxicity played an important part in the decline of the Roman Empire (Kobert, 1909; Gilfillan, 1965). Although the link between lead contamination and the decline of the Roman Empire is difficult to verify, there is little doubt that countless Roman citizens suffered from plumbism (Eisinger, 1984). That poisoning has been primarily attributed to lead contamination of food, based on the predilection of the Roman aristocracy to lead-sweetened Apician entrees and Columellan wine blends (Nriagu, 1983a, b, 1985b).

The principal sources of lead contamination to people in ancient Rome were grape syrups [sapa, defrutum, hepsema, car(o)enum, and siraeum (sireion)] with lead concentrations of 240–1000 mg/L (Eisinger, 1977). The syrups were specifically prepared in leaded pots, which produced fungicidal lead acetates (Nriagu, 1983a). Lead concentrations of wines fortified with the syrups (1 g/L) were high enough to cause plumbism (Eisinger, 1984). The syrups were also added to many foods and medicines, and the amount of lead in one teaspoon was high enough to cause plumbism (Nriagu, 1983a).

Roman foods were further contaminated by lead containers and vessels. Cooking and eating utensils were often made of lead or pewter, or were lead-lined. Some were decorated with lead-based glazes. Pots were either made of lead or lined with a lead alloy (stagnum), in order to give the foods a more agreeable taste and prevent formation of destructive copper sulfate verdigris (Nriagu, 1983a). Water was also collected in lead cisterns and transported in lead pipes. This was so common that the word plumbing was derived from the Latin word for lead, plumbum.

The uptake of contaminant lead from those ancient sources has been conservatively calculated (Nriagu 1983c). From the glutinous consumption of wine, the lead intake of the aristocracy is conservatively estimated to have been 50 μg/day. By comparison, wine consumption is estimated to have accounted for 50–60% of the average plebeian lead intake of 35 μg Pb/day, and the average intake of slaves, who had less access to lead-contaminated wines and foods, is estimated to have been 15 μg/day. The lead exposure for the Roman aristocracy was 5-fold greater than the current average uptake of individuals in North America (50 μg/day), and accounts for the widespread plumbism and gout among the aristocracy during the Roman Empire.

One of the tragedies of that epidemic was that there is evidence that some Romans recognized the toxicity of lead and did nothing to eliminate it. Nriagu (1985b) includes the following citation from Vitruvious in a review on the historical lead contamination in food:

> Water supply by earthen pipes has advantages. First, if any fault occurs in the work, anybody can repair it. Again, water is much more wholesome from earthenware pipes than from lead pipes. For it seems to be made injurious by lead because some white lead is produced from it; and this is said to be harmful to the human body. Thus, if what is produced by anything is injurious, it is not doubtful but that the thing is not wholesome in itself.

> We can take example by the workers in lead who have complexions affected by pallor. For when, in casting, the lead receives the current of air, the fumes from it occupy the members of the body, and burning then thereupon, robs the limbs of the virtues of the blood. Therefore it seems that water should not be brought in lead pipes if we desire to have it wholesome.

Nriagu then notes that Vitruvious regarded earthenware pipes as "cheap" and left no doubt of his preference for lead pipes in major water supply systems.

Although they were apparently aware of the effects of lead in drinking water, lead poisoning from the adventitious contamination of foods in ancient Rome (as well as in Europe and North America between the sixteenth and nineteenth centuries) was not immediately obvious to the physicians of the time, and could not have been accurately diagnosed. Consequently, the problems of lead contamination in food could not be fully recognized until the 1970s, when trace metal analytical techniques were developed to accurately establish the natural concentrations of lead in the environment and medical techniques were developed to identify sublethal lead toxicity.

6.   PRINCIPAL SOURCES OF LEAD CONTAMINATION IN FOOD

Industrial lead aerosols, lead-soldered cans, and lead plumbing are the principal sources of lead contamination in food (Figure 5). Their relative contributions, as well as those of other sources, depend on the food type and its history of exposure during production, harvesting, transport, packaging, preparation, and consumption.

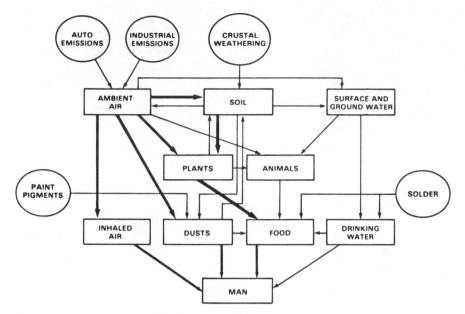

Figure 5. The pathways of lead exposure in the human environment, as evaluated by the Total Human Exposure Model for lead. The only natural input is from crustal weathering. In the baseline situation, those sources in ovals would be the minimum exposure in any reasonable North American residential situation (U.S. Environmental Protection Agency, 1986a).

6.1. Atmospheric Lead in Food

Anthropogenic lead aerosols account for approximately 40% of the lead in food (Elias, 1985; U.S. Environmental Protection Agency, 1986a). This estimate is based on calculations of the contributions of different sources to the total lead content of livestock (Penumarthy et al., 1980) and food crops (Wolnik et al., 1983, 1985), relative to natural concentrations of lead in the environment (Elias et al., 1982). The latter were determined with ultraclean measurements of lead concentrations and isotopic compositions in a remote terrestrial food chain. Those evidence a 10-fold increase in lead concentrations above natural levels from the deposition of atmospheric lead aerosols. The contribution of atmospheric lead to the total lead concentration in food crops and livestock is shown in Table 2.

The atmospheric lead content of crops correlates with their proximity to anthropogenic sources of lead. Numerous studies have shown that lead levels on the surfaces of vegetation are proportional to air lead concentrations, especially in the particle size range below 1 μm (Ratcliffe, 1975; Pilegaard, 1978). Other factors also contribute to the high degree of variability of lead collected on the plant surfaces. Earlier studies found differences related to plant species, meteorologic conditions, and local soil conditions (Dedolph et al., 1970; Motto et al., 1970; Schuck and Locke, 1970; Ter Haar, 1970; Rabinowitz, 1974; Welch

TABLE 2   Background Lead in Basic Food Crops and Meats

Crop	Natural Pb[a]	Indirect Atmospheric[a]	Direct Atmospheric[a]	Total[b]
Wheat	0.0015	0.0015	0.034	0.037
Potatoes	0.0045	0.0045	—	0.009
Field corn	0.0015	0.0015	0.019	0.022
Sweet corn	0.0015	0.0015	—	0.003
Soybeans	0.021	0.021	—	0.042
Peanuts	0.005	0.005	—	0.010
Onions	0.0023	0.0023	—	0.0046
Rice	0.0015	0.0015	0.004	0.007
Carrots	0.0045	0.0045	—	0.009
Tomatoes	0.001	0.001	—	0.002
Spinach	0.0015	0.0015	0.042	0.045
Lettuce	0.0015	0.0015	0.010	0.013
Beef (muscle)	0.0002	0.002	0.02	0.02[c]
Pork (muscle)	0.0002	0.002	0.06	0.06[c]

Source: U.S. Environmental Protection Agency (1986a).
[a]Data are from EPA (1986a).
[b]Data are from Wolnik et al. (1983, 1985).
[c]Data are from Penumarthy et al. (1980).
Units in μg Pb/g fresh weight.

and Dick, 1975). More recent studies have also shown lead concentrations are highest in leafy crops, which are subject to atmospheric deposition, and lowest in root crops (Nicklow et al., 1983; Nasralla and Ali, 1985). Although the total lead concentrations in preharvested food crops are primarily from atmospheric deposition, internal lead concentrations may be a combination of atmospheric and soil origin. The latter concentrations are most often a function of soil lead levels (Nicklow et al., 1983).

Lead contamination of livestock and poultry is also primarily derived from atmospheric lead. Grazing animals take up lead from forage and feed. Lead concentrations in forage are related to the atmospheric deposition rate, as in food crops. Forage grown adjacent to heavily utilized roads may contain more than 950 μg Pb/g (Graham and Kalman, 1974). Other factors that may contribute to the accumulation of lead in domesticated animals from both atmospheric and nonatmospheric sources include direct soil ingestion (Thorton and Abrahams, 1983), ingestion of processed food, and lead added to the meat during processing. However, the transfer of soil lead is generally quite small relative to the later two sources.

From harvest to packaging, lead concentrations in food increase by a factor of 2 to 12 (Table 3). Some of that contamination occurs between the field and processing of the food, although there are no definitive data on how much lead is added during those stages. The amount is obviously related to the type of crop, as

TABLE 3   Addition of Lead to Food Products[a]

Food	In the Field (A)	After Preparation for Packing (B)	After Packing (C)	After Kitchen Preparation (D)	Total Pb after Harvest (E)
Soft packaged					
Wheat	0.037	N/A[b]	0.065	—	—
Field corn	0.022	N/A	0.14	0.025	0.003
Potatoes	0.009	N/A	0.018	0.02	0.011
Lettuce	0.013	N/A	0.07	0.015	0.002
Rice	0.007	N/A	0.10	0.084	0.077
Carrots	0.009	N/A	0.05	0.017	0.008
Beef	0.01	N/A	0.07	0.035	0.025
Pork	0.06	N/A	0.10	0.06	—
Metal cans					
Sweet corn	0.003	0.04	0.27	0.28	0.28
Tomatoes	0.002	0.06	0.29	—	—
Spinach	0.045	0.43	0.68	0.86	0.82
Peas	N/A	0.08	0.19	0.22	0.14
Applesauce	N/A	0.08	0.24	0.17	0.09
Apricots	N/A	0.07	0.17	—	0.10
Mixed fruit	N/A	0.08	0.24	0.20	0.12
Plums	N/A	0.09	0.16	—	0.07
Green beans	N/A	0.16	0.32	0.16	—

Source: U.S. Environmental Protection Agency (1986a).
[a]This table summarizes the stepwise addition of lead to food products at several stages between the field and the dinner table. Data in column A are from Wolnik et al. (1983, 1985), columns B and C from National Food Processors Association (1982b), and column D from U.S. FDA (1985). Column E is calculated as column D − column A. Where data are not available in column A, the values in column B were used. For the most part, column C values closely approximate column D values, even though they are from separate studies, suggesting most of the lead in food production is added prior to kitchen preparation. Units in μg Pb/g fresh weight.
[b]N/A, data not available.

well as the procedures used in handling, transportation, and storage. For example, the mode in which a crop is transported and stored will influence the amount of atmospheric lead added, and the method and degree to which it is washed will influence the amount of contaminant lead retained. In cases in which the food product has not undergone extensive modification, (e.g., cooking, added ingredients), the added lead was probably derived from the atmosphere or from the machinery used to handle the product.

In food processing, where modification of the product has occurred, the most common ingredients added are sugar, salt, and water. In procedures involving cooking in water prior to packaging, the foods often absorb the lead in that water (Smart et al., 1981). A summary of available data (Table 3) indicates that about 30% of the total lead in canned goods is the result of prepackaging processes.

The procedures and materials used in food packaging contribute substantially to the lead levels in some foods. This includes the contamination from lead-soldered cans (Settle and Patterson, 1980; National Food Processors Association, 1982a).

The amount of lead solubilized from those cans also depends on the acidity and method of storage. One study (Capar, 1978) showed that lead in acidic foods that are stored refrigerated in open cans can increase by a factor of 2 to 8 in 5 days if the lead-soldered side seam is not protected by an interior lacquer coating.

The amount of lead added to food during kitchen preparation and storage has not been well studied either. Pathways of contamination include direct atmospheric deposition, contact with kitchen utensils that indirectly transfer atmospheric lead, contact and cooking with lead-contaminated water, and storage in lead-contaminated containers. The degree of contamination is dependent on the food type, the lead levels associated with the different sources, and the duration of contact with those sources. For example, vegetables can absorb up to 80% of the lead in the cooking water (Little et al., 1981). The amount of lead they absorb depends on the cooking water lead concentration, the type of vegetable, the water hardness, salting, and cooking time.

6.2.   Lead-Soldered Cans

Metal cans have been a principal source of lead contamination in foods since they were introduced in 1810 (Nriagu, 1985b). The contamination results from the dissolution of lead solder used to seal the side seam on three piece "sanitary cans," which are named after the Sanitary Can Company. Additional analyses of the lead concentrations of tuna from unsoldered cans (7 ng/g wet weight) by Settle and Patterson (1980) indicated that approximately 99.5% of the contamination in tuna in lead-soldered cans was derived from the solder (Table 4). In 1982 to 1983, lead solder accounted for 33–51% of the lead in children's dietary intake (Beloian, 1982) and approximately 40% of the lead in the average dietary intake in the United States (Elias, 1985; U.S. Environmental Protection Agency, 1986a).

TABLE 4   Prehistoric and Modern Lead Concentrations in Human Food from a Marine Food Chain[a]

Source	Prehistoric	Current
Surface seawater	0.0005	0.005
Albacore muscle, fresh	0.03	0.3
Albacore muscle from diepunched unsoldered can	—	7.0
Albacore muscle, lead-soldered can	—	1400
Entire anchovy from albacore stomach	—	21
Part of anchovy from lead-soldered can	—	4200

Source: Settle and Patterson (1980).
[a]Units in ng Pb/g fresh weight.

TABLE 5   Temporal Decrease in Lead Concentrations (μg/g) of Infant Foods

Food type	1971–1972	1971–1975	1976–1977	1979–1980	1980–1981
Evaporated milk	1.04	0.52	0.10	0.08	0.07
Canned infant formula (conc.)[a]	0.50	0.10	0.055	0.015	0.01
Infant juices	—	0.30	0.045	0.015	0.015
Glass-packed solid infant foods	—	0.15	0.05	0.03	0.02

Sources: Lamm et al. (1973); Schaffner (1981); Corwin (1982); Jelinek (1982).
[a]Dabeka and McKenzie (1987) reported lead concentrations of infant formulas in lead-free cans in Canada of 0.0012 μg/g on a "ready-to use" basis.

Several key studies provided the impetus for reducing the use of lead-soldered food cans. Investigators demonstrated that the lead leached from solder in evaporated milk (Lamm et al., 1973) and baby foods (Mitchell and Aldous, 1974) represented a significant health threat to small children. That information led to a rapid decrease in the lead concentrations of infant formula by a factor of 3 to 10 within 1 year (Lamm et al., 1973) and successive decreases in the lead content of other baby foods in following years (Table 5). A second catalyst for the elimination of lead solder in canned adult foods was provided by the study of Settle and Patterson (1980), which revealed that the lead concentration in canned tuna was more than 4000-fold greater than that of fresh fish.

The U.S. Food and Drug Administration has instituted a voluntary program for the phaseout of lead-soldered food cans. Side seam welding is becoming the common method of sealing three piece cans, and seamless two piece containers are now being used to can most beverages and many foods. Additionally, nonmetal containers are increasingly being utilized to package foods that were previously canned. Further decreases in the contamination of canned foods are projected to result from continued efforts to develop alternatives to lead-soldered cans.

However, lead-soldered cans still constitute an important source of lead contamination in food. Food is commonly, if not exclusively, packaged in lead-soldered cans in many countries. And the lead solder in that food continues to represent a significant health threat in those countries. It also continues to represent a potential health threat in countries, including the United States, that import food in lead-soldered cans.

6.3.   Lead in Drinking Water

Atmospheric lead aerosols are the predominant source of lead contamination in surface waters. The natural concentration of lead in surface water is approximately 0.02 μg/L, but atmospheric inputs of anthropogenic lead have increased that concentration to 5–30 μg/L (Patterson, 1980; U.S. Environmental Protection Agency, 1986a). The predominance of those inputs is revealed by the recent 40% decrease of lead in the Mississippi River, which correlates with the decrease in atmospheric emissions of industrial lead aerosols (Trefry et al., 1985). Most of this lead is removed during water treatment, so that the major contribution of lead to drinking water is lead within the distribution system. Atmospheric inputs accounted for approximately 15% of the lead in drinking water in 1982, but probably have decreased as atmospheric emissions have decreased.

Although atmospheric aerosols and lead based soldered cans are the primary source of lead contamination in food, there are numerous other sources. The relative importance of these other substances is increasing as atmospheric emissions of industrial lead are being reduced, with the elimination of lead alkyls in gasoline.

The adverse effects of water contaminated by lead plumbing are extensive (U.S. Environmental Protection Agency, 1986b). An estimated 42 million people in the United States are exposed to drinking water with lead levels exceeding 20 μg/L. Sublethal lead toxicities in 1988, resulting from that exposure, are projected to include 130,000 cases of hypertension in middle-aged males, 270,000 cases of diminished intelligence in children, and 680,000 cases of elevated pregnancy risk in women and their fetuses.

Lead plumbing was recognized as a source of contamination during the Roman Empire (Nriagu, 1985b), and is now considered a significant source of lead exposure in the United States (U.S. Environmental Protection Agency, 1986b; Cook, 1987). The lead-based solder, which has been commonly used in new residential plumbing with copper pipes, is probably the greatest single contributor to lead contamination of drinking water in the United States (Chin and Karalekas, 1984; AWWADVGM, 1985). The lead in connectors and pipes also contributes to the lead contamination in drinking water. One recent survey

TABLE 6   Lead Contamination in Tap Water by Age of Plumbing

Reference	Age of Housing	Mean Pb Level (μg/L)	Study Conditions
Sharrett	< 18 months	74	Median standing levels
et al.	< 5 years	31	No lead pipes
(1982)	> 5 years	4.4	
Nassau	Unoccupied	2690	Average, first flush
County...	< 2 years	540	
(1985)	2–10 years	60	
	> 10 years	10	
Philadelphia	< 2 years	90	Flushed
(1985)		5000	First flush
	> 2 years	60	Flushed
		500	First flush
	> 4.5 years	< 25	

of 153 public water systems found that almost 75% of the distribution systems contained lead service lines or connections, and that lead solder and flux were utilized in almost half of them (Chin and Karalekas, 1984).

The solubilization of lead from plumbing systems is a function of the amount of exposed material and the age of that material, the pH of the water, and the hardness of the water. A pH below 8 or alkalinity below 35 can increase the corrosive activity of the water and cause more lead to dissolve in the distribution system. Generally, lead leaches more readily in newer plumbing systems because the materials contain higher concentrations of leachable lead, and have not developed a protective build up of carbonate salts. Longer residence times in pipes between flushings can also substantially increase water lead concentrations (Moore, 1977). Lead contamination levels in tap water versus the age of the plumbing from several studies are listed in Table 6. These demonstrate that lead products used in plumbing will persist as the principal source of lead contamination in drinking water, even though the use of lead-based solder is currently restricted.

7. MODELS FOR DETERMINING DIETARY INTAKE OF CONTAMINANTS

Numerous model systems exist for determining the dietary intake of contaminants in food. The Committee on Food Consumption Patterns identified six capabilities needed to develop a model system to support and improve toxicological evaluations. These include the ability (1) to identify the primary patterns of food use and food components in the diet, (2) to identify extreme or unusual patterns of intake of foods or food ingredients including additives of incidental contaminants, (3) to identify the size and nature of populations at risk from use of certain foods or food products, (4) to determine the amount or number of food items in which a food additive may be permitted, (5) to determine the need to modify regulations in response to changes in food consumption, and (6) to determine intake of incidental contaminants and food additives.

Three parameters must be considered in estimating consumption of food contaminants by individuals. These are the kinds of foods consumed, the amounts of food consumed, and the concentration of the contaminant in each food. Beloian (1985) outlines a number of methods for obtaining data for the kinds and amounts of food consumed. These include food balance sheets (Hiemstra, 1968), food accounts (Pekkarinen, 1970; Young, 1981), food records, inventory records, and list recalls (Young, 1981; Beloian, 1985), food diaries and dietary records (Marr, 1971; Young, 1981; Beloian, 1985), dietary recall (Pekkarinen, 1970; Young, 1981), dietary history (Burke, 1947; Young, 1981), and food frequency record or recall (Marr, 1971; Morgan et al., 1978; Abrams, 1981).

There are many variations of the record and recall methods that provide consumption data on groups of people or on individuals. And when the various food consumption collection methods are evaluated, there are advantages and limitations to all of them (Young and Trulson, 1960). Different survey methods are likely to produce different values for estimates of mean intake. The accuracy of estimated intakes can also be influenced by a number of factors, such as sample size, survey period, and seasonality.

The collection of food samples for substance analyses generally involves one of two techniques, "duplicate portions" or "equivalent composites." Variations on the equivalent composite technique have been used in balance studies and also in food supply monitoring studies in various countries (Horuichi et al., 1956; Meranger and Smith, 1972; Kolbye et al., 1974; Mahaffey et al., 1975; Ministry of Agriculture, Fisheries and Food, 1975; Jelinek, 1982). For example, the U.S. Food and Drug Administration has used Total Diet Study programs to monitor chemical contaminants and nutrients in foods from representative diets for eight age/sex groups (Pennington, 1983).

The basic diets identified 234 typical food categories, and are the foundation for the U.S. Food and Drug Administration's revised Total Diet Study, which is often called the "market basket study." The diet information used here includes food, beverages, and drinking water for 2-year-old children, teenage males and females, adult males and females (25–30 years of age), and adult males and females (60–65 years of age). The 201 typical food categories that constitute the basic adult diets, and the 33 infant food categories, are an aggregation of 3500 categories of food actually consumed by participants in the two surveys that formed the basis of the Pennington study. Lead concentrations were determined for each of these 234 food categories by the U.S. Food and Drug Administration (1985). The U.S. Environmental Protection Agency (1986a) determined the probable source of lead in each food category and developed a model that can predict the relative contribution of lead from several sources to the total lead consumed in the human diet.

7.1. The Multiple Source Food Model

Ingestion of lead-contaminated food is the primary pathway of lead input to the body. Using the guidelines from the Committee on Food Consumption Patterns, a model has been developed that (1) identifies the primary patterns of food use and food components in the diet, (2) determines the amount of lead in each food category, (3) estimates the sources of lead in each food category, and (4) determines the intake of lead from incident consumption of food by category. The model does not attempt to identify the size and nature of the populations at risk, nor does it determine the need to modify regulations in response to changes in food consumption.

The Multiple Source Food Model uses the food consumption categories of Pennington (1983) and the individual food lead concentrations for each category. For source determinations, the food crop studies of Wolnik et al. (1983, 1985) are used to determine the original amount of lead present in each category of food at the time of harvest. Data from the National Food Processors Association (1982a) are used to estimate the amount of lead added during processing prior to packaging. The contribution of lead-based solder can be estimated from the difference between the canned and fresh product. Food that requires extensive kitchen preparation usually has more lead than similar food served fresh, so that the contribution for preparation can be estimated by this difference.

This model was originally developed for the U.S. Environmental Protection

TABLE 7   Recent Trends in Food and Beverage Consumption of Lead

Age/Sex Category	1982 (μg/day)	1983 (μg/day)	1984 (μg/day)	1985 (μg/day)
2 Year-old children				
Atmospheric Pb	10.9	7.8	7.2	4.9
Solder Pb	12.3	9.7	10.5	4.6
Other Pb	5.2	4.1	4.0	3.6
Total food and beverage Pb	28.4	21.6	21.7	13.1
Teenage females (14–16 yr)				
Atmospheric Pb	15.1	10.7	9.6	6.1
Solder Pb	14.9	9.7	13.0	4.3
Other Pb	7.6	5.3	5.0	4.3
Total food and beverage Pb	37.6	25.7	27.6	14.7
Teenage males (14–16 yr)				
Atmospheric Pb	21.8	15.4	14.2	8.6
Solder Pb	22.2	13.7	20.1	5.9
Other Pb	9.8	7.0	6.8	5.5
Total food and beverage Pb	53.8	36.1	41.1	20.0
Adult females (25–30 yr)				
Atmospheric Pb	15.2	10.5	10.1	6.1
Solder Pb	14.7	11.1	12.3	4.1
Other Pb	8.0	5.6	5.6	4.3
Total food and beverage Pb	37.9	27.2	28.0	14.5
Adult males (25–30 yr)				
Atmospheric Pb	23.5	15.0	15.3	8.6
Solder Pb	20.5	14.1	19.1	5.5
Other Pb	11.2	7.2	8.1	5.7
Total food and beverage Pb	55.2	36.3	42.5	19.8
Senior adult females (60–65 yr)				
Atmospheric Pb	13.8	10.1	10.0	5.9
Solder Pb	14.9	12.6	12.4	5.2
Other Pb	8.0	5.9	5.9	4.4
Total food and beverage Pb	36.7	28.6	28.3	15.5
Senior adult males (60–65 yr)				
Atmospheric Pb	17.9	12.7	12.9	7.5
Solder Pb	18.6	14.9	16.1	6.0
Other Pb	9.3	6.8	7.1	5.3
Total food and beverage Pb	45.8	34.4	36.1	18.8

Agency's Air Quality Criteria for Lead (1986a) in 1984, using food data for 1982–1983. A more extensive discussion of the methods was presented in that document. The model accounts for natural lead, indirect atmospheric lead that has been incorporated in soils, direct atmospheric lead added to food crops before harvest and during processing, and solder lead from food cans and drinking water. The lead that cannot be accounted for is considered lead of undetermined origin, although it is possible that some of this lead belongs in one of the previous categories. The results of the model are shown in Table 7 for the years 1982–1985. For simplicity, the categories natural lead, indirect atmospheric lead, and lead of undetermined origin have been combined into a miscellaneous category labeled "other."

The validation of any model lies in its ability to correctly predict trends in a dynamic situation. The food data for 1984 and 1985, using the same model parameters as for 1982–1983, show a decrease in the atmospheric lead contribution that corresponds in direction and degree to the actual change in atmospheric lead during this time (Figure 6). Those food categories with the greater atmospheric exposure decreased more than others. There was a corresponding decrease in the contribution of lead from solder in cans, although industry data are not available to document the actual reduction in the number of lead-soldered food cans used during this time period. Based on reductions of lead

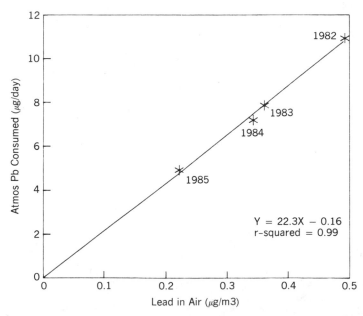

Figure 6. The relationship between lead in air and atmospheric lead in food (diet for 2-year-old children), as predicted by the Multiple Source Food Model. The model was based on data for 1982–1983 and was verified with data for 1984–1985. A similar relationship exists for solder lead in food, but there is not enough information available for quantification (Elias, 1987).

in canned foods, the model predicts that there was a 70% reduction in the use of lead soldered cans between 1982 and 1985.

The Multiple Source Food Model can also be used to project recent trends to some future date. Based on the assumption that lead in gasoline will reach 0.1 g/gal during 1988, the concentration of lead in air should reach an average of 0.05–0.1 μg/m^3 in the United States, except near point sources. Of course, there is always the possibility of increased emissions from municipal incinerators and coal-fired power plants. Nevertheless, at 0.05 μg/m^3, the atmospheric contribution of lead to food would be 5 μg/day for the diet of 2-year-old children. By the same token, if the projected decrease in the use of lead-based solder for food cans continues, the contribution of solder lead to food will reach a minimum during 1989. From this information, it is reasonable to conclude that by the end of 1989, the total lead in the diet of children in the United States will be 10–15 μg/day, and that most of this will be lead in drinking water and lead in dust inadvertently incorporated into food during processing and preparation. These are persistent forms of lead that will remain until drastic measures are taken for their removal.

Another feature of the Multiple Source Food Model is that lead exposure from unusual dietary sources can be evaluated. Such departures from the baseline diet might occur when the food source is from a contaminated area, such as seafood contaminated with lead-based bridge paint, or crops grown on land fertilized with sewage sludge. The model can also consider special diets for families with home gardens or who consume larger than normal amounts of ethnic or specialized foods.

7.2. The Total Human Exposure Model

The Total Human Exposure Model incorporates the results from the Multiple Source Food Model with the lead exposure by two other routes, inhaled air and ingested dust, to determine a baseline of human exposure to lead. Baseline exposure is the lowest achievable exposure to lead that does not require exceptional changes in life-style. It should not be confused with either natural levels or normal levels of lead exposure. A person exposed only to baseline levels of lead would eat a normal diet of supermarket food, live in an area, probably rural, that has less than 0.1 μg Pb/m^3 in the atmosphere, in a house with no lead-soldered pipes or history of lead-based paint, and engage in no hobbies, habits, or other activities, including smoking, that would increase lead exposure. Exposure above the baseline level is considered additive exposure. For each activity, the exposure in micrograms per day is added to the baseline exposure.

The data in Table 8 indicate a downward trend to total baseline exposure that is due largely to reductions of lead in air and food. There are no data that show a corresponding reduction of lead in dust. There is good evidence that children consume 100 mg of dust per day (LaGoy, 1987). This is a combination of housedust and street/playground dust. The evidence for the source of lead in these dusts comes from studies of dust with known origins. Housedust greater than 1000 μg Pb/g generally has a significant contribution from paint lead. Urban

TABLE 8   Recent Trends in Total Baseline Human Exposure to Lead

Age/Sex Category	1982 (μg/day)	1983 (μg/day)	1984 (μg/day)	1985 (μg/day)
2 Year-old children				
Food and beverage Pb	28.4	21.6	21.7	13.1
Air Pb	2.4	1.8	1.7	1.1
Dust Pb	20.1	20.1	20.1	20.1
Total food, air, and dust Pb	50.9	43.5	43.5	34.3
Teenage females (14–16 yr)				
Food and beverage Pb	37.6	25.7	27.6	14.7
Air Pb	4.8	3.5	3.4	2.2
Dust Pb	4.5	4.5	4.5	4.5
Total food, air, and dust Pb	46.9	33.7	35.5	21.4
Teenage males (14–16 yr)				
Food and beverage Pb	53.8	36.1	41.1	20.0
Air Pb	4.8	3.5	3.4	2.2
Dust Pb	4.5	4.5	4.5	4.5
Total food, air, and dust Pb	63.1	44.1	49.0	26.7
Adult females (25–30 yr)				
Food and beverage Pb	37.9	27.2	28.0	14.5
Air Pb	4.8	3.5	3.4	2.2
Dust Pb	4.5	4.5	4.5	4.5
Total food, air, and dust Pb	47.2	35.2	35.9	21.2
Adult males (25–30 yr)				
Food and beverage Pb	55.2	36.3	42.5	19.8
Air Pb	4.8	3.5	3.4	2.2
Dust Pb	4.5	4.5	4.5	4.5
Total food, air, and dust Pb	64.5	44.3	50.4	26.5
Senior adult females (60–65 yr)				
Food and beverage Pb	36.7	28.6	28.3	15.5
Air Pb	4.8	3.5	3.4	2.2
Dust Pb	4.5	4.5	4.5	4.5
Total food, air, and dust Pb	46.0	36.6	36.2	22.2
Senior adult males (60–65 yr)				
Food and beverage Pb	45.8	34.4	36.1	18.8
Air Pb	4.8	3.5	3.4	2.2
Dust Pb	4.5	4.5	4.5	4.5
Total food, air, and dust Pb	55.1	42.4	44.0	25.5

houses without interior lead-based paint usually have dust in the range of 300–1000 μg/g, most of which can be traced to the atmosphere (U.S. Environmental Protection Agency, 1986a). Current investigations of the persistency of lead have not yet revealed a downward trend that can be attributed to any particular source. It is likely that the slope of such a trend will be far more shallow than for air or food lead.

8. ADDITIVE LEAD EXPOSURE

Through the Total Human Exposure Model, such extraordinary factors as pica, location near point sources, lead-related hobbies (stained glass sculptures, soldering, target shooting), and secondary occupational exposure can be evaluated. Secondary occupational exposure occurs in households in which one or more members work in lead-related industries. These additive factors usually have a greater impact on the inhaled air and ingested dust routes than food and beverages, although housedust also influences lead added during food preparation.

There are several ways in which an individual may increase his or her exposure to lead above the baseline level. Smoking, living in an urban environment, excessive consumption of wine, or working in or living near a smelter are a few that are discussed in the Air Quality Criteria Document for Lead (U.S. Environmental Protection Agency, 1986a). We have chosen three topics that relate most directly to food consumption, and recognize that there are many other topics that could have been included.

8.1. Lead Exposure from Pica

Lead contamination in small children contrasts with that of others, in that the main source of lead in their diet is often from the ingestion of nonfood items through normal oral exploratory behavior (Charney et al., 1980) and normal hand-to-mouth activities (Lin-Fu, 1973; Roberts et al., 1974; Sayre et al., 1974; Sayre, 1978; Yaffe et al., 1983). Whereas it is normal for 2-year-old children to consume about 100 mg of dust per day (LaGoy, 1987), it is not unusual for children with pica to consume more than 1 g of dirt or paint chips per day. Lead-based paints (600–500,000 μg Pb/g) are the most important cause of high dose clinical lead toxicity in children (Centers for Disease Control, 1978). Other nonfood items (soil, ink, wallpaper) ingested by children may also be significant, especially soil (Landrigan et al., 1975; Lepow et al., 1975; Duggan and Williams, 1977; Centers, for Disease Control, 1978; Donner, 1979; Wesolowski et al., 1979; Roels et al., 1980; Yaffe et al., 1983).

Although there are very few data on the amount of paint or soil ingested by children, Boeckx's (1986) discussion of pica as a potential source of lead contamination in children illustrates the magnitude of this problem. He observes that the ingestion of just 200 mg of lead-containing paint chips per day can increase a child's total lead exposure by 2600 μg Pb/day. This is nearly 100-fold greater than the baseline lead exposure for children.

Lead-based paint as a source of lead contamination was first reported in Australia a century ago, yet "these observations were largely ignored or forgotten, to be rediscovered only in the early 1970's" (Lin-Fu, 1985). The severity of the problem in the United States was not recognized until a preliminary screening of inner-city infants revealed that 90% of them had blood lead concentrations above the current upper limit for normal children (Bradley et al.,

1956) and subsequent mass screenings revealed that almost 800, 000 infants and children had undue lead absorption (Chisholm and O'Hara, 1982). Estimates of the severity of this source of lead contamination in children have continued to increase as the recognized threshold of lead toxicity decreases (U.S. Dept of Health and Human Services, 1988), as discussed in the introduction to this chapter.

Despite restrictions on the lead concentrations of paints in 1979, pica remains a principal source of lead contamination. The lead concentrations in old paint, house dust, and soil in urban areas will remain high for a protracted period because of the limited mobility of lead. Lead concentrations of paints used on eating utensils, residential structures, toys, and furniture in the United States are now limited to a "safe" level of 600 μg/g. Although that concentration is significantly lower than previous concentrations, which were as high as 500, 000 μg/g, the consumption of 200 mg of this paint would increase a child's exposure to lead 4-fold above baseline.

8.2. Contamination from Applications of Sewage Sludge Fertilizers

The application of municipal sewage sludge fertilizers, with lead concentrations of approximately 500 μg/g (U.S. Environmental Protection Agency, 1976), may significantly increase the lead concentrations of both crops and groundwater. Average lead concentrations of food crops grown with sewage sludge (1.3–3.1 μg/g dry weight) are 10% higher than those (1.2–2.6 μg/g dry weight) grown in the same area without sludge (Table 9). Although those increases are relatively small, they represent increases of up to 500-fold above the natural lead concentrations in those crops. Sopper and Kerr (1980) observed groundwater concentrations of 90 μg/L 1 year following an application of sludge. This was 4-fold above background levels (20 μg/L). Tackett (1987) calculated that the

TABLE 9   Lead Concentrations of Vegetable Crops Grown with and without the Application of Municipal Sewage Sludge Fertilizer with a Lead Concentration of 530 μg/g[a]

Crop	Lead Concentration (μg/g dry weight)		Increase (%)
	Control	Sludge Application	
Lettuce	2.4	3.1	29.2
Broccoli	2.4	2.6	8.3
Potato	1.3	1.4	7.7
Tomato	1.6	1.7	6.3
Cucumber	2.6	2.6	0
Eggplant	1.2	1.3	8.3
String beans	2.5	2.7	8.0

Source: Tackett (1987).
[a]Data obtained from U.S. Environmental Protection Agency (1976).

combination of those increases may elevate the lead exposure of individuals by as much as 33%.

Contamination of crops with sewage sludge fertilizer has recently raised some concern within the National Food Processors Association (1982b), which has recommended that food crops that have been grown with sewage sludge fertilizer should not be purchased. Several major food companies have declined to buy food crops grown on fields where sludge has been applied, and it has been recommended that the utilization of sewage sludge fertilizers should be banned because of their high lead content (Tackett, 1987). However, sewage sludge fertilizers are still being applied to some agricultural fields.

8.3. Contamination from Lead Glazes

Lead glazes have been widely applied to cooking and eating vessels for several millennia (Nriagu, 1985b). A lead glaze is described in a Babylonian tablet from 1700 B.C., and applications of lead glazes may have been well established in China during the Chou Dynasty (1122–256 B.C.). Although the glazes were not commonly employed in western civilizations until the Islamic period, they became quite popular in the sixteenth century and have persisted since then. The U.S. Food and Drug Administration (1980) has placed a limitation on the solubility of lead (2.5–7 μg/mL) in kitchenware glazes, but because of difficulties of enforcement, they are still common on imported materials in the United States.

These glazes have remained a significant source of lead contamination since they were developed. Elevated bone lead concentrations in some ancient (1400 B.C.) North American Indians have been attributed to lead glazes on their pottery (Flegal et al., 1989). Lead glazing may have also been a principal factor in the epidemics of plumbism in ancient and Medieval periods, as well as the Devonshire colic and the West Indian dry gripes (Nriagu, 1985b).

For over 200 years, medical reports have recognized the ill effects of consuming foods stored in glazed containers, and the hazard had become commonly known by the end of the nineteenth century (Lin-Fu, 1985; Nriagu, 1985b). Lead glazes were also specifically cited in Accum's Treatise on the Adulteration of Foods, and Culinary Poison (1820). Reports of toxicity of lead glazes on pottery imported to the United States have continued to the present, both in the scientific literature (Spielholtz and Kaplan, 1980; Acra et al., 1981; Cubbon et al., 1981) and the press (Weisskopf, 1987). And the lead released from ceramic glazes into acidic foods stored in pottery is among the most frequently cited source of episodic cases of classic lead toxicity that are attributable to food with concentrations exceeding several parts per million (Mahaffey, 1978).

9. TOXIC EFFECTS OF LEAD

Estimates indicate that 3 million to 4 million preschool children have elevated blood lead levels above 15 μg/dL, which may impair hearing and heme formation (Hileman, 1987; U.S. Department of Health and Human Services, 1988). The ingestion of lead in food and water is a significant source of human lead contamination. This is primarily because, at the baseline level, the amount of lead exposure from the ingestion of contaminated food is substantial, relative to exposure from other sources (Table 8).

Approximately 15% of the lead ingested by adults and 50% of the lead ingested by children is metabolized (Hursh and Suomela, 1968; Harrison et al., 1969; Alexander et al., 1973; Rabinowitz et al., 1973; Blake, 1976; Chamberlain et al., 1978; Ziegler et al., 1978). Lead assimilated from ingested food therefore comprises a significant fraction of the total lead assimilated by the body. The ingestion of nonfood substances (e.g., dust, paint chips) also constitutes a significant fraction of the total lead metabolized, especially for children, even though the assimilation rate is much lower than for food or drinking water.

Lead is a nonessential element, with no known biological requirement. The uptake of lead is primarily from its binding to biomolecular substances, which are crucial to various physiological functions. This actuates a large number of toxicological actions, many of which did not become apparent until the latter part of this century. Lead compounds are now known to exert toxic effects on the central nervous, reproductive, renal, hematopoietic, and immune systems (Cantarow and Trumper, 1944; Pueschel et al., 1972; Rom, 1976; Meredith et al., 1978; Kerkvliet and Baecher-Steppan, 1982; Glickman et al., 1984; Weeden, 1984). This toxicity occurs at the subcellular level of organelle structures and processes, and continues up to the level of the overall general functioning and behavior of the organism (Needleman et al., 1979; Bull, 1980; U.S. Environmental Protection Agency, 1986c).

There are numerous, diverse effects of sublethal lead toxicity. The hematopoietic system appears to be especially sensitive to lead. Inhibition of components in the heme synthesis pathway has been observed in children with blood lead levels as low as 15 µg/dL (National Academy of Sciences, 1980; Boeckx, 1986). Symptoms of neurotoxicity are detectable in some human adults at blood lead levels as low as 40–60 µg/dL, and at even lower levels in children (Needleman et al., 1979; Mantere et al., 1982; Baker et al., 1983; Hogstedt et al., 1983). Sublethal concentrations of lead may also cause renal insufficiency, hypertension, and reproductive effects such as sterility, abortion, stillbirths, and neonatal deaths (Boeckx, 1986; U.S. Environmental Protection Agency, 1986c; Sharp et al., 1987).

Scarcely any studies of lead toxicity prior to 1980, with the possible exception of Needleman's (Needleman et al., 1979), dealt systematically with lower body lead levels (blood lead < 35 µg/dL) (Rutter, 1983). The Needleman study provided the most impressive evidence to date on the potentially damaging effects of lead levels in the range previously considered harmless (Rutter, 1980). Subsequent studies have further lowered the recognized level of sublethal lead toxicity to 10–15 µg/dL (U.S. Environmental Protection Agency, 1989). The thresholds for some deleterious effects have still not been established (Angle et al., 1982; Schwartz et al., 1986; U.S. Environmental Protection Agency, 1989), which substantiates the proposal that there may be no absolute threshold for lead toxicity (National Academy of Sciences, 1980).

10. SUMMARY

The anthropogenic production of lead has increased exponentially over the past 5 millennia to current rates of approximately 3×10^9 kg/yr. Emissions from that production to the environment now exceed those from natural sources by two orders of magnitude and have elevated lead concentrations throughout the biosphere by at least one order of magnitude. As a result of that contamination, natural lead concentrations in food must be derived from biogeochemical models. Many measurements of lead concentrations in food must be dismissed because they are invalid. Ultraclean methods of analysis have repeatedly demonstrated the inaccuracies of data collected by previously acceptable standard methods for measuring lead concentrations in the environment.

The principal sources of human lead exposure are atmospheric aerosols, solder, lead plumbing, and lead-based pigments. Of these, the predominant sources in food and beverages are atmospheric lead aerosols, and lead solder, with lead plumbing also a major problem in some regions. Atmospheric aerosols and paint pigments are major contributors to lead in dust, and can increase human exposure by direct ingestion or by incorporation of dust in food. The relative importance of these and other sources of lead in food depends on the food type and its history of exposure during production, harvesting, transport, packaging, preparation, and consumption.

Anthropogenic lead aerosols have elevated the lead concentrations in most organisms by an order of magnitude, and account for approximately 39% of the lead in food. This contamination is most prominent in leafy plants, which are most subject to atmospheric deposition. Lead aerosols are also the principal source of lead contamination in livestock and poultry, which accumulate the lead from contaminated forage and feed.

Lead solder accounts for approximately 40% of the dietary lead accumulated by adults, since it is a principal contaminant of both food and water. It may account for 99% of the lead in foods with lead-soldered cans. The health hazards of such foods are now recognized, and the utilization of lead-soldered cans for food is being phased out in several countries. However, lead-soldered cans are still commonly used in other countries both for local consumption and export.

Approximately 42 million people are exposed to drinking water with lead levels exceeding 20 μg/L, which has been proposed as the maximum concentration limit. Other studies indicate the exposure of lead from drinking water is even greater.

Pica is also a major source of lead contamination, especially in urban, preschool children. A child who ingests as little as 1 g of soil and 200 mg of lead paint chips in a day may be consuming a total of 2600 μg Pb/day and absorbing 550 μg Pb/day, which is significantly higher than baseline exposure of 35 μg Pb/day.

The contamination of acidic food stored in lead-glazed pottery is a common cause of episodic cases of classic lead toxicity. Lead concentrations in food often exceed several parts per million following contact with lead-glazed pottery. This has led to recent efforts to prohibit the use of lead glazes in kitchenware in several countries. However, lead glazes are still commonly used in other countries and imports from those countries continue to be a source of lead contamination.

Numerous models exist for determining the dietary intake of contaminant lead in food. These indicate that the ingestion of lead is approximately $50 \mu g$ Pb/day, and that this is the primary pathway of lead contamination in urban adults. Food accounts for about 66% of the baseline daily human consumption of lead.

The amount of lead consumed in food is of concern because of its toxicity. Lead is a pervasive element that is responsible for toxicological effects on the central nervous, reproductive, renal, hematopoietic, and immune systems. Inhibition of components in the hematopoietic system has been observed in children with blood lead levels as low as $15 \mu g/dL$. Other data suggest that there may be no threshold to lead toxicity in humans, and that any lead contamination in food may constitute a health threat.

REFERENCES

Abrams, I. J. (1981). In *Assessing Changing Food Consumption Patterns*. National Academy of Sciences, Washington D.C., pp. 119–134.

Accum, F. (1820). Treatise on the adulteration of foods and culinary poison. Longman, Hurst, Rees, Orne and Brown, London.

Acra, A., Dajani, R., Raffoul, Z., and Karahagopian, Y. (1981). "Lead-glazed pottery: A potential health hazard in the Middle East." *Lancet* 1(8217), 433–434.

Alexander, F. W., Delves, H. T., and Clayton, B. E. (1973). "The uptake and excretion by children of lead and other contaminants." In D. Barth, A. Berlin, R. Engel, P. Recht, and J. Smeets, eds., *Environmental Health Aspects of lead. Proceedings of International Symposium*, Amsterdam, The Netherlands, pp. 319–331.

American Water Works Association—DVGM (1985). *Internal Corrosion of Water Distribution Systems*. Cooperative Research Report. The AWWA Research Foundation.

Angle, C. R., McIntire, M. S., Swanson, M. S., and Stohs, S. J. (1982). "Erythrocyte nucleotides in children—Increased blood lead and cytidine triphospate." *Pediatr. Res.* 16, 331–334.

Baker, E. L., Feldman, R. G., White, R. F., and Harley, J. P. (1983). "The role of occupational lead exposure in the genesis of psychiatric and behavioral disturbances." *Acta Psychiatr. Scand. Suppl.* 67, 38–48.

Beloian, A. (1982). "Use of a food consumption model to estimate human contaminant intake." *Environ. Monit. Assess.* 2, 115–127.

Beloian, A. (1985). "Model for dietary survey data to determine lead exposure." In K. R. Mahaffey, ed., *Dietary and Environmental Lead: Human Health Effects*. Elsevier, Amsterdam, The Netherlands, pp. 109–155.

Biggins, P. D. E., and Harrison, R. M. (1979). "Atmospheric chemistry of automotive lead." *Environ. Sci. Technol.* 13, 558–565.

Blake, K. C. H. (1976). "Absorption of [203]Pb from gastrointestinal tract of man." *Environ. Res.* 11, 1–4.

Boeckx, R. L. (1986). "Lead poisoning in children." *Anal. Chem.* 58, 274A–287A.

Boutron, C., and Patterson, C. C. (1983). "The occurrence of lead in Antarctic recent snow, firn deposited over the last two centuries and prehistoric ice." *Geochim. Cosmochim. Acta* **47**, 1355–1368.

Boutron, C., and Patterson, C. C. (1986). "Lead concentration changes in Antarctic ice during the Wisconsin/Holocene transition." *Nature (London)* **323**, 222–225.

Boutron, C., and Patterson, C. C. (1987). "Relative levels of natural and anthropogenic lead in recent Antarctic snow." *J. Geophys. Res.* **92**, 8454–8464.

Boyle, E. A., Chapnick, S. D., and Shen, G. T. (1986). "Temporal variability of lead in the western North Atlantic." *J. Geophys. Res.* **91**, 8573–8593.

Bradely, J. E., Powell, A. E., Niermann, W., McGrady, K. R., and Kaplan, E. (1956). "The incidence of abnormal blood levels of lead in a metropolitan pediatric clinic: With observation on the value of coproporphyrinuria as a screening test." *J. Pediatr. (St. Louis)* **49**, 1–6.

Bruland, K. (1983). "Trace elements in sea water." In J. P. Riley, and R. Chester, eds., *Chemical Oceanography*, Vol. 8, Academic Press, New York, pp. 157–220.

Bull, R. J. (1980). "Lead and energy metabolism." In P. L. Singhal, and J. A. Thomas, eds., *Lead Toxicity*. Urban & Schwarzenberg, Baltimore, MD, pp. 119–168.

Burke, B. S. (1947). The dietary history as a tool in research. *J. Am. Diet. Assoc.* **23**, 1041–1046.

Burnett, M., Ng, A., Settle, D., and Patterson, C. C. (1980)." Impact of man on coastal marine ecosystems." In M. Branika and Z. Konrad, eds., *Lead in the Marine Environment. Proc. Int. Experts Discussion, Rovinj 1977*. Pergamon, New York, pp. 15–30.

Cantarow, A., and Trumper, M. (1944). *Lead Poisoning*. Williams & Wilkins, Baltimore, MD.

Capar, S. G. (1978). "Changes in lead concentrations of foods stored in their opened cans." *J. Food Saf.* **1**, 241–245.

Center for Disease Control (1978). "Preventing lead poisoning in young children. A statement by the Centers for Disease Control." *J. Pediatr.* **93**, 709–720.

Chamberlain, A. C., Heard, M. J., Little, P., Newton, D., Wells, A. C., and Wiffen, R. D. (1978). *Investigations into Lead from Motor Vehicles*. Harwell, United Kingdom, United Kingdom Atomic Energy Authority, Report No. AERE-R9198.

Charney, E., Sayre, J., and Coulter, M. (1980). "Increased lead absorption in inner city children: Where does the lead come from?" *Pediatrics* **65**, 226–231.

Chmiel, K. M., and Harrison, R. M. (1981). "Lead content of small mammals at a roadside site in relation to the pathways of exposure." *Sci. Total Environ.* **17**, 145–154.

Chin, D., and Karalekas, P. C. (1984). "Lead product use survey of public water supply distribution systems throughout the U.S." In *Proceedings of a Seminar on Plumbing Materials and Drinking Water Quality*. Cincinnati, Ohio.

Chisolm, J. J., Jr., and O'Hara, D. M., eds. (1982). *Lead Absorption in Children: Management, Clinical and Environmental Aspects*. Urban and Schwarzenberg, Baltimore, MD.

Cook, M. B. (1987). "Bad news, good news." *Environ. Sci. Technol.* **21**(6), 515.

Corwin, E. (1982). "On getting the lead out of food." *FDA Consumer* March, 19–21.

Cubbon, R. C. P., Roberts, W., and Marshall, K. (1981). "The extraction of lead from ceramic tableware by foodstuffs." *Trans. J. Br. Ceram. Soc.* **80**, 125–127.

Dabeka, R. W., and McKenzie, A. D. (1987). Lead, cadmium, and fluoride levels in market milk and infant formulas in Canada. *J. Assoc. Off. Anal. Chem.* **78**(4), 754–757.

Davis, A. O., Galloway, J. N., and Nordstrom, D. K. (1982). "Lake acidification: Its effect on lead mobility in the sediment of two Adirondack lakes." *Limnol. Oceanogr.* **27**, 163–167.

Dedolph, R., Ter Haar, G., Holtzman, R., and Lucas, H., Jr. (1970). "Sources of lead in perennial ryegrass and radishes." In *Symposium on Air Conservation and Lead*. Division of Water, Air and Waste Chemistry, 157th National Meeting, American Chemical Society. April 1969, Minneapolis, MN. *Environ. Sci. Technol.* **4**, 217–223.

Donner, H. E. (1979). "Soil lead and its potential hazards in California." In J. J. Wesolowski, ed., *The

Childhood Blood Lead Project. California Department of Health Services, Air and Industrial Hygiene Laboratory, SP 19.

Duggan, M. J., and Williams, S. (1977). "Lead in dust in city steets." *Sci. Total Environ.* **7**, 91–97.

Edgington, D. N., and Robbins, J. A. (1976). "Records of lead deposition in Lake Michigan sediments since 1800." *Environ. Sci. Technol.* **10**, 226–274.

Eisinger, J. (1977). "Lead and man." *Trends Biochem. Sci.* **2**, N147–N150.

Eisinger, J. (1984). "Lead in history and history in lead." Book Review. *Nature (London)* **307**, 573.

Elbaz-Poulichet, F., Hollinger, P., Huang, W. W., and Martin, J. M. (1984). Lead cycling in estuaries, illustrated by the Gironde Estuary, France. *Nature (London)*, **308**, 409–414.

Elias, R. W. (1985). "Lead exposures in the human environment." In K. R. Mahaffey, ed., *Dietary and Environmental Lead: Human Health Effects.* Elsevier, Amsterdam, The Netherlands, pp. 79–107.

Elias, R. W. (1987). "Recent changes in human lead exposure." In S. E. Lindberg and T. C. Hutchinson, eds., *Heavy Metals in the Environment.* International Conference on Heavy Metals in the Environment. CEP Consultants Ltd., Edinburgh, U.K., Vol. 2, pp. 197–202.

Elias, R. W., Hirao, Y., and Patterson, C. C. (1982). "The circumvention of the natural biopurification of calcium along nutrient pathways by atmospheric inputs of industrial lead." *Geochim. Cosmochim. Acta* **46**, 2561–2580.

Ericson, J. E., Shirahata, H., and Patterson, C. C. (1979). "Skeletal concentrations of lead in ancient Peruvians." *N. Engl. J. Med.* **300**, 946–951.

Facchetti, S., Geiss, F., Gaglione, P., Columbo, A., Garibaldi, G., Spallanzani, and Gilli, G. (1982). *Isotopic Lead Experiment.* Status Report. U.S. Environmental Protection Agency, Research Triangle Park, NC, p. 114.

Flegal, A. R. (1985). "Lead in a pelagic food chain." *Symp. Biol. Hung.* **29**, 83–90.

Flegal, A. R., and Patterson, C. C. (1983). "Vertical concentration profiles of lead in the central Pacific at 15°N and 20°S." *Earth Planet. Sci. Lett.* **64**, 19–32.

Flegal, A. R., and Patterson, C. C. (1989). Marine biogeochemistry of lead: Present and preindustrial fluxes through biological systems. Manuscript in preparation.

Flegal, A. R., and Smith, D. R. (1988). "Lead dispersion gradients in San Francisco Bay." *Proceedings of the 7th International Ocean Disposal Symposium.* Environment Canada, 468–483.

Flegal, A. R., Ericson, J. E., and Smith, D. R. (1989). "Skeletal concentrations of lead, cadmium, silver and zinc in North American Pecos Indians." *Lancet*, Manuscript in preparation.

Galloway, J. M., Thornton, J. D., Norton, S. A., Volchok, H. L., and McLean, R. A. N. (1982). "Trace metals in atmospheric deposition: A review and assessment." *Atmos. Environ.* **16**, 1677–1700.

Gilfillan, S. C. (1965). "Lead poisoning and the fall of Rome." *J. Occup. Med.* **7**, 53–60.

Glickman, L., Valciukas, J. A., Lilis, R., and Weisman, I. (1984). "Occupational lead exposure: Effects on saccadic eye movements." *Int. Arch. Occup. Environ. Health* **54**, 115–125.

Graham, D. L., and Kalman, S. M. (1974). "Lead in forage grass from a suburban area in northern California." *Environ. Pollut.* **7**, 209–215.

Hamelin, B., Shen, G. T., and Sholkovitz, E. R. (1987). "Anthropogenic Pb in sediments and corals: Isotopic contrast of Pb pollution from Europe, USA and Japan." *Eos* **68**, 1319.

Harrison, G. E., Carr, T. E. F., Sutton, A., and Humphreys, E. R. (1969). "Effect of alginate on the absorption of lead in man." *Nature (London)* **224**, 1115–1116.

Harrison, R. M., and Williams, C. R. (1982). "Airborne cadmium, lead and zinc at rural and urban sites in north-west England." *Atmos. Environ.* **16**, 2669–2681.

Hiemstra, S. J. (1968). *Food Consumption, Prices, Expenditures.* Agricultural Economic Report No. 138. U.S. Government Printing Office, Washington D.C., pp. 1–15.

Hileman, B. (1987). "Lead levels in water raises concerns anew." *Chem. Engin. News* **65**(51), 5.

Hogstedt, C., Hane, M., Agrell, A., and Bodin, L. (1983). "Neuropsychological test results and symptoms among workers with well-defined long-term exposure to lead." *Br. J. Ind. Med.* **40**, 99–105.

Horuichi, K., Yamamoto, T., and Tamori, E. (1956). Studies on the industrial lead poisoning: 1. Absorption, deposition, and excretion of lead. 2. A study on the lead content and daily food in Japan. *Osaka City Med. J.,* **3**, 84–113.

Huntzinger, J. J., Friedlander, S. K., and Davidson, C. F. (1975). "Material balance for automobile-emitted lead in Los Angeles basin." *Environ. Sci. Technol.* **9**, 448–507.

Hursh, J. B., and Suomela, J. (1968). "Absorption of ^{212}Pb from the gastrointestinal tract of man." *Acta Radiol.* **7**, 108–120.

Jelinek, C. F. (1982). "Levels of lead in the United States food supply." *J. Assoc. Off. Anal. Chem.* **65**, 942–946.

Kerkvliet, N. I., and Baecher-Steppan, L. (1982). "Immunotoxicology studies on lead: Effects of exposure on tumor growth and cell-mediated tumor immunity after syngeneic or allogeneic stimulation." *Immunopharmacology* **4**, 213–224.

Kobert, R. (1909). In P. Diergart, ed., *Beitrage asu der Geschichte der Chemie.* Franz Duticke, Liepzig, pp. 103–119.

Kolbye, A. C., Jr., Mahaffey, K. R., Fiorino, J. A., Corneliussen, P. C., and Jelinek, C. F. (1974). Food exposures to lead. *Environ. Health Perspect.* **7**, 65–74.

LaGoy, P. K. (1987). "Estimated soil ingestion rates for use in risk assessment." *Risk Anal.* **7**, 355–359.

Lamm, S., Cole, B., Glynn, K., and Ullmann, W. (1973). "Lead content of milks fed to infants—1971–1972." *N. Engl. J. Med.* **289**, 574–575.

Landrigan, P. J., Gehlbacj, S. H., Rosenblum, B. F., Shoults, J. M., Candelaria, R. M., Barthel, W. F., Liddle, J. A., Smrek, A. L., Staehling, N. W., and Sanders, J. F. (1975). "Epidemic lead absorption near an ore smelter: The role of particulate lead." *N. Engl. J. Med.* **292**, 123–129.

Lee, J. A., and Tallis, J. H. (1973). "Regional and historical aspects of lead pollution in Britain." *Nature (London)* **245**, 216–218.

Lepow, M. L., Brunkman, L., Gillette, M., Markowitz, S., Robino, R., and Kapish, J. (1975). "Investigation into sources of lead in the urban environment of urban children." *Environ. Res.* **10**, 415–426.

Lin-Fu, J. S. (1973). "Vulnerability of children to lead exposure and toxicity: Parts 1 and 2." *N. Engl. J. Med.* **289**, 1229–1233.

Lin-Fu, J. S. (1985). "Historical perspective on health effects of lead." In K. R. Mahaffey, ed., *Dietary and Environmental Lead: Human Health Effects.* Elsevier, Amsterdam, The Netherlands, pp. 43–63.

Lindberg, S. E., and Harriss, R. C. (1981). "The role of atmospheric deposition in an eastern U.S. deciduous forest." *Water, Air, Soil Poll.* **16**, 13–31.

Little, P., Fleming, R. G., and Heard, M. J. (1981). "Uptake of lead by vegetable foodstuffs during cooking." *Sci. Total Environ.* **17**, 111–131.

Livett, E. A., Lee, J. A., and Tallis, J. H. (1979). "Lead, zinc and copper analysis of British blanket peas." *J. Ecol.* **67**, 865–891.

Mahaffey, K. R. (1978). "Environmental exposure to lead." In J. O. Nriagu ed., *The Biogeochemistry of Lead in the Environment.* Elsevier/North Holland, Amsterdam, The Netherlands, pp. 1–39.

Mahaffey, K. R., Corneliussen, P. E., Jelinek, C. F., and Fiorino, J. A. (1975). "Heavy metal exposure from foods." *Environ. Health Perspect.* **12**, 63–69.

Mantere, P., Hanninen, H., and Hernberg, S. (1982). "Subclinical neurotoxic lead effects: Two-year follow-up studies with psychological test methods." *Neurobehav. Toxicol. Teratol.* **4**, 725–727.

Marr, J. W. (1971). "Individual dietary survey: Purposes and methods." *World Rev. Nutr. Diet.* **13**, 105–164.

Meranger, J. C., and Smith, D. C. (1972). "Heavy metal content of a typical Canadian diet." *Can. J. Public Health* **63**, 53–57.

Meredith, P. A., Moore, M. R., Campbell, B. C., Thompson, G. G., and Goldberg, A. (1978). "Delta-aminolaevulinic acid metabolism in normal and lead-exposed humans." *Toxicology* 9, 1–9.

Ministry of Agriculture, Fisheries and Food. (1975). Working party for the monitoring of foodstuffs for heavy metals." Fifth Report: *Survey of Lead in Food*. Her Majesty's Stationery Office, London, pp. 1–34.

Mitchell, D. G., and Aldous, K. M. (1974). "Lead content of foodstuffs." *Environ. Health Perspect.* 5, 59–64.

Moore, M. R. (1977). "Lead in drinking water in soft water areas: Health and hazards." *Sci. Total Environ.* 7, 109–115.

Morgan, R. W., Jain, M., Miller, A. B., Choi, N. W., Matthews, V., Munan, L., Burch, J. B., Feather, J., Howe, G. R., and Kelly, A. (1978). "Comparison of dietary methods in epidemiologic studies." *Am. J. Epidemiol.* 107, 488–498.

Motto, H. L., Daines, R. H., Chilko, D. M., and Motto, C. K. (1970). "Lead in soils and plants: Its relationship to traffic volume and proximity to highways." *Environ. Sci. Technol.* 4, 231–238.

Murozumi, M., Chow, T. J., and Patterson, C. (1969). "Chemical concentrations of pollutant lead aerosols, terrestrial dusts and sea salts in Greenland and Antarctic snow strata." *Geochim. Cosmochim. Acta* 33, 1247–1294.

Nasralla, M. M., and Ali, E. A. (1985). "Lead accumulation in edible portions of crops grown near Egyptian traffic roads." *Agr. Ecosys. Environ.* 13, 73–82.

Nassau County Department of Health (1985). "Report of investigation of drinking water contamination by lead/tin solder in Nassau County." New York, New York.

National Academy of Sciences, Committee on lead in the human environment. (1980). *Lead in the Human Environment*. National Academy of Sciences, Washington D.C.

National Food Processors Association, Can Manufacturers Institute, Inc. (1982a). "Comprehensive supplementary report covering further research, February 1980 through May 1982, on food-borne lead in the diet of children." Available from U.S. Department of Health and Human Services, Washington, D.C. FDA Docket No. 79N-0200.

National Food Processors Association (1982b). "Background document on the use of municipal sewage sludge in the production of foods for human consumption." Washington, D.C.

Needleman, H. L., Gunnoe, C. E., Leviton, A., Reed, R., Peresia, H., Maher, C., and Barnett, P. (1979). "Deficits in psychological and classroom performance of children with elevated dentine lead levels." *N. Engl. J. Med.* 300, 689–695.

Ng, A., and Patterson, C. (1981). "Natural concentrations of lead in ancient Arctic and Antarctic ice." *Geochim. Cosmochim. Acta* 45, 2109–2121.

Ng, A., and Patterson, C. C. (1982). "Changes of Pb and Ba with time in California offshore basin sediments." *Geochim. Cosmochim. Acta* 46, 2307–2322.

Nicklow, C. W., Comas-Haezebrouck, P. H., and Feder, W. A. (1983). "Influence of varying soil lead levels on lead uptake of leafy and root vegetables." *J. Am. Soc. Hortic. Sci.* 108, 193–195.

Nriagu, J. O. (1978). "Lead in the atmosphere." In J. O. Nriagu, ed., *The Biogeochemistry of Lead in the Environment. Part A: Ecological Cycles*. Elsevier/North-Holland, Biomedical Press, Amsterdam, The Netherlands, pp. 15–72.

Nriagu, J. O. (1979). "Global inventory of natural and anthropogenic emissions of trace metals to the atmosphere." *Nature (London)* 279, 409–411.

Nriagu, J. O. (1983a). *Lead and Lead Poisoning in Antiquity*. Wiley, New York.

Nriagu, J. O. (1983b). "Occupational exposure to lead in ancient times." *Sci. Total Environ.* 31, 105–116.

Nriagu, J. O. (1983c). "Saturnine gout among Roman aristocrats. Did lead poisoning contribute to the fall of the empire?" *N. Engl. J. Med.* 308, 660–663.

Nriagu, J. O. (1985a). "Cupellation: The oldest quantitative chemical process." *J. Chem. Educ.* 62, 668–674.

Nriagu, J. (1985b). "Historical perspective on the contamination of food and beverages with lead." In K. R. Mahaffey, ed., *Dietary and Environmental Lead: Human Health Effects*. Elsevier, Amsterdam, The Netherlands, pp. 1–41.

Nriagu, J. O., and Pacyna, J. M. (1988). "Worldwide contamination of the air, water and soils with trace metals—Quantitative assessment." *Nature (London)*, **333**, 134–139.

Palmer, K. T., and Kucera, C. L. (1980). "Lead contamination of sycamore and soil from lead mining and smelting operations in eastern Missouri." *J. Environ. Qual.* **9**, 106–111.

Patterson, C. C. (1980). "An alternative perspective—Lead pollution in the human environment: Origin, extent and significance." In *Lead in the Human Environment*. National Academy of Sciences, Committee on Lead in the Human Environment, Washington D.C.

Patterson, C. C. (1982). *Natural Levels of Lead in Humans*. Carolina Environmental Essay Series III. The Institute for Environmental Studies, University of North Carolina at Chapel Hill.

Patterson, C. C., and Settle, D. M. (1976). "The reduction of order of magnitude errors in lead analyses of biological materials and natural waters by evaluating and controlling the extent and sources of industrial lead contamination introduced during sample collecting and analysis." In P. LaFleur, ed., *Accuracy in Trace Analysis*. Natl. Bur. Stand. Spec. Publication 422, Washington D.C., pp. 321–351.

Patterson, C. C., and Settle, D. M. (1987). "Present and ancient sums of net global eolian fluxes of industrial and natural lead contained in seasalts, soil dusts, volcanic sulfate aerosols and anthropogenic carbonaceous aerosols to lands and seas at various geographic locations." *SEAREX Newslett.* **10**, 1–24.

Patterson, C. C., Shirahata, H., and Ericson, J. E. (1987). "Lead in ancient human bones and its relevance to historical developments of social problems with lead." *Sci. Total Environ.* **61**, 167–200.

Pekkarinen, M. (1970). Methodology in the collection of food consumption data." *World Rev. Nutr. Diet.* **12**, 145–171.

Pennington, J. A. T. (1983). "Revision of the Total Diet Study food list and diets." *J. Am. Diet. Assoc.* **82**, 166–173.

Penumarthy, L., Oehme, F. W., and Hayes, R. H. (1980). "Lead, cadmium, and mercury tissue residues in healthy swine, cattle, dogs, and horses from the midwestern United States." *Arch. Environ. Contam. Toxicol.* **9**, 193–206.

Pilegaard, K. (1978). "Airborne metals and SO_2 monitored by epiphytic lichens in an industrial area." *Environ. Pollut.* **17**, 81–92.

Pueschel, S. M., Kopito, L., and Schwachman, J. (1972). "Children with an increased lead burden: A screening and follow-up study." *J. Am. Med. Assoc.* **222**, 462–466.

Rabinowitz, M. B. (1974). "Lead contamination of the biosphere by human activity: A stable isotope study." Dissertation, Los Angeles, CA: University of California.

Rabinowitz, M. B., Wetherill, G. W., and Kopple, J. D. (1973). Lead metabolism in the normal human: Stable isotope studies. *Science (Washington, DC)*, **182**, 725–727.

Ragaini, R. C., Ralston, H. R., and Roberts, N. (1977). "Environmental trace metal contamination in Kellogg, Idaho, near a lead smelting complex. *Environ. Sci. Technol.* **11**, 773–781.

Ratcliffe, J. M. (1975). "An evaluation of the use of biological indicators in an atmospheric lead survey." *Atoms. Environ.* **9**, 623–629.

Robbins, J. A. (1978). "Geochemical and geophysical applications of radioactive lead." In J. O. Nriagu, ed., *The Biogeochemistry of Lead in the Environment*. Elsevier, Amsterdam, The Netherlands, pp. 285–293.

Roberts, T. M. (1975). "A review of some biological effects of lead emissions from primary and secondary smelters." In T. C. Hutchinson, S. Epstein, A. L. Page, J. Van Loon, and T. Davey, eds., *International Conference on Heavy Metals in the Environment*. Symposium Proceedings. Vol. 2, Part 2. Toronto, Canada. University of Toronto, Institute for Environmental Study, Toronto, Canada, pp. 503–532.

Roberts, T. M., Hutchinson, T. C., Paciga, J., Chattopadhyay, A., Jervis, R. E., Van Loon, J., and Parkinson, D. K. (1974). "Lead contamination around secondary smelters: Estimation of dispersal and accumulation by humans." *Science* **186**, 1120–1123.

Roels, H. A., Buchet, J. P., Lauwerys, R. R., Bruaux, P., Claeys-Thoreau, F., Lafontaine, A., and Verduyn, G. (1980). "Exposure to lead by the oral and the pulmonary routes of children living in the vicinity of a primary lead smelter." *Environ. Res.* **22**, 81–94.

Rom, W. N. (1976). "Effects of lead on the female and reproduction: A review." *Mt. Sinai J. Med.* **43**, 542–552.

Rutter, M. (1980). "Raised lead levels and impaired cognitive/behavioural functioning: A review of the evidence." *Dev. Med. Child. Neurol.* **22**, Supplement No. 42.

Rutter, M. (1983). "Low level lead exposure: Sources, effects and implications." In M. Rutter and R. R. Jones, eds., *Lead versus Health.* Wiley, New York, pp. 333–370.

Sayre, J. W. (1978). "Dust lead contribution to lead in children." *Second International Symposium on Environmental Lead Research*, Cincinnati, Ohio.

Sayre, J. W., Charney, E., Vastal, J., and Pless, Y. B. (1974). "House and hand dust as a potential source of childhood lead exposure." *Am. J. Dis. Child.* **127**, 167–170.

Schaffner, R. M. (1981). "Lead in canned foods." *Food Technol.* **35**, 60–64.

Schaule, B. K., and Patterson, C. C. (1981). "Lead concentrations in the northeast Pacific: Evidence for global anthropogenic perturbations." *Earth Planet. Sci. Lett.* **54**, 97–116.

Schaule, B., and Patterson, C. (1983). "Perturbations of the natural lead depth profile in the Sargasso Sea by industrial lead." In C. S. Wong, E. Boyle, K. W. Bruland, J. D. Burton, and E. D. Goldberg, eds., *Proc. NATO Adv. Res. Inst. on Trace Metals in Sea Water.* Plenum, New York, pp. 487–503.

Schell, W. R. (1986). "Deposited atmospheric chemicals." *Environ. Sci. Technol.* **20**, 847–853.

Schuck, E. A., and Locke, J. K. (1970). "Relationship of automotive lead particulates to certain consumer crops." *Environ. Sci. Technol.* **4**, 324–330.

Schwartz, J., Angle, C., and Pitcher, H. (1986). "The relationship between childhood blood lead and stature. *Pediatrics* **77**, 281–288.

Settle, D. M., and Patterson, C. C. (1980). "Lead in albacore: Guide to lead pollution in Americans." *Science* **207**, 1167–1176.

Sharp, D. S., Becker, C. E., and Smith, A. H. (1987). "Chronic low-level lead exposure: Its role in the pathogenesis of hypertension. " *Med. Toxic.* **2**, 210–232.

Sharrett, A. R., Carter, A. P., Orheim, R. M., and Feinleib, M. (1982). "Daily intake of lead, cadmium, copper and zinc from drinking water: The Seattle study of trace metal exposure." *Environ. Res.* **28**, 456–475.

Shen, G. T., and Boyle, E. A. (1987). "Lead in corals: Reconstruction of historic industrial fluxes to the surface ocean." *Earth Planet. Sci. Lett.* **82**, 289–304.

Shirahata, H., Elias, R. W., Patterson, C. C., and Koide, M. (1980). "Chronological variations in concentrations and isotopic compositions of anthropogenic atmospheric lead in sediments of a remote subalpine pond." *Geochim. Cosmochim. Acta* **44**, 149–162.

Smart, G. A., Warrington, M., and Evans, W. H. (1981). "Contribution of lead in water to dietary intakes. " *J. Sci. Food. Agr.* **32**, 129–133.

Sopper, W. E., and Kerr, S. N. (1980). Monitoring report: Stripmine Reclamation Project, Decker Site, Somerset County, Pa. Submitted to Pa. Dept. of Env. Res. (June 1980).

Spielholtz, G. I., and Kaplan, F. S. (1980). "The problem of lead in Mexican pottery." *Talanta* **27**, 997–1000.

Tackett, S. L. (1987). "Lead in the environment: Effects of human exposure." *Am. Lab.* July, 32–41.

Ter Haar, G. (1970). "Air as a source of lead in edible crops." *Environ. Sci. Technol.* **4**, 226–229.

Thorton, I., and Abrahams, P. (1983). "Soil ingestion—A major pathway of heavy metals into livestock grazing contaminated land." *Sci. Total Environ.* **28**, 287–294.

Trefry, J. H., Metz, S., Trocine, R. P., and Nelson, T. A. (1985). "A decline in lead transport by the Mississippi River." *Science* **230**, 439–441.

Turekian, K. K., Nozaki, Y., and Benninger, L. K. (1977). "Geochemistry of atmospheric radon and radon products." *Annu. Rev. Earth Planet. Sci.* **5**, 227–255.

U.S. Bureau of Mines. (1987). Lead. In: Minerals yearbook; Volume I. metals and minerals. Washington, DC: U.S. Department of the Interior.

U.S. Department of Health and Human Services. (1988). The nature and extent of lead poisoning in children in the United States: A report to Congress. Agency for Toxic Substances and Disease Registry, Public Health Service. July, 1988.

U.S. Environmental Protection Agency. (1976). "Applications of sewage sludge to cropland: Appraisal of potential hazards of the heavy metals to plants and animals." Report No. 64, Council for Agricultural Science and Technology, U.S. EPA 340/9-76-013. Washington, D.C.

U.S. Environmental Protection Agency. (1986a). "Air quality criteria for lead." Vol. II. U.S. EPA Environmental Criteria and Assessment Office, Research Triangle Park, NC. EPA/600/8-83/028bF.

U.S. Environmental Protection Agency. (1986b). "Reducing lead in drinking water: A benefit analysis." U.S. EPA Office of Policy Planning and Evaluation, Washington, DC. EPA-230-09-86-019.

U. S. Environmental Protection Agency. (1986c). "Air quality criteria for lead." Vol. IV. U.S. EPA Environmental Criteria and Assessment Office, Research Triangle Park, NC. EPA-600/8-83/028dF.

U.S. Environmental Protection Agency (1989). Supplement to the 1986 EPA Air Quality Criteria for Lead. Volume I Addendum. EPA/600/8-89/049A.

U. S. Food and Drug Administration. (1980). "Compliance program report of findings: FY 77 Total Diet Studies—Adult (7320.73)." U.S. Department of Health, and Human Services, U.S. Food and Drug Administration, Industry Programs Branch (HFF-326), Washington D.C.

U.S. Food and Drug Administration. (1985). Market basket survey: preliminary results for lead analysis. Washington, DC: U.S. Department of Health and Human Services.

Veron, A., Lambert, C. E., Isley, A., Linet, P., and Grousset, F. (1987). "Evidence of recent lead pollution in deep North-East Atlantic sediments." *Nature (London)* **326**, 278–281.

Wedeen, R. P. (1984). *Poison in the Post: The Legacy of Lead.* Southern Illinois University Press, Carbondale, IL.

Weisskopf, M. (1987). "Lead astray: The poisoning of America." *Discover* **8**, 68–75.

Welch, W. R., and Dick, D. L. (1975). "Lead concentrations in tissues of roadside mice." *Environ. Pollut.* **8**, 15–21.

Wesolowski, J. J., Flessel, C. P., Twiss, S., Stanley, R. L., Knight, M. W., Coleman, G. C., and DeGarmo, T. E. (1979). "The identification and elimination of a potential lead hazard in an urban park." *Arch. Environ. Health* **34**, 413–418.

Winder, C. (1984). *The Developmental Neurotoxicity of Lead.* MTP Press, Lancaster, England, 161 pp.

Wolff, E. W., and Peel, D. A. (1985). "The record of global pollution in snow and ice." *Nature (London)* **313**, 535–540.

Wolnik, K. A., Fricke, F. L., Capar, S. G., Braude, G. L., Meyer, M. W., Satzger, R. D., and Bonnin, E. (1983). "Elements in major raw agricultural crops in the United States: 1. Cadmium and lead in lettuce, peanuts, potatoes, soyabeans, sweet corn, and wheat." *J. Agr. Food Chem.* **31**, 1240–1244.

Wolnik, K. A., Fricke, F. L., Capar, S. G., Meyer, M. W., Satzger, R. D., Bonnin, E., and Gaston, C. M. (1985). "Elements in major raw agricultural crops in the United States: 3. Cadmium, lead and eleven other elements in carrots, field corn, onions, rice, spinach, and tomatoes." *J. Agr. Food Chem.* **33**, 807–811.

Yaffe, Y., Flessel, C. P., Weslows, J. J., Delrosar, A., Guirguis, G. N., Matias, V., Degarmo, T. E., Coleman, G. C., Gramlich, J. W., and Kelly, W. R. (1983). "Identification of lead sources in California children using the stable isotope ratio technique." *Arch. Environ. Health* **38**, 237–245.

Young, C. M. (1981). *Assessing Changing Food Consumption Patterns.* National Academy of Sciences, Washington D.C., pp. 89–118.

Young, C. M., and Trulson, M. F. (1960). Methodology for dietary studies and epidemiological surveys. II. Strengths and weaknesses of existing methods. *Am. J. Public Health.* **50**, 803–814.

Ziegler, E. E., Edwards, B. B., Jensen, R. L., Mahaffey, K. R., and Fomon, S. J. (1978). "Absorption and retention of lead by infants." *Pediatr. Res.* **12**, 29–34.

5

FOOD CONTAMINATION WITH ARSENIC IN THE ENVIRONMENT

Jerome O. Nriagu and Jose M. Azcue

National Water Research Institute, Burlington, Ontario, Canada

1. HISTORICAL

> It is an uncanny thought that this lurking poison [arsenic] is everywhere about us, ready to gain unsuspected entrance to our bodies from the food we eat, the water we drink and the air we breathe. (Karl Vogel, 1928).

Arsenic compounds were apparently known and employed in bronze alloys since about 3000 B.C. (Partington, 1935), and were used as medicaments well before the time of Hippocrates (about 400 B.C.). The use of such compounds as insecticides dates back to ancient times. Pliny (*Natural History*; cited by Whorton, 1969, p. 11), for example, noted that sandarach (realgar or red arsenic sulfide) was used in protecting grapes and the Chinese are believed to have employed arsenic compounds against garden insects as early as the tenth century A.D. (Whorton, 1969). During the Middle Ages, arsenic compounds found varying uses in agriculture and during the late 1700s some French farmers were substituting arsenic for lime as a treatment of grain blight and by 1846 a royal ordinance was issued to outlaw such applications of poisons on food items (Whorton, 1969).

The dawn of the era of arsenic insecticides began during the 1860s in the United States. The rapid spread of potato beetle from Colorado to Illinois and Wisconsin that began in 1860 helped to foster the interest in chemical insecticides as the existing pest control methods were found to be useless. In 1862, Rathvon reported the successful use of arsenic mixed with marshed potato or grated carrots against household insect pests. By 1868, Edwin Reynolds, a farmer in Fond du Lac, Wisconsin had reported that the potato beetle could be destroyed by applying a mixture of *Paris Green* (copper acetoarsenite, a common pigment of the period) and ashes to the foliage. Reynolds' claim immediately received a stiff rebuttal by the editors of *American Agriculture* (Vol. **27**, p. 321, 1868) who took issue with the "unsafe advice":

> The following is going the rounds of the press. "Sure death to potato bugs: take 1 lb Paris Green, 2 lbs pulverized lime. Mix together, and sprinkle the vines." We consider this unsafe, as there is no intimation of the fact, not generally known, that Paris Green is a compound of arsenic and copper, and a deadly poison... The poison would be very likely to kill the potato bugs, but how about the vines? [or the farmer who applied the mixture or the people who ate the potato treated with arsenic].

The bitter debate started by Reynold's use of chemical toxins as spray insecticides on fruit crops has continued unabated up to the present time. In the absence of alternative defense measures, Paris Green was enthusiastically accepted by the farmers and by 1875 well over 500 tonnes of the compound were being sold annually in New York market alone (Whorton, 1969).

The success of Paris Green prompted the trial of London Purple (a mixture of calcium arsenates and arsenites with some organic matter) as an insecticide. This by-product of the aniline dye industry had no use and was being dumped into the sea because of its high toxicity. Once it debuted in the 1870s, London Purple was quickly adopted by the agricultural industry because of its cheapness, greater adhesion to plant leaves, greater ease of application, and its more conspicuous color (Whorton, 1969).

The next episode in the saga of arsenical insecticides was linked to the rapid spread of the gypsy moth that was accidentally introduced into the United States in 1869 and soon was ravaging the forest and orchard trees throughout the New England states. Since both the Paris Green and London Purple are highly phytotoxic (the hardwood foliage was harmed as much by these insecticides as by the moth), they could not be used against the menacing moths. As a solution to this dilema, F. C. Moulton introduced lead arsenate, which is much less phytotoxic than its predecessors, as an insecticide in 1892. Lead arsenate was soon found to be effective against other insects (such as horn worm on tobacco crops) and could easily be applied by mechanical means. Calcium arsenate was subsequently introduced in 1906 primarily to control cotton pests in the sourthern parts of the United States. These two arsenic compounds were to remain the backbone of the insecticide industry until the widespread production of synthetic organic insecticides during and after the World War II. Their unassailed status at the turn of this century is reflected by the rapid increase in the agricultural application of lead arsenate (in the United States) from about 5.4 million kg in 1919 to over 13 million kg in 1929. During the same period, the use of calcium arsenate rose from about 1.4 million kg to over 13 million kg (Whorton, 1969). The insecticidal use of each of these compounds during the peak period of the 1930s and 1940s averaged about 23 million kg annually. Lead arsenate application peaked at about 40 million kg in 1944 while the calcium arsenate use exceeded 36 million kg. During the peak use period, lead and calcium arsenates were registered for use on 41 and 83 feed and food crops, respectively, in addition to many other nonfood uses on turf, ornamental, and shade trees (Alden, 1983).

Spray, farmers, spray with care,
Spray the apple, peach and pear;
Spray for scab, and spray for blight
Spray, O Spray, and do it right.

Spray the scales that's hiding there,
Give the insects all a share;
Let your fruit be smooth and bright,
Spray, O Spray and do it right.

Spray your grapes, spray them well,
Make first class what you've to sell,
The very best is none too good,
You can have it, if you would.

Spray your roses, for the slug,
Spray the fat potato bug;
Spray your cantaloupes, spray them thin,
You must fight if you would win. (E. G. Packard, 1906)

Packard's (1906) paean was symptomatic of the general attitude of farmers toward, and their growing heavy reliance on, lead and calcium arsenic insecticides at the beginning of this century.

Other inorganic arsenic compounds subsequently came into minor use, mostly as herbicides. Arsenic trioxide was found to be particularly effective as a soil sterilant. Sodium arsenite was applied in debarking trees, for controlling aquatic weeds, and as a defoliant prior to potato harvest. Today, inorganic arsenic compounds find occasional use, for example, in the control of bluegrass and other weeds on golf greens and fairways (Table 1). Mention, however, should be made of chromated copper arsenate (patented in 1938) and ammoniacal copper arsenate (patented in 1939), which are widely used as wood preservatives (Baldwin, 1983). By 1974, about 5 million kg of arsenic was used in wood preservation and this particular application continues to grow at the rate of about 10–25% per year (Woolson, 1983).

Although organic compounds of arsenic were being synthesized in 1907 by Paul Erlich, a German scientist, the commercialization of such compounds for agricultural purposes was a post-Word War II development. The use of herbicides containing monosodium methanoarsonate (MSMA), disodium methanoarsonate (DSMA), arsonic acid, and dimethylarsenic acid has grown rapidly since the mid-1970s and now represents one of the largest pesticides in modern use in terms volume (Wauchope and McDowell, 1984). Estimates of quantities used and acreages treated in the United States alone include 10 million kg and 10–12 million acres for MSMA + DSMA, 3.3 million kg and 2.1 million acres for arsenic acid, and 269,000 kg and 200,000 acres for cacodylic acid (Abernathy, 1983). The use of arsenic compounds, especially arsenilic acid, in poultry feed to control coccidiosis and promote chick growth came into practice during the 1930s (Luh et al., 1973; Anderson, 1983). Other current and historical uses of arsenic and its compounds are summarized in Table 1.

2. CONTAMINATION OF THE ENVIRONMENT WITH ARSENIC

The application of arsenic herbicides and pesticides represents a primary source of environmental contamination with arsenic. The high-pressure spraying techniques widely employed result in the contamination of not only the soils and vegetation but also the air and any surface waters nearby. The actual manufacturing of the arsenical pesticides often results in the discharge of arsenic whereas misplacement and leakage during transportation, distribution and

TABLE 1   Main Uses of Arsenic Compounds

Arsenic Compounds	Uses
Arsenic metal	In alloys (in combination with lead and copper)
	Automotive body solder (0.5% As)
	Industrial battery grid
	Electrophotography
High purity arsenic	Semiconductor applications
	Solar cells
	Optoelectronic devices
	Photoemissive surfaces
Cacodylic acid, MSMA, and DMSA[a]	Herbicides
	Defoliants in cotton fields
Calcium and lead arsenate	Pesticide on fruit crops[b]

TABLE 1 (*Continued*)

Arsenic Compounds	Uses
Arsenic acid	Desiccant or defoliant
	Wood preservative salts
	Feed additives
Sodium arsenite	Cattle and sheep dips
	Debarking trees
	Aquatic weed control
	Defoliant for potato
Phenylarsenic compounds	Animal feed additive
	Disease prevention
Arsenic trioxide and	Herbicides
arsenic salts	Soil sterilant
Refined arsenic trioxide	Decolorizer, and as oxidation/ reduction buffer in the production of glassware
Arsenic pentoxide	Wood preservative
Amorphous arsenic triselenoide	Electrophotography
Gallium arsenide	Light emitting diodes
	Solar cells
	Optoelectronic devices
Paris Green (copper acetoarsenite)	Sprays on apples and cherries[c], vegetables and small fruit[d], baits and mosquito larvae
Arsenic sulfide	Pyrotechnics
	Depilatory (leather industry)
Tertiary arsines	Catalysts or cocatalysts for polymerization of unsaturated compounds
	Ligands in coordination chemistry
Aromatic arseno compounds	Antisyphilitic drugs (such as salvasan and arsphenamine)
Melasoprol (arsobal) and trypansamide (tryponarsyl)	Treatment of trypanosomiasis
Carbarsone (Amebarsone) and glycobiarsol (Milibis)	Treatment of amebiasis
	Prevention of blackhead in turkey
Arsanilic acid	Feed additives
Arsonic and arsinic acids	Additives to motor fuel
	Corrosion inhibitors in steel
	Herbicides, fungicides, and bactericides
Roxarsone (3-nitro)	Feed additives

[a] MSMA, monosodium methanoarsonate; DSMA, disodium methanoarsonate.
[b] Until 1960.
[c] 1895–1920.
[d] 1895–1957.

storage can release some arsenic to the environment as well. In Texas, airborne concentrations of As are elevated from September to February, a period when cotton is ginned (Attrep et al., 1975). It has been estimated that 6 g of As is released during the ginning of a bale (500 lb) of cotton (Woolson, 1983). From the historical account above, it is obvious that, cummulatively, the quantity of pesticidal arsenic that has been released to the environment is clearly substantial. The arsenicals may be modified within the environment and/or dispersed by means of air and water to other ecosystems.

Many other sources also emit significant amounts of As to the atmosphere. Estimates by various authors (Table 2) suggest that 65% of the airborne As is derived, on a global basis, from the smelting of base metal ores (Nriagu and Pacyna, 1988). Most of the primary arsenic used commercially is obtained as a by-product of the beneficiation of gold and base metal ores (Carapella, 1979). Because arsenic trioxide, the main component of smelter flue dust, is volatile, a large fraction of the As in the ores is often vented to the atmosphere. It has been estimated that the Ronnskar smelter in northern Sweden used to release 50–115 tonnes of As per year (Beckman, 1978), a gold smelter in Yellowknife, Canada released 19–2600 tonnes of As per year, and the emission from the ASARCO smelter in Tacoma, Washington was about 7–152 tonnes (Woolson, 1983). A pyrite roasting plant in Barreiro/Seixal, Portugal has been known to release 1000–2000 tonnes of As annually to the atmosphere (Andreae et al., 1983). At such emission rates, the airborne concentrations of As around gold and base metal smelters can reach $2500 \, \mu g/m^3$ and the concentrations in surrounding soils can exceed $41,000 \, \mu g/g$ (Woolson, 1983).

It has been estimated that coal combustion contributes about 11% of the As emission to the global atmosphere (Nriagu and Pacyna, 1988). A halo of elevated As levels has been reported around many power plants (see Woolson, 1983). Other sources generally contribute $< 5\%$ of the estimated annual global anthropogenic As emission (Table 2).

On a local scale, significant air pollution with arsenic can stem from the burning of trash cotton plants sprayed with arsenic. For example, arsenic concentrations of $600–141,000 \, \mu g/m^3$ were reported at a distance of 46–91 m downwind of burning trash from a cotton gin (Woolson, 1983). Hutton and Symon (1986) believe that the buring of timber treated with cacodylic acid (CCA) is an important source of airborne As in the United Kingdom and that the disposal of the ashes of such timber contributes significantly to the arsenic burden of British soils. Microbial methylation processes can also volatilize some of the pollutant As in soils and sediments. It has been estimated that 17–35% of the total annual input of As into soils may be lost as methylarsines (Sandberg and Allen, 1975; Woolson, 1983).

The principal direct sources of arsenic pollution in surface waters include domestic and industrial wastewaters, electric power plants, base metal mining and smelting, and atmospheric fallout of contaminated aerosols (Table 3). Indirect sources of pollution include the leaching of slag heaps, ore bodies, and the residues of pesticides and fungicides from soils. It should also be noted that

TABLE 2 Worldwide Anthropogenic Arsenic Emissions to the Atmosphere ($\times 10^6$ kg/yr)

Sources	Nriagu and Pacyna (1988) [median (range)]	Chilvers and Peterson (1987) (mean)	Walsh et al. (1979) (mean)
Coal combustion			
Electric utilities	891 (232–1,550)	2,410	550
Industry and domestic	1,089 (198–1,980)	3,830	
Oil combustion			
Electric utilities	17.4 (5.8–29)		
Industry and domestic	39.6 (7.2–72)		
Pyrometallurgical nonferrous metal production			
Mining	60 (40–80)		13
Pb production	1,170 (780–1,560)	1,430	2,220[a]
Cu–Ni production	10,625 (8,500–12,750)	12,080[b]	13,000[b]
Zn–Cd production	460 (230–690)	780[c]	
Secondary nonferrous metal production			
Steel–iron	1,417 (355–2,480)	60	4,200
Refuse incineration			
Municipal	273 (154–392)	78	430
Sewage sludge	37 (15–60)		
Herbicides		3,440	1,920
Cement production	534 (178–890)		
Wood combustion	180 (60–300)	425	600
Miscellaneous	2,025 (1,250–2,800)	3,527	667
Total emissions	18,800 (12,000–25,630)	28,060	23,600

[a]Lead and zinc production.
[b]Copper production.
[c]Zinc production.

MSMA and CCA are highly soluble in water and can be washed away before they are stabilized in the soils.

Detailed reviews of the arsenic cycle in natural waters can be found, for example, in Ferguson and Gavis (1972), Fowler (1983), and Andreae (1983); no attempt is made to address such a topic here.

Highly elevated levels of As of natural origin have been reported in groundwater in many parts of the world. Arsenic poisoning due to excessive exposure to "natural" arsenic in drinking water has been documented in the Cordoba region of Argentina, the Antafagasta province of northern Chile, Coahuila–Durango (Mexico), the Tainan region of Taiwan, Lane County (Oregon), Fairbanks (Alaska), and so on (see Pershagen, 1983 for a good

TABLE 3   Inputs of Arsenic into the Aquatic Ecosystem

	As ($\times 10^6$ kg/yr)	
Source	Median	Range
Domestic wastewater		
Central	4.95	(1.8–8.1)
Noncentral	4.2	(1.2–7.2)
Steam electric	8.2	(2.4–14)
Base metal mining and	0.31	(0–0.75)
dressing		
Smelting and refining		
Nonferrous metals	7.0	(1.0–13)
Manufacturing processes		
Metals	0.88	(0.25–1.5)
Chemicals	3.8	(0.6–7.0)
Pulp and paper	2.28	(0.36–4.2)
Petroleum products	0.03	(0–0.06)
Atmospheric fallout	5.65	(3.6–7.7)
Dumping of sewage sludge	3.55	(0.4–6.7)
Total input, water	41	(12–70)

Source: Nriagu and Pacyna (1988).

overview). A detailed discussion of such naturally contaminated groundwaters is outside the objectives of this report. It needs to be noted that any added arsenic intake from anthropogenic sources can exacerbate the risks for populations in such contaminated areas.

Arsenic in soils can be derived either naturally from the weathering of arsenic-containing rocks and minerals or through the input of wastes from human activities. The principal anthropogenic sources include the application of arsenic insecticides (principal source), the addition of commercial fertilizers and manure, the discharge of municipal and industrial sewage, wastewaters, and slags, and the fallout of aerosols from ore smelters and fossil fuel combustion plants (see Table 4). In soils, the chemistry of arsenic is very similar to that of phosphorus and adsorption by soil constituents and the formation of insoluble salts with a number of cations often immobilize the two elements in soils.

Although arsenical insecticide is often applied directly to the plants, it is ultimately transferred to the soil, the principal sink for As in the environment. As most of the arsenical residues have low solubility and low volatility, they generally accumulate in the top soil layers (Hiltbold, 1975; Woolson, 1983; Adriano, 1986). The ultimate fate of the applied arsenical insecticide depends on the (1) absorption by inorganic and organic matter in the soil, (2) leaching and removal by runoff, (3) evaporation into the air and wind wafting of contaminated soil particles, (4) degradation and methylation by soil microorganisms, (5) biodegradation and photodecomposition of organoarsenic compounds in soils,

TABLE 4   Worldwide Emissions of Arsenic into Soils

Source	As ($\times 10^6$)(kg/yr)	
	Median	Range
Agricultural and food wastes	3.0	(0–6.0)
Animal wastes	2.8	(1.2–4.4)
Logging and other wood wastes	1.65	(0–3.3)
Urban refuse	0.4	(0.09–0.7)
Municipal sewage sludge	0.13	(0.01–0.24)
Miscellaneous organic wastes	0.13	(0–0.25)
Solid wastes, metal mfg.	0.11	(0.01–0.21)
Coal fly ash and bottom ash	21.9	(6.7–37)
Fertilizer	0.01	(0–0.02)
Peat (agricultural and fuel uses)	0.27	(0.04–0.5)
Wastage of commercial products	38.5	(36–41)
Atmospheric fallout	13.2	(8.4–18)
Total input soils	82	(52–112)
Mine tailing	9.1	(7.2–11)
Smelter slags and wastes	6.8	(4.5–9.0)
Total discharge on land	98	(64–132)

Source: Nriagu and Pacyna (1988).

and (6) translocation by means of biological systems to other environments. Some of these processes merely serve in translocating the arsenic whereas others can actually reduce its total concentration in soils.

Surface soils from 12 states of the United States that had a history of arsenic application averaged 165 μg/g (range 106–2550 μg/g) As compared to the average value of only 13 μg/g As in untreated soils (Adriano, 1986). A survey of treated agricultural soils in Ontario found mean As values of 8.9 μg/g in potato soils, 9.9 μg/g in peach soils, 23 μg/g in sour cherry and orchard soils, and 54 μg/g in soils from apple orchard compared to the value of 6.3 μg/g in the untreated soils (Frank et al., 1976a, b). The As concentrations showed a strong correlation with annual rates of arsenical application and with the duration of arsenical usage. In general, it has been found that long-term applications result in As concentrations in orchard soils that average 200–300 μg/g (Woolson, 1983). The contamination of soils with As generally results in the build-up of this toxin in the human food chain (see below).

3. TRANSFER OF ARSENIC TO PLANT MATERIAL

The contamination of vegetable crops with arsenic can occur either during spray application or from the root uptake of the pollutant arsenic in soils (Adriano, 1986). For many years, the main source of As in foods was the spray residues in fruits and vegetables (see above). In a chapter entitled *A Steady Diet of Arsenic and Lead*, Kallet and Schlink (1933) tried to reflect on what was a major public health issue of their day:

> A four-year old Philadelphia girl dead, in August, 1932, from eating sprayed fruit. With a background of cases like these, are you willing to have even small doses of arsenic, a deadly poison, administered to you and your children daily, perhaps several times daily? Willing or not, if you eat apples, pears, cherries and berries, celery, and other fruits and vegetables, you are also eating arsenic, and there is good reason to believe that it may be doing you serious, perhaps irreparable injury.
>
> The source of this dangerous poison is the lead arsenate which is sprayed on fruits and on some vegetables to protect them from the coddling moth and other insects destructive to plants. It is extensively used, especially in the Western States, which produce our most attractive and unblemished fruits. A residue of arsenic and lead remains on the fruits, and when you wash your apple or pear under the faucet you remove only a small part of the poison. The fruit grower, however, can, under government direction, remove the poison almost completely with a wash of dilute hydrochloric acid.... But the Federal Food and Drug Administration, proceeding on the unproved theory that arsenic in small quantities is not injurious to your health, permits growers to market fruit and vegetables contaminated with 12/1000 of a grain of arsenic, in the form of arsenic trioxide, per pound of fruit... with numerous fruit growers completely unequipped for removing the spray residue, with the staff of Government inspectors available for fruit inspection far too small to exercise more than a fraction of the necessary supervision, and with the Food and Drug Administration, in its usual fashion far more concerned about the economic interests of the growers than about the health of the public, one must be blind to suppose that a large part of the supply of apples and pears and many other fruits and vegetables is not contaminated with far more arsenic than is legally permitted. (Kallet and Schlink, 1933, pp. 47–48)

Because the spray of arsenical insecticides on food crops is now banned in most countries, this particular source of food contamination is no longer of concern. Nevertheless, the insecticide spray practice has left high levels of arsenic in agricultural soils of many parts of North America (Wauchope, 1983). Not only that, over 80% of the current As consumption is still in agriculture, and although the annual application rates for organoarsenical insecticides are much smaller than the loading rates for the inorganic insecticides, the long-term build-up of As in soils cannot be totally ignored.

The direct fallout of arseniferous aerosols and the uptake of arsenic from the contaminated soils often result in elevated levels of arsenic in vegetable crops around smelters (see *Environmental Impacts of Smelters* by Nriagu, 1984). The Anaconda incident dramatizes the environmental consequences of uncontrolled As emissions from smelters. The huge Washoe copper smelter at Anaconda, Montana was emitting enough arsenic to render the surrounding ecosystems harzardous to humans and farm animals. In a detailed classic study following the outbreak of cattle death at an alarming rate, Haywood (1904) found that some soils within a 8-km radius of the Washoe smelter contained hundreds of parts per million of arsenic and that the organs of the dead animals had highly elevated levels of arsenic. This outbreak of animal fatality associated with widespread environmental contamination with As, and the associated investigation, led to legislation forcing smelters to install effective dust collectors on their stacks (Whorton, 1969).

Plants can also assimilate the As in fly ash, sludge, and from manure dumped on land. Several studies have indeed demonstrated a good correlation between the *available* As in soils (usually defined as the fraction extractable with a mild acid) and the As concentration in whole plants (Walsh et al., 1977; Woolson, 1983). From the public health standpoint, the correlation may be somewhat misleading as plants tend to exclude As from fruits and seeds. In fact, the edible portions of plants grown on contaminated soils seldom accumulate dangerous levels of As because of (1) the phytotoxicity of As implying that the plant may be killed before it accumulates dangerous levels of this element in its edible parts; it should be noted, however, that plants vary considerably in their tolerance to arsenic (Table 5); and (2) the competition of As with the generally much more abundance phosphorus to gain entry into plant roots. The concentrations in arsenic ammended soils seldom exceed $0.5-1.0 \mu g/g$ and rarely exceed the tolerance limit set for As in edible plant materials (Woolson, 1983).

In his critical assessment of the available literature on plant uptake of As from soils, Woolson (1983) arrived at the following conclusions. (1) Arsenic levels in plants are low and various crops have differing capacities to accumulate this toxin. (2) The highest levels of As occur in the roots with the skin often containing higher concentrations than the inner flesh. The levels in the peel rise with increased As concentrations in soil. (3) At a given As concentration in soil, plants grown on sandy soils accumulate more As than those grown on silty or clayey soils. (4) Different As compounds are taken up at different rates and exercise differing toxicities to the plants. In spite of these reassuring conclusions, the contamination of agricultural soils with arsenic will remain a matter of some concern for years to come.

The unusual ability of tobacco crop to accumulate arsenic was noted in 1883 by A. M. Peters who found considerable residues of this element in tobacco leaves sprayed with Paris Green to prevent damage by tobacco worm (Whorton, 1969). The rough surface of the tobacco leaves apparently ensnared the sprays and reduced any losses by weathering. Subsequent studies have shown that tobacco crops also have a special ability for accumulating As from soils (Small and

TABLE 5   Comparative Sensitivity of Various Plants to Arsenic

Tolerant	Moderately Tolerant	Low Tolerant
Fruit Crops		
Apples	Cherries	Peaches
Pears	Strawberries	Apricots
Grapes		
Raspberries		
Dewberries		
Vegetables and Fruit Crops		
Rye	Beets	Peas
Mint	Corn	Onion
Asparagus	Squash	Cucumber
Cabbage	Turnips	Snap beans
Carrots	Radish	Lima beans
Parsnips		Soybeans
Tomato		Rice
Potato		Spinach
Swiss chard		
Wheat		
Oats		
Cotton		
Peanuts		
Tobacco		
Forage Crops		
Sudangrass	Crested wheat grass	Alfalfa
Bluegrass	Timothy	Bromegrass
Italian ryegrass		Clover
Kentucky bluegrass		Vetch
Meadow fescue		
Red top		

Source: Adriano (1986).

McCants, 1962; Walsh et al., 1977). The elevated levels in tobacco products have been shown to be an important route of arsenic exposure for smokers (Zielhuis and Wibowo, 1984).

3.1. Wine Contamination

The fact that the arsenic spray residues on grapes often resulted in substantial contamination of wines and grape juice was documented by the end of the last century (Whorton, 1969). In 1938, it was reported that 43% of the 336 samples of German wines contained over 3 mg/L As, the highest concentration observed being 14 mg/L (Koelsch, 1958). Even after the ban of arsenical sprays, the accumulation of elevated levels of arsenic in grapes grown in contaminated soils remained a problem. In 1976, over 50% of the samples of table wines sold in the United States contained over 0.05 mg/L As (Crescelius, 1977). It is believed that wine consumption is now one of the major routes of exposure of the general population to arsenic (Woolson, 1983). It would thus seem surprising that no data are currently available on the health effects of ingesting arsenic in wines (Pershagen, 1983).

4. ARSENIC IN SEAFOODS

Seafoods consumed by man come from many trophic levels, from algae to sea mammals. It has been known for a long time that aquatic fauna and flora accumulate both organic and inorganic arsenic from water, the first determination of this element in fish and other marine organisms having been made around 1900 (Lunde, 1977). The As content of edible seaweed, a common product in Japan and southeast Asian countries, has been shown to be 19–172 μg/g dry weight, the mean value being 112 μg/g (WHO, 1981). Similar As concentrations of 10–100 μg/g dry weight have been reported in marine algae from the coast of Norway (Lunde, 1973). Some seafood supplements prepared from kelp have been found to contain 0.6–58 μg/g As (WHO, 1981). In general, organoarsenic compounds (methanoarsonate, dimethylarsinate, dimethylarsenosugars, arseno-lipids, etc.) account for 60–99% of the arsenic in the marine algae (Andreae, 1986).

Lobsters and shrimps captured in Canadian waters contain 4–21 μg/g As (Lawrence et al., 1986). Canned clams and smoked oysters sold in Canada contain considerably more arsenic—16–46 μg/g (Luh et al., 1973). Although shellfish can contain up to 55 μg/g As, most of the observed concentrations typically fall between 5 and 20 μg/g on dry weight basis (Penrose et al., 1977; Woolson, 1975).

Pelagic fish typically contain 0.3–3.0 μg/g As (Woolson, 1975; Penrose et al., 1977). The concentrations of arsenic in marine fish (including haddock, halibut, cod, herring, mackeral, and sole) obtained throughout Canada were found to be 0.31–21 μg/g fresh weight (Lawrence et al., 1986). The distribution of the arsenic varies with the fish species and the sample location. The concentrations of As in Canadian freshwater fish of only 0.01–0.24 μg/g fresh weight were much lower than those of the marine fish (Lawrence et al., 1986). The higher concentrations of As in marine fish and shellfish can be attributed to lower As concentrations in freshwater and to the difference in the P:As ratios in the two aquatic ecosystems. In general, arsenobetaine and arsenocholine account for 80–90% of the arsenic in seawater fish whereas 70–85% of the arsenic in freshwater fish has been shown to be in the form of unidentified methanol-extractable arsenic compound (Lawrence et al., 1986). Studies on the distribution and speciation of arsenic in aquatic biota from other parts of the world are similar to the results obtained for the Canadian samples (Woolson, 1975; WHO, 1981; Anke et al., 1980; Cappon, Chapter 6, this volume).

Because of the high concentrations of As often reported, it has been assumed that this element is magnified in the aquatic food chain (WHO, 1981). The following As concentrations and the P:As ratios of the various trophic levels (reported by Andreae, 1986) clearly refute such an assumption and point to a food chain biodiminution of the arsenic.

Trophic Level	As[a]	P[a]	P:As Ratio
Surface ocean	0.0015	0.016	11
Marine algae	25	4,250	170
Crustacea	1.5	9,000	6,000
Pisces	1.7	18,000	10,600
Mammalia	0.2	43,000	215,000

[a]Concentrations are in μ/g for seawater and μg/g dry wt for the biomass.

The marked increase in P:As ratio suggests that the biochemical depletion of arsenic relative to phosphorus and its conversion to the relatively harmless forms represent a natural control on As toxicity up the food chain (Andreae, 1986).

5.   CONTAMINATION OF MEAT AND MEAT PRODUCTS WITH ARSENIC

It used to be a common practice to feed arsenicated fodder to cattle on the assumption that the arsenic served as a tonic, as with the Styrian arsenic eaters who were also known to feed arsenic to their horses. Concerns about arsenic poisoning from consumption of beef and milk with elevated levels of arsenic first surfaced as a result of arsenic poisoning of the livestock foraging in areas that were sprayed with the arsenical insecticides (Whorton, 1969). The concern about the health effects of eating contaminated meat and meat products became more pointed with the introduction of organoarsenic compounds as feed additives in the mid-1940s (Calvert, 1975). Arsenicated feeds have been shown to be effective in the control of coccidiosis in chicken, and in enhancing growth, better feathering, increasing egg production, improving feed conversion and pigmentation, and so on (Calvert, 1975; Anderson, 1983).

Animal feeds often contain significant amounts of arsenic. The analysis of 18 nonmedicated poultry feeds from 8 states in the United States showed an average As content of 0.39 μg/g (Anderson, 1983). Cattle feedlot diet without any known supplements has been shown to contain over 0.1 μg/g As (Capar et al., 1978). The medication of the feeds is actually designed to increase their arsenic contents to much higher levels. Four arsenicals are now widely used as animal feed additives, namely arsanilic acid, Roxarsone (3-nitro-4-hydroxyphenyl arsonic acid), Nitarsone (4-nitrophenyl arsonic acid), and Carbarsone (*p*-ureidobenzene arsonic acid). The recommended doses of arsanilic acid and sodium arsanilate are 50–100 μg/g feed (0.005–0.01%) in swine and poultry feeds for increasing the weight gains and feed efficiency and 250–400 μg/g in swine feed to be given for 5–6 days for the control of swine dysentry (Anderson, 1983). The doses for Roxarsone are 25–50 μg/g for chicken and turkey feed and 25–75 μg/g for swine feed, to enhance weight gains and feed efficiency. It is also recommended that the dose be increased to 200 μg/g for 5–6 days for the control of swine dysentry. Nitarsone at 170 μg/g and Carbarsone at 225–340 μg/g have been used mostly to prevent blackhead in turkeys (Anderson, 1983). It is clear that the medicated feeds contain high levels of the arsenic, which have even been known to poison the exposed animals (Ledet and Buck, 1978).

The arsenicals in feeds will result in elevated levels of arsenic residues in animal tissues. This point is well illustrated in Table 6, which shows elevated concentrations of As in most tissues of wethers fed arsanilic acid at levels of up to 273 μg/g diet for 28 days. Sharp increases in the levels of arsenic in most tissues have also been reported in poultry, swine, cattle, and sheep receiving arsenicated feeds (Calvert, 1975; Anderson, 1983; Anke et al., 1980). Nearly every study that has been reported, however, points to the fact the arsenic in animal tissues declines rapidly following the withdrawal from the exposure to arsenical feed (see the many reports in Anke et al., 1980). In other words, arsenic residues are not bioaccumulated in edible animal tissues. For the wether study (Table 6), it was shown that over 80% of the ingested As was excreted (Calvert, 1975) and that the

TABLE 6   Arsenic Accumulation in Tissues of Wethers Fed Arsanilic Acid for 28 Days and the Following Depletion of the Arsenic in the Liver[a]

	Exposure Dose (μg/g Feed)			
Tissue	0 (control)	27	144	273
Concentration after 28 days exposure				
Whole blood	<0.01	0.06	0.27	0.54
Liver	<0.01	3.1	27	29
Kidney	<0.01	3.2	12	24
Muscle	<0.01	0.2	1.1	1.2
Depuration rate from the liver				
Withdrawal time (days)				
0	<0.01	3.1	27	29
2	<0.01	4.9	15	27
4	<0.01	2.9	8.4	11
6	<0.01	1.9	3.5	5

[a]From Calvert (1975).

levels of As in the liver had declined to 5 $\mu g/g$ or less after a 6-day withdrawal time. In the United States, farmers are required to withdraw the arsenical feed at least 5 days before slaughter. Nevertheless, the disposition and excretion rate for the arsenic depends on the particular tissue, the method of feeding, the water intake, and health status of the animal (Calvert, 1975). Thus, arsenic is often detected in significant amounts in tissues of poultry and swine given medicated feeds and these meat items remain a major contributor of arsenic in human diet (see following).

6. ARSENIC RESIDUES IN FOODS AND DIETARY INTAKE OF ARSENIC

With the exception of coffee extracts, rice, spinach, parsley, and seafoods, the typical concentrations of arsenic in the major food items are often below 0.1 $\mu g/g$ (Table 7). Poultry meat and pork from animals raised on medicated feeds can contain up to 1.0 $\mu g/g$ but the As content of beef is rarely a matter of concern. Beef in most countries has been shown to contain <0.03 $\mu g/g$ and even the beef liver and kidney rarely contain over 0.15 $\mu g/g$ As (Vos et al., 1986). Furthermore, the As content of uncontaminated milk and diary products rarely exceed 0.05 $\mu g/g$.

As noted previously, seafood represents a major source of As in human diets. For some fishermen and local populations on some islands, the consumption of large amounts of seafoods may result in excessive intake of arsenic. For example, the 40,000 inhabitants of the Faeroe Island (Indian Ocean) have been known to ingest, on the average, about 197 g fresh weight of edible seafood tissues made up of 157 g of marine fish, 30 g of crustaceans, and 5.5 g of mollusks (Bernhard and Andreae, 1984). Assuming that the edible fish, crustacean and mullusk tissues contain 0.2, 1.0, and 3.0 $\mu g/g$ (fresh weight) of As, respectively, the intake of As by the Faeroe Island population is estimated to be 78 $\mu g/day$. This intake rate,

TABLE 7   Levels of Arsenic in Different Foods[a]

Food	Level (mg/kg)
Wine	0.005–0.03
Beer	0.005
Vegetables and fruit juices	0.01–0.03
Coffee extract	0.97
Spinach	0.23
Parsley	0.26
Asparagus	0.02
Potatoes	0.01–0.3
Carrots	0.007–0.08
Bush beans	0.007–0.03
Apple	<0.05–0.20

TABLE 7 (*Continued*)

Food	Level (mg/kg)
Sweet corn	<0.05–0.05
Cabbage	<0.05–0.1
Lettuce	<0.05–0.25
Tomato	<0.05–0.12
Onion	<0.05–0.12
Cucumber	<0.05–0.5
Soybeans	<0.1
Fatty oils	0.002–0.02
Margarine	0.002–0.01
Butter	0.01
Lard	0.005
Milk and milk products	0.003–0.025
Flour	0.01
Rice	0.22
Brown bread	0.003
Corn	0.05
Sugar	0.0001–0.04
Salt	0.03
Pork	0.003–0.03
Pig liver	0.02
Pig kidneys	0.01
Beef	0.001–0.065
Beef liver	0.005–0.07
Beef kidneys	0.02–0.132
Veal	0.002–0.012
Calf liver	0.02–0.04
Calf kidney	0.015–0.02
Poultry	0.001
Turkey liver	0.217
Eggs	0.01–0.5
Herring	0.8–1.43
Codfish	0.6–7.29
Codfish liver	0.8
Pollack	0.3–1.6
Cuttlefish	1.0
Shrimps	3.2–25.7
Shellfish	1.6–2.9
Lobster	1.5–122.0
Freshwater fish	0.01

[a]Based on compilations by Arnold (1988) and Shacklette et al. (1979).

equivalent to about $1.2\,\mu g/kg$ day, is below the FAO/WHO recommended maximum daily load for humans of $2\,\mu g/kg$ day (Codex Alimentarius Commission, 1984). Nevertheless, the long-term effects of constant exposure to such a high dose of arsenic must be of some concern. In Japan, the average daily seafood consumption is estimated to be 95 g fish, 5 g crustacea, and 12 g mollusks (Bernhard and Andreae, 1984) and results in an estimated arsenic intake, from this particular source, of $59\,\mu g/day$. In the United States, the average daily intake of As from seafoods is estimated to be $8.7\,\mu g$, from the consumption figures given by Nriagu (Chapter 3, this volume).

Although the arsenic in seafoods (known to occur mostly as organoarsenic compounds) is effectively taken up in the gastrointestinal tracts of the consumer, it is rapidly excreted without being broken down into the toxicologically active species. The rapid excretion via the kidney has been documented by concentrations of arsenic in urine of over $1.0\,mg/L$ observed after a fish meal (Pershagen, 1983). Apparently, the formation of very stable organoarsenic compounds in the aquatic biota provides some protection of the entire food chain, including the human consumer, against arsenic poisoning (Bernhard and Andreae, 1984; Andreae, 1986).

The mean arsenic concentrations in 10 food categories served in Canada range from $0.46\,\mu g/g$ in drinking water to over $60\,\mu g/g$ for meat and fish (Table 8). Using the data for the food categories, the daily As intake in this country is estimated to range from 2.6–101, with an average value of $17\,\mu g$ (Dabeka et al., 1987). Individual intakes are strongly influenced by the consumption of specific foods. On the average, meats and fish account for over 30% of the daily As intake while the grain and wheat products are responsible for 18% of the dietary As dose (Table 8). The dietary intake of arsenic in Canada of 2.6–101 $\mu g/day$, equivalent to 0.045–1.72 $\mu g/kg$ day, is well below the FAO/WHO recommended maximum acceptable consumption rate of 2.0 $\mu g/kg$ day.

Surveys of the total adult diets in the United States show that between 1978 and 1982, the average intake of As has been fairly constant at 45–50 $\mu g/day$ (Table 9). About 80% of the dietary As intake in 1981/1982 came from "meat, fish, and poultry" whereas grain and wheat products were responsible for about 17% of the dietary exposure. Thus, these two food groups accounted for over 95% of the dietary As intake by the United States population (Gartrell et al., 1985, 1986). Toddlers and infants consume considerably less arsenic in their meals. During 1981/1982, the average dietary intakes of As by toddlers and infants in the United States were 16 and 1.0 $\mu g/day$, respectively (Gartrell et al., 1985, 1986). The dietary intakes of arsenic in other countries (Table 9) are, in general, similar to rates in North America.

The sustained contamination of our air, water, and soils with arsenic will inevitably result in a build-up of this toxin in the human food chain. It is impossible, on the basis of the available information, to determine the fraction of our current dietary dose of arsenic that is "man made." Althought we do not know what the current margin of safety for dietary exposure of the general population to arsenic is, the erosion of this margin remains an on-going

TABLE 8 Description of Food Categories and Average Arsenic Concentration and Daily Intake in Canada[a]

Category	Foods description	Mean Food Intake (g)	Average Arsenic Concentration in Food Item (ng/g)	Average Arsenic Intake (µg/day)	Percentage of Intake from Different Categories
I	Cereals, porridge, bread, toast, buns, doughnuts, including bread from sandwiches	256	8.6	2.2	18.1
II	Water consumed directly, including duplicates from drinking fountains, ect.	319	0.46	0.15	1.4
III	Coffee, tea, beer, liquor, soft drinks (as consumed with or without milk, sugar, mixes, etc.)	780	1.0	0.78	8.6
IV	Fruit juices, fruits (canned or fresh)	297	3.8	1.13	6.4
V	Dairy products and eggs consumed directly; milk, eggs, omelette, yogurt, ice cream, milk shakes, cottage cheese, etc.	293	2.58	0.76	6.6
VI	Starch vegetables: rice, potatoes, yams, etc.	164	13.69	2.25	14.9
VII	Other vegetables: mushrooms, salads, vegetable-base soups, etc., including tomatoes and tomato juice	203	2.6	0.53	4.0
VIII	Meat, fish, poultry, meat-base soups, etc.	170	60.1	10.2	32.1
IX	Miscellaneous: pies, puddings other than rice, tarts, nuts, potatoes chips, chocolate bars, etc.	70	12.19	0.85	6.5
X	Cheese (other than cottage cheese)	17	8.46	0.14	1.4

[a]From Dabeka et al. (1987).

TABLE 9   Comparison of Dietary Arsenic Intakes by Adults in Different Countries

Country	Arsenic (μg/day)	Reference
Canada	16.7	Dabeka et al. (1987)
Canada	30	Smith et al. (1973)
United States	7[a]	Mahaffey et al. (1975)
United States	13–16	Kowal et al. (1979)
United States		
1976	50	Gartrell et al. (1985)
1977	54	Gartrell et al. (1985)
1978	45	Gartrell et al. (1985)
1979	47	Gartrell et al. (1985)
1980	48	Gartrell et al. (1986)
1981/1982	46	Gartrell et al. (1986)
UK	89	Food Additives and Contaminants Committee (1984)
UK	61	Evans and Sherlock (1986)
Belgium	12[b]	Buchet et al. (1983)
Holland	200[c]	de Vos et al. (1984)
	15[a]	
Austria	27	Pfannauser and Pechanek (1977)
New Zeland	55	Dick et al. (1978)
India	12	Krishna Murti (1987)

[a]Calculated using zero concentration for not detected.
[b]Median value.
[c]Calculated using detection limit values for not detected.

phenomenon. Contrary to what some people may believe, the so-called maximum permissible exposure dose has no relevance whatsoever in protecting the general population from the effects of long-term exposure to elevated levels of arsenic in the environment. As far as we know, there are few, if any, reliable early warning or biochemical indicators of subclinical arseniasis (Nriagu, 1988) and the maximum permissible dose pertains to observed symptoms of arseniasis.

REFERENCES

Abernathy, J. R. (1983). "Role of arsenical chemicals in agriculture." In W. H. Lederer and R. J. Fensterheim, eds., *Arsenic: Industrial, Biomedical, Environmental Perspectives.* Van Nostrand Reinhold, New York, pp. 57–62.

Adriano, D. C. (1986). *Trace Elements in the Terrestrial Environment.* Springer-Verlag, New York.

Alden, J. C. (1983). "The continuing need for inorganic arsenical pesticides." In W. H. Lederer and R. J. Fensterheim, eds., *Arsenic: Industrial, Biomedical, Environmental Perspectives.* Van Nostrand Reinhold, New York, pp. 63–70.

Anderson, C. E. (1983). "Arsenicals as feed additives for poultry and swine." W. H. Lederer and R. J.

Fensterheim, eds., *Arsenic: Industrial, Biomedical, Environmental Perspectives.* Van Nostrand Reinhold, New York, pp. 89–97.

Andreae, M. O. (1983). "Biotransformation of arsenic in the marine environment." In W. H. Lederer and R. J. Fensterheim, eds., *Arsenic: Industrial, Biomedical, Environmental Perspectives.* Van Nostrand Reinhold, New York, pp. 378–391.

Andreae, M. O. (1986). "Organoarsenic compounds in the environment." In P. J. Craig, ed., *Organometallic Compounds in the Environment.* Wiley, New York, pp. 198–228.

Andreae, M. O., Byrd, J. T., and Froellch, P. N. (1983). "Arsenic, antimony, germanium, and tin in the Tejo estuary, Portugal: modelling a polluted estuary." *Environ. Sci. Technol.* **17,** 731–737.

Anke, M., Gropped, B., Grun, M., Henning, A., and Meissner, D. (1980). "The influence of arsenic deficiency on growth, reproductiveness, life expectancy and health of goats." In M. Anke, H. J. Schneider, and Chr. Bruckner, eds., Book 3, Spurenelement—*Symposium Arsen*, GDR, pp. 25–32.

Arnold, W. (1988). "Arsenic." In H. Sigel and A. Sigel, eds., *Handbock of Toxicity of Inorganic Compounds.* Dekker, New York, pp. 79–93.

Attrep, Jr., Efurd, D. W., and Tribble, S. G. (1975). "Seasonal variation of atmospheric arsenic in a cotton growing region." *Texas J. Sci.* **26,** 549–552.

Baldwin, W. J. (1983). "The uses of arsenic as a wood preservative." In W. H. Lederer and R. J. Fensterheim, eds., *Arsenic: Industrial, Biomedical, Environmental Perspectives.* Van Nostrand Reinhold, New York, pp. 99–110.

Beckman, L. (1978). "The Ronnskar smelter—occupational and environmental effects in and around a polluting industry in northern Sweden." *Ambio* **7,** 226–231.

Bernhard, M., and Andreae, M. O. (1984). "Transport of trace metals in marine food chains." In J. O. Nriagu, ed., *Changing Metal Cycles and Human Health.* Springer, Berlin, pp. 143–167.

Buchet, J. R., Lauwerys, R., Vadervoorde, A., and Pycke, J. M. (1983). "Oral daily intake of cadmium, lead, manganese, copper, chromium, magnesium, calcium, zinc, and arsenic in Belgium: A duplicate meal study." *Food Chem. Toxicol.* **21,** 19–24.

Calvert, C. C. (1975). "Arsenicals in animal feeds and wastes." In E. A. Woolson, ed., *Arsenical Pesticides.* American Chemical Society, Washington D.C., pp. 70–80.

Capar, S. G., Tanner, J. T., Friedman, M. H., and Boyer, K. W. (1978). "Multielemental analysis of animal feed, animal wastes, and sewage sludge." *Environ. Sci. Technol.* **12,** 785–790.

Carapella, S. C. (1979). "Arsenic and arsenic alloys". In *Encyclopedia of Chemical Technology*, Vol. 3, Wiley, New York, pp. 243–265.

Chilvers, D. C., and Peterson, P. J. (1987). "Global cycling of arsenic." In T. C. Hutchinson and K. M. Meena, eds., *Lead, Mercury, Cadmium, and Arsenic in the Environment.* Wiley, New York, pp. 279–303.

Codex Alimentarius Comission. (1984). *Contaminants.* Joint FAO/WHO Food Standards Program. Codex Alimentarius, Vol. XVII, Edition 1.

Crescelius, E. A. (1977). "Changes in the chemical speciation of arsenic following ingestion by man." *Environ. Health Perspect.* **19,** 147–150.

Dabeka, R. W., McKenzie, A. D., and Lacroix, G. M. A. (1987). "Dietary intakes of lead, cadmium, arsenic and fluoride by Canadian adults: A 24-hour duplicate diet study." *Food Add. Contam.* **4,** 89–102.

de Vos, R. H., Dokkum, W., Olthof, P. D. A., Quirijas, J. K., Muys, T., and Poll, J. M. (1984). "Pesticides and other chemical residues in Dutch total diet samples (June 1976–July 1978)." *Food Chem. Toxicol.* **22,** 11–21.

Dick, G. L., Hughes, J. T., Mitchell, J. W., and Davison, F. (1978). "Survey of trace metals and pesticede residues in the New Zeland diet." *N.Z. J. Sci.* **21,** 57–69.

Evans, W. H., and Sherlock, J. C. (1986). "Relationships between elemental intakes within the United Kingdom total diet study and other adult dietary studies." *Food Add. Contam.* **4,** 1–9.

Ferguson, J. F., and Gavis, J. (1972). "A review of the arsenic in natural waters." *Water Res.* **6**, 1259–1274.

Food Additives and Contaminants Committee, (1984). *Report on the Review of the Arsenic in Food Regulations.* Ministry of Agriculture, Fisheries and Food. FAC/REP/39, HMSO, London.

Fowler, B. A. (ed.), (1983). *Biological and Environmental Effects of Arsenic.* Elsevier, Amsterdam.

Frank, R., Ishida, K., and Suda, P. (1976a). "Metals in agricultural soils of Ontario." *Can. J. Soil Sci.* **56**, 181–196.

Frank, R., Braun, H. F., Ishida, K., and Suda, P. (1976b). "Persistent organic and inorganic pesticide residues in orchard soils and vineyards of southern Ontario." *Can. J. Soil Sci.* **56**, 463–484.

Gartrell, M. J., Craun, J. C., Podrebarac, D. S., and Gunderson, E. L. (1985). "Pesticides, selected elements, and other chemicals in infant and toddler diet samples, October 1978–September 1979." *J. Assoc. Off. Anal. Chem.* **68**, 862–875.

Gartrell, M. J., Craun, J. C., Podrebarac, D. S., and Gunderson, E. L. (1986). "Pesticides, selected elements, and other chemicals in infant and toddler total diet samples, October 1980–March 1982." *J. Assoc. Off. Anal. Chem.* **69**, 123–145.

Haywood, J. K. (1904). USDA. Bureau of Chemistry, Bulletin **86**, p. 26.

Hiltbold, A. E. (1975). "Behaviour of organoarsenicals in plants and soils." In E. A. Woolson, ed., *Arsenical Pesticides.* American Chemical Society, Washington D.C., pp. 53–69.

Hutton, M., and Symon, C. (1986). "The quantities of cadmium, lead, mercury, and arsenic entering the U.K. environment from human activities." *Sci. Total Environ.* **57**, 129–150.

Kallet, A., and Schlink, F. J. (1933). *100,000,000 Guinea Pigs, dangers in Everyday Foods Drugs, and Cosmetics.* Vanguard, New York.

Koelsch, F. (1958). "Occupational arsenic lesions in vineyards and other places of employment." *Arch. Gewerbepathol. Gewerbehyg.* **16**, 405–438 (in German).

Kowal, N. E., Johnson, D. E., Kraemer, D. F., and Pahren, H. R. (1979). "Normal levels of cadmium in diet, urine, blood, and tissues of inhabitants of the United States." *J. Toxicol. Environ. Health* **5**, 995–1014.

Krishna Murti, C. R. (1987). "The cycling of arsenic, cadmium, lead, and mercury in India." In T. C. Hutchinson and K. M. Meena, eds., *Lead, Mercury, Cadmium, and Arsenic in the Environment.* Wiley, New York, pp. 315–333.

Lawrence, J. F., Michalik, Tam, G., and Conacher, H. B. S. (1986). "Identification of arsenobetaine and arsenocholine in Canadian fish and shellfish by high-performance liquid chromatography with atomic absorption detection and confirmation by fast atom bombardment mass spectrometry." *J. Agr. Food Chem.* **34**, 315–319.

Ledet, A. E., and Buck, W. B. (1978). "Toxicity of organic arsenicals in feedstuffs." In F. W. Oehme, ed., *Toxicity of Heavy Metals in the Environment.* Dekker, pp. 375–392.

Luh, M. D., Baker, R. A., and Henley, D. E. (1973). "Arsenic analysis and toxicity—a review." *Sci. Total Environ.* **2**, 1–12.

Lunde, G. (1973). "The synthesis of fat and water soluble arseno organic compounds in marine and limnetic algae." *Acta Chem. Scand.* **27**, 1586–1594.

Lunde, G. (1977). "Occurrence and transformation of arsenic in the marine environment." *Environ. Health Perspect.* **19**, 47–52.

Mahaffey, K. R., Corneliussen, P. E., Jelinek, C. F., and Fiorin, T. A. (1975). "Heavy metal exposure from foods." *Environ. Health Perspect.* **12**, 63–69.

Nriagu, J. O. (ed.), (1984). *Environmental Impacts of Smelters.* Wiley, New York.

Nriagu, J. O. (1988). "A silent epidemic of environmental metal poisoning?" *Environ. Poll.* **50**, 139–161.

Nriagu, J. O., and Pacyna, J. M. (1988). " Quantitative assessment of worldwide contamination of air, water and soils by trace metals." *Nature (London)* **333**, 134–139.

Packard, E. G. (1906). *Entomol. News* **17**, 256.

Partington, J. R. (1935). *Origins and Development of Applied Chemistry.* London, Longmans.

Penrose, W. R., Conacker, H. B. S., Meranger, R., Miles, J. C., Cunningham, W., and Squires, W. R. (1977). "Implications of inorganic/organic interconversions on fluxes of arsenic in marine food webs." *Environ. Health Perspect.* **19**, 53–59.

Pershagen, G. (1983). "The epidemiology of human arsenic exposue." In B. A. Fowler, ed., *Biological and Environmental Effects of Arsenic.* Elsevier, Amsterdam, pp. 198–232.

Pfannauser, W., and Pechanek, U. (1977). "The contamination of food in Austria with toxic heavy metals." *Lebensmit. Ernahrung* **30**, 88–92.

Pliny [The Elder] (1938–1962). *Natural History.* Translated by H. Rackham (Bks. 1–19), (33–35), W. H. S. Jones (Bks. 20–32), and D. E. Eichholz (Bks. 36, 37). LCL, 10 Vols.

Sandberg, G. R., and Allen, I. K. (1975). "A proposed arsenic cycle in an agronomic ecosystem." In E. A. Woolson, ed., *Arsenical Pesticides.* American, Chemical Society, Washington D.C., pp. 124–147.

Shacklette, H. T., Erdman, J. A., Harms, T. F., and Papp, C. S. E. (1979). "Trace elements in plant foodstuffs." In F. W. ed., *Toxicity of Heavy Metals in the Environment.* Dekker, New York, pp. 25–68.

Small, H. G., Jr., and McCants, C. B. (1962). "Residual arsenic in soils and concentrations in tobacco." *Tobacco Sci.* **6**, 34–36.

Smith, D. C., Leduc, R., and Charbonnean, C. (1973). "Pesticide residues in the total diet in Canada, III—1971." *Pest. Sci.* **4**, 211–214.

Vogel, K. (1928). "The significance of arsenic in the excretions." *Am. J. Med. Sci.* **176**, 215–224.

Vos, G., Hovens, J. P. C., and Delft, W. V. (1986). "Arsenic, cadmium, lead and mercury in meat, livers and kidneys of cattle slaughtered in the Netherlands during 1980–1985." *Food Add. Contam.* **4**, 73–89.

Walsh, L. M., Sumner, M. E., and Keeney, D. R. (1977). "Occurrence and distribution of arsenic in soils and plants." *Environ. Health Perspect.* **19**, 67–71.

Walsh, P. R., Duce, R. A., and Fasching, J. L. (1979). "Consideration of the enrichment, sources and flux of arsenic in the troposphere." *J. Geophys. Res.* **84**, 1719–1726.

Wauchope, R. D. (1983). "Uptake, translocation and phytotoxicity of arsenic in plants. "In W. H. Lederer and R. J. Fensterheim, eds., *Arsenic: Industrial, Biomedical, Environmental Perspectives.* Van Nostrand Reinhold, New York, pp. 348–377.

Wauchope, R. D., and McDowell, L. L. (1984). "Adsorption of phosphate, methanearsonate, and cacodylate by lake and stream sediments: Comparison with soils." *J. Environ. Qual.* **13**, 499–504.

Whorton, J. C. (1969). "Insecticide residues of foods as a public health problem: 1865–1938." Thesis, University of Wisconsin, Michigan.

Woolson, E. A. (1975). "Bioaccumulation of arsenicals." In E. A. Woolson, ed., *Arsenical Pesticides.* American Chemical Society, Washington D.C., pp. 97–107.

Woolson, E. A. (1983). "Man's perturbation on the arsenic cycle." In W. H. Lederer and R. J. Fensterheim, eds., *Arsenic: Industrial, Biomedical, Environmental Perspectives.* Van Nostrand Reinhold, New York, pp. 348–377.

World Health Organization (WHO), (1981). "Environmental Health Criteria 18. Arsenic." International Programme on Chemical Safety, Geneva, Switzerland.

Zielhuis, R. L., and Wibowo, A. A. E. (1984). "Standard setting and metal speciation: Arsenic." In J. O. Nriagu, ed., *Changing Metal Cycles and Human Health, Dahlem Konferentzen.* Springer, Berlin, pp. 323–344.

6

SPECIATION OF SELECTED TRACE ELEMENTS IN EDIBLE SEAFOOD

Chris J. Cappon

Environmental Health Sciences Center, University of Rochester Medical Center, Rochester, New York

1. **Introduction**
   1.1. Objective
   1.2. Importance of Seafood
      1.2.1. Dietary Importance
      1.2.2. Economic Importance
   1.3. Selected Trace Elements of Interest
   1.4. Seafood Types of Interest
2. **Mercury**
   2.1. Background
   2.2. Speciation
      2.2.1. Qualitative Speciation
      2.2.2. Quantitative Speciation
   2.3. Dietary Guidelines and Regulations
3. **Lead**
   3.1. Background
   3.2. Speciation
      3.2.1. Qualitative Speciation
      3.2.2. Quantitative Speciation
   3.3. Dietary Guidelines and Regulations
4. **Tin**
   4.1. Background
   4.2. Speciation
      4.2.1. Qualitative Speciation
      4.2.2. Quantitative Speciation
   4.3. Dietary Guidelines and Regulations

145

---

1. INTRODUCTION

During the second half of this century, there has been increasing concern about the presence and levels of numerous chemical contaminants in food used for human consumption. Trace elements are commonly cited as significant contaminants because of their general environmental ubiquity and ease of assimilation in the food chain. They are of unique concern because their environmental and toxicological impact is highly dependent on the different chemical forms in which they occur. Chemical forms of elements can be classified into three basic categories: inorganic compounds, inorganic and organic complexed ions, and element–organic compounds. Each chemical form of a given element has its own characteristic environmental distribution and interactive effects—essential or toxic—on living organisms. The qualitative and quantitative identification of elemental chemical forms (speciation) is critically important in assessing potential toxic effects of a specific element on susceptible organisms and ecosystems.

The recent interest in trace element speciation is threefold: (1) Toxicological research has documented both the dependency of element toxicity on chemical form and the occurrence of *in vivo* biotransformation processes in many animal organisms. For example, unlike ionic inorganic mercury, methylmercury is the most highly toxic form of the element and readily accumulates in specific body organs (liver, kidney, brain) (Clarkson, 1972). Inorganic arsenic (as arsenite or arsenate), on oral ingestion, is mainly converted to the less toxic dimethylarsinate, which is more rapidly excreted in the urine (Crescelius, 1977) (2) Organic forms of specific elements—notably mercury, lead, tin, and arsenic—are largely synthesized by various biotic and abiotic processes under environmental conditions. These processes involve complex interactions that include methylation, demethylation, hydrolysis, photolysis, and degradation (Zingaro, 1979). (3) Advances in state-of-the-art analytical technology, especially in the areas of gas–liquid chromatography and high-performance liquid chromatography, permit feasible and reliable measurement of specific element species at ultratrace (sub-ppb) levels in a variety of environmental and biological media. Because variation in speciation through biological and environmental processes can significantly alter an element's bioavailability and toxicity, quantitative information on trace element speciation in foods is of more interest than total element content.

1.1. Objectives

This chapter's main objective is a survey of documented speciation data for five selected trace elements—mercury (Hg), lead (Pb), tin (Sn), arsenic (As), and selenium (Se)—in edible seafood and seafood products. The survey will emphasize data published within the past 15 years, as most of the analytical speciation methodology for the above elements has been developed since the early 1970s. This survey is not intended to be exhaustive, especially since information on total element content is not emphasized. Earlier and more recent publications not covered are referenced in the many papers cited in this chapter. For each specific element, quantitative speciation data are outlined in tabular form. Appropriate dietary regulations and guidelines are also presented for comparison. The appropriate biological and physicochemical factors that influence the speciation behavior of each element in the aquatic environment and hence seafood will also be presented.

1.2. Importance of Seafood

1.2.1. Dietary Importance

A significant portion of the world's diet consists of seafood—fish and shellfish, either fresh or processed—from both freshwater and saltwater. Recently, the need to include more seafood, particularly fish, in the human diet has been emphasized, and a general trend toward increased per capita consumption has been realized in many industrialized nations, notably the United States (Tuna Research Foundation, Inc., 1978). The reasons for this are the relatively lower levels of saturated fat, cholesterol, and caloric intake when compared to other foods such as meat, poultry, and dairy products. The high availability of seafood in most countries also favors increased dietary consumption.

1.2.2. Economic Importance

Increased seafood consumption will have a positive economic impact on both marine and freshwater commercial fishing and the associated food processing industries. A secondary economic impact is reflected in regions in which increased sport fishing activity is a vital component of the regional economy. An excellent example is the multimillion dollar salmonid sports fishing in the Great Lakes—especially Lakes Michigan and Ontario.

The presence of persistent toxic chemicals in the aquatic food chain of important fishing areas can have a negative long-term impact on both local and regional economic health, as well as on the health of the general population consuming contaminated fish and seafood.

1.3. Selected Trace Elements of Interest

This chapter will focus on speciation of the following elements in seafood: Hg, Pb, Sn, As, and Se. There are four reasons for this: first, widespread industrial and commercial usage of these elements and their environmental discharge, ubiquity, persistence, and bioaccumulation in aquatic food chains; second, the known environmental diversity of their chemical forms; third, their widely investigated and established toxic impact on human health; and, finally, the existence of published speciation data, especially for Hg, Of these five elements, only Se is an essential trace nutrient for animals and probably man, but is highly toxic when present in excess of nutritional requirements (Zingaro and Cooper, 1974).

The general toxicity of species commonly identified in the environment and in living organisms is summarized for each element in Table 1. Corresponding discharge—terrestrial and atmospheric—sources into the aquatic environment are summarized in Table 2.

1.4. Seafood Types of Interest

The present survey will include edible seafood species and products from the following categories:

1. Marine seafood species—fish, shrimp, crab, lobster, clam, squid, octopus. This category has the majority of published speciation data, and represents the main dietary seafood source for most people.

2. Freshwater fish of regional commercial and recreational importance (e.g., Northern pike, salmonids, bass, perch).

3. Marine mammal species (e.g., seal, whale). There are limited data available because these species, as food sources, are restricted to specific population groups (e.g., Eskimo and Canadian Indian tribes). However, they will be included, where appropriate, for comparison purposes.

2. MERCURY

2.1. Background

Mercury is widely distributed in the environment both from natural geological activity and from industrial pollution. During the past 30 years, environmental

TABLE 1 Trace Element Speciation and Toxicity Relationships[a]

Element	Common Environmental Species	General Toxicity Relationships
Hg	Inorganic: Hg^0 (elemental), Hg^{2+} (ionic) Organic: $RHgX$, R_2Hg ($R = CH_3$, C_2H_5, C_6H_5)	Elemental $>$ ionic Organic $>$ inorganic; $R > R_2$: $CH_3 > C_2H_5 \gg C_6H_5 =$ inorganic
Pb	Inorganic: Pb^{2+} (ionic) Organic: R_4Pb, R_3Pb^+, R_2Pb^{2+} ($R = CH_3$, C_2H_5, or a mixture of these)	Organic $\gg$ inorganic; $R_4 > R_3 > R_2$; $C_2H_5 > CH_3$
Sn	Inorganic: Sn^{2+} (ionic) Organic: R_4Sn, R_3Sn^+, R_2Sn^{2+}, RSn^{3+} ($R = CH_3$, C_2H_5, C_4H_9, C_6H_{11}, C_8H_{17}, C_6H_5)	Organic $\gg$ inorganic; $R_4 = R_3 > R_2 > R$; $C_2H_5 > CH_3 > C_4H_9 > C_6H_5$
As	Inorganic: As^{3+}, As^{5+} (ionic); AsH_3 (arsine) Organic: $CH_3As(O)(OH)_2$, $(CH_3)_2As(O)OH$, $(CH_3)_2AsH$, $(CH_3)_3 As$; arsenobetaine, arsenecholine	Arsine $\gg +3 > +5$ Inorganic $>$ organic; however, organic arsines $> +3$ practically nontoxic
Se	Inorganic: Se^0 (elemental) Se^{4+}, Se^{6+} (ionic) Se^{2-} (as H_2Se) Organic: selenides (R_2Se, R_2Se_2, where $R = CH_3$, C_2H_5)	Practically nontoxic $+4 > +6$ highly toxic Highly toxic

[a]Reference: Cappon (1987).

149

TABLE 2   Major Trace Element Input Sources into the
Aquatic Environment

Element	Major Input Sources
Hg	Combustion of fossil fuels; mining, industrial (paper pulp, chloralkali), agricultural (fungicide) activities; volcanic emissions
Pb	Mining, industrial (battery, pigment, alloy) activities; automotive emissions
Sn	Mining, industrial (plating, alloys, chemicals) activities; pesticide usage (marine antifoulants)
As	Combustion of fossil fuels; mining, industrial (alloys, chemicals), agricultural (pesticide) activities; geological leaching
Se	Combustion of fossil fuels; mining, industrial (semiconductor, pigments, glass, chemicals) activities

Hg contamination has generated much worldwide concern, especially the presence of elevated Hg levels in edible seafood, which is the major human dietary source of this element. Since most of the Hg in the edible tissue is methylmercury, the most toxic chemical form, consumption of fish and seafood containing elevated Hg levels is potentially hazardous to human health. Human methylmercury poisoning resulting from ingestion of contaminated seafood was first established and well-documented by the Minamata and Niigata, Japan, epidemics (Kurland et al., 1960). Subsequent discoveries initially revealed elevated Hg contamination—predominately as methylmercury—of freshwater fish in Canada and the United States (Wobeser, 1970), Sweden (Westoo, 1966), and Finland (Hasanen and Sjoblon, 1968), which was in excess of maximum permissable levels in these countries.

These incidents and investigations have initiated several comprehensive analytical surveys of total Hg (Krenkel, 1973) and, more recently, methylmercury in several species of marine (Bebbington et al., 1977; Cappon and Smith, 1981, 1982a, b; Hight and Corcoran, 1987; James, 1983; Kamps et al., 1972; Luten et al., 1980; Rivers et al., 1972; Shimojo et al., 1975; Shultz and Crear, 1976; Shultz and Ho, 1979; Suzuki et al., 1973; Westoo, 1973) and freshwater (Bache et al., 1971; Buchanan, 1972; Cappon and Smith, 1981, 1982a; Cappon, 1984; Hattula et al., 1978; Linko and Terho, 1977; Luten et al., 1980; Westoo, 1967) fish and seafood (Cappon and Smith, 1982a; Hall, 1974; Hight and Capar, 1983; Uthe et al., 1972; Watts et al., 1976). Certain species of commercial marine fish—shark, swordfish, and tuna—have been of special interest in this respect as these species are known to concentrate significantly higher amounts of environmental Hg. Tuna has been the most frequently surveyed species (Cappon and Smith, 1982b) because of its extensive commercialization and per capita consumption, especially in the

TABLE 3 Regulations for Hg Content in Seafood and Related Dietary Consumption

Source	Date	Regulation	Reference
WHO	1972	*PTWI:* 5 μg/kg body weight, as total Hg, of which no more than 3.3 μg/kg body weight should be methylmercury (expressed as Hg). Based on high total Hg intake due to consumption of fish containing elevated methylmercury levels	WHO (1972)
United States FDA	1970	"Action Level" for total Hg in fish and shellfish: 0.5 ppm	FDA (1974)
	1978	"Action Level": 1.0 ppm (total Hg)	FDA (1979)
	1984	"Action Level": 1.0 ppm (Hg as methylmercury)	FDA (1984)
Finland	1970	0.5 mg/kg (safety limit, all fish species)	Hattula et al. (1978)
Canada (Food Directorate of Health and Welfare)	1970	0.5 ppm Hg (commercial fishing; guideline only)	Charlebois (1978)
Sweden	1967	> 1.0 mg/kg (commercial fish)	Birke et al. (1972)
	1968	< 1.0 mg/kg (health advisory: no more than one meal/week)	
Australia (NHMRC)	1976	0.5 ppm total Hg (statutory, commercial Seafood)	Bebbington et al. (1977)
Japan	1972	1.0 ppm, total Hg	Goldwater (1974)

151

United States. In addition, several human dietary and health studies have been conducted on specific population groups from selected regions in which consumption of Hg-contaminated fish occurs. These regions include Sweden (Birke et al., 1972; Skerfving, 1974), Japan (Kojima and Fujita, 1973; Suzuki et al., 1980), Canada (Charlebois, 1978), Peru (Turner et al., 1980), American Samoa (Clarkson et al., 1973), Finland (Sumari et al., 1972), United States (Alaska) (Galster, 1976), and the Seychelles (Matthews, 1983). Consequently, many nations have established maximum permissable Hg concentrations in fish used for human consumption, usually in the range from 0.4–1.0 μg Hg/g (ppm) (Watling et al., 1981). These concentrations are summarized in Table 3. Since the early 1970s, considerable effort has focused on the development of rapid and reliable analytical speciation methodology for Hg in seafood and other environmental and biological materials (Cappon, 1987; Cappon, 1988). Because of this, there is considerable quantitative data available on Hg speciation in seafood than for the other four elements surveyed in this chapter, even though only two predominant chemical forms—methylmercury and divalent inorganic Hg—have been identified.

2.2. Speciation

2.2.1. Qualitative Speciation

The predominant Hg species found in edible seafood are divalent inorganic (Hg^{2+}) and monomethylmercury (CH_3Hg^+). Most published analytical reports generally confirm that methylmercury, the most toxic form, exceeds 75% of the total Hg content in the edible fillet of marine and freshwater fish, although methylmercury percentages of less than 50 have been reported for a limited number of fish samples (Cappon and Smith, 1981; Shimojo et al., 1975). A notable example is the Pacific blue marlin (Shultz and Crear, 1976), the only pelagic species known to have a large percentage (usually > 80) of total muscle Hg present as inorganic Hg. In contrast, inorganic Hg predominates in shellfish (mollusks, crustaceans) and other nonfish species (e.g., octopus, squid). Other common commercial organomercurials, such as ethylmercury and phenylmercury, are generally not found in fish and shellfish. However, in one documented case involving direct contamination of a Japanese river with ethylmercury and inorganic Hg, accumulation of tissue ethylmercury to levels exceeding that for methylmercury was found for several fish species (Yamanaka and Ueda, 1975). Shafer et al. (1975) reported one trout sample suspected of containing ethylmercury. There has been one report of dimethylmercury [$(CH_3)_2Hg$] in freshwater fish (Koli and Carty, 1978), although no quantitative data were presented.

Methylmercury in seafood originates from both geological and anthropogenic [mainly as metallic Hg vapor (Hg^0) and Hg^{2+}] input into the aquatic environment. Biomethylation of Hg^{2+} is known to occur in sediment largely by a variety of bacterial and fungal organisms. This process is optimized by enrichment with organic substrates (plant or algal remains), oxygen and sulfide exclusion, and pH near 7 (Jensen and Jernelov, 1969). The main methylation pathway involves interaction of methylcobalamine (methyl-B_{12}) and soluble Hg^{2+} (Jernelov, 1970; Krenkel, 1973; Ridley et al., 1977) followed by bioaccumulation in the aquatic food chain. Total Hg and methylmercury levels are greatest for predatory fish species. Ascending the aquatic food chain, factors such as local habitat, diet, size, and age may influence the relative proportions of methylmercury and inorganic Hg in the edible tissue. However, metabolic processes involving Hg biotransformation, which may originate in other organ tissues (e.g., liver, kidney, intestine) may be more important. Suzuki et al. (1973) suggested the possibility of biotransformation of Hg in fish and that particular species may be capable of methylating ingested inorganic Hg, which is the predominant form in seawater and lower marine food chain organisms. Shultz and Crear (1976) suggested that ingested methylmercury is converted to the less toxic inorganic form in the liver and spleen of the Pacific blue marlin prior to subsequent accumulation in the flesh. Pan-Hon and Imura (1981) demonstrated *in vitro* demethylation of methylmercury by intestinal bacteria isolated from yellowfin tuna. However, Matsumura et al. (1975) reported *in vitro* methylation of inorganic Hg in tuna liver homogenates. Therefore, marine fish may also be capable of *in vivo* inorganic Hg methylation. Rudd et al. (1980) recently confirmed *in vitro* methylation by bacterial activity in the intestinal contents of several freshwater species. These investigations indicate only that Hg biotransformation and subsequent accumulation of inorganic Hg and methylmercury in fish are complex, species-dependent, and poorly understood.

More recently, acid rain has been cited as a factor that may play a key role in determining the level and form distribution of Hg in freshwater fish. Acid rain has been shown to increase methylmercury levels in the waters and fish of many rivers and lakes in the Northeastern United States, Southeastern Canada, and Northern Europe (Tomlinson, 1979). Tsai et al. (1975) and Jernelov (1980) demonstrated that low water pH leads to increased uptake of inorganic Hg by fish and other aquatic microorganisms. Jensen and Jernelov (1969) observed that a low water pH also enhances methylation–demethylation ratios by microorganisms. Thus, acid rain, coupled with localized discharges of Hg-contaminated industrial wastes, may be an important factor due to the ability of freshwater fish to methylate-ingested inorganic Hg. However, freshwater fish inhabiting acidified waterways may also accumulate significant amounts of methylmercury directly from the water. This methylmercury could originate from increased biomethylation of inorganic Hg at lower pH, or from absorption of airborne methylmercury by lower molecular weight organic material suspended in the water. Norstrom et al. (1976) found the efficiency of methylmercury uptake by fish from river water to be very high. In addition, Kramer and Neidhart (1975) observed the uptake of methylmercury from water by fish to be much greater than inorganic Hg. These findings suggest that water, along with food, may also contribute to higher methylmercury burdens in freshwater fish. Along with acid rain, impoundment of large bodies of water, usually for hydroelectric purposes, results in high methylmercury levels in fish (World Health Organization, 1976).

TABLE 4 Distribution of Hg and Se Chemical Forms in Edible Fish and Seafood Tissue

Category/Species	N	Water-Extractable Fraction (Percentage of Total Tissue Content)						Ref.
		Hg			Se			
		T^a	M^b	I^c	T^d	$-2, +4$	$+6$	
Fish, Freshwater								
Bass, largemouth		80.4	81.0	78.6	60.9	53.4	80.8	*f*
Bass, smallmouth		63.7	64.1	62.5	53.4	47.7	77.6	*e*
Bullhead		71.8	77.5	64.9	68.4	62.5	84.2	*e*
Muskellunge	2	59.7	58.4	58.6	67.0	62.7	59.9	*f*
Perch		70.4	70.0	71.3	65.2	62.2	85.7	*e*
Pike, Northern	7	58.7	57.2	71.2	63.1	63.5	62.3	*f*
Walleye		53.6	52.6	71.9	60.8	76.4	62.2	*e*
Fish, Marine								
NONPROCESSED								
Bluefish		31.8	17.2	60.3	58.7	57.6	64.0	*e*
Dolphin		38.0	27.0	46.6	59.2	64.6	49.3	*e*
Flounder		45.7	30.7	72.9	62.2	58.1	88.4	*e*
Mackerel		45.7	50.1	33.2	61.4	63.6	54.6	*g*
Red snapper		30.1	27.8	35.1	60.7	59.9	65.7	*e*
Salmon, pink		73.6	78.1	62.0	48.7	47.8	50.4	*g*
Shark, mako	2	33.1	29.0	50.6	61.3	58.4	67.5	*g*
Swordfish	2	29.0	28.3	32.4	63.5	58.7	51.1	*g*
Tuna	9	23.7	19.3	48.2	56.0	54.8	59.1	*g*
PROCESSED, OLD[h]								
Marlin, Pacific blue	2	32.2	23.1	34.5	12.4	9.1	45.9	*f*
Tuna	18	57.3	57.2	62.4	51.2	49.8	54.0	*g*
Processed, New[i]								
Mackerel	2	51.9	55.2	48.4	55.7	54.7	58.5	*f*
Salmon, pink		67.7	68.2	67.5	60.9	56.2	74.5	*f*
Tuna	12	32.5	29.6	45.4	55.4	51.7	64.7	*g*
Marine Organisms								*f*
MOLLUSKS (PROCESSED)								
Clam, minced		39.9	38.7	40.4	42.9	37.3	50.9	
Mussel, smoked		34.3	48.0	29.3	33.6	23.0	65.6	
Oyster		43.4	56.6	38.6	57.4	56.2	59.1	

TABLE 4   (*Continued*)

Category/Species	N	Hg			Se			Ref.
		T^a	M^a	I^c	T^d	$-2, +4$	$+6$	
CRUSTACEANS (PROCESSED)								
Crab		33.4	37.1	30.8	45.6	41.5	59.0	
Shrimp		42.6	25.0	62.8	41.4	39.0	48.0	
PODS (NONPROCESSED)								
Octopus		47.1	45.3	48.7	57.4	56.2	59.1	
Squid		63.1	66.8	60.6	56.7	56.6	47.8	

[a] Total Hg.
[b] Methylmercury.
[c] Inorganic mercury.
[d] Total Se.
[e] Cappon and Smith (1981).
[f] Cappon and Smith (1982a).
[g] Cappon and Smith (1982b).
[h] Sample (storage) age > 5 years.
[i] Sample (storage) age < 6 months.

Little is known about the exact chemical nature of inorganic Hg and methylmercury in seafood that may influence their relative bioavailability. Recent investigations (Cappon and Smith, 1981, 1981a, b) have examined the distribution of these two Hg forms in edible fish and seafood tissue. The overall results are summarized in Table 4. Different Hg distribution patterns were observed between nonprocessed and processed marine and freshwater fish. For freshwater and older processed (canned, ⩾ 5 years of storage age) marine categories, the majority (55–90%) of the total tissue Hg content was water extractable. For nonprocessed and newly processed (< 2 years of storage age) marine samples, only 22–47% was extractable. Water-extractable Hg is of interest because it represents Hg bound to lower molecular weight (< 100,000) tissue proteins or unbound neutral and ionic species. These species may be more readily absorbed from the gastrointestinal tract into the bloodstrean, thus having higher bioavailability. These specific Hg distribution patterns possibly result from differences in the source and form of Hg in marine and freshwater food chains. Unlike the marine environment, fish from inland waters subjected to acid rain and higher localized inputs of Hg waste discharges are exposed to higher methylmercury levels in the food chain and the water itself. More importantly, the nature of Hg metabolism in organs (e.g., liver, intestine) and subsequent transfer and binding to muscle tissue components (mostly proteins) may differ for each species category. The different distribution pattern between the new and older processed marine samples suggests that sample age (i.e., storage time) may be a more important determining factor than processing in tissue Hg form and distribution. Although processing may initiate a change in Hg distribution, possibly by altering tissue protein composition and chemical form of bound Hg, further changes can occur over an extended storage period. For canned tuna, there was no apparent effect of packing medium (water, oil) on the percentage of water-extractable Hg.

2.2.2. Quantitative Speciation

2.2.2.1. Marine Fish. Abundant Hg speciation data are available for this category, primarily because it includes several commercially important species (e.g., tuna, swordfish, salmon). Most of the data have been reported for fish taken from the North Atlantic and Pacific Oceans, primarily by investigators from the United States, Canada, Japan, and Western Europe. Table 5 presents representative results from selected surveys that provide data on total Hg content, along with methylmercury percentages.

For most marine species studied, the majority of total muscle Hg is present as methylmercury, which generally comprises more than 75% of the total Hg content. This percentage is higher for fish highest in total Hg, usually exceeding 0.5 ppm, notably larger pelagic species—shark, swordfish, and certain tuna species. The notable exception is Pacific blue marlin, where inorganic Hg is the predominant form (80–85% of the total muscle Hg content). In many cases, fish species having lower methylmercury percentages, particularly less than 60%, contain relatively low (< 0.2 ppm) tissue Hg levels. Methylmercury percentage in edible fish muscle may be related to size, age, or relative position in the food chain. For example, skipjack, a comparatively small variety of tuna, are notably lower in total Hg concentration and methylmercury percentage. Other marine species with lower reported methylmercury percentages were also of smaller size and lower in the marine food chain (i.e., nonpelagic species). These include mackerel, salmon, flounder, herring, sole, sardines, and anchovies.

The effect of packing medium on Hg content and form was investigated by the author for canned tuna and other processed seafood samples (Cappon and Smith, 1981, 1982a) (Table 6). Inorganic Hg was the major form in the packing medium. Although trace amounts of both Hg forms were detected in oil and water packings, total Hg content represents a very small fraction (< 0.1–7.6%) of the corresponding level in the total can contents. The oil or water may have contained endogenous Hg prior to the canning process or, at best, extracted insignificant amounts of methyl and inorganic Hg from the processed edible fillet, and this is reflected in the low concentration ratio (CR) values. Only samples of canned mackeral and salmon had significant levels of methylmercury in the packing water. Due to the higher CR and packing ratio (PR) values, a greater portion of muscle methylmercury may have become solubilized as a result of processing and storage. For the nonfish canned seafood samples studied, the packing water contained higher Hg levels (averaging 27.5% of the corresponding level in the total can contents), predominantly as the inorganic form. Since these samples were not packed as densely as tuna (i.e., higher average PR values,

TABLE 5 Hg Speciation in Marine Fish

Species	N	Total Hg, ppm[a]	Methylmercury[a] (%)	Remarks	Reference
			Nonprocessed		
See trout	69	0.218–0.296	81–95 (91)	Sweden: samples 1–2 yr of age	Westoo (1973)
Whiting	8	0.022–0.043	73–99	Baltic Sea	May et al. (1987)
Cod	5	0.021–0.040	91–96	Baltic Sea	May et al. (1987)
Dab	4	0.013–0.039	91–98	Baltic Sea	May et al. (1987)
Herring	4	0.020–0.030	88–92	Baltic Sea	May et al. (1987)
Haddock	2	0.033, 0.052	76, 83	Sweden	Westoo (1967)
Cod	2	0.026, 0.036	85.78	Sweden	Westoo (1967)
Herring	2	0.09, 0.95	66, 76	Canada	Uthe et al. (1972)
Tuna (skipjack, yellow fin)	15	0.086–1.96	65–113	Seychelles	Matthews (1983)
Bonita	5	0.805–1.286	77–98	Seychelles	Matthews (1983)
Kinfish	5	0.184–1.189	76–110	Seychelles	Matthews (1983)
Job	5	0.446–0.741	86–98	Seychelles	Matthews (1983)
Bourgeois	5	0.273–0.692	74–114	Seychelles	Matthews (1983)
Vielle platte	5	0.358–0.739	68–103	Seychelles	Matthews (1983)
Carangue balo	5	0.376–1.288	76–112	Seychelles	Matthews (1983)
Beume	5	0.370–1.33	92–105	Seychelles	Matthews (1983)
Tuna	17	1.296–1.950	71–99	Japan, ≥111 kg	Shimojo et al. (1975)
Tuna	17	0.550–0.981	49–98	Japan, <43 kg	Shimojo et al. (1975)
Barracuda	1	0.88	100	Senegal (W. Africa) Coast	Gras and Mondain (1982)
Sole	6	0.11	49	Senegal (W. Africa) Coast	Gras and Mondain (1982)
Tuna	5	0.22–0.25	59–88	Senegal (W. Africa) Coast	Gras and Mondain (1982)
Mullet	4	0.09	32	Senegal (W. Africa) Coast	Gras and Mondain (1982)
Carp	5	0.07	38	Senegal (W. Africa) Coast	Gras and Mondain (1982)
Sardine	10	0.11	45	Mediteranean	Aubert (1975)

(Continued)

TABLE 5 (*Continued*)

Species	N	Total Hg, ppm[a]	Methylmercury[a] (%)	Remarks	Reference
Bream, yellowfin	30	0.23	91	Australia	Beggington et al. (1977)
Tuna, yellowfin	20	0.38	97	Australia	Beggington et al. (1977)
Kingfish, yellowtail	20	0.18	94	Australia	Beggington et al. (1977)
Salmon	20	0.28	96	Australia	Beggington et al. (1977)
Snapper	30	0.33	97	Australia	Beggington et al. (1977)
Sea Mullet	30	0.03	100	Australia	Beggington et al. (1977)
Marlin, Pacific blue	29	0.35–14.0(4.78)	19.5	Hawaii	Rivers et al. (1972)
Tuna, yellowfin	22	0.24–1.32(0.54)	88.9	Hawaii	Rivers et al. (1972)
Skipjack	20	0.27–0.52(0.38)	92.7	Hawaii	Rivers et al. (1972)
Dolphin	10	0.17–0.31(0.25)	98.9	Hawaii	Rivers et al. (1972)
13 pelagic species	33	0.197	94.8	S. China Sea, Japan Sea	Suzuki et al. (1973)
Marlin, Pacific blue	35	0.13–16.80(4.35)	10.4	Hawaii	Shultz and Crear (1976)
Mackerel	6	0.05–0.70(0.33)	81.8	Netherlands	Luten et al. (1980)
Whiting, blue	15	0.03–0.29(0.15)	93.3	Netherlands	Luten et al. (1980)
Cod	53	0.03–0.49(0.16)	97.8	Netherlands	Luten et al. (1980)
Sole	5	0.02–0.29(0.13)	84.6	Netherlands	Luten et al. (1980)
Plaice	4	0.04–0.31(0.16)	87.5	Netherlands	Luten et al. (1980)
Marlin, Pacific blue	46	0.09–10.00(3.12)	12.8	Hawaii	Shultz and Ito (1979)
Tuna	3	0.41–1.11	46.7–6.92	American Samoa	Cappon and Smith (1981)
Tuna	6	0.23–6.88	60.3–92.5	Northern Peru	Cappon and Smith (1982a)
Bluefish	1	0.39	68.7	Retail	Cappon and Smith (1981)
Dolphin	1	0.47	38.4		Cappon and Smith (1981)
Flounder	1	0.20	65.0	Retail	Cappon and Smith (1981)
Red Snapper	1	1.36	70.4		Cappon and Smith (1981)
Shark, mako	2	1.45, 1.69	80.7, 82.3	Retail	Cappon and Smith (1982a)
Shark, mako	1	1.93	86.2		Hight and Corcoran (1987)

	n		Range		Reference
Swordfish	2	0.17, 0.99	76.8, 81.4	Retail	Cappon and Smith (1982a)
Swordfish	1	1.34	93.3		Hight and Corcoran (1987)
Salmon, pink	2	0.08, 0.14	60.1, 65.0	Retail	Cappon and Smith (1982a)
Swordfish	20	0.49–2.44	93–113	Frozen steaks	Kamps et al. (1972)
Swordfish	7	0.28–1.19	82.8–114.3	Frozen steaks	James (1983)
Mackerel	1	0.13	61.7	Retail	Cappon and Smith (1982a)
Carnivores, 9 species	23	0.03–0.12	77.0–91.9	Japan	Suzuki et al. (1980)
Omnivores, 4 species	18	0.01–0.05	81.6–97.2	Japan	Suzuki et al. (1980)

Processed

	n		Range		Reference
Tuna	11	0.06–0.54	67–125	White and light meat	Kamps et al. (1972)
Tuna	20	0.13–0.73	64–119	White and light meat	Hall (1974)
Tuna	13	0.04–0.69	57.1–100		James (1983)
Salmon	5	0.02–0.06	36–96	Pink, red sockeye, cohoe red	Hall (1974)
Herring, kippered	1	0.13	62		Hall (1974)
Sardines	1	0.04	55		Hall (1974)
Anchovies	1	**0.21**	43		Hall (1974)
Cat food	2	0.04	38, 54		Hall (1974)
Tuna	8	0.11–1.09	39.2–92.9		Cappon and Smith (1981)
Marlin, Pacific blue	2	0.36, 0.90	16.5, 21.0		Cappon and Smith (1982a)
Tuna	5	0.13–1.02	67.4–92.3 (78.9)	Old	Cappon and Smith (1982a)
Tuna	3	0.29–0.43	71.5–84.5 (77.1)	New	Cappon and Smith (1982a)
Mackerel	2	0.03, 0.12	53.2, 61.9		Cappon and Smith (1982a)
Tuna	13	0.10–0.55	87.7–91.7 (81.7)	New (<6 mo)	Cappon and Smith (1982b)
Tuna	13	0.07–0.47	57.4–94.7 (77.1)	Old (5–29 years)	Cappon and Smith (1982b)

[a]Values in parentheses represent averages of the corresponding ranges. Unless otherwise noted, total Hg consists of the inorganic and monomethyl forms.

159

TABLE 6 Hg and Se Content and Chemical Form in Seafood Packing Media[a]

Medium	Sample	N	Storage Age Category[b]	Total Hg[c] (ppb)	Methylmercury (%)	CR[d]	Total Se[c] (ppb)	Se^6+ (%)	CR[d]	PR[e]
Oil	Tuna	5	Old	0.09–0.47	15.8–81.2	0.013	0.06–0.22	8.5–18.2	0.046	0.20
	Tuna	3[f]	New	0.04–0.07	38.8–52.3	0.008	0.03–0.04	5.6–19.8	0.006	0.27
	Mackerel	1	New	0.10	17.2	0.034	0.07	7.6	0.022	0.20
Water[g]	Tuna	3	New	0.42–3.10	24.5–42.2	0.35	0.03–0.83	5.8–19.8	0.121	0.31
	Salmon, pike	1	New							
	Mackerel	1	New							
	Clam, minced	1	New	2.44–4.80	15.7–21.1	0.275	3.8–46.5	9.8–20.0	0.064	0.58
	Oyster	1	New							
	Crab	1	New							
	Shrimp	1	New							

[a]Data from Cappon and Smith (1982a).
[b]Old, ≥ 5 years; new, < 6 months.
[c]ng/mL oil, water.
[d]CR (concentration ratio): μg total Hg, Se in entire medium/μg total Hg, Se in total can contents.
[e]PR (packing ratio): g packing medium/g solid contents (net wt).
[f]Data from Cappon and Smith (1981).
[g]Values represent reported ranges for the canned fish and seafood groups.

0.58 vs. 0.31 for tuna), the samples were in greater contact with the packing medium, allowing a greater portion of tissue-bound Hg to solubilize during storage. The lower CR values for crab meat apparently resulted from the sample being more tightly packed.

2.2.2.2. Freshwater Fish. There have been fewer Hg speciation surveys of freshwater fish. This may be largely due to the less extensive commercial importance and corresponding dietary intake of this seafood source by the general population. However, recent attention has focused on fish—primarily larger predatory species (pike, walleye, salmonid)—in specific geographic regions experiencing industrial Hg pollution discharges and the effects of acid rain, and having localized commercial and sports fisheries and fish-eating populations.

The total Hg and methylmercury muscle levels in selected freshwater species (Table 7) are generally higher than those reported in most marine species, with the exception of Pacific blue marlin, shark, and contaminated tuna. Predatory species, notably Northern pike from regions having acidified lakes and water-ways, contain muscle Hg levels greatly exceeding 1 ppm. Methylmercury is also the predominant chemical form, but the corresponding percentages are signifi-cantly higher, usually greater than 90. As with marine species, the methylmercury percentage is highest in samples high in total Hg (> 1 ppm). This also includes other predatory species such as muskellunge, walleye, and largemouth bass. Bottom feeders and lower predatory species such as carp, bullhead, bream, and perch have lower methylmercury percentages. This is probably due to exposure to higher inorganic Hg levels from bottom sediment and detritus-feeding organisms. Size and age were shown to influence the methylmercury percentage in lake trout (Bache and Lisk, 1971). The percentage increased with age, exceeding 50% for fish beyond 2 years of age.

The effect of smoking on Hg speciation was recently examined in brown trout fillets (Cappon, 1984). Data presented in Table 8 reveal that there were similar ranges of total Hg and methylmercury content for hickory-smoked and unsmoked fillet samples. Westoo (1966) found that broiling or baking of fish meat did not alter the methylmercury content. Thus, the smoking process commonly used for salmonid and other species probably does not result in significant losses in total overall Hg content of the edible fillet. However, the smoked samples had a higher average methylmercury percentage. Some of the tissue-bound inorganic Hg may have been converted to methylmercury as a result of the elevated temperatures used in the initial smoking and subsequent baking steps. Whether or not the increased methylmercury percentage resulted solely from sample exposure to elevated temperature and smoke is speculative and requires further confirmation.

Freshwater invertebrates have been shown to accumulate high total and methylmercury levels in waterways contaminated by Hg from industrial discharge (Hildebrand et al., 1980).

2.2.2.3. Shellfish and Other Seafood. Quantitative Hg speciation data are more limited compared to fish. Selected Hg data for this category are presented in

TABLE 7 Hg Speciation in Freshwater Fish

Species	N	Total Hg (ppm)[a]	Methylmercury[a] (%)	Remarks	Reference
Perch	7	0.02–2.99	83–93	Sweden	Westöo (1967)
Pike, Northern	6	0.06–3.11	88–98	Sweden	Westöo (1967)
Pike, Northern	2	0.11, 0.88	100, 92	Canada	Uthe et al. (1972)
Whitefish	1	0.03	67	Canada	Uthe et al. (1972)
Salmon, coho	2	0.06, 0.22	50, 100	Canada	Uthe et al. (1972)
Trout, lake	2	0.06, 0.71	83, 82	Canada	Uthe et al. (1972)
Trout, rainbow	3	0.05–0.53	78–100	Canada	Uthe et al. (1972)
Walleye	1	0.65	100	Canada	Uthe et al. (1972)
Salmon	21	0.06–0.32	83–102	Ages 1–7 yr	Westöo (1973)
Trout, lake	27	0.19–0.66	30.8–101.8	Ages 1–12 yr	Bache and Lisk (1971)
Pike	14	0.11–2.42 (0.60)	96.7	Netherlands	Luten et al. (1980)
Perch	15	0.10–1.74 (0.82)	90.2	Netherlands	Luten et al. (1980)
Pike–perch	11	0.53–1.14 (0.85)	89.4	Netherlands	Luten et al. (1980)
Pike	64	0.23–3.96 (1.07)	100	Finland	Hattula et al. (1978)
Bream	19	0.05–0.81 (0.34)	64.7	Finland	Hattula et al. (1978)
Roach	27	0.11–1.13	98.0	Finland	Hattula et al. (1978)
Perch	47	0.10–4.68	84.0	Finland	Hattula et al. (1978)
Sunfish	16	0.08–0.27	52–86	United States	Buchanan (1972)
Pike, Northern	3	1.10–1.94	96.9–100	Sweden	Kamps et al. (1972)

	n				
Rock bass	10	0.22–0.70	78.9–99.5	United States	Hildebrand et al. (1980)
Hogsucker	10	0.19–0.46	84.6–103.8	United States	Hildebrand et al. (1980)
Bass, Largemouth	2	0.83	95.5	United States	Cappon and Smith (1981)
Smallmouth	1	0.42	68.2	United States	Cappon and Smith (1981)
Bullhead	1	0.17	58.8	United States	Cappon and Smith (1981)
Perch	1	0.24	69.4	United States	Cappon and Smith (1981)
Walleye	1	1.22	96.4	United States	Cappon and Smith (1981)
Muskellunge	2	0.45	76.6	Muscle smoked and cooked	Cappon and Smith (1982a)
		0.31	88.5	Nonsmoked	Cappon and Smith (1982a)
Pike, Northern	3	1.11–2.81	91.8–95.5	Ontario, Canada	Cappon and Smith (1982a)
Pike, Northern	4	0.77–1.68	84.2–94.1	Quebec, Canada	Cappon and Smith (1981)
Salmon, coho	7	0.26–0.80 (0.44)	70.7	United States (Lake Ontario)	Cappon (1984)
Salmon, chinook	6	0.26–0.43 (0.34)	71.3	United States (Lake Ontario)	Cappon (1984)
Trout, brown	4	0.14–0.37 (0.25)	68.2	United States (Lake Ontario)	Cappon (1984)
Trout, lake	7	0.16–0.29 (0.23)	71.0	United States (Lake Ontario)	Cappon (1984)
Dace, ayeu	75	0.04–4.65	15–87	Some samples contained up to 80% ethylmercury	Yananaka and Ueda (1975)
Eel	4	0.21–0.31 (0.26)	100	Netherlands	Luten et al. (1980)
Eel	2	0.19, 1.46	58, 81	Canada	Uthe et al. (1972)

[a] As per Table 5.

TABLE 8   Hg and Se Speciation in Smoked and Unsmoked Brown Trout Fillets[a]

Group	N	Total Hg (ppm)	Methylmercury (%)	Total Se (ppm)	Se[6+] (%)
Unsmoked	12	0.21	71.3	0.38	19.4
Smoked[b]	11	0.27	79.0	0.34	30.3

[a] Data from Cappon (1984). Values represent the means for each group.
[b] Fillets were hickory smoked for 10 h at 70–73°C then oven-baked for 10 min at 150°C before storage and analysis.

Table 9. The total tissue Hg concentrations are lower than those found in pelagic fish species, being usually less than 0.1 ppm. This reflects their lower position in the marine food chain and consequently, lower dietary Hg exposure. Limited speciation results indicate that these marine organisms, unlike fish, generally have significantly lower methylmercury percentages in the edible muscle. Methylmercury is the predominant form in shrimp, but other marine species have higher inorganic percentages. Hildebrand et al. (1980) found methylmercury to comprise approximately 50% of the total Hg in six species of freshwater benthic invertebrates in the vicinity of a chloralkali plant.

2.2.2.4.   Marine Mammals.   The principal species of interest in this category are whale and seal. They are included for two reasons. First, they represent a major food source, in addition to fish, for specific population groups, especially Inuit and Indian communities in coastal villages in arctic and subarctic regions of Canada and Greenland. Second, as these carnivorous species are part of the marine food chain, they are known to concentrate significant amounts of Hg in major edible tissues, muscle, and liver (Anas, 1974; Itano et al., 1984a, 1984b; Nagakura et al., 1974). The potential health hazard to these aboriginal population groups was recognized in the late 1960s when apparently dangerous levels of methylmercury were found in marine mammals (Charlebois, 1978).

Reported Hg speciation data for marine mammal species is presented in Table 10. Most of the Hg (approximately 80–95%) present in seal liver is inorganic despite the fact that the species' main food source, fish, contains mostly methylmercury. This is true for seals from marine, brackish, and freshwaters. These results support the hypothesis that ingested methylmercury is effectively converted to inorganic Hg by demethylation processes. It is not known whether these processes are of an enzymatic, bacteriological, or chemical nature. Reijnders (1980) found that total Hg and methylmercury tissue (brain, liver, kidneys) content in harbor seals increases with age, especially so for liver. Similar results (Gaskin et al., 1979) were seen in common porpoises in Canada, in which the methylmercury fraction in the liver varied from 7.4–41%, being lowest in liver with highest total Hg concentration. However, more than 80% of total muscle Hg was present as methylmercury. Itano et al. (1984b) observed a decrease in the high muscle methylmercury percentage in striped dolphins with age, from essentially

TABLE 9 Hg Specification in Shellfish and Other Seafood

Species	N	Total Hg (ppm)[a]	Methylmercury[a] (%)	Remarks	Reference
		Nonprocessed			
Shrimp	4	0.12–0.16 (0.14)	92.9	Netherlands. Each sample composite consisted of ~50–100 individuals	Luten et al. (1980)
Mussel	5	0.05–0.12 (0.07)	42.9	Each sample composite consisted of ~50–100 individuals	Luten et al. (1980)
Invertebrates, freshwater	60	0.02–3.75	29–83 (44)	United States. Six species	Hildebrand et al. (1980)
Lobster, dried meat	3	0.39	96.9	Japan	Suzuki et al. (1980)
Shellfish, meat	5	0.19	13	Japan	Suzuki et al. (1980)
Octopus	1	0.23	46.8	United States retail	Cappon and Smith (1982a)
Squid	1	0.08	34.8	United States retail	Cappon and Smith (1982a)
Squid, king	6	0.02–0.22	39–64	Japan	Nagakura et al. (1974)
		Processed			
Clam, minced	1	0.12	30.4		Cappon and Smith (1982a)
Mussel, smoked	1	0.07	28.6	Oil-packed	Cappon and Smith (1982a)
Oyster	1	0.04	31.2		Cappon and Smith (1982a)
Crab	1	0.17	47.3		Cappon and Smith (1982a)
Shrimp	1	0.08	55.4		Cappon and Smith (1982a)
Shrimp	1	0.01	78		Hall (1974)
Oyster	2	0.04	54		Hall (1974)
Clam, minced	2	0.05	59		Hall (1974)

[a]As per Table 5.

165

TABLE 10 Hg Speciation in Marine Mammals

Species	N	Total Hg (ppm)[a]	Methylmercury[a] (%)	Remarks	Reference
Whale, sperm	7	0.65–1.57	70.1	N. Pacific	Nagakura et al. (1974)
	6	0.54–1.48	71.1	Antarctic	Nagakura et al. (1974)
Sei	9	0.01–0.07	N.D.	N. Pacific	Nagakura et al. (1974)
Fin	8	0.01–0.03	N.D.	N. Pacific	Nagakura et al. (1974)
Seal, ringed (liver)	15	5.2–87.5 (26.2)	3.0	Canada, Northwest Territories	Eaton et al. (1980)
Seal, harbor (liver)	14	2.9–27.3, juveniles (9.3)	36.6	Wadden Sea	Reijnders (1980)
Seal, harbor (liver)	13	13.6–751, adults (254)	18.1	Wadden Sea	Reijnders (1980)
Porpoise			7.4–41.0	Canada (liver)	Gaskin et al. (1972)
			~100	(muscle)	
Seal, common (liver)	8	225–765	2–14, adult 16–78, juvenile	Netherlands	Koeman et al. (1973)
Seal, ringed bearded	454	0.32–27.5 (liver) 0.08–0.91 (muscle)	3.3–13.3 75	Canada	Smith and Armstrong (1978)
Dolphin, striped	26	1.27	96	Japan	Itro et al. (1984a)
	14	0.53–7.14	82–93	Japan	
Whale beluga	1	1.09	90	Canada	Uthe et al. (1972)
Seal	1	1.80	94	Meat	Uthe et al. (1972)
		387	0.7	Liver	Uthe et al. (1972)

[a]As per Table 5.

100% down to about 20%. Nagakura et al. (1974) found that Hg speciation in whale meat (dorsal and tail) was species-dependent, the differences reflecting feeding habits. For baleen whales, which feed on zooplankton and other early stage food chain organisms, methylmercury was undetectable (< 0.008 ppm) in meat. In contrast, sperm whales, whose diet consists of squid and fish in the upper portion of the food chain, had meat methylmercury percentages of about 70.

2.3.   Dietary Guidelines and Regulations

The long biological half-life for methylmercury in man, which averages about 70 days, raised the possibility that fish consumption might lead to development of toxic body burdens of methylmercury. As a result, guidelines were established by several nations for maximum permissible Hg concentrations in fish and seafood, usually in the range of 0.5–1.0 ppm. These guidelines were originally based on some of the observations made in the Hg poisoning outbreaks of Minamata and Niigata, Japan. Selected national and regional fish Hg guidelines and advisories are summarized in Table 3.

Current safety levels for dietary Hg exposure are often based on the level of Hg in whole blood. The first toxic effects are thought to appear in sensitive individuals at a blood concentration of 200 ng/mL (ppb) (Clarkson, 1972). A safety factor of about 10 is usually used to account for higher risk population groups, notably pregnant women and children. Thus, the provisional tolerable weekly intake (PTWI) of Hg recommended by the World Health Organization (WHO) (World Health Organization, 1972) is 200 μg which would correspond to a blood Hg level of 23 ppb. The WHO PTWI is probably not exceeded at present by the general population in any nation, but is usually exceeded by local fish-eating populations in some countries (e.g., Canada, Sweden, Peru, Japan, Seychelles).

The United States Food and Drug Administration (FDA) "Action Level" for total Hg in fish and shellfish in interstate commerce prior to 1978 was 0.5 ppm. This level, initially established in 1970, was derived from data on human fish consumption and methylmercury toxicity. Since 1978, this level has been 1.0 ppm (FDA, 1979). In 1984, the FDA changed the basis of this regulation from total Hg to Hg as methylmercury due to the presence of reliable analytical methodology for methylmercury. It was due to these regulations that a permanent ban on Pacific blue marlin and a temporary ban on swordfish were enacted during the 1970s. From the data presented in this chapter, larger marine and freshwater predatory fish (especially shark and Northern pike) usually contain total Hg concentrations in their edible tissue that exceed the FDA guidelines.

3.   LEAD

3.1.   Background

A great deal of environmental Pb data exist in the literature. However, unlike Hg, comparatively fewer data have been obtained on chemical speciation for Pb in fish and related seafood. There are two major reasons for this: (1) limited knowledge, until recently, concerning the significance and extent of biotransformations among various possible Pb species, as well as the ability of many organolead compounds, especially the tetraalkylleads, to degrade rapidly in the environment; and (2) the limited availability of suitable and reliable analytical speciation methodology for quantifying the more numerous environmental Pb species.

There have been no reported cases of human Pb poisoning due to ingestion of contaminated fish. However, initial concern about the possible effects of marine environmental contamination arose from the reports of high tetraethyllead concentrations in mussels collected near the 1976 S. S. Cavtat incident where a shipload of tetraethyllead was sunk in the Adriatic Sea (Harrison, 1977). Since that time, several key research investigations have greatly expanded the knowledge of environmental Pb chemistry and its speciation behavior. Also, specific chromatographic separation techniques have been developed and applied to Pb speciation in water, sediment, and fish.

Recently, tetraalkyllead toxicity has been linked to rapid metabolic dealkylation to the highly toxic ionic dialkyl and trialkyllead species (Botre et al., 1977; Grove, 1980). Biomethylation of inorganic Pb and ionic alkyllead can also occur (Schmidt and Huber, 1976; Wong et al., 1975).

3.2.   Speciation

3.2.1.   Qualitative Speciation

The major Pb species reported in edible seafood is the divalent inorganic form (Pb^{2+}). Recent speciation evidence has established the presence of a variety of organolead forms: molecular (tetraalkyl, R_4Pb) and ionic (R_3Pb^+, R_2Pb^{2+}), where R = methyl-, ethyl-. The ethyl derivatives are more common, although mixed alkylleads (methyl- and ethyl-) have been reported. Generally, the concentration ratio, alkyllead/total Pb, is less than 0.2.

The presence of several organolead species in fish is thought to arise from two input sources: (1) anthropogenic, mainly by atmospheric fallout, with surface runoff and industrial discharge being of secondary impact; and (2) natural alkylation of inorganic Pb by biotic and abiotic processes. Tetraalkyllead compounds are the predominant airborne organic species, but are capable of chemical breakdown to ionic dialkyl- and trialkyllead species, which can directly enter the aquatic environment by vapor phase washout, and therefore be available to marine life (Harrison and Radojevcic, 1985). Tetraalkyllead can also degrade once in the aquatic environment. Several investigators (Jarvie et al., 1975; Schmidt and Huber, 1976; Wong et al., 1975) have documented biological methylation of inorganic Pb and ionic organolead compounds in aquatic environments by a variety of microorganisms. The methylcobalamin coenzyme system is a key component of the biomethylation process (Wong et al., 1978). Experimental data obtained by Chau et al. (1979) indicate that Pb^{2+} methylation results in methylated Pb yields of $3.6–11.3 \times 10^{-4}\%$ in sediment media.

In aquatic environments, organolead species, particularly the tetraalkylleads, exist mainly in sediment, fatty tissue, and intestines of fish, and in the surface microlayer associated with hydrophobic materials. Chau et al. (1984) reported the first evidence for triethyl- and diethyllead compounds in fish, the respective concentration factors being 88 and 375 times over the corresponding water concentrations. It was postulated that this could be due to *in vivo* degradation of ingested tetraethyllead in fish, or from direct concentration of these compounds from water.

3.2.2.   *Quantitative Speciation*

Selected Pb speciation data for fish and seafood are summarized in Table 11. The majority of the data relate to freshwater varieties. The presence of tetraalkyllead compounds in marine fish and fish products was first reported by Sirota and Uthe (1977) who observed high alkyllead/total Pb concentration ratios in several of the fish products, ranging from 9.5–89.7% of total Pb content. The exact organic forms in this study were not identified because an atomic absorption technique was employed. More recent and specific Pb speciation data have been reported for freshwater fish, where chromatographic methodology was used as part of the investigation (Chau et al., 1980, 1985). Although the organolead/total Pb concentration ratios were lower (usually < 0.5), these values varied considerably between and within fish species. This strongly indicates that organolead levels in fish reflect localized fish feeding habits and pollution sources. Also, the presence of tetraethyllead may indicate that tetraalkyllead compounds may not be immediately metabolized by fish and may reside within the edible tissue in their original form for a long time, thus enhancing bioaccumulation. Another potential factor that may influence and possibly enhance quantitative uptake of organolead compounds by fish is low water pH, resulting from acid rain, similar to the case of Hg. However, there are currently no published data available to substantiate this effect.

3.3.   **Dietary Guidelines and Regulations**

National seafood dietary guidelines are currently based on data for total Pb content (Table 12). However, the occurrence of organolead compounds in fish poses questions as to the validity of existing dietary standards for Pb in food that do not specify chemical form. Health criteria for specific organolead forms are not yet available.

4.   **TIN**

4.1.   **Background**

The biogeochemistry of Sn is quite similar to that of Pb, although investigation into its environmental behavior has been more recent, beginning in the late 1970s. Like Pb, the recent investigations involving chemical speciation for Sn have been

TABLE 11 Pb Speciation in Fish and Seafood

Species	N	Total Pb (ppb)	RPb species % of Total	Pb	Remarks	Reference
Marine						
Cod	1	390		9.5	Japan, fresh liver	Sirota and Uthe (1977)
	1	520		24	Frozen liver	Sirota and Uthe (1977)
Flounder	1	5340		89.7		Sirota and Uthe (1977)
Mackerl	1	140		38.6		Sirota and Uthe (1977)
Dab	1	$0.01-0.62^a$		—	Only R_4Pb reported	Jermen et al. (1984)
Cod, liver	1	0.48^a		—	Only R_4Pb reported	Jermen et al. (1984)
					Freshwater	
Unspecified	4	55–273		8–38	Hexane-extractable	Cruz et al. (1979)
				2–29	Volatile	
				0.05–6.4	Tetraalkyl^b	
Trout, rainbow	4	0.26^a		$(CH_3)_4Pb$ identified only	Canada (Ganaraska R., Ontario)	Chau et al. (1979)
White sucker	8	772		15	Detroit and St. Clair Rivers	Chau et al. (1985)
Carp	2	183		56	Total alkyllead^c	
Perch, yellow	1	256		0		
Trout, brown	1	116		0		
Walleye	6	87		30		
Pike, northern	4	450		38		
Carp	2	6.91^a	d	e	St. Lawrence R.	Chau et al. (1989)
			18.6	f	Ontario	
		14.53^a	28.4	13.2		
				53.1		

Species	n						Notes
Pike	2	2.44[a]	42.6	8.8	48.6		Chau et al. (1980)
		2.72[a]	43.7	9.5	46.8		
White sucker	2	15.76[a]	22.1	50.1	27.8		
		11.31[a]	31.9	39.5	28.6		
Bass, small mouth	2	2.02[a]	12.6	15.6	71.8		
		3.40[a]	9.0	27.5	63.5		
			[g]	[h]	[i]		
Trout, rainbow	3	65–95	8.8–80.4	3.0–14.0	2.4–5.8		Ontario, Canada lakes and rivers
Salmon, coho	2	85	35.0	4.8	13.6		107 total samples;
Perch, white	1	185	—	2.3	0.3		17 samples contained
Bullhead	1	30	—	6.0	9.3		R_4Pb compounds
Crappie, black	1	115	—	2.3	1.3		
Perch, yellow	1	325	1.5	—	4.9		
Sucker	2	78	8.7	4.1	3.2		
Rockbass	1	215	0.8	1.8	1.6		
Sunfish	2	190	1.9	2.2	2.1		
Bass, smallmouth	1	120	15.0	1.3	1.3		
Bass, white	2	78	11.3	1.5	1.5		

[a] ppm.
[b] Tetramethyl, trimethylethyl, dimethyldiethyl, triethylmethyl, tetraethyl.
[c] R_4Pb, R_3Pb^+, R_2Pb^{2+} (R = ethyl); also triethylmethyl.
[d] Inorganic Pb.
[e] Ionic alkyllead; R_3Pb^+, R_2Pb^{2+} (R = methyl, ethyl).
[f] Tetralkyllead, as per b.
[g] Hexane-extractable (total organic Pb).
[h] Volatile Pb(R_4Pb).
[i] Tetralkyllead, as per b.

TABLE 12 Regulations for Pb, Sn, As and Se Content in Seafood and Related Dietary Consumption

Element	Source	Date	Regulation	Reference
Pb	WHO	1972	*PTWI*: 3 mg/person, equivalent to 0.05 mg/kg body weight. Not applicable to infants and children No hazard in short-term, only in long-term	WHO (1972)
	Great Britian	1974	Lead in Food Regulations: 2 ppm (all food, inorganic Pb) 0.5 ppm (canned baby food)	Waldron (1975)
Sn	West Germany	1971	Recommended daily intake from food: 0.0065 mg/kg body weight (R_2SnX_2 species)	Piver (1973)
As	United States FDA	1987	0.5 ppm, muscle meats	HSBD (1987)
	Great Britain	1969	1.0 mg/kg (statutory level for all food)	Schroeder and Balassa (1966)
	WHO	1967	50 mg/kg body weight, daily intake (adults) 2 mg/kg body weight, daily intake (adults)	WHO (1967)
		1984		Lawrence et al. (1986)
Se	United States	1980	Recommended dietary allowance: 200 mg/person/day	Food and Nutrition Board (1980)
		1976	60–120 mg/person/day	Food and Nutrition Board (1976)
		1977	50–100 mg/person/day as supplement for persons in low Se regions	Food and Nutrition (1977)
	Japan	1975	Proposed maximum acceptable daily intake: 500 mg/person	Sakurai and Tsuchiya (1975)

made possible by the development of suitable analytical technology. However, almost all of the ever-increasing speciation research activity has been concerned with the sediment, water column, and aquatic plant life of inland waterways.

4.2. Speciation

4.2.1. Qualitative Speciation

In addition to the predominant inorganic Sn, mainly as $Sn(OH)_4$, several organotin species have been identified in aquatic environments. These include methyl-, ethyl-, butyl-, cyclohexyl-, octyl-, and, less frequently, phenytin, usually as the di-, tri-, and tetraorganospecies. In fact, methylated Sn species are almost ubiquitous in natural waters—both sea- and freshwater—at the ppb level (Brinckman and Fish, 1981; Byrd and Andreae, 1982). These can exist as volatile tetraalkyltins (R_4Sn) and polar (ionic) alkyltin ($[R_nSn]^{(4-n)+}$) species. Studies on the environmental persistence of butyltin and phenyltin compounds indicate that abiotic and biological degradation occur by sequential dealkylation (Maguire, 1984) or dearylation (Soderquist and Crosby, 1980). However, abiotic and biotic methylation of inorganic Sn and methyltin compounds has been demonstrated in water and sediments (Guard et al., 1981).

Tin speciation in the aquatic environment is very complex. To the natural Sn^{2+} and Sn^{4+} and methyltin species, human activities add mainly butylated, octylated, phenylated, and even methylated derivatives. The most environmentally significant, due to their high toxicity and direct input, are the trisubstituted organic derivatives. The origin of environmental organotin is mainly antropogenic, biological methylation being of secondary importance. This is especially critical for fish species inhabiting inland lakes, rivers, and estuaries, since localized elevated levels of organotins are more likely to result from industrial wastewater discharges. Also, harbors and seaports, areas with large volumes of navigational activities, are places where significant leaching of organotins from marine antifoulant paints occurs.

Biomethylation of Sn is thought to be best represented by the reaction of methylcobalamin with Sn^{2+}, producing ionic mono- and dimethyl Sn(IV) species (Ridley et al., 1977). There is evidence that these organotin species are highly solvated, probably existing as $R_nSn(OH_2)_n^{(4-n)+}$, and showing strong lipophilic properties. The higher alkylated species, notably the butylated derivatives, are more lipophilic than the methyltins. Studies have suggested that butyltin species are persistent in the aquatic environment and that the tri- and dibutyl forms, along with inorganic Sn, are concentrated up to a factor of 10^4 in the surface microlayer of rivers and lakes (Maguire et al., 1985). Recent evidence has documented in vitro methylation of inorganic Sn and alkylated Sn species by sediment microbes isolated from Chesapeake Bay (Jackson et al., 1982), San Francisco Bay (Hodge et al., 1979), Great Bay (New Hampshire) (Donard et al., 1985), and several Canadian freshwater lakes (Maguire et al., 1982). For the marine environment, atmospheric transport of anthropogenic Sn species to seawater is more important than dissolved river input.

These factors strongly suggest that fish and other aquatic organisms inhabiting contaminated waterways may easily accumulate significant levels of inorganic and organotin compounds, mostly from the food chain but also from water and sediment. Another potential source could be *in vivo* biomethylation, as is the case for Hg, although there is no conclusive evidence for this in fish.

4.2.2. Quantitative Speciation

There is a general paucity of Sn speciation data for fish and other edible seafood. A partial reason for this is the relatively recent availability of reliable analytical methodology for organotin speciation. The major reason may be the recent application of this same methodology to evaluate Sn speciation in specific components—sediment, vegetation, water column—of freshwater and estuarine waterways. These waterways are not major sources of the more common varieties of commercial seafood. However, there are several regional areas [e.g., Chesapeake Bay, Ontario (Canada) lakes and rivers] in which commercial fisheries, aquatic food sources, and local fish-eating populations may be adversely affected in the future by the increasing presence and persistence of organotin pollution.

Pinel and Madies (1986) recently demonstrated that oysters can concentrate inorganic and organic Sn and that this concentration can become important with time. Oysters were exposed for several weeks to panels coated with antifouling paints containing tri-*n*-butyltin oxide (TBTO). High mortality rates, shell malformations, and organic Sn levels up to 2 ppm in the edible body tissue were observed. Piver (1973) reported an LD_{50} (24 hr) of 0.027 ppm for rainbow trout exposed to TBTO.

To date, the only available Sn speciation data for seafood were reported by Tugrul et al. (1983), for limpet and two species of fish from the Cilician basin of the Mediterranean Sea, and by Tsuda et al. (1987) for carp from Lake Biva, Japan. Although the majority of Sn was present as the inorganic form, there were significant amounts of methylated (mono-, di-, and tri-) and butylated Sn species detected in the edible muscle (Table 13). Compared to fish, limpets accumulated larger percentages of methylated Sn (35–75% versus 3–6% of the total Sn level).

TABLE 13 Sn Speciation in Fish and Seafood

Species	RSn species % of total Sn	Remarks	Reference
Fish[a]	3–6[b]	Mediteranean Sea	Tugrul et al. (1983)
Limpet	35–75[b]	Cilician basin	
Carp, crucian	0.013–0.016 ppm as tributyltin	Lake Biwa, Japan	Tsuda et al. (1987)

[a] Species not specified.
[b] Mono-, di-, and trimethylated Sn species.

4.3.   Dietary Guidelines and Regulations

Dietary regulations for Sn and its organic derivatives are scarce for two reasons. First, there have been few reported cases of human poisoning from ingestion of Sn-contaminated food. These cases resulted from contaminated food products, mainly fruit juices, packed in nonlacquered Sn-plated cans. However, storage of canned food, even in lacquered cans, can result in Sn contamination due to corrosion starting from defects in the lacquer surface. Schroeder et al. (1964) found relatively high concentrations in oil-packed fish from uncoated cans. Second, little information is available on particular sources and chemical forms of Sn in normal human diets, although estimated data are available on mean daily adult intake of Sn (Kehoe et al., 1940). These data reflect only total Sn levels, whereas no cases of human poisoning or dietary intake related to organic Sn have been reported. In comparison, the Sn levels found in fish, assuming a low percentage (< 10) of organic Sn, would probably not present any short-term risk to people exposed to normal fish consumption.

There is no established limit of safe dietary intake exposure for Sn or any of its compounds in the United States. To date, only West Germany has, since 1971, a recommended daily human food tolerance of 0.0065 mg/kg (body weight) for dialkyltin species (Piver, 1973) (Table 12).

5.   ARSENIC

5.1.   Background

In nature, As is widely but sparsely distributed. One exception to this is aquatic organisms, including most edible species. In general, compared to terrestrial animals, higher levels of total As are commonly found in commercially available marine seafood products. These levels are usually in the range of 1–45 ppm, and the numerous literature reports have been recently summarized by Penrose (1974) and Luten et al. (1982). More importantly, these levels are often in excess of established health limits. The initial concern over elevated As concentration in seafood resulted from the long history of associating As with poison, although recent toxicological research, together with speciation methodology, has moderated this concern. In addition, there are no known instances of poisoning by As that has accumulated naturally in food organisms.

Arsenic enters the aquatic environment indirectly from industrial and utility air emissions, and directly from localized effluent discharges. Open ocean water normally contains a low range of As concentration—1–8 ppb. The As content of fresh water varies considerably according to the location and pollution characteristics of the specific waterway. Levels of 0.4–10,000 ppb have been reported. The majority of natural and anthropogenic As in seawater is present as inorganic As, with approximately two-thirds as arsenite (As^{3+}). However, under aerated conditions, arsenate (As^{5+}) predominates, while As^{3+} predominates under anaerobic conditions. Like Hg, As is susceptible to methylation by microorganisms, mainly in sediment, forming volatile methylated arsines and nonvolatile organoarsenic acids (Lunde, 1977), with subsequent release to the water. Marine organisms can accumulate As by direct water contact and ingestion of food.

Although As levels found in marine animals exceed those found in their surroundings (water, sediment), there is a profound change in As speciation biomagnification pattern. Several investigators (del Riego, 1968; Penrose et al., 1977; Seydel, 1972) have shown that As concentrations are not magnified in the aquatic food web (sediment → phyto- or zooplankton → benthic and predatory species). Arsenic levels in pelagic fish (in the order of 0.3–3 ppm) are significantly lower than those of bottom feeders and shellfish (in the order of 1–55 ppm). Part of the reason for this is that As is removed from seawater by sedimentation, and is present in sediment at higher levels than water (2–20 ppm versus 1–8 ppb). The exact nature of As entry and subsequent biotransformation in marine organisms is unknown. It is quite possible that biotransformation of As in higher marine organisms may be initiated by intestinal flora.

5.2. Speciation

5.2.1. Qualitative Speciation

Present knowledge, along with recent investigations concerning As speciation in marine animals, reveals extremely low levels (expressed as a percentage of total As) of inorganic As and monomethylarsonic acid (MMAA) and dimethylarsinic acid (DMAA). Instead, practically all of the As found in edible seafood species is present either as small cationic organic molecules—arsenobetaine, arsenocholine—or as more complex As—phospholipid complexes. The predominance of these forms is significant because they are extremely stable, chemically and metabolically, and much less toxic.

Marine organisms have been shown to accumulate inorganic As and transform it into organoarsenicals. About 12 different organoarsenic compounds have been identified in algae (Andrae and Klumpp, 1979). Thus, it is likely that a major part of the As found in higher trophic level marine species originates from previously synthesized organoarsenicals from the lower stages of the aquatic food chain (e.g., algae, kelp). These substances are either water or lipid-soluble compounds, such as arsenosugars (Edmonds and Francesconi, 1977) and phospholipids (Cooney et al., 1978), which may be partially converted to other organic As species in higher tropic levels.

The most prevalent As compound found in marine animals is arsenobetaine. This cationic water-soluble organic species has been identified in the edible tissue of lobster (Edmonds and Francesconi, 1981a; Edmonds et al., 1977), shark (Cannon et al., 1979; Kurosawa et al., 1980), whiting (Edmonds and Francesconi, 1981b), plaice and crab (Luten et al., 1982), shrimp and flounder (Lawerence et al., 1986), and several fresh marine species (haddock, halibut, salmon, cod, herring, mackerel, sole, scallops (Lawrence et al., 1986). This compound was not found in freshwater fish (pike, bass, carp, pickerel, perch, whitefish) that contain a methanol-extractable As species that has not yet been identified (Lawrence et al., 1986). Other organic As compounds reported in marine organisms include arsenocholine in shrimp (Norin et al., 1983; Lawrence et al., 1986), dimethyl (ribosyl) arsine oxide, trimethylarsonium lactate, and O-glycerophosphoryltri-methylarsonium lactate in clam kidney (Benson and Summons, 1981). In addition, As-containing phospholipids, in which the nitrogen of the choline group is replaced by As, have been isolated from marine animals (Norin et al., 1983). Lunde (1977) reported the presence of polar lipid-soluble organic As compounds in marine oils. These lipid-soluble compounds have not yet been isolated in pure form. However, as isolation and analytical techniques improve, these and other complex organoarsenicals will be identified in marine organisms.

Recent human and animal metabolic studies strongly indicate that As present in seafood—all of it presumably as the specific organic compounds previously described—seems unable to mix with the inorganic As pool in the organism, and is quantitatively and rapidly eliminated from the body in the urine (Foa et al., 1984). Thus, As from seafood is apparently poorly absorbed by the body and the elevated dietary intake levels do not pose a serious hazard to human health.

5.2.2. Quantitative Speciation

Table 14 summarizes the rather limited current As speciation data for several commercial fish and shellfish species. Brooke and Evans (1981) reported very low percentages ($< 5\%$ of the total As content) of inorganic As in fish, shellfish, and fish products. Lawrence et al. (1986) recently presented the first conclusive speciation data on freshwater and marine fish and shellfish from Canada. Only the marine species contained arsenobetaine, and arsenocholine (shrimp and Pacific salmon only), as the major As constituents, ranging from 74–91% of the total As tissue concentration. This percentage showed no correlation to the total As levels, indicating that the As metabolic process is not species-dependent for marine animals. In shrimp samples containing both arsenobetaine (B) and arsenocholine (C), the concentration ratio of these species (B/C) varied according to type and harvest area. The same authors showed that freshwater fish contained no B or C. However, a more polar methanol-extractable As form not found in marine samples, was present in a similar percentage range of the total As level (71–85%). To date, the chemical structure and biological activity of this species are unknown.

5.3. Dietary Guidelines and Regulations

Legal limits for As in food were established during periods of public reaction to environmental poisoning incidents in which As was a suspected, but not proven, agent. These regulations (Table 12) were also based on data pertaining to total As content, and not on those typically found in seafood. For example, the only regulatory levels for As in foods in the United States are tolerances that have been established for residues in specified foods, resulting from the application of

TABLE 14 As Speciation in Fish and Seafood

Category/ Species	N	Totals As (ppm)	As Species, % of Total As	Remarks	References
			Marine Fish		
Herring	1	1.1	3.6	Inorganic As	Brook and Evans (1981)
Haddock	1	2.6	0.8	Inorganic As	Brook and Evans (1981)
Plaice	1	24.0	0.4	Inorganic As	Brook and Evans (1981)
Plaice	255	3–166	95.0	Netherlands: arsenobetaine	Luten et al. (1982)
Herring	2	1.1, 1.0	89, 86	Canada: arsenobetaine	Lawrence et al (1986)
Halibut	1	2.3	74	Canada: arsenobetaine	Lawrence et al (1986)
Haddock	2	3.5, 6.0	86, 78	Canada: arsenobetaine	Lawrence et al. (1986)
Cod	2	5.1, 7.4	80, 84	Canada: arsenobetaine	Lawrence et al. (1986)
Mackerel	1	0.55	80	Canada: arsenobetaine	Lawrence et al. (1986)
Sole	3	13.2, 5.2, 0.10	86, 85, 80	Canada: arsenobetaine	Lawrence et al. (1986)
Salmon	1	0.31	48	Arsenobetaine	Lawrence et al. (1986)
			42	Arsenocholine	Lawrence et al. (1986)
			Freshwater Fish		
Bass	1	0.12	71	Canada (Ontario and Alberta)	Lawrence et al. (1986)
Perch, yellow	2	0.055, 0.007	73, 85	Identity of organic As species unknown	Lawrence et al. (1986)

Perch, striped	1	0.24	71		Lawrence et al. (1986)
Pike	2	0.048, 0.023	71, 84		Lawrence et al. (1986)
Carp	1	0.18	72		Lawrence et al. (1986)
Pickerel	1	0.037	84		Lawrence et al. (1986)
Whitefish	1	0.024	79		Lawrence et al. (1986)
Marine Organisms					
Lobster	2	5.2, 4.7	87, 76	Canada; arsenobetaine	Lawrence et al. (1986)
Scallop	1	0.68	88	Canada; arsenobetaine	Lawrence et al. (1986)
Shrimp	2	20.8	76	Arsenobetaine	Lawrence et al. (1986)
			15	Arsenocholine	Lawrence et al. (1986)
		7.2	8	Arsenobetaine	Lawrence et al. (1986)
			72	Arsenocholine	Lawrence et al. (1986)
Crab species	3	4.0–8.6	79–85	Fresh; arsenobetaine	Brooke and Evans (1981)
Crab	1	1.5	5.3	Canned; inorganic As	Brooke and Evans (1981)
Lobster	1	3.6	1.9	Canned; inorganic As	Brooke and Evans (1981)
Shrimp	1	19	9.6	Boiled and frozen; arsenobetaine and arsenocholine	Norrin et al. (1983)

arsenical pesticides on food and feed crops and from animal feed additives. In addition, the regulatory tolerance level (1 ppm) in Great Britain, along with that of the United States, was established from experiments based on the assumption that the As was present as inorganic As.

The World Health Organization (WHO) has recently lowered the maximum tolerable daily As intake (from all dietary sources, including food and water) from 50 μg/kg body weight to 2 μg/kg with the recommendation that more research be performed to identify the nature of the arsenical compounds occurring in food and, in particular, seafood. This change reflected the concern over increased levels and, therefore, dietary intake, of the more toxic inorganic As. However, results of previously cited short-term metabolic studies, coupled with the known lack of toxicity for the major organoarsenicals identified in seafood, indicate that new As dietary tolerances may have to be established.

6. SELENIUM

6.1. Background

Of the five trace elements surveyed in this chapter, only Se is an essential micronutrient for plants and animals, and probably for man, but it is highly toxic when present in excess of nutritional requirements. Underwood (1977) reviewed this problem and considered the dietary concentration requirement for most animals to be in the order of 0.1–0.3 mg Se/kg food (dry weight basis), whereas the minimum toxic level might be close to 5 mg Se/kg. Thus, the margin between requirement and toxicity is relatively narrow.

Fish are known to contain higher levels of Se than most other foods and are thus a major dietary source of this element. This is important, since Se also has been shown to counteract both inorganic Hg and methylmercury toxicity in several animal species (Iwata et al., 1973; Parizek et al., 1974, 1976; Potter and Mattrone, 1974; Stoeswand et al., 1974). Selenium naturally present in the edible tissue of tuna (Ganther et al., 1972; Ohi et al., 1976; Stillings et al., 1974), swordfish (Friedman et al., 1978), and whale (Ohi et al., 1980), and in seal liver (Eaton et al., 1980), reduces methylmercury toxicity in rats fed diets containing these food sources. Marine mammals that feed exclusively on fish high in total Hg show no evidence of Hg intoxication (Koeman et al., 1973). More importantly, specific human population groups that consume large quantities of fish high in Hg (usually >1.0 ppm) have also shown no evidence of methylmercury poisoning. Because of these findings, it has been suggested that the Se content should also be considered in establishing future permissible Hg levels in fish and other commercially important seafood. Most of the present Hg dietary guidelines are based on data from Japanese ingesting freshwater fish highly contaminated with Hg and with lower Se levels, so that the effect of Se was discounted. However, different chemical forms and dietary sources of Se have different protective potencies against mercurial toxicity. For example, Se from animal origin (Ohi et al., 1976) and in the form of selenate (Se^{6+}) (Sumino et al., 1977) is less effective than Se from plant and inorganic sources in preventing methylmercury neurotoxicity in experimental animals. This is likely due to Se from plant foods, and in the form of selenite (Se^{4+}), having higher biological availability than that of animal origin (Yoshida et al., 1984). Although the protective effect of Se has been demonstrated in laboratory animals, it must be emphasized that its extension to humans has not been substantiated and therefore is a highly speculative matter. Data on Se levels and speciation in fish and other seafood are not as extensive as they are for Hg.

6.2. Speciation

6.2.1. Qualitative Speciation

Environmental input of Se into oceans and inland waterways arises from atmospheric deposition, terrestrial runoff, and wastewater discharge. The corresponding Se exposure to aquatic organisms is essentially inorganic—either as soluble Se^{4+} (selenite) or Se^{6+}. Elemental Se ($Se°$) is insoluble and becomes a major sink for Se, especially in sediment. In natural waters, the level of total Se is generally < 1 ppb except for certain freshwater bodies in regions of seleniferous soils. In most cases, total Se concentration in freshwater exceed those found in seawater. Several recent speciation surveys have reported Se^{6+} to be the predominant inorganic form in freshwater and in deeper layers of seawater. In the surface layer of seawater, Se^{4+} predominates. Where speciation of the Se^{2-}, $Se°$ forms was reported, river water contained higher percentages of these species than seawater. Burton et al. (1980) showed that Se^{4+} predominates in deeper waters and can originate from the oxidation of Se^{2-}, which is a common form in phytoplankton. Selenate was the predominate form (up to 90% of the total Se) in estuarian waters (Measures and Burton, 1978). More recently, Uchida et al. (1980), also found Se^{6+} to be most prevalent in seawater. The same authors reported Se^{2-} and Se^0 levels to be higher in river water than in seawater, whereas Se^{4+} in seawater was more prevalent than in river water. Thus, since many marine organisms, especially those at the lower end of the food chain, are likely to be exposed to Se mostly from water, the specific nature of the aquatic environment will influence the uptake of specific Se forms. For example, the markedly higher Se^{4+}/Se^{6+} concentration ratio in bottom seawater layers, as compared to surface water, may explain the preferential uptake of Se^{4+} by organisms, which has been observed in mussels (Fowler and Benayoun, 1976). The chemical nature of organic Se in water, reported by Sugimura et al. (1976) and Suzuki et al. (1981), has not been characterized. It may be specific inorganic Se species complexed to organic ligands, absorbed on the surface of suspended solids, or the element itself as part of an organic compound (e.g., amino acid, peptide, protein).

There is sufficient evidence for Se biomagnification in aquatic food chains. Adams and Johnson (1977) surveyed Se levels in water, sediment, zooplankton, and fish from the western basin of Lake Erie. A comparison of the Se concentration in various trophic levels indicated that Se progressively increased from water to sediment and from water to zooplankton to fish (e.g., common shiners, sheepshead, yellow perch, carp, walleye).

Selenium present in vegetation is incorporated into various amino acids on metabolism, whereas the chemical form of Se of animal origin is less known, although it is believed to be bound to glutathione and other proteins. For fish and other marine organisms, this may be true as biochemical fractionation studies on algal tissue showed Se to be largely associated with combined amino acids, one of the least toxic forms of Se. Thus, it is likely that fish and other edible seafood species may assimilate various organic and inorganic Se forms from various sources—sediment, water, and aquatic food chain organisms.

Selenium in edible seafood tissue exists as Se^{2-}, Se^{4+}, and Se^{6+}. To date, no attempts have been made to separately isolate and measure Se^{2-} in fish and seafood. Selenium that is metabolized by animals and becomes covalently bound to tissue proteins is believed to be Se^{2-}. However, the exact chemical forms of Se in animal selenoproteins have not been identified. Recent investigations have provided significant information on the distribution of Se species in fish and seafood (Cappon and Smith, 1981, 1982a,b). The results of this work are summarized in Table 4. Unlike what was observed in similar experimentation for Hg, the muscle Se distribution pattern was quite similar for all fish categories. Most of the total Se content, averaging 57.0%, was water-extractable. The only exceptions were canned marlin and fresh salmon. For marlin, an average of only 13.4% of the total Se was extractable. On an average percentage basis, Se^{6+} was more extractable than Se^{2-} and Se^{4+}, especially for the marlin, newly processed marine, and freshwater samples. These results agree with those of Itano et al. (1979) who found the majority (about 80%) of the total muscle Se in bream and tuna to be water-extractable. However, Soichi (1979) found only 40% of the Se in tuna muscle to be extractable by 0.05 M phosphate buffer. In a related study with canned tuna (Cappon and Smith, 1982b) the present author noted differences in the Se muscle distribution pattern (Table 4). For newly canned and older water-packed samples, the majority of the total Se content (averaging 55.6%) was water-extractable. On an average basis, Se^{6+} was more extractable than Se^{2-} and Se^{4+}. However, for older oil-packed samples, an average of only 48.2% of the total Se was extractable. These data suggest the possible influence of sample storage and packing medium, in conjunction with sample age, on the chemical form and distribution of Se in canned tuna. In contrast, canned mollusks and crustaceans, except oysters, revealed a Se distribution pattern different from that of fish. The majority—55.8%—of total Se was not water-extractable. The opposite was true for fresh octopus and squid. Like fish, Se^{6+} was the more extractable chemical form for all samples except squid. Lunde (1972) isolated a lipid-soluble Se compound from marine fish. An average of 60% of this compound was in the high-molecular-weight (MW > 5000) extract from liver and muscle tissue, indicating that part of this compound is bound to a lipoprotein.

6.2.2. Quantitative Speciation

6.2.2.1. Marine and Freshwater Fish. Although an increasing amount of data on the total Se content of fish and fish food products has become available in recent years, speciation data remain sparse. To date, Cappon and Smith (1978, 1981, 1982a) have investigated Se speciation in marine and freshwater fish and, more specifically, in canned tuna (1982b). The data from these studies are summarized in Table 15. All samples contained significant but variable percentages of Se^{6+}, ranging from 3.5 to 47.0% of the total tissue Se content. On the average, freshwater species contained higher Se^{6+} percentages than did marine samples (36.1 vs. 24.2%). For the canned tuna samples, there was no apparent effect of species, sample age, or packing medium on Se^{6+} percentage. As was found for shellfish, percentage Se^{6+} values were also independent of total tissue Se content.

A smaller percentage, averaging 12.1% of Se^{6+}, was present in the packing media from samples of canned fish and seafood (Table 6). Unlike Hg, the total Se levels represented more significant fractions (2.2–16.5%) of the corresponding levels in the total can contents. The CR values for the water-packed samples, averaged 2.5 times higher than those for the oil-packed samples, indicating the possibility of solubilization of tissue-bound Se by the packing water.

In one of the previous speciation investigations (Cappon and Smith, 1982a), a muskellunge sample that was smoked and cooked before analysis had a Se^{6+} percentage of 47.0, the highest value yet observed. In contrast, a fresh muskellunge sample had a corresponding value of 19.1%. The possible effect of smoking on Se speciation was recently examined in brown trout fillets (Cappon 1984), and the results are summarized in Table 8. Although smoked and unsmoked fillets and similar total Se concentration ranges, the smoked fillets had higher average Se^{6+} percentages, possibly due to oxidation of tissue-bound Se^{2-} to Se^{6+} at the elevated smoking and baking temperatures. Higgs et al. (1972) demonstrated that most common cooking techniques (baking, boiling) do not result in Se losses in seafood.

The significant Se^{6+} levels in fish and other edible seafood is of potential dietary significance regarding methylmercury intake and chronic toxicity. Sumino et al. (1977) demonstrated that added Se^{4+} is much more effective than Se^{6+} in releasing methylmercury from protein linkage in various animal tissues (blood, liver, kidney, tuna muscle) and can exert a greater influence in modifying its overall toxicity. There is no reported evidence of *in vitro* or *in vivo* Se^{6+} formation in animal tissue.

6.2.2.2. Shellfish and Other Seafood. Total Se content for this category is generally similar to the corresponding values reported for freshwater fish (Table 16). Limited speciation data obtained by Cappon and Smith (1982a) revealed significant Se^{6+} percentages (averaging 23.9% of the total Se content), but these values were highly variable and independent of the total Se content. For canned seafood samples, total Se levels in the packing water averaged 6.4% of the

TABLE 15 Se Speciation in Marine and Freshwater Fish

Species	N	Total Se (ppm)[a]	Se²⁻, Se⁴⁺ (%)	Se⁶⁺ (%)	Remarks	Reference
			Marine			
NONPROCESSED						
Tuna	3	0.25–0.62	80.1	19.9	American Samoa	Cappon and Smith (1981)
Tuna	6	0.24–0.64	77.2	22.8	Northern Peru	Cappon and Smith (1982a)
Bluefish	1	0.51	85.8	16.2	Retail	Cappon and Smith (1981)
Dolphin	1	0.21	72.8	27.2		Cappon and Smith (1981)
Flounder	1	0.17	83.4	16.6	Retail	Cappon and Smith (1981)
Red snapper	1	0.38	82.6	17.4		Cappon and Smith (1981)
Shark, mako	2	0.36, 0.24	66.7, 68.9	33.3, 31.1	Retail	Cappon and Smith (1982a)
Swordfish	2	0.48, 0.20	86.0, 74.5	14.0, 25.5	Retail	Cappon and Smith (1982a)
Mackerel	1	0.34	67.7	32.6	Retail	Cappon and Smith (1982a)
Salmon, pink	1	0.45	67.4	32.6	Retail	Cappon and Smith (1982a)
PROCESSED						
Tuna	8	0.49–0.83	70.1–85.1	16.9–29.3		Cappon and Smith (1981)
Marlin, Pacific blue	2	2.48, 4.15	94.1	5.9		Cappon and Smith (1982a)

Species	n	Range			Note	Reference
Tuna	5	0.32–0.77	71.1	28.9	Old	Cappon and Smith (1982a)
Tuna	3	0.37–0.89	79.9	20.1	New	Cappon and Smith (1982a)
Mackerel	2	0.40, 0.58	78.8	21.2		Cappon and Smith (1982a)
Tuna	13	0.47–1.33	70.1	29.9	New (<6 mo)	Cappon and Smith (1982b)
Tuna	13	0.36–1.00	72.2	27.8	Old (5–29 yr)	Cappon and Smith (1982b)
Freshwater						
Bass, largemouth	2	0.15, 0.47	67.6	32.4		Cappon and Smith (1981)
Bass, smallmouth	1	0.5	81.3	18.7		Cappon and Smith (1981)
Bullhead	1	0.26	78.8	21.2		Cappon and Smith (1981)
Perch	1	0.22	77.6	22.4		Cappon and Smith (1981)
Walleye	1	0.45	64.1	35.9		Cappon and Smith (1981)
Muskellunge	2	0.23	53.0	47.0	Muscle smoked and cooked	Cappon and Smith (1982a)
		0.32	80.9	19.1		Cappon and Smith (1982a)
Pike, Northern	3	0.42–0.58	81.4	18.6	Ontario, Canada	Cappon and Smith (1982a)
Pike, Northern	4	0.21–0.34	64.8	35.2	Quebec, Canada	Cappon and Smith (1981)
Salmon, coho	7	0.22–0.88 (0.45)	85.1	14.9	United States, Lake Ontario	Cappon (1984)
Salmon, chinook	6	0.22–0.65 (0.42)	71.7	28.3	United States, Lake Ontario	Cappon (1984)
Trout, brown	4	0.28–0.70 (0.48)	78.5	21.5	United States, Lake Ontario	Cappon (1984)
Trout, lake	7	0.16–0.33 (0.24)	72.5	27.5	United States, Lake Ontario	Cappon (1984)

[a]Values in parentheses represent averages of the corresponding ranges.

185

TABLE 16   Se Speciation in Shellfish and Other Seafood[a]

Species	N	Total Se (ppm)	Se^{2-}, Se^{4+} (%)	Se^{6+} (%)	Remarks
Nonprocessed					
Octopus	1	0.27	65.8	34.2	United States, retail
Squid	1	0.42	86.3	13.7	
Processed					
Clam, minced	1	0.16	56.4	43.6	Water-packed
Mussel, smoked	1	0.31	81.4	18.6	Oil-packed
Oyster	1	0.19	92.8	7.2	Water-packed
Crab	1	0.22	75.3	24.7	Water-packed
Shrimp	1	0.21	74.7	25.3	Water-packed

[a]Data from Cappon and Smith (1982a).

corresponding levels in the total can contents (Table 6). Although this was a significant fraction, it is considerably less than the corresponding average value for Hg, indicating that Se may have a lesser tendency to become solubilized by the packing water.

6.3.   Dietary Guidelines and Regulations

There are no current government regulations concerning human dietary intake or tolerable levels from seafood. This can be attributed to three factors. First, there have been no reported deficiency or toxicity symptoms in man from Se of food origin. Second, definite quantitative Se requirements for man are not known. Finally, the beneficial effects of Se on animal health have resulted in its widespread use as a livestock feed supplement, either as Se^{4+}, Se^{6+}, and selenoamino acids. In fact, this specific use of Se has resulted in recent regulations of permissible levels in domestic livestock feeds, both in the United States (FDA, 1974) and Canada (Stevenson, 1977).

Despite the lack of dietary regulations, dietary Se guidelines do exist, and these are summarized in Table 12. Earlier guidelines were based on data published prior to 1974. The World Health Organization (WHO) has considered Se, but has made no recommendations on dietary intake. Lo and Sandi (1980) noted that the only documented incidence of human Se toxicity was selenosis found in the seleniferous areas of South Dakota in the 1930s. This observation, in conjunction with normal levels recently reported in foods, lead these authors to suggest 500 μg/day as a maximum tolerable level. This would be achieved by a daily consumption of 500 g of seafood containing 1 ppm Se. However, the seafood dietary intake for the general population, at the Se levels reported in this survey (mostly <0.5 ppm) would result in a daily Se intake well below this suggested maximum level.

7. CONCLUSIONS AND RECOMMENDATIONS

7.1. Conclusions

The need to qualitatively and quantitatively assess the presence of numerous chemical forms of specific trace elements in fish and related seafood is essential to understanding not only the impact on aquatic environments, but also the long-term impact on human health. A key factor in this assessment is the availability and future technological advances in the specific analytical separation methodology used for trace element speciation. An essential component in the development and routine application of this methodology is the use of quality control procedures and programs for achieving method validation and ensuring accuracy of the speciation results. Although in most industrialized nations, environmental pollution control measures have been enacted along with dietary intake regulations, the environmental ubiquity and persistence of most trace element chemical species will demand continuous analytical monitoring and increased research efforts concerning fish and other seafood products to ensure adequate protection of human health.

7.2. Recommendations

There are four areas in which additional research efforts are needed to advance the knowledge of trace element speciation in fish and seafood:

1. Increased analytical monitoring programs that will provide an expanded data base for As, Pb, and Sn speciation.
2. Continuation of Hg speciation investigations, especially on freshwater species in regions of localized Hg pollution and acid rain fallout.
3. Expansion of the Se speciation data base, coupled with dietary intake and bioavailability studies with Hg in anticipation of reevaluating Hg dietary guidelines.
4. Employing strict quality control methods by use of appropriate standard reference materials and establishing participation in interlaboratory quality control programs.

REFERENCES

Adams, W. J., and Johnson, H. E. (1977). "Survey of the selenium contentent in the aquatic biota of Western Lake Erie." *J. Great Lakes Res.* **3**, 10–14.

Anas, R. E. (1974). "Residues in fish, wildlife and estuaries." *Pestic. Monit. J.* **8**, 12–14.

Andreae, M. O., and Klumpp, D. W. (1979). "Biosynthesis and release of organoarsenic compounds by marine algae." *Environ. Sci. Technol.* **13**, 738–741.

Aubert, M. (1975). "Le probleme du mercure eu Mediterranee." *Rev. Int. Oceanog. Med.* 37–38, 215–231.

Bache, C. A., Gutenmann, W. H., and Lisk, D. J. (1971). "Residues of total mercury and methylmercuric salts in lake trout as a function of age." *Science* **172**, 951–952.

Bebbington, G. N., Mackay, N. J., Chvojba, R., Williams, R. J., Dunn, A., and Autry, E. H. (1977). "Heavy metals, selenium and arsenic in nine species of Australian commercial fish." *Aust. J. Mar. Freshwater Res.* **28**, 277–286.

Benson, A. A., and Simmons, R. E. (1981). "Arsenic accumulation in great barrier reef invertebrates." *Science* **211**, 482–483.

Birke, G., Johnels, A-G., Plantin, L-O., Sjostrand, B., Skerfvirg, S., and Westermark, T. (1972). "Studies on humans exposed to methylmercury through fish consumption." *Arch. Environ. Health* **25**, 77–91.

Botre, C., Malizia, E., Melchiorri, P., Stacchini, E., Terayanti, G., and Zorsi, C. D. (1977). "Study and evaluation of organic lead levels in fishes and phytoplankton near Otrano." *Proc. Eur. Soc. Toxicol.*, 19th meeting, Copenhagen.

Brinckman, F. E., and Fish, R. H. (1981). "Environmental speciation and monitoring needs for trace metal-containing substances from energy-related processes." *Proc. DOE/NBS Workshop*, pp. 65–80.

Brooke, P. J., and Evans, W. H. (1981). "Determination of total inorganic arsenic in fish, shellfish and fish products." *Analyst* **106**, 514–520.

Buchanan, C. (1972). "Mercury analysis of fish in Monroe County." The Rochester Committee for Scientific Information, Bulletin 142.

Burton, J. D., Maher, W. A., Measures, C. I., and Statham, P. J. (1980). "Aspects of the distribution and chemical form of selenium and arsenic in ocean waters and marine organisms." *Thal. Jugoslav.* **16**, 155–164.

Byrd, J. T., and Andreae, M. O. (1982). "Tin and methyltin species in seawater: concentrations and fluxes." *Science* **218**, 565–569.

Cannon, J. R., Edmonds, J. S., Francesconi, K. A., and Langsford, J. B. (1979). "Arsenic in marine fauna." *Manag. Control Heavy Met. Environ., Intern. Conf., London*, pp. 283–286.

Cappon, C. J. (1984). "Content and chemical form of mercury and selenium in Lake Ontario salmon and trout." *J. Great Lakes Res.* **10**, 419–434.

Cappon, C. J. (1987). "GLC speciation of selected trace elements." *LC-GC* **5**, 400–418.

Cappon, C. J. (1988). "HPLC speciation of selected trace elements." *LC-GC* **6**, 584–599.

Cappon, C. J., and Smith, J. C. (1978). "Determination of selenium in biological materials by gas chromatography." *J. Anal. Toxicol.* **2**, 114–120.

Cappon, C. J., and Smith, J. C. (1981). "Mercury and selenium content and chemical form in fish muscle." *Arch. Environ. Contain. Toxicol.* **10**, 305–319.

Cappon, C. J., and Smith, J. C. (1982a). "Chemical form and distribution of mercury and selenium in edible seafood." *J. Anal. Toxicol.* **6**, 10–21.

Cappon, C. J., and Smith, J. C. (1982b). "Chemical form and distribution of mercury and selenium in canned tuna." *J. Appl. Toxicol.* **2**, 181–189.

Charlebois, C. T. (1978). "High mercury levels in Indians and Inuits (Eskimos) in Canada." *Ambio* **7**, 204–210.

Chau, Y. K., Wong, P. T. S., Bengert, G. A., and Kramar, O., (1979). "Determination of tetraalkyllead compounds in water, sediment, and fish samples." *Anal. Chem.* **51**, 186–188.

Chau, Y. K., Wong, P. T. S., Kramar, O., Bengert, G. A., Crug, R. B., Kinrade, J. O., Lye, J., and Van Loon, J. C. (1980). "Occurrence of tetraalkyllead compounds in the aquatic environment." *Bull. Environ. Contam. Toxicol.* **24**, 265–269.

Chau, Y. K., Wong, P. T. S., Bengert, G. A., and Dunn, J. L. (1984). "Determination of dialkyllead, trialkyllead, tetraalkyllead, and lead (II) compounds in sediment and biological samples." *Anal. Chem.* **56**, 271–274.

Chau, Y. K., Wong, P. T. S., Bengert, G. A., Dunn, J. L., and Glen, B. (1985). "Occurrence of alkyllead compounds in the Detroit and St. Clair Rivers." *J. Great Lakes Res.* **11**, 313–319.

Clarkson, T. W. (1972). "Recent advances in the toxicology of mercury with emphasis on the alkylmercurials." *CRC Crit. Rev. Toxicol.* **2**, 203–234.

Clarkson, T. W., Small, H., and Norseth, T. (1973). "Excretion and absorption of methylmercury after polythiol resin treatment." *Arch. Environ. Health* **26**, 173–176.

Cooney, R. V., Mumma, R. O., and Benson, A. A. (1978). "Arsonium phospholipid in algae." *Proc. Natl. Acad. Sci. U.S.A.* **75**, 4262–4264.

Crecelius, E. A. (1977). "Changes in the chemical speciation of arsenic following ingestion by man." *Environ. Health Perspect.* **19**, 147–150.

Cruz, R. B., Lorouso, C., George, S., Thomassen, Y., Kinrade, J. D., Butler, R. P., Lye, J., and Van Loon, J. C. (1980). "Determination of total, organic solvent extractable, volatile and tetraalkyllead in fish, vegetation, sediment and water samples." *Spectrochim. Acta* **35B**, 775–783.

del Riego, A. F. (1968). "Determinacion del arsenico en los organismos marinas." *Bol. Inst. Esp. Ocenaogr.* No. 134, 3–8.

Donard, O. F. X., and Weben, J. H. (1985). "Behavior of methylated tin compounds under simulated estuarian conditions." *Environ. Sci. Technol.* **19**, 1104–1110.

Eaton, R. D. P., Secord, D. C., Hewitt, P. (1980). "An experimental assessment of the toxic potential of mercury in ringed seal liver for adult laboratory cats." *Toxicol. Appl. Pharmacol.* **55**, 514–521.

Edmonds, J. S., and Francesconi, K. A. (1977). "Methylated arsenic from marine fauna." *Nature (London)* **265**, 436–439.

Edmonds, J. S., and Francesconi, K. A. (1981a). "Isolation and identification of arsenobetaine from the American lobster, *Homanus Americanas.*" *Chemosphere* **10**, 1041–1044.

Edmonds, J. S., and Francesconi, K. A. (1981b). "The origin and chemical form of arsenic in the school whiting." *Marine Pollut. Bull.* **12**, 92–96.

Edmonds, J. S., Francesconi, K. A., Cannon, J. R., Raston, C. L., Skelton, B. W., and White, A. H. (1977). "Isolation, crustal structure and synthesis of arsenobetaine, the arsenical constituent of the western rock lobster *Panularis longipes* cygnus George." *Tetrahedron Lett.* **18**, 1543–1546.

Foa, V., Colombi, A., and Maroni, M. (1984). "The speciation of the chemical forms of arsenic in the biological monitoring of exposure to inorganic arsenic." *Sci. Total. Environ.* **34**, 241–259.

Food and Drug Administration (1974). "Action level for mercury in fish, shellfish, crustaceans and other aquatic animals." *Fed. Reg.* **39**, 43738–43740.

Food and Drug Administration (1979). "Action level for mercury in fish, shellfish, crustaceans and other aquatic animals." *Fed. Reg.* **44**, 4012.

Food and Drug Administration (1984). FDA Compliance Policy Guide, sec. 7108.07.

Food and Nutrition Board (1976). "Selenium and human health." *Nutr. Rev.* **34**, 347–348.

Food and Nutrition Board (1977). "Are selenium supplements needed (by the general public)?" *J. Am. Diet. Assoc.* **70**, 249–250.

Food and Nutrition Board (1980). *Recommended Dietary Allowances,* 9th Rev. Ed. National Academy of Sciences, Washington, D.C.

Fowler, S. W., and Benayoun, G. (1976). "Influence of environmental factors on selenium flux in two marine invertebrates." *Mar. Biol.* **37**, 59–68.

Friedman, M. A., Eaton, L. R., and Caster, W. H. (1978). "Protective effects of freeze-dried swordfish on methylmercury chloride toxicity in rats." *Bull. Environ. Contam. Toxicol.* **19**, 436–441.

Galster, W. A. (1976). "Mercury in Alaskan Eskimo mother and infants." *Environ. Health Perspect.* **15**, 135–140.

Ganther, H. E., Gondie, C., Sunde, M. L., Kopecky, J. J., Wagner, P., Oh, S-H., and Hoekstra, W. B. (1972). "Selenium: Relation to decreased toxicity of methylmercury added to diets containing tuna." *Science* **175**, 1122–1124.

Gaskin, D. E., Ishida, K., and Frank, R., (1972). "Mercury in harbor porpoises from the Bay of Fundy region." *J. Fish Res. Board Can.* **29**, 1644–1646.

Gaskin, D. E., Stonefield, K. I., Sunda, P., and Frank, R. (1979). "Changes in mercury levels in harbor porpoises from the Bay of Fundy, Canada, and adjacent waters during 1969–1977." *Arch. Environ. Contam. Toxicol.* **8**, 733–736.

Goldwater, L. J. (1974) "Mercury: environmental consideration, Part II." *CRC Crit. Rev. Environ. Control* **4**, 251–339.

Gras, G., and Mondain, J. (1982). "Total mercury/methylmercury content of several species of fish of West African Coasts." *Toxicol. Eur. Res.* **4**, 191–195.

Grove, J. R. (1980). "Investigation into the formation and behavior of aqueous solutions of lead alkyls." In M. Branien and Z. Konrad, eds., *Lead Mar. Environ., Proc. Int. Experts Diss., 1977.* Pergamon Press, Oxford, pp. 45–52.

Guard, E. H., Cobet, A. B., and Coleman, III, W. M. (1981). "Biomethyltion of trimethytin compounds by estuarine sediments." *Science* **213**, 710–711.

HSDB (1987). Hazardous Substances Data Bank. Record for arsenic (computer printout). National Library of Medicine: Washington, D. C., April 9, 1987.

Hall, E. T. (1974). "Mercury in commercial canned seafood." *J. Assoc. Off. Anal. Chem.* **57**, 1068–1073.

Harrison, G. F. (1977). "The Cavtat Incident." In *Proceedings of the International Experts Discussion on "Lead-Occurrence, Fate and Pollution in the Marine Environment,"* Rovinj, Yogoslavia, pp. 45–52.

Harrison, R. M., and Radojevic, M. (1985). "Butylation of tetraalkyllead and ionic alkyllead for gas chromatography-atomic absorption analysis." *Environ. Technol. Lett.* **6**, 129–135.

Hasanen, E., and Sjoblom, V. (1968). "Mercury content of fish in Finland—year 1967." *Suom. Katatal.* **36**, 1–24.

Hattula, M. L., Sarkka, J., Janatuinen, J., Paasivirta, J., and Roos, A. (1978). "Total mercury and methylmercury contents in fish from Lake Paijanne." *Environ. Pollut.* **17**, 19–29.

Higgs, D. J., Morris, V. C., and Levander, O. A. (1972). "Effect of cooking on the selenium content of foods." *J. Assoc. Food Chem.* **20**, 678–680.

Hight, S. C., and Capar, S. G. (1983). "Electron capture gas-liquid chromatographic determination of methylmercury in fish and shellfish: Collaborative study." *J. Assoc. Off. Anal. Chem.* **66**, 1121–1128.

Hight, S. C., and Corcoran, M. T. (1987). "Rapid determination of methylmercury in fish and shellfish: Method development." *J. Assoc. Off. Anal. Chem.* **70**, 24–30.

Hildebrand, S. G., Strand, R. H., and Huckabee, J. W. (19880). "Mercury accumulation in fish and invertebrates of the North Fork Holston River, Virginia and Tennessee." *J. Environ. Qual.* **9**, 393–400.

Hodge, V. F., Seidel, S. L., and Goldberg, E. D. (1979). "Determination of tin (IV) and organotin compounds in natural waters, coastal sediments and macro algae by atomic absorption spectrometry." *Anal. Chem.* **51**, 1256–1260.

Itano, K., Sasaki, K., and Akehaski, H. (1979). "Distribution of selenium and mercury in the component of Skipjack and Red Sea bream tissues." *Shokuhim Eiseigaku Zasshi* **20**, 299–306.

Itano, K., Kawai, S., Miyazaki, N., Tatsukawa, R., and Fujiyama. T. (1984a). "Mercury and selenium levels in stripped dolphins caught off the Pacific Coast of Japan." *Agr. Biol. Chem.* **48**, 1109–1116.

Itano, K., Kawai, S., Miyazaki, N., Tatsukawa, R., and Fujiyama, T. (1984b). "Body burdens and distribution of mercury and selenium in striped dolphins." *Agr. Biol. Chem.* **48**, 1117–1121.

Iwata, H., Okamoto, H., and Ohsawa, Y. (1973). "Effect of selenium on methylmercury poisoning." *Res. Commun. Chem. Pathol. Pharmacol.* **5**, 673–680.

Jackson, J. A., Blair, W. R., Brinckman, F. E., and Iverson, W. P. (1982). "Gas-chromatographic speciation of methystannanes in the Chesapeake Bay using purge and trap sampling with a tin-selective detector." *Environ. Sci. Technol.* **16**, 110–119.

James, T. (1983). "Gas-liquid chromatographic screening method for determination of methylmercury in tuna and swordfish." *J. Assoc. Off. Anal. Chem.* **66**, 128–129.

Jarvie, A. W. P., Markall, R. N., and Potter, H. R. (1975). "Chemical alkylation of lead." *Nature (London)* **255**, 217–218.

Jermen, A., Delafortrie, A., Verdoot, D., Jacobs, T., and Dourte, P. (1984). "Determination of trace amounts of organic lead in fish." *Res. Agr. (Brussels)* **37**, 1025–1027.

Jernelov, A. (1970). "Release of methylmercury from sediment with layers containing inorganic mercury at different depths." *Limnol. Oceanogr.* **15**, 948–950.

Jernelov, A. (1980). "The effects of acidity on the uptake of mercury in fish." *Environ. Sci. Res.* **17**, 211–222.

Jensen, S., and Jernelov, A. (1969). "Biological methylation of mercury in aquatic organisms." *Nature (London)* **233**, 753–754.

Kamps, L. R., Clark, R., and Miller, H. (1972). "Total mercury monomethylmercury content of several species of fish." *Bull. Environ. Contam. Toxicol.* **8**, 273–279.

Koeman, J. H., Peeters, W. H. M., Koudstaal, C. H. M., Tijoe, P. S., and DeGoed, J. J. M. (1973). "Mercury-selenium correlation in marine mammals." *Nature (London)* **245**, 385–386.

Kojima, K., and Fujita, M. (1973). "Summary of recent studies in Japan on methylmercury poisoning." *Toxicology* **1**, 43–62.

Kramar, H. J., and Neidhart, B. (1975). "The behavior of mercury in the system water-fish." *Bull. Environ. Contam. Toxicol.* **14**, 699–704.

Krenkel, P. A. (1973). "Mercury: Environmental considerations, Part II." *CRC Crit. Rev. Environ. Control* **3**, 303–373.

Kurland, L. T., Faro, S. N., and Seidler, W. (1960). "Minamata disease." *World Neurol.* **1**, 370–390.

Kurosawa, S., Yasuda, K., Taguchi, M., Yamazaki, S., Toda, S., Morita, M., Vehiro, T., and Fuwa, K. (1980). "Identification of arsenobetaine, a water soluble organo-arsenic compound in muscle and liver of shark, *Prinace glaucus*." *Agr. Biol. Chem.* **44**, 1993–1994.

Lawrence, J. F., Michalik, P., Tam, G., and Conacher, H. B. S. (1986). "Identification of arsenobetaine and arsenocholine in Canadian fish and shellfish by high-performance liquid chromatography with atomic absorption detection and confirmation by fast atom bonbardment mass spectroscopy." *J. Agr. Food Chem.* **34**, 315–319.

Linko, R. R., and Terho, K. (1977). "Occurrence of methylmercury in pike and Baltic herring from the Turkeu Archipelago." *Environ. Pollut.* **14**, 227–235.

Lo, M.-T., and Sandi, E. (1980). "Selenium: Occurrence in foods and its toxicological significance—A review." *J. Environ. Pathol. Toxicol.* **4**, 193–218.

Lunde, G. (1972). "Location of lipid-soluble selenium in marine fish to the lipoproteins." *J. Sci. Food. Agr.* **23**, 987–994.

Lunde, G. (1977). "Occurrence and transformation of arsenic in the marine environment." *Environ. Health Perspect.* **19**, 47–52.

Luten, J. B., Ruiter, A., Ritskes, T. M., Rauchbarr, A. B., and Riekwel-Booy, G. (1980). "Mercury and selenium in marine and freshwater fish." *J. Food Sci.* **45**, 416–419.

Luten, J. B., Riekwel-Booy, G., and Rauchbaar, A. (1982). "Occurrence of arsenic in plaice (Pleuronectes platessa), nature of organo-arsenic compound present and its excretion by man." *Environ. Health Perspect.* **45**, 165–170.

Maguire, R. J. (1984). "Butyltin compounds and inorganic tin in sediments in Ontario." *Environ. Sci. Technol.* **18**, 291–294.

Maguire, R. J., Chau, Y. K., Bengert, G. A., Hale, E. J., Wong, P. T. S., and Kramar, O. (1982).

"Occurrence of organotin compounds in Ontario lakes and rivers." *Environ. Sci. Technol.* **16**, 698–702.

Maguire, R. J., Tkacz, R. J., and Sartor, D. L. (1985). "Butyltin species and inorganic tin in water and sediment of the Detroit and St. Claire Rivers." *J. Great Lakes Res.* **11**, 320–327.

Matsumura, F., Doherty, Y. G., Furakawa, F., and Bausch, G. H. (1975). "Incorporation of ^{203}Hg into methylmercury in fish liver. Studies on biochemical mechanisms in vitro." *Environ. Res.* **110**, 224–235.

Matthews, A. D. (1983). "Mercury content of commercially important fish of the Seychelles, and hair mercury levels of a selected part of the population." *Environ. Res.* **30**, 305–312.

May, K., Stoeppler, M., and Reisinger, K. (1987). "Studies on the ratio total mercury/methylmercury in the aquatic food chain." *Toxicol. Environ. Chem.* **13**, 153–159.

Measures, C. I., and Burton, J. C. (1978). "Behavior and speciation of dissolved selenium in estuarive waters." *Nature (London)* **273**, 293–296.

Nagakura, K., Arima, S., Kurihara, M., Koga, T., and Fujita, T. (1974). "Mercury content of whales." *Bull. Tokai Reg. Fish. Res. Lab.* **78**, 41–46.

Norin, H., Ryhage, R., Christakopoulous, A., Sandstrom, M., (1983). "New evidence for the presence of arsenocholine in shrimps (*Paudelein borealis*) by use of pyrolysis gas chromatography-atomic absorption spectrometry/mass spectrometry." *Chemosphere* **12**, 299–315.

Norstrom, R. J., McKinnon, A. B., and de Freitas, A. S. W. (1976). "A bioenergetics-based model for pollutant accumulation by fish. Simulation of PCB and methylmercury residue levels in Ottawa River yellow perch." *J. Fish Res. Board Can.* **33**, 248–260.

Ohi, G., Nishigaki, S., Seki, H., Tamura, Y., Maki, T., Konno, H., Ochiai, S., Yawada, H., Shimamura, Y., Mizaguchi, I., and Yagyer, H. (1976). "Efficacy of selenium in tuna and selenite in modifying methylmercury intoxication." *Environ. Res.* **12**, 49–58.

Ohi, G., Nishigaki, S., Seki, H., Tamura, Y., Maki, T., Minowa, K., Shimamura, Y., Mizoguchi, I., Inaba, Y., Takizawa, Y., and Kawanishi, Y. (1980). "The protective potency of marine animal meat against the neurotoxicity of methylmercury: Its relationship with the organ distribution of mercury and selenium in the rat." *Fd. Cosmet. Toxicol.* **18**, 139–145.

Pan-Hou, H. S. K., and Imura, N. (1981). "Biotransformation of mercurials by intestinal microorganisms isolated from yellow fin tuna." *Bull. Environ. Contan. Toxicol.* **26**, 359–363.

Parizek, J., Kalouskova, J., Babicky, A., Benes, J., and Paulik, L. (1974). In W. B. Hoekestra, J. W. Suthie, H. E. Ganther, and W. Merty, eds., *Trace Element Metabolism in Animals* University Park Press, Baltimore, Md, Vol. 2, pp. 119–128.

Parizek, J., Ostadolova, I., Kalouskova, J., Babicky, A., and Benes, J. (1976). In W. Mertz and W. E. Cornatzer, eds., *Newer Trace Elements in Nutrition.* Dekker, New York, pp. 96–105.

Penrose, W. R. (1974). "Arsenic in the marine and aquatic environments: Analysis, occurrence, and significance." *CRC Crit. Rev. Environ. Control* **4**, 465–482.

Penrose, W. R., Conacher, H. B. S., Black, R., Meranger, J. C., Miles, W., Cunninham, H. M., and Squires, W. R. (1977). "Implications of inorganic/organic interconversion on fluxes of arsenic in marine food webs." *Environ. Health Perspect.* **19**, 53–59.

Pinel, R., and Madiec, H. (1986). "Determination of "heavy" organotin pollution of water and shellfish by a modified hydride atomic absorption proceure." *Int. J. Environ. Anal. Chem.* **27**, 265–271.

Piver, W. T. (1973). "Organotin componds: Industrial applications and biological investigation." *Environ. Health Perspect.* **15**, 61–78.

Potter, S., and Mattrone, G. (1974). "Effect of selenite on the toxicity of dietary methylmercury and mercuric chloride in the rat." *J. Nutr.* **101**, 638-647.

Reijnders, P. J. H. (1980). "Organochlorine and heavy metal residues in harbour seals from the Wadden Sea and their possible effects on reproduction." *Netherlands J. Sea Res.* **14**, 30–65.

Ridley, W. P., Dizikes, L. J., and Wood, J. M. (1977). "Biomethylation of toxic elements in the environment." *Science* **197**, 329–332.

Rivers, J. B., Pearson, J. E., and Shultz, C. D. (1972). "Total and organic mercury in marine fish." *Bull. Environ. Contam. Toxicol.* **8**, 257–272.

Rudd, J. W. M., Furutani, A., and Turner, M. A. (1980). "Mercury methylation by fish intestinal contents." *Appl. Environ. Microbiol.* **40**, 777–782.

Sakurai, H., and Tsuchiya, K. (1975). "A tentative recommendation for the maximum daily intake of selenium." *Environ. Physiol. Biochem.* **5**, 107–118.

Schmidt, V., and Huber, F., (1976). "Aqueous chemistry of organolead and organothallium compounds in the presence of microorganisms." In *Organometals, Organometalloids: Occurrence Fate Environ.*, ACS Symp. Ser., Vol. 82, pp. 65–81.

Schroeder, H. A., and Balassa, J. J. (1966). "Abnormal trace metals in man: Arsenic." *J. Chron. Dis.* **19**, 85–96.

Schroeder, H. A., Balassa, J. J., and Tipton, I. H. (1964). "Abnormal trace metals in man: Tin." *J. Chron. Dis.* **17**, 483–502.

Schultz, C. D., and Crear, D. (1976). "The distribution of total and organic mercury in seven tissues of the Pacific blue marlin, *Makaira nigricans.*" *Pacific Sci.* **30**, 101–107.

Shultz, C. D., and Ho, B. M. (1979). "Mercury and selenium in blue marlin, Makaira nigricans, from the Hawaiian Islands." *Fish Bull.* **76**, 872–879.

Seydel, I. S. (1972). "Distribution and circulation of arsenic through water, organisms and sediments of Lake Michigan." *Arch. Hydrobiol.* **71**, 17–22.

Shafer, M. L., Rhea, V., Peeler, J. T., Hamilton, C. H., and Campbell, J. E. (1975). "A method for the estimation of methylmercuric compounds in fish." *J. Agr. Food Chem.* **23**, 1079–1083.

Shimojo, N., Yamaguchi, S., Kabu, S., Sun, K., Shiramizu, M., Hioda, Y., and Kurata, S. (1975). "Studies on total mercury, methylmercury and selenium in tunas." *Japan. J. Ind. Health* **17**, 240–241.

Sirota, G. R., and Uthe, J. F. (1977). "Determination of tetraalkyllead compounds in biological materials." *Anal. Chem.* **49**, 823–825.

Skerfving, S. (1974). "Methylmercury exposure, mercury levels in blood and hair, and health status in Swedes consuming contaminated fish." *Toxicology* **2**, 3–23.

Smith, T. G., and Armstrong, F. A. J. (1978). "Mercury and selenium in ringed and beared seal tissues from Arctic Canada." *Arctic* **31**, 75–84.

Soderquist, C. J., and Crosby, D. G. (1980). "Determination of triphenyltinhydroxide and its degradation products in water." *J. Agr. Food Chem.* **28**, 111–117.

Soichi, N. (1979). "Mercury and selenium in tuna muscle." *Chudokuhen Ho* **12**, 13–14.

Stevenson, C. L. (1977). "Selenium supplementation of livestock feeds." Trade Memorandum T-34, March 29, 1977, Agriculture Canada.

Stillings, B. R., Lagally, H., Bauersfield, P., and Soares, J. (1974). "Effect of cysteine, selenium, and fish protein on the toxicity and metabolism of methylmercury in rats." *Toxicol. Appl. Pharmacol.* **30**, 273–281.

Stoeswand, G. S., Bache, C.A ., and Lisk, D. J. (1974). "Dietary selenium protection of methylmercury intoxication of Japanese quail." *Bull. Environ. Contam. Toxicol.* **11**, 152–156.

Sugimura, Y., Suzuki, Y., and Migaka, Y. (1976). "The content of selenium and its chemical form in seawater." *J. Oceanograph. Soc. Jpn.* **32**, 235–239.

Sumari, P., Partanen, T., Hietala, S., and Heinonen, O. P. (1972). "Blood and hair mercury content in fish consumers. A preliminary report." *Work-Environ. Health* **9**, 61–65.

Sumino, K., Yamamoto, R., and Kitamura, S. (1977). "A role of selenium against methylmercury toxicity." *Nature* (London) **268**, 73–74.

Suzuki, T., Miyama, T., and Toyama, C. (1972). "The chemical form and bodily distribution of mercury in marine fish." *Bull. Environ. Contam. Toxicol.* **10**, 347–355.

Suzuki, T., Satoh, H., Yamamoto, R., and Kashiwazaki, H. (1980). "Selenium and mercury in food stuff from the locality with elevated intake of methylmercury." *Bull. Environ. Toxicol.* **24**, 805–811.

Suzuki, Y., Miyake, Y., Saruhashi, K., and Sugimura, Y. (1981). "A cycle of selenium in the ocean." *Pap. Meteorol. Geophys.* **31**, 185–189.

Tomlinson, G. H. (1979). "Acidic precipitation and mercury in Canadian lakes and fish." In *Scientific Papers from the Public Meeting on Acid Rain Precipitation*, May 1978 (Lake Placid, New York). Albany, New York, Science and Technology Staff, New York State Assembly.

Tsuda, T., Nakanishi, H., Akoi, S., and Takebayashi, J. (1987). "Determination of butyltin and phenyltin compounds in biological and sediment samples by electron-capture gas chromatography." *J. Chromatogr.* **387**, 361–370.

Tsai, S-C., Baush, G. M., and Matsumura, F. (1975). "Importance of water pH in accumulation of inorganic mercury in fish." *Bull. Environ. Contam. Toxicol.* **13**, 188–193.

Tugrul, S., Balkas, T. I., Goldberg, E. D., and Salihoglu. (1983). "The speciation of alkyltin compounds in the marine environment." *J. Etud. Pollut. Mar. Medit.* **6**, 497–504.

Tuna Research Foundation, Inc. (1978). *Tuna News: How Canned Tuna Came to Be.* Washington, D.C.

Turner, M. D., Marsh, D. A., Smith, J. C.. Englis, J., Clarkson, T. W., Rubio, L. E., Chiriboga, J., and Chiriboga, C. C. (1980). "Methylmercury in populations eating large quantities of marine fish." *Arch. Environ. Health* **35**, 367–378.

Uchida, H., Shimoishi, Y., and Toei, K. (1980). "Gas chromatographic determination of selenium (-II, 0), -(IV), and -(VI) in natural waters." *Environ. Sci. Technol.* **14**, 541–544.

Underwood, E. J. (1977). *Trace Elements in Human and Animal Nutrition*, 4th ed. Academic Press, New York, pp. 302–346.

Uthe, J. F., Solomon, J., and Grift, B. (1972). "Rapid semimicro method for the determination of methylmercury in fish tissue." *J. Assoc. Off. Anal. Chem.* **55**, 583–589.

Waldron, H. A. (1975). "Health standards for heavy metals." *Chem. Br.* **11**, 354–357.

Watling, R. J., McClurg, T. P., and Stauton, R. C. (1981). "Relation between mercury concentration and size in the mako shark." *Bull. Environ. Contam. Toxicol.* **26**, 352–358.

Watts, J. O., Boyer, K. W., Cortez, A., and Elkins, E. R., Jr. (1976). "A simplified method for the gas-liquid chromatographic determination of methylmercury in fish and shellfish." *J. Assoc. Off. Anal. Chem.* **59**, 1226–1233.

Westoo, G. (1966a). "Determination of methylmercury compounds in foodstuff. I. Methylmercury compounds in fish, identification and determination." *Acta Chem. Scand.* **20**, 2131–2137.

Westoo, G. (1966b). "Determination of methylmercury compounds in foodstuffs. II. Determination of methylmercury in fish, egg, meat, and liver." *Acta Chem. Scand.* **21**, 1790–1800.

Westoo, G. (1973). "Methylmercury as a percentage of total mercury in flesh and viscera of salmon and sea trout of various ages." *Science* **181**, 567–568.

Wobeser, F. M. S. (1970). "Mercury concentrations in muscle of fish from the Saskatchewan River." Presented to Canadian Committee on Freshwater Fisheries Research, Meeting No. 3, January, 1970, Ottawa, Ontario, Canada.

Wong, P. T. S., Chau, Y., and Luxan, P. L. (1975). "Methylation of lead in the environment." *Nature (London)* **253**, 263–264.

Wong, P. T. S., Silverberg, B. A., Chau, Y. K., and Hodson, P. V. (1978). "Lead and the aquatic biota." In J. Nriagu, ed., *Biogeochemistry of Lead.* Elsevier, New York, Ch. 17, pp. 279–342.

World Health Organization. (1967). "Specifications for the identify and purity of food additives and their toxicological evaluation: Some emulsifiers and stabilizers and certain other substances." *World Health Org. Tech. Rept.* Ser. 373, 10–15.

World Health Organization. (1972). "Evaluation of certain food additives." Joint FAO/WHO Expert Committee on Food Additives, 16th Report, WHO, Geneva, pp. 11–33.

World Health Organization. (1976). "Environmental Health Criteria. I. Mercury." WHO, Geneva, pp. 48–52.

Yamanaka, S., and Ueda, K. (1975). "High ethylmercury in river fish by man-made pollution." *Bull. Environ. Contam. Toxicol.* **14**, 409–414.

Yoshida, M., Iwami, K., and Yasumoto, K. (1984). "Determination of nutritional efficiency of selenium contained in processed skipjack meat by comparison with selenite." *J. Nutr. Sci. Vitaminol.* **30**, 395–400.

Zingaro, R. A. (1979). "How certain trace elements behave." *Environ. Sci. Technol.* **13**, 282–287.

Zingaro, R. A., and Cooper, W. C. (1974). *Selenium.* Van Nostrand-Reinhold, New York.

7

RESIDUES OF DITHIOCARBAMATE FUNGICIDES IN FRUITS, VEGETABLES, AND SOME FIELD CROPS

Nigel A. Smart

Harpenden Laboratory, Agricultural Development and Advisory Service, Ministry of Agriculture, Fisheries and Food, Hatching Green, Harpenden, Herts, United Kingdom

1. INTRODUCTION

Dithiocarbamates are wide-spectrum nonsystemic fungicides widely used in agriculture and horticulture over several decades. Some dithiocarbamates are formulated as seed treatments and all have been used, in some way, for disease control on growing crops. The range of crops treated with dithiocarbamates is very wide and includes cereals, potatoes and other field crops, soft and top fruit, tropical fruit, and protected crops. A large number of formulations are marketed throughout the world, some of them as mixtures with other pesticides and others intended for use in tank mixes at the time of spraying.

Dithiocarbamate fungicides are classified into two main groups: dimethyldithiocarbamates and ethylenebisdithiocarbamates (EBDCs). All, except thiram, are salts or complexes with a metal ion. Dimethyldithiocarbamates include ferbam, thiram, and ziram; thiram is a dimethyldithiocarbamate disulfide and can be regarded as a subgroup. Ethylenebisdithiocarbamates include mancozeb, maneb, nabam, zineb, and a zineb–polyethylene thiuram disulfide complex. Propylenebisdithiocarbamates (propineb) form a third, smaller group.

Dithiocarbamates act on the surface of the seed, plant, or fruit, and have nonspecific fungicidal activity. There is little or no evidence of any of the pathogens they control developing resistance in the field. Consequently they have proved to be very reliable and have been used on a large scale. Extensive research and development on their use have been carried out and information on residues in crops, and to a lesser extent in the daily diet, is available for many situations and for many countries. Results of these studies are summarized and drawn together in this chapter.

Some of the information has been collated, in a different form, in successive Evaluations of the Joint Food and Agriculture Organization (FAO)/World Health Organization (WHO) Meeting on Pesticide Residues (Food and Agriculture Organization, 1965, 1968, 1971, 1975, 1978, 1981, 1986a). I have drawn on this material as well as using world literature and UK sources.

2. PHYSICAL PROPERTIES OF DITHIOCARBAMATES

The dithiocarbamate fungicides used in agriculture are solids. They have low volatility, although some degradation products (Section 4) may be more volatile.

Dithiocarbamates tend to be unstable to heat, light, and moisture over longer periods of time so that formulations will often have lost a significant percentage of active ingredient after storage for 2 or 3 years. They degrade on the surface of seed, plant, or fruit over a period of weeks, although complete degradation can take some months.

Solubility in water varies: nabam is very soluble at about 200 g/L, ferbam and thiram much less so at about 130 and 30 mg/L, respectively, and zineb is regarded as almost insoluble (in terms of control of disease with its solution) at about 10 mg/L.

3. USE OF DITHIOCARBAMATES IN AGRICULTURE

3.1. Extent of Use

The 1985 Food and Agriculture Organization Yearbook (1986b) gave the extent of use of dithiocarbamates in some countries in recent years. Tonnages are abstracted in Table 1. It should be noted that the amounts used are generally given in terms of the active ingredients although Hungary, Italy, Jordan, and Oman have submitted the weights of formulated materials.

TABLE 1   Annual Usage of Dithiocarbamate Pesticides in Agriculture in Some Countries[a]

Continent/Country	Dithiocarbamates (tonnes) Used in Year[b]		
	1982	1983	1984
Africa			
Kenya		198	250
Libya		73.3	71.4
North/Central America			
Mexico	3,400	3,305	3,715
United States	5,000		
South America			
Argentina	1,317.8	1,277.7	826.4
Uruguay	111.4	166.8	266.8
Asia			
Brunei Darus	0.2	0.2	
Cyprus	153.8	235.7	
India	1,713	1,559	
Israel	337	358	
Jordan	2,874.8		
Korea Republic	1,823.3		
Oman	6.2	12	39.2
Pakistan	88.1		
Turkey	934.6		
Europe			
Australia	232.2	220.7	
Czechoslovakia	650.1	942.2	
Denmark	1,148.5	1,374.7	1,580.4
Hungary	3,193.2	4,341.5	3,734.6
Italy	9,723.8	11,265.5	
Norway	37.2	28.5	
Poland	1,410.2	1,241.7	486.3
Portugal	759.2		
Sweden	438	307.5	383

[a] Abstracted from Food and Agriculture Organization (1986b).
[b] The tonnages are of active ingredients, except for Hungary, Italy, Jordan, and Oman, where the weights of formulated materials are given.

TABLE 2   Annual Usage of Dithiocarbamate Pesticides in the
United Kingdom, 1980–1983[a]

Type of Crop	Year of Enquiry	Tonnes of Active Ingredient
Soft fruit	1980	19.95
Protected crops	1981/1982	11.52
Hardy nursery stock	1981	5.21
Vegetables, bulbs	1981/1982	22.34
Cereals	1982	207.43
Other arable	1982	332.83
Grass	1982	9.37
Orchard	1983	11.61
Hops	1983	19.79
Total	1980–1983	640.05

[a]Simplified from Sly (1986).

The average total amount of dithiocarbamates used yearly in England and Wales during 1980–1983 was about 640 tonnes (active ingredients) (Sly, 1986). This tonnage was spread over about 426,000 ha. The distribution among types of crops is given in Table 2. During 1976–1979 the average total usage of dithiocarbamates each year had been about 1006 tonnes, applied over about 610,000 ha (Sly, 1981).

The amounts of dithiocarbamates used worldwide has probably decreased in recent years.

3.2. Pattern of Use

Dithiocarbamates are used to control disease on most fruit, vegetable, and field crops in one country or another. Compounds are formulated in a number of ways for effective use on the different crops and in differing conditions; thiram, however, is used essentially in horticulture only and as a seed treatment. Formulations are usually wettable powders, although dusts are also marketed. Mixtures with other fungicides, such as metalaxyl, are common. Ultralow volume formulations are available both for ground and aerial application. Many applications (even up to 16) may be made on some crops during a single growing season, although registration authorities may limit this number.

Rates of application are generally in the range of 1–5 kg of active ingredient (a.i.)/ha for all dithiocarbamate fungicides. Some crops, such as onions and cucurbits, are treated at a concentration at the lower end of this range. Diseases in stone fruit are controlled at concentrations in the range 1.8–3.4 kg a.i./ha, whereas olives may need spraying at 8 kg a.i./ha. Rates of application recommended and registered for use vary from country to country.

Similarly, recommended (or compulsory) intervals between last applications to edible crops and harvest for consumption vary from country to country and from crop to crop. Intervals are relatively short for potatoes, ranging from 7–10 days. Intervals for lettuce vary between 2 and 6 weeks, depending on the type of culture and the pesticide. Other protected crops need preharvest intervals of 1–3 weeks. A number of regulatory authorities have increased the preharvest interval in recent years so as to prevent unnecessary residues reaching the consumer.

4. DEGRADATION FOLLOWING APPLICATION

The degradation pathways of dimethyldithiocarbamates and EBDCs (together with propylenebisdithiocarbamates) are similar, involving successive oxidations and hydrolyses. Bisdithiocarbamates pose more of a problem as they degrade through the corresponding thiourea. Ethylenethiourea (ETU) (and propylene-thiourea, PTU) may be formed during the manufacture or storage of bisdithio-carbamates, on plants following application of bisdithiocarbamate formul-ations, or in food containing bisdithiocarbamate residues during cooking or processing procedures.

Degradation pathways of EBDCs in soils and other environments as well as in crops are outlined in the monograph of the Commission on Pesticide Terminal Residues of the Applied Chemistry Commission of IUPAC (1977). Ethylenethio-urea and propylenethiourea residues require special consideration.

However, residues analyses usually determine a "total" residue from liberation of carbon disulfide during acid digestion of the analytical sample. This will include dithiocarbamate, hydantoin, thiourea, plant constituents, and so on (see Section 6). The core "dithiocarbon" residue is therefore determined, irrespective of the source or identity of the moiety degraded during acid digestion.

Several detailed studies have shown that a substantial portion of EBDC parent residues may be removed from raw agricultural crops by a simple washing procedure (IUPAC, 1977; Phillips, 1976; Phillips et al., 1977).

Residues of dithiocarbamates on crops stored at room temperature or in a freezer before analysis may degrade (Panel on Determination of Dithiocarb-amate Residues, 1981). Residues of dithiocarbamates in crops that are prepared and cut up for subsampling and stored before analysis (for other pesticides) are known to degrade. Hence it is advisable always to analyze a sample for dithiocarbamate residues on its arrival at a laboratory. The degradation product ETU may be found in heat-processed food that had contained appreciable residues of EBDCs at a higher level (of ETU) than had been in the fresh commodity (Commission of the European Communities Scientific Committee on Pesticides, 1985).

5. TOXICOLOGY AND STANDARDS FOR RESIDUES IN FOODS

5.1. Toxicology of Dithiocarbamates

Toxicity data for most of these pesticides are summarized in the Evaluations of the Joint FAO/WHO Meeting on Pesticide Residues (JMPR) for 1965, 1967,

1970 (mancozeb only), 1974, 1977, and 1980 (Food and Agriculture Organization,1966, 1968, 1971, 1975, 1978, 1981).

The European Communities Scientific Committee on Pesticides (1985) has reviewed residues of dithiocarbamates, and associated compounds, and accepted in principle the conclusions reached in the FAO/WHO evaluations.

The sources quoted by the Food and Agriculture Organization and the European Communities should be consulted for further information.

5.2. Acceptable Daily Intakes

The JMPR has, where sufficient information was available, estimated acceptable daily intakes (ADIs) for pesticides. The acceptable daily intake of a chemical is the daily intake that, during an entire lifetime, appears to be without appreciable risk on the basis of all known facts at that time. It is expressed in milligrams of the chemical per kilogram of body weight.

The ADI given for EBDCs by the JMPR is 0.05 mg/kg body weight, with not more than 0.002 mg/kg body weight present as ethylenethiourea (Food and Agriculture Organisation, 1987). The ADI for ferbam and ziram (dimethyldithiocarbamates) is 0.02 mg/kg body weight. The JMPR withdrew temporary ADIs for thiram and propylenebisdithiocarbamates in 1985, pending further toxicological data.

5.3. Maximum Residue Limits

5.3.1. Codex Alimentarius Commission Limits

The Codex Committee on Pesticide Residues (CCPR) recommends, on the advice of the JMPR, maximum residue limits (MRLs) for pesticides in foodstuffs, where these are necessary and where sufficient information is available. The MRLs are published by the Codex Alimentarius Commission (CAC) as a fixed standard. In general, a Codex MRL refers to the residue resulting from the use of a pesticide under circumstances designed to protect the crop or food commodity against attack by pests, according to good agricultural practice, which may vary from country to country, or region to region, because of differences in local pest control requirements. Therefore, residues in food, particularly when pesticides are applied to crops close to harvest, will also vary. In recommending Codex limits, these variations in residues caused by differences in "good agricultural practice" are taken into consideration as far as possible on the basis of available data. They take into account the minimum quantities necessary to achieve adequate control of pests and diseases, and which are applied in a manner that will leave the smallest practical residue which must also be toxicologically acceptable.

CAC MRLs for dithiocarbamates are determined and expressed as milligrams of carbon disulfide per kilogram and refer separately to the residues arising from any or each of the groups of dithiocarbamates: (1) dimethyldithiocarbamates, resulting from the use of ferbam or ziram, and (2) EBDCs, resulting from the use of mancozeb, maneb, or zineb (including zineb derived from nabam plus zinc sulfate). They are

> 5 mg/kg for celery, currants (black, red, white), grapes, and lettuce (head)
>
> 3 mg/kg for apple, peach, pear, strawberry, and tomato
>
> 1 mg/kg for banana, cherries, endive, melon (except watermelon), and plums
>
> 0.5 mg/kg for carrot, common bean, and cucumber
>
> 0.2 mg/kg for wheat
>
> 0.1 mg/kg for potato (Food and Agriculture Organisation, 1987)

These limits do not cover residues from thiram or propylenebisdithiocarbamates, as ADIs have been withdrawn (Section 5.2); guideline levels have not been recommended for residues of these chemicals. The limit for lettuce has recently been increased from 1 mg/kg.

5.3.2. European Community Limits

Maximum levels have also been fixed for residues of pesticides by the European Communities (EC) and are derived somewhat differently. The 1976 Directive on pesticide residues in or on fruits and vegetables (European Economic Community, 1976) fixed a maximum level of 3.8 mg/kg thiram (only) on strawberries and grapes and 3.0 mg/kg in other fruits and vegetables. As it is common to screen for total dithiocarbamates in terms of total carbon disulfide liberated on acid digestion, and as the methodology for determining thiram only is meager, this limit is rarely used in practice. The EC have, at the time of writing, no limits for residues of dithiocarbamates, as a group, or for any other dithiocarbamate. EC have no limits for dithiocarbamates in cereals.

5.3.3. National Limits

Many countries, although accepting CAC and/or EC limits, nevertheless have maximum limits of their own. Although similar to the internationally agreed levels, there are differences both in the classification of commodities and in the levels set. Thus, in the case of lettuce, whereas the CAC limit is 5 mg/kg, proposed in 1985, a limit of 4 mg/kg is used in France and The Netherlands, 3 mg/kg in Australia and the German Democratic Republic, 2 mg/kg in Austria, Belgium, Czechoslovakia, Denmark, Federal Republic of Germany, Italy, and Ireland, 1 mg/kg in Finland and Sweden, and 0.5 mg/kg in Israel; the United States has varied limits depending on the dithiocarbamate pesticide applied. It is not possible to list here the differences in limits for other commodities that countries have chosen to use, although a similar range is common.

6. ANALYSIS OF DITHIOCARBAMATE RESIDUES

The carbon disulfide evolution procedure is still the method of choice for determining residues of all dithiocarbamates. Residues expressed in terms of carbon disulfide will include dimethyldithiocarbamates, EBDCs, and propylenebisdithiocarbamates. "Dithiocarbon" degradation products (Section 4) will be included in the analytical results, although these results may include carbon disulfide from other sources and natural plant constituents (for example, in allium species) in the sample.

Originally, and for some years, the carbon disulfide liberated by acid digestion, and after cleanup in wash solutions, was complexed with a copper salt and an aliphatic amine to give a yellow complex [copper N, N-bis(2-hydroxyethyl) dithiocarbamate], the intensity of absorption of the complex being measured at (say) 435 nm (Keppel, 1969, 1971; Arbeitsgruppe "Pestizide," 1977). Collaborative study did not fully confirm this procedure as somewhat low results were obtained with some dithiocarbamates and erratic results with others (Keppel, 1971).

In another colorimetric procedure carbon disulfide was adsorbed in potassium hydroxide solution and converted to potassium xanthate. This product is measured at 302 nm, with baseline correction.

In more recent headspace methods, the carbon disulfide gas liberated within closed reaction bottles containing the acidified sample is withdrawn by syringe and directly injected into a gas chromatograph fitted with flame photometric detection (McLeod and McCully, 1969; Panel on Determination of Dithiocarbamate Residues, 1981; Hill and Edmunds, 1982). Collaborative study on lettuce gave mean recoveries of added zineb, maneb, mancozeb, or thiram of 75–93% at 3 or 9 mg/kg levels. Overall coefficients of variation averaged about 25%. Lesage (1980) and Hill and Edmunds (1982) have indicated that copper, for example from Bordeaux mixture, can interfere in the analysis of residues of EBDCs by carbon disulfide evolution procedures.

To enforce MRLs reliably in crops, following application of one or several dithiocarbamates for disease control, it may be necessary to determine the following separately:

1. total dithiocarbamate
2. ETU (PTU) precursors, or EBDCs, or ETU (PTU) itself
3. thiram, or demonstrate its absence
4. propineb.

The difference between the total dithiocarbamate determination and the EBDCs, propineb, and thiram, taken together, would be empirically ascribed to dimethyldithiocarbamates. Such a series of analyses would be costly and will neglect other "dithiocarbon" degradation products (except ETU, PTU) and naturally interfering compounds.

The method of Greve and Hogendoorn (1978) attempts to determine ETU precursors separately from other dithiocarbamates in terms of the 1,2-bis(pentafluorobenz) derivative of liberated ethylenediamine. Earlier, Rohm and Haas (unpublished) developed method for determining residues as ethylenediamine.

Approaches to HPLC determination of thiram residues, as an entity separate from other dimethyldithiocarbamates, are being examined in a number of laboratories.

Methods for determining ETU (and PTU) in foodstuffs have been reviewed elsewhere (Bottomley et al., 1985).

7.   RESIDUES FOUND IN SUPERVISED TRIALS

Supervised trials in which the crop husbandry, and crop protection in particular, are closely managed often provide data for registration purposes. Usually only the pesticide under scrutiny is applied.

7.1.   General Remarks

Supervised trials carried out in the 1940s and 1950s, when dithiocarbamates were first notified to registration authorities for use as pesticides, involved formulations and application equipment different from those in current practice. The earlier Evaluations of the JMPR (Food and Agriculture Organization, 1965, 1968) summarize these. Subsequent work is reviewed here, chronologically.

7.2.   Mancozeb Residues in Potato; United States, 1969

Residues of mancozeb (determined as ethylenediamine and reported as mancozeb) in potato were reported to the 1970 JMPR (Food and Agriculture Organization, 1971). Over 300 samples were collected in 1969, grown in 11 states in the United States. Treatments had ranged from 1 to 3 lb/acre with total quantities ranging from 8 to 36 lb/acre, depending on the number of treatments (13 treatments was the maximum). No residue (< 0.05 mg/kg) was found in only 5% of the samples. In a group having the highest treatment rate and an 8-day harvest interval, results ranged up to 0.60 mg/kg, averaging 0.17 mg/kg. The highest level found was 0.78 mg/kg. No ETU was detected at a sensitivity of 0.05 mg/kg.

7.3.   Mancozeb Residues in Banana; Honduras, 1972

One application of a mancozeb formulation at 6.75 or 10.5 kg a.i./ha was made at six sites. Residues, as carbon disulfide, ranged up to 31 mg/kg for both 7-day and 14-day preharvest intervals. Residues of ETU were $\leqslant 0.05$ mg/kg (Food and Agriculture Organization, 1975).

7.4.   Mancozeb Residues in Celery; United States, 1972

One application of manozeb to celery at 1.6 kg a.i./ha, at two sites, gave residues of 1.0–4.4 mg/kg, as carbon disulfide, for a 7-day preharvest interval and

0.7–1.8 mg/kg for a 14-day preharvest interval. Residues were ⩽ 0.08 mg/kg after 21 days. Residues of ETU were ⩽ 0.02 mg/kg (Food and Agriculture Organization, 1975).

7.5. EBDC Residues in Wheat; Denmark, 1973–1976

Residues data were reported by Lauridsen et al. (1980) for fungicidal treatments to the standing crop both at normal rates and at high rates. Residues in harvested grains were, with few exceptions, less than 0.3 mg/kg with 36 days between treatment and sampling. Applications of the fungicides 2 days before harvest could leave residues of 4 mg/kg, when four treatments had been used in all on the crop.

7.6. Zineb Residues in Lettuce; Denmark, 1975–1976

Residues data were reported by Lauridsen et al. (1980) for fungicidal treatments to glasshouse crops both at normal rates and at higher rates. In lettuce grown in the winter, residues at harvest 91 and 63 days after a treatment with zineb at normal dosage carried out 11 and 39 days after planting were less than 0.6 and 8 mg/kg, respectively.

7.7. EBDC Residues in Apple; Canada, 1975 and 1976

Ross et al. (1978) have reported studies of residues of EBDCs (and ETU) in apple and apple products. Residues of EBDCs in fresh apples harvested 42 days after the last of nine sprays, each of either mancozeb or metiram, were 1.7 and 0.5 mg/kg (as EBDC), respectively, in 1975. Similar trials in 1976 showed EBDC residues of less than 1 mg/kg in fruit that had received one, two, three, or four cover sprays of the compound.

7.8. Residues of Mancozeb, Maneb, and Zineb in Various Crops

The 1977 JMPR considered information on EBDC residues from supervised trials from various sources (Food and Agriculture Organization, 1978). (Residues in this report were determined as total carbon disulfide and expressed as the parent compound.)

Apple contained residues in the range 0.6–3.1 mg/kg (3.1 mg/kg after a 28-day preharvest interval). Carrot contained residues at or below 0.7 mg/kg in two trials. Grape, however, could contain residues up to 7.5 mg/kg at a 14-day preharvest interval, averaging about 4 mg/kg over a month between last application and harvest. One of two trials on spinach led to high residues averaging 54 mg/kg. Five other trials with tomato are summarized in the Evaluations, residues ranging up to 2.3 mg/kg, although averaging 1–2 mg/kg or less. Cucurbits contained lower residues. Residues in wheat (grain) were up to 0.4 mg/kg at harvest 28–47 days after application of fungicide.

7.9. Propineb

The 1977 JMPR reviewed information on propineb residues in a wide range of crops (Food and Agriculture Organization, 1978). These Evaluations set out the results in some detail. Residues in currants were higher than might have been expected.

7.10. EBDC Residues in Lettuce, Endive, and French Bean; Italy

Trials on lettuce, endive, and French bean were carried out to examine the amount of dithiocarbamate residue corresponding to various preharvest intervals. Three plots of each vegetable were treated 10, 20, and 30 days before harvest with one of five dithiocarbamates at 160 g a.i./ha. Residues analyses showed that a preharvest interval of about 30 days guaranteed no detectable residues in the vegetables examined. At 20 days residues up to 2.5 mg/kg could be found in endive (Di Muccio et al., 1981).

7.11. Mancozeb Residues in Lettuce; Poland, 1979

Czarnik et al. (1980) reported residues in lettuce grown in an unheated foil tent and treated twice, at an 8-day interval, with a mancozeb formulation. Mancozeb (determined and reported as carbon disulfide) was 4.4 mg/kg 7 days after the first application, 21.6 mg/kg 5 days after the second, 2.4 mg/kg 9 days after the second, and 14 days after the second application, and thereafter residues were less than 0.05 mg/kg. Residues of ETU were also determined.

7.12. Mancozeb Residues in Lettuce; United Kingdom, 1983

Eight residue trials were carried out in the United Kingdom in February–April, 1983, to provide residues data for lettuce grown under cover, under conditions of low light and temperature. The plots were treated with a mancozeb/metalaxyl mixture at 1.44 kg a.i./ha (a rate lower than that used with EBDCs alone). At a preharvest interval of 21 days, residues ranged from 0.5 to 9.5 mg/kg and, at an interval of 28 days, they ranged from 0.4–5.3 mg/kg (Food and Agriculture Organization, 1986a).

7.13. EBDC Residues in Lettuce; France, 1980–1982

Data were obtained from two areas in France in which EBDCs (maneb, mancozeb, thiram) were applied to glasshouse lettuce at normal rates of about 1.6 kg a.i./ha and with recommended preharvest intervals of 4–6 weeks. Of a total of 134 samples, 22 contained residues less than 1 mg/kg, as carbon disulfide, 21 contained residues in the 1–2 mg/kg range, 69 contained residues in the 2–4 mg/kg range, and 22 contained residues above 4 mg/kg (Food and Agriculture Organization, 1986a).

7.14.   Trials to Study Degradation in Processing of the Crop

Several reports of such trials have been described: Ripley et al. (1978) of mancozeb on grapes, Ripley and Cox (1978) of maneb or mancozeb on tomato, Ripley and Simpson (1977) of zineb on pear and Blundstone (Campden Food Preservation Research Association, private communication) of several EBDCs on spinach, celery, blackcurrant, apple, and Brussels sprouts. These reports contain incidental information on residues at harvest following normal applications and preharvest intervals.

8.   RESIDUES IN RETAIL-TYPE SAMPLES OF KNOWN TREATMENT HISTORY OBTAINED FROM GROWERS

These samples were taken from situations in which dithiocarbamates were known to have been used and so all could possibly contain residues. Other pesticides may also have been applied.

8.1.   Protected Lettuce, United Kingdom

8.1.1.   General, 1974–1979

A survey of pesticides in retail lettuce in the United Kingdom during July 1974–June 1975 indicated an unduly high level of dithiocarbamate residues (Hatfull, 1976). As a result of this and other work, Government Departments in the United Kingdom tightened recommendations for safe use of these pesticides on lettuce in 1978, allowing fewer applications during heading-up and increasing the preharvest interval, depending on the type of culture of the crop. Analysis of samples taken from commercial growers at harvest showed that dithiocarbamate fungicides are more persistent in crops marketed during the winter and early spring. During 1979 monitoring of crops of lettuce (and tomato) sprayed with zineb and thiram at recommended rates continued. Analysis of crops for the retail market at harvest, and whose histories of pesticide treatment were known, indicated that the 1978 regulations had achieved the intended reduction in residues of dithiocarbamates (Agricultural Science Service, 1981).

8.1.2.   Country-Wide Survey, 1979–1980

During 1979–1980, 101 samples of protected lettuce taken at or very near to harvest were analyzed for residues of dithiocarbamates, by a carbon disulfide headspace procedure. Of these 59% contained less than 3 mg/kg (as total carbon disulfide), 30% were in the 3–50 mg/kg range, and residues in excess of 100 mg/kg were found in only two samples (Agricultural Science Service, 1982).

8.1.3.   Country-Wide Survey, 1983–1984

During the year autumn 1983 to autumn 1984, a survey of pesticide residues in protected crops was carried out to obtain data on the wide range of such commodities. Eighty-one samples of lettuce with relevant pesticide treatment data were received from growers throughout the country. Of these 1 mg/kg (carbon disulfide) was exceeded in seven samples and 5 mg/kg was exceeded in only one sample (Agricultural Science Service, 1987).

8.2. Soft Fruit, United Kingdom, 1982

Soft fruits of known treatment histories were obtained directly from growers at harvest in 1982 and analyzed for a wide range of pesticides, including dithiocarbamates. A total of 53 samples were analyzed for total carbon disulfide by the head space procedure. Of these, 26 samples contained residues of dithiocarbamates above the limits of determination of 0.02 mg/kg. One sample only, of currants, contained residues above 5 mg/kg (Agricultural Science Service, 1984).

8.3. Tomato, United Kingdom

8.3.1. *Single Holding, 1979*

Dithiocarbamates, and ETU, were determined in tomato from a crop planted in May and sprayed twice with zineb in mid- and late June. Samples for market were taken at weekly intervals between early August and early October. Dithiocarbamate residues, determined as carbon disulfide, fell from an initial value of 1.1 mg/kg to 0.02 mg/kg by late September (Ministry of Agriculture, Fisheries and Food, 1982).

8.3.2. *Country-Wide Survey, 1983–1984*

As a part of a survey of pesticides in protected crops (see Section 8.1.3), 101 samples of tomatoes of known treatment history were analyzed for dithiocarbamate by the headspace procedure. None of these contained residues in excess of 3 mg/kg (Agricultural Science Service, 1987).

8.4. Potato, Denmark

Samples of potato from six commercial fields sprayed with maneb at 1.75 a.i./ha on two to four occasions and harvested in October were analyzed for total dithiocarbamate residues. Residues, determined by a colorimetric procedure, were ≤ 0.2, the limit of determination, in five cases and 0.31 mg/kg in the sixth (Lauridsen et al., 1980).

8.5. Blackcurrant, Denmark, 1975–1976

Eight crops of blackcurrant treated with mancozeb, and one treated with propineb, at 2–2.4 kg a.i./ha were analyzed for total dithiocarbamate residues. Residues were 2–50 mg/kg, depending on timing and number of treatments (Lauridsen et al., 1980).

9. RESIDUES FOUND IN RETAIL MONITORING

Residues found in samples of unknown origin and of crops in which dithiocarb-amate may or may not have been used are described in this section. Results obtained in various countries are described for each country, in turn, and then subgrouped in commodities, as appropriate. It is not convenient to present the wide range of commodities in terms of large number subheadings.

9.1. Belgium, Lettuce, 1983

Galoux and Bernes (1983) reported pesticide residues determined in 110 samples of Belgian lettuce intended for export. Only two samples (2.7 and 2.9 mg/kg) contained residues above the 2 mg/kg (as carbon disulfide) limit used in Belgium, the Federal Republic of Germany, and Switzerland.

9.2. Denmark, All Commodities, 1984–1985

The results of Danish monitoring of retail produce for pesticide residues by the National Food Institute have been published in reports edited by Andersen (1980, 1981, 1983) for 1976–1978, 1978–1979, 1980–1981, and by Orbaek (1985, 1987) for 1982–1983 and 1984–1985. All contain data on dithiocarbamate residues (determined and expressed as carbon disulfide) in both home-produced and imported foodstuffs. The figures for 1984 are summarized in Table 3 and for 1985 in Table 4. Only five samples contained dithiocarbamate residues in excess of Danish maximum limits.

9.3. Federal Republic of Germany, Fruits and Vegetables, 1976–1982

A considerable amount of screening of commodities for pesticide residues is carried out in the Federal Republic, although much of it is not reported in the literature. Kampe (1977, 1983) has reported series of results.

In 1976, 143 samples of fruits and vegetables, taken in Hessen and Rheinland-Pfalz, were analyzed for pesticide residues. Apple, blackcurrant, lettuce, and tomato were analyzed for dithiocarbamates, using the Keppel procedure. Four of 12 samples of apples contained dithiocarbamate, averaging 0.65 mg/kg of residue. Four of 6 samples of blackcurrant contained the pesticides, averaging 0.98 mg/kg. One of 12 samples of lettuce contained a residue, of 0.44 mg/kg. Two of 12 samples of tomato contained dithiocarbamates, averaging 0.60 mg/kg. None exceeded the Federal Republic maximum limits.

Analysis of samples taken in 1982 showed low residues in apples (0.2–1 mg/kg) and a few high residues in lettuce (overall range 0.2–6.7 mg/kg).

9.4. Finland, Apple, Banana, Lettuce, and Tomato, 1979–1981

Hemminki et al. (1982) tabulate some examples of residues analysis of pesticides by the Customs Laboratory in Finland in 1979–1981. Only median and mean

TABLE 3 **Dithiocarbamate Residues in Crops in Denmark, 1984 (Retail Monitoring)**[a]

Commodity	Total Number of Samples Analyzed	Home-Produced		Imported	
		Number of Positive	Range	Number of Positive	Range
Apple	149	23	0.10–0.72	57	0.10–1.13
Asparagus	8	0	—	—	—
Banana	7	—	—	0	—
Bean	2	0	—	—	—
Blackberry	6	0	—	—	—
Blackcurrant	26	8	0.18–2.23	—	—
Brussels sprouts	13	3	0.15–0.32	—	—
Cabbage	19	3	0.25–0.86	0	—
Cauliflower	32	1	0.72	1	0.18
Celeriac	16	0	—	—	—
Celery	19	1	0.24	2	0.10–0.21
Cherry	32	5	0.11–0.50	0	—
Chinese cabbage	34	1	0.50	0	—
Clementine	37	—	—	19	0.10–0.78
Cucumber	60	0	—	6	0.10–0.88
Elderberry	9	0	—	—	—
Grapefruit	40	—	—	1	0.11
Grape	46	—	—	17	0.10–0.95
Iceberg lettuce	7	0	—	1	0.10
Kale	9	2	0.12–0.18	—	—
Leek	21	0	—	0	—
Lemon	36	—	—	6	0.10–2.06
Lettuce	49	2	0.11–0.14	5	0.10–0.20
Melon	5	0	—	2	0.28–0.85
Mushroom	2	0	—	—	—
Nectarine	6	—	—	3	0.10–0.18
Onion	24	1	0.25	1	0.14
Orange	66	—	—	10	0.11–0.43
Peach	24	—	—	18	0.19–1.88
Pear	55	3	0.14–0.36	3	0.29–0.81
Peas	24	1	0.10	—	—
Pepper	21	0	—	2	0.14–0.23
Plum	21	1	0.17	1	0.11
Pumpkin	8	0	—	—	—
Radish	14	0	—	—	—
Raspberry	6	1	0.23	—	—
Red cabbage	12	2	0.15–0.36	—	—
Redcurrant	4	1	0.58	—	—
Spinach	2	0	—	—	—
Spring cabbage	13	1	0.27	—	—
Strawberry	36	0	—	1	0.10
Tomato	52	2	0.11–0.19	13	0.10–0.77

[a]Prepared from Orbaek (1987).

TABLE 4 Dithiocarbamate Residues in Crops in Denmark, 1985 (Retail Monitoring)[a]

Commodity	Total Number of Samples Analyzed	Home-Produced		Imported	
		Number of Positive	Range	Number of Positive	Range
Apple	122	9	0.12–0.80	32	0.10–0.84
Asparagus	1	0	—	—	—
Banana	10	—	—	0	—
Bean	4	0	—	—	—
Blackberry	1	0	—	—	—
Blackcurrant	24	8	0.17–4.19	—	—
Brussels sprouts	14	5	0.12–0.18	—	—
Cabbage	26	0	—	0	—
Cauliflower	32	2	0.12–0.24	—	—
Celeriac	18	3	0.11–0.17	—	—
Celery	17	1	0.23	2	0.15–0.24
Cherry	26	4	0.13–0.59	0	—
Chinese cabbage	32	0	—	0	—
Clementine	36	—	—	6	0.10–0.33
Cress	8	7	0.16–2.35	—	—
Cucumber	42	0	—	3	0.15–0.28
Grapefruit	39	—	—	1	0.11
Grape	53	—	—	30	0.10–2.06
Kale	12	4	0.11–0.17	—	—
Leek	18	0	—	0	—
Lemon	35	—	—	2	0.10–0.15
Lettuce	65	4	0.21–2.95	10	0.10–0.96
Melon	7	0	—	0	—
Mushroom	8	0	—	—	—
Nectarine	8	—	—	3	0.25–0.46
Onion	8	1	0.13	0	—
Orange	66	—	—	5	0.10–0.16
Peach	22	—	—	5	0.14–0.60
Pear	38	14	0.12–0.92	6	0.14–0.57
Peas	15	0	—	—	—
Pepper	25	0	—	1	0.56
Plum	19	2	0.10–0.11	—	—
Potato	4	0	—	—	—
Radish	11	0	—	—	—
Raspberry	15	3	0.13–4.47	—	—
Red cabbage	6	0	—	—	—
Redcurrant	14	2	0.47–1.90	—	—
Spinach	2	0	—	—	—
Spring cabbage	13	0	—	0	—
Strawberry	42	0	—	0	—
Tomato	39	0	—	6	0.10–0.16

[a]Prepared from Orbaek (1987).

values are given and not ranges. The estimated mean residues of dithiocarbamates in apple, banana, lettuce, and tomato are 0.55, 0.09, 0.20, and 0.17 mg/kg, respectively, for analyses of 1522, 56, 378, and 121 samples.

9.5.   France, Fruits and Vegetables, 1967–1977

Cassanova et al. (1979) described an analysis of results from screening over 500 samples of fruits and vegetables for pesticides from 1967 to 1977. Only two samples, of fruits, exceeded the maximum limit for dithiocarbamates.

9.6.   Italy, Selected Fruits, and Vegetables, 1978–1982

Table 5 combines information given by Lorusso et al. (1984) and by Di Muccio

TABLE 5   Dithiocarbamate Residues in Some Italian Retail Produce, 1978–1982[a]

Commodity	Year	Samples Examined	Number of	
			Positive Residues (>0.4 mg/kg)	Residues Exceeding 2 mg/kg
Apple	1978	7	3	0
	1980	5	0	0
	1982	38	10	2
Artichoke	1980	5	0	0
Aubergine	1980	5	0	0
Beetroot	1980	5	0	0
Celery	1978	8	4	3
	1980	5	3	3
	1982	30	8	4
Courgette	1980	7	0	0
Cucumber	1980	7	0	0
Fennel	1980	5	0	0
Grape	1979	68	8	0
	1980	10	0	0
	1982	22	4	2
Lettuce	1978	11	3	0
	1980	5	3	3
	1982	40	10	1
Peach	1978	12	5	2
	1982	34	6	0
Pepper	1980	7	0	0
Plum	1980	7	0	0
Tomato	1979	19	1	0
	1980	8	5	3
	1982	59	19	2

[a]Prepared from Lorusso et al. (1984) and Di Muccio et al. (1982).

et al. (1982) on apple, artichoke, aubergine, beetroot, celery, courgette, cucumber, fennel, grape, lettuce, peach, pepper, plum, and tomato. About 0.7% of retail samples contained residues in excess of a 2 mg/kg limit. Foschi et al. (1985) determined dithiocarbamate residues in 48 samples of commercial pears. Dithiocarbamates were absent in 14 samples and exceeded legal limits in three samples.

9.7. The Netherlands, Fruits and Vegetables

1031 samples of various fruits and vegetables were analyzed specifically for thiram residues in the mid-1970s. In only 14 instances was the tolerance of 3.8 mg/kg exceeded. Of these, 12 were of strawberry (ranging from 4.2 to 14.6 mg/kg) and the other two were one of lettuce (out of 590 samples) and one of currant (Food and Agriculture Organization, 1978).

9.8. Poland, Fruits and Vegetables, 1979–1982

Dabrowski (1984) and Wegorek and Dabrowski (1985) have discussed results of residues analyses of samples of agricultural crops in Poland in this period. About 10,000 analyses of fruits and vegetables for dithiocarbamates were made in the 5 years 1976–1980 and about 5000 in the 2 years 1981–1982. Residues were determined and expressed as total carbon disulfide. The results are summarized in Table 6. About 1% were above the Polish limit. Results of analyses for some commodities normally grown under glass include a small number from samples of field grown crops.

TABLE 6   Dithiocarbamate Residues in Fruits and Vegetables in Poland 1979–1982[a,b]

	1979–1980			1981–1982		
Commodity	Number of Samples Analyzed	Number in Range 0.5–2.0 mg/kg	Number > 2.0 mg/kg	Number of Samples Analyzed	Number in Range 0.05–2.0 mg/kg	Number > 2.0 mg/kg
Lettuce	664		10	355	46	10
Cucumber	373	0	0	986	126	8
Tomato	414	19	4	1542	276	12
Celery	46	3	1	273	105	11
Onion	263	28	5	308	90	5
Bean				90	2	0
Currant	1297	50	23	307	56	0
Gooseberry	269	9	0	41	2	0
Apple	776	2	0	993	86	0
Plum	92	0	0	104	7	0

[a] Prepared from Dabrowski (1984) and Wegorek and Dabrowski (1985).
[b] Limit of determination 0.05 mg/kg.
Residues determined and expressed as carbon disulfide.

9.9. Sweden, Fruits and Vegetables, 1981–1984

Andersson (1986) has described monitoring of pesticide residues in fruits and vegetables in Sweden, 1981–1984. Samples were mainly taken at wholesalers before distribution to retailers. Dithiocarbamate residues were determined as carbon disulfide. There is an anomaly in that results were expressed as zineb from January 1981 to February 1983 and as carbon disulfide from March 1983. Results for dithiocarbamates are summarized in Table 7. Over 3% were above maximum limits. Swedish maximum levels, as carbon disulfide, are 1.0 mg/kg for all fruits and vegetables, except 0.5 mg/kg for carrot and 0.1 mg/kg for potato (but expressed as zineb from January 1981 to February 1983).

TABLE 7   Dithiocarbamate Residues (as Carbon Disulfide) in Crops in Sweden, 1981–1984 (Retail Monitoring)[a]

	Home-Produced				Imported			
	Number of Samples Analyzed for Dithio-carbamates	Number of Samples within Given % Range of MRL			Number of Samples Analyzed for Dithio-carbamates	Number of Samples within Given % Range of MRL		
Commodity		20–50	51–100	> 100		20–50	51–100	> 100
Apple	45				367	39	27	18
Cabbage	—				16	1	—	—
Carrot	1				16	—	—	—
Cherry	—				37	3	3	1
Chinese cabbage	24				48	1	2	—
Cucumber	9				61	2	1	1
Grapefruit	—				5	—	1	—
Grape	—				104	8	5	1
Iceberg lettuce	32				136	1	—	6
Lettuce	53		1		94	4	1	3
Nectarine	—				6	2	—	—
Onion	15				48	—	—	—
Peach	—				104	7	12	5
Pear	26				220	24	16	18
Pepper (sweet)	—				21	1	2	—
Plum	—				32	2	1	—
Potato	107				105	—	—	—
Strawberry	31				78	1	3	9
Tomato	17				117	3	—	5

[a] Prepared from Andersson (1986).

9.10.   United Kingdom, Fruits and Vegetables, 1974–1984

9.10.1.   Lettuce

Following earlier reports of dithiocarbamate pesticide residues in lettuce, six members of the Association of Public Analysts (Hatfull, 1976) sampled retail outlets in 1974–1975. Analyses were done by colorimetric determination of carbon disulfide liberated by acid digestion and had a reporting limit of 0.5 mg/kg. Results (for this series of analyses) were expressed as ethylenebisdithiocarbamic acid. During the summer period, July to September 1974, 120 samples were examined and 10 contained dithiocarbamates, in amounts ranging from 0.5 to 4.2 mg/kg. During the period October to December 1974, 140 samples were examined and 58 contained dithiocarbamates in amounts up to about 200 mg/kg. During the winter period January to March 1975, 148 samples were examined and 82 contained residues above the reporting limit; several residues were greater than 100 mg/kg. During April to June 1975, 103 samples were examined and only 16 contained residues above the reporting limit. The pattern of higher residues in winter months is consistent with that noted in Section 8.1.1. In most cases it was not known whether the lettuce had been grown in this country or abroad.

The Association of Public Analysts continued their surviellance of dithiocarbamate residues in 1981 and 26 samples of lettuce were examined. Six samples contained residues: three were in the 0.002–6.9 mg/kg range (as ethylenebisdithiocarbamic acid) (Hatfull, 1983). During 1983, the Association examined 32 samples of retail lettuce and found 4 to have residues of dithiocarbamates. The levels found were 24.0, 0.13, 0.10, and 0.05 mg/kg (as dithiocarbamate) (Nicolson, 1984).

Other workers analyzed retail lettuce during 1982 for a range of pesticide residues, including dithiocarbamates (Agricultural Science Service, 1983). A headspace procedure having a limit of detection of 0.1 mg/kg was used. Of 21 samples of home-grown lettuce, 10 contained detectable residues and 8 exceeded the (then) CAC MRL of 1 mg/kg. Of 32 samples of imported lettuce analyzed, two contained residues above the reporting limit, both of which also exceeded the CAC MRL. The 1982 work was incorporated in a more general survey described by the Working Party on Pesticide Residues (in preparation).

9.10.2.   Soft Fruit

In the 1983 survey of the Association of Public Analysts (Nicolson, 1984), gooseberry and raspberry were examined for dithiocarbamate residues. Of 11 samples of gooseberry, only one contained residues above the detection limit of 0.02 mg/kg, at 0.03 mg/kg (expressed as ethylenebisdithiocarbamic acid). One residue, of 0.023 mg/kg, was found in 17 samples of raspberry. Also in 1983, other workers examined 59 samples of home-produced soft fruit and found 19 samples to contain residues of dithiocarbamates above the limit of determination of 0.02 mg/kg, although below the CAC MRL (Agricultural Science Service, 1984).

Eighteen samples of imported strawberry were also analyzed for dithiocarbamates and five contained residues above the CAC MRL of 2 mg/kg; three others contained residues above the 0.02 mg/kg limit of determination but below the MRL.

9.10.3. Other Crops

During the 1981 survey of the Association of Public Analysts (Hatfull, 1983), bean, grape, mushroom, peach, pear, and plum were also analyzed for dithiocarbamate pesticide residues. No detectable residues were found in 2 samples of french bean, 5 samples of pear, or 2 samples of plum. Two of 12 samples of grape contained residues of 0.09 and 0.06 mg/kg (as ethylenebisdithiocarbamic acid). Three of 20 samples of mushroom contained significant dithiocarbamate residues but the levels were below 1 mg/kg. Of 31 samples of peach 14 contained significant residues, the highest of which was 1.0 mg/kg.

A few determinations in beetroot, carrot, and potato were made in the 1983 survey of the Association of Public Analysts (Nicolson, 1984). Four samples of carrot and five of potato all contained less than 0.1 mg/kg dithiocarbamate. One sample of beetroot contained 0.63 mg/kg dithiocarbamate. Edmunds (in preparation) describes results of retail monitoring of U.K. produce for pesticide residues during 1981–1984. This chapter contains results for dithiocarbamate residues that overlap the specific information given above for lettuce and soft fruit. Dithiocarbamate residues in mushroom (42 samples), peach (30 samples), and tomato (51 samples) were all 0.3 mg/kg or below. Of 12 samples of celery examined, 5 contained residues in the 0.1–1 mg/kg range and two contained 1.4 and 2.0 mg/kg.

9.11. Switzerland, Fruits and Vegetables, 1980–1983

Schüpbach (1986) has discussed pesticide residues in this 4-year period in Swiss fruits and vegetables by the official food control of the City of Basle. About 850 samples were examined but only two samples of fruit and one of vegetables contained dithiocarbamates in excess of the maximum allowable. Details of residues found in commodities are not reported.

10.   RESIDUES IN TOTAL DIET SAMPLES

10.1.   Finland, 1980

Elinkeinohallitus (1982) calculated the average exposure to dithiocarbamates in Finland in 1980 as in vegetables, 0.74, in fruit, 8.9, in berries, 0.83, and in total 10.5 mg per person per year.

10.2.   France

Deschamps and Hascoet (1983) have assessed the annual intake of dithiocarbamate pesticides in a national survey of the food quality in France. They took account of the limit of detection of 0.3 mg/kg in two ways: as zero, and as positive and integrating 90% of that limit into the intake calculation. Using a zero limit of detection, vegetables (average intake of 144.8 kg per person) contained a mean residue of 1.19 mg/kg dithiocarbamate, giving an annual intake of 171.8 mg per person. When a positive limit of detection was used for calculation, the mean residue in fruits was 0.322 mg/kg, giving an annual intake of 23.0 mg dithiocarbamates per person.

For fruits (average intake 71.5 kg per person), the mean residue was 0.095 mg/kg, using a zero limit of detection, giving an annual intake of 67.9 mg per person. When a positive limit of detection was used for calculation, the mean residue in fruits was 0.322 mg/kg, giving an annual intake of 23.0 mg dithiocarbamates per person.

The total intake of dithiocarbamates from products of plant origin was 178.6 mg, on the basis of a zero limit of detection and 222.3 mg, on the basis of a 0.3 mg/kg limit of detection.

10.3. The Netherlands, 1976–1978

De Vos et al. (1984) have described a comprehensive study of pesticides and other chemical residues in Netherlands total diet samples from June 1976 to July 1978. Thiram and dithiocarbamates were determined in the food classes grain and cereal products, potato, vegetables and garden fruits, root vegetables, legume vegetables, and fruits by the method of Keppel (detection limit 0.2 mg/kg as carbon disulfide). However, dithiocarbamates were not detected (the method gave recovery figures outsides the 80–120% range and so was not as precise as had been hoped). Using the proportions by weight of the commodity groups in the total diet, the maximum total annual intake appears to be certainly less than 70 mg dithiocarbamates per person (190 μg/day).

10.4. Switzerland, 1981–1983

Wüthrich et al. (1985) gave results for determinations of dithiocarbamates, among other pesticide residues, in the Swiss diet. Five diet groups were examined twice over the two years. A colorimetric method, having a limit of detection of 0.05 mg/kg, as carbon disulfide, was used to determine dithiocarbamate residues. The mean daily intake overall was 34 μg (as carbon disulfide) per day, with a maximum of 51 μg.

Dithiocarbamates were found only in the fruits, vegetables and salads group, having a mean of 62 mg/kg and a range of up to 203 mg/kg from the limit of detection.

10.5. United Kingdom, 1979–1981

In two total diet studies (Ministry of Agriculture, Fisheries and Food, 1982, 1986), fruit and vegetable groups were examined for total dithiocarbamate residues, using a headspace procedure having a limit of determination of 0.1 mg/kg (as carbon disulfide). About twenty diets were examined in each study. Any residues of dithiocarbamates were at a or below the limit of determination.

11. CONCLUSION

The recent monitoring data for residues of dithiocarbamates are summarized, as far as information is adequate, in Table 8. About 1–2% overall of the 18,000 retail samples included contained residues above a limit. This is a similar proportion to that found for residues of organophosphorus pesticides (Smart, 1986) or of pesticides more generally in retail samples (Ministry of Agriculture, Fisheries and Food, 1986).

Two general considerations need to be taken into account when considering monitoring data. First, as discussed in Section 5.3.3, limits used in different countries vary and the international CAC recommendation is not always used. The numbers of samples exceeding limits are, therefore, somewhat arbitarily

TABLE 8   Summary of Proportion of Retail Samples of Fruits and Vegetables Found to Contain Dithiocarbamate Residues

Country	Year(s)	Commodity	Number of Samples Analyzed	Proportion of Samples above a Maximum Limit
Belgium	1983	Lettuce	110	2% above 2 mg/kg
Denmark	1984–1985	All Fresh commodities	2069	0.3% above Danish limits (0.5, 1.0, 2.0 mg/kg)
FRG	1976	All	143	None
	1979–1982	Potato, lettuce, carrot, apple	720	< 0.5% lettuce, above 2 mg/kg
France	1967–1977	All	about 450	0.4%
Italy	1978–1982	Range of 13 commodities	436	5.7% above Italian limit (2 mg/kg)
The Netherlands	Mid-1970s	Varied	1031	1.4% above 3.8 mg/kg (thiram only) (The Netherlands limit)
Poland	1979–1980	Range of 9 commodities	4194	1.0% above 2.0 mg/kg Polish limit
	1981–1982	Range of 10 commodities	5099	0.9% above 2.0 mg/kg Polish limit
Sweden	1981–1984	Many fresh commodities	1975	3.4% above Swedish limits
United Kingdom	1981—1983	Lettuce	111	10.8 above 1 mg/kg
	1983	Soft fruit	103	2% above CAC limit
	1983	Celery, mushroom peach, Tomato	135	None
Switzerland	1980–1983	Range of commodities	850	0.2%

determined, although over 10 countries some consensus proportion over an average limit is probably arrived at. Second, sampling and analytical procedures will not give a precise mean value of the residue in a consignment. Some samples taken will give results that are lower than the mean (or "true") value and some results that are higher, the mathematical form of the distribution being unknown, although commonly assumed to approximate to normal. Hence, sampling of some consignments having mean residues just below the limit will give analytical results in excess of the limit and, conversely, some consignments having mean residues just above the limit will appear to contain less than the limit. This is normal in sampling foodstuffs and an overall small percetage of retail monitoring results above maximum limits can be accepted.

Another, more specific, consideration is the underlying reasoning leading to recommendation of maximum residue limits. CAC limits emphasize the need for good agricultural practice and indicate a level that is necessary for pest or disease control while well within toxicological requirements. CAC limits are based on ADIs (Section 5.2) arrived at on the basis of a day-to-day intake in the foodstuff for a lifetime. It is clear that many crops treated with dithiocarbamates are not eaten in such a way, except by an extremely small percentage of the population on a specialized diet, and over any year sources will vary, perhaps widely. CAC MRLs are not health standards but considered recommendations, using all the scientific information available, based on the practical usage to control pests, and set within, and sometimes well within, the level signified on the basis of ADI alone. Residues in lettuce, in particular, were high in some countries in the 1970s, as evaluated alongside the (then) 1 mg/kg CAC limit. Horticultural practices have since been modified and CAC currently recommend a 5 mg/kg limit. Registration authorities have lengthened intervals between last application and harvest and restricted the overall dosage that can be used for disease control.

Appreciable residues of EBDCs in commodities that are heat processed, for canning or other purposes, can sometimes lead to high residues of the decomposition product ETU. The subject has been considered by the Commission of the European Commission Scientific Committee on Pesticides (1984) and some further information is needed.

Information from total diet studies is not clearcut and studies in several countries do not support the total annual intake figure of 200 mg dithiocarbamates reported by Deschamps and Hascoet (1983) in France. Work in four other countries indicates intakes of less than half this figure. As the JMPR has currently given ADIs for EBDCs, ferbam, and ziram but not for thiram or propylenebis-dithiocarbamates, it is impossible to further evaluate the toxicological situation.

Although residues of dithiocarbamates can occur in many foodstuffs there is little evidence to suggest that these trace levels pose a threat to health.

REFERENCES

Agricultural Science Service (1981). *Pesticide Science 1979*. Ministry of Agriculture, Fisheries and Food, Reference Book 252 (79). HMSO, London, p. 46.

Agricultural Science Service (1982). *Pesticide Science 1981.* Ministry of Agriculture, Fisheries and Food, Reference Book 252 (81). HMSO, London, pp. 51–55.

Agricultural Science Service (1983). *Pesticide Science 1982.* Ministry of Agriculture, Fisheries and Food, Reference Book 252 (82). HMSO, London, p. 29.

Agricultural Science Service (1984). *Pesticide Science 1983.* Ministry of Agriculture, Fisheries and Food, Reference Book 252 (83). HMSO, pp. 23–26.

Agricultural Science Service (1987). *Research and Development Report 1985.* Ministry of Agriculture, Fisheries and Food, Reference Book 257 (85). HMSO, London, pp. 93–94.

Andersen, K. S. (ed.). (1980). *Pesticide Residues in Danish Food, 1976–1978.* Statens Levnedsmiddelinstitut. Publication no. 46, Søborg.

Andersen, K. S. (ed.). (1981). *Pesticide Residues in Danish Food, 1978–1979.* Statens Levnedsmiddelinstitut. Publication no. 54, Søborg.

Andersen, J. R. (ed.). (1983). *Pesticide Residues in Danish Food, 1980–1981.* Statens Levnedsmiddelinstitut. Publication no. 78, Søborg.

Andersson, A. (1986). "Monitoring and biased sampling of pesticide residues in fruits and vegetables. Methods and results, 1981–1984." *Vår Föda.* Suppl. 1/86, 8–55.

Arbeitsgruppe "Pestizide" (1977), rapporteur H-P. Thier. "Zur Rückstandsanalytik der Pestizide in Lebensmitteln. 5. Mitteilung: Analytik von Dithiocarbamat-Rückständen. *Lebensm. gerichtl. Chem.* **31**, 25–27.

Bottomley, P., Hoodless, R. A., and Smart, N. A. (1985). "Review of methods for the determination of ethylenethiourea (imidazolidine-2-thione) residues." *Residue Rev.* **95**, 45–89.

Casanova, M., Duval, E., and Guichon, R. (1979). "Les résidus de pesticides dans les fruits et légumes: résultats d'enquêtes." *Phytiat-Phytopharn.* **28**, 193–198.

Commission of the European Communities Scientific Committee on Pesticides (1985). "The residues of dithiocarbamates and their associated compounds." In *Reports of the Scientific Committee on Pesticides (Second Series)*, Luxembourg, 1985 (EUR 10211 EN), pp. 33–43.

Czarnik, W., Dabrowski, J., Sadlo, S., Pelz, A., Srzedzinska, A., and Pilecka, D. (1980). "Effects of growth conditions on the dynamics of mancozeb and thiuram disappearance in lettuce." *Mater. Ses. Nauk. Inst. Ochr. Rosl. (Poznan)* **20**, 307–314.

Dabrowski, J. (1984). "Determination of pesticide residues in domestic agricultural produce." *Przem. Chem.* **63**, 431–433.

Deschamps, P., and Hascoet, M. (1983). "National survey of the food quality in France." In R. Greenhalgh and N. Drescher, eds., *Pesticide Chemistry Human Welfare and the Environment*, Vol. 4. Pesticide Residues and Formulation Chemistry. Pergamon, Oxford, pp. 147–152.

De Vos, R. H., van Dokkum, W., Olthoff, P. D. A., Quirijns, J. K., Muys, T., and van der Poll, J. M. (1984). "Pesticides and other chemical residues in Dutch total diet samples (June 1976–July 1978)." *Fd. Chem. Toxic.* **22**, 11–21.

Di Muccio, A., Camoni, I., and Rubbiani, M. (1981). "Dithiocarbamate Fungicides. I. Residue decay in field tests." *Nuovi Ann. Ig. Microbiol.* **32**, 278–288.

Di Muccio, A., Rubbiani, M., and Camoni, I. (1982). "Dithiocarbamate pesticides. 2. Residue levels in farm products on the market." *Nuovi Ann. Ig. Microbiol.* **33**, 67–71.

Elinkeinohallitus (1982). "Torjunta-aineiden jäämät elintarvikkeissa vuosina 1977–1980." *Sarja* A2/1982, Helsinki (Finnish).

European Economic Community (1976). "Council Directive of 23 November 1976 relating to the fixing of maximum levels for pesticide residues in and on fruit and vegetables" (76/895/EEC). *Off. J. Eur. Comm.* no. L340, 9.12.76, 26–31.

Food and Agriculture Organization (1965). *Evaluations of the Toxicity of Pesticide Residues in Food.* Joint Meeting of the FAO Working Party of Experts and the WHO Expert Committee on Pesticides, Rome, 15–22 March, 1965, pp. 117–119, 142–143, 151–152, 181–184, 190–193.

Food and Agriculture Organization (1968). *1967 Evaluations of Some Pesticide Residues in Food. The Monographs.* Joint Meeting of the FAO Working Party of Experts and the WHO Expert Committee on Pesticide Residues, Rome, 4–11 December. FAO/PL: 1967/M/11/I. Rome, pp. 156–161, 175–186, 197–199, 226–237.

Food and Agriculture Organization (1971). *1970 Evaluations of Some Pesticide Residues in Food. The Monographs.* Joint Meeting of the FAO Working Party of Experts and the WHO Expert Committee on Pesticide Residues, Rome, 9–16 November. AGP: 1970/M/12/1. Rome, pp. 269–284, 407–422.

Food and Agriculture Organization (1975). *1974 Evaluations of Some Pesticide Residues in Food. The Monographs.* Joint Meeting of the FAO Working Party of Experts and the WHO Expert Committee on Pesticide Residues, Rome, 2–11 December. AGP: 1974/M/11. Rome, pp. 261–264, 451–470.

Food and Agriculture Organization (1978). *Pesticide Residues in Food: 1977 Evaluations. The Monographs.* Joint Meeting of the FAO Panel of Experts on Pesticide Residues in Food and the Environment and the WHO Expert Committee on Pesticide Residues, Geneva, 6–15 December. FAO Plant Production and Protection Paper. 10 sup. Rome, pp. 163–172.

Food and Agriculture Organization (1981). *Pesticide Residues in Food: 1980 Evaluations. The Monographs.* Joint Meeting of the FAO Panel of Experts on Pesticide Residues in Food and the Environment and the WHO Expert Committee on Pesticide Residues, Rome, 6–15 October. FAO Plant Production and Protection Paper. 26 sup. Rome, pp. 180–194.

Food and Agriculture Organization (1986a). *Pesticide Residues in Food: 1985 Evaluations. The Monographs.* Joint Meeting of the FAO Panel of Experts on Pesticide Residues in Food and the Environment and the WHO Expert Committee on Pesticide Residues, Geneva, 23 September–2 October. FAO Plant Production and Protection Papers 72/1, pp. 101–103 and 269, and 72/2, pp. 153–170, Rome.

Food and Agriculture Organization (1986b). *1985 Production Yearbook*, Vol. 39. FAO Statistics Series no. 70. Rome, p. 306.

Food and Agriculture Organization (1987). *Guide to Codex Recommendations concerning Pesticide Residues.* Part 2. Maximum Limits for Pesticide Residues. CAC/PR 2—1987, Rome.

Foschi, Taccheo, B. M., Bagarolo, L., Paroni, S., Pratella, G. C., Avancini, D., Stringani, G., Camoni, L., De Stanchina, G., and Fadanelli, L. (1985). "Monitoring of dithiocarbamate residues in pears." *Inf. Fitopatol.* **35**, 39–40.

Galoux, M., and Bernes, A. (1983). "Residues of plant-protection agents on winter lettuce. Results of a survey conducted in 1983." *Note Tech. Cent. Rech. Agron. Etat, Gembloux*, **9/37**, 20 pp.

Greve, P. A., and Hogendoorn, E. A. (1978). "Determination of residues of ethylene bis(dithiocarbamates) (ETU precursors) as 1,2-bis(pentafluorobenzamido)-ethane." *Med. Fac. Landbouww. Rijksuniv. Gent* **43/2**, 1263–1268.

Hatfull, R. S. (1976). "Surveys of pesticide residues in lettuce 1 July, 1974–30 June 1975. Report prepared on behalf of the Association of Public Analysts." *J. Assoc. Publ. Analysts* **14**, 75–86.

Hatfull, R. S. (1983). "Survey of pesticide residues in foodstuffs, 1981. A Report on behalf of the Association of Public Analysts." *J. Assoc. Publ. Analysts* **21**, 19–24.

Hemminki, K., Vaino, H., Rosenberg, C., and Kiviranta, A. (1982). "Pesticide residues in food: How much and how safe?" *Regul. Toxic. Pharmacol.* **2**, 229–231.

Hill, A. R. C., and Edmunds, J. W. (1982). "Headspace methods for the analysis of dithiocarbamate pesticide residues in foodstuffs." *Anal. Proc.* **19**, 433–435.

IUPAC (1977). "Monograph on ethylenethiourea." *Pure Appl. Chem.* **49**, 675–689.

Kampe, W. (1977). "The problem of plant protectant residues, as shown by studies on farm products." *Kali-Briefe* **12**, 9 pp.

Kampe, W. (1983). "Plant protectant residues." *Landwirtsch. Forsch. Sonderh.* **40**, 174–181.

Keppel, G. E. (1969). "Modification of the carbon disulphide evolution method for dithiocarbamate residues." *J. Assoc. Off. Anal. Chem.* **52**, 162–167.

Keppel, G. E. (1971). "Collaborative study of the determination of dithiocarbamate residues by a modified carbon disulphide evolution method." *J. Assoc. Off. Anal. Chem.* **54**, 528–532.

Lauridsen, M. G., Dahl, M. H., Hansen, K. E., and Hansen, T. (1980). "Investigations for residues of dithiocarbamates and ethylenethiourea in different crops." *Tidsskr. Pleavl.* **84**, 245–256.

Lesage, S. (1980). "Effect of cupric ions on the analysis of ethylenebis(dithiocarbamate) residues in tomato juice." *J. Assoc. Off. Anal. Chem.* **63**, 143–145.

Lorusso, S., Camoni, I., Di Muccio, A., and Chiaccherini, E. (1984). "Analytical techniques and instrumentation applicable for the control of contamination of pesticide residues in food." *Latte* **9**, 984–1000.

McLeod, H. A., and McCulley, K. A. (1969). "Head space gas procedure for screening food samples for dithiocarbamate pesticide residues." *J. Assoc. Off. Anal. Chem.* **52**, 1226–1230.

Ministry of Agriculture, Fisheries and Food (1982). *Report of the Working Party on Pesticide Residues (1977–1981)*. Food Surveillance Paper No. 9. HMSO, London, 38 pp.

Ministry of Agriculture, Fisheries and Food (1986). *Report of the Working Party on Pesticide Residues (1982 to 1985)*. Food Surveillance Paper No. 16. HMSO, London, 50 pp.

Nicolson, R. S. (1984). "Association of Public Analysts Surveys of Pesticide Residues in Food, 1983." *J. Assoc. Publ. Analysts* **22**, 51–57.

Orbaek, K. ed. (1985). *Pesticide Residues in Danish Food, 1982–1983.* Statens Levnedsmiddelinstitut. Publication no. 107, Søborg.

Orbaek, K. ed. (1987). *Pesticide Residues in Danish Food 1984–1985.* Levnedsmiddelstyrelsen, Publication no. 147, Søborg.

Panel on Determination of Dithiocarbamate Residues (1981) of the Committee for Analytical Methods for Residues of Pesticides and Veterinary Products of the Ministry of Agriculture, Fisheries and Food. "Determination of residues of dithiocarbamate pesticides in foodstuffs by a headspace method." *Analyst* **106**, 782–787.

Phillips, W. F. (1976). "A residue study of Munzate-D and ethylenethiourea." Du Pont de Nemours & Co. (Inc) RPAR Dec. 22, 1977, Vol. 12, Ap D 12 pp.

Phillips, W. F., Grady, M. D., and Freudenthal, R. (1977). "Effects of good processing on residues of two ethylenebisdithiocarbamates (EBDC) fungicides and ethylenethiourea." U.S. EPA Environ. Health Res. Ser. EPA-600/1-77-021, 181 pp.

Ripley, B. D., and Cox, D. F. (1978). "Residues of ethylenebis(dithiocarbamate) and ethylenethiourea in treated tomatoes and commercial tomato products." *J. Agr. Fd. Chem.* **26**, 1137–1143.

Ripley, B. D., Cox, D. F., Wiebe, J., and Frank, R. (1978). "Residues of Dikar and ethylenethiourea in treated grapes and commercial grape products." *J. Agr. Fd. Chem.* **26**, 134–136.

Ripley, B. D., and Simpson, C. M. (1977). "Residues of zineb and ethylenethiourea in orchard-treated pears and commercial pear products." *Pestic. Sci.* **8**, 487–491.

Ross, R. G., Wood, F. A., and Stark, R. (1978). "Ethylenebisdithiocarbamate and ethylenethiourea residues in apples and apple products following sprays of mancozeb and metiram." *Can. J. Plant Sci.* **58**, 601–604.

Schüpbach, M. R. (1986). "Pesticide residues in fruits and vegetables." *Deutsch. Lebensm. Rund.* **82**, 76–80.

Sly, J. M. A. (1981). "Review of usage of pesticides in agriculture, horticulture and forestry in England and Wales 1975–1979." Reference Book 523. Ministry of Agriculture, Fisheries and Food, London.

Sly, J. M. A. (1986). "Review of usage of pesticides in agriculture, horticulture and animal husbandry in England and Wales, 1980–1983." Reference Book 541. Ministry of Agriculture, Fisheries and Food, London.

Smart, N. A. (1986). "Organophosphorus pesticide residues in fruits and vegetables in the United Kingdom and some other countries of the European Community since 1976." *Rev. Environ. Contam. Toxic.* **98**, 99–160.

Wegorek, W., and Dabrowski, J. (1985). "Contamination of crops by pesticide residues." *Arch Ochr. Srodowiska* **1/2**, 97–110.

Wüthrich, C., Müller, F., Blaser, O., and Marek, B. (1985). "Pesticides and other chemical residues in Swiss Diet samples." *Mitt. Geb. Lebensm. Hyg.* **76**, 260–276.

8

RESIDUES IN FOODSTUFFS FROM BROMOMETHANE SOIL FUMIGATION

Nigel A. Smart

*Harpenden Laboratory, Agricultural Development and Advisory Service,
Ministry of Agriculture, Fisheries and Food, Hatching Green,
Harpenden, Herts, United Kingdom*

1. INTRODUCTION

The insecticidal action of bromomethane (methyl bromide) gas as a space fumigant was recognized as early as 1932 (Le Goupil) and initial records of its fungicidal and nematicidal action were made in the late 1930s and early 1940s (Roughan, 1981). It was found that soil could be successfully fumigated for control of a wide spectrum of pests and diseases: insects, nematodes, fungi, viruses, and weed seeds. In the late 1960s following developments in the Netherlands, bromomethane soil sterilization by contractors was readily available in a number of countries and became a standard practice in protected cropping. Control is effective and, after a short airing period, there is minimal risk of phytotoxicity to subsequent crops. Bromomethane fumigation of field soil for crops such as strawberries is carried out in countries having a warm temperate western margin (Mediterranean) climate.

The practice of bromomethane soil fumigation and standards for any residues arising in foodstuffs are described in this chapter. Information on residues that have occurred in crops following soil fumigation is presented in some detail, protected crops being the main concern. The review does not cover the dynamics of bromomethane residues in soils and fumigant action; neither does it cover residues from postharvest treatments of cereal stores, flour warehouses, or other premises storing foodstuffs, with bromomethane. Reference is made to naturally occurring "background" levels of total bromine in foodstuffs.

In the literature, residues are expressed in terms of "total bromine," "bromide," "bromide ion," or "inorganic bromide," and I have generally accepted the terminology used by workers to whom I refer. Some confusion appears and although "total bromine" or "bromine" usually refers to determinations following complete dry ashing of samples (thus including naturally occurring bromine as well as pesticide residue), "inorganic bromide" or "bromide ion" may refer not only to determinations by aqueous/solvent extraction but also to determinations by rigorous dry ashing.

2. PROPERTIES OF BROMOMETHANE

2.1. Physical Properties

Bromomethane is a colorless gas, the liquid having a boilingpoint of 3.6°C at normal pressure. The low boiling point and high vapor pressure allow a deep penetration into the soil and rapid dispersion of the gas when released under plastic sheeting spread over a soil surface. It is available commercially in small cans or large cylinders as a liquid under pressure. Bromomethane is odorless and an additive, such as chloropicrin (2%), is sometimes included in its formulation to give warning of any substantial leakage during use.

2.2. Toxicological Properties

Toxicity data for bromomethane have been reviewed by Alexeeff and Kilgore (1983) and by the Commission of the European Communities Scientific Committee on Pesticides (1984). After fumigations, bromomethane in soils, or foodstuffs, gradually decomposes to give bromide ion, among other products. The toxicology of bromide ion is different from that of bromomethane itself and has also been detailed in the literature (Food and Agriculture Organization, 1967; Commission of the European Communities Scientific Committee on Pesticides, 1984). A symposium in Bilthoven in 1982 discussed bromide in toxicology (Poulsen, 1983) and, more recently, a workshop in Avignon has considered toxicological aspects of bromomethane fumigation.

3. AGRICULTURAL PRACTICE

3.1. Outline of the Present Process

The chemical is applied to the soil in glasshouses at intervals that may be as short as 1 year. Gas is released from several points at soil level beneath a sealing sheet of plastic, usually polyethylene, the edges of which are buried in the soil. Normally, under glass, gas is introduced beneath the sheet from perforated lay-flat polyethylene tubing connected to a thermal vaporizer that is coupled to the cylinder of bromomethane. In open fields, liquid bromomethane is injected at an appropriate depth from hollow tines by an applicator which simultaneously lays polyethylene sheeting to seal the soil. The soil remains covered for 2–4 days, or more, to allow the bromomethane to penetrate the soil and act. When the sheeting is removed, any remaining gas escapes into the atmosphere. The soil is normally left undisturbed for 24 hr after which it may be cultivated. A further airing period of 48 hr or more is allowed when there is then little risk of residual gas at a phytotoxic level in the soil. The soil may need to be leached before sensitive crops are planted out.

Bromomethane is currently applied at rates of up to $125 \, g/m^2$, the amount depending on the national regulations and agreement between grower and fumigator. Application is made by trained and licensed contractors because of the high toxicity of bromomethane gas to humans and other mammals. Codes of practice are issued by responsible departments in most countries.

3.2. Current Trends

Concern over an undue proportion of retail samples of protected crops monitored for total bromine residues having levels in excess of maximum limits (Section 4.2), and the toxic nature of the fumigant itself, has tended to reduce the advantages of bromomethane fumigation. Approaches to combat the problems have varied. Extensive leaching of glasshouse soils has been used to reduce residues in lighter soils in the Netherlands. In France, fumigation is carried out in August and a tomato crop is grown before lettuce; a water drench is applied to the soil, after cleaning the tomato crop, before planting out the lettuce. Other fumigants that were common in the 1960s, such as metham-sodium, are used in the United Kingdom as well as bromomethane, despite longer unproductive periods for fumes to clear. Although many possible compounds have been researched, it is clear that no new generally effective fumigant has been marketed to supercede bromomethane. Steam sterilization is still practiced and improved techniques attempting to reduce the high costs have been suggested. Grafting of tomatoes and more selective herbicide control in lettuce are also alternatives to bromomethane fumigation. Solar sterilization is being studied for regions and climate belts with appropriate sunshine intensity and duration. The newer approaches are considered as part of an integrated pest control strategy.

3.3. Extent of Use

The Food and Agriculture Organization (FAO) Production Yearbooks, in the FAO Statistics Series, do not give any specific information on the extent to which bromomethane is used as a soil fumigant or as a space fumigant in countries of the world. There is little information available in the literature generally on the extent to which bromomethane fumigation in soil is used.

Alexeeff and Kilgore (1983) give the amounts used in California from 1970–1981; in 1981 the figure was about 2800 tonnes.

Wegman et al. (1983) state that in the Netherlands in 1979 the yearly use in glasshouses was about 2000 tonnes.

In 1972 use in England and Wales covered 250 ha and in 1976 the extent was 570 ha (Sly, 1981). In 1981 use of bromomethane in England and Wales covered 434 ha of glasshouse soil; 282 tonnes of bromomethane was used in treatments (Sly, 1986). By 1985, use on edible glasshouse crops in the U.K. had declined further and was about 170 tonnes (Longland et al., 1988).

3.4. Degradation following Application

3.4.1. Interaction with and Residues in the Soil

Once the gas is in the soil reversible processes include physical adsorption–desorption, chemisorption and desorption, and solution and loss in water. Irreversible processes such as chemical reaction and decomposition also take place. Quantitative descriptions of these processes during and after soil fumigation are few and of limited application.

The exact form in which bromine derived from bromomethane exists in the soil is unclear. In contrast to bromine occurring naturally in the soil mineral structure, bromine derived from bromomethane is initially only slightly absorbed onto soil particles and is thus mobile in the soil. It is available for uptake by plants and can also be leached. Bromide is reported to be leached readily in light soils, but leaching from heavy clay soils is difficult.

The methyl moiety of the pesticide is commonly assumed to degrade to methanol although the evidence is scanty and superficial. The recent report of the Commission of the European Communities Scientific Committee on Pesticides (1984) indicates that, when applied to soil, bromomethane reacts chemically in methylation of polymeric organic material and to a lesser extent in hydrolysis, resulting in formation of inorganic bromide. Whether small organic molecules in soil become methylated is unknown.

3.4.2. Degradation and Interaction within Plants

Many workers have demonstrated the presence of increased total bromine levels in plants grown on bromomethane-fumigated soils that could be related to precrop fumigation. The levels of total bromine in various crops are the main subject of this review and are discussed in detail in Sections 7 to 10.

If small organic species in soils are methylated (or brominated) they may be taken up by growing plants. Alternatively, any larger methylated (or brominated) molecules formed in soils may be unsuitable to enter the plant phloem, either immediately, or after subsequent transformations. (It should be noted that direct fumigation of wheat, cocoa beans, and other products with bromomethane has led to methylation of some amino acids and proteins (Asante-Poku et al., 1974; Swamy and Reddy, 1974; Afifi et al., 1982) resulting in enzyme inhibition (Price et al., 1981). Some work has also suggested that N-methylation of various amino acids by bromomethane may yield precursors of carcinogenic N-nitroso compounds (Dunkelberg, 1980).

4. STANDARDS FOR RESIDUES IN FOODS

4.1. Acceptable Daily Intakes

The Joint Food and Agriculture Organization (FAO) World Health Organization (WHO) Meeting on Pesticide Residues (JMPR) has, where sufficient information is available, estimated acceptable daily intakes (ADIs) for pesticides. The ADI of a chemical is the daily intake that, during an entire lifetime, appears to be without appreciable risk on the basis of all known facts at that time. It is expressed in milligrams of the chemical per kilogram of body weight (b.w.).

The ADI given by the JMPR for inorganic bromide is 1 mg/kg b.w. (Food and Agriculture Organization, 1987).

Bromomethane itself is not currently cleared toxicologically by by the JMPR (further information is needed) and no ADI is available.

4.2. Maximum Residue Limits

4.2.1. Codex Alimentarius Commission Limits

Where necessary and sufficient information to do so is available the Codex Committee on Pesticide Residues (CCPR) recommends, on the advice of the JMPR, maximum residue limits (MRLs) for pesticides in foodstuffs. The MRLs are published by the Codex Alimentarius Commission (CAC) as a food standard. In general, a Codex MRL refers to the residue resulting from use of a pesticide under circumstances designed to protect the crop or food commodity against attack by pests, according to good agricultural practice, which may vary from country to country, or region to region, as a result of differences in local pest control requirements. Therefore, residues in food, particularly when pesticides are applied to crops close to harvest, may also vary. In recommending Codex limits, these variations in residues resulting from differences in "good agricultural practice" are taken into consideration as far as possible on the basis of available data. Codex residue limits take into account the minimum quantities of pesticide necessary to achieve adequate control, and which are applied in a manner that will leave the smallest practical residue which must also be toxicologically acceptable.

CAC MRLs for inorganic bromide (determined and expressed as total bromide from all sources) in fruit and vegetables that may have been grown on bromomethane-fumigated soils are (Food and Agriculture Organization, 1987) as follows:

	MRL (mg/kg)
Cabbages, head	100
Celery	100
Cucumber	50
Lettuce, head	100
Strawberry	30
Tomato	75

The 1983 JMPR had recommended 300 mg/kg for celery.

When an ADI is not given by the JMPR, because of incomplete or uncertain toxicological data, only guideline levels (GLs) for the highest acceptable residues can be given. Guideline levels are given by the CAC for bromomethane (as a space fumigant) in cereal grains, cereal products, dried fruit, nuts, and cocoa foodstuffs but no GLs have been recommended for fruits and vegetables that may have been grown on bromomethane-fumigated soils (as distinct from inorganic bromide residues).

4.2.2. European Community Limits

Maximum levels have also been fixed for residues of pesticides by the European Communities (EC) and are derived somewhat differently. The amending Directive (European Economic Community, 1982) to the 1976 Directive on pesticide residues in or on fruits and vegetables fixed a maximum level of 0.1 mg/kg bromomethane in fruits and vegetables (potatoes and sugar beet are

TABLE 1   National Maximum Levels for Residues of Inorganic Bromide and Bromomethane

Country	Commodities	Limit (mg/kg)
Inorganic Bromide		
Australia	Strawberry	20
	Brassica, lettuce, tomato, curcurbits	20
Austria	Cucumber	50
	Lettuce, tomato	30
Belgium	Leaf vegetables, roots and tubers	50
	Tomato	30
	Strawberry	20
Finland	Fresh fruits and vegetables	30
Federal Republic of Germany	Lettuce	50
	Other vegetables, except lettuce	30
	Strawberry	20
Israel	Strawberry	30
The Netherlands	Strawberry	20
	Leaf vegetables, root vegetables	50
	Other vegetables	30
Sweden	Fruit and vegetables, except potato	30
United States	Brassica and leafy vegetables	100
	Curcurbit vegetables (except watermelon group)	250
	Celery	50
	Strawberry	60
	Watermelon group	100
Bromomethane		
Federal Republic of Germany	All vegetable foodstuffs	0.1
The Netherlands	Fruit, vegetables (except potato)	Zero
United States	Brassica (cole), leafy vegetables, bulb vegetables, curcurbits, fruiting vegetables, root and tuber vegetables	0.1
	Melon	0.5

not included). The EC have no maximum limit for total inorganic bromide in fruits and vegetables, only a limit for cereals.

4.2.3. National Limits

A number of countries, while accepting CAC and/or EC limits, nevertheless, have maximum limits of their own. Although similar to the internationally agreed levels, there are differences both in the classification of commodities and in the levels set. Examples, covering the fruit and vegetable commodities relevant to this review, are for inorganic bromide (from soil fumigation) and for bromomethane itself are given in Table 1. Most countries do not have a national limit for bromomethane in these commodities.

5. METHODS OF ANALYSIS

5.1. General Remarks

A number of approaches to determining total bromine or bromide ion in crops grown on fumigated soil have been described in the literature; some have been more thoroughly tested and documented than others. Analyses for bromomethane itself in crops grown on fumigated soil have very rarely been carried out.

5.2. Dry Ashing to Determine Total Bromine

Samples have been dried and ground before treatment with alcoholic sodium hydroxide and ashing in a muffle furnace (500–600°C). The residue was taken up in dilute acid for reaction with ethylene oxide to form 2-bromoethanol for gas chromatography (GC) with electron-capture detection (Greve and Grevenstuk, 1976; Roughan et al., 1983). Shiga et al. (1986) determined bromide ion by high-performance liquid chromatography (HPLC) after ashing.

The residue following ashing has also been oxidized by chloramine to hypobromite forming bromophenol blue by reaction with phenol red, which was determined spectrophotometrically (Orbaek, 1987).

5.3. Wet Extraction of Bromide Ion

Samples of commodities having a high water content and thin outer surface have been macerated with an aqueous medium, sometimes at above room temperature, to extract bromide ion (Stijve, 1981; Andersson, 1986) usually for reaction with ethylene oxide for GC with electron-capture detection.

5.4. Other Determinative Steps

Electrochemical detection of bromide ion in this context has been described by Stijve (1981) and Nangniot et al. (1984).

High-performance liquid chromatography (HPLC) separations with UV detection (205 nm) have been reported for determining inorganic bromide in vegetables by Van Wees et al. (1984) and Shiga et al. (1986).

Nazer et al. (1982) have used X-ray fluorescence to determine bromine in various crops.

These and other approaches were discussed by Declercq at the European Workshop on methyl bromide as a soil fumigant at Avignon in 1986 (Scudamore, private communication).

5.5. Interlaboratory Studies

At least four interlaboratory studies of methods have been described.

Greve and Grevenstuk (1979) have reported using lettuce with incurred bromide residues ranging from 5 to 200 mg/kg to examine their procedure. Coefficients of variation for repeatability and reproducibility ranged from 4.0 to 9.6% and from 8.3 to 18.4%, respectively. An interlaboratory study of GC and electrochemical procedures was carried out on NBS spinach (about 50 mg/kg bromine, dry weight) and a composite dried lettuce sample (about 5000 mg/kg bromine, dry weight) by laboratories in the U.K. already doing work on fresh crops. Results were acceptable, having an overall standard deviation of about 12% (Agricultural Science Service, 1981b). Nangniot et al. (1984) have compared determination of bromide in vegetables by GC and ionemetry. They suggest that an ion-selective electrode method is preferable to GC for determining bromide in crops.

Beernaert and Vandezande (1986) have reported an interlaboratory study on the efficacy of the method of Stijve (1981). Results for tomato compared well with those using neutron activation analysis; an interlaboratory standard deviation of 2.29 mg/kg was obtained at about 19 mg/kg level (wet weight). Maximum deviations of about 20% were obtained with lettuce, corn, and celery samples. Although acceptable it was possible to improve the method.

5.6. Confirmation of Identity

Confirmation of identity has been achieved by using propylene oxide (instead of ethylene oxide) to form 1-bromo-2-propanol and 2-bromo-1-propanol for GC determination (Stijve, 1985). Mass spectrometry has been used to confirm bromoethanol from normal derivatization (Roughan et al., 1983).

5.7. Repeatability

Coefficients of variation associated with series of determinations on the same substrate and in the same laboratory are usually in the 2–10% range (see, for example Section 5.5), depending on the levels of bromine being determined; the higher coefficients are associated with low background levels.

Differences between replicate determinations on similar subsamples are larger than this. Coefficients of variation of results on samples taken from single heads of lettuce from some pot experiments were 7–22%. Variation in levels in heads of lettuce taken from commercial glasshouses have been about 60%.

6. BACKGROUND LEVELS

Bromine is widely distributed in the environment. Total bromine in soils does not normally exceed 5 mg/kg dry weight, except for coastal soils that can attain levels of 100 mg/kg. Rainwater and surface water do not normally exceed a level of 1 mg/L (seawater contains much higher levels).

Naturally occurring total bromine levels in fresh fruits and vegetables can generally be expected to be below 10 mg/kg fresh weight and will not normally exceed 50 mg/kg. Table 2 gives some main values for "natural" total bromine in relevant crops, derived from some 19 literature sources (Roughan, private communication). Similar values are given by Pendias and Pendias (1984).

There is some evidence that plants selectively absorb bromide ion relative to chloride as Br/CI ion ratios are generally larger in plants than in the soil and soil water. The level of nitrogen in soil can affect the extent of bromide uptake.

TABLE 2   "Natural" Bromine in Crops

Group	Commodity	Moisture Content (%)	Mean Bromine, Wet Weight (Several Sources of Data) (mg/kg)
Root and tuber	Carrot		1.75
vegetables	Potato	78	2.73
	Radish	91	1.68
Bulb vegetables	Onion	88	2.27
Stem vegetables	Asparagus	93	2.58
	Celery		2.57
	Leek		1.96
Fruiting	Cucumber	96	0.58
vegctables	Melon		0.68
	Pepper		0.53
	Tomato	94	1.40
	Watermelon		0.43
Leafy	Chicory		(15.73)
vegetables and	Endive		2.63
salads	Lettuce	96	1.28
	Spinach		2.5
Brassicas	Brussels sprouts		1.32
	Cabbage		1.49
	Cauliflower		0.89
	Kale		2.4
	Kohlrabi		1.4

Results from determinations of bromine in crops grown on fumigated soils are viewed in terms of these background levels.

7. RESIDUES IN SUPERVISED TRIALS

Supervised trials in which the crop husbandry and the crop protection in particular are closely managed often provide data for registration purposes. Usually only the pesticide under scrutiny is applied.

7.1. Lettuce

7.1.1.   1978 and 1979 Protected Crops, United Kingdom

A multifactorial glasshouse unit fumigated annually with bromomethane at about $100 \, g/m^2$ was used (Kempton, 1978). Forty-five samples of lettuce from a crop grown during 1978 following 1977 fumigation were analyzed for bromide content. These contained 40–165 mg/kg (fresh weight) and only 16% contained less than 50 mg/kg bromide. There were no significant differences in residues between varieties.

Lettuce grown in 1979 following the additional 1978 fumigation were also analyzed (Kempton, 1979). Twelve samples contained residues ranging from 125 to 225 (mean 210) mg/kg bromide (fresh weight).

7.1.2.   1980 Protected Crops Trials on Six Sites, United Kingdom

During 1980 supervised trials were carried out to investigate the effect of leaching following soil fumigation with bromomethane (75 or $100 \, g/m^2$) (Food and Agriculture Organization, 1985). Glasshouse soils having different textures and different histories of bromomethane fumigation (all having been treated recently) were used, so as to obtain results from different soil types. Residues data are presented in Table 3 that show, in the first column giving residue results, the "normal" residue figures for crops grown in soils not having any leaching. Soil residues at sites A and B were low and residues in lettuce were unaffected by leaching. On the other sites there was a positive but diminishing response to increasing rates of leaching. The very high residues at site F were probably due to the high organic matter (51%) and the previous history of fumigation.

7.1.3.   1981 Protected Crops on Three Sites, Denmark

Soil in three glasshouses was treated with bromomethane at $100 \, g/m^2$ according to normal practice (Helweg and Rasmussen, 1982). Leaching of the soil decreased the bromide content of the lettuce grown in the soil from about 400 mg/kg to generally below 30 mg/kg, when harvested 3 months after fumigation. Table 4 gives the mean residues results for the three sites.

7.1.4.   1977–1978 Trial on an Open Plot, United Kingdom

To assess the uptake of bromine ion by crops from open field fumigation with bromomethane, a level plot (clay + loam, 12% organic matter) was fumigated at

TABLE 3 Bromide Residues in Lettuce Grown under Protection on Soil Fumigated with Bromomethane and Leached with Water before Planting[a]

Site	Soil Texture	Soil Organic Matter (%)	Number of Applications in Previous Years	Interval between Planting and Harvest (weeks)	Bromide Residues in Lettuce (mg/kg Fresh Weight) for the Water Application Rates (mm)				
					0	100	200	300	400
A	Fine sandy loam	4	2	16	81	102	92	150	142
B	Sandy loam	12	1	12	62	70	232	93	114
C	Sandy loam	13	2	13	307	215	98	157	117
D	Sandy loam		1	24	765	427	392	284	250
E	Sandy loam	8	4	21	676	335	328	164	134
F[b]	Loamy peat	51	4	14	1958	1534	1001	—	—
Mean sites A–E					378	230	228	170	151

[a]Summarized from Food and Agriculture Organization (1985).
[b]At F, the leaching treatments were applied after planting the lettuce crop.

239

TABLE 4   Bromide Residues in Lettuce Grown in Untreated Soil and in Soil Treated with Bromomethane[a]

| Treatment | Bromide Residues in Lettuce (mg/kg, Fresh Weight): Days after Treatment | | |
	31 (First Crop)	66 (Second Crop)	99 (Third Crop)
Untreated	6.0	1.8	1.9
100 g/m² bromomethane with no leaching	385.0	123.0	23.9
100 g/m² bromomethane with leaching	24.2	12.3	5.5

[a]Reproduced, with permission, from Helweg and Rasmussen (1982).

$100\,g/m^2$ with bromomethane and left uncovered for 5 days (Agricultural Science Service, 1981a). Three days after the end of fumigation the plot and an adjacent untreated plot were marked off into microplots. The plots were then planted with crops, at commercial densities, for the next 18 months.

Lettuce harvested approximately 12 weeks after fumigation contained between 146 and 458 mg/kg bromide (fresh weight), with a mean value of 335 mg/kg. Controls ranged from 3 to 7 mg/kg with a mean value of 4 mg/kg. Lettuce planted 1 year after fumigation contained about 21 mg/kg at harvest.

It has been shown that washing heads of lettuce grown on bromomethane-fumigated soil and containing bromide residues of about 300 mg/kg removes very little of the residue, even after a considerable period of time.

7.2.   Tomato (Protected Cropping)

7.2.1.   1978 Trial, United Kingdom

A multifactorial glasshouse unit fumigated annually with bromomethane at about $100\,g/m^2$ was used (Kempton, 1978). Forty samples of tomato fruit taken over a 10-week period following the 1978 lettuce crop (Section 7.1.1) showed residues of bromide ranging from 11 to 312 mg/kg (fresh weight). Nineteen samples contained residues above 50 mg/kg.

7.2.2.   1980–1981 Trials on Seven Holdings, United Kingdom

During 1981 the uptake of bromine ion by tomato fruit from plants grown on soils previously treated with bromomethane at 75 or $100\,g/m^2$ was investigated (Roughan and Roughan, 1984b). Where possible the fruit from each truss of three selected and labeled plants on each holding were sampled and analyzed for total bromine content. Abridged results are presented in Table 5. The bromide levels

TABLE 5   Bromide Residues in Tomato Fruits Grown under Protection on Soil
Previously Fumigated with Bromomethane[a]

| Site | Dates of Soil Fumigation | Date of Planting Tomato Crop | Plant Number | Bromide Residues[b] (mg/kg Fresh Weight) | |
				Plants Mean	Site Mean
1	Each October, 1973–1980 (8 fumigations)		1 2 3	39 18 17	25
2	Each October, 1977–1980 (4 fumigations)		1 2 3	54 41 42	46
3	Before tomato crops 1977–1980 (4 fumigations)		1 2 3	158 187 111	152
4	April 1977, 1980, 1981 (3 fumigations)	15 April 1981	1 2 3	39 55 32	42
5	November 1977, 1978, and December 1980 (3 fumigations)	27 Feb. 1981	1 2 3	31 23 38	31
6	November 1978, 1980 (2 fumigations)	26 Feb. 1981	1 2 3	30 32 26	29
7	October 1978, 1979 (2 fumigations)	Approx. 15 April 1981	1 2 3	15 14 6	12

[a]Simplified from Roughan and Roughan (1984b).
[b]Sampled summer 1981.

exceeded 30 mg/kg (fresh weight) in tomato fruit throughout the major part of the
season for the six holdings treated with bromomethane in 1980/1981. The overall
mean was 57 mg/kg for all samples from the six sites. The highest residue found
was 326 mg/kg from the site having soil with an organic matter content in excess
of 50%.

7.2.3.   1979–1980 Trial in Plastic House

Soil in a plastic house was treated with bromomethane fumigant at rates of 0, 450,
900, and 1350 kg/ha and, after airing, tomatoes were planted out (Nazer et al.,
1982). Samples of fruit were taken at 3-week intervals and analyzed for total
bromine. The highest residue was 13.7 mg/kg. There was a decrease in bromine
residues with time.

7.3. Cucumber (Protected Cropping)

7.3.1. 1978 Trial, United Kingdom

A multifactorial glasshouse unit fumigated annually with bromomethane at about $100 \, g/m^2$ was used (Kempton, 1978). Thirteen samples of cucumber fruit were taken in 1978, following 1977 fumigation, and contained residues of bromide ranging from 11 to 26 mg/kg (fresh weight). Four times as much residue was in the skins as in the inner flesh.

7.3.2. 1979–1980 Trial in a Plastic House

Soil in a plastic house was treated with bromomethane fumigant at rates of 0, 450, 900, and 1350 kg/ha and, after airing, cucumbers were planted out (Nazer et al., 1982). Samples of fruit were taken at 3-week intervals and analyzed for total bromine. The highest residue was 19.4 mg/kg well below United States and CAC limits. There was a decrease in bromine residues with time.

8.   RESIDUES IN RETAIL-TYPE SAMPLES OF KNOWN TREATMENT HISTORY, OBTAINED FROM GROWERS

These samples were taken from situations in which bromomethane was known to have been used and so all could possibly contain bromine residue. Other pesticides may also have been applied.

8.1. Lettuce (Protected Cropping)

8.1.1. Initial Investigations, United Kingdom, 1972

Kempton and Maw (1972) investigated fumigated glasshouses in the north of England and found bromide residues in lettuce ranging from 80 to 550 mg/kg (fresh weight).

Williams also in 1972, found 45 to 360 mg/kg bromide (fresh weight) in lettuce grown in soils fumigated with bromomethane just before planting out (Roughan and Roughan, 1984a).

8.1.2. Survey, United Kingdom, November 1977–April 1978

Samples of lettuce were collected direct from growers in England and Wales during this period (Roughan and Roughan, 1984a); data on bromomethane usage were recorded on an accompanying questionnaire. Fifty-seven samples were analyzed for bromide ion content. The results, set out in Table 6, show 67% of the samples had exceeded 100 mg/kg bromide (fresh weight).

8.1.3. Survey, United Kingdom, October 1979–September 1980

During this period, samples of protected lettuce were collected directly from a wide selection of holdings in England and Wales (Roughan and Roughan, 1984a). Sampling was biased toward those on which bromomethane had been used for

TABLE 6 Bromide Residues in Lettuce Grown under Protection in Soil Previously Fumigated with Bromomethane and Sampled Direct from the Grower[a]

Survey Date	Total Number of Samples Analyzed	Range of Bromide Residues in Lettuce (mg/kg, Fresh Weight)												
		0–50		51–100		101–200		201–500		501–1000		1001–2000		2001–3000	
		No.	%	No.	%	No.	%	No.	%	No.	%	No.	%	No.	%
November, 1977–April, 1978[b]	57	9	16	10	18	12	21	20	35	5	9	1	2	0	0
October, 1979–September 1980[c]	684	135	20	64	9	111	16	174	25	117	17	68	10	15	2
October, 1979–September, 1980[d]	144	143[e]	99	1	1	0	0	0	0	0	0	0	0	0	0

[a]Simplified from Roughan and Roughan (1984a).
[b]From bromomethane-fumigated soil.
[c]From soil fumigated with bromomethane since 1974 (1 to 7 applications).
[d]From soil fumigated with bromomethane before 1974 or not at all.
[e]Of the 143 samples in the 0–50 mg/kg range, 124 (i.e., 86%) were in the range 0–10 mg/kg.

soil sterilization during any of the preceding 6 years. When each sample was collected, a questionnaire was completed giving bromomethane treatments, cropping and cultural details, and often data on pesticide usage. A total of 828 samples were analyzed from 470 individual holdings. Table 6 shows that, of the 684 samples taken from crops grown on land fumigated with bromomethane, 71% exceeded 100 mg/kg bromide (fresh weight) and approximately 30% contained in excess of 500 mg/kg.

The data were analyzed in terms of interval between last fumigation and planting the sampled crop, number of fumigations, and a broad classification of soil type. No firm conclusions could be drawn except that the bromide residue declines with increasing interval between fumigation and planting out. Even where the soil has received only one application, residues can remain above background levels for over 3 years. Averaging all results from different monthly intervals between fumigation and planting-out for general soil types, and plotting these against a 3-year time scale, indicates that the heavier the soil the higher the residues in lettuce (very heavy clay soils form a small somewhat anomalous group). Variations in bromide content of lettuce grown on different sites having the same general soil type and the same interval between fumigation and planting-out are very great. The higher the organic matter content of the soil the greater the probability of retention of bromomethane is soil and consequent high bromide ion residues in lettuce.

8.2.   Cucumber (Protected Cropping), United Kingdom 1981 Survey

Twenty-five commercial producers (with a total of 30 separate houses) within the three major growing regions (for cucumbers) in England and Wales were identified by the Agricultural Service (Roughan and Roughan, 1984b). The selection of holdings was biased toward those on which bromomethane had been used during or since the autumn of 1980, generally at 100 g/m^2 rate. Samples were collected during September and October of 1981. The mean interval between the most recent soil fumigations and planting was 28 weeks. The mean bromide ion level in cucumber was 27 mg/kg (fresh weight), with a range of 1–109 mg/kg.

TABLE 7   Bromide Residues in Cucumbers and Celery Grown under Protection on Soil Previously Fumigated with Bromomethane and Sampled Direct from the Grower[a]

Survey (Date)	Total Number of Samples Analyzed	Range of Bromide Residues (mg/kg, Fresh Weight)					
		0–15	16–49	50–74	75–99	100–299	300–1000
Cucumber (1981)	30	16	8	3	1	2	0
Celery (1982)	38	11	8	3	4	8	4

[a]Summarized from Roughan and Roughan (1984).

Results are set out in Table 7. There were insufficient samples to show any correlation between bromide residues, treatment data, and cultural regime.

8.3. Celery (Protected Cropping), United Kingdom 1982 Survey

Samples were obtained from 35 commercial producers of self-blanching celery (with a total of 38 separate houses) within the south of England (Roughan and Roughan 1984b). The selection of holdings was biased toward those on which bromomethane had been used during the previous 2 years, generally at the $75 \, g/m^2$ rate. Samples were collected during April to July 1982. Results are set out in Table 7. The mean interval between the most recent fumigation and planting the celery was 51 weeks and the mean bromide ion level in the celery was $104 \, mg/kg$.

9. RESIDUES IN RETAIL SAMPLES OF UNKNOWN TREATMENT HISTORY

The crops may not have been grown under protection nor necessarily treated with bromomethane. Other pesticides appropriate to the commodity may have been present.

9.1. Lettuce

9.1.1. Belgium, 1983

110 samples of lettuce ready for export from Belgium were taken during the period January to April and analyzed for residues of pesticides (Galoux and Barnes, 1983). Of these 19.1% (21 samples) contained total bromide residues above the maximum limit of $50 \, mg/kg$ permissible in Belgium. Four samples contained more than $100 \, mg/kg$, namely, 871, 387, 278, and $234 \, mg/kg$ total bromide.

9.1.2. Denmark, 1978–1985

190 samples of lettuce, both home-produced and imported, were analyzed for bromine residues in this period (Andersen, 1981; Orbaek, 1985, 1987). Only a few contained residues above $50 \, mg/kg$ total bromine. More detail is given in Table 8. No determinations were made in 1980–1981.

9.1.3. Sweden, 1981–1984

Twenty-four samples of lettuce were analyzed for extractable bromide in this period (Andersson, 1986). Five contained residues in excess of $100 \, mg/kg$. Further detail is given in Table 9.

9.1.4. United Kingdom, Imported from The Netherlands, 1981–1983

The United Kingdom had noted that bromide residues in lettuces in The Netherlands were reported to be less than in the United Kingdom, as a result of

TABLE 8 Bromide Residues in Fruits and Vegetables in Retail Monitoring in Denmark, 1978–1985[a]

Commodity	Year	Total of Samples Analyzed	Home-produced		Imported	
			Number of Positive	Range (mg/kg)	Number of Positive	Range (mg/kg)
Lettuce	1978	69	41	2.01–71.6	6	2.23–21.2
	1979	27	11	2.12–29.8	10	3.33–58.4
	1982	15	—		6	8.11–25.4
	1983	20	—		7	6.37–36.0
	1984	22	4	3.00–17.5	10	5.40–40.0
	1985	37	6	3.30–42.0	18	5.90–71.0
Iceberg lettuce	1984	2	—		1	7.9
Chinese cabbage	1982	9	—	—	3	3.05–3.81
	1983	10	—			
	1984	26	2	3.40–4.80	2	5.00–5.10
	1985	23	—	—	5	3.30–4.40
Spinach	1982	1	—		1	3.05
	1984	2	—		1	7.50
	1985	1	—		—	—

[a]Summarized from Andersen (1981) and (Orbaek (1985, 1987).

TABLE 9 Bromide Residues in Fruits and Vegetables in Retail Monitoring in Sweden, 1981–1984[a]

Commodity	Home-produced				Imported			
	Number of Samples analyzed for Br	Number of Samples within Given % Range of MRL			Number of Samples Analyzed for Br	Number of Samples within Given % Range of MRL		
		20–50	51–100	>100		20–50	51–100	>100
Chinese cabbage	—				13	2	2	1
Lettuce	1				23	1	—	5
Tomato	3				18	5	6	2
Chicory leaves	—				1	—	—	—
Cucumber	—				—			
Celery	—				—			

[a]Simplified from Andersson (1986).

246

use of lower bromomethane application rates. So as to examine the pattern of total bromine residues in Netherlands lettuce, samples were specifically taken in three successive years (Smart, 1986). Twenty analyses in 1981 ranged from 1 to 53 mg/kg (mean 22 mg/kg) bromide (fresh weight). Ten analyses in 1982 ranged from 2 to 61 mg/kg (mean 20 mg/kg) bromide (fresh weight). In 1983 samples were obtained between February and July from the London wholesale market and the levels found in individual heads of lettuce ranged from 2 to 61 mg/kg (mean 25 mg/kg from 28 analytical results). The general level of bromide residues average 20–25 mg/kg but, as the lettuce were sampled over different times in each year, the figures should be interpreted with caution.

9.1.5. United Kingdom, 1981–1982

Samples of lettuce from Belgium, Cyprus, France, Israel, Spain, and the United States were also analyzed for bromide residues (Roughan and Roughan, 1984b). Results are summarized in Table 10. No residues exceeded 100 mg/kg.

TABLE 10   Bromide Residues in Normally U.K.-Produced Fruits and Vegetables Imported into U.K. (Excluding Dutch Lettuce), 1981–1982[a]

Crop	Country of Origin	Total Number of Samples	Range of Bromide Residues (mg/kg, Fresh weight)	Mean Bromide (mg/kg, Fresh Weight)
Aubergine	The Netherlands	4	2 23	11
Celery	Israel	4	7–14	10
	Spain	16	2–8	4
	United States	1	4	
Cucumber	Canary Islands	15	0.3–10	3
	The Netherlands	7	0.1–14	7
	Spain	3	0.2–10	4
Green pepper	The Netherlands	7	0.4–5	2
Lettuce	Belgium	1	5	5
	Cyprus	5	1	1
	France	9	0.2–19	4
	Israel	6	1–4	2
	Spain	11	0.4–4	2
	United States	12	0.1–2	1
Onion	The Netherlands, Israel	3	0.4–1	1
Radish	The Netherlands	12	0.1–48	13
	Israel	2	5	5
	United States	1	1	1
Tomato	Canary Islands	8	1–5	4
	The Netherlands	14	1–39	11
	Spain	15	1–7	3

[a]Abstracted from Roughan and Roughan (1984b).

9.1.6.   United Kingdom, Home-Produced, 1977–1978

Samples of lettuce grown in the United Kingdom were collected from retail outlets within four areas of England from July 1977 to June 1979 (Roughan and Roughan 1984a). A summary of analytical results is given in Table 11. Over 20% exceeded 100 mg/kg bromide.

9.1.7.   United Kingdom, Home-Produced, 1981–1982

As part of a survey of total bromine residues in fruits and vegetables, samples of home-produced lettuce were purchased from shops locally during the period July 1981 to April 1982 (Roughan and Roughan, 1984a). A summary of analytical results is given in Table 11. A smaller percentage of samples contained residues above the recommended limit than in 1977–1978.

9.2.   Tomato

9.2.1.   Sweden, 1981–1984

21 samples of tomatoes were analyzed for extractable bromide in the period (Andersson, 1986). Two contained residues in excess of 100 mg/kg and six residues in excess of 100 mg/kg, fresh weight. Further detail is given in Table 9.

9.2.2.   United Kingdom, Home-Produced, 1979

102 samples of English tomatoes were obtained from retail outlets from March to November 1979 (Roughan and Roughan, 1984b). Analytical results are summarized in Table 12. A number of residues were above 75 mg/kg bromide.

9.2.3.   United Kingdom, Home-Produced, 1981–1982

As part of a survey of bromide residues in fruits and vegetables, samples of English tomatoes were analyzed (Roughan and Roughan, 1984b). Results are summarized in Table 12. No samples contained residues in excess of 75 mg/kg.

9.2.4.   United Kingdom, Imported, 1981–1982

Samples of imported tomatoes were also analyzed for bromide content during 1981–1982 (Roughan and Roughan, 1984b). Results are summarized in Table 10. None exceeded 75 mg/kg bromide.

9.3.   Cucumber

9.3.1.   United Kingdom, Home-Produced, 1979

Samples of English cucumbers were obtained from retail outlets from March to November (Roughan and Roughan, 1984b). Results of analyses of the 107 samples are summarized in Table 12. Two samples contained residues in excess of 50 mg/kg.

9.3.2.   United Kingdom, Home-Produced, 1981–1982

As part of a survey of bromide residues in fruits and vegetables, samples of English cucumbers were analyzed. Results are summarized in Table 12 (Roughan

TABLE 11 Bromide Residues in Retail Samples of U.K.-Produced Lettuce 1978–1982[a]

Date of Survey	Total Number of Samples Analyzed	Range of Bromide Residues (mg/kg, Fresh Weight)												
		0–50		51–100		101–200		201–500		501–1000		1001–2000		2001–3000	
		No.	%	No.	%	No.	%	No.	%	No.	%	No.	%	No.	%
July 1978–Sept. 1979	384	274	71	30	8	25	7	37	10	14	4	3	1	1	0.3
July 1981–	69	65	94	1	1	1	1	2	3	0	0	0	0	0	0

[a] Abstracted from Roughan and Roughan (1984a).

TABLE 12 Bromide Residues in English Cucumber and Tomato from Retail Outlets, 1979–1982[a]

Year	Commodity	Total Number of samples	Range of Bromide Residues (mg/kg, Fresh Weight)				
			0–5	6–10	11–30	31–100	101–200
March–November 1979	Cucumber	107	102	1	2	0	2
	Tomato	102	63	3	16	15	5
June 1981–July 1982	Cucumber	36	26	1	7	2	0
	Tomato	33	19	3	6	5	0

[a] Abstracted from Roughan and Roughan (1984b).

249

and Roughan, 1984b). Two of the 36 samples contained residues in excess of 50 mg/kg.

9.3.3. United Kingdom, Imported, 1981–1982

Twenty-five samples of imported cucumbers were also taken for analysis during 1981–1982 (Roughan and Roughan, 1984b). Results are summarized in Table 10. None exceeded 50 mg/kg.

9.4. Other Crops

9.4.1. Denmark, 1978–1985

Chinese cabbage and spinach were sampled and analyzed for total bromine (Andersen, 1981; Orbaek, 1985, 1987). Results are summarized in Table 8. All residues of bromine were low.

9.4.2. United Kingdom, Home-Produced, 1981–1982

As a part of the survey of total bromine residues in fruits and vegetables in the United Kingdom, 13 samples of celery were obtained locally and analyzed (Roughan and Roughan, 1984b). Results ranged from 1 to 178 mg/kg bromide (fresh weight), having a mean of 27 mg/kg.

9.4.3. United Kingdom, Imported, 1981–1982

Results of analyses of other relevant commodities of imported produce in 1981–1982 are also found in Table 10 (Roughan and Roughan, 1984b).

9.5. Overall View

Pyysalo (1983) surveyed reports of pesticide residues in some foodstuffs in international trade and suggested that high residues of bromine "often" occurred in lettuce, celery, and tomato, among protected and field crops (at the end of 1981). Residues were only occasionally to be found in cucumber and strawberry.

10. BROMINE IN TOTAL DIET STUDIES

10.1. The Netherlands

de Vos et al. (1984) examined the pesticides and other chemical residues in Netherlands total diet samples for the period June 1976 to July 1978. For the relevant commodity groups, bromine (from all sources) was: vegetables and garden fruits, mean 2.1 mg/kg, range 0.9–5.8 mg/kg; root vegetables, mean 2.0 mg/kg, range 0.36–3.2 mg/kg; fruits, mean 0.82 mg/kg, range 0.07–4.9 mg/kg. The mean daily intake of bromine was 9.4 mg.

Two other concurrent studies to estimate the daily total intake of bromine by the consumer in the Netherlands were reported by Greve (1983). The first study, in May–June 1976, gave a mean bromine content in the total diet of 3.6 ± 1.4 mg/kg and a mean bromine of 7.8 ± 2.8 mg per person per day. The second study in January–March, 1978 gave a mean bromine content in the total diet of 3.2 ± 1.2 mg/kg and a mean bromine intake of 7.6 ± 3.0 mg per person per day.

Coosemans and Van Assche have also analyzed two typical diets in the Netherlands and found that the average bromide content per day was about 8.5 mg (private communication).

10.2. Switzerland

Wüthrich et al. (1985) have estimated the average daily intake of inorganic bromide, among other agrochemical residues, to the Swiss consumer for the years 1982 and 1983. The mean daily intake of bromide in the total diet was 2.98 mg. The mean daily intake in the fruit, vegetables, and salad group was 2.24 mg.

10.3. United Kingdom

Cross et al. (1978) have reported a mean daily intake of 9 mg of total bromine for Glasgow, Scotland.

Five sets of total diet samples obtained (in a wider context) in 1978 to 1979 were analyzed for total bromine. Results for bromine in the relevant food groups are: fruit and sugars, mean 2 mg/kg, range 1–5 mg/kg; root vegetables, mean 3 mg/kg, range 2–4 mg/kg; other vegetables, mean 19 mg/kg, range 6–40 mg/kg (Working Party on Pesticide Residues, 1986). The mean intake in the total diet per person per day was less than 8.4 mg.

Six sets of total diet samples obtained between April and December 1982 were analyzed for total bromine. Results for bromine in relevant food groups were: green vegetables, mean 16 mg/kg, range 3–73 mg/kg; potato, mean 4 mg/kg, range 4–5 mg/kg; other vegetables, mean 5 mg/kg, range 4–7 mg/kg; canned vegetables, mean 3 mg/kg, range 2–4 mg/kg; fresh fruit, mean 2 mg/kg, range 1–2 mg/kg.

One sample in the green vegetable group contained 73 mg/kg; mean bromine concentration in the other 5 samples was 4 mg/kg (range 3–6 mg/kg) (Working Party on Pesticide Residues, 1986). The mean intake in the total diet per person per day was less than 7.8 mg.

10.4. United States

Duggan and Corneliussen (1972) reported a mean total bromine intake of 16.3 mg/day in a study in the United States in 1970.

11.   CONCLUSIONS

Recent monitoring of samples of individual crops having a likelihood of being grown on bromomethane-fumigated soil show only a very few percent to have residues above CAC internationally recommended limits (MRLs). The high figures are mainly for lettuce. It is clear that the proportion of samples having residues above MRLs has declined since the late 1970s, when in some countries residues in lettuce could be as high as 500–1000 mg/kg. Leaching of treated soils, attention to timing of application and integrated pest control have all helped to reduce such residue levels.

The proportion of samples having residues in excess of MRLs is understandably greater when taken from trials, or from holdings where bromomethane fumigations were known to have been made, than during retail monitoring. Not all growers use bromomethane to control pests and it is likely that, in the U.K. for example, less than half have ever done so (and the proportion is declining). Consequently, most of the lettuce, tomato, cucumber, celery, and other relevant commodities on the retail market may not have been grown on soil having had (recent) treatments of bromomethane.

The MRLs for bromide ion in protected crops were recommended by the JMPR in 1983 and endorsed by CCPR only in 1984, so that these food standards, which have not as yet been finally accepted by participating governments, are subsequent to much of the work reviewed. It is important to realize that MRLs for bromide ion (often determined as total bromine) are not health standards; they do not imply that residues above the limit are deleterious to health. CAC limits emphasize the need for good agricultural practice and indicate a level that is necessary for pest control while well within toxicological requirements. CAC limits take account of acceptable daily intakes (Section 4.1) arrived at on the basis of a day-to-day intake in the foodstuff for a lifetime. It is clear that toxicological evaluation of the intake of crops discussed in this chapter will be less stringent as the crops are not eaten in such a way, except by an extremely small percentage of the population on a specialized diet. Also, over any year the sources of lettuce, for example, will vary widely. Although in many countries a few commodity samples are identified each year as containing residues of bromide at above CAC limits, levels in crops are diminishing overall.

General sampling considerations indicate that, as procedures do not give a precise mean value of the residue in a consignment, a very few consignments having mean residues just below the limit will give analytical results in excess of the limit. Over a large number of consignments this is acceptable.

Total diet studies in a number of countries show that the mean intake of bromine by various groups of people is not a problem, as evaluated on current toxicological information. The contribution of total bromine through such commodities as protected crops and strawberries to the total diet is modest. Bromine-containing compounds are otherwise deliberately added to a number of food items during processing for the retail market. It is evident that use of bromomethane to fumigate soils in growing such crops does not lead to significant food contamination.

REFERENCES

Afifi, F. A., El-Rafai, S., and El-Ballal, A. S. I. (1982). "Comparative amino acids alternations in commercial rice grains under effect of fumigants and radiation treatments." *Egypt J. Food. Sci.* **10**, 81–94.

Agricultural Science Service (1981a). *Pesticide Science, 1979.* Ministry of Agriculture, Fisheries and Food, Reference Book 252(79) HMSO, London, p. 46.

Agricultural Science Service (1981b). *Pesticide Science, 1980.* Ministry of Agriculture, Fisheries and Food, Reference Book 252(80) HMSO, London, pp. 45–46.

Alexeeff, G. V., and Kilgore, W. W. (1983). "Methyl bromide." *Residue Rev.* **88**, 101–153.

Andersen, K. S. (ed.) (1981). *"Pesticide Residues in Danish Food, 1978–1979."* Statens Levnedsmiddelinstitut. Publication No. 54, Søborg.

Andersson, A. (1986). "Monitoring and biased sampling of pesticide residues in fruits and vegetables. Methods and results, 1981–1984." *Vår Föda,* Suppl. 1/86, 8–55.

Asante-Poku, S., Aston, W. P., and Schmidt, D. E. (1974). "Site of decomposition of methyl bromide in cocoa beans." *J. Sci. Fd. Agr.* **25**, 285–291.

Beernaert, H., and Vandezande, A. (1986). "Gas-chromatographic determination of inorganic bromide in glasshouse-cultivated vegetables. A collaborative study." *Meded. Fac. Landbouwwet. Rijksuniv. Gent.* **51**(21A), 191–197.

Commission of the European Communitics Scientific Committee on Pesticides (1984). "The use of methyl bromide as a fumigant of plant growing media." In *Reports of the Scientific Committee for Pesticides (Second Series),* Luxembourg, 1985 (EUR 10211 EN), pp. 15–32.

Cross, J. D., Dale, I. M., Smith, H., and Smith, L. B. (1978). "Dietary bromine in the Glasgow area." *Radiochem. Radioanal. Lett.* **35**(6), 291–300.

de Vos. R. H., van Dokkum, W., Olthof, P. D. A., Quirijins, J. K., Muys, T., and van der Poll, J. M. (1984). "Pesticides and other chemical residues in Dutch Total Diet Samples (June 1976–July 1978)." *Fd. Chem. Toxic.* **22**(1), 11–21.

Duggan, R. E., and Corneliussen, P. E. (1972). "Dietary intake of pesticide chemicals in the United States (III) June 1968–April 1970." *Pestic. Monit. J.* **5**, 331–341.

Dunkelberg, H. (1980). "On the problems in respect to the formation of N-nitroso compound precursors when using alkylating agents in the fumigation of foodstuffs. II. N-Methylation of various amino acids by the action of methyl bromide." *Zbl. Bakt. Hyg., I Abt. Orig. B.* **171**, 48–54.

European Economic Community (1982). "Council Directive of 19 July 1982 (82/528/EEC)." *Off. J. Eur. Comm. No. L* 234, 9.8.82, p. 1.

Food and Agriculture Organization (1967). *1966 Evaluations of Some Pesticide Residues in Food. The Monographs.* Joint Meeting of the FAO Working Party and the WHO Expert Committee on Pesticide Residues," Geneva, 14–21 December, 1966. Rome, PL CP. 15, pp. 112–125.

Food and Agriculture Organization (1985). *Pesticide Residues in Food: 1983 Evaluations. The Monographs.* Joint Meeting of the FAO Panel of Experts on Pesticide Residues in Food and the Environment and the WHO Expert Committee on Pesticide Residues, Geneva, 5–14 December, Rome. FAO Plant Production and Protection Paper 61, pp. 59–72.

Food and Agriculture Organization (1987). *Guide to the Codex Recommendations concerning Pesticide Residues.* Part 2, Maximum Limits for Pesticide Residues. Rome (CAC/PR. 2–1986).

Galoux, M. and Bernes, A. (1983). "Residues of plant-protection agents on winter lettuce. Results of a survey conducted in 1983." *Note Tech. Cent. Rech. Agron. Etat, Gembloux,* **9/37**, 20 pp.

Greve, P. A. (1983). "Bromide-ion residues in food and foodstuffs." *Fd. Chem. Toxic.* **21**, 357–367.

Greve, P. A., and Grevenstuk, W. B. F. (1976). "Optimization studies on the determination of bromide residues in lettuce." *Meded. Rijksfac. Landbouww. Gent.* **41**, 1371–1381.

Greve, P. A., and Grevenstuk, W. B. F. (1979). "Gas-liquid chromatographic determination of bromide-ion in lettuce: Interlaboratory studies." *J. Assoc. Off. Anal. Chem.* **62**, 1155–1159.

Helweg, A., and Rasmussen, A. N. (1982). "Influence of soil fumigation with methyl bromide in bromide content in soil and in lettuce grown in the soil." *Tidsskrift for Planteavl. (Copenhagen)* **86**, 461–469.

Kempton, R. J. (1978). "Bromide residues in glasshouse soils and crops." *Rep. Glasshouse Crops Res. Inst.* **1978**, pp. 79–80.

Kempton, R. J. (1979). "Bromide residues in glasshouse soils and lettuce." *Rep. Glasshouse Crops Res. Inst.* 1979, 80–81.

Kempton, R. J., and Maw, G. A. (1972). "Soil fumigation with methyl bromide: Bromide accumulation by lettuce plants." *Ann. Appl. Biol.* **72**, 71–79.

Le Goupil, M. (1932). "Les proprietes insecitide du bromure de methyle." *Rev. Pathol. Veg. Entomol. Agr. Fr.* **19**, 169–172.

Longland, S., Chapman, P. J., and Cole, D. B. (1988). "Edible Glasshouse Crops, 1985." Reference Book 562. Ministry of Agriculture, Fisheries and Food, London.

Nangniot, P., Agneessens, R., Zenon-Roland, L., and Berlemont-Frennet, M. (1984). "Comparative study of the determination of bromide in vegetables by gas chromatography and ionemetry." *Analusis* **12**(4), 197–200.

Nazer, I. K., Hallak, A. B., Abu-Gharbieh, W. I., and Saleh, N. S. (1982). "Bromine residues in the soil and fruits of certain crops after fumigation with methyl bromide." *J. Radioanal. Chem.* **74**, 113–116.

Orbaek, K. ed. (1985). *Pesticide Residues in Danish Food, 1982–1983.* Statens Levnedsmiddelinstitut. Publication No. 107, Søborg.

Orbaek, K. ed. (1987). *Pesticide Residues in Danish Food, 1984–1985.* Statens Levnedsmiddelstyrelsen. Publication No. 147, Søborg.

Pendias, A. K., and Pendias, H. (1984). *Trace Elements in Soils and Plants.* CRC Press, Boca Ratan, FL, pp. 217–218.

Poulsen, E. (1983). "International symposium on residues and toxicity of bromide: Summary and conclusions." *Fd. Chem. Toxic.* **21**, 421–422.

Price, N. R., Bunyan, P. J., and Loveder, C. J. (1981). "The effect of methyl bromide on esterase activity in wheat." *J. Sci. Fd. Agr.* **32**, 17–20.

Pyysalo, H. (1983). "General approaches to the identification of pesticide residues in samples of unknown origin." In R. Greenhalgh and N. Drescher, eds., *Pesticide Chemistry: Human Welfare and the Environment*, Vol. 4. Pesticide Residues and Formulation Chemistry. Pergamon, Oxford, pp. 123–128.

Roughan, J. A. (1981). "A review of the use of methyl bromide in horticulture." In *Pesticide Science 1980.* Ministry of Agriculture, Fisheries and Food Reference Book **252**(80). HMSO, London, pp. 46–53.

Roughan, J. A., and Roughan, P. A. (1984a). "Pesticide residues in foodstuffs in England and Wales. Part I: Inorganic bromide ion in lettuce grown in soil fumigated with bromomethane." *Pestic. Sci.* **15**, 431–438.

Roughan, J. A., and Roughan, P. A. (1984b). "Pesticide residues in food stuffs in England and Wales. Part II: Inorganic bromide ion in cucumber, tomato and self-blanching celery grown in soil fumigated with bromomethane, and the 'natural' bromide ion content in a range of fresh fruit and vegetables." *Pestic. Sci.* **15**, 630–636.

Roughan, J. A., Roughan, P. A., and Wilkins, J. P. G. (1983). "Modified gas-liquid chromatographic method for determining bromide/total bromine in foodstuffs and soils." *Analyst* **108**, 742–747.

Shiga, N., Shimamura, Y., Matano, O., and Gota, S. (1986). "Analytical method for total bromine in crops by high-performance liquid chromatography." *Nippon Noyaku Gakkaishi* **11**, 585–589.

Sly, J. M. A. (1981). *Review of Usage of Pesticides in Agriculture, Horticulture and Forestry in England and Wales 1975–1979.* Reference Book 523. Ministry of Agriculture, Fisheries and Food, London.

Sly, J. M. A. (1986). *Review of Usage of Pesticides in Agriculture, Horticulture and Animal Husbandry in England and Wales, 1980–1983.* Reference Book 541. Ministry of Agriculture, Fisheries and Food, London.

Smart, N. A. (1986). In *Agricultural Science Service 1984: Research and Development Report.* Ministry of Agriculture, Fisheries and Food. Reference Book 257 (84). HMSO, London, p. 113.

Stijve, T. (1981). "Gas chromatographic determination of inorganic bromide residues—a simplified procedure." *Dtsch. Lebensm. Rundsch.* **77**, 99–101.

Stijve, T. (1985). "Inorganic bromide—a simple method for the confirmation of residue identity." *Dtsch. Lebensm. Rundsch.* **81**, 321–324.

Swamy, P. M., and Reddy, S. B. (1974). "Effect of methyl bromide fumigation on changes in nitrogenous constituents of groundnut seedlings." *Curr. Sci.* **43**, 595–597.

Van Wees, A. M. P., Rijk, M. A. H., Rijnaars, M. W., and de Vos, R. H. (1984). "Chromatographic methods for the determination of inorganic bromide in vegetables." *Anal. Chem. Symp. Ser. 21* (*Chromatogr. Mass. Spectrom. Nutr. Sci. Food Saf.*), 19–25.

Wegman, R. C. C., Hamaker, P., De Heer, H. (1983). "Bromide-ion balance of a polder district with large-scale use of methyl bromide for soil fumigation." *Fd. Chem. Toxic.* **21**, 361–367.

Working Party on Pesticide Residues. (1986). *Report of the Working Party on Pesticide Residues (1982–1985).* Food Surveillance Paper No. 16, Ministry of Agriculture, Fisheries and Food. HMSO, London.

Wüthrich, F., Müller, F., Blaser, O., and Marek, B. (1985). "Pesticides and other chemical residues in Swiss diet samples." *Mitt. Geb. Lebensm. Hyg.* **76**, 260–276.

9

PESTICIDE CONTAMINATION OF FOOD IN THE UNITED STATES

Arun P. Kulkarni and Mitra Ashoke

Florida Toxicology Research Center, Department of Environmental and Occupational Health, College of Public Health, University of South Florida, Tampa, Florida

1. INTRODUCTION

"What's for Dinner?" is a common, everyday question heard in every household all over the world. Today health-conscious person planning a nutritious meal for the family worries about how to cut down on red meat, increase fiber, and plan a diet that is low in sodium with no cholesterol. Considering this, a menu of fish, baked potatoes, home baked bread, and fresh, crisp garden salad for dinner and strawberries or an apple for desert sounds just right. What could be wrong with such an apparently wholesome, healthy dinner? Perhaps a lot. It is possible that each item is contaminated with residues of pesticides that may pose serious future health problems.

These days the concept of good food has changed considerably. Good food is not only expected to look fresh and taste good but it must be good for you. The presence of pesticide residues in food products is a clear violation of this simple, reasonable expectation. A recent poll conducted by the Food Marketing Institute indicates that 75% of today's consumers consider pesticides a "serious hazard" in supermarket food. In the February 1987 report entitled "Unfinished Business," the U.S. Environmental Protection Agency (EPA) ranked pesticides in food as one of the nation's most serious health and environmental problems. The time has come to reexamine this issue and to seek answers to questions such as what pesticides are found in food, what are the potential health hazards associated with the pesticide contamination of food, what level of pesticide residue is safe, what federal regulatory agencies are doing or ought to be doing about it, and what the consumer can do. In this chapter we have attempted to address these questions.

2.   PESTICIDE PRODUCTION AND USE

2.1.   Definition of Pesticide

In a strict sense, pesticides are chemical agents that are expected to selectively kill target organisms or pests and be nonlethal to nontarget organisms including humans. Most of the pesticides used today do not exhibit such a high degree of selectivity and, by this definition, fail to qualify as pesticides.

Under the Federal Environmental Pesticide Control Act, pesticides are defined as including

(i) any substance or mixture of substances intended for preventing, destroying, repelling, or mitigating any pest (insect, rodent, nematode, fungus, weed, other forms of terrestrial or aquatic plant or animal life, or viruses, bacteria, or other micro-organisms except viruses, bacteria, or other micro-organisms on or in living human or other animals, which the Administrator declares to be a pest) and (ii) any substance or mixture of substances for use as plant regulator, defoliant or desiccant.

2.2.   Pesticide Production

In assessing the extent to which pesticides may pervade the human food supply, the first consideration, of course, is to determine how much of these chemicals are produced. Pesticide manufacture is a multibillion dollar industry in United States. The production volume for various pesticides peaked in 1975. Following a decline in 1976, it has been hovering around the 1.5 billion pounds per year mark. At present there are more than 1800 basic chemicals that are employed as active ingredients of pesticides dispersed in approximately 33,600 formulations. Of the total number of pesticide formulations registered with U.S. Environmental Protection Agency (EPA), about 49% are insecticides, 23% are industrial or household pesticides, 15% are herbicides, 9% are fungicides, and 3% are rodenticides.

2.3.   Pesticide Usage

Little over half of the domestic consumption of pesticides in the United States goes for crop production with the remaining being used by government, industry, and homeowners. The land area treated by pesticides represents about 10–15% of the total in 50 states. In 1984, farmers spent about $5 billion on pesticides. This accounted for only 4% of all the crop production costs incurred by the farmers. Available statistics for the late 1970s indicate a leveling off in domestic pesticide consumption between 0.8 and 1.0 billion pounds. This roughly translates into a potential exposure of every American to residues from approximately 4–5 lb of pesticides per year. To some extent, the actual level of exposure to pesticide residues varies (1) geographically, being higher in agricultural states such as California and Florida, as well as (2) during different months within a year, being highest in summer, the crop growing season.

3. NEED FOR PESTICIDES

Agriculture is a basic industry that supplies food, one of the most fundamental necessities of life. People all around the world depend on this industry for substenance, but in many underdeveloped countries productivity is too low to maintain an adequate food supply. According to a 1967 estimate, the total crop loss in the world caused by various pests was in the order of $75,000 million annually or about about 35% from the potential production, enough to feed approximately 700 million people for 1 year without additional land or cultivation.

Whereas house flies and other bugs can be tolerated, insects and arthrapods that cause unnecessary suffering and loss of human lives through the spread of diseases such as malaria and typhoid are certainly not acceptable. During the past 30 years pesticides such as DDT have prevented millions of deaths from malaria alone. Considering the scientific and technological maturity of that era, these actions appear, even today, both appropriate and justified. The past dramatic success in crop protection combined with a lack of effective alternatives means that our current use of pesticides to protect farm animals such as cattle, pigs, and poultry as well as household pets, such as cats and dogs, from diseases and parasitic infections seems inevitable. It is obvious that there will be a need for pesticides until such time as we discover how to control pests without the use of chemicals.

4. REGULATION OF PESTICIDES

According to Reed et al. (1987) the use of pesticides (which includes insecticides, herbicides, fungicides, fumigants, and other chemical agents) has increased considerably in the past 30 years in the United States as well as around the world and this increase is expected to continue in the immediate future. Consequently, the question arises as to the safety of food for human or animal consumption. In other words, are the residues remaining in or on the food materials for consumption low enough so as not to cause any acute or chronic pathophysiological conditions in humans and animals. To resolve such an apparently simple question is not an easy task, as in reality a multifaceted complex issue involved. In recent years, this issue has turned into a battleground for biomedical scientists, economists, politicians, and the general public.

Americans first became aware of the possible hazards to the ecosystem and human health from widespread pesticide use with the publication in 1962 of *Silent Spring* by Rachel Carson. In this book, which became a national best seller, Carson argued that many pesticides in use on farms and timberlands had unknown and cumulative toxic effects that could be gauged only after many years of tests. Because so little was known about the effects of pesticides on plants, animals, and humans, Carson said their use should be curtailed. She also criticized the U.S. Department of Agriculture's endorsement of increased pesticide use and contended that farmers often exceeded permitted tolerances without detection. After *Silent Spring* raised the issue, public interest, awareness, and concern have increased regarding the human health safety of pesticide residues in food.

4.1. Delaney Amendment

Today, much of the debate over food safety has centered on the 1958 Delaney Clause, which prohibits any food additive that induces cancer in humans or animals. This law requires that a newly proposed substance must undergo rigorous testing designed to establish the safety of its intended use before it is marketed.

The EPA is entrusted with the responsibility for registration of pesticides for use in the United States under the statutes of the Federal Insecticide, Fungicide and Rodenticide Act (FIFRA). The EPA, governed by the Federal Food, Drug and Cosmetic ACT (FDCA) is responsible for setting "acceptable" residue limits of different pesticides that are used on human and animal food. The Food and Drug Administration (FDA) is responsible for determining and enforcing these limits; for residues in meat and poultry the responsibility is with the U.S. Department of Agriculture (USDA).

The FDA requires information on the chemical composition of the substance, its manufacturing process, and the analytical methods used to detect and measure its presence in the food product at the levels of expected use. The analytical method must be sensitive enough to determine compliance with the regulations. The data must establish how the intended purpose will be achieved. Finally, data must be provided establishing the safety of the product after its intended use. This usually requires extensive testing of the product. Although the aim of the Delaney amendment is excellent, serious problems exist in the evaluation of data, the protocols for carcinogenicity testing, data interpretation, and the extrapolation of animal data to humans. Regulatory agencies such as the FDA and the EPA have embarked on the task of ascertaining and maintaining the quality and safety of our food. The FDA has developed a well-organized pesticides monitoring program to determine the "acceptable" limits of pesticide residues in human and animal food. It should be mentioned that the FDA recognizes that such a monitoring system cannot ensure the public that foods will be free of all pesticide residues.

The procedure of determining "acceptable" limits of a pesticide on a particular commodity involves an appreciable amount of systematic screening, which is strictly followed by the EPA and FDA. It is not possible within the scope of this chapter to describe the details of the process. For more details of the procass, the reader is referred to the excellent review by Reed et al. (1987). A brief outline of the process, however, is included here.

1. Before an "acceptable" limit is established by the EPA, the applicant for a particular pesticide registration must submit his application with adequate

residue data from trial field applications of the compound. This is to be supplemented by the results from a variety of toxicological tests that will help EPA reviewers to make predictable safety judgments.

2. The review of such information by the EPA leads to the determination of an acceptable daily intake (ADI) that is, the level at which a chemical, consumed daily during an entire lifetime, does not appear to pose appreciable risk. The "acceptable" residue of a pesticide on a specific food item is determined by comparing the ADI for that particular compound with the possible daily dietary intake contributed by all other food items. Since the ADI is based on the no-effect level in animals, a safety factor (10–10,000) is incorporated in the formula to arrive at the final ADI value for humans.

5.   PESTICIDE RESIDUES IN FOOD

5.1.   History of Pesticide Monitoring

Prior to the existence of the EPA, the FDA was the sole authority setting tolerance values and monitoring and enforcing residue limits of pesticides on food. In the late 1930s and early 1940s the FDA monitored only residues of pesticides containing copper, mercury, lead, and/or arsenic. The major pesticides used at that time were compounds containing lead arsenate and the FDA concentrated its monitoring activities on this compound (Jelinek, 1985). In the last 30 years and with the establishment of the EPA in 1970, there has been a tremendous upsurge in the production and application of different pesticides on crops for pest control. The primary intention is, of course, to produce healthy, nutritious food materials for human and animal consumption that will be available all year. With increased awareness of toxicological properties, farmers and agricultural workers are now more cautious in the handling and use of various pesticidal chemicals. The use of agricultural chemicals certainly has helped to improve the quality of food on the table over what it was 20 or 30 years ago. The FDA, in 1961, started a Total Diet Program (also called the Market Basket Study), and the data from such studies, in general, show that the dietary exposure to certain selected pesticide residues are consistently well below their ADI values (Duggan and McFarland, 1967). In the beginning, the program primarily investigated pesticide residues in the adult diet (representing the basic 2-week diet of a 16- to 19-year-old male, statistically confirmed to be America's most voracious eater). From August 1974, infant (6-month-old) and toddler (2-year-old) diets were also included in the program (Johnson et al., 1979). The procedure to determine pesticide residues in adult, infant, and toddler diets is briefly described in the following.

The market baskets are collected from four different geographical regions: northeast, south, west, and northcentral. Foods are prepared as they are consumed normally and are then itemized into 12 composite classes (Table 1). Each food composite (the number of items in each composite for every food class

TABLE 1   Food Composite Classes Analyzed for Pesticides, August 1975– March 1982

Class	Food Composite
1	Dairy products
2	Meat–fish–poultry
3	Grain and cereal products
4	Potatoes
5	Leafy vegetables
6	Legume vegetables
7	Root vegetables
8	Garden fruits
9	Fruits
10	Oils, fats, and shortening
11	Sugar and adjuncts
12	Beverages

may vary from year to year, but usually is not less than 20), one from each market basket, is prepared. Prepared food composites with similar characteristics are analyzed for organochlorine, organophosphorous, carbamate insecticides, and herbicides by analytical procedures that are usually modified from time to time to allow for quantitation at levels that are 5–10 times lower than those of the tolerance values.

The procedure implemented in the Total Diet Study for infants and toddlers is essentially the same as that of the adult Total Diet Study, with minor modifications. For infants and children, drinking water, milk, dairy products, and dairy substitutes other than milk were included; there was no food class for root

TABLE 2   Food Composite Classes for Infants and Toddlers Analyzed for Pesticides, October 1978–March 1982

Class	Food Composite
1	Drinking water
2	Whole milk
3	Other dairy products and dairy substitutes
4	Meat, fish, and poultry
5	Grain and cereal products
6	Potatoes
7	Vegetables
8	Fruit and fruit juices
9	Oils and fats
10	Sugar and adjuncts
11	Beverages

vegetables, leafy vegetables, or legume vegetables. Instead vegetables are categorized as a class in itself as they constitute only 7% of the infant and 5% of the toddler diet (Gartrell et al., 1986a). The food classes for infants and toddlers are shown in Table 2.

5.2.   Common Pesticide Residues in Food

A wide variety of synthetic chemicals possessing pesticidal action have been used in agriculture since the early 1940s. The major families of chemicals employed for this purpose include chlorinated hydrocarbons (or organochlorines), organophosphorous pesticides, carbamates, a newer class known as the pyrethroids, and a few others. The most commonly noted residues in food, in decreasing order, are those of organochlorines, organophosphates, and carbamate pesticides. Other less commonly encountered pesticide residues include those of fungicides, herbicides, and fumigants used for various purposes.

5.2.1.   Organochlorine Pesticides

The organochlorine pesticides represent the first group of synthetic chemicals used successfully as insecticides in agriculture. Besides 2, 2-bis(p-chlorophenyl)-1, 1, 1-trichloroethane (p, p'-DDT), kelthane, methoxychlor, benzene hexachloride, lindane, endosulfan, mirex, and kepone, the other members in this family of pesticides include cyclodienes such as aldrin, dieldrin, chlordane, heptachlor, toxaphene, and endrin. From the early 1940s to the early 1970s the organochlorines were widely used in agriculture. These compounds as a class are regarded as having low acute toxicity, but possess a greater potential for chronic toxicity when compared to the organophosphates or carbamates. They are sometimes referred to as "hard" pesticides because of their long persistence in the environment. In view of their propensity to accumulate in man, animals, and the environment, the use of several organochlorine pesticides is either restricted or banned in the United States as well as in few other countries.

5.2.2.   Organophosphorous Pesticides

The first organophosphate insecticide, tetraethylpyrophosphate (TEPP), was developed in Germany during World War II. However, due to its high mammalian toxicity and instability it was soon replaced by less toxic and more stable compounds. Of the numerous triesters of phosphoric acid synthesized and tested for insecticidal activity, parathion was the first commercial compound from this series introduced in 1944. Since then this insecticide has become very popular with farmers and continues to be used extensively throughout the world. Potential food contaminants due to this group of pesticides include those of methamidophos, dimethioate, parathion, methyl parathion, guthion, chlorpyrifos, malathion, diazinon, ethion, fenitrothion, and a few others.

5.2.3.   Carbamate Pesticides

The insecticidal activity of various esters of carbamic acid was discovered in the 1950s. The carbamates, unlike the organophosphates, have relatively low dermal toxicities, with the exception of aldicarb. The mechanism of toxic action is similar to the organophosphates (i.e., inhibition of cholinesterase), but the important difference is that the enzyme inhibition due to the carbamates is reversible whereas with the organophosphates it is irreversible. The most common insecticidal compounds of this class that are detected as a residue in food are carbaryl, aldicarb, baygon, and zectran.

The carbamates, having a bulky substitution on nitrogen rather than oxygen, exhibit different biocidal properties. They are poor inhibitors of the enzyme cholinesterase and, therefore, lack insecticidal properties. However, they possess herbicidal activity and are used commercially. Examples include propham, asulam, swep, phenmedipham, chlorbufam, and chloropam.

Another group of carbamates that exhibits fungicidal activity includes dithiocarbamates such as ziram, ferbam, maneb, zineb, and nabam, whereas metham-sodium has nematocidal activity. Small amounts of residues of these carbamate pesticides are commonly found in some fresh fruits and vegetables.

5.3. Factors Influencing Pesticide Residues in Food

5.3.1. Factors Influencing Pesticide Residues in Plant Products

Most commonly pesticides are applied directly to the leaves of plants. Some are absorbed by the foliage and translocated to the other parts of the plant. When pesticide is applied to soil, it is absorbed by the roots and then distributed to all parts of the plant. Various factors that determine the levels of pesticide residues left in or on the harvested plant material include the (1) nature of the pesticide, (2) pesticide formulation, (3) dosage, (4) number of applications, (5) time from application to harvest, (6) photodegradation and metabolism by the plant and microbes, (7) weather (8) migration in soil, (9) volatilization, (10) quality of the irrigation water, (11) method of harvest, (12) postharvest application, and (13) methods and conditions of storage. Prior knowledge of these factors, though helpful, does not allow precise estimation of the pesticide residue content of the final product.

5.3.2. Factors Influencing Pesticide Residues in Animal Products

The levels of residues in animal products intended for human consumption are governed by the pesticide residue content of the animal feed and by those factors that affect the absorption, distribution, metabolism, excretion, and storage of pesticides in the animal body. The magnitude of pesticide residues in human tissues is also governed by these factors. In commercial food of animal origin, the occurrence of organochlorine pesticide residues is most common. This originates primarily from the contamination of pasture land by pesticides that drift from adjacent fields. Alternately, absorption of pesticide and their residues by the roots from contaminated soils adulterates hay and fodder.

During the past 10 years, left over pulp and other waste products generated by the orange juice, sugar, and gasohol industries are being used as animal feed. These products have been shown to be potential sources of pesticide residues in beef and pork (Ober et al., 1987a). More recently, the products manufactured from recycling of animal manure are being used to a limited extent as feed ration ingredients for cattle, sheep, pigs, poultry, and other animals. Mowafy et al. (1987) reported the dieldrin content of animal feed, manure, and commercial products C-I and C-II prepared by recycling of manure.

5.3.2.1. Biomagnification. The organochlorine pesticides such as p, p'-DDT and their metabolites are lipophilic compounds and tend to concentrate in membrane lipids and fat depots. This build-up of pesticide residues in the different segments of our environment, and especially in the aquatic life food web, is well documented (Edwards, 1970). Some pesticides move up to the food chain through a process generally referred to in the literature as biomagnification, in which minute amounts of pesticides stored in organisms low in the food chain are transferred and stored at higher concentrations in the tissues of predators higher in the food chain. The magnitude of biomagnification depends on (1) the nature of the pesticide, (2) the amount of pesticide available to the organism by contact or ingestion, (3) the rate of pesticide intake by the organism, and (4) the rate of biotransformation and elimination of the pesticides by the organism in terms of time and quantity.

5.4. Pesticide Residues in Domestic versus Imported Food Products

According to the U.S. General Accounting report of April 1987, annual food consumption in the United States is in excess of 290 billion pounds, with imported food accounting for a significant and increasing portion of the total. In 1985, 43 billion pounds of food was imported. This represents about 25% of all fresh fruits and 5–6% of all vegetable products consumed. Between 1979 and 1985 the FDA analyzed 101,191 food samples for pesticide residues and found that 4028 or 4% of the total of the samples were contaminated. The violation rate of 6.1% noted for imported food was twice that of domestically grown food. Monitoring of pesticide residues in imported food is complicated by inadequate information about the pesticide used on each food item in the exporting country. Based on the data collected during 1983–1985, the violation rate was in general higher for pineapples, peppers, mangoes, cabbage, okra, tangerines, and blackberries; bananas, onions, garlic, plums, pumpkin, papaya, and a few other food commodities did not show any pesticide contamination.

5.5. Pesticide Residues in Processed Food

The effects of various processes employed in the preparation of ready-to-eat food are not fully understood. Some processed foods contain higher levels of residues than allowed by the raw commodity tolerance because residues concentrate during processing. Benomyl, for example, concentrates in processed tomato products. Although dehydration, on the one hand, can increase concentration, increased volatilization during high-temperature processing may decrease the amount of pesticide residue in the food. Also some pesticides are degraded due to enhanced hydrolysis during processing.

Daily consumption of food products such as jams, jellies, syrups, and candy add very little to the body burden of pesticide residues. The maximum amount of residues in a day's ration from this source includes 0.001 mg of DDT, 0.004 mg of herbicides, and 0.005 mg of carbaryl (Duggan and Weatherwax, 1967). Beverages, in general, are not considered to contain significant amounts of pesticide residues.

5.6.   Pesticide Residues in Adult Diet

Johnson et al. (1981) reported that during the period of August 1975 to July 1976, 240 food composites were examined and 1039 residues of 47 different compounds (pesticides and industrial chemicals) were found. These results represented an 8% increase in the number of residues reported for an earlier period in the same number of food composites. Bearing in mind the concern in this article with pesticide residues in food, the frequency of occurrence of commonly encountered pesticide residues within each food class is represented in Table 3. A comparison of 1975–1976 data with that of 1980–1982 is also included. Table 4 compares the average concentrations (expressed in ppm) of the different classes of pesticides in the food composite of 1975–1976 and 1980–1982.

From Tables 3 and 4 it is evident that organochlorine residues are found in two main food classes: dairy products and meat, fish, and poultry. The remaining organochlorine residues were distributed among the other food classes with garden fruits and leafy vegetables containing half of them. The organophosphates were detected mainly in dairy products, and potatoes. Malathion was the most frequently detected in this class of pesticides in grain and cereal products. Residues of the carbamate pesticide, carbaryl, were found on garden fruits at levels ranging from 0.002 to 0.004 ppm.

5.7.   Pesticide Residues in Infant and Toddler Diet

In the infant diet, pesticide residues are found most frequently in the grains and cereals and oils and fats food classes. The most frequently encountered class of compounds is the organochlorines, with p, p'-DDE being present in traces in almost every category of food except drinking water, sugar and food additives, and beverages. Organophosphates are not as commonly detected as the organochlorines. However, malathion residues were detected in 13 samples of grain and cereal products.

In the toddler diet pesticides were found most frequently in oils and fats, meat, fish, and poultry, and fruit and fruit juices. The most common compounds detected were p, p'-DDE, dieldrin, heptachlor (mainly in milk and dairy substitutes), lindane (sugar and adjuncts) of the organochlorine group, and diazinon and malathion (mainly in grain and cereal products and oils and fats) of the organophosphate group.

The frequency of occurrence and levels of commonly encountered pesticide

TABLE 3 Frequency of Occurrence of Common Pesticide Residues by Food Class in Adult Food Composites (240 Composites from 20 U.S. Cities)[a]

Pesticide	Food Class											
	1	2	3	4	5	6	7	8	9	10	11	12

Data for August 1975–July 1976

ORGANOCHLORINES

Pesticide	1	2	3	4	5	6	7	8	9	10	11	12
Chlordane	0	0	1	1	0	0	0	0	0	0	0	0
p,p'-DDE	14	20	0	3	8	3	3	2	0	1	0	0
p,p'-DDT	0	15	0	0	1	0	0	0	0	0	0	0
Dieldrin	15	19	1	4	2	0	1	13	2	3	0	0
HCB	5	11	0	0	0	0	0	0	3	0	0	0
Heptachlor epoxide	13	15	0	2	0	0	0	0	0	0	0	0
Lindane	2	7	1	0	1	0	0	5	0	1	7	0
Octachlor epoxide	5	12	0	0	0	0	0	0	0	0	0	0
Toxaphene	0	0	0	0	1	0	0	0	0	0	0	0

ORGANOPHOSPHATES

Pesticide	1	2	3	4	5	6	7	8	9	10	11	12
Diazinon	0	0	5	0	2	0	0	2	1	0	0	0
Ethion	0	1	0	0	0	0	1	1	3	0	0	0
Malathion	0	0	19	0	0	0	0	0	0	7	3	0
Parathion	0	0	0	0	1	2	1	1	0	0	0	0

CARBAMATES

Pesticide	1	2	3	4	5	6	7	8	9	10	11	12
Carbaryl	0	0	0	0	0	0	0	2	3	0	0	0

FUNGICIDES

Pesticide	1	2	3	4	5	6	7	8	9	10	11	12
Captan	0	0	0	0	0	0	0	0	2	0	0	0

Data for October 1980–March 1982

ORGANOCHLORINES

Pesticide	1	2	3	4	5	6	7	8	9	10	11	12
Chlordane	0	0	0	3	0	0	0	1	0	0	0	0
p,p'-DDE	12	23	0	8	13	1	17	3	1	1	1	0
p,p'-DDT	0	0	0	0	4	0	3	0	0	0	0	0
Dieldrin	16	24	1	8	4	0	6	19	2	5	0	0
HCB	4	16	0	0	0	0	0	0	0	22	4	0
Heptachlor epoxide	13	22	0	3	0	1	1	3	0	1	0	0
Lindane	0	9	0	0	0	0	0	3	0	0	13	0
Octachlor epoxide	3	15	0	1	0	0	0	1	1	0	0	0
Toxaphene	0	1	1	0	2	0	2	1	0	1	0	0

TABLE 3   (*Continued*)

Pesticide	Food Class											
	1	2	3	4	5	6	7	8	9	10	11	12
ORGANOPHOSPHATES												
Diazinon	0	0	16	0	3	0	3	4	3	3	4	0
Ethion	0	0	0	0	0	0	2	3	14	0	0	0
Malathion	0	1	27	0	0	0	1	1	3	17	9	0
Parathion	0	0	0	0	5	2	6	0	4	0	0	0
CARBAMATES												
Carbaryl	0	0	0	0	0	0	0	0	3	0	0	0
FUNGICIDES												
Captan	0	0	0	0	0	0	0	0	2	0	0	0

*a*Data adapted from Johnson et al. (1981) and Gartrell et al. (1986b).

residues in infant and toddler diet samples are represented in Tables 5, 6, 7, and 8. In 1986, the H. J. Heinz Company announced that the plant products treated with any one or more of the 13 pesticides that are currently under EPA review would not be used in the manufacture of baby food (Mott and Snyder, 1987).

5.8. Pesticide Residues in Fruits

Duggan and Weatherwax (1967) reported that fruits contribute about 10% of the total dietary intake of organochlorine pesticide residues. Trace amounts of organophosphates, carbamates, bromides, and arsenic may also be found.

   More recent data on the pesticide contamination of common fruits are given in Table 9. Depending on the fruit crop in question, the number of pesticides allowed vary from 30 for bananas and watermelons to 110 for apples. In contrast, the FDA's routine methods for monitoring pesticide residues on fruits can detect residues of only 50–60% of these pesticides. Monitoring thus far has not revealed, except for one occasion, the presence of pesticide residues in bananas. This is probably because the peel acts as an effective barrier. The residues of 30 to 40 pesticides can be detected in most fruits, the highest number being 43 for apples. The percentage contamination was found to be about 20% for grapefruits, oranges, and pears; over half the samples of cherries, peaches, and strawberries showed pesticide residues. Most commonly observed residues include those of carbaryl and organophasphate insecticides. The fungicide Captan was detected in apples, cherries, grapes, peaches, and strawberries (Mott and Snyder, 1987).

   Aldicarb, a carbamate insecticide, is widely used in fruit (citrus groves) and vegetable (e.g., potatoes) crops. It is readily taken up by roots and distributed into the stem, leaves, and fruits of the plant. This suggests that serious contamination levels, which may lead to serious health problems and even death, are possible from its injudicious use (Anonymous, 1986). Several minor foodborne aldicarb

TABLE 4 Levels of Pesticide Residues by Food Class in Adult Food Composites (240 Composites from 20 U.S. Cities)[a]

Data for August 1975–July 1976

Pesticide	Food Class											
	1	2	3	4	5	6	7	8	9	10	11	12
ORGANOCHLORINES												
Chlordane	0.002											
p,p'-DDE		0.01	T[b]	T	0.003	T	T	T		T		
p,p'-DDT		0.002			T							
Dieldrin	T		T	T	T		T		T	T		
HCB	T	T								T		
Heptachlor epoxide	T	T		T								
Lindane	T	T	T		T			T		T		
Octachlor epoxide	T	T									T	
Toxaphene					T							
ORGANOPHOSPHATES												
Diazinon			T		T		T	T				
Ethion		T					T	T	T			
Malathion			0.02							0.003	T	
Parathion						T	T	T	T			
CARBAMATES												
Carbaryl								0.002	T			
FUNGICIDES												
Captan									0.003			

Data for October 1980–March 1982

	1	2	3	4	5	6	7	8	9	10	11
ORGANOCHLORINES											
Chlordane	0.0015	0.003								T	T
p,p'-DDE	T	T	T	0.0024	T	0.005	T	T	T	T	T
p,p'-DDT	T	T		T		T		T		T	
Dieldrin	T	0.001	T	T		0.002	T	T	0.002	T	
HCB	T	T			T	T	T	0.002	0.002	T	
Heptachlor epoxide	T	T		T	T	T		T	T		
Lindane	T							T		T	T
Octachlor epoxide	T	T		T			T	T			
Toxaphene		T	0.002	0.002		0.006	T	0.002		0.002	
ORGANOPHOSPHATES											
Diazinon		0.001		T		T		T	T	T	T
Ethion				0.002		T		T	T		T
Malathion		T	0.004	0.02		T	T	T	0.001		
Parathion				T		T	T	T	T		
CARBAMATES											
Carbaryl				0.004							
FUNGICIDES											
Captan				0.005							

[a]Modified from Johnson et al. (1981) and Gartrell et al. (1986b). Values are average concentrations expressed in ppms of the frequency of occurrences in Table 2.
[b]T, trace residues or well below detection limits.

TABLE 5   Frequency of Occurrence of Common Pesticide Residues by Food Class in Infant Food Composites (120 Composites from 13 U.S. Cities)[a]

Pesticide	1	2	3	4	5	6	7	8	9	10	11

Data for October 1980–March 1982

ORGANOCHLORINES

Pesticide	1	2	3	4	5	6	7	8	9	10	11
p,p'-DDE	0	7	7	8	0	3	3	0	1	0	0
Dieldrin	0	4	6	5	0	3	1	0	0	0	0
HCB	0	2	0	3	0	0	0	0	3	0	0
Heptachlor epoxide	0	5	6	1	0	1	0	0	0	0	0
Lindane	0	0	1	0	2	0	0	0	0	0	0
Octachlor epoxide	0	1	0	3	0	0	0	0	0	0	0
Toxaphene	0	0	0	0	0	0	0	0	3	0	0

ORGANOPHOSPHATES

Pesticide	1	2	3	4	5	6	7	8	9	10	11
Diazinon	0	0	0	1	6	0	0	0	1	0	0
Ethion	0	0	0	0	0	0	0	2	0	0	0
Malathion	0	0	0	0	13	0	0	0	3	0	0
Parathion	0	0	0	0	0	0	0	2	0	0	0

CARBAMATES

Pesticide	1	2	3	4	5	6	7	8	9	10	11
Carbaryl	0	0	0	0	0	0	0	2	0	0	0

FUNGICIDES

Pesticide	1	2	3	4	5	6	7	8	9	10	11
Captan	0	0	0	0	1	0	0	0	0	0	0

Data for August 1974–July 1975

ORGANOCHLORINES

Pesticide	1	2	3	4	5	6	7	8	9	10	11
p,p'-DDE	0	5	4	4	0	0	2	0	0	0	0
Dieldrin	0	4	4	7	1	2	0	0	2	0	0
HCB	0	2	1	3	0	0	0	0	1	0	0
Heptachlor epoxide	0	2	1	3	0	0	0	0	0	0	0
Lindane	0	0	0	0	2	1	0	0	0	0	0
Octachlor epoxide	0	0	0	1	0	0	0	0	0	0	0
Toxaphene	0	0	0	0	0	0	0	0	3	0	0

ORGANOPHOSPHATES

Pesticide	1	2	3	4	5	6	7	8	9	10	11
Diazinon	0	0	0	0	4	0	0	0	0	0	0
Ethion	0	0	0	0	0	0	0	0	0	0	0
Malathion	0	0	0	0	0	10	0	0	0	0	0
Parathion	0	0	0	0	0	0	1	0	0	0	0

[a]Data adapted from Johnson et al. (1979) and Gartrell et al. (1986a).

TABLE 6 Frequency of Occurrence of Common Pesticide Residues by Food Class in Toddler Food Composites (120 Composites from 13 U.S. Cities)[a]

Pesticide	Food Class										
	1	2	3	4	5	6	7	8	9	10	11

Data for October 1980–March 1982

ORGANOCHLORINES

Chlordane	0	0	0	0	0	0	0	0	1	0	0
p,p'-DDE	0	7	11	11	0	6	1	1	7	1	0
p,p'-DDT	0	0	0	1	0	0	0	0	0	0	0
Dieldrin	0	4	12	9	0	3	5	0	2	0	0
HCB	0	2	8	6	0	0	0	0	12	1	0
Heptachlor epoxide	0	5	12	4	0	1	0	0	4	0	0
Lindane	0	0	0	2	0	0	0	0	0	10	0
Octachlor epoxide	0	1	6	6	0	0	0	0	0	0	0

ORGANOPHOSPHATES

Diazinon	0	0	0	2	7	0	3	2	4	2	0
Ethion	0	0	0	0	0	0	0	5	0	0	0
Malathion	0	0	0	0	13	0	0	2	8	5	1
Parathion	0	0	0	0	0	0	2	1	0	0	0

CARBAMATES

Carbaryl	0	0	0	0	0	0	2	2	0	1	0

FUNGICIDES

Captan	0	0	0	0	0	1	0	1	0	0	0

Data for August 1974–July 1975

ORGANOCHLORINES

p,p'-DDE	0	5	4	4	0	0	2	0	0	0	0
Dieldrin	0	4	4	7	1	2	0	0	2	0	0
Heptachlor epoxide	0	2	1	3	0	0	0	0	0	0	0
Lindane	0	0	0	0	2	1	0	0	0	0	0
Octachlor epoxide	0	0	0	1	0	0	0	0	0	0	0
Toxaphene	0	0	0	0	0	0	0	0	3	0	0

ORGANOPHOSPHATES

Diazinon	0	0	0	0	4	1	1	0	0	1	0
Malathion	0	0	0	0	0	10	0	0	0	0	0
Parathion	0	0	0	0	0	0	1	0	0	0	0

[a]Data adapted from Johnson et al. (1979) and Gartrell et al. (1986a).

TABLE 7 Level of Pesticide Residues by Food Class in Infant Food Composite (120 Composites from 13 U.S. Cities)[a,b]

Data for October 1980–March 1982

Pesticides	Food Class										
	1	2	3	4	5	6	7	8	9	10	11
ORGANOCHLORINES											
p,p'-DDE		0.002	T[c]	0.002					T		
Dieldrin		T	T	T		T	T				
HCB		T	T	T					0.002		
Heptachlor epoxide		T	T	T		T	T				
Lindane			T		T		T				
Octachlor epoxide		T	T	T							
Toxaphene									0.03		
ORGANOPHOSPHATES											
Diazinon				T	0.005				T		
Ethion								T			
Malathion					0.04				0.004		
Parathion								T			
CARBAMATES											
Carbaryl								0.008			
FUNGICIDES											
Captan						T					

Data for August 1974–July 1975

ORGANOCHLORINES							
p,p'-DDE			0.002			0.004	0.001
Dieldrin		0.002		0.001		0.001	T
Heptachlor epoxide						T	T
Lindane				T	0.002		
Octachlor epoxide						T	
Toxaphene		0.35					
ORGANOPHOSPHATES							
Diazinon	T		T	T	T	T	
Malathion				0.017			
Parathion			T	T			

[a]Levels are expressed as the mean values in ppm.
[b]Data adapted from Johnson et al. (1979) and Gartrell et al. (1986a).
[c]T, trace residues that were well below the detection limits for that compound.

275

TABLE 8 Level of Pesticide Residues by Food Class in Toddler Food Composites (120 Food Composites from 13 U.S. Cities)[a,b]

Data for October 1980–March 1982

Pesticide	Food Class										
	1	2	3	4	5	6	7	8	9	10	11
ORGANOCHLORINES											
Chlordane								T[c]			
p,p'-DDE	0.001	0.003	0.002	0.002	T	T	T	0.001	T		
p,p'-DDT		T	T								
Dieldrin	T	0.001	T		T	T		T			
HCB	T	T	T		T			0.002	T		
Heptachlor epoxide	T	T	T		T			T			
Lindane	T	T	T						0.002		
Octachlor epoxide	T	T	T								
ORGANOPHOSPHATES											
Diazinon		T		0.003		T	T	T	T		
Ethion							T				
Malathion				0.02			T	0.03	T	T	
Parathion						T	T				
CARBAMATES											
Carbaryl						0.008	0.008		0.004		
FUNGICIDES											
Captan					T		T				

Data for August 1974–July 1975

ORGANOCHLORINES							
p,p′-DDE	0.001	0.004	0.012		T		T
p,p′-DDT		T	0.001				
Dieldrin	T	0.004	0.002	0.001	T	T	
Heptachlor epoxide	T	T	T		T		
Lindane		T	T	T			0.002
Octachlor epoxide			T		T		
Toxaphene						0.061	
ORGANOPHOSPHATES							
Diazinon	T	T	T	T	T		T
Malathion				0.009		0.048	0.003
CARBAMATES							
Carbaryl	T				T		

[a]Values are expressed as mean residues in ppm.
[b]Data adapted from Johnson et al. (1979) and Gartrell et al. (1986a).
[c]T, trace residues that are well below detection levels for that compound.

TABLE 9   Pesticide Contamination of Fruits[a]

| | | Pesticides | | | |
| | | Regulation | | Contamination | |
Fruit	Consumption (lb/yr/person)	Permitted Number	Detection (%)	Observed Number	Contamination (%)
Apples	22	110	50	43	33
Bananas	11	30	50	0	0
Cherries	—	15	60	25	50
Grapefruit	—	80	60	15	20
Grapes	—	80	60	30	33
Oranges	7	90	50	30	20
Peaches	10	100	55	36	50
Pears	6	100	50	27	20
Strawberries	—	70	50	39	60
Watermelons	—	30	50	7	4

[a]Modified from Mott and Snyder (1987)

poisoning incidences involving a few people have been reported in the literature (Marshall, 1985; Goes et al., 1980). Green and co-workers (1987) reported 264 cases with 61 definite cases of aldicarb poisoning from Oregon and northern California. The largest recorded outbreak of illness caused by pesticide contamination in food in the United States was due to aldicarb-contaminated watermelons (Green et al., 1987). It is important to note that aldicarb is not registered for use on watermelons. It is also classified as a highly toxic carbamate pesticide [acute oral LD_{50} in rats is 0.9 mg/kg, (Hansen and Speigel, 1983)]. The residue levels in watermelons ranged from 0.01 to 6.3 ppm. The possible reasons for such an incidence as discussed by Green and co-workers (1987) could be (1) deliberate misapplication of a pesticide not registered for use on melons or (2) soil persistence of the pesticide from a previous season when it was used on crops for which it was registered. Evidence indicates that the second reason could be a more likely possibility. The major metabolite of aldicarb is aldicarb sulfoxide (AS). The half-life of AS is 360 days at 15°C and pH 7.5 (Hansen and Speigel, 1983). Maitlen and Powell (1982) reported residues in vegetables such as potatoes, alfalfa, mint, mustard greens, and radishes 406 to 456 days after soil treatment with aldicarb. However, the soil samples from the same sites had no detectable residues. In the watermelon incidence six melons did not have detectable aldicarb or AS residues. Both aldicarb and AS were detected in nine persons with symptoms of poisoning. The possible explanation put forth was that the HPLC method for detecting AS was insensitive as compared to other carbamates and their metabolites. This issue exemplifies the necessity of developing more sensitive analytical methods for detecting pesticide residues in food. Despite the aldicarb incidence, the overall level of pesticide contamination appears to be very low in watermelons as compared to many other fruits (Table 9).

5.9. Pesticide Residues in Vegetables

Pesticide contamination of vegetables is of great concern as these, like fruits, are perishable products and are consumed by the general public before the contamination can be detected or the adulterated lot can be removed from the market. The information on this subject for selected common vegetables is presented in Table 10.

Americans, on average, consume 54 lb of potatoes per person per year. Root and tuber crops such as potatoes, carrots, and sweet potatoes absorb pesticide residues from soil. It is therefore not uncommon to detect the residues of organochlorine pesticides such as DDT, BHC, aldrin, dieldrin, and chlordane in these vegetables. Of these contaminants, BHC taints the flavor of potatoes and some other foods. According to Duggen and Weatherwax (1967), less than 0.001 mg of residues of these organochlorine pesticides is noted in the daily ration of potatoes in the American diet. Small amounts of residues of organophosphates such as parathion and dizinon, herbicides such as trifluralin, and fungicides such as dicloram and chlorpropham are also found in potatoes, carrots, and sweet potatoes.

Currently, the use of 20 to 40 pesticides is permitted on different vegetable crops. Residues of 50–70% of these can be detected by the FDA analytical methods. Except for corn, residues of 20 to 40 pesticides have been detected on various vegetables sold in supermarkets (Table 10). Relatively less pesticide contamination of corn may be due to the protection provided by the husk. Among different vegetables consumed in the United States special consideration must be given to tomatoes, cucumbers, and bell peppers for two essential reasons. First, to retain moisture, to preserve their appearance, and to prevent spoilage during transport and storage, these vegetables are coated with wax containing fungicides. Second, large quantities of tomatoes, cucumbers, and

TABLE 10 Pesticide Contamination of Common Vegetables[a]

| | | Pesticides | | | |
| | | Regulation | | Contamination | |
Vegetables	Consumption (lb/yr/person)	Permitted Number	Detection (%)	Observed Number	Contamination (%)
Bell pepper	—	70	60	39	50
Cabbage	5	60	70	36	20
Carrots	8	50	50	50	25
Corn	11	80	60	8	1
Green beans	11	60	60	32	25
Lettuce	11	60	60	40	33
Onions	5	50	60	18	10
Potatoes	54	90	55	38	20
Tomatoes	24	100	55	42	50

[a]Modified from Mott and Snyder (1987)

bell peppers are imported into the United States from 40–50 different countries. Residues of up to 39 pesticides can be found on bell peppers. About half of the samples collected from U.S. markets showed pesticide contamination, whereas a contamination rate of 80% was noted in imported bell peppers. Essentially the same was true for cucumbers and tomatoes (Mott and Snyder, 1987). The residues of insecticides most commonly detected in different vegetables include endosulfan, methamidophos, chlorpyriphos, dimethoate, and acephate.

5.10.   Pesticide Residues in Grains and Other Plant Products

To protect from the losses caused by insects and fungi, grains are fumigated several times during storage. Pesticides used for this purpose include dibromoethane, methyl bromide, and, more recently malathion, pyrethrum, and other fumigants. Both malathion and pyrethrum are degraded in grain with time and during cooking and, therefore, are believed not to pose any serious health problem. During 1983–1984 high levels of dibromoethane were detected in flour and cake mixes. In view of its known mutagenic and/or carcinogenic properties and reproductive toxicities in animals, EPA banned all agricultural uses of this chemical in 1984. It is comforting to know that milling contaminated grain yields flour with less residues, but the bran and other related products contain high levels of pesticides.

With the exception of treated olives, plant oils such as soybean oil, peanut oil, or cottonseed oil, with or without hydrogenation, contain little contamination of organochlorine insecticides. Residue levels are further lowered in refined oils as compared to crude oils.

5.11   Pesticide Residues in Milk

5.11.1.   Pesticide Residues in Animal Milk

It has long been known that pesticides are secreted in the milk if cows are exposed to relatively high concentrations of pesticides. Controlled studies have indicated that, in general, pesticides are excreted rapidly via milk at first and then more slowly at a rate similar to that usually noted for storage loss of pesticides from other tissues. About 15 years ago the pesticides detected in cow milk and other dairy products were almost entirely chlorinated hydrocarbon insecticides. The concentration of residues of these chemicals in dairy products usually corresponds to the butterfat content of the product. Subsequent to the ban on the use of several organochlorine insecticides, the level of pesticide contamination in cow milk has dropped significantly. It is difficult to predict when the current level (0.1 ppm) of milk contamination by the residues of organochlorine insecticide will reach zero.

Heptachlor, a chlorinated cyclodiene pesticide, was first isolated from technical chlordane. It was registered in 1952 to be used commercially as a pesticide. The EPA canceled its registration in 1976 for uses other than termite control through subsurface soil treatment. However, in January 1986 heptachlor and its primary metabolites were detected in raw, unprocessed milk in Arkansas, Missouri, and Oklahoma (Stehr-Green et al., 1988). The contamination was the result of feeding cattle grain mash left over from alcohol production in an Arkansas gasohol plant. This mash had high levels of heptachlor and its metabolites (heptachlor epoxide, transnonachlor, and oxychlordane), which were also identified in raw milk samples from dairy cows in these farms. Heptachlor and its principal metabolite, heptachlor epoxide, were detected in levels that were as high as 89.2 ppm in milk (seven times higher than the acceptable levels) on a fat basis. Approximately 140 dairy herds in Arkansas, Oklahoma, and Missouri were quarantined due to contamination by the banned insecticide heptachlor. Most dairy products from eight states were subject to recall. Three persons responsible for this contamination were sentenced to prison terms (Mott and Snyder, 1987).

The serum levels of heptachlor epoxide, oxychlordane, and transnonachlor in the exposed farm-family members were compared with the national population data collected in the Second National Health and Nutrition Examination Survey (NHANES II). It was found that the farm-family members had statistically significant high levels of serum heptachlor epoxide, serum oxychlordane, and serum transnonachlor as compared with the NHANES II data (Stehr-Green et al., 1986). However, the authors were unable to detect any evidence of acute or subacute effects of heptachlor and its metabolites in the exposed population with such high serum pesticide residues. Except for this isolated incidence, contamination of cow milk, in general, with residues of organochlorine pesticides has been very low and declining in recent years and the presence of residues of carbamates, organophosphates, and other pesticides such as herbicides and fungicides is a rarity.

Saxena and Siddiqui (1982) reported residues of six organochlorine pesticides in the milk of buffalo and goat. This study also found that the concentration of these chemicals in human milk was 12–13 times greater than levels detected in buffalo and goat milk.

5.11.2. Pesticide Residues in Human Milk

Human breast milk is known to contain residues of insecticides, most commonly those of organochlorine pesticides (Curley and Kimbrough, 1969). These pesticide residues most probably originate indirectly from the residues in food consumed or as a result of direct exposure. Woodard et al. (1976) reported residues of DDT to be much greater in the milk samples of rural blacks than in those of urban residents. Separate studies conducted in different parts of the world clearly indicate this to be a global problem. Fortunately, over the years, the concentrations are declining in countries in which the use of organochlorine pesticides is more restricted or banned (Coulston, 1985). Thus, the reports from Kenya (Kanja et al., 1986), Australia (Hornabrook et al., 1972), France (Luquet et al., 1975), Japan (Tojo et al., 1986), Canada (Davies and Mes, 1987), Italy (Dommarco et al., 1987), India (Saxena and Siddiqi, 1982), and the United States (Harris and Highland, 1977) have documented the presence of DDT and its principal metabolite, DDE, in human milk. Bakken and Seip (1976) analyzed 50

TABLE 11   DDT and DDE Residues in Human Milk in Different Countries

Study Region	Mean (ppm)[a]		N	Reference
	DDT	DDE		
Australia	18.10	9.59	1235	Hornabrook et al. (1972)
Kenya	27.63	24.13	302	Kanja et al. (1986)
India	0.125	0.314	71	Saxena and Siddiqi (1982)
Italy	7.0	40.0	130	Dommarco et al. (1987)
Canada	6.1	75.9	18	Davies and Mes (1987)
United States	52.9	352.1	1400	Harris and Highland (1977)

[a]All values expressed as ppm in milk fat.

samples of human breast milk in Norway and found that all the samples were contaminated with DDT, benzene hexachloride, and hexachlorobenzene. The concentrations of residues of these pesticides varied from very small to 11 times the WHO recommended maximum for cow's milk. Selected data are given in Table 11. The data on the levels of pesticide residues in samples of human colestrum are sparse in the literature. The analysis of results obtained in a Yugoslavian study (Vukavic et al., 1986) indicated the presence of six organochlorines and the concentration of the residues was found to exhibit seasonal variation. Pediatricians recommend breast-feeding of newborns, which means that babies are exposed to small quantities of this pesticide from the very first day of life. Other organochlorines that have been detected in traces in human milk and reported in these studies include chlordane, heptachlor epoxide (principal metabolite of heptachlor), hexachlorobenzene, hexachlorocyclohexane, dieldrin, and lindane.

5.12.   Pesticide Residues in Fish

5.12.1.   Organochorine Pesticide Residues in Fish

Fish is one of the principal food constituents of a health-conscious American's diet. It should be realized that these animals live in a media that is continuously being polluted by dangerous man-made chemicals. Prior to the existence of the EPA, chemical and other related industries used rivers, streams, and even oceans as dumping grounds. The EPA since 1970 has established severe restrictions on such activities. But the pollution or its delayed effects still continue to be detected in edible seafood. Pesticides, as a group, represent one of the most common pollutants in the water. The literature abounds with information on the extent of such contamination. The most common method used for detection of pesticide contamination is residue analysis of fish tissues. For the benefit of the reader, a summary of some important studies is presented in Table 12. It is obvious from Table 12 that of the eight chlorinated hydrocarbons, p, p'-DDT was present in all the samples gathered in different regions of the world. This further substantiates literature reports cited earlier in this chapter that p, p'-DDT is still persistent in

TABLE 12 Pesticide Residues in Fish from Different Parts of the World[a]

Fish	1[b]	2	3	4	5	6	7	8	Country	Reference
Finfish	70				4			120	United States	Eisenberg and Topping (1985)
Seafish[c]		8[d]							Iraq	Al-Omar et al. (1986)
Freshwater fish[e]	0.8	0.7		0.9					Poland	Zamojski et al. (1986)
Trout	28	821		10		7	8		Spain	Teran and Sierra (1987)
Freshwater fish[f]	140	14	43		3005	900	116		Chile	Ober et al. (1987b)
Indian shad[g] and cyprinid		72[d]		18	20	14	5	26	Iraq	Dou-Abul et al. (1987)
Roach and perch		23	1	1					West Germany	Schuler et al. (1985)

[a]All values expressed as the mean in ppb.
[b]1, p,p'-DDE; 2, p,p'-DDT; 3, HCB; 4, HCH; 5, dieldrin; 6, heptachlor and its epoxide; 7, lindane; 8, chlordane.
[c]Mean content of pesticide residues in 14 different kinds of sea fish.
[d]Value represents total DDT (DDE + DDT + TDE).
[e]Mean content of pesticides in the meat of four different kinds of fish.
[f]Mean content of pesticide in seven fish and four shellfish species.
[g]Mean content of pesticides in fish collected from the Shatt al-Arab river throughout the year.

283

the environment, in this case the aquatic environment. In the Chilean study dieldrin was present in very high levels (3005 ppb), which is approximately 30 times the FAO/WHO acceptable daily intake. In a recent National Pesticide Monitoring Program (NPMP) of organochlorine residues in freshwater fish it was concluded that p, p'-DDE (the most persistent metabolite of DDT) constituted approximately 70% of the p, p'-DDT residues since 1974 (Schmitt et al., 1985). After the ban of DDT in 1972 in the United States, the use of toxaphene increased considerably. NPMP reported high residue of this insecticide in fish and the environment following its use. The EPA recently canceled the registration of toxaphene for most uses (USEPA, 1982) and this has resulted in a decline in toxaphene concentrations in fish in several areas of the Great Lakes and in the southern United States. Incidentally, these areas had earlier reported high levels of this insecticide in different species of fish (Schmitt et al., 1985).

5.12.2. Organophosphate Pesticide Residues in Fish

In terms of acute and chronic toxicity, the organophosphates are considered more toxic than the organochlorines or carbamates. The main reason for such a difference is the irreversible inhibition of cholinesterase, which leads to neurological problems. But environmental persistence of these chemicals is not as prolonged as the organochlorines. Studies indicate that methylparathion accumulation occurred in carp fish when added in a carp-rearing pond within the first week. However, levels declined to 85% of the original level at the end of the second week (Sabharwal and Belsare, 1986).

5.13. Pesticide Residues in Meat Products

In a survey conducted by Duggan and Weatherwax (1967), DDT was the compound found in highest concentration (0.39 ppm on a fat basis) in the fat of red meat. In addition, residues of other 11 organochlorine pesticides were present in amounts exceeding 0.01 ppm in about 3% of the samples. These workers also reported the presence of residues of DDT and eight other organochlorine pesticides in poultry. The occurrence of pesticide residues in meat products has been noted in countries other than the United States as well. In a recent study from Nigeria (Atuma, 1985), the presence of organochlorine pesticides was reported in the meat of cows, goats, antelopes, as well as chickens. Similarly, the analyses of lamb and beef samples in Iraq revealed contamination by lindane, dieldrin, heptachlor, and chlordane (Al-Omar et al., 1985).

Cattle ear tags filled with insecticides such as pyrethroids and organophosphates have been used to provide economic control of the horn fly in the United States. But such processes provided excellent sources of contamination for the meat and fat of these animals. In a 14-week study of cattle treated with cypermethrin and chlorpyrifos ear tags, it was found that chlorpyrifos residues were detectable in perinatal fat in ranges of 0.01–0.128 ppm at 10–12 weeks posttreatment. In the fat, muscle tissue, and other organs, the values of chlorpyrifos ranged from 0.004–0.021 ppm and those for cypermethrin none detected to 0.011 ppm, respectively. These values were recorded at the time of slaughter 14-weeks after treatment. Cypermethrin residues were not detectable in any of the perianal fat, muscle, and other tissues (Byford at al., 1986). The EPA tolerances for cypermethrin and chlorpyrifos in cattle fat, meat, and meat products are 0.05 and 2.0 ppm, respectively. Although the detected residues of chlorpyrifos were well below the EPA tolerance values, human exposure through consumption of such contaminated meat is possible.

6.   HEALTH EFFECTS OF PESTICIDE RESIDUES IN FOOD

6.1.   Health Effects of Acute Pesticide Exposure

Although reports of mass poisoning due to pesticide contamination of food products are not common in the United States, the recent case of aldicarb contamination of watermelons makes it abundantly clear that the possibility for acute pesticide poisoning of epidemic proportions exists in the future.

6.2.   Health Effects of Chronic Pesticide Exposure

Perhaps the major concern today revolves around the potential danger of chronic exposure to pesticide residues. From the public health and safety point of view, this is the least understood aspect. Animal data and the results of various *in vitro* tests strongly suggest that a number of pesticides are carcinogenic and/or mutagenic. The committee on Scientific and Regulatory Issues Underlying Pesticide Use Patterns and Agricultural Innovation (1987) published a report on pesticides' potential to induce cancer. According to this report, of the 289 pesticides, the EPA found 53 active ingredients that are oncogenic or potentially oncogenic. This figure represents about 18% of all pesticides used on foods and includes 19 insecticides, 17 herbicides, 14 fungicides, and 3 others. Unfortunately, the data supporting this for many pesticides are incomplete. For some, especially insecticides, most registered uses on food have been canceled. Several organo-phosphorous insecticides or their oxygen analogs have been found to be teratogenic in rodent or chick embryo assays (Kimbrough and Gaines, 1968; Roger et al., 1969; Byrne and Kitos, 1983). Among insecticidal carbamates, extensively used carbaryl exhibits teratogenicity (Moscioni et al., 1977) whereas fungicidal ethylenebisdithiocarbamates break down in the environment or during cooking to ethylenethiourea that causes liver, lung, and thyroid cancer, genetic mutation, and birth defects in animals and is considered a probable human carcinogen (USGAO, 1987; Mott and Snyder, 1987).

6.3.   Health Effects of Contaminants in Pesticides

Most chemical reactions employed in the manufacture of active ingredients of pesticides also generate a wide array of impurities. Some of these impurities are invariably present in the final pesticide formulation. Toxicological data on these chemicals are almost nonexistent mainly due to the lack of interest on the part of manufacturers and the lax approach taken by the government regulatory agencies. Although the human health effects associated with such impurities are largely unknown, the importance of the subject has become abundantly clear from the well-publicized case of 2, 3, 7, 8-tetrachlorodibenzo-p-dioxin (TCDD) contamination in the herbicidal formulations of Agent Orange and its health effects on veterans resulting from its use in the Vietnam war. Another example is the presence of small amount (0.1–0.2%) of impurities such as O, O, S-trimethyl phosphorothioate (TMP) and isomalathion in technical grade malathion, a commonly used pesticide that is known to have a potent insecticidal activity and low mammalian toxicity (March et al., 1956). TMP causes unusual acute delayed toxicity and a number of reports have described the signs of this toxicity, including weight loss, red staining around the mouth and nose, a transient decrease in the weights of the spleen and thymus, and an increased incidence of pneumonia secondary to delayed toxicity (Mallipudi et al., 1979; Hammond et al., 1982; Rodgers et al., 1985; Thomas and Imamura, 1986). Besides causing immunosuppression, TMP significantly potentiates the acute toxicity of malathion presumably by inhibiting malathion carboxylesterase (Thomas and Imamura, 1986). Other examples of pesticide contamination include the presence of sulfotepp in diazinon, DDT in chlorobenzilate and dicofol, hexachlorobenzene in quintozene, tecnazene, and chlorothalonil, and nitrosamine in trifluralin (Matt and Snyder, 1987).

6.4. Health Effects of Inert Ingredients in Pesticides

The public health issue of pesticides exposure is further complicated by the presence of impurities in so-called inert ingredients such as solvents, wetting agents, and emulsifiers that are used in various pesticide formulations. More than 1200 such chemicals are currently employed as inert ingredients in different pesticide formulations. Of these approximately 50 are known to cause cancer, neurotoxicity, and birth defects in animals and another 50 chemicals are suspected of producing adverse health effects based on their structural similarity to proven toxicants.

7. WHAT THE CONSUMER CAN DO

1. *Variety in Diet*: It is generally accepted that a varied diet is the best means of ensuring adequate amounts of essential nutrients. A varied diet is also an excellent means of avoiding ingestion of harmful levels of pesticide residues from food. However, this may not always be possible. For example, for babies, milk is the primary source of nourishment.

2. *Extra Care*: Simple washing of fruits and vegetables will remove to some extent residues of certain insecticides, herbicides, and fungicides located on the

surface. Similarly, peeling of carrots, potatoes, and fruits, discarding the outer leaves of cabbage and lettuce, using flour with less bran, and removing fat from meat before cooking, or discarding it during cooking and avoiding it in a prepared meal will all help in reducing the exposure to pesticide residues.

3. *Organic Food*: Another alternative available to the consumers is to buy or produce their own organic fruits and vegetables. Crop cultivation methods without the use of pesticides have been known for quite some time. Several commercial farms in different states are currently producing organically grown fruits and vegetables. Some states have defined organic food and the certification programs verify that food sold as organic is really organic. In 1978, about $1 million worth of organic food was sold in California. In 1988 this increased to $40 million. Experts project that by 1997, organic food will capture approximately 7% of the market share, worth $500 million.

8.   WHAT THE GOVERNMENT CAN DO

1. The EPA should employ higher standards during pesticide reregistration and set low tolerance values. The agency should also discourage the practice of pesticide usage for cosmetic purposes.

2. Currently, the EPA estimation of exposure to pesticide residues is based on the estimates of per capita consumption of the food product. The established avarage food intake values need to be periodically revised to take into account any change in the dietary pattern. For example, fresh fruit and vegetable consumption in the United States has increased recently because of their increased availability in supermarkets. Unique dietary patterns of certain age groups should also be considered. For example, milk is usually the only source of nourishment for newborns. Certain products such as apples and bananas are consumed by children more than adults. On a per kilogram weight basis, a typical toddler consumes 19 times more noncitrus fruit juice, 13 times more milk, 7 times more apples, and 6 times more bananas than the typical adult woman (Mott and Snyder, 1987). In addition, tolerance values should be based on total exposure. At present domestic pesticide exposure via air and water is not taken into account. The current policy of treating "raw versus processed food, new versus old pesticides" differently in setting tolerances should be abandoned as these issues are irrelevant from the consumer protection point of view.

3. The EPA relies heavily on animal testing for determination of potential health effects of pesticide exposure in humans. Such an extrapolation of animal data to humans always introduces an element of uncertainty. At present, this is compensated by the incorporation of a safety factor in the calculations. It is expected that this approach would lead to either an underestimation or overestimation of the true health risk for humans associated with the dietary exposure of certain chemicals. Although for many reasons the use of animals for this purpose may seem inevitable, the unreliability of the resulting data cannot be

ignored. A number of laboratories are currently studying drugs and other chemicals using human tissues and, therefore, gathering human data for pesticides is clearly possible. Both the FDA and EPA should encourage research activity in this area if tragedies similar to that of thalidomide are to be avoided in the future. In light of this, the EPA's use of terms such as "probable human carcinogen or possible human carcinogen" based on animal data for certain pesticides appears meaningless.

4. The FDA is charged with the responsibility of enforcing the pesticide residue tolerances established by the FDA for all domestic and imported food. In reality, the FDA annually tests less than 1% of the food being consumed in the United States. Furthermore, the FDA is overlooking several pesticides with moderate to high health risk potential. This results from the FDA's reliance on multiresidue testing methods that cannot detect certain pesticides. As discussed earlier, ethylenebisdithiocarbamates produce ethylenethiourea, a probable human carcinogen. Crops that are known to have been treated with these chemicals in the United States that have not been tested by the FDA include beans, melon, pears, carrots, and peppers. During 1978–1987 not a single sample of imported food was tested for these compounds (U.S. GAO, 1987). Obviously, this is not an acceptable situation. In order to discharge its lawful duty of protecting the public from unsafe food, the FDA must have an effective monitoring system. This calls for an expansion of the monitoring activity using more effective, reliable, and quick methods.

5. To maximize the deterrent capabilities of the residue monitoring, the congress should give the FDA appropriate authority so that the producers of adulterated food not removed from the market are penalized.

6. We understand very little about the underlying biochemical mechanisms responsible for toxicity of pesticides in humans. Virtually nothing is known regarding the outcome of interactions between different pesticides and hundreds of other chemicals to which the general population is exposed on an every day basis. Immediate attention of the EPA and FDA to these long neglected research needs is critical. Scientific evidence gathered from such studies will provide a sound foundation necessary for intelligent decision making in the areas of reregistration of pesticides and in setting tolerances for them in the future.

7. The record indicates that for the past few years, the FDA's routine inspection of imported food has been very spotty and sporadic. In view of this the current practice of exporting the pesticides banned in the United States to third world countries is dangerous and must be stopped so long as the United States continues to import food products from these countries. This is important if the import of food contaminated by the same pesticides is to be avoided in the future.

8. Exploration of alternatives to chemical pest control has received much less attention than it deserves. Advances in plant breeding to develop disease and pest-resistant crop varieties, genetic engineering, innovations in biological, cultural, and integrated pest management, and so on certainly offer some promise for nonchemical pest control.

9. CONCLUSIONS

Successful pest management, triggered by the discovery of the miracle pesticide DDT, was a blessing for millions of people in the tropics. The subsequent introduction of other members of the family of organochlorine pesticides in the agricultural industry played a major role in the "green revolution" that we witnessed in many parts of the globe. Unfortunately, it appears that this blessing may turn into a curse. The unwise and indiscriminate use of billions of pounds of these pesticides between 1945 and the late 1970s has unquestionably left our ecosystem polluted. The air we breaths, the water we drink, and the food we eat all contain traces of these unwanted residues. Experts believe that DDT residues in the environment will be detectable even in the early part of the next century.

At present there is confusion, fear, and uncertainty. There is no doubt that the food quality today in terms of pesticide contamination has improved vastly over the past few years. Evidence is clear that since the ban of highly persistent organochlorine pesticides, there has been a decline in their residues both in food as well as in animal and human tissues. However, although the present public demands and the regulatory climate tend to require zero tolerance and tacitly imply absolute safety, the use of pesticides seems inevitable and the total elimination of pesticide residues in food appears to be essentially impossible in the near future. There are two main reasons for this. First, it must be realized that pesticides hold a unique position among the environmental contaminants as their distribution in the environment is the way they are used. Second, analytical technology, which is undergoing dramatic changes today, if adopted, will detect the presence and identity of trace amounts of pesticide residues in food. In fact our ability to chase down the vanishing zero has outrun our ability to test human health effects of such low level exposure to pesticides. However, the fact that pesticide residues are found in each food product and the incidence of contamination by organophosphate and carbamates, which replaced the organochlorines, is on the rise is alarming and clearly attests to the fact that the present measures are not adequate. On numerous occasions, questions have been raised, answered, and debated. It appears that the time has come to take some action, accept the responsibility, and reach a realistic understanding among all the segments of our society. Because "risk to human health" is the central issue, it is imperative that rather than considering a "perceived risk" we must establish the "real risk." A total ban on pesticides, based on the perceived risk of cancer and other health problems, would result in starvation and death by diseases such as malaria. Contrary to the data gathered from experiments on animals, epidemiological surveys have not yet clearly identified any pesticide as a human mutagen, carcinogen, or teratogen at the present exposure levels. This does not necessarily mean that pesticides are harmless to humans. Considering the limitations in epidemiological methods, and indications that the years of debate over the validity of extrapolation of animal data to human may never be resolved, a logical solution appears to be to develop sensitive, nonevasive biomarkers that will detect preneoplastic changes, reproductive toxicities, and other undesirable effects in human organs and tissues. Obvious prerequisites for this include a full understanding of pesticide metabolism in human tissues and the biochemical mechanisms of their toxicity. Biomedical scientists must accept this challange with appropriate support and encouragement from government regulatory agencies. Several years of neglect in research in this area has resulted in scores of laws based on perceived rather than on proven risk. These laws have failed to weed out the harmful pesticides or to punish offenders and are not providing adequate protection to the general public from the potential health hazards of pesticide residues in food.

A future free of fear from pesticide contamination of food is an achievable goal only if we act now and bring about the necessary changes in societal attitudes, laws, and scientific research. The solution(s) of the problem must involve scientists, those who produce food, and those who eat.

REFERENCES

Al-Omar, M., Al-Bassomy, M., Al-Ogaily, N., and Shebl, D. A. D. (1985). "Residue levels of organochlorine insecticides in lamb and beef from Baghdad." *Bull. Environ. Contam. Toxicol.* **34**, 509–512.

Al-Omar, M. A., Al-Ogaily, N. H., and Shabil, D. A. (1986). "Residues of organochlorine insecticides in fish from polluted water." *Bull. Environ. Contam. Toxicol.* **36**, 109–113.

Anonymous. (1986). "Aldicarb food poisoning from contaminated melons—California." *J. Am. Med. Assoc.* **256**, 175–176.

Anonymous. Committee on Scientific and Regulatory issues Underlying Pesticide Use Patterns and Agricultural Innovation (1987). *Regulating Pesticides in Food.* National Academy Press, Washington D.C.

Atuma, S. S. (1985). "Residues of organochlorine pesticides in some Nigerian food samples." *Bull. Environ. Contam. Toxicol.* **35**, 735–738.

Bakken, A. F., and Seip, M. (1976). "Insecticides in human milk." *Acta Paediat. Scand.* **65**, 535–539.

Byford, R. L., Lockwood, J. A., Smith, S. M., Harmon, C. W., Jahnson, C. C., Luther, D. G., Morris, H. F., Jr., and Penny, A. J. (1986). "Insecticide residues in cattle treated with a cypermethrin, chlorpyrifos, piperonyl butoxide-impregnated ear tag." *Bull. Environ. Cont. Toxicol.* **37**, 692–697.

Byrne, D. H., and Kitos, P. A. (1983). "Teratogenic effects of cholinergic insecticides in chick embryos—IV. The role of tryptophan in protecting against limb deformities." *Biochem. Pharmacol.* **32**, 2881–2890.

Coulston, F. (1985). "Reconsideration of the dilemma of DDT for the establishment of an acceptable daily intake." *Reg. Toxicol. Pharmacol.* **5**, 332–383.

Curley, A., and Kimbrough, R. (1969). "Chlorinated hydrocarbon insecticides in plasma and milk of pregnant and lactating women." *Arch. Environ. Health* **18**, 156–164.

Davies, D., and Mes, J. (1987). "Comparison of the residue levels of some organochlorine compounds in breast milk of the general and indigenous Canadian populations." *Bull. Environ. Contam. Toxicol.* **39**, 743–749.

Dommarco, R., DiMuccio, A., Camoni, I., and Gigli, B. (1987). "Organochlorine pesticide and polychlorinated biphenyl residues in human milk from Rome (Italy) and surroundings." *Bull. Environ. Contam. Toxicol.* **39**, 919–925.

Dou-Abul, A. A. Z., Al-Omar, M., Al-Obaidy, S., and Al-Ogaily, N. (1987). "Organochlorine pesticide residues in fish from the Shatt-Al-Arab river, Iraq." *Bull. Environ. Contam. Toxicol.* **38**, 674–680.

Duggan, R. E., and McFarland, F. J. (1967). "Residues in food and feed. Assessments include raw food and feed commodities, market basket items prepared for consumption, meat samples taken at slaughter." *Pestic. Monit. J.* **1**, 1–5.

Duggan, R. E., and Weatherwax, J. R. (1967). "Dietary intake of pesticide chemicals: Calculated daily consumption of pesticides with food are discussed and compared with currently accepted values." *Science* **157**, 1006–1010.

Edwards, C. A. (1970). "Insecticide residues in soils." *Res. Rev.* **13**, 83.

Eisenberg, M., and Topping, J. J. (1985). "Organochlorine residues in finfish from Maryland Waters 1976–1980." *J. Environ. Sci. Health.* **B20**, 729–742.

Gartrell, M. J., Craun, J. C., Pondrebarac, D. S., and Gunderson, D. L. (1986a). "Pesticides, selected elements and other chemicals in infant and toddler total diet samples, October 1980-March 1982." *J. Assoc. Off. Anal. Chem.* **69**, 123–145.

Gartrell, M. J., Craun, J. C., Pondrebarac, D. S., and Gunderson, D. L. (1986b). "Pesticides, selected elements and other chemicals in adult total diet samples, October 1980-March 1982." *J. Assoc. Off. Anal. Chem.* **69**, 146–159.

Goes, E. A., Savage, E. P., Gibbons, G., Aaronson, M., Ford, S. A., and Wheeler, H. W. (1980). "Suspected food borne carbamate pesticide intoxications associated with ingestion of hydroponic cucumbers." *Am. J. Epidemiol.* **111**, 254–260.

Green, M. A., Heumann, M. A., Wehr, M. H., Foster, L. R., Williams, L. P., Jr., Polder, J. A., Morgan, C. L., Wagner, S. L., Wanke, L. A., and Witt, J. M. (1987). "An outbreak of watermelon-borne pesticide toxicity." *Am. J. Pub. Health* **77**, 1431–1434.

Hammond, P. S., Braunstein, H., Kennedy, J. M., Badwy, Sm. A., and Fukuto, T. R. (1982). "Mode of action of delayed toxicity of O,O,S-trimethyl phosphorothioate in the rat." *Pestic. Biochem. Physiol.* **18**, 77–82.

Hansen, J. L., and Speigel, M. H. (1983). "Hydrolysis studies of aldicarb, aldicarb sulfoxide and aldicarb sulfone." *Environ. Toxicol. Chem.* **2**, 147–153.

Harris, S. G., and Highland, J. H. (1977). "Birthright denied." Environ. Defense Fund, Washington D.C.

Hayes, W. J. (1982). *Pesticides Studied in Man.* Williams & Wilkins, Baltimore, pp. 172–208.

Hornabrook, R. W., Dyment, P. G., Gomes, E. D., and Wiseman, J. S. (1972). "DDT residues in human milk from New Guinea natives." *Med. J. Aust.* **1**, 1297–1300.

Jelinek, C. F. (1985). "Control of chemical contaminants in foods: Past, present and future." *J. Assoc. Off. Anal. Chem.* **68**, 1063–1068.

Johnson, R. D., Manske, D. D., New, D. H., and Pondrebarac, D. S. (1979). "Pesticide and other chemical residues in infant and toddler diet samples(I)—August 1974–July 1975." *Pestic. Monit. J.* **11**, 116–131.

Johnson R. D., Manske, D. D., and Pondrebarac, D. S. (1981). "Pesticide, metal, and other chemical residues in adult total diet samples—(XII)—August 1975–July 1976." *Pestic. Monit. J.* **15**, 54–69.

Kanja, L., Skare, J. U., Nafstad, I., Maitai, C. K., and Lokken, P. (1986). "Organochlorine pesticides in human milk from different areas of Kenya, 1983–1985." *J. Toxicol. Environ. Health* **19**, 449–464.

Kimbrough, R. D., and Gaines, T. B. (1968). "Effect of organic phosphorus compounds and alkylating agents on the rat fetus." *Arch. Environ. Health* **16**, 805–808.

Luquet, F. M., Goursaud, J., and Casalis, J. (1975). "Pollution of human milk in France by organochlorine insecticide residues." *Path. Biol., Paris* **23**, 45–49.

Maitlen, J. C., and Powell, D. M. (1982). "Persistence of aldicarb in soil relative to the carry-over residues into crops." *J. Agr. Food Chem.* **30**, 589–592.

Mallipudi, N. M., Umetsu, N., Toia, R. F., Talcott, R. E., and Fukuto, T. R. (1979). "Toxicity of O,O,S-trimethyl and triethyl phosphorothioate in rats." *J. Agr. Food Chem.* **27**, 463–473.

March, R. B., Fukuto, T. R., Metcalf, R. L., and Maxon, M. G. (1956). "Fate of P^{32}-labeled malathion in the laying hen, white mouse and American cockroach." *J. Econ. Entomol.* **49**, 185–195.

Marshall, E. (1985). "The rise and decline of temikTM." *Science* **229**, 1369–1371.

Moscioni, A. D., Engel, J. L., and Casida, J. E. (1977). "Kynurenine formamidase inhibition as a possible mechanism for certain teratogenic effects of organophosphorus and methylcarbamate insecticides in chicken embryos." *Biochem. Pharmacol.* **26**, 2251–2258.

Mott, L., and Snyder, K. (1987). *Pesticide Alert. A Guide to Pesticides in Fruits and Vegetables.* Sierra Club Books, San Francisco, 179 pp.

Mowafy, L. I., Marzouk, M. A., El-Ahraf, M., and Willis, W. V. (1987). "Determination of dieldrin concentration in recycled cattle feed and manure by liquid chromatography." *Bull. Environ. Contam. Toxicol.* **38**, 396–403.

Ober, A. G., Santa Maria, I., and Carmi, J. D. (1987a). "Organochlorine pesticide residues in animal feed by cyclic steam distillation." *Bull. Environ. Contam. Toxicol.* **38**, 404–408.

Ober, A., Valdivia, M., and Santa Maria, I. (1987b). "Organochlorine pesticide residues in Chilean fish and shellfish Species." *Bull. Environ. Contam. Toxicol.* **38**, 528–533.

Reed, D. V., Lombardo, P., Wessel, J. R., Burke, J. A., and McMahon, B. (1987). "The FDA pesticides monitoring program." *J. Assoc. Off. Anal. Chem.* **70**, 591–595.

Roger, J. C., Uphall, D. G., and Casida, J. E. (1969). "Stucture activity and metabolism studies on organophosphate teratogens their alleviating agents in developing hen eggs with special emphasis on bidrin. *Biochem. Phamacol.* **18**, 373–392.

Rodgers, K. E., Imamura, T., and Devens, B. H. (1985). "Effects of subchronic treatment with *O, O, S*-trimethyl phosphorothioate on cellular and humoral immune response system." *Toxicol. Appl. Pharmacol.* **81**, 310–318.

Sabharwal, A. K., and Belsare, D. K. (1986). "Persistence of methyl parathion in a carp rearing pond." *Bull. Environ. Contam. Toxicol.* **37**, 705–709.

Saxena, M. C., and Siddiqui, M. K. J. (1982). "Pesticide pollution in India: Organochlorine pesticides in milk of woman, buffalo, and goat." *J. Dairy Scie.* **65**, 430–434.

Schmitt, C. J., Zajicek, J. L., and Ribick, M. A. (1985). "National pesticide monitoring program: Residues of organochlorine chemicals in freshwater fish 1980–81." *Arch. Environ. Contam. Toxicol.* **14**, 225–260.

Schuler, W., Brunn, H., and Manz, D. (1985). "Pesticides and polychlorinated biphenyls in fish from the Lohn river." *Bull. Environ. Contam. Toxicol.* **34**, 608–616.

Slade, R. E. (1945). "The gamma-isomer of hexachlorocyclohexane (Gammexane). An insecticide with outstanding properties." *Chem. Ind.* **1945**, 314–319.

Stehr-Green, P. A., Schilling, R. J., Burse, V. W., Steinberg, K. K., Royce, W., and Denny Donnell, H. (1986). "Evaluation of persons exposed to dairy products contaminated with heptachlor." *J. Am. Med. Assoc.* **256**, 3350–3351.

Stehr-Green, P. A., Wohlleb, J. C., Royce, W., and Head, S. L. (1988). "An evaluation of serum pesticide residue levels and liver function in persons exposed to dairy products contaminated with heptachlor." *J. Am. Med. Assoc.* **259**, 374–377.

Teran, M. T., and Sierra, M. (1987). "Organochlorine insecticides in trout, *Salmo trutta fario* L., taken from four rivers in Leon, Spain." *Bull. Environ. Contam. Toxicol* **38**, 247–253.

Thomas, I. K., and Imamura, T. (1986). "Immunosuppressive effect of an impurity of malathion: Inhibition of murine T and B-lymphocyte responses by *O, O, S*-trimethyl phosphorothioate." *Toxicol. Appl. Pharmacol.* **83**, 456–464.

Tojo, Y., Wariishi, M., Suzuki, Y., and Nishiyama, K. (1986). "Quantitation of chlordane residues in mother's milk." *Arch. Environ. Contam. Toxicol.* **15**, 327–332.

U.S. Environmental Protection Agency (EPA). (1982). "Toxaphene decision document." Office of Pesticide Programs, washington D.C.

U.S. Environmental Protection Agency (EPA). (1987). "Unfinished business: A comparative assessment of environmental problems."

U.S. General Accounting Office Testimony. (1987). "Federal regulation of pesticide residues in food."

Vukavic, T., Pavkov, S., Cusic, S., Roncevic, N., Vojinovic, M., and Tokovic, B. (1986). "Pesticide residues in human colustrum: Seasonal variation, Yugoslavia." *Arch. Environ. Contam. Toxicol.* **15**, 525–528.

Woodard, B. T., Ferguson, B. B., and Wilson, D. J. (1976). "Insecticide levels in milk of rural indigent blacks." *Am. J. Dis. Child.* **130**, 400–403.

Zamojski, J., Smoczynski, S., Skibriewska, K. A., and Amarowicz, R. (1986). "Organochlorine insecticides and heavy metals in fish from Mutek Lake, N.E. Poland." *Bull. Environ. Contam. Toxicol.* **37**, 587–592.

10

FOOD CONTAMINATION WITH INSECTICIDE RESIDUES FROM NONAGRICULTURAL ENVIRONMENTAL SOURCES

Parm Pal Singh and Balwinder Singh

Department of Entomology, Punjab Agricultural University, Ludhiana, India

1. INTRODUCTION

Due to the inherently toxic nature of insecticide residues, there is worldwide concern over their presence in human food. Consequently, many studies have to be done and much data have to be produced before a chemical is considered safe for the consumer and is allowed to be used for pest control. However, most safety studies are limited to preharvest insecticidal applications on crops and do not take into consideration the incidental routes through which the foodstuffs may acquire insecticide residues before they reach the dining table of the consumer. Several studies have revealed that there are many diffuse sources of contamination that are normally not considered when discussing food safety, but can produce inadvertent contamination of the foodstuffs with residues of insecticides and related compounds at toxicologically significant levels. Moreover, although it is easy to regulate residues from an unambiguous route of contamination such as the use of insecticides on food commodities, contamination with unintentional residues is generally difficult to control. In addition, because of degradation, residues from field treatment of crops are depleted before the produce reaches the public. In contrast, most inadvertent contamination involves persistent compounds and occurs shortly before the food reaches the ultimate consumer; thus there is less chance for insecticide residues to become degraded before consumption of the foodstuffs by consumers.

The limited literature on this topic indicates that the level of residues in foods arising from unconventional sources of contamination is generally low. However, the available information is sufficient to warrant further study and illustrates the need for the persons engaged in the regulation and estimation of insecticide residues in foods to be aware of the possibility of incidental contamination of food commodities.

2. FOOD CONTAMINATION FROM PUBLIC HEALTH USAGE OF INSECTICIDES

Several vector-transmitted diseases prevalent in tropical and subtropical areas are estimated to threaten more than one-third of the world population (Büchel, 1983). One factor in the strategy to suppress the vector-borne diseases is systematic control of the transmitting vector by the use of chemicals. As a result, in addition to agricultural applications, there has been intensive use of insecticides in the public health programs in developing countries. In spite of the availability of a number of alternate compounds, persistent organochlorine compounds such as DDT and BHC have been used and are still being employed in large quantities to control vectors in many Third World nations, primarily because of their low cost. In this context, it has been maintained that although outdoor use of DDT should be avoided as far as possible, indoor spraying of DDT in routine antimalarial operations does not involve a significant risk to humans (Fontaine, 1978; WHO, 1971; Spindler, 1983). However, studies evaluating the impact of these compounds on human health and the environment in developing nations are very rare. Farvar (1979) concluded that after more than a quarter of a century of public health use of pesticides and only a few biological monitoring programs and residue impact studies, an enormous empirical experiment had been conducted with no experimental design to assess the outcome and little care as to its deletereous effects on the people of developing countries.

Application rates of 1 g DDT/m² twice a year or the technical BHC equivalent to 0.2 g γ-isomer/m² three times a year in antimalarial spraying operations are more than 10 times higher than agricultural application rates. Compared to outdoor applications, such indoor applications may result in a thousand to a million times higher ambient residues (Lewis and Lee, 1976); the residues, not subject to environmental degradation or dispersal, remain confined to the immediate vicinity of human beings for a long time. Reports indicate that such organochlorine residues are present on walls, in household goods, in housedust, and in the indoor environment and may lead to considerable nondietary as well as dietary human exposure by contamination of bovine milk, human milk, wheat grains, straw, and so on. Kalra and Chawla (1985) pointed out that vector control programs, which are generally evaluated only in terms of disease control, should also be assessed with respect to their insecticide residue implications.

2.1. Bovine Milk

In India, bovine milk and its products have been reported to be contaminated with excessive residues of DDT and BHC (Kalra and Chawla, 1981; Kalra et al., 1983; Dhaliwal, this volume). However, the sources of contamination of these commodities are not clearly known. Studies done to ascertain the contribution of indoor residual spraying of DDT and BHC for mosquito control to the contamination of milk and milk products have indicated a relationship between public health usage of these insecticides and the presence of their residues in bovine milk.

Kapoor et al. (1980) collected 54 buffalo milk samples from February to April 1979 from rural houses that had recently received different antimalarial sprays. The analysis of these samples showed the presence of residues of both DDT and BHC in all the samples (Table 1). However, average DDT residues in DDT-sprayed districts were about 2.5 times higher than the levels encountered in samples from BHC-sprayed area. Contamination of milk with BHC residues was more than 7 times higher in BHC-sprayed localities than in DDT-sprayed areas. Thus, the results indicated some relationship between the use of DDT or BHC for mosquito control and environmental contamination with the corresponding residues.

Whereas Kapoor et al. (1980) compared DDT and BHC residues in bovine milk from rural dwellings that had been treated with one of these insecticides for malaria control on the basis of a single sampling, Singh et al. (1986) assessed the overall situation by monitoring these insecticide residues in milk from randomly selected houses in areas receiving application of different insecticides for public health purposes. The study involved two villages (Harnampura and Jaspal Bangar) in the Ludhiana district sprayed with DDT under the National Malaria Eradication Program (NMEP) and two villages (Phulewal and Kup Kalan) in the

TABLE 1   Mean and Range of DDT and BHC Residues (mg/kg, Whole Milk Basis) in Buffalo Milk Samples Collected from Areas Sprayed with These Insecticides for Malaria Control[a]

Insecticide Used for Mosquito Control	District (Number of Samples)	DDT Residues (Range)	BHC Residues (Range)
DDT	Ludhiana (14)	0.35 (0.07–0.72)	0.05 (0.01–0.19)
	Jalandhar (10)	0.39 (0.05–1.57)	0.04 (0.02–0.10)
	Total DDT area (24)	0.37 (0.05–1.57)	0.05 (0.01–0.19)
BHC	Ferozepur (16)	0.17 (0.05–0.49)	0.39 (0.05–2.06)
	Sangrur (14)	0.14 (0.08–0.24)	0.38 (0.13–0.91)
	Total BHC area (30)	0.15 (0.05–0.49)	0.38 (0.06–2.06)

[a] Based on Kapoor et al. (1980).

Sangrur district sprayed with BHC (Figure 1). Ten bovine milk samples were collected from different houses from each village at various intervals (Table 2). The inquiries revealed that although some dwellings were completely covered under NMEP, others were partially treated or not treated with insecticides during the sampling year.

Median values of DDT residues in bovine milk samples from DDT-sprayed areas were three to five times higher than the corresponding samples from BHC-sprayed areas (Table 2). During the last two samplings, the level of BHC residues found in samples from BHC-sprayed localities was three to four times higher than that in the samples from DDT-sprayed localities. However, during the first sampling, a few samples from DDT-sprayed area contained exceptionally high BHC residues. This inconsistency in the case of BHC residues may be due to agronomic use of this insecticide. In India, about 60% of BHC consumption is in agriculture while DDT is mainly used in public health programs.

Statistical analysis of the data revealed a direct correlation between indoor application of DDT in malaria control and its residue levels in bovine milk throughout the study area. A similar trend was observed for BHC for the last two samplings. The difference in DDT and BHC residues in bovine milk samples from the villages sprayed with the same insecticide was always nonsignificant. More than 40% of milk samples collected during this study contained DDT residues at levels exceeding the extraneous residue limit of 0.05 mg/kg (fresh weight basis).

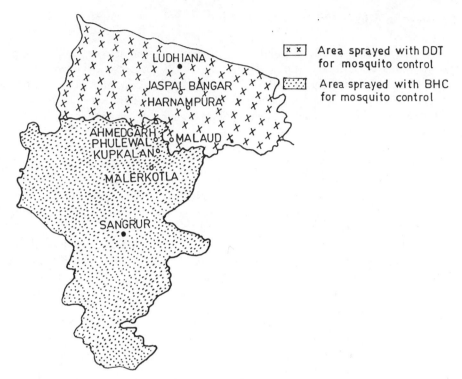

Figure 1. Location of sampling sites in the Ludhiana and Sangrur districts of Punjab, India.

TABLE 2   DDT and BHC Residues (mg/kg, Whole Milk Basis) Found in Bovine Milk Samples Collected from DDT- and BHC-Sprayed Areas[a]

	DDT Residues Median (Range)		BHC Residues Median (Range)	
Date	DDT Area	BHC Area	DDT Area	BHC Area
May 9, 1984	0.21 (0.09–0.91)	0.04 (ND–0.12)[b]	0.46 (Trace–6.04)[c]	0.27 (0.01–0.73)
August 2, 1984	0.17 (0.06–0.51)	0.05 (0.01–0.14)	0.04 (0.02–0.24)	0.15 (0.03–1.62)
January 18, 1985	0.18 (0.03–0.44)	0.04 (0.02–0.36)	0.02 (0.01–0.13)	0.07 (0.02–0.60)

[a]Based on Singh et al. (1988).
[b]ND, not detected (less than 0.005 mg/kg).
[c]Trace, less than 0.01 mg/kg.

No limit has been prescribed for BHC residues, but their levels in a number of samples were also excessive. Because spraying of DDT and BHC under NMEP is carried out exclusively by Government agencies, the presence of residues at levels exceeding the regulatory limit was considered to be relevant in the regulation of insecticide residues in milk and milk products in India (Singh et al., 1986).

2.2. Human Milk

As children are considered a population group highly sensitive to xenobiotics and mothers' milk is a primary source of infant nutrition, the presence of insecticide residues in breast milk is viewed with concern throughout the world. A few surveys carried out in developing countries have shown the presence of DDT and BHC residues in human milk at levels much higher than in western countries (Farvar, 1979; Kalra and Chawla, 1981). While discussing the occurrence of organochlorine compounds in biological specimens from human population of Guatemala, Farvar (1979) noted that indoor spraying of DDT for malaria control seemed to be the main cause of high accumulations of DDT residues in human milk, whereas agricultural use of insecticides appeared to be a secondary source of contamination. Other investigations on human milk carried out in developing countries also support this hypothesis.

Hornabrook et al. (1972) conducted a study in seven districts of New Guinea and reported that breast milk samples obtained from donors living in districts sprayed with DDT for malaria control averaged 0.242 mg/kg DDT, whereas milk samples from donors living in unsprayed districts of New Guinea had about 20 times less (mean level 0.011 mg/kg) DDT residues.

A survey carried out by the Centre for the Biology of Natural Systems in Guatemala noted high concentrations of DDT residues (Figure 2) in samples of mother's milk from donors in three rural Pacific Coast communities (Olszyna-Marzys et al., 1973). Of these, La Bomba had not been subjected to heavy agricultural spraying of the insecticides, whereas Cerro Colorado and El Rosario had received heavy applications of organochlorine compounds on cotton crops. However, all three communities had received about 13 years of DDT spraying for malaria control. As there was no significant difference among the DDT levels in mother's milk from these areas (p less than 0.01), it appeared that indoor antimalarial use of DDT was a more important source of accumulation of DDT residues in human milk than the agronomic use of this insecticide.

This observation was supported by a follow-up study in the same three areas and the analysis of human milk samples from five other communities in Guatemala (Winter et al., 1976). During these investigations, a pronounced difference in the level of contamination of human milk with DDT, which seemed to be closely associated with residential spraying of this insecticide by the National Malaria Eradication Service, was observed. The highest average DDT level of 0.864 mg/kg was found in Livingston, a community in which all homes had been treated with DDT up to 1974 (Figure 2). The authors noted that there was no large-scale agricultural use of DDT around Livingston.

In the next four communities with abnormally high levels of DDT residues, the malaria control service had discontinued all pesticide sprays in 1970 in Asuncion Mita and switched to propoxur in 1970 in Carro Colorado and El Rosario and in 1972 in La Bomba. The authors also compared the residue levels observed in 1974 with those reported for three communities in 1970 by Olszyna-Marzys et al. (1973) and noted a drop in average DDT contamination of human milk after this

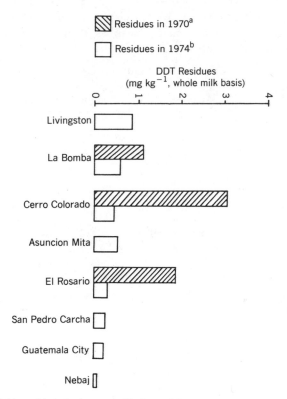

Figure 2. Insecticide residues in human milk from different locations in Guatemala. [a]Based on Winter et al. (1976). [b]Based on Olszyna-Marzys et al. (1973).

insecticide was no longer used for mosquito control (Figure 2). The three communities (San Pedra Carcha, Guatemala, and Nebaj) found to have minimum contamination of human milk with DDT residues lie in nonmalarial areas and were not receiving antimalarial sprays.

Joia et al. (1985) analyzed human milk samples collected in 1984 from three villages (Ahmedgarh, Kup Kalan, and Malerkotla) in the Sangrur district of Punjab, India sprayed with BHC under NMEP and two villages (Jaspal Bangar and Malaud) in the adjoining district of Ludhiana sprayed with DDT (Figure 1). The average DDT residue level (0.70 mg/kg) in human milk samples collected from DDT-sprayed areas, when compared by t test with the mean level (0.44 mg/kg) in samples from BHC-sprayed area, was found to be significantly higher at the 5% significance level (Table 3). However, the difference in BHC residues in samples from two areas (0.21 and 0.29 mg/kg) was not significant at this level. This may be due to the use of BHC for purposes other than mosquito control in DDT-sprayed area, as about 60% of BHC use in India is in agriculture whereas DDT is used primarily in public health programs.

A comparison of data from villages situated close to the boundary between the districts of Ludhiana and Sangrur with that from villages located in the interior of

TABLE 3   Residues of DDT and BHC (mg/kg, Whole Milk Basis) in Human
Milk Samples Collected from Various Localities Treated with Different
Insecticides for Malaria Control[a]

	DDT Residues		BHC Residues	
Locality	Mean (Range)	Median	Mean (Range)	Median
BHC-Sprayed District				
Kup Kalan	0.93	0.41	0.42	0.38
(n = 12)	(0.02–4.25)		(0.05–1.18)	
Ahmedgarh	0.49	0.37	0.41	0.31
(n = 18)	(0.04–1.88)		(0.04–1.28)	
Malerkotla	0.27	0.17	0.18	0.14
(n = 39)	(0.02–1.54)		(0.03–1.30)	
Total	0.44	0.21	0.29	0.17
(n = 69)	(0.02–4.25)		(0.03–1.30)	
DDT-Sprayed District				
Jaspal Bangar	0.54	0.30	0.10	0.08
(n = 13)	(0.08–2.47)		(0.03–0.34)	
Malaud	0.76	0.68	0.26	0.19
(n = 39)	(0.22–2.14)		(0.05–0.80)	
Total	0.70	0.52	0.21	0.14
(n = 52)	(0.08–2.47)		(0.03–0.80)	

[a]Based on Joia et al. (1985).

these two districts revealed that human milk samples from border sites had
relatively more residues of the insecticide being employed in the adjoining district
for mosquito control. For example, samples collected from Kup Kalan and
Ahmedgarh in the Sangrur district, which are situated in a region adjoining a
DDT-sprayed district, had higher DDT residues (median levels 0.41 and
0.37 mg/kg, respectively) as compared to DDT residue levels observed for
Malerkotla (median value 0.17 mg/kg) located in the interior of the district
(Table 3, Figure 1). In a similar way, both Jaspal Bangar and Malaud are in the
Ludhiana district in which DDT is used in antimalarial operations. However, the
median level of BHC residues in human milk samples from Jaspal Bangar located
in the interior of the district was 0.08 mg/kg, whereas the median level of BHC
residues in samples from Malaud situated in the fringe area of the Ludhiana
district was 0.19 mg/kg. The atmospheric transport of persistent organochlorine
compounds is well documented and may be responsible for these differences.

Considering that an infant weighs 3 kg and consumes 600 mL of breast milk
daily, DDT present at median values of 0.53 and 0.21 mg/kg in human milk
samples from DDT-and BHC-sprayed areas was calculated to represent an
intake level of 0.11 and 0.04 mg/kg/day, respectively. These figures are about 20

and 8 times higher than the acceptable daily intake (ADI) of 0.005 mg/kg/day for DDT. No ADI has been prescribed for BHC, but Joia et al. (1985) considered that their high levels in comparison with values reported from other regions of the world (Farvar, 1979; Jensen, 1983) demand concern.

Human milk has unique nutritional and immunological advantages over alternate feeding methods for infants. As the above-discussed reports suggest, there is a relationship between indoor application of DDT and BHC for mosquito control and their residues in human milk, and it would seem necessary to identify the routes of contamination and take suitable measures to reduce human ingestion of these insecticides.

2.3. Wheat Grains and Straw

Several monitoring studies conducted in Punjab, India revealed widespread contamination of wheat grains and flour with DDT and BHC at levels generally less than 1 mg/kg (Joia et al., 1978; Kalra et al., 1986). These insecticides are not recommended for use on wheat crops or stored grains. It has been suggested that these residues might be the result of addition of these chemicals in stored commodities to prevent insect infestation (ICAR, 1967). However, the levels of contamination observed in most of the samples in monitoring studies are too low to occur from an admixture of the insecticides with grains and could be arising from routes other than direct treatment. In India, large quantities of harvested wheat and straw are stored in rural premises, many of which receive antimalarial DDT or BHC application. A few studies have investigated the possible relationship between insecticide use for public health and the presence of insecticide in stored food and feed.

A simulated trial investigation of the contamination of stored wheat grains following the indoor residual spraying of DDT and BHC was reported by Kalra and Chawla (1983). The analysis of representative samples found that levels of DDT and BHC up to 1.22 and 13.8 mg/kg, respectively, were acquired by stored wheat during the experimental period.

A study of the relationship between usage of DDT and BHC for malaria control and contamination of wheat grains and straw was conducted by Singh et al. (1988). At three intervals, 20 wheat grain samples having no history of insecticide usage for insect control were collected randomly from fields or houses from each of the two adjoining districts of Ludhiana, where DDT is used for mosquito control, and Sangrur, where BHC is used (Figure 1, Table 4). At harvest time, wheat samples were collected from fields while the crop was being harvested. During later samplings, wheat grains stored in houses for home consumption were obtained. In addition, 20 samples of wheat straw of a crop harvested in April 1985 were also collected from rural houses of each district in April 1986.

All 120 samples of wheat grains analyzed during this study contained residues of both DDT and BHC. At harvest time, wheat grains from DDT- as well as BHC-sprayed areas had similar low levels of these compounds representing background contamination due to the ubiquitous presence of these chemicals

TABLE 4 DDT and BHC Residues (mg/kg) in Wheat Grains and Straw Samples Collected from Premises Treated with These Insecticides for Mosquito Control[a]

	DDT Residues				BHC Residues			
	DDT-Sprayed Area		BHC-Sprayed Area		DDT-Sprayed Area		BHC-Sprayed Area	
Date of Sampling	Mean ± SD (Range)	Median	Mean ± SD (Range)	Median	Mean ± SD (Range)	Mean	Mean ± SD (Range)	Mean
Wheat Grains								
April, 1985 (Harvest time)	0.02 ± 0.01 (0.01–0.05)	0.02	0.02 ± 0.02 (Trace[b]–0.05)	0.02	0.01 ± 0.01 (Trace–0.03)	0.01	0.01 ± 0.01 (0.01–0.03)	0.01
December, 1985	0.14 ± 0.25 (0.01–1.10)	0.07	0.04 ± 0.05 (Trace–0.24)	0.03	0.04 ± 0.04 (0.01–0.24)	0.02	5.18 ± 14.31 (0.02–60.21)	0.21
April, 1986	0.51 ± 0.78 (0.06–2.40)	0.16	0.01 ± 0.01 (Trace–0.01)	0.01	0.04 ± 0.05 (Trace–0.19)	0.02	2.14 ± 7.77 (0.11–35.06)	0.28
Wheat Straw								
April, 1986	1.35 ± 2.58 (0.05–10.94)	0.46	0.06 ± 0.07 (Trace–0.28)	0.03	0.05 ± 0.04 (0.02–0.18)	0.03	0.24 ± 0.35 (0.02–1.49)	0.13

[a]Based on Singh et al. (1988).
[b]Trace, less than 0.01 mg/kg.

(Table 4). At the time of sampling, made after 8 and 12 months of storage of grains in rural dwellings, median levels of DDt residues in samples from DDT-sprayed districts were greater than those in samples from BHC-sprayed areas by factors of 2 and 16, respectively. On the other hand, median BHC residues in wheat grains from BHC-sprayed district were 10 and 14 times higher than in the corresponding samples from DDT-sprayed villages. During this time, DDT levels in wheat grain samples from BHC-sprayed districts and BHC residues in samples from DDT-sprayed districts remained similar to background level observed at harvest time.

The analysis of 40 wheat straw samples also showed that samples from DDT-sprayed areas had acquired about 15 times higher DDT residues (median value 0.46 mg/kg) as compared to samples from BHC-treated area (median value 0.03 mg/kg). In contrast, BHC residues in straw from BHC-sprayed area were about four times higher than those in samples from DDT-sprayed districts (Table 4).

Therefore, the study of Singh et al. (1988) revealed a relationship between indoor use of DDT and BHC for mosquito control and contamination of wheat grains and straw stored in the treated premises. The authors considered that there is a possibility of substantial dietary intake of these insecticide residues by consumption of wheat grains contaminated at levels encountered in the study. As wheat straw in the area of investigation is commonly used as cattle feed, it was considered that residues in this commodity are likely to become bioconcentrated and might be contributing toward the high level of DDT and BHC observed in milk and milk products of the region (Kalra et al., 1983).

3. FOOD CONTAMINATION FROM HOUSEHOLD AND RELATED USES OF INSECTICIDES

As most people spend the majority of their time indoors, use of insecticides inside the premises is of special concern. However, until now, little emphasis has been placed on the problem of residues of insecticides resulting from their use in dwellings. While describing toxic hazards of insecticides in his book on the control of domestic pests, Busvine (1980) elaborated only an environmental residues and precautions for persons occupationally engaged in pest control and did not discuss the residue hazards from the household use of insecticides. The possibility of an accident or hazardous situation developing from the home use of pesticides was revealed by a survey of 196 urban families in Charleston, South Carolina (Keil et al., 1969). It was found that 89% of houses covered in the study used pesticides, with one-third of them applying these chemicals at least weekly during the entire year. Private pest control operators regularly visited about 42% of the homes. The chemicals were often stored in unlocked areas (88%), within the reach of small children (66%), and near food or medicines (54%). Two-thirds of users neither wore gloves while applying pesticides nor washed their hands after application of the pesticides.

There are several reports correlating greater human body burden of insecticide residues with their greater use in households. Radomski et al. (1968) reported that individuals using insecticides extensively in homes had levels of DDT and DDE in their adipose tissues three to four times higher than persons using small amounts of insecticides in their homes.

Davies et al. (1975) demonstrated the possibility that insecticide residues in house dust resulting from domestic use of insecticides may be an important source of human exposure. Takahashi et al. (1981) observed that organochlorine residue levels in breast milk were higher among donors living in homes treated by professional pest control operations than in donors living in untreated homes. However, the extent of human exposure resulting from oral intake by contamination of foodstuffs present in the vicinity of treated surfaces is not clearly known.

The presence of pesticides and their analogues in the air inside residential, office, and industrial buildings has been discussed by several workers (Lewis and Lee, 1976; Seiber et al., 1983). Contamination of the air of rooms and food preparation areas following the routine application of pesticides for preventive pest control has also been reported (Wright and Leidy, 1978, 1980; Wright et al., 1981). The studies conducted on insecticide residues on nontarget areas of the rooms and kitchens by Wright and Jackson (1971, 1975) and Wright et al. (1984) indicate the potential of indoor applications of insecticides for contaminating food commodities.

3.1.   Penetration of Insecticides through Food Packages

Experiments have shown that commonly used packaging materials do not form an impermeable barrier to insecticides and packaged foodstuffs can be contaminated when they are stored in contact with or in close proximity to the insecticide-treated surface or environment.

Queen (1953) observed that protection of foods and surfaces from vaporized lindane is not achieved by covering with or packaging in common wrapping materials except for metal. Siakotos (1956) found that no commercial package (box board, kraft paper, cellophane, glassine, or aluminium foil) completely resists the penetration of lindane vapors. The pinholes which do occur in certain grades of aluminium foil were considered to lead to the permeability of the foil to lindane vapors.

Highland et al. (1966) studied the migration of piperonyl butoxide from multiwall kraft bags treated on the outside surface into four commodities. Piperonyl butoxide was present in the stored commodities after only 1 month of storage in the treated bags. The highest mean residues in composite samples after 18 months of storage were 6.28 mg/kg in flour, 5.50 mg/kg in polished rice, 4.74 mg/kg in nonfat dry milk, and 0.57 mg/kg in navy beans. The higher residues were found in foods that contained more fat. Polyethylene, which was used as a liner in dry milk packages, did not act as a barrier to the piperonyl butoxide.

Yeo and Bevenue (1969) found that insecticide-treated shelf papers available in retail stores for household use can contaminate the food commodities with appreciable amounts of insecticide residues. The sorption effect of insecticides with flour was found to vary with the mode of exposure and type of bag used. When bags were kept in direct contact with a shelf paper impregnated with chlordane, flour packages in polyethylene bags contained the maximum level of chlordane residues (28.9 mg/kg). For single thickness and double thickness kraft paper bags, these values were 24.4 and 18.3 mg/kg, respectively. When the bags were stored in area containing chlordane-treated shelf paper but not in direct contact with it, the amounts of residues found in single and double thickness kraft paper bags were 0.4 and 0.3 mg/kg, respectively, whereas samples in polyethylene bags sorbed only 0.06 mg/kg chlordane. The flour in sealed commercial bags as purchased from grocery stores when kept in contact with chlordane-treated shelf paper acquired residue levels of 2.6 and 5.9 during storage intervals of 21 and 50 days, respectively.

3.2. Residues Arising from the Use of Dichlorvos Polyvinyl Strips

Dichlorvos (DDVP) is a contact and stomach insecticide with a fumigant action. The major use of this insecticide is in the form of Vapona Pest Strips, a slow release formulation in which dichlorvos is contained in a poly(vinyl chloride) matrix from which it slowly and continuously evaporates to give a vapor concentration that is effective for pest control in enclosed spaces. In a series of trials carried out in the United Kingdom, Australia, and France to determine levels of dichlorvos that occur in air when Vapona Pest Strips are used in domestic conditions, residue values up to 0.24 μg/L were observed (Elgar and Steer, 1972). When such slow release strips are used to control insects in kitchens and food shops, there is a likelihood that foodstuffs kept in these places will acquire insecticide residues.

Elgar et al. (1972a) conducted experiments in the United Kingdom and France to determine the residues of dichlorvos that occur in food prepared in kitchens in which Vapona Pest Strips had been placed. The housewives were asked to prepare food and drinks for one extra adult at each meal or whenever refreshments were taken. Each sample of food was left exposed until the equivalent course was consumed by the family members, then deep frozen and analyzed for dichlorvos residues. The results showed that residue concentrations from the samples collected 1, 6, and 10 days after hanging the strips were 0.03, 0.03, and 0.02 mg/kg for the homes in the United Kingdom and 0.02, 0.02, and less than 0.01 mg/kg for the French trial (Table 5).

In a similar experiment conducted in Modesto, California separate breakfast and dinner meals were secured from 15 homes in which Vapona Strips had been installed (Collins and DeVries, 1973). Of the 174 meals collected during a 13-week interval, only 18 breakfasts and 18 dinners had discernible dichlorvos residues (Table 5). Of these 36 samples, 24 contained dichlorvos at 0.02 mg/kg, while residue levels exceeding 0.03 and 0.04 mg/kg were found in seven and five samples, respectively.

As Vapona strips are also used for pest control in food shops, experiments

TABLE 5   Dichlorvos Residues in Meals Prepared in Domestic Kitchens Equipped with Dichlorvos Polyvinyl Chloride Strips

Country	Number of Samples	Time Interval after Hanging Dichlorvos Strip (Weeks)	Dichlorvos Residues (mg/kg)	
			Mean	Range
U.K.[a]	14	1	0.032	0.02–0.09
		6	0.032	0.02–0.05
		10	0.016	Less than 0.01–0.03
France[a]	14	1	0.019	Less than 0.01–0.07
		6	0.016	0.01–0.03
		10·	Less than 0.01	Less than 0.01–0.02
United States[a]	15	1	0.02	Less than 0.02–0.03
		8	Less than 0.02	All less than 0.02
		13	Less than 0.02	All less than 0.02

[a]Based on Elgar et al. (1972).
[b]Based on sum of breakfast and dinner values reported by Collins and DeVries (1973),

were carried out in the United Kingdom and France under practical conditions to estimate dichlorvos residues resulting from this practice (Elgar et al., 1972b). To find the maximum exposure of the consumers, the emphasis was placed on foodstuffs that are displayed unwrapped in shops and are eaten without further processing. In the United Kingdom and France, the 17 and 20 shops were covered and each foodstuff was purchased from 7 to 11 and 10 shops, respectively. About 78% of samples from the United Kingdom and 66% of samples from France contained dichlorvos residues at concentrations less than the limit of detection of 0.05 mg/kg. High fat food items such as fruit pastry and cheese contained relatively higher residues. The high level of dichlorvos found in apples (0.19 mg/kg) in the United Kingdom and 0.30 mg/kg in France) was considered to be due to retention of the insecticide in their waxy skin and longer exposure time as compared to other foods such as bread. The foodstuffs from the United Kingdom were also analyzed for dichloroacetaldehyde, a primary hydrolysis product of dichlorvos. However, its concentration in all the samples was less than its limit of determination of 0.03 mg/kg.

3.3.   Residues Arising from Insecticidal Lacquers and Paints

Besides their other uses in buildings, insecticides incorporated in lacquers and paints have also been used to combat crawling insects in homes, hospitals, kitchens, ships, and various other places. Studies have shown that use of such formulations in close proximity of foodstuffs can constitute a toxic hazard. Dyte (1960) reported that malathion lacquers emit vapors of the insecticide that were absorbed by the flour. Samples of flour and offal that had been in contact with an endrin lacquer in a British flour mill were found to acquire residue concentrations varying from 79 to 323 mg/kg (Dyte and Tyler, 1960). Substantial contamination was produced even by storage of flour in containers treated about 18 months before, and could result from as little as 3 hr contact with the lacquer.

3.4. Residues of Aryl/Alkyl Phosphates and Synergists in Foods

During the screening of composites of the Food and Drug Administration's Total Diet Program of chemical contaminants, Daft (1982) observed, in some samples, phosphorus-containing residues with gas chromatographic retention times longer than the pesticide residues normally encountered. Further studies identified these compounds as esters of phosphoric acid that are used as flame retardants in the manufacture of various packaging materials such as paper goods, adhesives, and plastics. The assay of a margarine stick along with its tissue-lined wrapper revealed the presence of 2-ethylhexyldiphenyl phosphate at a level of 50 mg/kg. The margarine stick itself contained this compound at a concentration of 20 mg/kg. Similarly, plastic bread bags were found to contain this ester up to 400 mg/kg, whereas the bread samples had the residue below 0.5 mg/kg. In both cases, the incurred ester appeared to have migrated from the wrapper to the food commodities. Daft (1982) noticed that phosphate ester diffusion appeared to occur in foods that were moist or lipoid in nature or had long contact with a contaminated wrapping. Some of the phosphate esters used as flame retardants have a toxicity similar to that of organophosphate insecticides in test systems with fish (Sasaki et al., 1981). However, little is known about toxicity of these chemicals to human beings.

During the analysis of human milk samples for insecticide residues in Japan, Miyazaki et al. (1980) observed an unknown additional peak. The application of alternate gas chromatographic columns and mass spectroscopic analysis identified the unknown compound as 1, 1'-oxybis(2, 3, 3, 3-tetrachloropropane) or S-421 (Miyazaki et al., 1981; Miyazaki, 1982). This compound is used as a synergist for pyrethroid insecticides against insects such as mosquitoes, houseflies, and cockroaches. The average S-421 residues in human milk samples from the Tokyo metropolitan area were 2.0, 1.5, and 0.5 ng/g on a whole milk basis in 1978, 1979, and 1980, respectively. The maximum concentration observed was 3.8 ng/g in a sample collected in 1979. The analysis of mosquito coils and aerosols used to control mosquitoes and houseflies also revealed the presence of this compound. Miyazaki (1982) considered that the presence of S-421 residues in human milk may be due to the use of commercial insecticide formulations containing this synergist. The toxicological significance of S-421 in human milk at levels encountered in this study is not clearly known (Miyazaki et al., 1981).

4. RESIDUES IN FOODS ARISING FROM INSECTICIDE USE FOR STRUCTURAL PEST CONTROL

Termites, commonly known as white ants, are common structural pests in almost all the warmer parts of the world. As termites are insidious insects operating inconspicuously, their presence in dwellings and other buildings is generally noticed only when they have become established and are in a position to cause considerable damage. Because their control at this stage involves substantial labor and money, prophylactic and remedial antitermite measures are taken in buildings as a routine manner. As a preventive measure, aqueous emulsions of organochlorine insecticides are used as a soil drench to form an insecticidal barrier between the soil and woodwork of the building. The follow-up termite control measures consist of application of insecticides to soil, wall voids, chimney bases, termite-attacked woods, under basement floors, and other relevant sites.

To avoid frequent applications of the insecticides, chemicals such as aldrin, dieldrin, chlordane, and heptachlor are commonly employed for termite control. These compounds are notorious for their persistence and retain their toxicity for years. When Metcalf (1975) devised a pest management rating of insecticides to provide a rationale for the intelligent selection of insecticides, he segregated such chemicals into a group that had little place in pest management and suggested that these insecticides may be used for the control of structural pests only. However, the residues of insecticides used for termite control have been detected in air samplings of homes receiving such treatments and some studies have indicated that even such use of these chemicals can pose environmental problems and cause contamination of the foods.

Of 498 ground-floor apartments sample in 1980, Livingston and Jones (1981) found levels of chlordane ranging from not detectable to 37.8 μg/m^3. In view of the observation that the majority (77%) of the apartments contained insecticide concentrations above the detection limit, it was considered that the presence of chlordane vapors may be quite widespread in dwellings that have been treated with this insecticide for termite control. Wright and Leidy (1982) monitored ambient air for 1 year after insecticide treatment given by pest control firms to six houses infested with active subterranean termites. In three houses treated with chlordane, the highest average concentration (5.01 μg/m^3) was observed after 12 months of application. In the other three houses treated with formulations containing chlordane as well as heptachlor, the highest mean concentrations (5.81 and 1.80 μg/m^3, respectively) were found 6 months after application.

In a survey conducted in 1979-1980 in Western Australia to evaluate the effectiveness of restrictions placed on the use of organochlorine pesticides, Stacey et al. (1984) noted decreases in the concentrations of hexachlorobenzene (0.025–0.008 mg/kg) and total DDT (0.078–0.046 mg/kg) in samples of human milk. However, dieldrin level showed an increase from 0.005–0.009 mg/kg. Statistical analysis of these data suggested a correlation between dieldrin levels in breast milk and insecticidal treatment of the houses for the control of termites. As pesticide consumption patterns indicated that aldrin was the main insecticide used for that purpose at that time, the authors considered that its use might have contributed toward the higher levels of dieldrin, a conversion product of aldrin, in human milk. Considering that breast milk forms an important part of the diet of an infant, the presence of residues of a toxic compound such as dieldrin in it warrants concern. Assuming that a neonate weighing 3 kg consumes 600 mL of mother's milk in a day, dieldrin present at an average level of 0.009 mg/kg would result in an intake of 1.8 μg/kg body weight/day. This level is 18 times higher than the acceptable daily intake of dieldrin of 0.001 mg/kg body weight/day.

To confirm that the use of insecticides in houses to control termites leads to an increase in their level in human milk, Stacey and Tatum (1985) studied 14 houses located in the Perth metropolitan area that had been treated during the previous 12 months with chemicals for termites. Aldrin had been used in only three houses covered in this study and results showed that levels of dieldrin in the human milk of residents continued to increase up to about 8 months after treatment. The breast milk from donors living in houses that had received aldrin treatments in the past (more than a year ago) also contained dieldrin at levels equal to or greater than the mean values of the previous survey by Stacey et al. (1984). This suggests that aldrin treatment for termite-proofing of the houses can lead to contamination of human milk for years and the decline in dieldrin concentrations in human milk would take a long time, even after stopping the use of aldrin.

In contrast, changes in levels of both chlordane and heptachlor residues in human milk after treatment of houses with these insecticides were more rapid. For both chemicals, their levels or the levels of their metabolites began to rise soon after treatment, attaining maximum concentrations after about 1 month and declining to near pretreatment values by the fourth month (Figure 3).

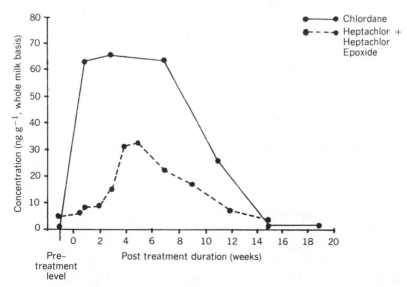

Figure 3. Effect of treatment of dwellings with chlordane and heptachlor for termites on their residue levels in human milk of the inhabitants. Based on Stacey and Tatum (1985).

TABLE 6   Uptake of Insecticide Residues by Carpet Swatch Samples (15.2 × 7.6 × 1.3 cm) Kept for 24 hr in Houses before and after Antitermite Treatments[a]

Insecticidal Treatment	Residue Level[b]	
	Pretreatment	Posttreatment
1% chlordane	0.12	2.25
	(0.01–0.36)	(1.45–3.62)
0.5% chlordane and	0.01	1.15
		(0.96–1.38)
0.25 heptachlor	0.01	0.39
		(0.15–0.71)

[a]Based on data of Wright and Leidy (1982).
[b]Microgram per swatch, mean of three replicates (range).

Stacey and Tatum (1985) considered that, as houses were treated in such a way that the inhabitants had no chance of coming in contact with treated surfaces, the increase in levels of insecticides in human milk in the posttreatment period might have occurred through inhaling the vapors over a period of time. However, Wright and Leidy (1982) observed that carpet swatches (15.2 × 7.6 × 1.3 cm) placed on floors of houses for 24 hr after antitermite treatment absorbed 0.15–3.62 µg of the insecticide used (Table 6). As discussed in earlier parts of this section, a number of studies have detected airborne residues of insecticides in houses even long after receiving antitermite applications. So there is a possibility that edible food items kept in such dwellings might be absorbing insecticide residues from the ambient air, thus resulting in their oral intake through contaminated food. However, as Stacey and Tatum (1985) did not analyze air or food samples from the treated houses, no conclusion regarding the route of contamination of human milk can be drawn.

5. TRANSFERENCE OF INSECTICIDE RESIDUES TO FOODS FROM FREIGHT CONTAINERS MADE FROM TREATED TIMBER

The economics of present day world trade depend heavily on the transport of various commodities including food items by cargo ships. Because of its favorable strength-to-weight ratio and physical properties that enable it to withstand shock loads, wood has been extensively used in the construction of freight containers as flooring or protective claddings on the metal walls. In recent years, increasing amounts of plywood are also being used for this purpose. The high humidity of the marine environment makes timber and plywood highly susceptible to insect attacks. Moreover, quarantine services of some countries require that to qualify for immediate release of food commodities on import, timber used for the construction of freight containers must have been treated against attack. Otherwise, the freight containers must be unpacked to allow inspection for insect infestation. To avoid this costly and time-consuming procedure, preservatives that remain effective for years have been employed to immunize the wood used in freighters against attack. Wood preservatives containing several organochlorine compounds such as aldrin, dieldrin, chlordane, and lindane are approved for the treatment of timber under the Australian Quarantine Regulations. As other international recommendations on this matter are not available, treatments recommended by the Australian authorities have been adopted widely as a means of protecting timber from attacks by pests (Fishwick and Rutter, 1982; Paton, et al., 1984). However, some recent studies have indicated that foodstuffs stored on or near treated wood in cargo ships can pick up substantial insecticide residues, which in some situations may exceed the maximum residue limits (MRLs) set by the Codex Alimentarious Commission.

Attention concerning the possibility of contamination of foodstuffs with pesticide residues during their transport in freight containers made from treated wood was drawn by Rutter (1978) when a consignment of imported carobs, *Ceratonia siligua* (L.), was found to contain about 22 mg/kg dieldrin along with small amounts of lindane, aldrin, and endrin. This combination of residues seemed unlikely to have arisen from common agricultural practices and inquiries from the region of origin of consignment revealed that dieldrin had not been used on the carobs. As dieldrin had been widely used in the United Kingdom during industrial pretreatment of lumber, consideration was given to the possibility of the transfer of these residues from timber used for construction of freighters.

To confirm this, Rutter (1978) kept samples of two adsorbent materials (charcoal and peat) in well-used freight containers having wooden floors. In 2 weeks, both the materials had taken up dieldrin residues greater than 2 mg/kg. The analysis of scrappings of the floor timber revealed the presence of about 10 mg/kg dieldrin. Further tests performed by storing a range of food commodities on or near the wooden surfaces of a simulated container demonstrated that dieldrin residues up to 94 mg/kg along with lesser amounts of endrin, aldrin, and lindane could result from such exposure. The storage of several food commodities in two unused commercial containers was also found to result in a similar uptake of dieldrin residues (Table 7).

Fishwick and Rutter (1982) noted that a large proportion of freight containers being constructed in recent times are lined with plywood that is treated with organochlorine insecticides by adding them to the resin used to glue the thin layers of wood being bonded. The extent to which the insecticides from treated plywood can migrate to foodstuffs stored nearby was investigated by using a simulated freight container having a floor made from plywood impregnated with a commercial formulation preservative understood to contain dieldrin as one of its active ingredients. The analysis of the plywood revealed that in addition to dieldrin at a level of 42 mg/kg, it also contained aldrin at about 350 mg/kg and lindane at 52 mg/kg. Samples of several foodstuffs in polyethylene bags were kept inside the container either on the floor or suspended in the container about 1 m

TABLE 7 Sorption of Dieldrin Residues (mg/kg) by Some Food Commodities by Their Storage in Two Commercial Containers with Floors Made from Dieldrin-Treated Wood[a]

| | Food Commodity | | | | | | | |
| | Carobs | | Soya Beans | | Groundnuts | | Flour | |
Mode of Exposure	C-1[b]	C-2[c]	C-1	C-2	C-1	C-2	C-1	C-2
In tray on floor	1.01	1.5	0.7	1.5	0.7	1.6	1.6	2.8
Sealed polyethylene bag on floor	1.8	13.0	3.4	26.0	3.2	17.8	12.0	59.0
Sealed polyethylene bag suspended 2 m above the floor	0.1	0.6	0.1	5.8	0.2	4.4	0.5	2.4

[a]Based on data of Rutter (1978).
[b]C-1 Storage in a commercial container for 1 week.
[c]C-2 Storage in another commercial container for 2 weeks.

above the floor. After an exposure of 2 weeks, the samples were removed and analyzed for pesticide residues (Table 8). The experiments were also performed with a well-used commercial freight container with a floor made from insecticide-treated plywood.

In the case of the simulated container that had flooring made from freshly treated plywood, aldrin, dieldrin, and lindane residues absorbed by foods stored on the floor ranged from 0.26 to 1.10, 0.04 to 0.15, and 0.09 to 0.32 mg/kg, respectively (Table 8). For the well-used freight container, aldrin, dieldrin, and lindane residues up to 0.31, 0.29, and 0.20 mg/kg were acquired. The food commodities suspended in polyethylene bags and not in direct contact with

TABLE 8 Insecticide Residues (mg/kg) Acquired by Some Food Commodities after Their Storage for 2 Weeks in Polyethylene Bags on Floors of Freight Containers[a]

| | Carobs | | Flour | | Groundnuts | | | |
Insecticide Residue	SC[b]	CC[c]	SC	CC	SC	CC	Rice SC	Tea SC
Aldrin	0.26	0.07	0.40	0.15	0.28	0.31	0.70	1.10
Dieldrin	0.04	0.23	0.10	0.06	0.08	0.29	0.10	0.15
Lindane	0.09	0.03	0.13	0.20	0.09	0.04	0.12	0.32

[a]Based on data of Fishwick and Rutter (1982).
[b]SC, simulated freight container having floor made from plywood recently treated with dieldrin.
[c]CC, well-used commercial container having insecticide-treated plywood floor.

plywood generally accumulated comparatively less, but still substantial amounts of residues. This suggested that transfer of insecticides can take place exclusively via the vapor phase and might be making a significant contribution toward food contamination.

Fishwick and Rutter (1982) also conducted a limited survey of organochlorine residues in foodstuffs taken from commercial loads carried in freight containers to assess the nature and magnitude of residues that can be expected under actual transport conditions. In 45 samples of cocoa beans analyzed, the highest dieldrin and lindane concentrations encountered were 0.08 and 0.029 mg/kg, respectively. The estimation of insecticide residues in 75 samples of other commodities such as groundnuts, almonds, rice, lentils, beans, and sugar revealed the presence of various compounds at low levels. The maximum concentrations of aldrin, dieldrin, and lindane observed in these miscellaneous food items were 0.04, 0.012, and 0.026 mg/kg, respectively. Chlordane was not detected in bulk samples of the various commodities. However, floor sweepings taken from one freight container and consisting chiefly of whole groundnuts were found to contain chlordane at 0.80 mg/kg. In general, these residue levels were considered unlikely to present a health hazard to consumers.

A series of trials conducted by the Australian Quarantine Service to investigate the sorption of organochlorine insecticides by flour stored on or near treated plywood or laminated timber used in freight containers have been reported by Paton et al. (1984). A test was conducted using 1- to 2-year-old empty freight containers (internal volume 36.2 m³) selected from a group that had recently arrived from overseas. Five of these had plywood floors treated with chlordane whereas three containers had sawn timber floors impregnated with dieldrin. Samples of flour (300 g) were kept either directly on the floor or suspended in polyethylene bags 1 m above the floor in the middle of the container. The analysis of samples after 14 days exposure revealed that flour kept on the floor as well as in polyethylene bags had absorbed appreciable amounts of insecticide residues even though the containers had been in commercial service for up to 2 years (Table 9).

TABLE 9   Range of Insecticide Residues (mg/kg) Absorbed by Flour Kept in Commercial Containers for 14 days[a]

	Type of Flooring	
Mode of Exposure	Chlordane-Treated Plywood[b]	Dieldrin-Treated Sawn Timber[c]
Storage on floor	0.2–0.6	0.4–6.8
Polyethylene bags suspended in middle	0.1–0.3	0.1–0.3

[a]Based on Paton et al. (1984).
[b]Chlordane residues.
[c]Dieldrin residues.

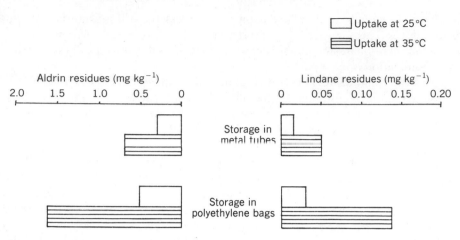

Figure 4.   Effect of temperature on uptake of aldrin and dieldrin residues by wheat flour. Based on Paton et al. (1984).

When Paton et al. (1984) performed experiments with similar small-scale containers with floors constructed from dieldrin-treated laminated sawn timber, residue levels up to 125 mg/kg dieldrin were observed in stored flour. As this trial indicated that foodstuffs stored near freshly treated laminated timber floors of freight containers could result in residues exceeding the Maximum Residue Limits set by the Codex Alimentarius Commission for international food legislation, the authors reported that the Australian Plant Quarantine Service has withdrawn the approval for the use of organochlorine insecticides for immunizing laminated sawn timber.

In view of the variations in volatilization rates of insecticides with temperature changes, Paton et al. (1984) also studied the uptake of insecticide residues by flour at two temperatures. Ten small-scale metal containers ($20 \times 20 \times 50$ cm) with floors of plywood treated with aldrin and lindane were constructed and 50 g flour samples were kept in these either in metal tubes placed vertically on the floor or in polyethylene bags suspended in the middle. The containers were then sealed and a set of five was maintained at 25°C and 40% relative humidity (RH) and another similar set was kept at 35°C and 40% RH. The residue analysis of flour samples after 40 days exposure period showed that flour at the higher temperature had taken up two to five times more residues (Figure 4). The authors concluded that as internal temperatures up to 50°C commonly occur inside the closed freight containers, the intensity of heat can have a significant effect on the magnitude of contamination of the foodstuffs.

6. CONCLUSIONS

The occurrence of insecticide residues in food commodities from nonagricultural usage of insecticides has received little attention until now. However, it is apparent from the available literature that the potential exists for the inadvertent transfer of significant amounts of insecticide residues from treated surfaces to foods in a variety of situations. Though the levels of residues reported in several studies do not seem to be hazardous, their presence in human food cannot be ignored and deserves greater attention in the future.

The literature reviewed suggests that the probability of unintentional food contamination is highest for chemicals with relatively high stability, such as organochlorine compounds. Of the different sources of incidental contamination, indoor use of insecticides for mosquito control in antimalarial programs should merit comparatively more attention. Although the most intensive use of insecticides in developing countries has been in public health programs, there are very few reports on the impact of such usage on human health and the environment. So detailed studies to delineate the extent and routes of contamination of various commodities by this mode of insecticide use seem necessary. Among the various food items, especially high insecticide residues have been observed in bovine and human milk. Due to the significance of human milk in infant nutrition, efforts to reduce contamination of this substrate demand priority.

The various aspects of unconventional routes of contamination of foods discussed in this review are not exhaustive but only indicative. The personnel involved in the regulation and determination of insecticide residues in foodstuffs must remain attentive and carefully observant toward the possibility of more and more routes of incidental contamination coming to light with further studies.

REFERENCES

Büchel, K. H. (1983). "Political, economic and philosophical aspects of pesticide use for human welfare." In J. Miyamoto and P. C. Kearney, eds., *Pesticide Chemistry: Human Welfare and the Environment*, Pergamon Press, Oxford, Vol. 1, pp. 3–19.

Busvine, J. R. (1980). *Insects & Hygiene*. Chapman & Hall, London, pp. 125–133.

Collins, R. D., and DeVries, D. M. (1973). "Air concentrations and food residues from use of Shell's No-Pest insecticide strip." *Bull. Environ. Contam. Toxicol.* **9**, 227–233.

Daft, J. L. (1982). "Identification of aryl/alkyl phosphate residues in foods." *Bull. Environ. Contam. Toxicol.* **19**, 221–227.

Davies, J. E., Edmundson, W. F., and Raffonelli, A. (1975). "The role of housedust in human DDT pollution." *Am. J. Public Health*, **65**, 53–57.

Dyte, C. E. (1960). "Preliminary tests of an insecticidal lacquer containing malathion." *Pest Technol.* **2**, 98–99.

Dyte, C. E., and Tyler, P. S. (1960). "The contamination of flour by insecticidal lacquers containing endrin and dieldrin." *J. Sci. Fd. Agr.* **11**, 745–750.

Elgar, K. E., and Steer, B. D. (1972). "Dichlorvos concentrations in the air of houses arising from the use of Dichlorvos PVC strips." *Pestic. Sci.* **3**, 591–600.

Elgar, K. E., Mathews, B. L., and Bosio, P. (1972a). "Dichlorvos residues in food arising from the domestic use of Dichlorvos PVC strips." *Pestic. Sci.* **3**, 601–607.

Elgar, K. E., Mathews, B. L., and Bosio, P. (1972b). "Vapona strips in shops—Residues in foodstuffs." *Environ. Qual. Safety* **1**, 217–221.

Farvar, M. T. (1979). "Collection of human biological specimens in developing countries for monitoring organochlorine compounds." In A. Berlin, A. H. Wolff, and Y. Hasegawa, eds., *The Use of Biological Specimens for the Assessment of Human Exposure to Environmental Pollutants.* Nijhoff, The Hague, pp. 155–163.

Fishwick, F. B., and Rutter, I. (1982). "Transfer of insecticide residues to foods stored on or near plywood impregnated with organochlorine insecticides." *Pestic. Sci.* 13, 263–268.

Fontaine, R. E. (1978). *House Spraying with Residual Insecticides with Special Reference to Malaria Control.* World Health Organization, WHO/VBC/78, 704; WHO/VBC/78.904.

Highland, H. A., Jay, E. G., Phillips, M., and Davis, D. F. (1966). "The migration of piperonyl butoxide from treated multiwall kraft bags into four commodities." *J. Econ. Ent.* 59, 543–545.

Hornabrook, R. W., Dyment, P. G., Gomes, E. D., and Wiesman, J. S. (1972). "DDT residues in human milk from New Guinea natives." *Med. J. Aust.* 25, 1297–1300.

ICAR (1967). *Report of Special Committee on Harmful Effects of Pesticides.* Indian Council of Agricultural Research, New Delhi, 93 pp.

Jensen, A. A. (1983). "Chemical contaminants in human milk." *Residue Rev.* 79, 1–128.

Joia, B. S., Chawla, R. P., and Kalra, R. L. (1978). "Residues of DDT and HCH in wheat flour in Punjab." *Ind. J. Ecol.* 5, 120–127.

Joia, B. S., Singh, P. P., and Battu, R. S. (1985)." Organochlorine insecticides in human milk in Punjab, India." *International Conference on Pesticide Toxicity, Safety and Risk Assessment, October 27–31, 1985.* Industrial Toxicology Research Centre, Lucknow, India.

Kalra, R. L., and Chawla, R. P. (1981). "Monitoring of pesticide residues in the environment." In B. V. David, ed., *Indian Pesticide Industry—Facts and Figures.* Vishwas Publications, Bombay, pp. 251– 285.

Kalra, R. L., and Chawla, R. P. (1983). *Studies on Pesticide Residues and Monitoring of Pesticidal Pollution.* Punjab Agricultural University, Ludhiana, pp. 180–190.

Kalra, R. L., and Chawla, R. P. (1985). "Pesticidal contamination of foods in the year 2000 AD." *Proc. Ind. Natl. Sci. Acad.* B.52, 188–204.

Kalra, R. L., Chawla, R. P., Sharma, M. L., Battu, R. S., and Gupta, S.C. (1983). "Residues of DDT and HCH in butter and ghee in India." *Environ. Pollut.* 6B, 195–206.

Kalra, R. L., Chawla, R. P., Battu, R. S., Singh, P. P., Udeaan, A. S., and Joia, B. S. (1986). "Magnitude and sources of DDT and HCH residues in wheat in Punjab." *National Conference on Short-Term and Long-Term Hazards of Pesticides and Strategies for their Safe Use,* February 24–26, 1986, New Delhi.

Kapoor, S. K., Chawla, R. P., and Kalra, R. L. (1980). "Contamination of bovine milk with DDT and HCH residues in relation to their usage in malaria control programme. *J. Environ. Sci. Health* B15, 545–547.

Keil, J. E., Winklea, J. F., Pietsch, R. L., and Godsden, R. H. (1969). "A pesticide use survey of urban households." *Agr. Chem.* 24, 10–12.

Lewis, R. G., and Lee, R. E. Jr. (1976). "Air pollution from pesticides: Sources, occurrence and dispersion." In R. E. Lee, Jr., ed., *Air Pollution from Pesticides and Agricultural Processes.* CRC Press, Cleveland, pp. 30–32.

Livingston, J. M., and Jones, C. R. (1981). "Living area contamination by chlordane used for termite treatment." *Bull. Environ. Contam. Toxicol.* 27, 406–411.

Metcalf, R. L. (1975). "Insecticides in pest management." In R. L. Metcalf and W. Luckman, eds., *Introduction to Insect Pest Management.* Wiley Interscience, New York, pp. 265–266.

Miyazaki, T. (1982). "Residues of the synergist S-421 in human milk collected from the Tokyo Metropolitan area." *Bull. Environ. Contam. Toxicol.* 29, 566–569.

Miyazaki, T., Akiyama, K., Kaneko, S., Horii, S., and Yamagishi, T. (1980). "Chlordane residues in human milk." *Bull. Environ. Contam. Toxicol.* 25, 518–523.

Miyazaki, T., Kaneko, S., Horii, S., and Yamagishi, T. (1981). "Identification of the synergist bis(2, 3, 3, 3-tetrachloropropyl) ether in human milk." *Bull. Environ. Contam. Toxicol.* **26**, 420–423.

Olszyna-Marzys, A. E., deCampos, M., Farvar, M. T., and Thomas, M. L. (1973). "Residues de plagincidas clorados en leche humana y de vaca de Guatemala." *Bol. Of. Sanitaria Panamericana*, 75. (Cited in Winter et al., 1976.)

Paton, R., Luke, B. G., and Roerts, G. (1984). "Studies on the sorption of organochlorine insecticides by flour stored on or near treated laminated timber or plywood as used in freight containers." *Pestic. Sci.* **15**, 624–629.

Queen, W. A. (1953). "Distribution characteristics and adsorption of characteristics of vaporized lindane." *Bull. Assoc. Fd. Drug Off. U.S.* **17**, 127–139.

Radomski, J. L., Deichman, W. B., Clizer, E. E., and Rey, A. (1968). "Pesticide concentrations in the liver, brain and adipose tissue of terminal hospital patients." *Fd. Cosmet. Toxicol.* **6**, 209–220.

Rutter, I. (1978). "Pesticide residues in food arising from transport in freight containers." *Chem. Indust.* July 15, 531–532.

Sasaki, K. M., Takeda, M., and Uchiyama, M. (1981). "Toxicity, absorption and elimination of phosphoric acid triesters by killfish and goldfish." *Bull. Environ. Contam. Toxicol.* **27**, 775–782.

Seiber, J. M., Kim, Y., Wehner, T., and Woodrow, J. E. (1983). "Analysis of xenobiotics in air." In J. Miyamoto and P. C. Kearney, eds., *Pesticide Chemistry: Human Welfare and the Environment*, Pergamon Press, Oxford, Vol. 4, pp. 3–12.

Siakotos, A. N. (1956). "Package exposure to continuously vaporized lindane." *J. Econ. Ent.* **49**, 481–484.

Singh, P. P., Battu, R. S., Joia, B. S., Chawla, R. P., and Kalra, R. L. (1986). "Contribution of DDT and HCH used in malaria control programme towards the contamination of bovine milk." *Proc. Symp. Pesticide Residues and Environmental Pollution*, S. D. College, Muzaffarnagar, pp. 86–92.

Singh, P. P., Battu, R. S., and Kalra, R. L. (1988). "Insecticide residues in wheat grains and straw arising from their storage in premises treated with BHC and DDT under malaria control program." *Bull. Environ. Contam. Toxicol.* **40**, 696–702.

Spindler, M. (1983). "DDT: Health aspects in relation to man and risk/benefit assessment based thereupon." *Residue Rev.* **90**, 1–34.

Stacey, C. I., and Tatum, T. (1985). "House treatment with organochlorine pesticides and their levels in human milk—Perth, Western Australia." *Bull. Environ. Contam. Toxicol.* **35**, 202–208.

Stacey, C. I., Perriman, W. S., and Whitney, S. (1984). "Organochlorine pesticide residue levels in human milk, Western Australia 1970–80." *Arch. Environ. Health* **40**, 102–108.

Takahashi, W., Saidin, D., Takei, G., and Wong, L. (1981). "Organochlorine pesticide residues in human milk in Hawaii, 1979–80." *Bull. Environ. Contam. Toxicol.* **27**, 506–511.

WHO (1971). *The Place of DDT in Operations against Malaria and Other Vector-Borne Diseases*. Official records of World Health Organization, Geneva, No. 190, App. **14**, pp. 176–182.

Winter, M., Thomas, M., Wernick, S., Levin, S., and Farvar, M. T. (1976). "Analysis of pesticide residues in 290 samples of Guatemala mother's milk." *Bull. Environ. Contam. Toxicol.* **16**, 652–657.

Wright, C. G., and Jackson, M. D. (1971). "Propoxur, chlordane and diazinon on porcelain China saucers after kitchen cabinet spraying." *J. Econ. Ent.* **64**, 457–459.

Wright, C. G., and Jackson, M. D. (1975). "Insecticide residues in nontarget areas of rooms after two methods of crack and crevice application." *Bull. Environ. Contam. Toxicol.* **13**, 123–138.

Wright, C. G., and Leidy, R. B. (1978). "Chlorpyrifos residues in air after application to crevices in rooms." *Bull. Environ. Contam. Toxicol.* **20**, 340–343.

Wright, C. G., and Leidy, R. B. (1980). "Insecticide residues in the air of buildings and pest control vehicles." *Bull. Environ. Contam. Toxicol.* **24**, 582–589.

Wright, C. G., and Leidy, R. B. (1982). "Chlordane and heptachlor in the ambient air of houses treated for termites." *Bull. Environ. Contam. Toxicol.* **28**, 617–623.

Wright, C. G., Leidy, R. B., and Dupree, H. E., Jr. (1981). "Insecticides in the ambient air of rooms following their application for control of pests." *Bull. Environ. Contam. Toxicol.* **26**, 548–553.

Wright, C. G., Leidy, R. B., and Dupree, H. E., Jr. (1984). "Chlorpyrifos and diazinon detection on surfaces in dormitory rooms." *Bull. Environ. Contam. Toxicol.* **32**, 259–264.

Yeo, C. Y., and Bevenue, A. (1969). "The adsorption of chlordane by wheat flour from chlordane-treated shelf paper." *J. Stored Product Res.* **5**, 325–336.

11

ENVIRONMENTAL CONTAMINATION OF FOOD DERIVED BY NITROSAMINES FROM PESTICIDES

Ali Khayat

Beatrice/Hunt-Wesson, Inc., Fullerton, California

1. INTRODUCTION

It has been over 30 years since Magee and Barnes (1956) discovered that nitrosodimethylamine was an animal carcinogen. Since then, an additional 150 or more N-nitroso compounds have been identified as carcinogens (Magee et al., 1976). Animal studies with nitrosamines have shown that these compounds are carcinogenic, although there has been no record that nitrosamines cause human cancer.

Foods have been the potential link between nitrosamines and human carcinogenesis (Crosby and Sawyer, 1976), especially those foods that have been preserved by nitrites and nitrates. Nitrosamine contaminants in certain pesticide formulations, especially dinitroaniline and some acidic herbicides, have been firmly established. These contaminants arise from either chemical synthesis, as with dinitroaniline, or from nitrosation of amines and amides, as with certain halobenzoic acid formulations.

Ross et al. (1976) first reported the powerful carcinogen N-nitrosodimethylamine (NDMA) and N-nitrosodi-n-propylamine (NDPA) were found at ppm levels in commercial herbicide formulations. Since the discovery of N-nitroso contaminants in certain pesticides, a wide variety of products have been analyzed in response to a request from the United States Environmental Agency (Cohen et al., 1978; Bontoyan et al., 1976; Zweig et al., 1980; Kearney, 1980). Products include dinitroanilines, dimethylamine and ethylamine salts, phenoxyalkanoic acids, quaternary salts, amides, carbamates, organophosphates, triazines, urea derivatives, and many other pesticides. The potential magnitude of the N-nitroso problem has been indicated by a list of 51 nitrosated pesticides reviewed by Kearney (1980). However, the EPA analysis survey reveals that for positive detection N-nitroso compounds in pesticide formulations is limited to dinitroanalines, amines, and ethylamine salts of acidic herbicides and quaternary salts.

Although the nitrosamine contaminants occur at low levels (parts per million), many have been found to cause cancer in laboratory animals and may present a hazard to pesticide users. In pesticides nitrosamines are formed by the action of nitrosating agents on secondary amines in manufacturing processes, by the use of nitrites, nitrate as container corrosion inhibitors, or as impurities in amine reagents used in synthesis.

Occupational exposure to dimethylnitrosamine is reported to cause jaundice in humans and is shown to be hepatotoxin and a hepatocarcinogen in rats. Many homologous series of these nitrosamines have been synthesized and most have been reported to be carcinogenic to experimental animals. In 1962 an epidemic in stricken sheep in Norway was traced to the consumption of nitrate-treated fish meal (Edner et al., 1964). Dinitrosamine formation in nitrate-treated fish raised the possibility that a similar event might occur in meat and other dietary products treated with nitrites or nitrates.

Formation of N-nitroso compounds has been observed from all types of amines, amides, and quaternary ammonium compounds and many other nitrogen-containing compounds (Mirvish, 1975). The presence of volatile nitrosamines in pesticide formations was first reported by Ross (1976). Of the pesticides tested by Ross et al. (1976), three formulations with dimethylamine salts of acidic herbicides were found to contain 0.3–640 ppm of *N*-nitrosodimethylamine (NDMA). One formulation of the dinitroaniline, trifluralin, was found to contain 154 ppm *N*-nitrosodipropyl amine (NDPA). Ross et al. (1977) observed no detectable nitrosamines in the air, water, or crops as the result of treflan-containing NDPA application. The nitrosamines are either simple nitrosodialkyl amines or the nitroso derivative of the parent pesticides. Formation of nitrosated pesticides in soil or water would not appear to be a major problem.

In vivo nitrosation of pesticides and dialkyamines has been studied. Trace amounts of nitrosated carbamates and low yields of nitrosated dialkyamines were observed as a result of feeding rats and guinea pigs the parent compound and nitrite. No evidence of nitrosoatrazine in stomach contents, tissue, and milk was found when atrazine, and nitrite, was fed to rats and goats. Nitrosamines are photolabile and the volatile members of the class are partially dissipated by volatilization and subsequent photodecomposition in air. Most nitrosamines are stable to hydrolysis in aqueous solution at the pHs usually found in ground and surface waters and are photodegraded in solutions. The nitrosodialkyamines are rapidly dissipated in soil. When crops were exposed to ^{14}C-labeled nitroso compounds, no nitrosated products were identified in either the foliage or grain.

Uptake of radioactivity into plants has been demonstrated by ^{14}C-labeled nitroso compounds. However, no nitrosated products have been identified in either the foliage or grain of the crops studied (Kearney et al., 1977).

Among the pesticides carrying dimethylated amino groups, the dithiocarbamates, dimethylureas and phosphine, form NDMA. Among the drugs carrying tertiary amino groups, amino pyrine became the best known to produce NDMA. Malignant tumors of livers and lungs in rats given amino pyrine together with nitrates have been observed. In the presence of nitrite at low pH, nitrosocarbaryl is produced which is mutagenic to *Escherichia coli, Haemophilus influenzae*, and *Saccharomyces cerevisiae* and is carcinogenic to rats. Other monomethyl carbamate pesticides, such as aldicarb, buturon, carbofuran, linuron, methomyl, and propoxur, also form N-nitroso derivatives.

Dimethylnitrosamine was formed in samples of soil, municipal sewage, and lake water, supplemented with 250 ppm of dimethylamine and 100 ppm of nitrite. The toxic compound was formed in similar quantities in soil and sewage samples sterilized by gamma radiation and in lake water sterilizer by autoclaving. No nitrosamine was produced from the precursors when the soil was freed of organic matter by ignition (Kearney, 1980).

Incubation of the ignited soil with glucose and an inoculum of microorganisms led to the formation of nonleachable organic matter and the return of the nitrosating capacity.

Fine et al. (1977) reported NDMA levels in dimethylamine formulations of 2, 3, 6-trichlorobenzoic acid to be as high as 0.06%. Volatile N-nitrosamines have been found to be present at the 200–600 mg/L level in amine base pesticide formulation that have been tested. These products are applied as a fine aqueous mist and it must be assumed that drifting winds could expose herbicides applicators to levels in excess of 100 μg/day.

Reliable knowledge about potential human health risks due to exposure to environmental N-nitroso compounds is of relatively recent origin. After the formation of carcinogenic nitrosamines in tobacco smoke was first suggested by Druckrey and Preussmann (1962), Ender et al. (1964) provided the first evidence of occurrence of N-nitrosodimethylamine in nitrite treated fish meal. Sakshaug et al. (1965) and Sander (1967) were the first to provide unequivocal nitrosamine formation from precursors after an earlier negative results in an animal experiment.

The present knowledge of human exposure to nitrosamines is illustrated in Table 1 (Preussmann et al., 1983). The total exposure is subdivided into exposure to preformed N-nitroso compounds, exogenous exposure, which is subdivided into life-style and occupational exposure and endogenous exposure, *in vivo* formation of such compounds from precursors, for example, nitrosable amino groups and nitrosating agents.

Confirmation of the observations in regards to nitrosamine formations and evaluation of the potential exposure to man and his environment and the reduction of the impurity were pursued vigorously. Since the original reports on nitrosamines associated with pesticides, improvements in pesticide manufacturing processes and formulations have reduced the concentration of nitrosamine in most pesticide formulations. Extensive monitoring studies have also been conducted on environmental residues resulting from nitrosamine impurities in pesticide formulations.

Exposure to nitrosamine from tobacco and tobacco smoke has been reported by Hoffman et al. (1982) and Hecht et al. (1983). It is evident from this report that the tobacco smoke is responsible for the highest nonoccupational nitrosamine

TABLE 1   Total Exposure to Nitrosamines

External Exposure		Internal Exposure	
Living Habits	Work Exposure	Precursor's Uptake	Precursor's Formation
Cigarette smoke Drugs	Chemical industry Rubber, leather, metal industry	Nitrite Nitrous gases	Nitrite from nitrate (gastric juices, saliva)
Food Indoor air	Fish factory Pesticides	Nitrosable amino compounds	
Household commodities			

exposure known so far. Data reported by Preussmann et al. (1983) indicate that the nitrate concentration in tobacco is of crucial importance for nitrosamine yields in tobacco smoke. Tobacco smoke nitrosamines arise from nitrosation of tobacco alkaloids either during tobacco processing or during burning. Nitrosodiethylamine contamination of tobacco and its smoke results from the use of diethanolamine salt of maleic hydrazide. The exposure of smokers to total nitrosamines may be estimated to be 16 and 86 μg/day for a smoker of 20 cigarettes per day (U.S. commercial cigarettes and French nonfilter cigarettes, respectively).

In this review the mechanism of nitrosamine formation, the chemistry, environmental concerns, toxicological properties, scope, and limitation of nitrosamine removal from pesticides will be discussed.

2. DETERMINATION OF NITROSAMINES

Several methods have been used to detect nitrosamines.

2.1. Colorimetric Methods

Daiber and Preussmann (1964) have developed methods for the colorimetric determination of nitrosamines and nitrosamides. An improved colorimetric method has been published by Eisenbrand and Preussmann (1970). The nitrosamine is reacted with HBr/glacial acetic acid to form nitrosylbromide (BrNO) and the corresponding amine hydrobromide. The liberated BrNO is reacted with sulfanilic acid and the resulting diazonium derivative is coupled with N-(1-naphthyl)ethylenediamine. The sensitivity of the method is calculated to be 2–3 μg/kg. Alternatively, the liberated amine can be reacted with heptafluorobutyryl chloride and the derivative determined by gas chromatography (Eisenbrand, 1972).

2.2. Gas Chromatography

Many gas chromatographic methods have been developed for determination of nitrosamines using a wide variety of very nonpolar to highly polar packed and capillary columns. Detectors used for quantitative determination include FID, AFID, EC, MS, Coulometric, and Coulson detector. Nitrosamines are determined either directly or after derivatization.

2.3. Thermal Energy Analyzer

Fine et al. (1973) and Fine and Rufeh (1974) developed a new analytical instrument (thermal energy analyzer, TEA, or thermoluminous analyzer), highly specific and extremely sensitive for both volatile and nonvolatile N-nitroso compounds. The N-nitroso compound, usually dissolved in dichloromethane, is injected into a flash catalytic pyrolyzer, where the N—NO bonds are ruptured to form nitrosyl radicals (NO°), which are swept into a connected reaction chamber. Ozone, developed by electric discharge, also enters the chamber and reacts with the nitrosyl radicals giving excited NO_2^*. The excited molecules rapidly decay to their ground state with characteristic emission in the near infrared. The light emission is measured with an IR-sensitive photomultiplier response, which is directly proportional to the number of decays and thus to the number of moles of the N-nitroso compound, is amplified and displayed on a chart recorder.

N-Nitroso compounds are divided into two distinct categories, nitrosamines and nitrosamides. The nitrosamines consist of compounds of the general structure R^1R^2N—N=O where R^1 and R^2 can be alkyl or aryl groups, or parts of a ring. The nitrosamides are more diverse and less accurately named, but consist primarily of compounds with a carbon heteroatom double bond adjacent to the nitrogen, RN(NO)CY. Most frequently X=O as with nitrosamides (X=O, Y=alkyl, aryl), nitrosocarbamates (X=O, Y=OR^1), and nitrosoureas X=O, Y=NH_2, NHR, NR_2, etc.). For convenience, nitrosoguanidines (X and Y=NH and/or NR) can also be included in this second category, along with nitrososulfonamides [RN(NO)SO_2R^1]. Since pesticides encompass all of these classes, the variety of possible nitrosated pesticides is considerable.

3. PROPERTIES AND SYNTHESIS

Nitrosamines are rapidly photodecomposed in dilute acid solutions, but are relatively stable to light in neutral solutions (Chow, 1967). Burns and Alliston (1971) and Polo and Chow (1976) indicate that nitrosamines are photolabile in aqueous solutions (pH 7–10) but are more readily photolyzed at lower pH values. There are apparently no pH effects in the ranges of 3–9 normally observed for natural waters (Saunders and Mosier, 1979). It is generally believed that nitrosamines are rapidly decomposed by light in the vapor phase (Bamford, 1939;

TABLE 2   Pesticides Converted to Nitroso Derivatives

Acephate (O, s-dimethyl-N-acetylphosphoramidothioate, Seiler, 1977)
Aldicarb [2-methyl-2-(methylthio)propionaldehyde-o-(methylcarbamoyl)oximine, Seiler, 1977]
Antu (1-naphthylthiourea, Seiler, 1977)
Atrazine [(2-chloro-4-ethylamino)-6-(isopropyl)-s-triazine, Eisenbrand et al., 1975; Kearney et al., 1977]
Bassa (2-sec-butylphenyl-N-methylcarbamate, Uchiyama et al., 1975)
Benomyl [methyl-1-(butycarbamoyl)-benzimidazole-2-yl-carbamate, Seiler, 1977]
Benzthiazuron [1-(2-benzothiazolyl)-3-methylurea, Eisenbrand et al., 1975; Seiler, 1977]
Butralin [4-(1, 1-dimethylethyl)-N-(1-methylpropyl)-2, 6-dinitrobenzenamine, Oliver and Kontson, 1978]
Burturon [3-(4-chlorophenyl)-1-methyl-1-(1-methyl-2-propynyil)-urea, Seiler, 1977]

TABLE 2 *(Contd.)*

Carbaryl (1-naphthylmethylcarbamate, Egert and Greim, 1976;
  Eisenbrand et al., 1975; Elispuru et al., 1973; Seiler, 1977; Uchiyama et al., 1975)
Carbendazine (methyl-2-benzimidazole carbamate, Seiler, 1977)
Carbofuran (2, 3-dihydro-2, 2-dimethylbenzofuran-7-yl-methylcarbamate, Seiler, 1977)
Chlorobromuron [3-(4-bromo-3-chlorophenyl)-1-methoxy-1-methylurea, Seiler, 1977]
Chlorothiamid (2, 6-dichlorothiobenzamide, Seiler, 1977)
Chlorotoluron [3-(3-chloro-*p*-tolyl)-1, 1-dimethylurea, Seiler, 1977]
Chloroxuron (3[*p*-(*p*-chlorophenoxy)phenyl]-1, 1-dimethylurea, Egert and Greim, 1976)
Cychuron (3-cyclooctyl-1, 1-dimethylurea, Egert and Greim, 1976)
Cypendazole [methyl 1-(5-cyanopentylcarbamoyl-2-(methoxycarbonylamino)-
  benzimidazole. Seiler, 1877]
Daminozide (succinic acid 2, 2-dimethylhydrazide, Seiler, 1977)
Dimethoate [*O, O*-dimethyl-*S*-(*N*-methylcarbamoylmethyl)phosphorodithioate,
  Seiler, 1977]
Dimefox (*N, N, N', N'*-tetramethylphosphorodiamidic fluoride, Egert and Greim, 1976)
Dioxacarb (*O*-1, 3-dioxolan-2-yl-phenyl methylcarbamate, Seiler, 1977)
Diuron [3-(3, 4-dichlorophenyl)-1, 1-dimethylurea, Seiler, 1977]
Dodine (dodecylguanidine acetate, Seiler, 1977)
Ethiofencarb (2-ethylthiomethylphenyl methylcarbamate, Egert and Greim, 1976;
  Seiler, 1977)
ETU (ethylenethiourea, Seiler, 1977)
Ferbam (ferric dimethyldithiocarbamate, Sen et al., 1974)
Fluometuron [1, 1-dimethyl-3-(3-trifluoromethyl)phenylurea, Seiler, 1977]
Formelanate [3-(dimethylamino)methylene-aminophenyl methylcarbamate, Seiler, 1977]
Glyphosate [*N*-(phosphonomethyl)glycine, Seiler, 1977; Khan and Young, 1977]
Hopcide (2-chlorophenyl-*N*-methylcarbamate, Uchiyama et al., 1975)
Linuron [3-(3, 4-dichlorophenyl)-1-methoxy-1-methylurea, Seiler, 1977]
Maqbarl (3, 5-xylyl *N*-methylcarbamate, Seiler, 1977)
Meobal (3, 4-xylyl *N*-methylcarbamate, Seiler, 1977)
Methabenzthiazuron [1-(2-benzothiazolyl)-1, 3-dimethylurea, Uchiyama et al., 1975]
Methomyl [*S*-methyl-*N*-(methylcarbamoyl-oxy)-thioacetimidate, Seiler, 1977]
Metoxuron [3-(3-chloro-4-methoxyphenyl)-1, 1-dimethylurea, Seiler, 1977]
Mipsin (2-isopropylphenyl *N* methylcarbamate, Seiler, 1977)
Monolinuron [3-(*p*-chlorophenyl)-1-methoxy-1-methylurea, Seiler, 1977]
Monuron [3-(*p*-chlorophenyl)-1, 1-dimethylurea, Seiler, 1977]
Phenmedipham [3-methoxycarbonylaminophenyl *N*-(3-methylphenyl) carbamate,
  Seiler, 1977]
Prometryne [2, 4-bis(isopropylamino)-6-(methylthio)-3-triazine, Egert and Greim, 1976]
Propham (isopropylcarbanilate, Seiler, 1977)
Propoxur (*O*-isopropoxyphenyl methylcarbamate, Eisenbrand et al., 1975)
Propyzamid [*N*-(1, 1-dimethylpropynyl)-3, 5-dichlorobenzamide, Seiler, 1977]
Simazine [2-chloro-4, 6-bis(ethylamino)-3-triazine, Eisenbrand et al., 1975]
Suncide [methyl-*N*-(3, 4-dichlorophenyl) carbamate, Seiler, 1977]
Swep [methyl-*N*-(3, 4-dichlorophenyl) carbamate, Seiler, 1977]
Thiram [bis(dimethylthiocarbamoyl)disulfide, Egert and Greim, 1976; Sen et al., 1975]
Tsumacide (3-tolyl *N*-methylcarbamate, Seiler, 1977)
Ziram (zinc dimethyldithiocarbamate, Eisenbrand et al., 1975)

Hanst et al., 1977; Pitts, 1978). The half-life of NDPA in air is estimated to be about 20 min under cloudy conditions and 10 min in bright sunlight. In addition, nitrosamines are subject to oxidation, reduction, alkylation, condensation, and other types of reactions with appropriate reagents.

Several pesticides have been converted to their nitroso derivatives under laboratory conditions for a variety of reasons. Table 2 lists many of these pesticides (Kearney, 1980).

4. MECHANISMS OF NITROSAMINE FORMATION

Any class of reduced nitrogen compound can serve as the nitrosatable precursor of an N-nitroso compound. Also, every nitrogen coordination state from primary to quaternary has been converted to a nitrosamine. As shown in Table 3, a wide variety of N-nitroso compounds might be expected to be encountered in our complex environment. Any of the higher oxidation states of nitrogen can serve as a nitrosating agent. Nitrosation may involve an accelerator.

Nitrosation is affected by agents related to nitros acid having structure ONX, where $X = OAlk$, NO_2, NO_3, halogen, tetrafluroborate, hydrogen sulfate, or

TABLE 3 Nitrosatable substrates, nitrosating agents and accelerators of nitrosation

State	Nitrosatable Substrates
Primary, secondary, tertiary, or quaternary	
Amides	
Amines	
Ammonium compounds	
Carbamates	
Cyanamides	
Guanidines	
Hydrazines	
Hydroxylamines	
Ureas	

Nitrosating Agents	Oxidation State of Nitrogen	Reference
Coordinated NO_3	+5	Croisy et al. (1980)
Nitrogen dioxide	+4	Challis et al. (1978)
Nitrites, nitrosamines, nitro compounds	+3	Douglass et al. (1978)
Nitric oxide	+2	Ragsdale et al. (1965)

Accelerators of Nitrosation	Reference
Nucleophiles	Fan and Tannenbaum (1973)
Electrophiles	Croisy et al. (1980)
Physical stimuli	Challis et al. (1980)
Microbes (dead or alive)	Yang et al. (1977)

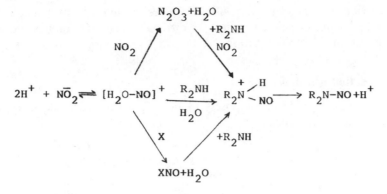

Scheme 1. N-nitrosation mechanisms according to Ridd (1961).

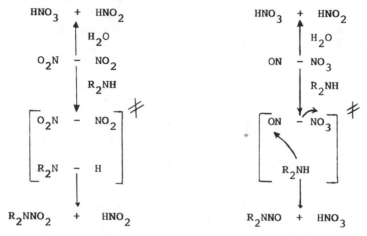

Scheme 2. Mechanisms of N-nitration and N-nitrosation by N_2O_4 according to Chillis et al. (1978).

$^+OH_2$. The nitrosonium cation is present at high concentrations only at high acidities (see schemes 1 and 2). In the presence of an active nucleophilic agent the nitrosonium cation is converted into nitrous acid and further into nitrite ions such as

$$ON^+ + OH^- \rightleftharpoons HNO_2 \rightleftharpoons H^+ + NO_2^-$$

At pH $\geqslant 7$, the equilibrium in the reaction is completely displaced to the right. In its simplest form the nitrosation of amines includes electrophilic attack by the nitrosating species on the lone pairs of electrons of the nitrogen atom and subsequent deprotonation of the alkylnitrosoammonium cation, such as

$$\begin{array}{c} R^1 \\ R \end{array}\!\!\!> \!NH + ON^+ - X \rightleftharpoons \begin{array}{c} R^1 \\ R \end{array}\!\!\!> \!N\!\!<\!\!\begin{array}{c} H \\ NO \end{array} \xrightarrow{-H^+} \begin{array}{c} R^1 \\ R \end{array}\!\!\!> \!N\!-\!NO$$

The reaction of primary amines with aldehydes in the presence of alcohols and nitrites under mildly acidic reaction conditions forms nitrosamines bearing an ether group in the alpha position.

4.1.   Reaction with Inorganic Acids

N-Nitrosodialkylamine decomposes on heating with hydrochloric acid into dialkylamine hydrochloride.

$$R_1R_2N{-}N{=}O \underset{H_2O}{\overset{HCl}{\rightleftharpoons}} R_1R_2NH{\cdot}HCl + HNO_2$$

Hydrogen chloride has been more pronounced in denitrosating activity. The carbonate group has also a significant influence on the results of the denitrosation. Acid cleavage of aliphatic nitrosamines perhaps goes through an equilibrium state between nitrosamine and its protonated form leading to separation of the nitroso group such as

$$R^1RN{-}N{=}O + H^+ \rightleftharpoons R^1RHN{-}\overset{+}{N}{=}O \overset{slow}{\rightleftharpoons} R^1RNH + NO^+$$

The reaction takes place much more readily in hydrochloric acid than in sulfuric or perchloric due to nucleophilic properties of hydrochloric acid.

$$\underset{}{>}\overset{+}{N}H{-}N{=}O \overset{Cl^-}{\longrightarrow} \underset{}{>}\overset{+}{N}H{-}N \underset{O^-}{\overset{Cl}{<}} \overset{}{\longrightarrow} \underset{}{>}NH + NOCl$$

Fridman et al. (1971), however, believed that a more correct mechanism of denitrosation of aliphatic nitrosamines involves electrophilic attack on the oxygen atom of the nitroso group that will be eliminated as the result of the generation of partial positive charges at the nitrogen atom such as

$$\underset{}{>}\overset{\delta+}{N} \cdots\cdots \overset{\delta+}{N} \cdots\cdots \overset{\delta-}{O} \cdots\cdots H \overset{HA}{\longrightarrow} \underset{}{>}NH{\cdot}HA + \overset{+}{N}O$$

As it has been observed, nitrosation pathways are very diverse and it is comforting to note that only a few combinations of circumstances have been implicated to environmental nitrosamine formation.

The most common interaction is the reaction of di- or trisubstituted ammonium derivatives with a nitrite ion for the secondary amines under acidic conditions.

4.2.   Nitrosamine Formation in Pesticides

Herbicides originally found by Ross et al. (1977) to be contaminated were formulated as amine salts of carboxylic acids. Usually the formulations were contaminated due to the fact that it was placed in metal cans together with nitrite to keep the container from corroding and this promoted formation of NDMA. Prevention of NDMA formation was achieved by not using nitrite to control corrosion of the container. Most dimethylamine phenoxyalkanoic acid products now contain no detectable NDMA (Zweig et al., 1980).

Treflan herbicide was reported to be contaminated with over 100 ppm of NDPA (Ross et al., 1977). Nitrosamine contamination was due to a small amount of nitrite produced during the manufacturing of this compound by nitric acid acting as an oxidizer, which could combine under nitrosation conditions with the nitric acid to produce N_2O_4 that then could react with dipropylamine and convert to nitrosamine (see Figure 1).

Scrubbing of the mixture with sodium carbonate under aeration at 70°F for 1 hr reduced the level of nitrosamin considerably (see Figure 2).

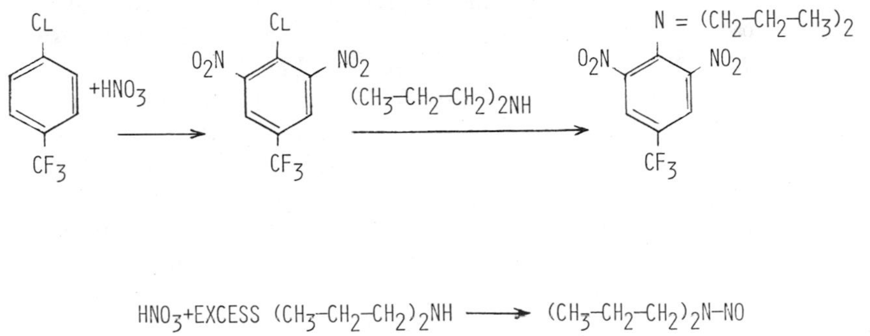

$$HNO_3 + EXCESS \ (CH_3-CH_2-CH_2)_2NH \longrightarrow (CH_3-CH_2-CH_2)_2N-NO$$

Figure 1. Synthesis of Treflan and the probable source of contamination by NDPA.

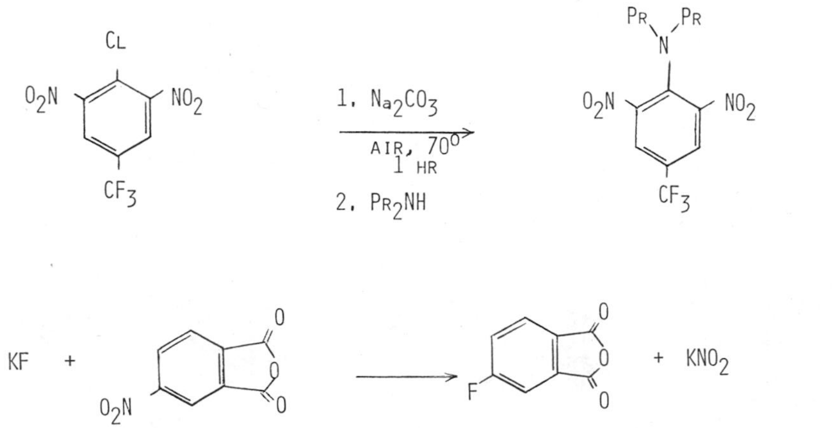

Figure 2. Reduction of nitrosamines by scrubbing with sodium carbonate and nudeophilic displacement of nitrite from a nitro compound.

Trifluralin, 4-chlorobenzotrifluoride (PCBT), is dinitrated to yield 4-chloro-3,5-dinitro-α,α,α-triflurotoluene (Dinitro PCBT), which in turn is aminated with n-dipropylamine to yield the product (Probst et al., 1975). Residual nitrosating agents are present as impurities in amine causing the problem in trifluraline formulation (see Figure 3).

Nitrosating reagents would include residual nitrate—nitrite or oxides of nitrogen. The active nitrosating agents, in this case oxides of nitrogen, were found to be the key components and are indicated along with possible precursors (Ingold, 1953).

In the manufacturing process, great care should be devoted to the purification and removal of nitrogen oxides resulting from nitration prior to the amination step (Cannon and Eizember, 1978). Therefore modification of the process was directed to the purification of Dinitro PCBT to remove nitrogen oxides. In the case of the synthesis of compounds, such as pendimethaline and butralin (see Figure 4), which are secondary amines, a major side reaction product is nitroso compound formation. Nitration of the aniline can yield 12–62% N-nitroso-N-(1-ethylpropyl)-3,4-dimethyl-2,6-dinitroaniline (Diehl et al., 1979).

N-Nitroso derivative must be denitrosated to make the process acceptable as well as avoid the presence of the N-nitroso derivative in the product. Denitrosation of N-(1-ethylpropyl)-3,4-dimethylaniline utilizes a nitration mixture of 35–53% water by weight, nitric acid and sulfuric acid in molar ratios of 3.25:1 and 2.25:1 of the aniline compound. The reaction proceeds for 2 hr at 35–70°C. Denitrosation is effected by adding hydrochloric and sulfamic acids to the mixture, maintaining a temperature of 70–100°C over a period of 1–6 hr, then recovering the N-(1-ethylpropyl)-2,6-dinitro-3,4-dimethylaniline product formed. Hydrochloric acid and ferrous chloride may be used for denitrosation, too.

Prevention of nitrosamine formation in other important commercial dinitroanilines such as benefin, trifluralin, profluralin, isopropalin, ethalfluralin and

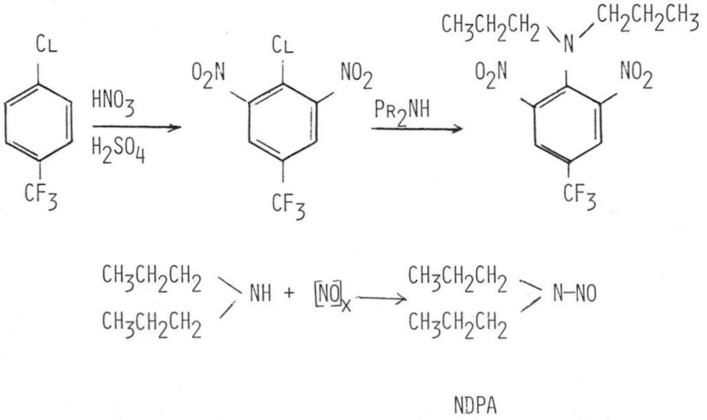

NDPA

Figure 3. Trifluraline synthesis.

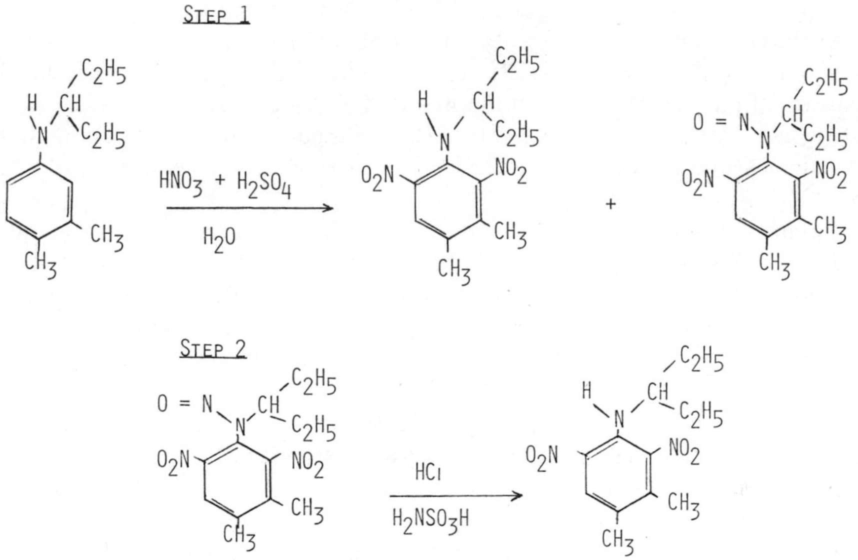

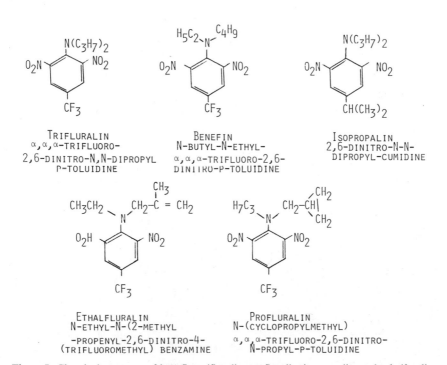

Figure 4. Pendimethalin synthesis.

Figure 5. Chemical structure of benefin, trifluralin, profluralin, isopropalin, and ethalfuralin.

333

other tertiary amines is accomplished by eliminating sources of nitrosating agents from the reaction mixture prior to amination (see Figure 5).

Cannon and Eizember (1978) suggested the following procedure for the removal of nitrosating agents in Dinitro PCBT. A nonreactive gas, such as air, nitrogen, or carbon dioxide, is bubbled through the reaction mixture containing an aqueous solution of base such as sodium carbonate, at a temperature of 50–100°C.

Removal of nitrosating agents can be carried out by several different routes, working the reaction mixture with base prior to sparging with gas, causing reduction of nitrate and nitrite ions, or simultaneous base treatment and gas sparging followed by base treatment leading to the reduction of nitrogen oxides. Such processes will result in 95% reduction of nitrosamines.

5.   REDUCTION OF N-NITROSO COMPOUNDS IN FORMULATIONS

Certain commercial pesticides, such as the phenoxy or benzoic acids, have been formulated as secondary amine or ethanol amine salts (see Figure 6). N-Nitrosodimethylamine (NDMA) content in formulated 2,3,6-trichlorobenzoic acid (Benzac) stored in metal containers was reported to be 640 ppm (Ross et al.,

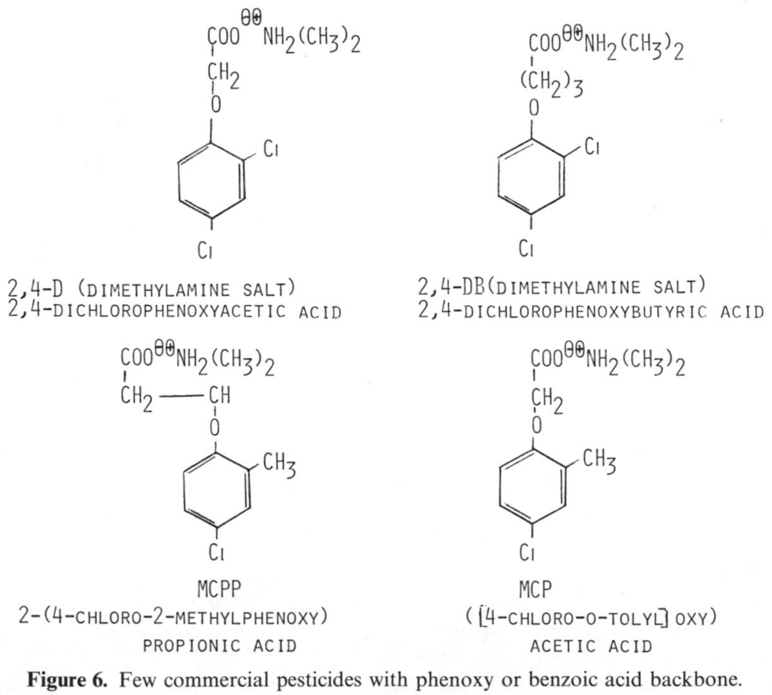

Figure 6. Few commercial pesticides with phenoxy or benzoic acid backbone.

TABLE 4   Nitrosamine Content of Technical Trifluralin[a] Sampled in Glass and Tin Containers

Lot Sample Number	Total Volatile Nitrosamine (ppm)	
	Sampled in Glass	Sampled in Tin Can
6683	0.17	7.68
6684	0.16	3.30
6685	0.20	6.86
6697	0.23	3.20
6719	0.25	4.11

[a]Storage: 1 month at ambient temperature.

1976). Sodium nitrite had been used as a corrosion inhibitor, which reacted with components of the mixture to yield nitrosating species.

Addition of 0.1–0.5% sodium nitrite to dimethylamine salts of 2, 4-D resulted in the formation of NDMA in 50–60 days in glass or metal containers (Bontoyan et al., 1976). Other corrosion inhibitors that do not produce nitrosating agents should be applied. The use of tin-plated containers for commercial pesticides susceptible to nitrosamine formation is unsatisfactory. Grove (1979) reported analytical results on nitrosamine formation (NDPA) in glass and tin containers from samples of trifluralin that contained less than 1 ppm nitrosamine (Table 4).

In a matter of a few days, increased amounts of NDPA in trifluralin were observed in the tin containers stored at ambient temperatures. The tin containers did not have corrosion inhibitors. However, container manufactures use a flux in the tinning process; many contain either nitrite or nitrite salts. Archer and Wishnok (1976) demonstrated the formation of nitrosamines from constituents of polymeric liners of metal cans. For pesticide formulations prone to be nitrosated, container specifications are vital.

Another source of N-nitroso contamination of pesticides is commercially synthesized amines. As much as 25 ppm NDMA has been found in different batches of dimethyl amines (Cohen et al., 1978). Scavengers inhibit nitrosamine formation by competing with the amine for the nitrosating agent (see Figure 7). Ascorbic acid is a typical scavenger.

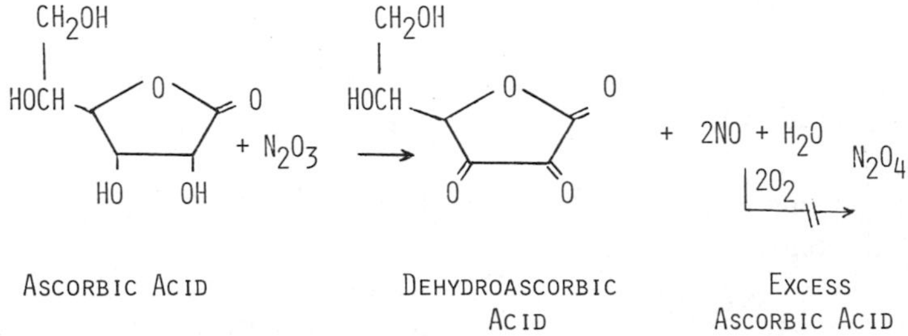

Figure 7. Scavenger activity of ascorbic acid.

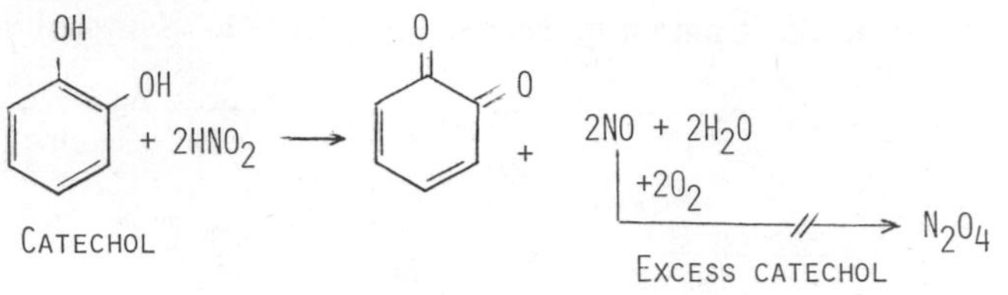

Figure 8. Inhibition of nitrosamin formation by phenols.

It is known that commercially available amines have impurities. In order to avoid unwanted impurities in final formulations of pesticides rigid specifications are needed.

The use of scavengers for the reduction of oxides of nitrogen that are present in the atmosphere is highly recommended. Bassow (1976) has shown that about 50 ppb of nitrous oxide and nitrogen dioxide is present in the atmosphere of cities. Ammonia is converted to nitrite by microorganisms in soil and natural water. As a result of such a wide distribution of nitrosating agents in the environment, various techniques should be considered to reduce the level of nitrosamines and their rate of formations. Addition of reducing compounds such as ascorbic acid to the formulation is recommended. Extensive review by Douglass et al. (1978) was carried out. Phenols also can inhibit nitrosamine formation (see Figure 8).

However, phenols under other conditions catalyze nitrosamine formation. 4-Methylcatechol catalyzes the nitrosation of dimethylamine and piperidine in the presence of excess nitrite. Sulfur compounds inhibit nitrosamine form- ation. Bisulfite, sulfur dioxide, or sulfamic acid reduces nitrite (Douglass et al., 1978).

$$SO_2 + 2HNO_2 \rightarrow 2NO + H_2SO_4$$

$$SO_2 + 2NO + H_2O \rightarrow N_2O + H_2SO_4$$

$$NaNO_2 + H_2NSO_3H \rightarrow NaHSO_4 + N_2 + H_2O$$

Nitrite is reduced to molecular nitrogen in the presence of sulfamate. Sulfamate serves as a denitrosating agent in the synthesis of pendimethalin (Diehl et al., 1979).

6. INTRAMOLECULAR REARRANGEMENT AND FORMATION OF NITROSAMINE

Herbicides having backbone of dinitroaniline have unusual properties; they can act as amine contributors as well as nitrosating agents. When trifluralin is heated to 70°C N-nitrosodipropylamine (NDPA) is formed. The quantity of NDPA is increased as the time of heating is extended (Grove, 1979). This is illustrated in Figure 9.

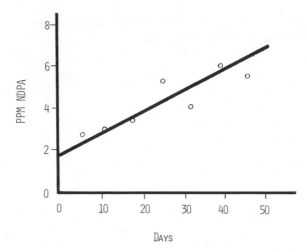

Figure 9. Nitrosamin generation from technical trifluralin heated at 70°C.

The mechanism of the reaction is not understood at the present time, but is presumed to be an internal rearrangement effecting a nitro group.
The reaction occurs with technical material only, and, once formulated, the nitrosamine content is not altered as a function of time at ambient formulation temperatures (see Figure 10). This intramolecular rearrangement requires careful attention to conditions in the formulation of dinitroaniline herbicides.

The nitrosating agent is formed by the action of a proton on one nitro group that then will react with traces of residual dipropylamine present in the mixture. Nitrosamine content once formed is not changed as long as the material is stored at ambient temperature and it is formed only in technical grade material. As a result, this intramolecular rearrangement requires careful attention to conditions in the formulations of dinitroaniline herbicides.

Nitrosamines are considered to be very stable compounds and once formed are difficult to destroy. The following reaction conditions were used to treat Trifluraline for the removal of nitrosamines. Trifluralinc was treated in a neat or an appropriate solvent, at 50–140°C for 5–8 hr.

The amount of acid required for nitrosamine destruction is dependent on the level of the nitrosamine impurity, the dinitroaniline being treated, the organic

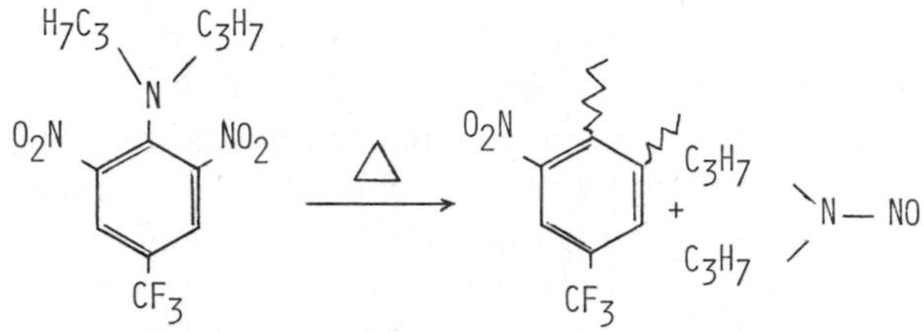

Figure 10. Hypothetical intramolecular rearrangement of dintroanilin herbicides.

TABLE 5   Treatment of Trifluraline by Different Acids for the Removal of NDPA[a]

Acid	NDPA (ppm)
50% Sulfuric acid	22
70% Sulfuric acid	<1
85% Sulfuric acid	<1
10% Hydrochloric acid	81
33% Hydrochloric acid	<1
37% Hydrochloric acid	<1
Hydrogen chloride gas[b]	<1
50% Formic acid	74
98% Formic acid	58
70% Acetic acid	58
Oxalic acid[c]	9
40% Phosphoric acid	90
48% Hydrobromic acid	<1
Ascorbic acid[d]	85

[a]Treatment conditions were time, 20 min; temperature, 70°C; and amount of acid 20% w/w of acid to trifluralin. Untreated trifluralin contained 69 ppm of NDPA.
[b]Hydrogen chloride gas flow was 35 mL/min.
[c]Time, 2 hr.
[d]Time, 3 hr.

solvent used, temperature, and time. Reduction of preformed nitroso compounds in pesticides is not an easy task, especially because of the fact that they are present in trace amounts. Detailed work of Eizember (1978), Eizember et al. (1979), and Eizember and Vogler (1980) on reduction of nitrosamines in a wide variety of dinitro analogs has shown it is possible to achieve 25 to 30-fold reduction. Typical reaction parameters along with some results are illustrated in Table 5. As observed, the acidic reagents vary widely in their ability to reduce the levels of NDPA in trifluralin. Obviously, the concentration of the acid is critical to produce the desired effect.

Hydrochloric acid at 10% concentration, 40% phosphoric acid, and ascorbic acid prompted additional formation of nitrosamine. Hydrogen chloride gas was the most efficient at destroying NDPA impurities.

Loss of NDPA as a function of time during treatment of trifluralin is illustrated in Figure 11. The reaction mechanism with either hydrochloric or sulfuric acid is postulated in Figure 12.

A charged intermediate of nitrosamine is formed after protonation. In the presence of hydrogen chloride the protonated intermediate is rapidly attached by strong chloride anion generating nitrosyl chloride and secondary amine. The nitrosyl chloride is removed by hydrogen chloride gas flow or reacts with water.

Table 6 illustrates the effect of halogens and halogen-releasing agents under similar laboratory conditions as reported on removal of NDPA from trifluralin (Probst et al., 1975).

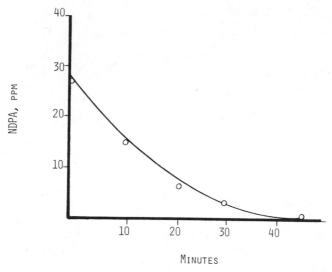

Figure 11. Removal of NDPA from impure trifluralin with gaseous hydrogen chloride.

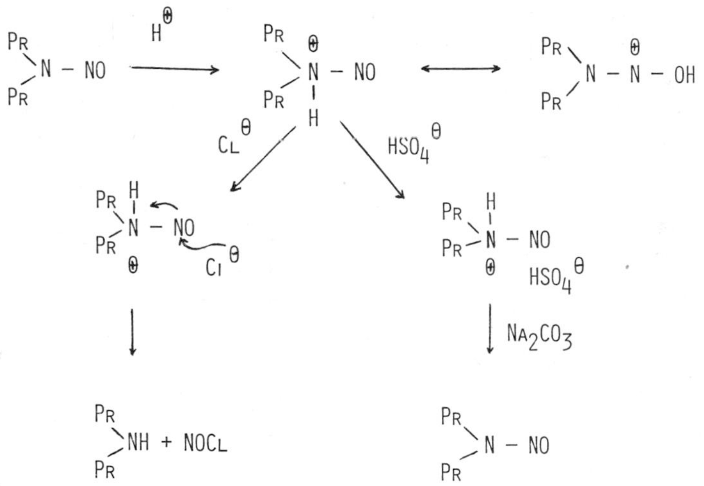

Figure 12. Reaction mechanism of nitroso compounds by either hydrochloric or sulfuric acid.

   Chlorine is less efficient than bromine in removing NDPA from trifluralin and requires higher reaction temperature and longer time. Also, N-bromosuccinimide is effective in removing nitrosamines from dinitroanilines. A variety of inorganic halides have been examined by Eizember and Vogler (1980) for the removal or destruction of nitrosamine. Inorganic halides such as phosphorus trichloride, phosphorus tribromide, thionylchloride, and sulfonyl chloride can be used to destroy nitrosamines in dinitroaniline pesticides, which

TABLE 6   NDPA Removal from Trifluralin[a]

	Grams Used	Temperature (°C)	Time (min)	NDPA (ppm)
Bromine	0.2	70	20	<1
Bromine	0.1	90	15	<1
Chlorine	35 mL/min	70	30	16
Chlorine	15 mL/min	110	120	22
N-Bromosuccinimide	0.5	70	60	1.7
Iodine	0.1	70	60	78

[a]Each reaction used 30 g of trifluraline containing 68 ppm NDPA.

should be effective in destroying nitrosamines in a wide variety of inert or relatively unreactive solvents or diluents.

Rigid control of denitrosation reactions is needed in order to avoid modification of dinitroaniline and ethylfluralin by the addition of HCl to the double bonds (see Figure 13). This is achieved by accurate control of time and temperature of the reaction.

Denitrosation of dialkylnitrosamines such as N-nitrosodimethylamine and aryl-alkyl nitrosamines such as N-nitroso-N-(1-ethylpropyl)-3,4-dimethyl-2,6-dinitroaniline was accomplished according to the procedure described by Ross and Chiarello (1979). Nitrosamine contamination was reduced by treating the mixture with a ketone or an aldehyde in the presence of strong acid such as hydrochloric or hydrobromic acid under pressurized condition at 110°C for 1–2 hr (see Figure 14).

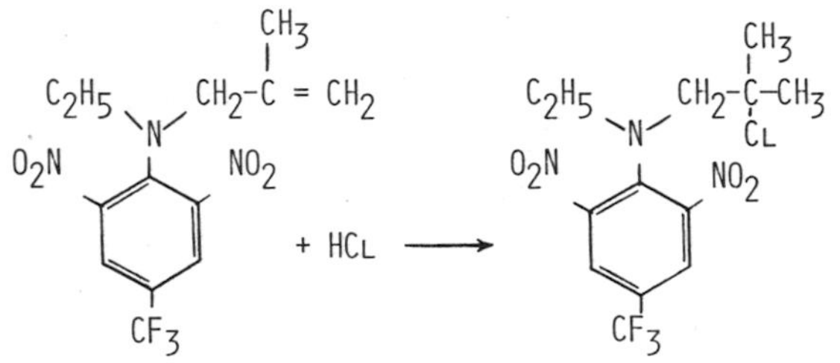

Figure 13. Addition of HCl to double bonds of dinitroaniline type compounds.

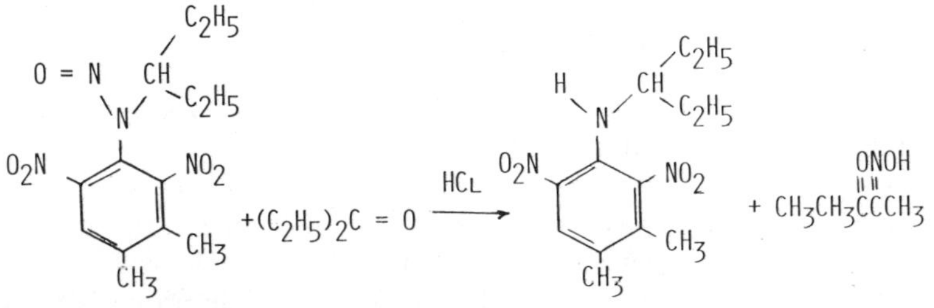

Figure 14. Nitrosamin reduction by treatment with ketones or aldehydes.

TABLE 7   Degree of Denitrosation in the Above Reaction

Sample	Sample Taken (Hours after Reaction Began)	Found	
		%A[a]	%B[a]
1	0	91.2	1.3
2	1	90.6	0.06
3	2	91.2	0.006
4	3	90.1	0.003
5	4	91.3	<0.003

[a]A, N-(1-ethylpropyl)-2,6-dinitro-3,4-xylidine; B, N-(1-ethylpropyl)-N-nitroso-2,6-dinitro-3,4-xylidine.

Denitrosation was studied by Biggs and Williams (1976) extensively. Table 7 shows degree of dinitrosation of N-(1-ethylpropyl)-3,4-dimethyl-2,6-dinitroaniline (40% by weight) and N-(1-ethylpropyl)-N-nitroso-3,4-dimethyl-2,6-dinitroaniline (15% by weight) in the presence of diethyl ketone and solvent of ethylene chloride. After addition of HCl and sealing of the vessel, the reaction mixture is heated at 90°C. Nitroso derivative was reduced to less than 30 ppm after 5 hr of heating.

Eizember and Vogler (1980) have shown many inorganic halides are effective in destruction of nitrosamine impurities during the synthesis of dinitroanilines. The following reaction illustrates synthesis of oryzalin which is accompanied by destruction of nitrosamine (see Figure 15).

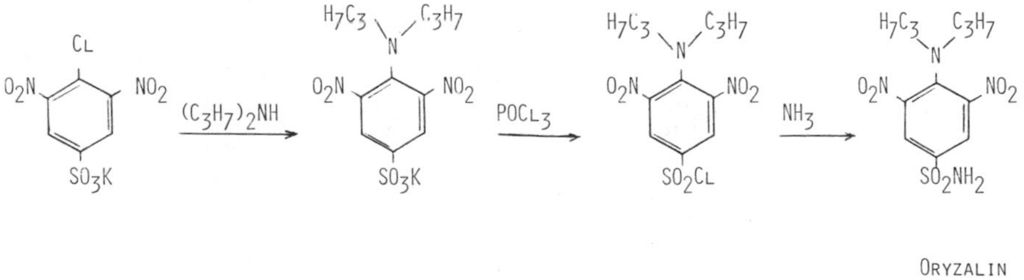

Figure 15. Synthesis of oryzalin.

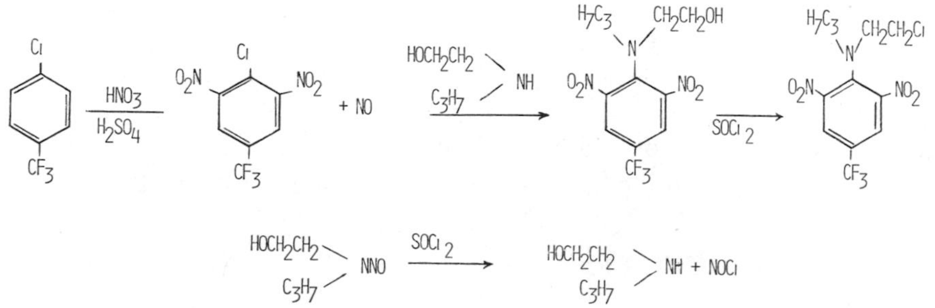

Figure 16. Destruction of nitroso compound by thionyl chloride.

Similarly, thionyl chloride, an inorganic halide, has been emphasized in the synthesis of fluchloraline, N-(2-chloroethyl)-α, α, α-trifluoro-2, 6-dinitro-N-propyl-p-toluidine yielding fluchloraline (Zweig et al., 1980). The N-2-hydroxyethyl-N-propylnitrosamine formed during the amination reaction from nitrosating agents is effectively destroyed by chloro substitution process (see Figure 16). Nitrosamine content is reported for fluchloralin to be 0.5–1.7 ppm.

7. TOXICITY OF NITROSAMINE

The main acute action of the dialkyl and cyclic nitrosamines is on the liver, resulting in hemorrhagic centrilobular necrosis (Barnes and Magee, 1954; Magee and Barnes, 1967). Other organs are much less severely affected; the main features are peritoneal and sometimes pleural exudates that may contain a high proportion of blood and a tendency to hemorrhage into the lungs and other organs. In protein-deficient rats there may be detectable necrosis of some renal tubules (Hard and Butler, 1970a) and of the tests following treatment with N-nitrosodimethylamine. Chronic administration of many nitrosamines induces tumors of the liver and other organs. The livers of rats and other species chronically exposed to nitrosamines show various pathological changes, including biliary hyperplasia, fibrosis, nodular prenchymal hyperplasia, and the formation of enlarged hepatic parenchymal cells with very big nuclei.

The nitrosamides include some of the most powerful known chemical mutagens, for example N-methyl-N'-nitrosoguanidine (Magee and Barnes, 1967). As with acute toxicity, there is a clear distinction between the mutagenic actions of the nitrosamines and of the nitrosamides. The latter compounds are powerfully mutagenic in all the usual microbial test systems and in Drosophila, but the nitrosamines are active only in the insect. This difference seems to be related to the lack of nitrosamine metabolism in the microorganisms and the probable capacity of Drosophila to perform the enzyme-catalyzed reactions necessary for their biological activation. This conclusion is supported by the demonstration that N-nitrosodimethylamine is mutagenic when incubated with bacteria in the presence of rat liver microsomes (Malling, 1971) and the fact that several nitrosamines are actively mutagenic when tested by the host-mediated assay (Legator and Malling, 1971).

The teratogenic actions of the nitroso compounds have been much less extensively investigated. As with transplacental carcinogenesis, the nitrosamides are active while the nitrosamines are not except when administered late in pregnancy (Pielsticker et al., 1967). This is probably related to the capacity for direct action of the former compounds and the lack of sufficient amounts of the activating enzymes in the fetus until the last few days of pregnancy.

8. CARCINOGENIC EFFECTS

Liver tumors developed in mice and rats following chronic treatment with a variety of N-nitroso compounds. In the 11 strains of adult mice in which N-nitrosodimethylamine was tested, hemangiomatous tumors are the most frequent liver tumors, and parenchymal cell tumors developed less frequently. N-Nitrosodiethylamine induced mainly parenchymal cell tumors in the seven strains tested except in the BALB/c and NMRI strains where hemangiosarcomas and hemangioendotheliomas were predominant (Clapp et al., 1971; Kunz et al., 1969). Newborn and young adult mice appear more likely to develop parenchymal liver cell tumors (Toth et al., 1964; Toth and Shubik, 1967; Terracini et al., 1966; Vesselinovitch, 1969; Frei, 1970; Gargus et al., 1969; Turusov, 1973; Wood et al., 1970). Both types of tumors were induced with other nitrosamines such as N-nitrosodibutylamine, N-nitrosopiperidine, di-N-nitrosopiperazine, N-nitrosomorpholine, and a single dose of N-methyl-N-nitrosourea produced parenchymal cell tumors in newborn BC3 F_1 mice but not in adults (Terracini and Testa, 1970).

8.1. Formation of Nitrosamines in Soil

Nitrosamines in soil can arise from a pesticide metabolite such as dimethylamine or from the parent pesticide. Nitrosodimethylamine could be generated in raw but not in autoclaved sewage with dimethylamine (Verstraete and Alexander, 1971). It has been demonstrated by Ayanaba and Alexander (1973) that extract from Cryptococcus species catalyzed nitrosamine synthesis at pH 7.5. Other investigators such as Mills and Alexander (1976) have indicated that nitrosamines can be formed in some soils through nonenzymatic reactions at neutral pHs. However, organic matter is of importance in this synthesis.

The systems described so far were all amended with high levels of amines and nitrites or nitrates. When high levels of amines and nitrites or nitrates are present in soil, nitrosation takes place easily.

Formation of nitrosodimethylamine has been observed by Pancholy (1976) after addition of both nitrite and dimethylamine to soil. Concentration of nitrosodimethylamine was followed and it was noticed that significant increase for 12–15 days was observed, which was followed by a decrease to near zero in 30 days.

Autoclaved soils did not form any nitrosamines. It was also observed that in polluted and fertilized soil when supplemented with additional amines (10 ppm) after incubation contained 0.1–0.5 ppm of nitrosamines. Apparently formation of nitrosamines is favored by anaerobic conditions. Although nitrosodimethylamine was produced from both tri- and dimethylamines, no nitrosamine was detected from several other organic compounds containing dimethylamino groups including the herbicide monuron.

In soils with high levels of sodium and nitrite herbicides [^{14}C]atrazine (Kearney et al., 1977) and [^{14}C]butralin (Oliver and Kontson, 1978) were found to form nitrosamines. When ammonium nitrite was substituted for sodium nitrite, no nitrosation took place. Interestingly, the nitroso atrazine formed but then rapidly disappeared; nitroso butraline also formed rapidly but was still detectable after 6 months.

Similarly, Tate and Alexander (1974) reported formation of trace quantities of nitrosamine in soils in the presence of nitrites from soils containing dimethylamine and diethylamine from dimethyldithiocarbamate and diethyldithiocarbamate. Khan and Young (1977) detected nitrosoglyphosate in several soils using a method for measuring nonvolatile nitrosamines. Although the herbicide Glyphosate could theoretically lose the phosphoro groups to yield sarcosine, which could then form a nitroso compound, neither sarcosine nor nitrososarcosine was found in soil incubated with nitrites (Khan and Young, 1977).

Miyamoto (1978) reported that no N-nitrosocarbamates could be detected in three soils containing sodium nitrite at 100 and 1000 ppm and 20 ppm of three insecticides, carbaryl, carbofuran, or propoxur.

Nitrosodipropylamine, a contaminant trifluraline, was analyzed by Fine et al. (1976) in the air and in irrigation water from tomato fields in the Sacramento Valley, California, before, during, and after application of the herbicide trifluraline. No nitrosamine was detected.

8.2. Nitrosamine Formation and Uptake in Plants

Herbicides in their formulations contain many ingredients that are nitrosable substances. In addition they are used in plants that contain many more nitrosable compounds. Several studies were designed to investigate this possibility (Sander et al., 1975; Dressel, 1967, 1976, 1977; Raymond and Alexander, 1976). The results of such studies yielded negative findings. Also, it was found that plants that are ingested by humans could not take up and store potentially carcinogenic N-nitroso compounds formed in the plant environment. Such findings are understandable since most of the data on the disposition of nitrosamine in plant parts relies on the use of compounds such as NDMA that have a short half-life in plants. In addition it is expected that relatively low levels of nitrosamines are found in plants due to the presence of low levels of basic amines (Magee et al., 1976).

Nitro compounds can give rise to free nitrosating agents in several ways: by heterolytic cleavage to nitrite (Schmeltz and Wenger, 1979; Markenzich et al., 1977); by photolysis (Lippert and Kelm, 1978), to nitrogen dioxide, which can dimerize to N_2O_4 or potentially even recombine with the aryl radical from which it was homolytically generated to form an aryl nitrite ester, an as yet apparently uncharacterized compound type that should be a powerful nitrosating agent (see Figure 17); or by autoxidation (Franck, 1970) or microsomal oxidation (Ullrich et al., 1978) to nitrous acid (see Figure 18). Schmeltz et al. (1977) showed that tobacco grown in fields on which the antisuckering agent, MH-30 (which is maleic hydrazide compound as the diethanolamine salt), was used had relatively high concentrations of nitrosodiethanolamine (NDElA), which is the result of *in vivo* nitrosation of the diethanolamine in the plant (see Figure 19).

Analysis of barley, wheat, spinach, lettuce, carrots, and tomatoes for the presence of nitrosodimethyamine and nitrosodiethylamine in soils treated with 0–123 kg/ha nitrogen by Dressel (1976) showed absence of such compounds. Sander et al. (1975) found no nitrosamines in treated fields of wheat with heavy

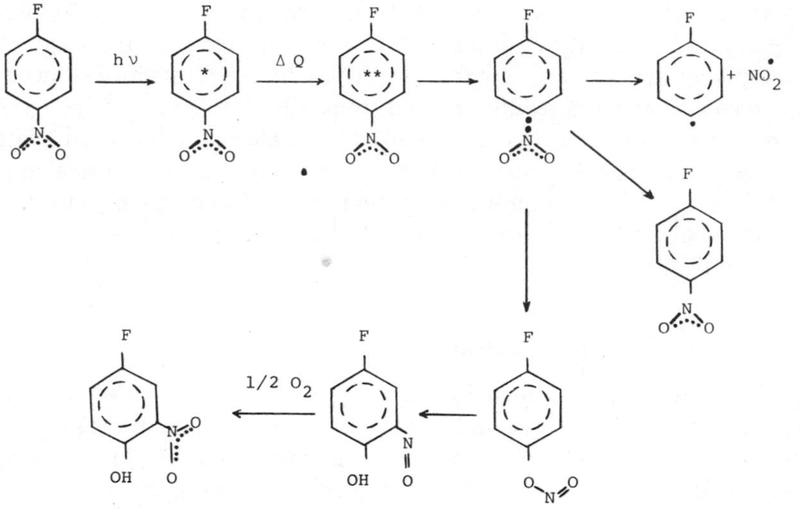

Figure 17. Photolytic generation of nitrogen dioxide and an aryl nitrite ester from a nitro compound (Lippert and Kelm, 1978).

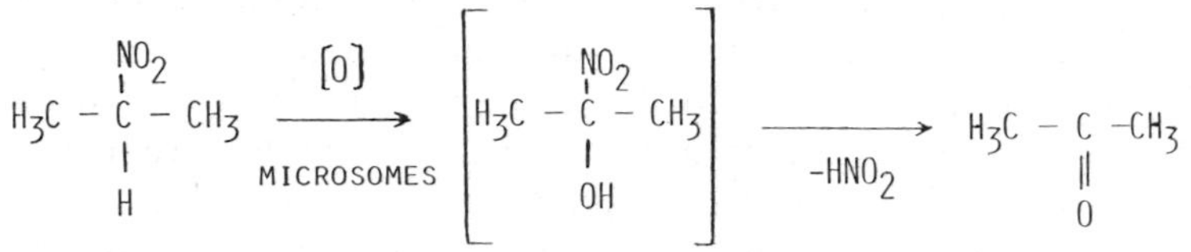

Figure 18. Metabolic generation of nitrite from a nitro compound *in vitro* (Ullrich et al., 1978).

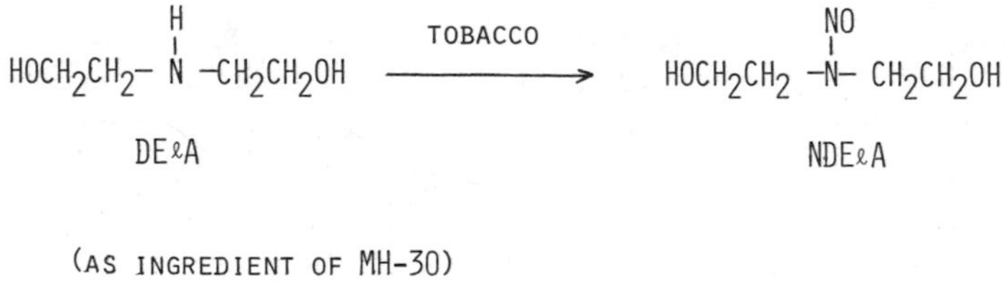

DEℓA

NDEℓA

(AS INGREDIENT OF MH-30)

Figure 19. Mechanism of contamination of tobacco by NDELA (Schmeltz et al. 1977).

doses of nitrogen fertilizers and secondary amines such as dimethylamine, *N*-methylethylamine, and *N*-methyl-*n*-benzylamine. Kearney et al. (1977) studied the uptake of NDPA and *N*-nitrosopendimethalin by soybeans in the fields. It was found that concentrations of 0.1, 1, 10, and 100 ppb in soils resulted in no measurable residues of nitrosamines in soybean seeds after 110 days.

Sander et al. (1975) reported that uptake of nitrosamines from water by plants, cress, decreased rapidly when the nitrosamine contamination of water was eliminated, concluding that the nitrosamines did not accumulate in green plant material. The accumulation of NDMA into lettuce and spinach was shown by Raymond and Alexander (1976). These investigators also showed that the same compound was readily leached from soil by water.

Nitrosodiethanolamine was detected in cured tobacco that had been treated in the field with the growth regulator MH-30 (maleic hydrazide formulated as its diethanolamine salt). Nitrosamines were lacking in tobacco on which the growth regulator had not been used. It was concluded that the nitrosation had occurred in the plant (Schmeltz et al., 1977).

8.3. Formation of Nitrosamines in Animals

Nitroso derivatives are formed in mammalian organisms by various amino compounds, especially from secondary amines of low or moderate basicity, in the presence of nitrite. This is presumed to be formed primarily in the stomach. Nitrosamine formation has been illustrated indirectly by induction of various tumors, by concomitant administration of the amines with nitrite to experimental animals. N-Methylaniline, N-methylbanzylamine, morpholine, and piperazine are examples (Sander et al., 1975).

The influence of nitrite concentration in tumor formation was demonstrated by feeding studies with 2500 ppm N-methylbenzylamine and 800 ppm nitrite. When the nitrite concentration was reduced to 600 ppm or below, no tumors were found. The lowest concentrations of ethylurea and nitrite that produced tumors were around 500 ppm in feed or in drinking water.

When insecticidal carbamate compounds and nitrite were fed to rats, nitrosamines were detected *in vivo* in the stomachs of rats and guinea pigs. When 16.2 mg ring-labeled m-cresyl N-methylcarbamate was orally administrated to male Sprague–Dawley rats with a 4-fold excess of sodium nitrite, less than 0.1% of the nitrosocarbamate was detected after 15 and 60 min (Miyamato and Hosokawa, 1977). Marco et al. (1976) reported that rats fed atrazine with nitrite produced no detectable nitrosoatrazine in stomach contents, stomach wall, or excretion. Nitrosamine can be formed through dialyklamines when combined with high nitrite concentration and acidic conditions. Sen et al. (1974) fed the fungicides thiram, ziram, and ferbam to guinea pigs with an excess of nitrite and obtained very low levels of NDMA in the stomach. Oral feeding of mice at maximal tolerated doses of various pesticides (benzthiazuron, carbarbyl, carbofuran, dimethoate, ethiofencarb, formetanate, linuron, maneb, methanbenzthiazuron, propham, or propoxur) together with nitrite developed no increased micronuclei in bone marrow erythrocytes. In contrast, ETU produced a significant increase of micronucleated polychromatic erythrocytes under similar conditions (Seiler, 1977). Similar pesticides can form nitrosamines as a result of their metabolism to dialkylamines when combined with high nitrite concentration and acidic conditions. Eisenbrand et al. (1974) administered ziram with a 40-fold molar excess of nitrite to rats and obtained an average yield of NDMA of 0.9%.

9. DEGRADATION AND METABOLISM OF NITROSAMINES

9.1. Degradation of Nitrosamines in Water and Air

It has been demonstrated by Preussman et al. (1975) that ethylnitrosourea is decomposed in aqueous solution in the presence of Cu^{2+} and OH^- ions. Ni^{2+} ions have also shown similar but less pronounced effects. NDMA, NDEA, and NDPA were not degraded in lake water during a 3.5-month period as was shown by Tate and Alexander (1975).

N-Methyl-N'-nitro-N-nitrosoguanidine, a compound known to be relatively stable in aqueous solution, was strongly decomposed in the presence of Cu^{2+} ion. However the stability of N-methyl-N-nitrosourethane was not influenced by the presence of heavy metal ions. Wolfe et al. (1967) studied the sunlight photolysis of N-nitrosoatrazine in the aquatic environment. N-Nitrosoatrazine was stable toward hydrolysis over 3 weeks in water at pH 5.5 and 8.0 or in river water at pH 7.1. However, the compound was degraded by sunlight yielding desethylatrazine and atrazine. The half-life of this compound was about 10 min throughout the year in the United States.

Aliphatic nitrosamines are generally volatile and are expected to enter the atmosphere rather rapidly. NDMA vapor has been shown to be unstable to ultraviolet light. NDPA was transformed with a half-life of less than 7 days into N,N-dipropylnitramine, which degraded to several other compounds including N-dipropylpropionamide (Crosby et al., 1978).

9.2. Degradation of Nitrosamines in Soil

Stability of nitrosodialkylamines in soil has been studied by many investigators and the literature in this regard is somewhat conflicting. It has been reported by Ayanaba et al. (1973) that over 90% of the NDMA formed from dimethylamine disappears in about 9 days after reaching a maximum concentration of 1.2 ppm. However, Tate and Alexander (1975) reported that NDMA was not degraded in flooded soil or in microbial enrichments from bog sediments. Also NDEA and NDPA were not metabolized by enrichment cultures from soil or sewage. It was reported by Tate and Alexander that microbial degradation of NDMA, NDEA, and NDPA was rather slow and somewhat around 50% of these compounds remain after 14 days in sewage undergoing microbial degradation.

Oliver et al. (1979b) studied degradation of $[^{14}C]$NDPA in aerobic soil. Initially ^{14}C losses by volatilization competed with losses with $^{14}CO_2$ production. However, after a few days $^{14}CO_2$ accounted for all of the additional ^{14}C trapped.

Sterilization of soil inhibited $^{14}CO_2$ production but extended the time period over which NDPA evaporation volatilization was observed. The rate of $^{14}CO_2$ production was the same whether the NDPA was labeled at carbon 1, 2, or 3, and the half-life was estimated to be about 3 weeks. It was concluded that degradation was at least partly microbial and once the degradation began, the reaction proceeded rapidly all the way to CO_2. Field studies indicated that more than 90% of the NDPA incorporated in the top 10 cm of soil in a 30-cm cylinder that had dissipated within 3 weeks. Losses were most likely due to volatilization, degradation, and leaching. Surface application of moist soil with NDMA and NDPA showed nearly 80% of the NDMA was lost in a few hours and volatilization of NDPA was slight and rather slow.

Incorporation of nitrosamines into the soil reduced both the rate and extent of volatilization. However, volatilization of certain pesticides is not as rapid as low-molecular-weight nitrosamines.

Disappearance of N-nitrosocarbamates from soil that was exposed to sunlight was extremely rapid with a half-life of 5–25 min (Miyamoto, 1978).

The rate of disappearance was nitrosoporpoxur < nitrosocarbofuran < nitrosocarbaryl. Only 12% of N-nitrosoatrizine was recovered from aerobic Matapeake loam after 1 month, and the recovery was less than 1% after 3 and 4 months from the soil. N-Nitroso derivatives from two secondary amines, dinitraniline herbicides, butralin and pendimethalin were found to be relatively stable in aerobic soil and significant portions could be recovered after 6 months. Metabolism of N-nitrosopendimethalin was found to be carried out by streptomyces culture isolated from an aerobic soil (Lusby et al., 1978) and the reduction of a nitro group and hydroxylation of a ring methyl seemed to be the major reactions. This compound was rapidly degraded in flooded and anaerobic soil.

9.3. Degradation of Nitrosamines in Plants

Information in regards to nitrosamines in plants is rather scarce in literature. It has been shown by Marco et al. (1976) that corn grown to maturity in the greenhouse with a soil treated with N-nitrosoatrazine or N-nitrosohydroxyatrazine and fertilizers containing nitrate and nitrite did not contain either NNA or NNHA in the stalks or grain. Soybeans grown to maturity in soil treated with [^{14}C]NDPA or [^{14}C]nitrosopendimethalin showed no radioactive uptake of either compound in the mature beans (Kearney et al., 1977).

9.4. Metabolism of Nitrosamines in Animals

A significant amount of research has been done on nitrosodimethylamine and low-molecular-weight nitrosodialkyamines. Also a substantial amount of work has been reported on cyclic nitrosamines.

Magee (1956) studied the metabolism distribution in the body and excretion of nitrosodimethylamines in rats, rabbits, and mice. Nitrosoamines show a distinct specificity in their carcinogenicity. The actual carcinogen is a metabolite of the original nitrosamine. It has been shown that the liver is the main site for the metabolism of NDMA. It has been shown that the rate of metabolism is quite rapid and only 34% of the NDMA could be recovered from a rat 8 hr after oral dosing (Magee, 1972; Dutton and Heath, 1956). Only 1.7% of the NDMA was recovered from the urine in 24 hr and none in the feces.

Dutton and Heath (1956) and Heath and Dutton (1958) studied the rate of disappearance of $[^{14}C]$NDMA in the mouse and rat. It was observed that 44–66% of the ^{14}C was eliminated as a $^{14}CO_2$ within 6 hr with 5% in the urine of both species. The proposed metabolism of NDMA is presented in Figure 20.

It is believed that the initial step of metabolism involved an enzyme-catalyzed demethylation via an unstable α-hydroxy compound that decomposes and leads to the formation of formalydehyde (Druckrey et al., 1967; Brouwers and Emmelot, 1960; Magee and Hultin, 1962).

It has been shown that methyl groups from NDMA are metabolized to formate, which eventually ends up in lipids, specifically in 3-*sn*-phosphati-

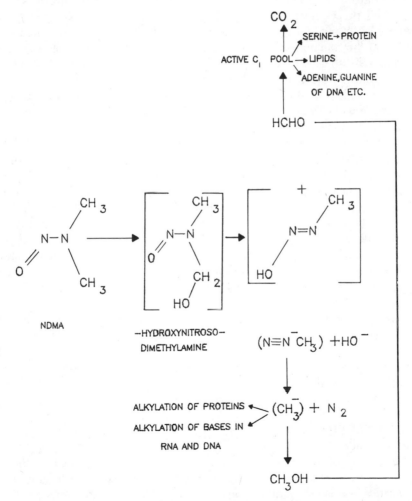

Figure 20. Proposed metabolism of NDMA (Heath and Dutton, 1958).

dylcholine. *In vitro* most of the carbon atoms in NDMA can be accounted for as methanol and formaldehyde (Lake et al., 1976).

The fate of nitrogen has been studied by Heath and Dutton (1958). Traces of methylamine, hydroxylamine, and nitrite were detected in liver and/or urine. Most of the amino nitrogen was converted into ammonia and both nitrogen atoms in NDMA became evenly distributed in the nitrogen constituents of the body.

Methylation of proteins and nucleic acids as studied by Magee and Hultin (1962) and Magee and Farber (1962) were postulated to be carried out by methyldiazonium ion or a carbonium ion derived from it. Methylation of proteins include methylation at 1 and 3 positions of histidine carried out by ^{14}C NDMA (Magee and Hultin, 1962). [^{14}C]NDMA activity was also detected in the C-3 atom of serine, pointing out that the carbon atoms from NDMA had entered the C_1 metabolic pool. Nucleic acids is at the N-7 position of guanine in RNA and DNA as reported by Magee and Farber (1962). Proteins and nucleic acids are the main reason for carcinogenicity of nitrosamines, especially the position of O-6 of guanine in nucleic acids (Loveless, 1969). Methylase activity and its inhibitors were studied by Friedman et al. (1976) and the results indicated that nitrososarcosin and nitrosodiethylamine suppressed the enzyme activity in rat liver. The work *in vivo* has involved high and toxic doses and it would be important to consider some work at lower dosages.

NDMA is metabolized rapidly, mainly in the liver, and is demethylated to C-1 intermediates. Most of these are oxidized to carbon dioxide or used in the normal metabolism of the body. In rats most of the ^{14}C[1-C] is converted to CO_2. The percentage of the dose excreted was unchanged in urine and the evidence is that the 7-position of guanine was being ethylated but not methylated (Kruger, 1972). The products of such reactions have been identified by giving oral doses of NDEA (Blattmann and Preussmann, 1973). Metabolism is similar to that of NDMA with additional occurrence of β-oxidation with the urinary elimination of oxidation products still containing intact nitrosamine group.

It has been demonstrated by Kruger (1971, 1972) that when rats were given 1-[^{14}C]NDPA led to the formation of 7-[^{14}C]*n*-propylguanine and 7-[^{14}C]methylguanine in the RNA of rat liver but no 7-[^{14}C]methylguanine was detected when 2-[^{14}C]NDPA was administered. This indicated that during metabolism the C-1 and C-2 atoms split with the C-1 atom involved in the subsequent methylation. The splitting was considered to follow β-hydroxylation resembling fatty acid metabolism.

10. CONCLUSION

With the advent of more sensitive and reliable analytical tools such as thermo energy analyzer for the detection of nitrosamine, the presence of such compounds as impurities in pesticide formulations was an unsuspected phenomenon. Trace contamination of pesticide formulations with nitrosamine was shown to be a real phenomenon. Reduction or removal of such trace components was achieved by prevention of nitrosation either by direct chemical destruction or purification steps. The ease of nitrosation reactions with appropriate amines demands careful attention to possible side reactions that take place during the synthetic process. The reduction of nitrosamine contaminants in pesticides should be considered an individual and unique situation in the manufacture of a specific pesticide and can be controlled by careful investigation of the sources.

Areas that should be considered for further research are systematic studies on occupational exposure to nitrosamines in order to obtain an adequate data base for further neurological studies. Investigation in regards to nitrosated peptides and evaluation of their carcinogenic potential require attention. Also identification of relevant amine precursors in indogenous nitrosation and clarification of the role of high nitrate exposures need to be further studied.

It is evident from animal studies that nitrosamines are carcinogens and mutagens that occur in the environment and can be formed in the body from noncarcinogenic precursors. The possible role of nitrosamines in the causation of human cancer is not known and is under intensive investigation throughout the world. Experiments in relation to volatility of low-molecular-weight nitrosamines including NDPA have indicated that these compounds are somewhat volatile and disappear rather rapidly after application to the surface of soil. However incorporation into the soil decreases both the rate and extent of volatilization. No uptake of ^{14}C into the plants was observed when soybeans were grown in soils treated with high levels of $[^{14}C]NDPA$ or N-nitrosopendimethaline.

The results of several studies showed that plants that are ingested by humans could not take up and store potentially carcinogenic N-nitroso compounds formed in the plant environment.

Analysis of barley, wheat, spinach, lettuce, carrots, and tomatoes for the presence of NDMA and NDEA in soils treated with 0–123 kg/ha nitrogen showed absence of such compounds. Studies on the uptake of NDPA and N-nitrosopendimethalin by soybeans in the fields treated at 0.1, 1, 10, and 100 ppb level showed an absence of measurable quantities of nitrosamines in the plant. From such studies one can conclude that the food supply is safe and wholesome and the chances of being contaminated with nitroso compounds is very minute.

REFERENCES

Archer, M. C., and Wishnok, J. S. (1976). *J. Environ. Sci. Health, All* **583**, 10411.

Ayanaba, A., and Alexander, M. (1973). *Appl. Microbiol.* **25**, 862.

Ayanaba, A., Verstraete, W., and Alexander, M. (1973). *Proc. Soil. Sci. Soc. Am.* **37**, 565.

Bamford, C. H. (1939). *J. Chem. Soc.* Part I, **12**, 85.

Barnes, J. M., and Magee, P. N. (1954). *Br. J. Ind. Med.* **11**, 167.

Bassow, H. *Air Pollution Chemistry. An Experimenter's Source Book.* Hayden Book Company, Rochell Park, NJ, 1976.

Biggs, I. D., and Williams, D. L. H. (1976). *J. Chem. Soc. Perkin II* 601 ff.

Blattmann, L., and Preussmann, R. (1973). *Z. Krabsforsch* **79**, 3.

Bontoyan, W., Wright, D., Jr., and Law, M. W. (1979). *J. Agric. Food Chem.* **27**, 631.

Brouwers, J. A. J., and Emmelot, P. (1960). *Exp. Cell. Res.* **19**, 467.

Burns, D. T., and Alliston, G. V. (1971). *J. Food Technol.* **6**, 433.

Cannon, W. N., and Eizember, R. F. (1978). U.S. Patent 4,120,905, Oct. 17.

Challis, B. C., Edwards, A., Hunma, R. R., Kyrtopoulos, S. A., and Outram, J. R. (1978). In E. A. Walker, L. Griciute, M. Castegnaro, and R. C. Lyle, eds., *Environmental Aspects of N-Nitroso Compounds*. IARC Scientific Publications No. 19, International Agency for Research on Cancer, Lyon, 127.

Challis, B. C., Outram, J. R., and Shuker, D. E. G. (1980). In E. A. Walker, L. Griciute, M. Castegnaro, and M. Borzsonyi, eds., *N-Nitroso Compounds: Analysis, Formation and Occurrence*. IARC Scientific Publications No. 31, International Agency for Research on Cancer, Lyon, 43.

Chow, Y. L. (1967). *Can. J. Chem.* **45**, 53.

Clapp, N. K., Tyndall, R. L., and Otten, J. A. (1971). *Cancer Res.* **31**, 196.

Cohen, S. Z., Zweig, G., Law, M. W., Wright, D., Jr., and Bontoyan, W. R. (1978). *IARC Sci. Publ.* **19**, 333.

Croisy, A. F., Fanning, J. C., Keefer, L. K., Slavin, B. W., and Uhm, S. J. (1980). In E. A. Walker, L. Griciute, M. Castegnaro, and M. Borzsonyi, eds., *N-Nitroso Compounds: Analysis, Formation, and Occurrence*. IARC Scientific Publications No. 31, International Agency for Research on Cancer, Lyon, 83.

Crosby, D. G., Humphrey, R., and Moilanen, K. W. (1978). Unpublished results.

Crosby, N. T., and Sawyer, R. (1976). *Adv. Food Res.* **22**, 1.

Daiber, D. and Preussmann, R. (1964). *Z. Anal. Chem.* **206**, 344.

Diehl, R. E., Levy, S. D., and Gastrock, W. H. (1979). U. S. Patent 4,136,117, Jan. 23.

Douglass, M. L., Kabacoff, B. L., Anderson, G. A., and Cheng, M. C. (1978). *J. Soc. Cosmet. Chem.* **29**, 581.

Dressel, J. (1967). 85, 191315 U. *Qual. Plant, Plant Foods Nut.* **25**, 381.

Dressel, J. (1976). *Qual. Plant-Plant Foods Hum. Nutr.* **25**, 381; cf. *CA* **85**, 191315.

Dressel, J. (1977). *Z. Lebensm. Unters.-Forsch.* **163**, 11.

Druckery, H., and Preussmann, R. (1962). *Naturwissenschaften*, **49**, 498.

Druckery, H., Preussmann, R., Ivankovic, S., and Schmahl, D. (1967). *Z. Krebsforsch.* **69**, 103.

Dutton, A. H., and Heath, D. F. (1956). *Nature (London)* **178**, 644.

Edner, F., Havre, G. N., Helgebostad, A., Coppang, N., Matson, R., and Ceh, L. (1964). *Naturwissenschaften* **51**, 637.

Egert, G., and Greim, H. (1976). *Mut. Res.* **37**, 179.

Eisenbrand, G. (1972). *IARC Sci. Publ.* **3**, 64.

Eisenbrand, G., and Preussmann, R. (1970). *Arzenimittel-Forsch.* **20**, 1513.

Eisenbrand, G., Ungerer, O., and Preussman, R. (1974). *Fd. Cosmet. Toxicol.* **12**, 229.

Eisenbrand, G., Janzowski, C., and Preussmann, R. (1975). *J. Chromatog.* **115**, 602.

Eizember, R. F. (1978). U. S. Patent 4,127,610, Nov. 28.

Eizember, R. F., and Vogler, K. R. (1980). U. S. Patent 4,185,035, Jan. 22.

Eizember, R. F., Vogler, K. R., Souter, R. W., Cannon, W. N., and Wege, P. M., II. (1979). *J. Org. Chem.* **44**, 784.

Elespuru, R. K. and Lijinsky, W. (1973). *Fd. Cosmet. Toxicol.* **12**, 229.

Fan, T-Y., and Tannenbaum, S. R. (1973). *J. Agric. Food Chem.* **21**, 237.

Fine, D. H., and Rufeh, F. (1974). *IARC Sci. Publ.* **9**, 40.

Fine, D. H., Rufeh, F., and Gunter, G. (1973). *Anal. Lett.* **6**, 731.

Fine, D. H., Ross, R., Fan, S., Rounbehler, D. P., Silvergleid, A., Song, L., and Morrison, J. (1976). Presented at the 172nd Am. Chem. Soc. Natl. Mtg., San Francisco, CA, Sept. 2.

Fine, D. H., Ross, R., Rounbehler, D. P., Silvergleid A., and Song, L. (1977). *Nature* (*London*) **265**, 753.

Franck, B., Conrad, J., and Misbach, P. (1970). *Angw. Chem. Intl. Ed. Engl.* **9**, 892.

Frei, J. V. (1970). *Cancer Res.* **30**, 11.

Friedman, M. A., Sander, V., and Woods, S. (1976). *Toxicol. Appl. Pharmacol.* **36**, 395.

Gargus, J. L., Reese, W. H., Jr., and Rutter, H. A. (1969). *Toxicol. Appl. Pharmacol.* **15**, 92.

Grove, D. R. (1979). Eli Lilly and Company, unpublished data.

Hanst, P. L., Spence, J. W., and Miller, M. (1977). *Environ. Sci. Tech.* II, 403.

Hard, G. C., and Butler, W. H. (1970a). *Cancer Res.* **30**, 2796.

Hard, G. C., and Butler, W. H. (1970b). *J. Pathol.* **102**, 201.

Heath, D. F., and Dutton, A. H. (1958). *J. Biochem.* **70**, 619.

Hecht, S. S., Castonguay, A., Rivenson, A., Mu, B., and Hoffman, D. (1983). *J. Environ. Sci. Health* 1.

Hoffmann, D., Adams, J. D., Brunnenmann, K. D., and Hecht, S. S. (1982). In R. A. Scanlan and S. R. Tannenbaum, eds. *N-Nitroso Compounds*. ACS Ser. No. 174. American Chemical Society, Washington D.C., 247.

Ingold, C. K. (1953). *Structure and Mechanism in Organic Chemistry*. Cornell University Press, Itheca, NY.

Kearney, P. C. (1980). *Pure Appl. Chem.* **52**, 499.

Kearney, P. C., Oliver, J. E., Helling, C. S., Isensee, A. R., and Knotson, A. (1977). *J. Agric. Food Chem.* **25**, 117.

Khan, S. U., and Young, J. C. (1977). *J. Agric. Food Chem.* **25**, 1430.

Kruger, F. W. (1971). *Z. Krebsforsch.* **76**, 145.

Kruger, F. W. (1972). *Topics Chem. Carcinogenesis, Proc. 2nd Int. Symp.*, p. 213.

Kunz, W., Schaude, G., and Thomas, C. (1969). *Z. Krebsforsch.* **72**, 291.

Lake, B. G., Minski, M. J., Phillips, J. C., Gangolli, S. D., and Lloyd, A. G. (1976). *Life Sci.* **17**, 1599.

Legator, M. S., and Malling, H. V. (1971). In A. Hollaender, ed. *Chemical Mutagens: Principles and Methods for Their Detection*. Plenum, New York and London, 1971.

Lippert, E., and Kelm, J. (1978). *Helv. Chim. Acta* **61**, 279.

Loveless, A. (1969). *Nature* (*London*) **223**, 206.

Lusby, W. R., Oliver, J. E., and Kearney, P. C. (1978). Presented at the 176th Am. Chem. Soc. Natl. Mtg. Miami, FL, Sept. 14.

Magee, P. N. (1956). *Biochem. J.* **64**, 676.

Magee, P. N. (1972). *Topics Chem. Carcinogenesis, Proc. 2nd Int. Symp.* **259**.

Magee, P. N., and Barnes, J. M. (1956). *J. Cancer* **10**, 114.

Magee, P. N., and Barnes, J. M. (1967). *Adv. Cancer. Res.* **10**, 163.

Magee, P. N., and Farber, E. (1962). *Biochem. J.* **83**, 111.

Magee, P. N., and Hultin, T. (1962). *Biochem. J.* **83**, 106.

Magee, P. N., Montesano, R., and Preussmann, R. (1976). In C. E. Searle, ed., *Chemical Carcinogens*. ACS Monograph 173. American Chemical Society, Washington, D.C., 491.

Malling, H. V. (1971). *Mut. Res.* **13**, 425.

Marco, G. J., Boka, G., Cassiday, J. E., Ryskiewich, D. P., Simoneaux, B. J., and Summer, D. D. (1976). Presented at the 172nd Am. Chem. Soc. Natl. Mtg., San Francisco, CA.

Markenzich, R. L., Zamek, O. S., Donahue, P. E., and Williams, F. J. (1977). *J. Org. Chem.* **42**, 3435.

Mills, A. L., and Alexander, J. (1976). *Environ. Qual.* **5**, 437.

Mirvish, S. S. (1975). In C. G. King and J. J. Burns, eds., *Second Conference on Vitamin C, Anal. Sci.* N.Y. Acad. Sci., New York, Vol. 258, p. 175.

Miyamoto, J. (1978). Unpublished results.

Miyamoto, J., and Hosokawa, M. (1977). Unpublished results.

Oliver, J. E., and Kontson, A. (1978). Presented at the 174th Am. Chem. Soc. Natl. Mtg., Anaheim, CA.

Oliver, J. E., Kearney, P. C., and Kontson, A. (1979a). *J. Agric. Food Chem.* **27**, 887.

Oliver, J. E., Lusby, W. R., and Smith, R. H. (1979b). Presented at the 177th Am. Chem. Soc. Natl. Mtg., Honolulu, Hawaii, Apr. 2.

Pancholy, S. K. (1976). Agronomy Abstract, 1976 Ann. Meeting, Houston, TX. *Pest. Toxic Chem. News*, May.

Pielsticker, K., Mohr, U., and Klemm, J. (1967). *Naturwissenschaften* **54**, 340,

Pitts, J. N., Grosjean, D., Van Canwenberghe, K., Schmid, J. P., and Fritz, D. R. (1978). *Environ. Sci. Technol.* **12**, 946.

Polo, J., and Chow. Y. L. (1976). *J. Natl. Cancer. Inst.* **56**, 997.

Preussmann, R., Deutsch-Wenzel, R., and Eisenbrand, G. (1975). *Z. Krebsforsch.* **84**, 75.

Preussmann, R., O'Neill, I. K., Eisenbrand, G., Spiegelhalder, B., and Bartsch, H., eds. (1983). *Environmental Carcinogens: Selected Methods of Analysis*, Vol. 6, *N-Nitroso Compounds*. International Agency for Research on Cancer, Lyon.

Probst, G. W., Golab, T., and Wright, W. L. (1975). In P. C. Kearney and D. D. Kaufman, eds., *Dinitroanilines. Herbicides, Chemistry, Degradation and Mode of Action*. Dekker, New York, Vol. I, Chap. 9, p. 453.

Ragsdale, R. O., Karstetter, B. R., and Drago, R. S. (1965). *Inorg. Chem.* **4**, 420.

Raymond, D. D., and Alexander, M. (1976). *Nature (London)* **262**, 394.

Ridd, J. H. (1961). *Q. Rev.* **15**, 418.

Ross, L. J., and Chiarello, G. A. (1979). U. S. Patent 4,134,917, Jan. 16.

Ross, R. D., Morrison, J., Rounbehler, D. P., Fan, S., Fine, D. H. (1976). Presented at the Division of Pesticide Chemistry 172nd National Meeting of the American Chemical Society, San Francisco, CA, Sept.

Ross, R. D., Morrison, J., Rounbehler, D. P., Fan, S., and Fine, D. H. (1977). *J. Agric. Food Chem.* **25**, 1416.

Sakshaug, J., Sogmen, E., Hansen, M. A., and Koppeny, N. (1965). *Nature (London)* **206**, 1261.

Sander, J. (1965). *Arch. Hyg. Bakteriol.* **151**, 22.

Sander, J., Schweinsberg, F., La Bar, J., Burkle, G., and Schweinsberg, E. (1975). *Gann Monograph Cancer Res.* **17**, 145.

Saunders, D. G., and Mosier, J. W. (1979). *J. Agric. Food Chem.* **27**, 584.

Schmeltz, I., and Wenger, A. (1979). *Fd. Cosmet. Toxicol.* **17**, 105.

Schmeltz, I., Abidi, S., and Hoffmann, D. (1977). *Cancer Lett.* **2**, 125.

Seiler, J. P. (1977). *Mut. Res.* **48**, 225.

Sen, N. P., Donaldson, B. A., and Chrabonneau, C. (1974). *IARC Sci. Publ.* **9**, 75.

Tate, A. L., III, and Alexander, M. J. (1975). *J. Natl. Cancer Inst.* **54**, 327.

Tate, R. L., and Alexander, M. (1974). *Soil Sci.* **118**, 317.

Terracini, B., and Testa, M. C. (1970). *Br. J. Cancer* **24**, 588.

Terracini, B., Palestro, G., Ramella Gigliardi, M., and Montesano, R. (1966). *Br. J. Cancer* **20**, 871.

Toth, B., and Shubik, P. (1967). *Cancer Res.* **27**, 43.

Toth, B., Magee, P. N., and Shubik, P. (1964). *Cancer Res.* **24**, 1712.

Turusov, V., Tomatis, L., Guibert, D., Duperray, B., and Pacheco, H. (1973). *IARC Sci. Publ.* **4**, 84.

Uchiyama, M., Takeda, M., Suzuki, T., and Yoshikawa, K. (1975). *Bull. Environ. Contam. Toxicol.* **14**, 389.

Ullrich, F., Hermann, G., and Weber, P. (1978). *Biochem. Pharmacol.* **27**, 2301.

Verstraete, N., and Alexander, M. (1971). *J. Appl. Bacteriol.* **34**.

Vesselinovitch, S. D. (1969). *Cancer Res.* **29**, 1024.

Wolfe, H. R., Durham, W. F., and Armstrong, J. F. (1967). *Arch. Environ. Health* **14**, 622.

Wood, M., Flaks, A., and Clayson, D. B. (1970). *Eur. J. Cancer* **6**, 433.

Yang, H. S., Okun, J. D., and Archer, M. C. (1977). *J. Agric. Food Chem.* **25**, 1181.

Zweig, G., Selim, S., Hummel, R., Mittelman, A., Wright, D., Jr., Law, S. C., Jr., and Regelman, E. (1980). *IARC Sci. Publ.* **31**, 555.

12

PESTICIDE CONTAMINATION OF MILK AND MILK PRODUCTS

G. S. Dhaliwal

Section of Ecology,
College of Agriculture,
Punjab Agricultural University
Ludhiana, India

1. INTRODUCTION

Pesticides are vital in meeting the increasing food needs of the growing population and in containing vector-borne diseases, particularly in developing countries. However, the use of pesticides in agriculture and the public health sector is not free from adverse effects on human health and the environment. Although the per hectare use of pesticides in developing countries is less than that in many developed countries (Table 1), the problems caused by their unregulated use can be equally severe. Moreover, the present use of less than 0.5 kg/ha of pesticides in developing countries is likely to increase at least by 3-fold, even when integrated pest management programs are implemented (Pathak and Dhaliwal, 1986). The use of pesticides in agriculture in India may increase from the current 75,000 tonnes to over 200,000 tonnes by the year 2000 (Mehrotra, 1986). Hence, extreme precautions would be needed on the part of scientists and policymakers so that this prospective increased use of pesticides does not further degrade the quality of our environment.

The contamination of food materials with pesticides is one of the major problems confronting humans. The consumer runs the greatest risk of exposure to pesticides through the contaminated food. Since the degradation of chlorinated insecticides is slow, lactating animals excrete them as such in milk over a long period of time and such lipophilic residues may become further concentrated in high fat milk products. Hence, the pesticidal contamination of milk and milk products, because of their special significance in the human diet, should be viewed with great concern.

2. NATURE AND MAGNITUDE OF CONTAMINATION

2.1. Milk

Monitoring surveys have been conducted in different countries that reveal widespread contamination of milk with DDT [1, 1, 1-trichloro-2, 2-bis(chlorophenyl)ethane] analogues and HCH (hexachlorocyclohexane) isomers.

TABLE 1 Pesticide Consumption in Selected Developing and Developed Countries

Country	Quantity (kg/ha)	Country	Quantity (kg/ha)
Bangladesh	0.20	Philippines	0.70
Pakistan	0.30	United States	1.49–3.00
Thailand	0.40	West Europe	1.87–3.00
Indonesia	0.40	West Germany	3.00
India	0.57	Japan	10.79–11.80

Source: Pesticides Information **9**, 45 (1983).

2.1.1. DDT Analogues

The work on screening of bovine milk for DDT contamination was started in India during the mid-1960s. However, the early studies did not take into consideration the different analogues. Tripathi (1966) analyzed five milk samples collected from Pantnagar (Uttar Pradesh) and found four samples to be contaminated with DDT. However, the level of contamination was below the maximum residue limit (MRL) of 1.25 ppm prescribed for milk on fat basis (0.05 ppm on whole milk basis) by FAO/WHO (Anonymous, 1979). On the other hand, out of 17 samples collected from Delhi during 1965, 13 were contaminated with DDT and 9 (5.3%) contained DDT above the MRL (Agnihotri et al., 1974). When the monitoring in Delhi was resumed in 1972, out of 14 samples, 13 contained DDT residues above MRL and there was about a 3.3-fold increase in concentration of DDT in milk (Agnihotri et al., 1974). The results of another survey of milk carried out in Hyderabad (Andhra Pradesh) revealed that 58 (80%) of the 72 samples contained DDT above the MRL (Lakshaminaryana and Krishna Menon, 1975). The maximum levels of DDT in milk collected from Delhi and Hyderabad were 2 and 5 ppm, respectively.

The first attempt to analyze different analogues of DDT in the milk in India was made in 1976 by Dhaliwal and Kalra (1977). DDT residues were found in the form of p, p'-TDE [1, 1-dichloro-2, 2-bis(p-chlorophenyl)ethane], p, p'-DDE, and p, p'-DDT (Table 2). The milk supplied by the Punjab Dairy Development Corporation (PDDC) which came from many villages and hence was considered to be representative of the large area, contained 0.26 ppm DDT, about 71.7% of which consisted of p, p'-TDE. Out of 60 samples, 44 (73%) contained DDT residues above the MRL. Subsequent studies in 1977 revealed that 40 of 42 samples obtained from the depots of PDDC milk plants situated in the cities of Amritsar, Bathinda, Chandigarh, and Ludhiana contained DDT residues above

TABLE 2   DDT Contamination of Milk Collected from Different Localities in Punjab, India[a]

Location	Period of Sampling[b]	Contamination Level (ppm Whole Milk Basis)			
		p, p'-DDE	p, p'-TDE	p, p'-DDT	Total DDT
Rural areas	A	0.04	0.39	0.13	0.56
	B	0.04	0.16	0.03	0.23
Punjab Dairy Development Corporation	A	0.03	0.25	0.07	0.35
	B	0.02	0.13	0.03	0.18
Ludhiana city dairies	A	0.01	0.06	0.02	0.09
	B	0.01	0.05	0.003	0.06
PAU milk supply	A	0.003	0.03	0.006	0.04
	B	0.006	0.02	0.007	0.03

[a]Dhaliwal and Kalra (1977).
[b]A, September–October 1976; B, January–February 1977.

the MRL; the average level varied from 0.15 to 0.29 ppm on a whole milk basis (Kalra et al., 1978). Some samples showed total DDT levels as high as 10–17 times the MRL. DDT residues in milk were found mainly in the form of p,p'-TDE, p,p'-DDE, and p,p'-DDT. Traces of o,p'-DDT [1,1,1-trichloro-2-(p-chlorophenyl)-2-(o-chlorophenyl)ethane] and o,p'-TDE [1,1-dichloro-2-(p-chlorophenyl)-2-(o-chlorophenyl)ethane] were also detected. However, p,p'-TDE constituted more than 50% of DDT residues in the majority of the samples. The high level of TDE in milk suggests that the intake of DDT in cattle occurs mainly through contaminated feed (Kalra et al., 1986). Thus, the knowledge of the TDE–DDT relationship is considered very useful in predicting the sources of contamination of milk.

Almost the same picture of DDT contamination of milk in Punjab persisted during 1979, 1980, and 1981 (Table 3). Kapoor et al. (1980) collected 54 milk samples from preselected rural areas, which had been sprayed with either DDT or HCH for the control of malaria. DDT was detected in all the samples, the concentrations ranging from 0.047 to 1.568 ppm. Singh et al. (1986) also detected DDT in 111 of 112 samples of milk from rural areas of Punjab and the magnitude of contamination varied from traces to 0.91 ppm.

Of 44 milk samples collected from Bombay, 12 were contaminated with DDT, the level (whole milk basis) ranging from 2.8 to 10.8 ppm (average 5.00 ppm) in samples from government booths and from 4.8 to 6.3 ppm (average 5.6 ppm) in samples from local vendors (Khandekar et al., 1981). Chauhan et al. (1982) found 35 of 105 milk samples collected from Hisar (Haryana) to be contaminated with DDT ranging from traces to 7 ppm. The milk samples from Lucknow were found to be contaminated with DDT ranging from 0.015 to 0.071 ppm with an average of 0.045 ppm for buffalo milk and from 0.011 to 0.073 ppm with an average of 0.042 ppm for goat milk (Saxena and Siddiqui, 1982). In another study reported from Lucknow, the average level of total DDT in bovine milk was found to be 0.218 ppm (Kaphalia et al., 1985). In Marathwada region of Maharashtra, buffalo milk contained 0.014–1.749 ppm of DDT residues and cow milk contained 0.003–1.415 ppm (Jadhav, 1986). The higher level of contamination of buffalo than of cow milk might be due to the higher fat content in buffaloes than in cows and their differences in metabolism.

A recent survey sponsored by the Food and Agriculture Organization (FAO) through the Ministry of Health, Government of India, revealed that 60% of the bovine milk samples were contaminated with DDT above the MRL. Throughout India, of 980 samples from Andhra Pradesh, Delhi, Maharashtra, and Punjab, 95% were contaminated with DDT, the levels ranging from 0.19 to 216 ppm on fat basis (Kalra and Chawla, 1986). These results clearly indicate widespread contamination of bovine milk with excessive residues of DDT in India. In general, the milk samples from rural areas contain higher residues than the urban milk supply.

Unlike India, cow milk in Japan showed DDT contamination levels much below the MRL. The first survey of milk for pesticidal contamination in Japan was started in 1969 and the average level of total DDT contamination was

TABLE 3 DDT Contamination of Bovine Milk in India

Year of Sampling	Locality	Source of Milk Supply	Range (ppm Whole Milk Basis)				Average Level of DDT (ppm)
			p,p'-DDE	p,p'-TDE	p,p'-DDT	Total DDT	
1977[a]	Amritsar	PDDC[f]	0.016–0.055	0.03–0.33	0.02–0.17	0.07–0.56	0.26
	Bathinda	PDDC	0.010–0.08	0.002–0.38	0.001–0.15	0.004–0.61	0.24
	Chandigarh	PDDC	0.008–0.042	0.022–0.14	0.009–0.08	0.037–0.25	0.15
	Ludhiana	PDDC	0.011–0.06	0.04–0.35	0.05–0.22	0.10–0.63	0.29
1979[b]	Ferozepur	RA	0.02–0.15	BDL[g]–0.26	0.01–0.09	0.05–0.50	0.17
	Jalandhar	RA	0.02–0.31	0.01–0.80	0.02–0.40	0.05–1.51	0.39
	Ludhiana	RA	0.02–0.14	0.04–0.42	0.01–0.28	0.07–0.71	0.35
	Sangrur	RA	0.03–0.09	0.03–0.10	0.01–0.05	0.08–0.24	0.14
1980[b]	Ferozepur	UA	0.01–0.03	Tr–0.03	Tr–0.02	0.01–0.08	0.02
	Ludhiana	RA	0.01–0.06	0.01–0.27	0.02–0.22	0.05–0.55	0.16
	Ludhiana	UA	0.01–0.05	0.01–0.32	BDL–0.07	0.02–0.44	0.12
	Sangrur	UA	0.01–0.04	BDL–0.03	BDL–0.05	0.01–0.10	0.05
1980–1981[c]	Ludhiana	UA	0.02–0.06	0.07–0.24	0.03–0.09	0.14–0.36	0.24
1980–1981[d]	Lucknow	UA	0.008–0.016	0.006–0.028	BDL–0.022	0.015–0.071	0.045
	Lucknow	UA(GM)	0.007–0.052	BDL–0.018	0.002–0.021	0.011–0.073	0.042
1980–1981[e]	Lucknow	UA	0.047	0.114	0.038	0.218	0.22

[a]Kalra et al. (1978).
[b]Kalra and Chawla (1983).
[c]Singh (1982).
[d]Saxena and Siddiqui (1982). Year of sampling not mentioned.
[e]Kaphalia et al. (1985). Year of sampling not mentioned, only average values of contaminants reported.
[f]PDDC, sale depots of Punjab Dairy Development Corporation; RA, rural areas; UA, urban areas; UA(GM), urban areas (goat milk).
[g]BDL, below detectable limits; Tr, traces.

TABLE 4   Pesticidal Contamination of Milk Collected from Four Localities around Baghdad, Iraq[a]

Pesticide	Contamination Level (ppm)			
	Abu-Ghraib	Fidhelia	Musaib	Tarmia
p,p'-DDE	0.15	0.17	0.23	0.23
p,p'-TDE	0.18	0.12	0.17	0.17
o,p'-DDT	0.37	0.57	0.36	0.38
p,p'-DDT	0.02	0.01	0.01	0.05
Total DDT	0.75	0.88	0.78	0.85
Lindane	0.03	0.05	0.06	0.04
Aldrin	0.02	0.04	0.06	0.03
Dieldrin	0.02	0.01	0.01	0.01
Endrin	0.06	0.05	0.03	0.10

[a]Al-Omar et al. (1985).

0.023 ppm (Uyeta et al., 1970). The parent compound of DDT was the major contaminant, comprising 48% of total DDT residues. The highest levels of 0.08 ppm of total DDT were reported from Miyagi and Okayama (Tanabe, 1972; Tomizawa, 1977). The cow milk in three Central American countries, namely El Salvador, Guatemala, and Honduras, contained DDT residues varying from 0.30 to 32.31 ppm with an average of 4.22 ppm (Mazariegos, 1976).

The results obtained from the analysis of 52 cow milk samples taken from 12 commercial dairies in Israel revealed an average level of contamination of 0.29 ppm of total DDT on a fat basis (Veirov et al., 1977). However, the proportion of DDE was higher than the combination of DDT and TDE. The residues in goat milk were much lower (0.02–0.07 ppm) and consisted of only DDE. On the other hand, the DDT residues in goat milk from India revealed the presence of p,p'-DDE, p,p'-TDE, and p,p'-DDT and the total DDT contamination level was similar to that of buffalo milk (Saxena and Siddiqui, 1982). Thirty-nine milk samples collected directly from milk tankers from four different localities of Baghdad (Iraq) were contaminated with an average total DDT level of 0.81 ppm on a fat basis (Al-Omar et al., 1985). The major analogue detected in milk was o,p'-DDT, followed by p,p'-DDE/p,p'-TDE and p,p'-DDT, respectively (Table 4). The differences in DDT analogues might be due to different feeding practices and to differences in metabolism.

2.1.2.   HCH Isomers

There are several reports that document the contamination of milk with HCH isomers. Lakshminarayana and Krishna Menon (1975) found 25% of the samples of bovine milk (37 of 127) in Hyderabad (Andhra Pradesh) to be contaminated with HCH in concentrations ranging from traces to 5 ppm. Chauhan et al. (1982) detected HCH in only 3 of 105 samples of milk collected from Hisar (Haryana) and the level of contamination varied from traces to 7 ppm. However, Kapoor et al. (1980) detected HCH in all the 54 milk samples collected from rural areas of

Punjab and the level of contamination varied from 0.014 to 2.057 ppm. The low incidence of contamination in the former two studies might be due to the employment of less sensitive thin-layer chromatographic method.

The milk samples collected from rural and urban areas of four districts of Punjab (Kalra and Chawla, 1983) revealed widespread contamination of milk with HCH in the range of 0.01 to 2.06 ppm (Table 5). In another study, the level of HCH in milk from urban area varied from 0.014 to 0.053 ppm with an average of 0.041 ppm (Singh, 1982). Singh et al. (1986) found all the 112 samples of bovine milk collected from rural areas to be contaminated with HCH and the level ranged from traces to 6.04 ppm. HCH contamination was detected mainly in the form of α- and β-isomers, although small amounts of γ- and δ-isomers were also detected. However, in the milk samples analyzed from Lucknow, α-isomer was most pronounced, followed by γ- and β-isomers, the average levels being 0.242, 0.060, and 0.011 ppm, respectively (Kaphalia et al., 1985). These differences in the proportions of HCH isomers reflect the different routes of exposure of dairy animals.

The buffalo milk in Lucknow showed total HCH contamination of 0.313 ppm (Kaphalia et al., 1985). However, an earlier study from Lucknow found buffalo milk to be contaminated with 0.02–0.098 ppm of HCH as compared to 0.002–0.068 ppm of HCH detected in goat milk (Saxena and Siddiqui, 1982). On the other hand, buffalo milk from Marathwada region of Maharashtra contained 0.003–0.475 ppm of HCH as compared to 0.005–0.217 ppm in cow milk (Jadhav, 1986). Thus, buffalo milk, being richer in fat than cow and goat milk, reveals higher levels of contamination. Among the 980 samples of milk analyzed from different parts of India, 90% of samples have been found to be contaminated with HCH, the level varying from 0.12 to 40 ppm on fat basis (Kalra and Chawla, 1986). Although there is no prescribed residue limit for HCH, these results reveal contamination of milk at quite high levels in India.

Severe contamination of market cow milk by HCH isomers was discovered in 1969 in Japan (Uyeta et al., 1970). The level of β-HCH constituted 66.8% followed by α-, δ-, and γ-isomers in the ratio of 26.9, 4.3, and 2.0%, respectively (Table 6). Based on analytical results of cow milk from January to February 1970, the level of β-HCH ranged from 0.001 to 2.68 ppm, whereas that of γ-HCH ranged from BDL (below detectable level) to 0.05 ppm (Kanazawa, 1983). The contamination level of α-HCH was less than that of β-HCH, whereas δ-HCH was at par with γ-HCH. The higher levels of β-HCH seem to be due to its greater stability, lower volatility, and higher capacity to accumulate in lipids (Tomizawa, 1977). The contamination levels of HCH in southwestern parts were generally higher than those in the northeastern parts. The highest total HCH of 0.976 ppm was reported from Osaka, out of which 0.753 ppm was β-HCH (Kanazawa, 1983).

The milk samples collected from 12 commercial cow dairies in Israel revealed highest contamination of 0.38 ppm of α-HCH followed by 0.28 ppm of γ-HCH (Veirov et al., 1977). The β-isomer was detected at a level less than 0.05 ppm. The milk samples collected from stores also contained the highest residues of 0.25 ppm of α-HCH followed by 0.19 ppm of γ-HCH. The contamination levels of

TABLE 5 HCH Contamination of Bovine Milk in India

Year of Sampling	Locality	Source of Milk Supply	Range (ppm Whole Milk Basis)					Average Level of Total HCH (ppm)
			α-HCH	β-HCH	γ-HCH	δ-HCH	Total HCH	
1979[a]	Ferozepur	RA[e]	0.01–0.54	0.03–1.40	0.005–0.14	Tr–0.13	0.06–2.06	0.40
	Jalandhar	RA	0.007–0.01	0.002–0.02	0.003–0.02	0.005–0.05	0.02–0.10	0.03
	Ludhiana	RA	0.001–0.04	0.002–0.12	0.002–0.014	Tr–0.02	0.005–0.19	0.05
	Sangrur	RA	0.03–0.25	0.08–0.56	0.01–0.06	0.01–0.07	0.13–0.91	0.38
1980[a]	Ferozepur	UA	0.004–0.016	0.002–0.029	0.003–0.020	0.002–0.009	0.011–0.064	0.04
	Ludhiana	RA	0.001–0.009	0.004–0.089	0.001–0.011	0.001–0.004	0.012–0.098	0.03
	Ludhiana	UA	0.003–0.028	0.002–0.049	0.001–0.035	BDL–0.025	0.006–0.088	0.02
1980–1981[b]	Ludhiana	UA	0.005–0.027	0.005–0.023	0.003–0.008	0.001–0.011	0.014–0.053	0.04
1980–1981[c]	Lucknow	UA	—	—	0.005–0.018	—	0.020–0.098	0.06
	Lucknow	UA (GM)	—	—	0.001–0.039	—	0.002–0.068	0.04
1980–1981[d]	Lucknow	UA	0.242	0.011	0.060	—	0.313	0.31

[a]Kalra and Chawla (1983).
[b]Singh (1982).
[c]Saxena and Siddiqui (1982). Year of sampling not mentioned.
[d]Kaphalia et al. (1985). Year of sampling not mentioned, only average values of contaminants reported.
[e]RA, rural areas; UA, urban areas (GM), urban areas (goat milk); BDL, below detectable levels; Tr, traces.

TABLE 6   HCH Contamination of Cow Milk in Japan[a]

	Contamination Level (ppm)				
Commodity	α-HCH	β-HCH	γ-HCH	δ-HCH	Total HCH
Raw milk	0.088	0.255	0.009	0.017	0.369
Market milk	0.097	0.236	0.007	0.016	0.356
Market milk	0.066	0.220	0.006	0.011	0.303
Market milk	0.078	0.140	0.005	0.011	0.234
Market milk	0.059	0.129	0.003	0.008	0.199
Market milk	0.022	0.033	0.002	0.002	0.059

[a]Adapted from Tomizawa (1977).

α-, β-, and γ-isomers in goat milk varied from 0.01 to 0.02, 0.02 to 0.05 and 0.03 to 0.05 ppm, respectively.

These variations in the ratio of various isomers in milk from India, Japan, and Israel might be due to the different routes of exposure and to the differences in metabolism. However, contamination of milk by HCH, particularly the β-isomer, is quite alarming.

2.1.3.   Other Pesticides

A number of other chlorinated hydrocarbons such as aldrin, dieldrin, chlordane, endrin, endosulfan, and heptachlor, have been detected by several workers in the milk. However, none of these insecticides has been detected in milk samples analyzed in Punjab (Kalra and Chawla, 1983). Chauhan et al. (1982) detected endosulfan in 3 of 105 milk samples collected from Hisar (Haryana), the level ranging from traces to 2.5 ppm. The milk samples from Lucknow were contaminated with aldrin, the average level being 0.022 ppm (range 0.001–0.042 ppm) in buffalo and 0.005 ppm (BDL–0.014 ppm) in goat milk (Saxena and Siddiqui, 1982). Khandekar et al. (1981) found that 12 of 23 milk samples collected from the local vendors in Bombay had alarmingly high dieldrin level, which varied from 38 to 126 ppm (average 76 ppm) on a fat basis. This level is more than 500 times higher than the MRL of 0.15 ppm recommended by FAO/WHO (Anonymous, 1979). Such a high level of contamination is a matter of grave concern and further monitoring of milk over large areas should be carried out to ascertain the recurrence of such high levels.

The contamination of cow milk with dieldrin has also been reported from Japan, although the levels were low, varying from traces to 0.01 ppm (Uyeta et al., 1970; Tomizawa, 1977). In another report, the highest dieldrin residue in cow milk was reported to be 0.019 ppm (Tanabe, 1972), although no aldrin or endrin was detected in any of the samples. The dieldrin contamination of cow milk has also been reported from Israel with an average level of 0.01 ppm (Veirov et al., 1977). Other contaminants detected at low levels in a few milk samples from Israel included aldrin, heptachlor, and heptachlor epoxide. Varying levels of aldrin

(0.02–0.06 ppm), dieldrin (0.01–0.02 ppm), and endrin (0.03–0.10 ppm) have also been detected in milk samples (Table 4) around Baghdad in Iraq (Al-Omar et al., 1985).

2.2. Butter

Butter is often contaminated with several organochlorine insecticides in India.

2.2.1. *DDT Analogues*

Only two of the four butter samples collected from Pantnagar (Uttar Pradesh) were found to be contaminated with DDT (Tripathi, 1966). However, all eight samples of butter from Delhi were excessively contaminated with DDT (Agnihotri et al., 1974). The residue in branded butter varied from 1.1 to 8.0 ppm, and had a mean of 4.1 ppm; the unbranded local butter revealed a contamin-

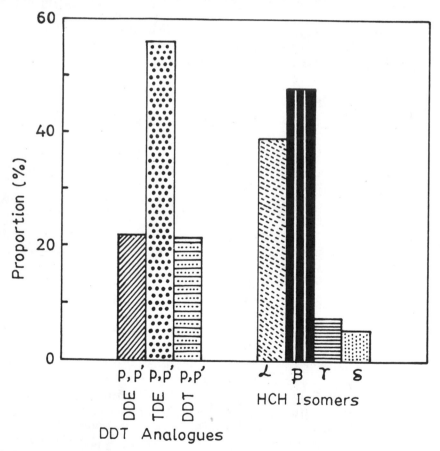

Figure 1. Composition of DDT analogues and HCH isomers in butter in India (Dhaliwal and Kalra, 1978; Kalra et al., 1983; Takroo et al., 1985).

ation level of 2.8 to 3.8 ppm with an average of 3.3 ppm. These two studies did not take into consideration the different analogues of DDT.

The analyses of DDT analogues in butter in India were initiated by Dhaliwal and Kalra (1978) and were subsequently continued by Kalra et al. (1983) and Takroo et al. (1985). DDT contamination of butter existed in the form of p,p'-DDE, p,p'-TDE, and p,p'-DDT. In a few instances, o,p'-DDE, o,p'-TDE and DDMU [1-chloro-2,2-bis(p-chlorophenyl)ethylene] were also detected. However, the major portion of DDT contamination of butter samples was in the form of p,p'-TDE (Figure 1). Among the five brands of butter analyzed by Takroo et al.

TABLE 7   Contamination of Butter with DDT in Different States of India

Year of Sampling	State	Contamination Level (ppm)			
		p,p'-DDE	p,p'-TDE	p,p'-DDT	Total DDT
1977[a]	Delhi	0.25	1.55	0.35	2.15
	Gujarat	0.94	4.28	1.22	6.44
	Haryana	0.59	2.84	0.95	4.41
	Punjab	0.47	3.13	0.67	4.30
	Rajasthan	0.55	2.95	0.71	4.21
1978–1979[b]	Andhra Pradesh	0.44	0.97	0.28	1.69
	Chandigarh	0.96	3.17	1.22	5.35
	Delhi	0.28	1.03	0.38	1.72
	Gujarat	0.64	2.08	0.54	3.26
	Haryana	1.11	3.94	1.73	6.79
	Karnatka	0.14	0.60	0.23	0.97
	Madhya Pradesh	0.53	0.69	0.43	1.63
	Punjab	0.87	3.88	1.44	4.84
	Rajasthan	0.97	3.07	1.23	5.28
	Tamil Nadu	0.58	0.99	0.22	1.80
	Uttar Pradesh	0.35	0.99	0.32	1.66
	West Bengal	0.41	1.78	0.27	2.47
1980[b]	Delhi	0.72	1.54	1.02	3.28
	Gujarat	0.73	1.23	0.43	2.35
	Punjab	0.58	1.25	0.70	2.53
1981[b]	Gujarat	2.02	3.36	1.03	6.33
	Haryana	0.24	0.23	0.08	0.55
	Himachal Pradesh	0.12	0.31	0.31	0.74
	Madhya Pradesh	0.11	0.15	0.06	0.32
	Punjab	0.71	1.87	0.80	3.42
1981[c]	Gujarat	3.59	3.14	3.25	10.80
	Uttar Pradesh	1.98	2.12	0.76	5.31

[a]Dhaliwal and Kalra (1978).
[b]Kalra et al. (1983).
[c]Takroo et al. (1985). Year of sampling not mentioned.

(1985), two contained higher amounts of p, p'-TDE and three contained higher amounts of p, p'-DDE. This disparity in the proportion of different metabolites seems to be due to different routes of exposure of dairy animals to pesticides in different areas.

The data for India as a whole (Table 7) show considerable variations in DDT contamination of butter, which is expected in view of marked differences in the pattern of pesticide usage in different parts of India. The butter samples from Gujarat, Punjab, Haryana, Rajasthan, and Chandigarh showed quite severe contamination with DDT. Even within Gujarat state, samples of butter obtained from Palanpur showed consistently lower levels of contamination than the samples from other parts of the state (Kalra and Chawla, 1983). The average level of 4.35 ppm of DDT contamination of butter samples in India (Table 7) is higher than even the maximum level of DDT reported in butter from 12 countries to the International Dairy Federation (Downey, 1972). A recently concluded study on monitoring of food commodities in India sponsored by FAO revealed that 66% of the butter samples were contaminated with DDT above the MRL. The median and 90th percentile values of DDT in butter were 2.5 and 8.5 mg/kg, respectively (Kalra and Chawla, 1986). These values are far in excess of corresponding values reported from many other countries. However, the values from cotton-growing regions of Guatemala have been reported to be 9.0 and 15.0 mg/kg, respectively (Anonymous, 1982). The contamination level of more than 2 mg/kg of total DDT has been found in 80 of the 145 samples of butter analyzed in India (Kalra and Chawla, 1986). However, only 5 of 1141 samples of dairy products analyzed in the United States have been found to be contaminated above the level of 2 mg/kg (Duggan and Duggan, 1973). Low levels of DDT contamination in butter have also been reported from Tehran region of Iran (Hashemy-Tonkbabony and Assadi-Hangaroodi, 1979).

2.2.2.   HCH Isomers

The HCH contamination of butter in India is made up of α-, β-, γ-, and δ-isomers (Table 8). On an all-India basis, the β-isomer constitutes the highest percentage (40%) followed by the α-isomer with 39% (Figure 1). However, all five brands of butter analyzed in Lucknow (Takroo et al., 1985) and samples from West Bengal (Kalra et al., 1983) contained higher levels of α-HCH than β-HCH. Other isomers were found only in small quantities.

Although α- and γ-isomers are degraded more readily than the β-isomer (Brooks, 1976), the presence of high levels of α-HCH in butter indicates continuous exposure of animals to HCH. The combined contamination with α- and β-HCH in the majority of butter samples is quite excessive and needs to be viewed seriously. However, there are no maximum residue limits prescribed for these isomers. Such a limit has been prescribed only for the γ-isomer, which is 0.1 ppm according to the Food and Agriculture Organization and World Health Organization (FAO/WHO, 1973) and 0.2 ppm according to Codex Alimentarious Commission (Anonymous, 1979). The limit of 0.1 ppm was exceeded in 38% of butter samples and 0.2 ppm in 9.5% of samples (Kalra et al., 1983).

TABLE 8 Contamination of Butter with HCH in Different States of India

Year of Sampling	State	α-HCH	β-HCH	γ-HCH	δ-HCH	Total HCH
1978–1979[a]	Andhra Pradesh	1.52	2.52	0.10	0.08	4.17
	Chandigarh	0.86	1.04	0.13	0.06	2.10
	Delhi	0.25	0.48	0.03	0.03	0.79
	Gujarat	0.54	0.86	0.09	0.05	1.54
	Haryana	0.52	1.11	0.15	0.07	1.84
	Karnatka	0.86	0.63	0.15	ND[c]	1.64
	Madhya Pradesh	0.28	0.57	0.04	0.08	1.00
	Punjab	0.81	1.53	0.12	0.07	2.52
	Rajasthan	0.74	0.98	0.08	0.05	1.88
	Tamil Nadu	0.15	1.03	0.03	0.04	1.25
	Uttar Pradesh	0.83	1.02	0.12	0.06	2.04
	West Bengal	2.06	1.73	0.34	0.20	4.34
1980[a]	Delhi	0.08	0.17	0.03	0.02	0.30
	Gujarat	0.21	0.53	0.09	0.06	0.94
	Punjab	0.10	0.15	0.03	0.02	0.30
1981[a]	Gujarat	0.88	1.30	0.12	0.18	2.48
	Haryana	1.42	1.70	0.45	0.50	4.07
	Himachal Pradesh	0.21	ND	0.01	ND	0.22
	Madhya Pradesh	0.53	0.10	0.02	0.07	0.72
	Punjab	0.55	0.35	0.06	0.09	1.06
1981[b]	Gujarat	0.70	0.12	0.37	—	1.30
	Uttar Pradesh	0.59	0.12	0.37	—	1.10

[a]Kalra et al. (1983).
[b]Takroo et al. (1985). Year of sampling not mentioned.
[c]ND, not detected.

The HCH contamination levels in butter were generally less than those of DDT. This may be attributed to the decreased stability of its major components (α- and γ-isomers) compared to that of DDT in different areas of the environment. The butter samples from Andhra Pradesh and West Bengal showed quite excessive levels of contamination; a level of total HCH above 2 mg/kg has been detected in 48 of the 145 samples of butter analyzed; the median and 90th percentile values were 1.5 and 5.5 mg/kg, respectively (Kalra and Chawla, 1986). The average level of contamination on all-India basis was 1.63 ppm (Table 8), which is much higher than the levels reported from many developed countries (Downey, 1972; Anonymous, 1982).

2.2.3. Other Pesticides

Although aldrin, dieldrin, heptachlor, and heptachlor epoxide have been frequently detected in Europe and the United States (Downey, 1972), none of these pesticides was detected in India in any of the butter samples, even using the method applicable to acid-labile compounds (Dhaliwal and Kalra, 1978; Kalra et al., 1983).

2.3. Ghee

Ghee (clarified butterfat) is also excessively contaminated with DDT and HCH in India. All of the five ghee samples analyzed at Ludhiana were found to be contaminated with DDT above the MRL (Kalra et al., 1983). The major metabolite in ghee was also p, p' TDE. The total DDT contamination varied from 2.53–4.87 ppm (Table 9). The analysis of 22 ghee samples collected from Sitapur and 20 from Lucknow revealed an average total DDT contamination of 9.86 and 4.47 ppm, respectively (Table 9). However, ghee samples from Sitapur contained a maximum level of 28.8 ppm of total DDT compared to 20.4 ppm in samples from Lucknow (Lata et al., 1984). The major metabolite from ghee samples from both Sitapur and Lucknow was p, p'-DDT as compared to p, p'-TDE in ghee samples from Punjab. This suggests different sources of contamination in Punjab and the two places in Uttar Pradesh.

The HCH level in ghee samples collected from Punjab varied widely, from 0.30–6.65 ppm with an average of 2.49 ppm (Kalra et al., 1983). However, the average levels of total HCH in ghee samples from Lucknow and Sitapur districts were 1.29 and 1.42 ppm, respectively (Table 9). The maximum total HCH level from Lucknow was 1.8 ppm, compared to 4 ppm in the ghee sample collected from Sitapur district (Lata et al., 1984). Kalra et al. (1983) detected α-, β-, γ-, and

TABLE 9   Pesticidal Contamination of Ghee in Two States of India

	Contamination Level (ppm)						
	Punjab					Uttar Pradesh	
		Local Brand[a]		Popular Brand[a]			
Contaminant	Popular Brand[a]	I	II	I	II	Sitapur District[b]	Lucknow Market[b]
p, p'-DDE	1.03	0.50	0.50	0.90	0.66	2.29	1.27
p, p'-TDE	2.64	1.44	1.50	1.90	1.50	—	—
o, p-TDE	—	—	—	—	—	2.63	1.18
p, p'-DDT	1.20	0.72	0.53	0.85	0.56	4.75	1.10
Total DDT	4.87	2.66	2.53	3.65	2.72	9.86	1.49
α-HCH	0.14	4.33	0.15	1.90	2.65	—	—
β-HCH	0.26	1.20	0.05	2.55	1.60	—	—
γ-HCH	0.10	0.70	0.06	1.35	0.15	—	—
δ-HCH	0.08	0.42	0.04	0.60	0.12	—	—
Total HCH	0.58	6.65	0.30	6.40	4.52	1.42	1.29

[a]Kalra et al. (1983).
[b]Lata et al. (1984).

TABLE 10 Contamination of Infant Formula with DDT and HCH in India

Year of Sampling	Sample Number	DDT					HCH				Total HCH
		p,p'-DDE	p,p'-TDE	p,p'-DDT	o,p'-DDT	Total DDT	α-HCH	β-HCH	γ-HCH	δ-HCH	
1977[a]	1	0.33	1.76	0.63	—	2.72	—	—	—	—	—
	2	0.25	1.04	0.40	—	1.69	—	—	—	—	—
	3	0.36	1.03	0.26	—	1.65	—	—	—	—	—
	4	0.17	1.02	0.33	—	1.52	—	—	—	—	—
1981[b]	1	0.070	0.105	0.046	Tr[c]	0.221	0.119	0.164	0.022	0.022	0.327
	2	0.062	0.160	0.080	Tr	0.302	0.108	0.102	0.040	0.044	0.294
	3	0.012	0.019	0.015	Tr	0.046	0.024	0.040	0.008	0.011	0.083
	4	0.071	0.250	0.180	Tr	0.501	0.056	0.144	0.025	0.030	0.255
	5	0.170	0.280	0.011	0.008	0.469	0.112	0.180	0.037	0.025	0.354
	6	0.084	0.140	0.090	Tr	0.314	0.044	0.130	0.015	0.030	0.219
	7	0.136	0.170	0.080	Tr	0.386	0.080	0.184	0.021	0.034	0.319
	8	0.072	0.130	0.045	Tr	0.247	0.037	0.130	0.021	0.040	0.228
	9	0.0145	0.030	0.032	Tr	0.077	0.030	0.132	0.021	0.060	0.243
	10	0.019	0.021	0.008	Tr	0.048	0.027	0.037	0.020	0.031	0.115
	11	0.018	0.021	0.034	Tr	0.073	0.016	0.018	0.020	0.020	0.074
	12	0.018	0.053	0.009	Tr	0.080	0.024	0.016	0.040	0.007	0.087
	13	0.006	0.014	0.018	Tr	0.038	0.010	0.011	0.008	Tr	0.029

Contamination Level (ppm)

[a]Dhaliwal and Kalra (1978); figures are based on fat basis.

[b]R.P. Chawla (personal communication); figures are based on total weight basis.

[c]Tr, traces.

371

δ-isomers in ghee samples, whereas Lata et al. (1984) could not detect the δ-isomer. In general, the nature of DDT analogues and HCH isomers in ghee is similar to that detected in butter, suggesting no preferential loss during additional processing involved in its preparation.

2.4. Infant Formula

All of the four popular brands of infant formula originating from Gujarat, Maharashtra, and Punjab were found to be contaminated with DDT above the MRL (Dhaliwal and Kalra, 1978). The concentration of DDT on fat basis varied from 1.52–2.72 ppm, and averaged 1.9 ppm (Table 10). More than 60% of DDT contamination was in the form of p, p'-TDE. The analysis of 13 more samples of infant formula from large commercial houses (R. P. Chawla, personal communication) revealed widespread contamination with DDT and HCH (Table 10). The contamination level of total DDT on a total weight basis varied from 0.04–0.47 ppm and averaged 0.22 ppm, whereas that of total HCH varied from 0.03–0.35 ppm with an average of 0.20 ppm. Thus, the spray drying process used in the manufacture of infant formula does not cause substantial reduction in DDT contamination.

2.5. Cheese

The cheese samples collected from Ludhiana were contaminated with DDT and HCH (Table 11). The mean level of contamination with DDT was 0.78 ppm, whereas the maximum level detected was 1.42 ppm (Singh, 1982). Similarly, the mean HCH level was 0.14 ppm and the maximum HCH detected was 0.27 ppm. The average level of contamination of cheese with DDT in three Central American countries, namely El Salvador, Guatemala, and Honduras is quite high, being 3.30 ppm (Mazariegos, 1976). However, the overall results from most of the European countries and Canada indicate that the levels of DDT, HCH, aldrin/dieldrin, and heptachlor/heptachlor epoxide in cheese in these countries were negligible (Downey, 1972).

TABLE 11   Contamination of Cheese with DDT and HCH in Punjab, India[a]

Contaminant	Level (ppm)	Contaminant	Level (ppm)
p, p'-DDE	0.139	Total DDT	0.783
p, p'-TDE	0.483	α-HCH	0.040
p, p'-DDT	0.121	β-HCH	0.035
o, p'-DDE	0.017	γ-HCH	0.053
o, p'-TDE	0.010	δ-HCH	0.015
o, p'-DDT	0.013	Total HCH	0.144

[a]Singh (1982).

3.   DIETARY INTAKE OF PESTICIDES

The contamination of milk is to be viewed with concern as it is consumed in substantial quantities by infants and the sick. The dietary intake of pesticides through milk and milk products has been calculated by several workers (Table 12). Dhaliwal and Kalra (1977) estimated DDT intake by a person through milk contaminated with 0.26 ppm to be about 117 μg. This is higher than the total dietary intake of DDT of 2.35 μg in the United States (Gartell et al., 1985) and 5.0 μg in England (Tincknell, 1980). The situation is particularly serious for an infant. This milk, if taken by a 3-month-old child (average weight 5 kg) at the rate of 875 mL (5 × 175 mL) per day, would result in daily intake of 225 μg of DDT. This is nine times higher than the acceptable daily intake of 0.005 mg/kg body weight (25 μg for a body weighing 5 kg). Similarly, the consumption of infant milk formula contaminated with DDT at the 1.90 ppm level (Dhaliwal and Kalra, 1978) by a 3-month-old child at the normal feeding rate of 135 g/day would result in a daily intake of 47 μg of DDT. This level is about twice the acceptable daily intake of 0.005 mg/kg of body weight. As children are considered to be much more susceptible than adults, they are prone to a much greater health risk.

The consumption of a mixture of 75% milk and 25% curd results in dietary intake of total DDT and total HCH with an average levels of 116 and 20.0 μg/person day whereas the consumption of contaminated cheese results in intake of DDT and HCH at average levels of 36.8 and 6.8 μg, respectively (Singh, 1982). Even when the rate of consumption of contaminated milk is low, the associated intake of DDT and HCH is quite high, the average level being 17.4 and 25.0 μg/person day, respectively (Kaphalia et al., 1985). Although the acceptable daily intake for HCH has not been fixed, these levels seem to be unacceptably high. The dietary intake of 20–25 μg of HCH through milk alone in India is higher than the total HCH dietary intake of 1 μg in the United States (Gartell et al., 1985) and 6.9 μg in the United Kingdom (Tincknell, 1980).

4.   SOURCES OF CONTAMINATION

The different routes of contamination of milk with pesticides have been shown in Figure 2. Furthermore, the knowledge of DDT analogues and HCH isomers in milk can prove useful in predicting the sources of contamination of milk in the stores. Several studies have revealed that TDE is the predominant product of excretion into milk of animals given DDT orally whereas DDT is predominant when animals are treated dermally (Witt et al., 1966; Kapoor, 1985; Kalra et al., 1986). Similarly, α-HCH and β-HCH are the major isomers in milk in animals orally fed with technical-grade HCH, whereas α- and γ-HCH are major isomers when animals are treated dermally (Veirov et al., 1977; Kapoor, 1985).

The relative significance of different sources of contamination may vary from one country to another and even from one region to another in a country. The high level of TDE in milk and milk products in India (Dhaliwal and Kalra, 1977,

TABLE 12 Some Estimates of Dietary Intake of DDT through Milk and Milk Products

Commodity	Contamination Level (ppm)	Consumer	Consumption of Commodity (g or mL/day)	Calculated Daily Intake of DDT (μg)	Proportion of Acceptable Daily Intake (%)	Reference
Milk	0.26	Adult	450	117	470	Dhaliwal and Kalra (1977)
Milk	0.26	Infant	875	225	900	Dhaliwal and Kalra (1977)
Infant formula	1.90	Infant	135	47	190	Dhaliwal and Kalra (1978)
Milk and milk products[a]	0.24	Adult	491	116	465	Singh (1982)
Cheese	0.78	Adult	47	37	150	Singh (1982)
Milk	0.22	Adult	80	17.4	70	Kaphalia et al. (1985)

[a]Comprises 75% milk and 25% curd.

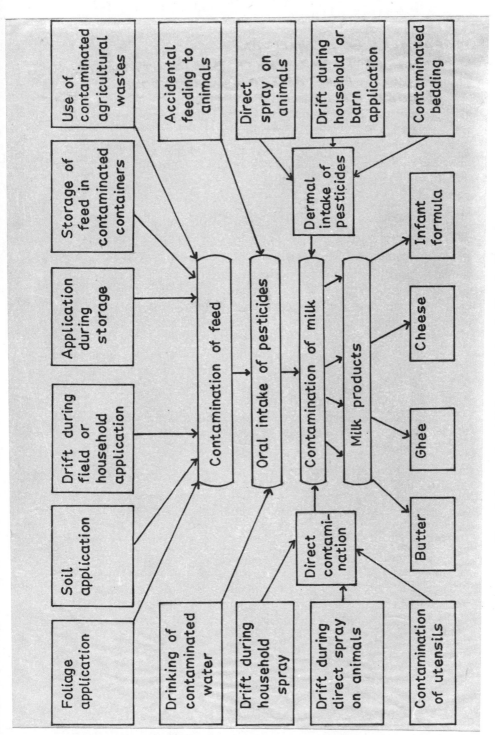

Figure 2. Sources of pesticidal contamination of milk and milk products.

375

1978; Kalra et al., 1983) suggests that the cattle ingest DDT mainly through contaminated feed. Since there is a very small amount of p, p'-DDE in milk obtained from buffalo fed p, p'-DDT (Kalra et al., 1986), higher amounts of p, p'-DDE detected in milk, butter, and ghee suggest its intake by lactating animals through aged DDT residues. Similarly, high levels of α- and γ-HCH in cow milk in Israel (Veirov et al., 1977) suggest that the contamination results from the dermal application of HCH. The high level of β-HCH detected in cow milk in Japan (Kanazawa, 1983) suggests that the contamination is the result of oral ingestion of HCH through contaminated feed. In view of the above findings, the role of the following sources of pesticidal contamination of dairy products has been clearly established.

4.1. Intradomicilary Spray of DDT and HCH

The high level of TDE in bovine milk led Dhaliwal and Kalra (1977, 1978) to suggest that indoor spraying of DDT might contaminate stored feed and thereby contribute partly toward the exposure of cattle to DDT. In addition to their agricultural usage, both DDT and HCH are extensively used for the control of mosquitoes under the National Malaria Eradication Programme (NMEP). For this purpose either DDT at a dosage of 1 g/m^2 twice a year or HCH at a dosage of 1.5 g/m^2 three times a year is sprayed inside the rural houses, cattle sheds, and other dwellings. As a matter of fact, the higher level of DDT contamination of milk from rural areas in September–October 1976 than in January–February 1977 (Table 2) was attributed to indoor spraying of DDT in August 1976 (Dhaliwal and Kalra, 1977).

Kapoor et al. (1980) carried out extensive studies to determine the DDT and HCH contamination of bovine milk in four districts of Punjab, out of which two

TABLE 13   Pesticidal Contamination of Milk in Different Localities Treated with DDT and HCH for the Control of Malaria[a]

Contaminant	Level (ppm Whole Milk Basis)			
	Ludhiana[b]	Jalandhar[b]	Ferozepur[c]	Sangrur[c]
p, p'-DDE	0.069	0.096	0.046	0.054
p, p'-TDE	0.186	0.188	0.066	0.061
p, p'-DDT	0.099	0.104	0.036	0.021
Total DDT	0.354	0.388	0.168	0.136
α-HCH	0.012	0.009	0.083	0.090
β-HCH	0.023	0.009	0.263	0.227
γ-HCH	0.007	0.009	0.030	0.022
δ-HCH	0.011	0.010	0.015	0.037
Total HCH	0.052	0.037	0.391	0.375

[a]Kapoor et al. (1980).
[b]Areas sprayed with DDT.
[c]Areas sprayed with HCH.

districts were sprayed with DDT and two with HCH under NMEP. They found a direct correlation between the usage of these insecticides for malaria control and contamination of bovine milk. The mean level of DDT contamination in milk in DDT-sprayed localities was two to three times higher than the level encountered in HCH-sprayed localities. The contamination of milk with HCH was 7–10 times more in HCH-sprayed localities than that in DDT-sprayed localities (Table 13). Similar studies were conducted by Singh et al. (1986) who found DDT contamination of milk from DDT-sprayed areas to be three to five times higher than the corresponding levels from HCH-sprayed areas. The level of HCH contamination from HCH-sprayed areas was three to four times higher than that from DDT-sprayed areas. These studies clearly demonstrate a direct correlation between intradomicilary application of DDT and HCH for malaria control and levels of contamination of bovine milk.

4.2. HCH Spray on Rice

The origin of HCH contamination in cow milk in Japan has been attributed to the rice straw, used as feed, obtained from paddy crop to which HCH has been applied for the control of stem borers and leafhoppers (Tanabe, 1972). In Japan, rice straw left after harvest of rice grains is used as roughage for livestock, especially in the southern part of the country. There was a direct correlation between the degree of HCH contamination of cow milk and its concentration in rice straw (Table 14). When rice straw was withdrawn as feed to cows, there was considerable reduction in HCH contamination of milk (Table 15). The contamination levels of milk from cows fed rice straw as roughage were higher than those fed on other feeds (Kanazawa, 1983). Also, the HCH contamination of cow milk from different areas almost paralleled the pesticide use. The correlation between the time of application of HCH and levels of contamination of rice straw was also

TABLE 14   Contamination of Cow Milk and Rice Straw with HCH at Two Locations in Japan[a]

| | Contamination Level (ppm) | | | |
| | Rice Straw | | Cow Milk | |
HCH Isomers	Location I[b]	Location II[c]	Location I[b]	Location II[c]
α-HCH	0.418	0.156	0.172	0.034
β-HCH	0.307	0.115	0.191	0.060
γ-HCH	0.108	0.072	0.007	0.002
δ-HCH	0.174	0.081	0.018	0.003
Total HCH	1.007	0.424	0.388	0.099

[a]Tanabe (1972).
[b]Areas showing high HCH contamination.
[c]Areas showing less HCH contamination.

TABLE 15   Changes in HCH Contamination of Milk from Cows Fed on Fodders without Rice Straw[a]

Year	Month	α-HCH	β-HCH	γ-HCH	δ-HCH	Total HCH
				Contamination Level (ppm)		

Year	Month	α-HCH	β-HCH	γ-HCH	δ-HCH	Total HCH
1970	January	0.065	0.069	0.002	Tr[b]	0.136
	February	0.024	0.023	0.003	Tr	0.050
	May	0.035	0.038	0.002	0.002	0.077
	September	0.024	0.023	0.002	0.002	0.051
	October	0.011	0.022	0.001	0.003	0.037
1971	January	0.004	0.003	Tr	0.001	0.008
	March	0.010	0.015	0.002	0.002	0.029

[a]Tanabe (1972).
[b]Tr, traces.

established. When HCH was applied at later stages of crop growth to control rice leafhoppers, the level of contamination of rice straw was about 10 times higher than when pesticide was applied in the early stages of growth to control stem borers (Kanazawa, 1983). These results demonstrate that the main route of HCH contamination of cow milk in Japan is the use of rice straw as feed from the crop to which HCH has been applied.

4.3.   HCH Spray on Cows

The main source of contamination of cow milk in Israel has been attributed to the spraying of animals with Hexalone (a formulation of HCH) for insect control, which leads to an increase in α- and γ-HCH in milk (Veirov et al., 1977). The level of HCH contamination of milk varied considerably in different dairies. In three dairies where the cows had not been sprayed with HCH for at least 1 year, the levels of α- and γ-HCH were much lower and did not exceed 0.02–0.05 ppm. Milk from other dairies that were sprayed with HCH was contaminated with 1.4–1.6 ppm of γ-HCH. The rapid increase in α- and γ- HCH in milk was actually demonstrated by spraying the individual cows with HCH. The seasonal fluctuations in the concentration of HCH in milk were also attributed to the use of HCH for spraying of cows.

4.4.   DDT Application on Cotton

DDT contamination of cow milk has been found to be higher in cotton-growing areas of Central America than in non-cotton-growing areas. About 90% of all pesticides used in Central America is sprayed on cotton. During September to October, when most pesticides are sprayed on cotton, DDT contamination of milk has been found to be the highest (Mazariegos, 1976). The monitoring of milk samples in three different zones revealed that DDT contamination of milk was highest in milk from the farms in the middle of cotton areas, the levels being 3.47 ppm for whole herd and 5.87 ppm for an individual cow as compared to 0.35 ppm in noncotton areas for both whole herd and individual cow. However, the milk nearer to the cotton area contained 1.13 and 0.93 ppm DDT for whole herd and individual cow, respectively. The farms about 45 km away from the cotton area had only 0.3 ppm of DDT in the milk of cows from the whole herd. The DDT levels in milk closely followed the time of application of insecticides in zones nearest to the cotton-growing areas. Thus, the role of DDT application in contamination of milk has been clearly established.

5. PESTICIDE USAGE VIS-À-VIS CONTAMINATION

The problem of pesticidal contamination of milk and milk products in a country depends on the consumption of pesticides, their usage pattern, and the efficacy of regulatory measures. The persistent organochlorine insecticides, the use of which has been severely restricted in many developed countries, are still being extensively used in developing countries for the control of insect pests in agriculture and public health programs. Therefore, the pesticidal contamination of milk and milk products in most of the developed countries has shown a downward trend following restrictions on their usage (Downey 1972; Anonymous, 1982; Steffey et al., 1984). The continuous decline in the contamination levels in these countries has been attributed to the use of less persistent pesticides. Moreover, the situation was not so alarming even when large amounts of persistent chlorinated hydrocarbons were used in these countries (Büchel, 1983). However, in most of the developing countries, the pesticidal contamination of milk and milk products continues to be serious due to lack of pesticide regulations.

There has been a sharp increase in the use of pesticides in agriculture and public health in India, particularly since 1975. DDT and HCH constitute more than 60% of the total pesticide use. Since its introduction in 1948 in India, 270,000 tonnes of DDT has been used out of which 220,000 tonnes in public health and 50,000 tonnes in agriculture. At present the annual consumption of DDT is about 12,000 tonnes—10,000 tonnes in public health and 2,000 tonnes in agriculture (Mehrotra, 1985a). The total consumption of HCH since 1949 when it was introduced in India is estimated to be about 575,000 tonnes out of which 500,000 tonnes were in agriculture and 75,000 tonnes in public health. The present annual consumption of HCH is about 36,000—30,000 tonnes in agriculture and 6,000 tonnes in the public health sector (Mehrotra, 1985b). Despite so much application of these persistent insecticides, the incidence of malaria has shown a dramatic increase during the mid-1970s (Figure 3).

The contamination of milk and milk products with DDT and HCH is the direct outcome of their use and their high environmental persistence. The contamination level of milk with DDT and HCH in Punjab showed little decline from 1977 to 1981 (Tables 3 and 5). The analysis of popular brands of butter

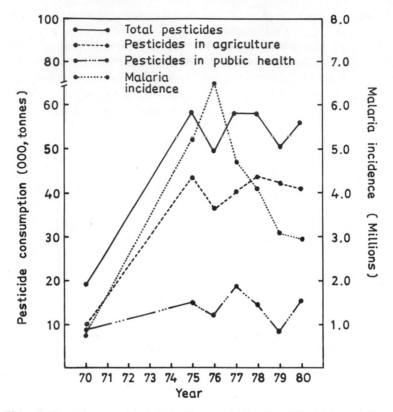

Figure 3. Pesticide use and malaria incidence in India, 1970–1980 (Mehrotra, 1986).

produced in Gujarat and Punjab (India) at the wholesale level revealed that contamination of milk with DDT and HCH continued to persist at high levels during 1977–1981 (Figure 4). From a total of 980 milk samples obtained from different parts of India, the presence of DDT, HCH, dieldrin, and endrin has been reported to be 95, 90, 1.0, and 0.2% of the samples, respectively (Kalra and Chawla, 1986). The lower incidence of dieldrin and endrin corresponds to their reduced use in agriculture.

The contamination levels of cow milk in Japan with HCH in different areas almost paralleled pesticide use in 1969–1970. The levels in southwestern parts, especially Osaka and Nagasaki, were higher than the levels in the northeastern parts including Hokkaido, Tohoku, and Kanto (Tanabe, 1972). There were severe outbreaks of insect pests in rice in southern Japan leading to frequent HCH application that was reflected in the contamination of milk (Table 16). The use of HCH in Japan was prohibited on feed crops in January 1970 when Uyeta et al. (1970) reported high levels of contamination of cow milk with HCH. This led to an eventual ban of HCH use in 1971 (Tanabe, 1972). The dairy farmers were also educated not to feed rice to their cattle (Kojima and Araki, 1975). These steps led to a considerable decrease in contamination of cow milk (Figure 5). Generally the

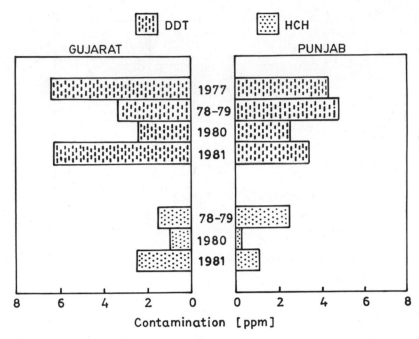

Figure 4. Variations in contamination of butter with DDT and HCH in Gujarat and Punjab, India, 1977–1981 (Dhaliwal and Kalra, 1978; Kalra et al., 1983).

reduction in contamination in the early stage was rapid in highly contaminated areas but took much more time in later stages to reach lower levels. By 1971, β-HCH levels in cow milk were reduced to about 0.01 ppm. The total amount of HCH produced in Japan until 1971 was 40,000 tonnes, DDT was 15,000 tonnes, and the sum of aldrin and endrin was only 3,500 tonnes. The amount of HCH

TABLE 16   Changes in HCH Contamination of Cow Milk in Japan with Geographical Latitude[a]

Geographical Latitude[b]	Contamination Level (ppm)				
	α-HCH	β-HCH	γ-HCH	δ-HCH	Total HCH
1 (South)	0.069	0.211	0.005	0.010	0.295
2	0.069	0.161	0.009	0.013	0.252
3	0.033	0.123	0.003	0.014	0.173
4	0.063	0.201	0.006	0.015	0.285
5	0.034	0.081	0.002	0.004	0.121
6	0.014	0.028	0.003	0.001	0.046
7	0.029	0.052	0.002	0.007	0.090
8 (North)	0.010	0.009	0.001	—	0.020

[a]Tomizawa (1977).
[b]Locality is divided into eight blocks from south (1) to north (8) latitude.

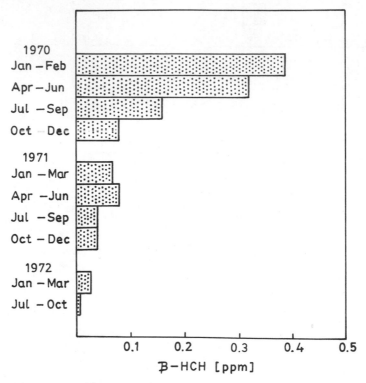

Figure 5. Reduction in β-HCH contamination of cow milk in Japan following restrictions imposed on the use of HCH; data based on average of eight locations (Kojima and Araki, 1975).

used in rice totalled 33,000–40,000 tonnes each year (Tomizawa, 1977). Hence, low levels of DDT and dieldrin detected in milk correspond to their low usage.

6. CONCLUSIONS

The contamination of milk and milk products with organochlorine pesticides is quite alarming in developing countries, particularly in India. The major contaminants are DDT and HCH, which are extensively used for pest control in agriculture and the public health sector without proper care and regulations. The results suggest an urgent need for restricting the use of these pesticides in order to lower their levels in milk.

DDT is mainly used for the control of malaria and has decreased the number of cases of malaria from about 100 million in 1947 to about 0.1 million in 1965. Since then, the incidence of malaria has shown an upward trend despite heavy use of this and other insecticides. At present, there are more than 10 million cases of malaria in India. The low effectiveness of DDT might be due to the development of resistance by the vectors and low coverage by the insecticide. Therefore, there is an urgent need to reconsider the continuing use of DDT for the malaria control program. This has become more essential in view of the direct correlation between intradomicilary spraying of DDT and contamination of milk.

HCH is mainly used for pest control in agriculture. Among the various isomers, only the γ-isomer is insecticidal and also less persistent and rapidly degraded by animals. Other isomers, though noninsecticidal, pose a greater hazard to human health. The acceptable daily intakes of α-, β-, and γ-HCH are 0.005, 0.001, and 0.0125 mg/kg, respectively. Therefore, the technical-grade HCH should either be replaced with γ-HCH (lindane) or other safer pesticides. In fact, both DDT and HCH should be phased out in such a manner so that their levels in different components of the environment can be reduced to a minimum.

ACKNOWLEDGMENTS

Thanks are due to Dr. Parm Pal Singh, Pesticide Residue Laboratory, Department of Entomology, Punjab Agricultural University, Ludhiana for going through the manuscript and making many valuable suggestions.

REFERENCES,

Agnihotri, N. P., Dewan, R. S., Jain, H. K., and Pandey, S. Y. (1974). Residues of insecticides in food commodities from Delhi-II. High fat content food materials. *Indian J. Ent.* **36**, 203–208.

Al-Omar, M. A., Tameesh, A. H., and Al-Ogaily, N. H. (1985). Dairy product contamination with organochloride insecticide residues in Baghdad district. *J. Biol. Sci. Res.* **16**, 133–144.

Anonymous (1979). *Summary of Acceptances of Recommended Worldwide and Regional Codex Standards and Recommended Codex Maximum Limits for Pesticide Residues.* Joint FAO/WHO Food Standards Programme, CAC/Acceptances/Rev. 1, Rome, 12.45 pp.

Anonymous (1982). *Summary and Assessment of Data received from FAO/WHO Collaborating Centres for Food Contamination Monitoring.* UNEP/FAO/WHO National Food Administration, Uppsala, Sweden, 82 pp.

Brooks, G. T. (1976). *Chlorinated Insecticides, Volume I, Technology and Application.* CRC Press, Cleveland, Ohio, 249 pp.

Büchel, K. H. (1983). Political, economic and philosophical aspects of pesticide use for human welfare. In J. Miyamoto and P. C. Kearney, eds., *Pesticide Chemistry: Human Welfare and the Environment.* Pergamon Press, New York, Vol. 1, pp. 3–19.

Chauhan, R., Singh, Z., and Dahiya, B. (1982). Organochlorine insecticides as food contaminants. *First Int. Conf. Food Sci. Technol.*, Bangalore, 23–26. May, 1982, Abstract No. 9, Section II, pp. II.5.

Dhaliwal, G. S., and Kalra, R. L. (1977). DDT residues in milk samples from Ludhiana and surrounding areas. *Indian J. Ecol.* **4**, 13–22.

Dhaliwal, G. S., and Kalra, R. L. (1978). DDT residues in butter and infant formula in India, 1977, *Pestic. Monit. J.* **12**, 91–93.

Downey, W. L. (1972). *Pesticide Residues in Milk and Milk Products.* International Dairy Federation Bulletin, 51 pp.

Duggan, R. E., and Duggan, M. B. (1973). Pesticide residues in food. In C. A. Edwards, ed., *Environmental Pollution by Pesticides.* Plenum Press, London, pp. 334–364.

FAO/WHO (1973). *Pesticide Residues in Food*. Report of the 1972 joint meeting of the FAO working party of experts on pesticide residues and the WHO expert committee on pesticide residues, WHO Tech. Rep. Series No. 525, FAO Agricultural Series No. 90, Rome, 47 pp.

Gartell, M. J., John, C. C., David, S. P., and Gunderson, E. L. (1985). Pesticides, selected elements, and other chemicals in adult total diet samples, October 1979–September 1980. *J. Assoc. Off. Anal. Chem.* **68**, 1184–1197.

Hashemy-Tonkbabony, S. I., and Assadi-Hangaroodi, F. (1979). Chlorinated pesticide residues in butter from the Tehran region, Iran. *J. Fd. Prot.* **42**, 202–203.

Jadhav, G. D. (1986). DDT and BHC residues in milk samples from Marathwada region. In S. C. Goel, ed., *Pesticide Residues and Environmental Pollution*. Sanatan Dharam College, Muzaffarnagar, India, pp. 72–85.

Kalra, R. L., and Chawla, R. P. (1983). *Studies on Pesticide Residues and Monitoring of Pesticidal Pollution*. Final Technical Report, PL 480 Project, IN-ARS-65, Punjab Agricultural University, Ludhiana, India, 230 pp.

Kalra, R. L., and Chawla, R. P. (1986). Pesticidal contamination of foods in the year 2000 A.D. *Proc. Indian Natl. Sci. Acad.* **B52**, 188–204.

Kalra, R. L., Chawla, R. P., Dhaliwal, G. S., and Joia, B. S. (1978). DDT and HCH residues in foodstuffs in the Punjab. *Proc. Symp. Nuclear Techniques in Studies of Metabolism, Effects and Degradation of Pesticides*, Department of Atomic Energy, Bombay, India, pp. 19–30.

Kalra, R. L., Chawla, R. P., Sharma, P. L., Battu, R. S., and Gupta, S. C. (1983). Residues of DDT and HCH in butter and ghee in India 1978–1981. *Environ. Pollut.* (Ser. B), **6**, 195–206.

Kalra, R. L., Chawla, R. P., Joia, B. S., and Tiwana, M. S. (1986). Excretion of DDT residues into milk of the Indian Buffalo, *Bubalus bubalis* (L.) after oral and dermal exposures. *Pestic Sci.* **17**, 128–134.

Kanazawa, J. (1983). Pesticide residue problems in Japan. *International Symposium on Pesticide Use in Developing Countries—Present and Future*. Tropical Agriculture Research Series No. 16, Yatabe, Tsukuba, Ibaraki, Japan, pp. 179–187.

Kaphalia, B. S., Siddiqui, F. S., and Seth, T. D. (1985). Contamination levels in different food items and dietary intake of organochlorine pesticide residues in India. *Indian J. Med. Res.* **81**, 71–78.

Kapoor, S. K. (1985). *Excretion of DDT—analogues and HCH-isomers in Buffalo Milk after Oral Administration and Dermal Application*. Ph.D. Thesis, Punjab Agricultural University, Ludhiana, India, 143 pp.

Kapoor, S. K., Chawla, R. P., and Kalra, R. L. (1980). Contamination of bovine milk with DDT and HCH-residues in relation to their usage in malaria control programme. *J. Environ. Sci. Health* **B15**, 545–557.

Khandekar, S. S., Noronha, A. B. C., and Banerji, S. A. (1981). Organochlorine pesticide residues in eggs and milk available in Bombay markets. *Sci. Cult.* **47**, 137–139.

Kojima, K., and Araki, T. (1975). Recent status of organochlorine pesticide residues in foods in Japan. In F. Coulston and F. Korte, eds., *Environmental Quality and Safety*. Georg Thieme, Stuttgart, Academic Press, New York, pp. 74–79.

Lakshminarayana, V., and Krishna Menon, P. V. (1975). Screening of Hyderabad market samples of food-stuffs for organochlorine insecticide residues. *Indian J. Plant Prot.* **3**, 4–19.

Lata, S., Siddiqui, M. K. J., and Seth, T. D. (1984). Chlorinated pesticide residues in Desi ghee. *J. Fd. Sci Technol.* **21**, 94–95.

Mazariegos, F. (1976). An environmental and economic study of the consequences of pesticide use in Central American cotton production. *Impact Monitoring of Agricultural Pesticides*. United Nations Environmental Programme, AGP:1976/M/4, Rome, pp. 15–18.

Mehrotra, K. N. (1985a). Use of DDT and its environmental effects in India. *Proc. Indian Natl. Sci. Acad.* **B51**, 169–184.

Mehrotra, K. N. (1985b). Use of HCH (BHC) and its environmental effects in India. *Proc. Indian Natl. Sci. Acad.* **B51**, 551–595.

Mehrotra, K. N. (1986). Pest control strategies for 2000 A.D. *Proc. Indian Natl. Sci. Acad.* **B52**, 10–16.

Pathak, M. D., and Dhaliwal, G. S. (1986). Insect control. In M. S. Swaminathan and S. K. Sinha, eds., *Global Aspects of Food Production.* Tycooly International, Oxford, pp. 357–386.

Saxena, M. C., and Siddiqui, M. K. J. (1982). Pesticidal pollution in India: Organochlorine pesticides in milk of women, buffalo and goat. *J. Dairy Sci.* **65**, 430–434.

Singh, P. P. (1982). *Insecticide Residues in Food in the Punjab—A Typical Dietary Survey.* Ph.D. Thesis, Punjab Agricultural University, Ludhiana, India, 123 pp.

Singh, P. P., Battu, R. S., Joia, B. S., Chawla, R. P., and Kalra, R. L. (1986). Contribution of DDT and HCH used in malaria control programme towards the contamination of bovine milk. In S. C. Goel ed, *Pesticide Residues and Environmental Pollution.* Sanatan Dharam College, Muzaffarnagar, India, pp. 86–92.

Steffey, K. L., Mack, J., Macmonegle, C. W., and Petty, H. B. (1984). A 10-year study of chlorinated hydrocarbon insecticide residues in bovine milk in Illinois (USA), 1972–1981. *J. Environ. Sci. Health* **B19**, 49–66.

Takroo, R., Kaphalia, B. S., and Seth, T. D. (1985). Chlorinated pesticide residues in diferent brands of butter. *J. Fd. Sci. Technol.* **22**, 57–59.

Tanabe, H. (1972). Contamination of milk with chlorinated hydrocarbon pesticides. In F. Matsumura, G. M. Boush, and T. Misato, eds., *Environmental Toxicology of Pesticides.* Academic Press, New York, pp. 239–256.

Tincknell, R. C. (1980). Pesticides: an industry view of safety. *Pestic. Inf.* **6**, 22–25.

Tomizawa, C. (1977). Past and present status of residues of pesticides manufactured in Japan. *Jpn. Pestic. Inf.* **30**, 5–42.

Tripathi, H. C. (1966). *Organochlorine Insecticide Residues in Agricultural and Animal Products in Terai Area.* M.Sc. Thesis, Uttar Pradesh Agricultural University, Pantnagar, India.

Uyeta, M., Taue, S., and Nishimoto, T. (1970). Residues of BHC isomers and other organochlorine pesticides in fatty foods of Japan. *J. Fd. Hyg. Soc.* **11**, 256–263.

Veirov, D., Aharonson, N., and Alumos, E. (1977). Residues of HCH isomers and DDT derivatives in Israel milk and their seasonal fluctuations. *Phytoparasitica* **5**, 26–33.

Witt, J. M., Whiting, F. M., Brown, W. H., and Stull, J. W. (1966). Contamination of milk from different routes of animal exposure to DDT. *J. Dairy Sci.* **49**, 370–380.

13

PASTEURIZED DAIRY PRODUCTS: THE CONSTRAINTS IMPOSED BY ENVIRONMENTAL CONTAMINATION

John D. Phillips and Mansel W. Griffiths

The Hannah Research Institute
Ayr, Scotland

---

1. INTRODUCTION

Milk and milk products will support the growth of a wide variety of bacteria. The many types of bacteria associated with milk have been well documented, and a comprehensive list of the common genera isolated from milk products has been given by Gilmour and Rowe (1981).

The numbers and types of bacteria found in raw milk reflect the hygienic

TABLE 1 Microflora Change in Milks of varying Quality[a]

Group	% Isolates Found in Raw Milk within Count Range of				
	<5000	5000–20,000	>20,000–200,000	>200,000–1,000,000	>1,000,000
Micrococci	69	52	30	21	19
Streptococci	13	14	17	7	21
Asporogenous Gram-positive rods	10	10	15	13	4
Coli-aerogenes	0	1	9	26	3
Gram-negative rods	6	7	22	29	51
Aerobic spore formers	1	12	4	1	0
Streptomycetes	0	1	1	1	1
Unclassified	1	3	2	2	1

[a]Data from Thomas et al. (1963).

conditions under which the milk was produced. A number of workers including McKenzie (1962) and Thomas et al. (1963) have shown that the microflora changes dramatically as the quality of the milk deteriorates (Table 1). About 70% of the flora of low count raw milk is composed of micrococci, but the incidence of Gram-negative rod-shaped organisms increases with increase in count.

Before describing in detail the steps involved in the production and processing of milk, it is useful to outline briefly the major groups of bacteria commonly found in milk and milk products.

The bacterial flora of raw milk can be divided, for all practical purposes, into two groups. These are (1) bacteria that are sensitive to heat and are destroyed by pasteurization, and (2) the so-called thermoduric organisms that are heat resistant and can survive laboratory pasteurization of 63°C for 30 min. However, the heat resistance of bacteria in milk products can be affected by a number of factors such as total solids concentration (Daemen, 1981) and water activity (Senhaji and Loncin, 1975). The latter has been used to explain the apparent protective effect of fat in the heat treatment of milk products (Senhaji and Loncin, 1975), but other workers (Luedecke and Harmon, 1966) have shown that fat concentration does not exert a significant effect on the thermoresistance of bacteria.

1.1. Bacteria Not Surviving Heat Treatment

The heat-sensitive bacteria are mainly Gram-negative organisms of several genera (Table 2). Many of these bacteria are capable of growth at refrigeration temperatures although their optimum growth temperature lies around 21°C. The term "psychrotroph" was coined by Eddy (1960a) to describe these bacteria. A detailed review of the literature suggests that it is inappropriate to define a precise growth rate for psychrotrophic bacteria in raw milk. At any fixed temperature there is a large variation in growth rate that is species related (Cousins et al. 1977). Muir and Phillips (1984) have shown that at temperatures of 4, 6, and 8°C there is a broad range of apparent generation times for organisms growing in farm bulk tank and creamery silo milk. Greene and Jezeski (1954) observed generation times of 26–38, 11–15, and 4–7 hr at temperatures of 0–2, 4–6 and 10°C, respectively, for two strains of *Pseudomonas* spp. and a strain of *Enterobacter aerogenes* in milk. Seven pseudomonads had average generation times of 28, 7.1, and 4.1 hr at 0, 7, and 15°C respectively (Patel et al., 1983). Other *Pseudomonas* spp. were shown to have a mean generation time of 8.7 hr at 4°C (Richard, 1981), whereas *Ps. fluorescens* P26 had generation times of 10, 5.5, and 4.5 hr at 4, 10, and 21°C, respectively (Mayerhofer et al., 1973). Phillips and Griffiths (1987b) have related growth parameters to temperature for a number of psychrotrophic bacteria commonly found in milk and milk products.

Although it is generally recognized that most Gram-negative psychrotrophs are destroyed by pasteurization procedures, there have been reports in the literature that certain *Pseudomonas* spp. can survive pasteurization if present

TABLE 2 Gram-Negative, Heat-Sensitive Bacteria Commonly Isolated from Raw Milk

Genus	Source	Reference
Achromobacter (now classified as *Alcaligenes* or *Pseudomonas*)	Soil	Thomas (1966)
Acinetobacter	Soil and water supplies, milking equipment surfaces	Bergey (1974) Cousins and Bramley (1981)
Aeromonas	Water supplies	Bergey (1984)
Alcaligenes	Soil	Thomas (1966)
	Water supplies and milking equipment surfaces	Cousins and Bramley (1981)
Chromobacterium	Soil and water supplies	Bergey 1974
Enterobacter	Mastitic infections	Tolle (1980)
	Water supplies and milking equipment surfaces	Cousins and Bramley (1981)
Escherichia	Bedding of housed lactating cattle	Bramley and Neave (1975)
	Mastitic infections	Tolle (1980)
	Water supplies and milking equipment surfaces	Cousins and Bramley (1981)
Flavobacterium	Soil	Thomas (1966)
Klebsiella	Sawdust bedding and mastitic infections	Tolle (1980)
	Water supplies and milking equipment surfaces	Cousins and Bramley (1981)
Pseudomonas	Soil	Thomas (1966)
	Streak canal of udder	Tolle (1980)
	Water supplies and milking equipment surfaces	Cousins and Bramley (1981)
Serratia	Soil and water supplies	Bergey (1974)
	Milking equipment surfaces	Cousins and Bramley (1981)

initially at levels in excess of 1×10^5 cfu/mL (Macaulay et al., 1963; Weckbach and Langlois, 1977).

With modern methods of milk production and handling it is the psychrotrophic bacteria that have assumed greatest importance due to their ability to grow rapidly at the low temperatures used for milk storage. The importance of these organisms has been extensively reviewed by Witter (1961) and Cousin (1982).

In a survey of the quality of raw milk in southwest Scotland, Muir et al. (1979) found that over 70% of isolates obtained from creamery silo and farm bulk tank milks were *Pseudomonas* spp. Only the groups Enterobacteriaceae, *Acinetobacteria*, and Gram-positive organisms accounted for more than 5% of the psychrotrophic bacteria isolated. Similar results were reported by Orr et al. (1964).

1.2.   Thermoduric Organisms

Bacteria that are able to survive laboratory pasteurization procedures of 63°C for 30 min are usually referred to as thermodurics. The common thermoduric bacteria isolated from milk are Gram-positive, but not all Gram-positive organisms found in milk are capable of surviving pasteurization (Table 3). There is one species of Gram-negative bacteria that is considered thermoduric, namely *Alcaligenes tolerans*.

TABLE 3   Gram-Positive Bacteria Isolated from Raw Milk

Genus	Source	Strains Capable of Surviving Pasteurisation (65°C/30 min)
Arthrobacter	Teat surfaces[a]	None
Bacillus	Bedding material used for housing lactating cows. Pasture[b]	All species survive as spores
Clostridium	Cow's feces, fodder, bedding and teats Water supplies[a]	All species survive as spores
Corynebacterium	Streak canal[c]	*Corynebacterium lacticum, C. liquefaciens*
Lactobacillus	Cow's feces, silage[d]	None
Microbacterium	Milking equipment[a]	*Microbacterium lacticum*
Micrococcus	Milking equipment, teat skin[a] Udder and streak canal[c]	*Micrococcus luteus, M. varians*
Sarcina	Soil, mud[c]	None
Staphylococcus	Teat skin[a] Udder and streak canal[c]	None
Streptococcus	Teat skin[a] Udder and streak canal[c]	*Streptococcus thermophilus, Strep. bovis, Strep. durans, Strep. faecalis*

[a] Cousins and Bramley (1981).
[b] McKinnon and Pettipher (1983).
[c] Tolle (1980).
[d] Bergey (1986).

The importance of thermoduric bacteria in farm milk supplies became apparent as a result of the work of Hussang and Hammer (1931) in the United States and Anderson and Meanwell (1933) in the United Kingdom. Surveys have shown that over 70% of milk samples have thermoduric counts below 1,000 cfu/mL (Thomas et al., 1966a, b, 1967) and thermoduric contamination is negligible when statisfactory cleansing and disinfection procedures are used. However, there is also evidence that with the decreasing use of steam sterilization on farms and the consequent increase in the use of chemical disinfectants, the levels of thermoduric bacteria in milk are increasing.

There are four main groups contributing to the thermoduric flora of milk.

1.2.1. Spore-Forming Bacteria

Although anaerobic spore-forming bacteria of the genus *Clostridium* can be isolated from milk, they are unable to proliferate due to the high redox potential. It is thought that their incidence is associated with silage feeding of cattle (Goudkov and Sharpe, 1965).

By far the most important group of spore-forming bacteria found in milk is of the genus *Bacillus*. The most common species found in raw milk are *B. licheniformis, B. pumilus, B. brevis, B. megaterium,* and *B. subtilis* (Martin, 1981; Phillips and Griffiths, 1986b; Waes, 1976), but strains of *B. cereus, B. circulans,* and *B. coagulans* among others are also found (Coghill and Juffs, 1979b; Grosskopf and Harper, 1974; Jayne-Williams and Franklin, 1960; Mikolajcik, 1970). Many of these *Bacillus* spp. are able to adapt to growth at refrigeration temperatures (Grosskopf and Harper, 1974) and there are numerous reports of psychrotrophic spore formers being isolated from milk (Chung and Cannon, 1971; Coghill and Juffs, 1979b; Cousin, 1982; Credit et al., 1972; Grosskopf and Harper, 1969; Johnston and Bruce, 1982; Mikolajcik and Simon, 1978; Shehata and Collins, 1971). There is even one report in the literature of the presence of a psychrotrophic strain of *Clostridium* in milk (Bhadsavle et al., 1972), but this remains unconfirmed.

By far the greatest number (86%) of psychrotrophic, thermoduric organisms isolated from raw milk is *Bacillus* spp. (Johnston and Bruce, 1982). Of these, the species present in greatest numbers were *B. cereus, B. licheniformis,* and *B. coagulans.*

Although there are reports that suggest that the spore load in milk varies with season, being higher in winter than in summer (Billing and Cuthbert, 1958; Franklin et al., 1956; Ridgway, 1954; Stewart, 1975), other workers claim little seasonal variation (Phillips et al., 1981b; Waes, 1976). However, McKinnon and Pettipher (1983) found higher psychrotrophic spore counts in raw milks during the summer, and this was confirmed by Phillips and Griffiths (1986b).

Some of the strains of *Bacillus* isolated may have generation times of about 6 hr in milk at 4°C (Magdoub et al., 1983; Shehata et al., 1971). There is a long lag period before growth commences at these lower temperatures and they are usually outgrown by any Gram-negative psychrotroph that may be present. The generation times of a number of psychrotrophic *Bacillus* have been studied in milk at various temperatures and the results are presented in Table 4.

TABLE 4 Generation Times (Hours) of Psychrotrophic *Bacillus* Species

Strain	Temperature of Growth (°C)							Reference
	0	2	4	6	7.2	8	10	
B. cereus HRM044	n.d.[a]	n.g.	n.d.	23	n.d.	n.d.	4	
B. cereus HRM045		n.g.		19			3	
B. cereus v. mycoides HRM042		n.g.		15			n.d.	
B. cereus/thuringiensis HRM067		n.d.		12			n.d.	
B. circulans MRM054		36		10			5	
B. circulans MRM056		30		11			n.d.	
B. circulans MRM064		26		13			9	
B. circulans/lentus MRM184		19		11			n.d.	Griffiths and Phillips
B. circulans/polymyxa HRM080		n.g.		12			n.d.	(unpublished results)
B. lentus MRM305		20		7			3	
B. mycoides HRM068		n.g.		22			4	
B. mycoides/thuringiensis MRM223		n.g.		22			n.d.	
B. polymyxa MRM304		31		9			6	
B. pumilus KRM029		26		17			12	
B. thuringiensis MRM218		n.g.		12			3	
B. subtilis RH22								
B. circulans RH3								
B. coagulans TS3 Group A	30	13–19	7.5–14	6–9	n.d.	5–6	4–4.5	Shehata et al. (1971)
B. coagulans TS4								
B. laterosporus								
B. circulans F7								
B. coagulans F8								
B. licheniformis Group B	n.d.	n.d.	n.d.	24–36	7	16–22	8–14.5	
B. pumilus								
B. brevis								
B. megaterium								
B. cereus								
B. coagulans	n.d.	24–30	n.d.	n.d.	n.d.	n.d.	n.d.	Grosskopf and Harper (1969)

[a] n.d., not determined; n.g., no growth.

1.2.2.   Corynebacteria

Corynebacteria are rod-shaped organisms, and members of this group such as *Microbacterium lacticum* can form a substantial part of the bacterial population of milk (Thomas et al., 1967; Turbutt et al., 1960). A few authors consider microbacteria to be psychrotrophic (Coghill, 1982; Collins, 1981; Johnston and Bruce, 1982; Stanley and Rose, 1967). They have been shown to form up to 10% of the micobial flora of laboratory pasteurized milk after incubation at 7°C (Johnston and Bruce, 1982). Pure cultures of *M. lacticum* in sterile milk had generation times of from 40 hr to 10 days at 8°C (Seiler et al., 1984). It is likely that the organisms survive heat treatment and are then conserved by cold storage. This view is strengthened by the observations of Williams (1956) who showed that *M. lacticum* had little effect on the quality of pasteurized milk at 15°C. Only in rare cases are thermoduric microbacteria found to be responsible for defects in dairy products (Credit et al., 1972; Gillies, 1971; Mourgues and Auclair, 1973; Washam et al., 1977).

1.2.3.   Micrococci

Micrococci often predominate in the microflora of raw milk (Thomas et al., 1962), but only a small percentage (about 10%) is thermoduric (Lakshminarasim and Iya, 1955; McKenzie, 1965). Milks from farm bulk tanks had similar thermoduric microflora consisting of aerobic spore-forming bacilli (about 40%) and micrococci (about 40%), irrespective of whether their thermoduric count was low or high (Thomas et al., 1967). However, after storage at 7°C thermoduric micrococci formed only 1% of the population, being easily outgrown by strains of *Bacillus* spp. (Johnston and Bruce, 1982). This inability to grow in raw milk was to some extent due to the presence of inhibitory substances, lactenins, in milk (Thomas et al., 1967).

1.2.4.   Streptococci

Strains of *Streptococcus thermophilus, Strep. bovis,* and *Strep. faecalis* can survive pasteurization of milk. As with micrococci, they do not grow well in refrigerated milk although they may grow rapidly in milk at ambient temperatures to produce acid. Phillips et al. (1981b) isolated streptococci from spoiled samples of pasteurized double cream (48% fat) stored at 10°C, but not from samples kept at 6°C. It is uncertain whether these streptococci were thermoduric or had entered the cream after pasteurization. Recently, in our laboratory, *Streptococcus* spp. have been isolated from pasteurized milks stored at 4°C, although again these may have been postpasteurization contaminants.

2.   SOURCES OF BACTERIAL CONTAMINATION OF RAW MILK

Having briefly considered the types of bacteria found in milk, we should now consider their origin. The production, processing, and marketing of pasteurized dairy products arc multifacetted operations lasting several days. The chain of events from cow to consumer (represented diagrammatically in Figure 1) thus provides many opportunities for the introduction of bacterial contamination. Some of these contaminants may grow during the periods of storage encountered on the farm, at the factory, in the food store, and, not least, at the home of the consumer. The importance of the contamination to the quality of the final

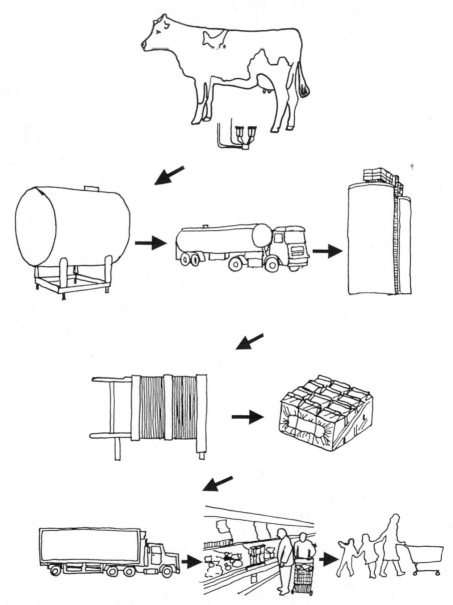

Figure 1.   Opportunities for bacterial contamination of milk during its passage from cow to consumer.

product depends on the point in the manufacturing cycle that it is introduced as well as on the nature of the contaminating organism. This review will attempt to provide a study of the steps involved in providing the public with their daily 'pinta" (or gallon in the case of the United States). The survey is presented in the context of the U.K. dairy industry, but most of the problems encountered are universal.

2.1.   Contamination of Milk at Its Source–The Cow

Under normal conditions, milk emerging from a cow's udder is essentially sterile. Bacteria may be present as a result of udder infection or as a result of environmental contamination of the udder from bedding material, foodstuffs, feces, etc.

By far the most common source of udder infection is mastitis, which can be the result of intramammary infections (mainly due to *Staphylococcus aureus* and *Streptococcus agalactiae*) or extramammary sources (*Escherichia coli* and *Streptococcus uberis*) (Hill, 1986). The causative agents of mastitis have been described by Tolle (1980). Mastitic infection is not always easy to detect, as, in addition to the clinical form of the disease during which changes in the milk or udder are readily observed, there exists a subclinical form. In the latter case, both milk and udder appear normal and the animals will not be isolated. Thus, milk from these cows will enter the bulk tank and increase the bacterial content dramatically (Cousins and Bramley, 1981). Wilson and Richards (1980), in a survey of the incidence of mastitis in the United Kingdom, showed that about 30% of dairy cows were affected by microbial infection in one or more udder quarters. The significance of udder infection as a factor increasing the bacterial count of herd bulk milk was studied by Bramley et al. (1984). These workers detected *Staph. aureus* or mastitis streptococci in 86% of milk samples analyzed. However, the numbers were below 1×10^4 cfu/mL in 90% of these cases. In herds with a high incidence of *Strep. uberis* infection, bulk tank milk counts reached 1×10^5 cfu/mL, albeit in a small number of samples.

Mastitis prevention in Britain involves extensive use of teat dip procedures and dry cow therapy with antibiotics. Methods for the control of mastitis have been reviewed by Bramley and Dodd (1984) and Hill (1986). The levels of mastitic milks entering the supply to processing sites has declined in the United Kingdom over the past few years as indicated by the fall in somatic cell counts of farm bulk tank milks (Longstaff, 1985). This was mainly due to the introduction of a centralized testing scheme for ex-farm milks that monitors both somatic cell count and total viable bacterial count (Harding, 1987). Farmers are paid for their milk on the basis of these results. It is recognized that the improvement in quality is not due to a decrease in the levels of mastitis, but rather to farmers being more prudent in their additions of mastitic milks to the bulk supply (Prentice and Neaves, 1986).

In addition to the mastitis-causing bacteria, other bacteria may infect the mammary gland, and many of these are pathogenic to man. These include

Mycobacterium bovis, Brucella abortus, Salmonella spp., *Listeria monocytogenes,* and *Coxiella burnettii* (Tolle, 1980).

The organisms introduced into the milk supply through intramammary infection are probably of little relevence to the quality of pasteurized products. However, it is in everyone's best interest that these organisms are eliminated from the milk supply in the interest of safety. Mastitis pathogens can survive in refrigerated raw milk, but they do not multiply at these temperatures. They will proliferate at 15°C and above although there is evidence that *Strep uberis* may be able to grow at temperatures below this (Cousins and Bramley, 1981). Any of these organisms that have survived in the raw milk will be destroyed by pasteurization.

Of potentially greater importance to the processor are the bacteria introduced into milk by the exterior surface of the udder. Dung, mud, bedding materials, soil, vegetation, and several other materials are likely to come into contact with the cow between milkings. All these are a rich source of microorganisms and can lead to surface contamination of the teats. Counts in milk from unwashed udders have been reported to range from 1×10^4 to 1×10^5 cfu/mL compared with 5×10^3 or less from washed udders (Johns, 1962). Thomas et al. (1971) have shown that even after washing, the numbers of organisms on teat surfaces can be high.

Arguably the greatest contribution to the contamination of the udder is afforded by bedding materials. All the common sources of bedding (such as sawdust, sand, shavings, and straw) have high bacterial counts and easily adhere to skin surfaces. A reduction in the levels of contamination on teats occurs in the summer when cows are turned out to graze (Cousins and Bramley, 1981; McKinnon and Pettipher, 1983) and this can result in a noticeable decrease in bulk tank milk counts during summer periods. McKinnon and Pettipher (1983) found that total counts for all types of bedding were similar (about 5×10^9/g) whereas that of pasture was considerably less (about 8×10^7/g).

Micrococci, including coagulase-negative staphylococci, are the dominant organisms found on teats of cows during winter housing. This group of bacteria is present at concentrations about 1×10^4 cfu/teat out of a total population of between 1×10^5 and 1×10^7 cfu/teat. Streptococci are also present although Gram-negative organisms are less numerous (Cousins, 1978; Cousins and Bramley, 1981; Thomas et al., 1971).

Of widespread concern to the processor is the occurrence of spore-forming bacteria in raw milk supplies. Some of these can survive pasteurization, germinate, and cause off-flavors in the final product (Cox, 1975; Griffiths et al., 1981a). It has been estimated that 90% of the spore count of bulk milk is contributed from teats (McKinnon and Pettipher, 1983; Underwood et al., 1974), and these are almost entirely *Bacillus* spp.

As far as the production of pasteurized products is concerned, by far the most important of the spore-forming bacteria associated with raw milk are those capable of growth at refrigeration temperatures. Psychrotrophic strains of *Bacillus* spp. were first isolated from raw milks by Grosskopf and Harper (1969). Since then several workers have reported on their incidence in raw milk (Table 5).

TABLE 5 Occurrence of Psychrotrophic Spore Formers in Raw Milk Samples

Location of Study	Source of Raw Milk	Number of Samples Examined	Samples Containing Psychrotrophic Spore Formers (%)	Reference
Ohio, United States	Individual producers	Not known	25.0	Grosskopf and Harper (1969)
Alabama, United States	Individual producers	18	83.0	Chung and Cannon (1971)
California, United States	Individual producers	97	30.0	Shehata and Collins (1971)
Ohio, United States	Individual producers	32	31.3	Grosskopf and Harper (1974)
Ohio, United States	Tanker loads	20	90.0	
Ohio, United States	Individual producers and tanker loads	109	39.0	Mikolajcik and Simon (1978)
Queensland, Australia	Tanker loads	30	23.3	Coghill and Juffs (1979b)
West Scotland	Individual producers	1040	22.6	Johnston and Bruce (1982)
England	Individual producers	8	100.0	McKinnon and Pettipher (1983)
Karnal, India	Individual producers	51	92.2	Sharma et al. (1984)
Oregon, United States	Individual producers	559	24.9	Bodyfelt (1986)
West Scotland	Individual producer	34	32.4	Phillips and Griffiths
West Scotland	Creamery silo	34	44.1	(unpublished results)

399

The most common psychrotrophic *Bacillus* spp. isolated from raw milks are *B. cereus, B, circulans, B. coagulans,* and *B. licheniformis* (Johnston and Bruce, 1982; Phillips and Griffiths, 1986b). The principal source of psychrotrophic spore formers in milk appears to be the upper layer of soil in pasture land and not bedding (McKinnon and Pettipher, 1983). Milk from cows at grass contained a higher proportion of psychrotrophic spore formers than milk from cows on bedding. This may account in part for seasonal variations in levels of psychrotrophic *Bacillus* spp. in raw milk (McKinnon and Pettipher, 1983; Phillips and Griffiths, 1986b).

Spores of *Clostridium* spp. can be introduced into milk from bedding and feedstuffs, especially silage (Carini et al., 1985; Donnelly and Busta, 1981). Although clostridia may cause spoilage problems in cheese, the evidence is that they do not multiply in raw milk because of its high redox potential (Goudkov and Sharpe, 1965). This, coupled with the fact that they are unable to grow at low temperatures, suggests that they are of strictly limited importance in the spoilage of pasteurized dairy products. Indeed, there has only been one account of a psychrotrophic strain of clostridia that was claimed to cause spoilage of pasteurized milk (Bhadsavle et al., 1972). Even then spoilage was not detected until after 46 days at 7.2°C.

The wide variety of bacteria present in the cow's environment that can be transferred into milk via teat surfaces can be controlled to some degree by preventing heavy soiling of the udder. The teats should also be washed and dried before milking. Control of the bacteriological contamination of milk from the surface of the teat and udder has been reviewed by Joergensen (1980).

2.2. Contamination of Milk from Milking Equipment

The only major source of bacteria in milk in the period between leaving the udder and collection is from inadequately disinfected milking equipment. Although there are no legally defined standards for cleaned and disinfected milking equipment surfaces in the United Kingdom, a suggested satisfactory figure would be 5×10^5 cfu/m^2 (British Standards Institution, 1982). However, milk of an acceptable bacteriological quality can be produced where rinses of the milking equipment are 1×10^7 to 1×10^9 cfu/m^2 (Cousins and McKinnon, 1979). In their survey of 10 farms, Bramley et al. (1984) showed that there was no correlation between the bacteriological quality of herd milk and the cleanliness of the milking machine and pipeline as assessed by plant rinses. Furthermore, a survey by Panes et al. (1979) showed that in spite of 25% of milking machines having high rinse counts of $> 1 \times 10^8$ cfu/m^2, only 13.5% of milk samples had total initial counts of $> 1 \times 10^5$ cfu/mL. Notwithstanding this, there is ample evidence that bacteria can proliferate rapidly in milk residues in badly cleaned milking plant (Thomas and Thomas, 1978b). This is especially true for difficult to clean areas such as crevices, joints, and dead ends. Rubber components of milking machines are also notorious for harboring potential milk contaminants (Thomas and Thomas, 1977b).

A wide variety of bacterial types can be introduced into milk from milk mineral deposits present in milking equipment, and these have been extensively reviewed in a series of articles by Thomas and Thomas (1977a, b, c, d, 1978a, b). Poorly cleaned milking equipment is a major source of psychrotrophic contamination of milk (Cousins, 1977; Marshall, 1985; Thomas and Thomas, 1977b). Gram-negative bacteria, particularly *Acinetobacter* spp., predominate among the microflora that adhere to stainless-steel milk transfer pipeline (Lewis and Gilmour, 1987).

Considerable variation in the microflora of milking equipment occurs from farm to farm due to differences in cleaning regimes and the magnitude of bacterial contamination (Druce and Thomas, 1972). Adequate cleaning of pipeline milking machines affords the only protection against the introduction of bacteria into milk during milking (Campbell and Marshall, 1975; Flake et al., 1972). Variations in temperature and cleaning procedure are known to affect the attachment of bacteria to stainless steel (Stone and Zottola, 1985). Generally, in-place cleaning, using hot detergent-disinfectant solutions or acidified boiling water, is the main regime employed. The effectiveness of cleaning procedures depends to a large extent on the design of plant and on factors such as the hardness of the water supply, which itself can give rise to deposit formation on milking equipment (Palmer, 1980).

The relative contributions to the bacterial load of raw milk from the udder and milking plant have been studied by Kurzweil and Busse (1973). These authors found that the udder microflora were predominant in low-count milks, but with increasing count there was an increase in milking machine flora. Thus, problems with high-count milks are mainly associated with the introduction of contaminants from milking machines.

2.3. Other Sources of Contamination of Raw Milk

The risk of bacterial contamination of milk from milking personnel and from aerial contamination is slight with machine milking, provided adequate precautions are taken (Cousins and Bramley, 1981; Palmer, 1980). It is also unlikely that contaminated water is a major source of bacteria in milk. Nevertheless, all water used in the production of milk should be of potable quality (Cousins and Bramley, 1981; Palmer, 1980). Where water is supplied from a storage tank, there is a danger of bacterial contamination from rodents, birds etc. unless it is properly protected.

2.4. Contamination of Milk during Refrigerated Storage on the Farm

Farm bulk tanks do not as a rule contribute greatly to the bacterial load of raw milk. Their design makes them easy to clean and they have a much lower bacterial content than pipeline milking plants (Druce and Thomas, 1972). Bacteria may be introduced into the milk from outlet plugs or cocks that tend to be difficult to clean. There is speculation that poorly cleaned farm bulk tanks may be a source of contamination by psychrotrophic spore formers, but this needs to be clarified.

Although they do not directly contribute much to the overall contamination of raw milk, prolonged storage in the tanks can have a profound effect on milk quality. During storage psychrotrophs, introduced into the milk at earlier stages of production, may proliferate. There is an increasing use of alternate (every other) day collection of bulk tank milks. As a consequence, milk may be stored on the farm for 48 hr or more. The effect of alternate day collection on the bacteriological quality of bulk tank milk has been reviewed by Thomas and Druce (1971). Most of the surveys carried out suggest that alternate day collection has little or no effect on the bacterial count of farm bulk tank milk providing the milk was rapidly cooled to 4°C or below before addition to the tank. Bockelmann (1974) has concluded that the critical time for milk storage at refrigeration temperatures lies between 60 and 72 hr. This is supported to some extent by the work of Griffiths et al. (1987a) who showed little change in bacterial counts of farm bulk tank milk stored at 2°C for 48 hr. Bulk tank milk stored at 6°C showed, on average, a 2 log cycle increase in growth after the same storage period. Similar results were achieved by Muir et al. (1978b). During the latter study ex-farm milk stored at 4°C for 48 hr experienced a 1 log cycle increase in psychrotroph count. Thus, it is imperative that the temperature of milk stored on the farm should be kept as low as possible and should certainly not exceed 4°C.

The initial level of bacteria present in milk prior to bulk storage is also of great importance to the length of time the milk can be stored without encountering gross contamination (Gehriger, 1980). There is a good linear relation between initial psychrotroph count and safe storage time for raw milks (Griffiths et al., 1987a). Even at low temperatures, milks produced under poor hygienic conditions, which have high levels of psychrotrophs present, have a poor keeping quality.

Although milks stored on farm for comparatively long periods of time do not exhibit a marked increase in bacterial numbers, there will be a detrimental effect on the length of time the milk can be held safely at the processing site. The length of the lag phase of bacterial growth in farm bulk tank milk at temperatures between 2 and 4°C appears to be about 48 hr (Griffiths et al., 1987a; Muir et al., 1978b). Alternate day collection will, therefore, result in milk arriving at the factory containing bacteria in the exponential phase of growth, and hence capable of rapid multiplication. Orr et al. (1964, 1965) observed no apparent deterioration in the quality of bulk collected milk due to alternate day collection, but when the alternate day collected milk was held for a further 24 hr at the processing site at 5°C, there was an appreciable increase in bacterial numbers.

The quality of farm bulk tank milk in the United Kingdom is acceptable, and has improved dramatically as a result of payment on the basis of total bacterial count. At present, 76% of the milk produced in England and Wales has a total count of less than 2×10^4 cfu/mL with 98% less than 1×10^5 cfu/mL (Prentice and Neaves, 1986). The weighted average lies around 1.9×10^4 cfu/mL (Longstaff, 1985). Similar results are found for the Scottish Milk Marketing Board area where approximately 36% of producers supply milks with total counts less than 1.5×10^4 cfu/mL and 95% of producers' milk has a count less than 5.0×10^4 cfu/mL. The bacterial content of bulk tank milk in Britain is similar to the standard suggested for ex-farm milk produced under good hygienic conditions (Solberg et al., 1974) and shows what can be achieved by diligence on the part of farmers coupled with constant monitoring of the milk supply.

2.5. Transportation of Milk to Creameries

All milk supplies in the United Kingdom are transported to creameries in collection tankers, although on occasion it may be transferred to large trunker tankers for longer journeys. Again the main causes of increased bacterial count during this stage are contamination due to inadequately cleaned vehicles and growth of bacteria already present in the milk. The latter will be governed by the temperature of the milk and the length of time taken to transport it. The temperature at which the milk arrives at the processing site is dependent to some extent on the initial temperature of the farm bulk tank supply.

Milk from tankers on arrival at the creamery usually has a slightly higher bacterial content than that of farm bulk tank milk sampled at collection (Thomas, 1974). A 2-fold increase in count is common. The increase is mainly the result of proliferation of psychrotrophic and coliform bacteria (Leali, 1965; Thomas, 1974). Pastore (1968) suggests that there is little evidence that contamination with thermoduric organisms occurs at this stage of milk processing.

Inclusion of milk from a single farm with high bacterial counts can have a significant effect on the total colony count of tanker milk (Møller-Madsen, 1965; Orr et al., 1960).

2.6. Storage of Milk at the Processing Site

The importance of obtaining an extended storage life for raw milk at creameries is increasing as manufacturing practices change. The introduction of a 5-day working week coupled with the adoption of a quota system for milk production within the EEC have resulted in bulking and extended storage of milk at creameries.

Muir et al. (1978a) surveyed the bacteriological quality of creamery silo milk at a number of sites in southwest Scotland. The average total bacterial count of the milks sampled was 1.7×10^5 cfu/mL, with the mean psychrotroph count at 1.3×10^5 cfu/mL. The majority of bacteria present (70.2%) belonged to the genus *Pseudomonas* (Muir et al., 1979), but among other groups isolated were Enterobacteriaceae (7.7%), Gram-positive bacteria (6.9%), and miscellaneous Gram-negative rod-shaped organisms (12.6%). On further storage of the silo milks for 48 hr at 6°C, the psychrotroph count increased by 2 log cycles to 1.3×10^7 cfu/mL (Muir et al., 1978a). Other work has also shown marked increases in bacterial counts for raw milks stored for 48 hr at 5°C at the creamery (Crawford, 1967). Poor quality milk may only be stored for 24 hr (McLarty and

Robb, 1968). The increase in count on storage at the creamery is almost entirely due to psychrotrophic growth. Dommett et al. (1986), in a study of storage conditions at a factory, concluded that psychrotrophic growth patterns were similar whether milk was selected according to initial total counts or not, and whether storage was in large air-agitated silos or small paddle-agitated vats. Growth rates were higher during filling of the silos. These workers found that final bacterial numbers in the vats were dependent only on initial levels and storage time, with storage time being the most significant factor affecting the final quality. The temperature of the milk throughout did not rise above 6°C.

Bockelmann (1970) stated that thermoduric bacteria do not grow in cold-stored creamery silo milk, but recent work (Griffiths et al., 1987a) has shown that Gram-positive bacteria grow in raw milk at 6°C. There was even slow growth of these organisms at 2°C. It was not clear, however, whether these Gram-positives were heat resistant.

3. CONSEQUENCES OF PSYCHROTROPHIC BACTERIAL GROWTH IN RAW MILKS

Although most psychrotrophs are destroyed by pasteurization, they produce heat-stable, exocellular enzymes that can survive pasteurization and even UHT temperatures (Cogan, 1977; Cousin, 1982; Law, 1979). The enzymes that are of most relevence to the processor are the proteases that can degrade caseins and, to a lesser degree, whey proteins (Fairbairn and Law, 1986), and lipases that hydrolyze milk fat (Stead, 1986). Other enzymes that may be of importance are phospholipases (Griffiths, 1983) and glycoprotein glycosidases (Marshall, 1982).

The proteases and lipases produced by psychrotrophs of a number of genera retain some 60–70% of activity after heating at 77°C for 17 sec (Griffiths et al., 1981b). Muir et al. (1979) reported that 53.4% of isolates of psychrotrophic bacteria from creamery silo milks were capable of degrading milk protein and/or fat. The levels of exocellular enzymes synthesized by these organisms depends on a number of environmental factors. Thus, the conditions under which milk is stored could have a profound effect on the rate of enzyme synthesis and, hence, milk degradation.

3.1. Lipase Production by Psychrotrophic Bacteria during Storage

Lipase probably occurs in most raw milk (Adams and Brawley, 1981), but much of the enzyme activity is associated with a native milk lipoprotein lipase. This enzyme activity seems to be destroyed by HTST pasteurization (Shipe and Senyk, 1981). The effect of endogenous lipase on milk products has been the subject of a number of reviews (Anderson, 1983; Chilliard and Lambert, 1984; Deeth and FitzGerald, 1983; Downey, 1980) and is outside the scope of this article.

The role of microbial lipase in the spoilage of dairy products has been reviewed by Lawrence (1967a, b) and more recently by Stead (1986). The main factors governing their synthesis in milk during storage are temperature and degree of aeration. A reduction in temperature from 25 to 4°C progressively increases lipase production in whole milk with 25% more enzyme produced at 4°C than at 10°C (Bucky et al., 1986). Other workers have also shown increased lipase synthesis by psychrotrophic bacteria at temperatures below the optimum for growth (Alford and Elliott, 1960; Andersson, 1980; Khan et al., 1967; Lawrence et al., 1967; Nashif and Nelson, 1953a). Recent work in our laboratory suggests that whereas lipase may be produced by a number of psychrotrophs at 6°C, there is very little enzyme synthesized by cells grown at 2°C (Figure 2). The optimum temperature for lipase production by a *Pseudomonas fluorescens* strain is 8°C (Andersson, 1980).

The effect of aeration on lipase production is more difficult to assess. In some cases high aeration depresses the synthesis of lipase by a number of pseudo-monads (Alford and Elliott, 1960; Lu and Liska, 1969; Nadkarni, 1971; Nashif and Nelson, 1953a; Te Whaiti and Fryer, 1978). High aeration was needed for production of high lipase activity for two *Ps. fluorescens* strains growing in skim milk at low temperatures (4–7°C) (Dring and Fox, 1983; Fox and Stepaniak, 1983). This increased enzyme activity may be caused by lipase production profiles shifting toward earlier stages of growth (Bucky et al., 1986) or higher population densities (Stepaniak et al., 1987b). Opposite results were obtained by Rowe and Gilmour (1982). These workers showed that lipase production by *Ps. fluorescens* grown on simulated milk medium at 7°C was immediately preceeded by a large decrease in O_2 tension. When O_2 tension was forcibly decreased there was a concomitant increase in lipase synthesis. Griffiths and Phillips (1984a) showed little difference in lipase synthesis by the natural microflora of raw milk (80–100% *Pseudomonas* spp.) when the O_2 tension was maintained at a high level by aeration.

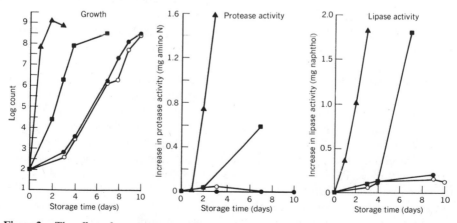

Figure 2. The effect of temperature on the growth and enzyme production of an *Aeromonas* sp. grown in milk at 2°C (○), 6°C (●), 10°C (■), and 21°C (▲).

3.2.   Protease Production by Psychrotrophic Bacteria during Storage

The proteases produced by psychrotrophic bacteria have been extensively reviewed in a recent article by Fairbairn and Law (1986). As with lipase production, storage conditions can have a marked effect on protease synthesis by psychrotrophic bacteria.

It has been reported that protease synthesis by a psychrotrophic strain of *Ps. fluorescens* was immediately preceeded by a fall in the O_2 tension of the medium (Rowe and Gilmour, 1982). This effect was confirmed by a study of the natural psychrotrophic flora growing in aerated and nonaerated milk (Griffiths and Phillips, 1984a). The amount of protease produced by these organisms was about 50% less under aeration than nonaeration. On the other hand, Keen and Williams (1967) reported an increase in protease production during aeration of a *Ps. lachrymans* culture. Aeration also increased protease synthesis by *Ps. fluorescens* (O'Donnell, 1975) and other pseudomonads (O'Donnell, 1975; Stepaniak et al., 1987b). However, little or no protease synthesis occurred during aeration of a *Ps. fragi* culture (O'Donnell, 1975). Conflicting results have also been obtained with two pseudomonads grown in skim milk (Te Whaiti and Fryer, 1978). Flushing of raw milk with N_2 resulted in a lack of O_2 and prevented the accumulation of proteases from psychrotrophic bacteria (Murray et al., 1983). Similar results were obtained by other workers (Skura et al., 1986). Fairbairn and Law (1986) point out that the results of Murray et al. (1983) are open to criticism because of the low bacterial numbers present. Also no account was taken of the effect N_2 flushing had on the microflora. For instance, maximal production of protease in mixed cultures is about half that found in corresponding single strain cultures (Stepaniak et al., 1987b).

Temperature of storage may exert an effect on protease production by psychrotrophs growing in milk. Levels of protease produced per unit growth are higher at refrigeration temperatures for a number of *Pseudomonas* spp. (Fairbairn and Law, 1986; Juffs et al., 1968; McKellar, 1982; Peterson and Gunderson, 1960). We have some evidence that storage of milk at 2°C prevents synthesis of protease by a number of psychrotrophic bacteria, including an *Aeromonas* sp. (Figure 2).

The conflicting evidence on the effect of the physical environment on the production of exocellular enzymes by psychrotrophic bacteria suggests that there is no universal method of storage that can limit their production. In the vast majority of studies carried out it has been found that appreciable enzyme synthesis occurs only when the organisms have reached late log phase or stationary phase of growth and bacterial numbers are above 5×10^6 cfu/mL (Table 6). It may be that existing methods of assay for these enzymes are not sensitive enough to detect low levels of enzyme that may be synthesised earlier in the growth cycle. However, Stepaniak et al. (1987b) using a sensitive ELISA assay system for the detection of protease and lipase showed that population levels for a strain of *Ps. fluorescens* must reach at least 1×10^6 cfu/mL before these enzymes could be detected. Thus, the most effective way of limiting the presence of these degradative enzymes in pasteurized products is to keep growth of psychrotrophs

TABLE 6 Bacterial Levels Required in Raw Milk to Bring about an Effect on Final Product

Level in Raw Milk before Defect Encountered in Final Product (cfu/mL)	Organism Responsible	Product Studied	Reference
Lipase			
$>1 \times 10^6$	Natural flora	Raw milk	Suhren et al. (1975)
$>5 \times 10^6$	Natural flora	Raw milk	Muir et al. (1978b)
$>1 \times 10^7$	Natural flora	Raw milk	Overcast and Skean (1959)
1.9×10^7	Natural flora	Pasteurized milk	Bandler et al. (1981)
$>5 \times 10^6$	*Pseudomonas* P46	UHT milk	Bucky, Robinson and Hayes (1987a)
$>1 \times 10^7$	*Pseudomonas fragi*		
$>1 \times 10^6$	*Pseudomonas fluorescens*	UHT milk	Shelley et al. (1986)
$>1 \times 10^7$	Natural flora	Pasteurized cream	Tekinson and Rothwell (1974)
3.6×10^6	Natural flora	Pasteurized cream	Griffiths et al. (1981a)
$3\text{–}5 \times 10^6$	*Alcaligenes viscolactis*	Dutch cheese	Driessen and Stadhouders (1971)
	Pseudomonas fluorescens	Cheddar cheese	Law (1979)
Protease			
3×10^6	Natural flora	Raw milk	Pâquet et al. (1987)
$>1 \times 10^6$	Natural flora	Raw milk	Juffs (1975)
$>1 \times 10^8$	*Pseudomonas fluorescens*	Pasteurized skim milk	Stepaniak et al. (1982a)
$>1 \times 10^6$	Natural flora	Pasteurized milk	Juffs (1975)
5×10^7	*Pseudomonas* sp.	UHT milk	Bergtsson et al. (1973)
5×10^7	*Pseudomonas fluorescens*	UHT milk	Law et al. (1977)
$>1 \times 10^6$	Natural flora	Pasteurized cream	Juffs (1975)
$>1 \times 10^7$	Natural flora	Pasteurized cream	Griffiths et al. (1981a)
$>1 \times 10^7$	*Pseudomonas* sp.		
	Acinetobacter sp.	Cheddar cheese	Law et al. (1979)

in raw milk to below 1×10^6 cfu/mL. A method for predicting the safe storage life of raw milk based on initial psychrotroph counts has been proposed (Muir and Phillips, 1984).

4. CONTROL OF GROWTH OF PSYCHROTROPHIC BACTERIA IN RAW MILK

There are several methods available to both the producer and processor for limiting the growth of psychrotrophic bacteria in raw milk. These have been briefly reviewed by Cousin (1982) and Fairbairn and Law (1986).

4.1. Control of Growth of Psychrotrophs by "Deep Cooling"

It is well established that decreasing the storage temperature of raw milk results in a decrease in growth rate of bacteria present. The predominant factor controlling storage life is probably the length of the lag phase of the growth cycle (Stadhouders, 1982). Decreasing the storage temperature of farm bulk tank milk from 6 to 2°C markedly affects the length of the lag phase and results in a 48 hr extension of the time taken for bacterial counts to reach 1×10^6 cfu/mL (Griffiths et al., 1987a). Creamery silo milk behaved differently in that no lag phase was evident at either 2 or 6°C due to the age of the milk. Decreasing the storage temperature to 2°C still resulted in an average 29 hr increase in storage life. The storage life was strongly dependent on the initial psychrotroph count of the raw milk.

Storage of raw milk at temperatures around 2°C may adversely affect the processing properties due to dissociation of casein micelles (Reimerdes, 1982). Work has shown no detrimental effect on the properties of a number of products, including cheese, manufactured from raw milk stored at 2°C (Banks et al., 1987; Griffiths et al., 1987b). The soluble β-casein content of the milk did increase with decreasing storage temperature, but reaggregation of the casein micelles occurred on pasteurization.

4.2. Control of Growth of Psychrotrophs by Thermization

Thermization is the term given to a heat treatment at subpasteurization temperatures applied to milk either on receipt at the creamery or on the farm itself (Zall, 1980). The process is used in Europe, particularly in the Netherlands (Foley and Buckley, 1978), but, as yet, has not been widely adopted in Britain.

Griffiths et al. (1986a) studied a number of time–temperature combinations and found that temperatures of 65°C or more were most effective. Thermization at 65°C for 15 sec resulted in a 2 to 3 log cycle decrease in psychrotroph count, with the result that milk could be stored for up to 5 days at 6°C before counts reached 1×16^6 cfu/mL. Similar results have been achieved by other workers (Bjorgum et al., 1978; Coghill et al., 1982; Gilmour et al., 1981; Stadhouders,

1982). The efficiency of the thermization was strongly correlated with the initial bacterial load of the raw milk.

There were no adverse effects on the yield or quality of Cheddar cheese manufactured from thermized milk (Banks et al., 1986; Johnston et al., 1987), nor was the production of dried skim milk affected (West et al., 1986).

4.3.   Other Methods of Preventing Psychrotrophic Bacterial Growth

4.3.1.   By the Addition of CO_2

The addition of CO_2 to milk to inhibit the growth of psychrotrophic bacteria has been described by several researchers (Duthie et al., 1985; Gill and Tan, 1979; King and Mabbitt, 1983; Shipe et al., 1982). The storage life of poor quality milk could be increased by 3 days at 4°C with no adverse effects on fermented products made from the treated milk (Law and Mabbitt, 1983). However, there are indications that the addition of CO_2 may increase the heat resistance of spores of a number of *Bacillus* spp. (Guirguis et al., 1984).

4.3.2.   By the Lactoperoxidase System

Other methods of preserving milk include activation of natural inhibitory systems in milk such as lactoperoxidase (Bjork et al., 1975; Reiter and Marshall, 1979). This technique has been reviewed by a number of authors (J. G. Banks, et al., 1986; Bjork, 1980; Korhonen, 1980; Reiter, 1981). It is doubtful whether the lactoperoxidase system has any commercial application for large-scale processors.

4.3.3.   By the Addition of Lactic Acid Bacteria

Lactic acid bacteria have been added to raw milk to prevent psychrotrophic bacterial growth (Juffs and Babel, 1975). Although some encouraging results have been reported by Honer (1981), problems with flavor defects may limit the use to cultured products (Mikolajcik, 1979).

5.   SOURCES OF BACTERIAL CONTAMINATION OF PROCESSED MILK

At spoilage, or after extended periods of storage in the range 4–6°C, the organisms most frequently found in pasteurized products are Gram-negative, psychrotrophic bacteria with strains of *Pseudomonas* spp. being predominant (Table 7) (Phillips et al., 1981b; Schröder et al., 1982; Schultze and Olson, 1960; Tekinsen and Rothwell, 1974; Thomas and Druce, 1969). The microflora of these spoiled products are little different from those of the raw milk from which they were manufactured (Muir et al., 1979; Phillips et al., 1981b). This is unexpected as most of the Gram-negative psychrotrophs associated with spoilage of milk and milk products are susceptible to pasteurization (Table 8). Cogan (1977) quoted *d*-values at 71.7°C of between 0.0001 to 0.3243 sec for a number of pseudomonads.

TABLE 7   Microflora (%) of Pasteurized Milks and Creams at Spoilage or after Extended Periods of Storage in the Range 4–6°C

Microorganisms Present	a	b	c	d	e	f
Gram-negative						
Pseudomonas	60	70	63	62	88	40
Achromobacter	20	10	10	36	—	—
Alcaligenes	15	10	10	—	2	—
Acinetobacter	—	—	—	—	5	17
Coli-aerogenes	—	8	15	2	1	13
Flavobacterium	—	2	1	—	1	3
Aeromonas	—	—	—	—	—	11
Gram-positive						
Bacillus	—	—	—	—	—	12
Coryneform group	—	—	—	—	—	2
Lactobacillus	—	—	—	—	—	2
Micrococci	—	—	—	—	2	—
Yeast	5	—	1	—	1	—
Total number of samples	—	—	—	3	21	35
Storage temperature (°C)	4 ± 1	4	4	6	5	6

[a] Milk; Overcast and Skean (1959).
[b] Milk; Schultze and Olson (1960).
[c] Whipping cream; Schultze and Olson (1960).
[d] Milk; Dempster (1968).
[e] Cream; Tekinsen and Rothwell (1974).
[f] Cream; Phillips et al. (1981b).

Thus, even the most heat resistant of these bacteria should be reduced by much more than the 10 log cycles required by an effective HTST pasteurization procedure. Despite reports of heat-resistant strains of *Pseudomonas* (Macaulay et al., 1963; Weckbach and Langlois, 1977), it seems likely that the Gram-negative psychrotrophs isolated from pasteurized products are reintroduced into the product after heat treatment. This assumption is supported by the fact that aseptic collection of product at a point as close to the heat exchanger as possible results in a marked increase in shelf-life (Table 9) (Phillips et al., 1981a; Schröder and Bland, 1983). The problem of postpasteurization contamination was the subject of a review by Griffiths et al. (1985a) in which its importance, sources, detection, and elimination were discussed. Nevertheless, a brief commentary on this problem is considered useful.

5.1.   Importance of Postpasteurization Contamination

Post-heat treatment contamination is widespread in the U.K. dairy industry, with 92% of milk samples and more than 80% of cream samples analyzed showing

TABLE 8 Heat Resistance of Psychrotrophic and Other Micro-organisms in Milk

Organism	Temperature (°C)	d-value[c] (minutes)	Z-value[d] (°C)	$d_{71.0}$ (seconds)	References
Pseudomonas fluorescens	48.8	2.6	4.4	0.0010[a]	Chaudhary et al. 1960; Cogan 1977
Pseudomonas fluorescens	52.0	1.3	n.d.	n.d.	Lawton and Nelson 1955
Pseudomonas fragi	48.8	8.9	6.9	0.2563[a]	Chaudhary et al. 1960; Cogan 1977
Pseudomonas fragi	48.1	7.1	7.6	0.3243[a]	Luedecke and Harmon 1966
Pseudomonas fragi D	48.9	21.9	n.d.	n.d.	Collins 1961
Pseudomonas mephitica	48.1	2.6	3.7.	0.0001[a]	Kaufmann and Andrews 1954; Cogan 1977
Pseudomonas viscosa	48.8	2.3	4.9	0.0029[a]	Chaudhary et al. 1960; Cogan 1977
Pseudomonas viscosa A	48.9	3.7	n.d.	n.d.	Collins 1961
Salmonella binza	68.3	0.52	n.d.	0.30[a]	Read et al. 1968
Salmonella seftenberg 775W	71.7	0.02	n.d.	1.20	Read et al. 1968
Salmonella typhimurium	68.3	0.9	5.3	0.22[a]	Bradshaw et al. 1987
Escherichia coli	76.7	0.004	6.5	n.d.	Evans et al. 1970
Campylobacter jejuni	56.0	0.3	5.1	0.6[b]	Waterman 1982
Yersinia enterocolitica	68.3	0.09	5.1	1.17[a]	Lovett et al. 1982
Listeria monocytogenes	71.7	0.015	6.3	0.9	Bradshaw et al. 1985
Staphylococcus aureus	76.7	0.004	5.11	n.d.	Evans et al. 1970
Streptococcus bovis	62.8	2.6	n.d.	n.d.	Ienistea et al. 1970; Cogan 1977
Streptococcus durans	62.8	7.5	n.d.	n.d.	Ienistea et al. 1970; Cogan 1977
Sterptococcus faecalis	62.8	10.3	n.d.	n.d.	Ienistea et al. 1970; Cogan 1977
Streptoccus faecium	62.8	3.5	n.d.	n.d.	Ienistea et al. 1970; Cogan 1977

n.d. = not determined;
[a]Extrapolated values;
[b]$d_{63.5}$ value;
[c]d-value is the time of heating, at a particular temperature, to reduce the number of bacterial cells to 10% of its initial value;
[d]Z-value is the change in temperature yielding a 10-fold change in *d*-value.

411

TABLE 9 Increase in Shelf Life by Elimination of Postpasteurisation Contamination

Product	Shelf Life (Days to Reach 10^6 cfu/mL at 5–6°C)		Improvement in Shelf Life	Reference
	ppc Present	ppc Absent		
Milk				Schröder and
Bottled	7,1	30 7	4.3 ×	Bland (1983)
Cartoned	9.3	22.7	2.4 ×	
Cream	7.6	13.2	1.7 ×	Phillips et al. (1981a)

the presence of Gram-negative bacteria (Phillips et al., 1981a; Schröder, 1984). The position has changed little over the past 40 years since Crossley (1948) stated that postpasteurization contamination had a major effect on the shelf life of cream. Elimination of post-heat treatment contaminants led to a 15-day increase in keeping the quality of milk at 5°C (Schröder et al., 1982), and a 6.5-day increase in the shelf life of cream at 6°C (Phillips et al., 1981a).

There are signs that the dairy trade in the United Kingdom is becoming more aware of the problem, as evidenced by the introduction, in Scotland, of a pasteurized milk quality testing scheme. Retail samples of milk from participating dairies are tested weekly and those meeting the required standards will be allowed to display a quality mark on their packaging.

5.2. Detection of Postpasteurization Contamination

The test used to assess the quality of the milks participating in the scheme described above was developed in our laboratory (Griffiths et al., 1984a, b; Griffiths and Phillips, 1986; Phillips and Griffiths, 1985). It relies on the fact that Gram-negative psychrotrophs can be selectively grown by incubation at 21°C for 25 hr in the presence of an inhibitor system (comprising crystal violet, penicillin, and nisin) that prevents the growth of Gram-positive bacteria. The latter interfere with the performance of the test (Phillips et al., 1983b). The level of bacterial growth after this preincubation correlates well with the keeping quality of a number of pasteurized products (Griffiths and Phillips, 1986; Phillips and Griffiths, 1985). The selective enrichment is required because numbers of Gram-negative psychrotrophs present in commercially pasteurized products are too small to be detected by conventional plate counting methods. Initial psychrotroph counts of freshly pasteurized milks are usually of the order of 50 cfu/L or less (Schröder et al., 1982). Other traditional tests for estimating bacterial quality of milk such as standard plate count and coliform count are of little value in assessing the quality of pasteurized products (Blankenagel, 1976; Griffiths and Phillips, 1986; Martins et al., 1982). This is because organisms enumerated by these techniques do not necessarily grow at refrigeration temperatures and vice versa. Methods that are meaningful in terms of keeping quality, for example, the

Moseley test (Coghill and Juffs, 1979a; Felmingham and Juffs, 1977; Lück et al., 1980; Moseley, 1958; Mull et al., 1974; Randolph et al., 1965; Thomas, 1969) take 8–10 days to perform. Results are not obtained until after the effective shelf life of the product and are of historical interest only.

The preincubation test can be combined with rapid methods of bacterial estimation such as bioluminescence, DEFT (Griffiths et al., 1984a; Griffiths and Phillips, 1986), impedance monitoring (Griffiths and Phillips, 1984b, 1986), or catalase estimation (Phillips and Griffiths, 1987a) to provide a useful indication of expected shelf life within 26 hr of production.

Other inhibitor systems for preventing the growth of Gram-positive bacteria have been studied (Phillips and Griffiths, 1986a), and recent results suggest that a better correlation with shelf life can be obtained after preincubation with cetrimide–fucidin–cephaloridine. This inhibitor system allows selective growth of pseudomonads (Mead, 1985), although other Gram-negative bacteria may also grow (Phillips and Griffiths, 1986a).

Many different combinations of preincubation conditions have been described (Griffiths et al., 1985a) and methods of assessment of dairy product quality and potential shelf life are the subject of an extensive review by Bishop and White (1986).

5.3. Sources of Postpasteurization Contaminants

5.3.1. Problems Related with the Heat Exchanger

Contaminants are able to enter the pasteurized product at any stage of the processing chain after the heating unit. The regeneration, cooling sections, and related pipelines of the pasteurizer itself can contribute to recontamination (Anas, 1977; Sing, 1972). During a period of extended operation of a plate heat exchanger, there can be a build up of bacteria on the surface of the pasteurized milk side of the regenerator section. These bacteria are mainly thermoduric strains of streptococci such as *Streptococcus thermophilus* (Bouman et al., 1982; Stadhouders, 1982). They may not influence the keeping quality of refrigerated milk, but can lead to unacceptably high plate counts.

Contamination of the pasteurized product can also arise from the bacteria concentrated along the rubber gaskets of the heating unit (Lück et al., 1980). The specific defect encountered most often is the accumulation of soil in the cooling section or in dead-end fittings (Sing, 1972). Potentially the most dangerous form of contamination contributed by the heat exchanger is the adulteration of pasteurized milk with raw milk by leakage in the regeneration unit due to corrosion (Berg, 1986).

5.3.2. Problems Associated with Pipelines and Holding Tanks

Subsequent manufacturing steps can contribute to contamination of pasteurized commodities. Pipelines and storage tanks sited between the flow diversion valve of the heat exchanger and the filling machines can contribute to increases in bacterial load (Marshall and Appel, 1973; Schröder, 1984). One possible cause of pollution of heat-treated products in storage tanks is due to the introduction of airborne bacteria (Wainess, 1981). As tanks are emptied, the contents are replaced with air, which, unless the tanks have adequately designed air filtration systems, is an extremely high potential source of contamination (Heldman et al., 1964). Mechanical ventilation and air filtration are now considered essential even in the smallest dairies. There is some conflict in the importance attached to airborne contamination in limiting the shelf life of pasteurized milk. Cannon (1970) showed no relation between the two factors, but Wainess (1981) stated that some dairies have improved the keeping quality of milk by installing high efficiency particulate air filters.

5.3.3. Contamination during Packaging

Despite all these possible sites in the processing chain at which bacteria can be introduced, the step that has the greatest influence on the keeping quality of heat-treated dairy products is the filling operation (Moseley, 1980). Filling machines are an important source of contamination of pasteurized milk (Griffiths and Phillips, 1986; Schröder, 1984), albeit at a low level (10–500 psychrotrophs/L). Pasteurized milks containing 500 psychrotrophic bacteria/L normally have shelf lives of less than 130 hr at 6°C. Improvements in the design of filling machines have resulted in a 100% increase in the keeping quality of milk at 7°C (Schröder and Bland, 1983). This new generation of packaging machines incorporates features such as exclusion of air from the filler (by using bellows instead of pistons) and carton sterilization by UV light and hydrogen peroxide. Indirect aerial contamination may also occur at the filling stage from condensation formed on the machines as well as from smearing of products by moving parts of filling valves (Mrozek, 1970).

A significant, but decreasing, proportion of milk sold in the United Kingdom is bottled. Bottle fillers are notoriously difficult to clean (Schröder, 1984). Thus, shelf lives of bottled milk are substantially less than those of cartoned milks produced from the same raw milk supply (Griffiths and Phillips, 1986). As well as contamination by fillers, improperly sanitized bottles can also severely limit keeping quality (Schröder et al., 1982; Thomas and Druce, 1969). Other packaging materials, such as properly prepared plastic and laminated plastic materials, are not considered an important source of bacteria (Lück, 1981). Packaging materials may contribute to contamination indirectly due to the permeability of the material to bacteria. For instance, contamination can be introduced into bottled milk via lids that are not hermetically sealed (Schröder et al., 1982).

5.4. Elimination of Postpasteurization Contamination

It is evident that postpasteurization contamination can be introduced at any stage after heat treatment. The problem is entirely due to inadequate sanitary practices at the creamery. Consequently, ensuring adequate cleansing and sterilizing procedures will substantially reduce the dilemma. Olson (1964) showed that in the absence of sanitization, the shelf life of pasteurized milk decreased from 31 to 9 days. Good keeping quality was reattained immediately after the introduction of proper cleansing procedures. Similar effects were observed by Griffiths and Phillips (1986). The use of in-line sampling methods coupled with an appropriate technique for assessing postpasteurization contamination can lead to dramatic improvement in shelf life (Ellicker et al., 1964; Griffiths and Phillips, 1986).

Clean-in-place procedures have been reviewed by Barron (1984), and a general code of hygienic practice for the dairy industry has been drawn up (Anon, 1984). Particular attention should be paid to rinse water used in plant cleaning. This should contain $5 \mu L$ available chlorine/L. A common fault is the indiscriminate use of untreated water in phases of operations after heat treatment of product. Considerable regard should also be focused on the design and construction of equipment employed in dairy plants. These can have a striking effect on the hygienic quality of pasteurized products (Wainess, 1982).

6. IMPORTANCE OF HEAT-RESISTANT BACTERIA IN PASTEURIZED PRODUCTS

On elimination of postpasteurization contamination, the microflora of heat-treated products consist almost entirely of Gram-positive bacteria that have survived pasteurization. These mainly comprise streptococci, micrococci, coryneforms, and aerobic spore-forming bacteria of the genus *Bacillus* (Thomas et al., 1967; Davies, 1975a). Heat-resistant bacteria capable of growth at refrigeration temperatures have been isolated from pasteurized milk, and these include strains of *Arthrobacter, Microbacterium, Streptococcus, Corynebacterium*, and *Bacillus* (Washam et al., 1977). For pasteurized milk free of post-heat treatment contamination, the shelf life is limited by the development of psychrotrophic strains of spore-forming bacteria (including *B. cereus, B. licheniformis*, and *B. circulans*) and *Microbacterium* spp. originating from the raw milk (Langeveld et al., 1973; Morgues and Auclair, 1973). The sources and types of these psychrotrophic bacilli have been discussed earlier. Some strains of *Bacillus* sp. (notably *B. circulans*) isolated from both raw and pasteurized milk can grow at temperatures as low as 2°C (Table 4).

A correlation exists between the keeping quality of pasteurized milk and the thermoduric count of raw milk from which it was manufactured (Auclair, 1986; Mourgues et al., 1983; Schröder and Bland, 1984). Organoleptic defects in milk stored at 7.2°C for 6 days were observed when thermoduric counts reached $3-4 \times 10^6$ cfu/mL (Tinuoye and Harmon, 1975). Other workers have isolated appreciable numbers of *Bacillus* spp. from spoiled pasteurized milk (Coghill and Juffs, 1979b; Collins, 1981; Cox, 1975; Credit et al., 1972; Gillies, 1971) and creams (Phillips et al., 1981b). Their levels in pasteurized milk vary between 0 and 60 spores/L while counts of between 1 and 1600 spores/L have been quoted for raw milk (Cannon, 1972; McKinnon and Pettipher, 1983). There is evidence that problems associated with heat resistant psychrotrophic bacteria are increasing in the United States with 20–25% of shelf life defects attributable to these organisms (Bodyfelt, 1980).

Growth of *Bacillus* spp. in pasteurized milk products can lead-to a number of defects. For example, *B. cereus* is the causative organism for the fault known as "bitty cream," which is characterized by fat destabilization when cream is added to a hot beverage (Cox, 1975; Franklin, 1969; Stone and Rowlands, 1952). This bacterium is also associated with a sweet-curdle defect in pasteurized dairy produce (Mikolajcik, 1978). Heat resistant psychrotrophs cause other "off" flavors in pasteurized milks such as bitterness. Fruity, rancid, and yeasty tastes have been ascribed to these bacteria as well (Collins, 1981). The first manifestation of a taint caused by the growth of thermoduric psychrotrophs is the formation of a sweet curd or pellicle on the bottom of the container. In the early spring about 25% of milk samples examined after 12 days of storage at 7°C exhibited this spoilage phenomenon (Collins, 1981). The incidence of spoilage of pasteurized cream by psychrotrophic spore formers was also highly seasonal (Phillips et al., 1981b). The greatest number of *Bacillus* spp. were isolated from the cream from June to October. This seasonality was unrelated to the spore or thermoduric count of the original raw milk. Experiments suggest that it is due to many factors (Phillips and Griffiths, 1986b). Arguably the most important contributary influence is the seasonal variation in levels of psychrotrophic spore formers isolated from raw milk. These are most abundant in raw milk from June to October (McKinnon and Pettipher, 1983; Phillips and Griffiths, 1986b; Stewart, 1975). Therefore, more of these organisms will gain entry into the pasteurized product at this time of year.

Spores of some strains of psychrotrophic *Bacillus* can exist as either fast or slow germinating types depending on their origin (Hutchinson, 1974; Labots and Hup, 1964; Labots et al., 1965; Stadhouders et al., 1980; Stewart, 1975; Wilkinson and Davies, 1973). In general, spores isolated from soil and dung germinate more rapidly than those isolated from milk and milking equipment. Strains isolated from pasteurized milk are mainly of the fast germinating variety (Davies, 1975b), and the proportion of fast to slow germinating types present in milk may have a bearing on the rate of spoilage (Franklin, 1967; Labots et al., 1965). If the principal source of psychrotrophic spore formers found in milk is soiled teat surfaces (McKinnon and Pettipher, 1983), then the ratio of fast to slow germinating spores may vary with season. Equivocal evidence was produced to show that fast germinating spores predominated in raw milk during the summer–autumn period (Phillips and Griffiths, 1986b). This could have a bearing on the seasonal variation in spoilage of pasteurized products by *Bacillus* spp. Furthermore, a substance that promotes germination of spores of some psychrotrophic bacilli has been isolated from milk (Hutchinson, 1974, 1975; Wilkinson and Davies, 1973, 1974). Its activity was related to somatic cell count (Davies, 1977). Hence, the concentration of this germinant can change with season (Davies and Wilkinson, 1977) and is highest from June to August (Phillips and Griffiths, 1986b).

Evidence has been presented to show thermoduric, psychrotrophic bacteria enter pasteurized goods after the heat treatment step (Coghill, 1982; Davies, 1975b; Phillips and Griffiths, 1986b) and are introduced in greater numbers in the summer and autumn. There is, however, little variation in initial numbers of psychrotrophic spore formers in pasteurized milk with high and low levels of postpasteurization contamination; average counts of these bacteria were 16 and 31 spores/L respectively.

There is little information concerning the spoilage of pasteurized milk by heat-resistant psychrotrophs other than *Bacillus* spp. Samples of commercially pasteurized milks stored at 6°C are often found to contain appreciable numbers of micrococci and streptococci. Milks of one producer studied by our laboratory frequently contain psychrotrophic streptococci that can cause spoilage within 7 days at 6–7°C. At spoilage, they account for 90% of the microflora. It is unclear whether these organisms are thermoduric, and are derived from the raw milk, or are present as contaminants introduced after processing. Seiler et al. (1984) observed no significant growth of *Microbacterium lacticum* that had survived pasteurization in milks stored at 5, 8, or 11°C.

6.1. Elimination of Heat-Resistant Psychrotrophic Bacteria from Pasteurized Milk Products

Various methods have been tried to control the growth of heat-resistant bacteria in pasteurized dairy products. To date, no really effective method to achieve this goal has been devised. Some of the techniques used are outlined below.

6.1.1. Bactofugation

Bactofugation, a process designed to separate microorganisms from milk by centrifugation, was originally intended to prolong the shelf life of market milk. Although more than 100 of these units are in operation, mainly in Europe, very few are used for this purpose. The increase in shelf life of bactofuged market milk is less than expected due to rapid growth of the remaining flora and recontamination of the milk. It is claimed that about 95% of the aerobic spore-forming population of raw milk can be removed by bactofugation (Sillén, 1987). No evidence has been produced to show the effect of bactofugation on spores of psychrotrophic species.

6.1.2. Antibacterial Agents

Several additives are active against spore-forming bacteria, and a number have been used experimentally to prolong the shelf life of pasteurized milk. Nisin, for example, is a polypeptide antibiotic that is a highly effective inhibitor of spore formers in foods including processed cheese, evaporated milk and clotted cream (Fowler, 1979). It is not a permitted additive for pasteurized dairy products. Pilot plant studies have shown that addition of nisin at concentrations of 25 U/g or above to pasteurized cream (48% fat) prior to pasteurization resulted in no significant bacterial growth in creams over a 38-day period at 10°C (Phillips et al., 1983a). These creams were essentially free of contamination with Gram-negative bacteria. Unfortunately, if Gram-negative organisms were present there was a decrease in shelf life of the cream. This was caused by stimulation of the growth of Gram-negatives by the nisin. The average generation time for four pseudomonads was reduced by more than 2 hr in the presence of nisin. This makes the use of nisin impractical for pasteurized products unless the absence of post-heat treatment contamination can be assured.

The use of other inhibitors has been suggested, but the "fresh" image of these foodstuffs may be compromised by their incorporation.

6.1.3.   Use of Naturally Occurring Compounds in Milk

Some compounds that occur naturally in milk may have a role in controlling the growth of bacteria, including psychrotrophic bacteria in milk (Banks et al., 1986). Lysozyme is present in raw milk (Chandan et al., 1964) and survives pasteurization temperatures (Griffiths, 1986). It is an enzyme that breaks down the cell wall of Gram-positive bacteria. Lysozyme has been used to prevent the late blowing of cheeses by clostridial spores (Carini et al., 1985). Its use in extending the shelf life of pasteurized products has not been investigated fully, but its application may be limited by economic considerations. The enzyme is reportedly active only when the spore population is low (about 3 spores/10 mL of milk) (Stadhouders et al., 1985).

The composition of milk could play a part in determining the keeping quality of pasteurized goods manufactured from it. We have conducted experiments that involved addition of spores of a *Bacillus cereus* strain to raw milks with high (37.2%) and low (27.8%) unsaturated fatty acid composition. These milks were obtained by manipulation of the diet of cows. The shelf life of cream manufactured from milk high in unsaturates was 2.5 days longer at 8°C than for cream low in unsaturated fatty acids and no deleterious flavors were observed in the former. The effect of varying milk composition by feeding on the microflora has not been studied. The presence of a factor(s) in milk that promotes germination of spores of some *Bacillus* spp. has been mentioned in Section 6 of this review. Studies in our laboratory indicate that augmentation of milk containing spores with this factor will accelerate germination. The germinated cells lose their heat resistance and can be killed by normal pasteurization procedures.

6.1.4.   Novel Heat Treatments

Spores may be induced to germinate by sublethal heat treatments (Keynan and Evenchik, 1960). Several workers have noted that increasing the temperature of pasteurization of both milk (Franklin, 1969; Galesloot, 1953, 1955; Kessler and Horak, 1984; Schröder and Bland, 1984) and cream (Brown et al., 1980) resulted in a decrease in keeping quality. The effect is most noticeable at temperatures of 78°C and above (Burton, 1986). The reduction in keeping quality is primarily due to the activation of spores of *Bacillus* spp. at the higher temperatures (Wilkinson and Davies, 1973). These then germinate and grow. Other factors such as the destruction of naturally occurring inhibitor systems at the higher temperatures may also be important (Auclair, 1954; Schröder and Bland, 1984). Researchers have tried to capitalize on this heat activation phenomenon to devise processing conditions that are optimal for the activation of spores routinely found in milk. Following a period of time to allow for germination, the germinated cells can be destroyed by a further heat treatment. Mikolajcik and Koka (1968) found a 2-hr interval suitable for the germination of heat-shocked spores of *B. cereus* in pasteurized milk. Laboratory studies on several strains of *Bacillus*, including some psychrotrophic strains, indicate that temperatures between 105 and 125°C for very short holding times (approximately 1 sec) are effective in activating spores of the majority of strains tested (Guirguis et al., 1983; Martin et al., 1966). Guirguis et al. (1983) recommended an optimum activation temperature of 115°C for 1 sec. This activated 8 of the 11 strains of *Bacillus* tested. Repetition of these experiments on a pilot plant gave different results (Griffiths et al., 1986b, c). Instead of activation at 115°C, there was an extremely effective "kill" of the spores at this temperature. Care should therefore be taken when extrapolating results achieved in the laboratory to a commercial system.

A modified Tyndallization process was ineffective in improving the shelf life of either milk or cream (18% fat) (Brown et al., 1979). The procedure involved two heat treatments at 72°C for 15 sec interspersed with anaerobic storage for 6 hr to allow germination of heat-activated spores. However, the activation temperature used in this study was too low to produce a significant effect on *Bacillus* spores.

Germination of heat-activated spores of *B. cereus* in skim milk was stimulated by previous growth of *Ps. fluorescens* and *Ps. fragi* (Overcast and Atmaram, 1974). Counts of pseudomonads in excess of 2.5×10^4 cfu/mL are required before these germinants are produced (Mikolajcik and Simon, 1978). No relation was found between psychrotrophic growth in raw milk and subsequent growth in pasteurized milk (Overcast and Adams, 1966; Watrous et al., 1971). High levels of *Ps. fragi* in milk, far from having a stimulatory effect, in fact, had an inhibitory effect on subsequent growth of psychrotrophs (Overcast and Adams, 1966).

7.   IMPORTANCE OF HEAT-RESISTANT BACTERIAL ENZYMES IN PASTEURIZED PRODUCTS

It has been stated earlier in this chapter that if the count of Gram-negative psychrotrophs in milk exceeds 5×10^6 cfu/mL detectable amounts of exocellular enzymes can be produced. These enzymes are extremely thermostable (Tables 10A and 10B), and, if present in raw milk, will survive pasteurization remaining active in the finished product. Indirect evidence that heat-resistant enzymes can be involved in the deterioration of short shelf life dairy commodities was produced by Patel and Blankenagel (1972). Flavor defects were encountered in pasteurized milk if bacterial counts in the raw milk were greater than 5×10^6 cfu/mL. However, it was not determined whether the taints were present in the raw milk prior to heating. In most cases, raw milk is processed before synthesis of enzymes becomes apparent. Nevertheless, there will be occasions

TABLE 10A Heat Resistance of Bacterial Extracellular Enzymes: Lipases

Organism	Heating Medium	d-Value[a] (min) for		Temperature (°C)	$Z_{d\text{-value}}$[b] (°C)	Range	Reference
		First Inact.	Second Inact.				
Pseudomonas fluorescens							
AFT36	Synthetic milk salts buffer	7.50	Absent	70	42.3	100–150	Fox and Stepaniak (1983)
22F	Skim milk	0.23[c]	298.9[c]	74	4.8[d]	50–63	Driessen (1983)
31H	Skim milk	1.67	66.7	74	—	—	Driessen and Stadhouders (1974)
AFT29	Synthetic milk salts buffer	0.14	Absent	70	38.0	80–150	Dring and Fox (1983)
Pseudomonas putrefaciens							
R48	Skim milk	0.74	Absent	74	—	—	Driessen and Stadhouders (1974)
Pseudomonas fragi							
14-2	Skim milk	0.27	17.5	74	—	—	Driessen and Stadhouders (1974)
Pseudomonas spp.							
21B	Skim milk	54.0[c]	Absent	74	55.2	—	Kishonti (1975)
MC50	Distilled water	40.0	Absent	100	36.0	100–150	Adams and Brawley (1983)
Alcaligenes viscolactis							
23a1	Skim milk	0.56	Absent	74	—	—	Driessen and Stadhouders (1974)
Alcaligenes sp.							
23a2	Skim milk	1.6	16.4	74	—	—	Driessen and Stadhouders (1974)
Achromobacter sp.							
230	Skim milk	33.0	Absent	74	—	—	Driessen and Stadhouders (1974)

[a] d-value is the time of heating, at a particular temperature, to reduce activity to 10% of its initial value.
[b] $Z_{d\text{-value}}$ is the change in temperature yielding a 10-fold change in d-value.
[c] Calculated and/or extrapolated.
[d] First activation rate.

TABLE 10B Heat Resistance of Bacterial Extracellular Enzymes: Proteinases

Organism	Enzyme	Heating Medium	d-Value[a] (mins)	Temperature (°C)	$Z_{d\text{-value}}$[b] (°C)	Range	Reference
Pseudomonas fluorescens							
P26	Protease	Whey	611.0	71.4	n.d.	—	Meyerhofer et al. (1973)
AFT36	Proteinase	Synthetic milk salts buffer	219.0	70.0	31.9	70–150	Stepaniak and Fox (1983)
OM2	Proteinase	SMUF[d]	41.0	74.0	62.5[c]	74–140	Mitchell et al. (1986)
OM41	Proteinase	SMUF	157.0	74.0	39.0[c]	74–140	
OM82	Proteinase	SMUF	190.0	74.0	32.5[c]	74–140	
OM186	Proteinase	SMUF	38.5	74.0	36.5[c]	74–140	
OM227	Proteinase	SMUF	0.02	74.0	n.d.	—	
OM228	Proteinase	SMUF	13.5	74.0	32.0[c]	74–140	
Pseudomonas spp.							
21B	Protease	Skim milk	160.0	74.0	44.5[c]	—	Kishonti (1975)
MC60	Protease	Casein solution	304.0[c]	74.0	32.5	110–150	Adams et al. (1975)
AFT21	Proteinase I	SMUF	149.0	70.0	32.0[c]	70–150	Stepaniak and Fox (1985)
	Proteinase II	SMUF	118.0	70.0	27.0[c]	70–150	
	Proteinase III	SMUF	239.0	70.0	30.3	70–150	
Serratia marcescens							
OM1191	Proteinase	SMUF	13.0	74.0	37.0[c]	74–140	Mitchell et al. (1986)
OM1192	Proteinase	SMUF	5.5	74.0	30.0[c]	74–140	

[a–c]See Table 10A.
[d]SMUF, synthetic milk ultrafiltrate (Jenness and Koops, 1962).

when raw material containing milk-degrading enzymes of bacterial origin will be processed.

7.1.   Proteases in Pasteurized Products

Most of the work on the effects of protease on liquid milk has been carried out on UHT product (Fairbairn and Law, 1986). UHT milks appear to be more susceptible to proteolysis than pasteurized milks (McKellar, 1981) and little is known about the importance of residual protease in pasteurized goods. The enzyme produced by several *Pseudomonas* spp. is active at refrigeration temperatures. *Ps. fluorescens* strains AR11, P1, AFT36, and OM82 synthesized proteases that retained 33, 31, 16, and 6% of their maximum activity respectively at 4°C (Alichandis and Andrews, 1977; Mitchell et al., 1986; Stepaniak et al., 1982b, 1987a). The proteases of several other pseudomonads retained about 20–40% of maximum activity at 4–7°C (Gebre-Egziabher et al., 1980; Patel et al., 1983; Stepaniak and Fox, 1985). The proteases of pseudomonads are not denatured by prolonged refrigerated storage at 2°C, and, allowing for the reduced reaction rate, casein is hydrolyzed to the same degree at 2 as at 30°C (Juffs and Doelle, 1968). Not all proteases produced by psychrotrophic pseudomonads maintain significant activity at low temperatures. Malik and Mathur (1984) showed that only about 2.5% of activity remained at 5°C for the enzyme from *Pseudomanas* sp. B-25.

Milk proteins are usually attacked by bacterial proteases in the sequence κ-casein > β-casein > α-casein (Fairbairn and Law, 1986) but other degradation patterns are not uncommon (Suhren, 1983). Proteolysis of milk leads to gelation (Fairbairn and Law, 1986; McKellar, 1981) and the production of unclean and bitter off flavors in liquid milk (Mayerhofer et al., 1973; Torrie et al., 1983) and cream (Griffiths et al., 1981a). A correlation exists between the flavor score of pasteurized milk and the psychrotroph count of the raw milk from which it is manufactured (Janzen et al., 1982). These workers also showed a linear relation between relative protease activity and flavor score for pasteurized skim and whole milk. White and Marshall (1973) conducted experiments in which they added 7.9 units/mL of protease to milk 12 hr before and immediately after pasteurization. They also added 1×10^5 cells of *Ps. fluorescens* to the milk 12 hr before heat treatment. Proteolysis in the milks containing enzyme or pseudomonads was greater than in control milks to which no additions had been made. Neither samples with protease or bacteria added before pasteurization could be differentiated from the negative control by organoleptic assessment. The taste panel was able to distinguish control samples from milks with added protease after a week's storage at 7°C. McKellar (1981) found that proteolysis could be detected by assay of free amino acids before organoleptic defects arose. The amount of purified protease required to produce flavor defects in milk may be as low as 0.2 units/mL (Mayerhofer et al., 1973). However, taints were not apparent until after 30 days of storage at 4°C. The period of storage of pasteurized products is too short to result in noticeable off-flavors induced by residual protease

(Driessen, 1983) because advanced proteolysis is required to create the small peptides responsible for bitterness (Visser et al., 1975).

Methods for assaying protease activity in milk have been reviewed (Mottar and Driessen, 1987; Seminega et al., 1985), and, recently, a sensitive ELISA system has been described (Stepaniak et al., 1987b). The immunological properties of proteases from pseudomonads differ (Jackman et al., 1983; Symons et al., 1985), which will make the adoption of an ELISA technique for routine assay difficult. There is still no ideal method for quantifying protease activity in dairy products.

7.1.1. Removal of Protease Activity from Pasteurized Products

The proteases of a number of Gram-negative psychrotrophs are more susceptible to heat denaturation at temperatures around 55°C than at those adopted during HTST pasteurization (Barach et al., 1976; Dalaly and Abbo, 1982; Driessen, 1983; McKellar and Cholette, 1983; Patel et al., 1983; Stepaniak et al., 1982b). This low temperature inactivation appears to be a two-step process involving a conformational change followed by aggregation with casein micelles to form an inactive enzyme complex (Barach et al., 1978). Autolysis of the enzyme may be implicated as well (Barach et al., 1978; Richardson, 1981; Stepaniak and Fox, 1983; West et al., 1978). The use of a treatment of 55°C for 1 hr to eliminate proteases from processed milk has been proposed (West et al., 1978), and incorporation of this heating step into a UHT operation resulted in a substantial increase in shelf life. Unfortunately, proteases from Gram-negative psychrotrophs vary considerably in their sensitivity to this treatment (Griffiths et al., 1981b; Kocak and Zadow, 1985; Marshall and Marstiller, 1981). Also, the order in which the heat treatments are carried out is important for inactivation of some enzymes (Griffiths et al., 1981b; Stepaniak and Fox, 1983). This variability in response makes commercial exploitation of the technique untenable.

7.2. Lipases in Pasteurized Products

Although there are very few reports on the effects of lipases on the quality of pasteurized dairy products, as with proteases, several psychrotrophic bacteria produce heat-resistant lipolytic enzymes. These may remain active after pasteurization (Law, 1979; Stead, 1986). Lipases from *Pseudomonas fluorescens* (Andersson, 1980; Christen and Marshall, 1984a; Dring and Fox, 1983; Fox and Stepaniak, 1983; Stepaniak et al., 1987a). *Ps. fragi* (Alford and Pierce, 1961; Nashif and Nelson, 1953b), other *Pseudomonas* spp. (Alford and Elliott, 1960), and *Acinetobacter* spp. (Breuil and Kushner, 1975) all retain activity at refrigeration temperatures. A relation exists between psychrotrophic counts of raw milk and residual enzyme activity in UHT milks manufactured from it (Mottar, 1981). When lipase from *Ps. fluorescens* was added to UHT milk at levels commonly found in raw milk (0.3 units/mL), the heat-treated milk became rancid within 5–8 days of storage at 8°C.

Creams prepared from milk containing high concentrations of lipase due to the growth of psychrotrophs developed rancidity on storage (Davis, 1981). Addition of lipase from *Ps. fragi* to cream resulted in a significant pH drop caused by lipolysis (Nashif and Nelson, 1953b). The problem of lipolysis may be exacerbated because of the accumulation of lipase in the fat phase on separation of milk (Downey, 1980; Kishonti and Sjöström, 1970; Stead, 1983). The action of lipase may also be enhanced by the presence of glycosidases, phospholipases, and proteases produced by psychrotrophs (Alkanhal, 1985; Chrisope and Marshall, 1976; Griffiths, 1983; Marin et al., 1984; Owens, 1978). The combined action of these enzymes can disrupt the fat globule membrane, making the fat more accessible to lipase action. However, only native milk lipase and not *Ps. fluorescens* lipase was affected by this synergistic action (Griffiths, 1983). FitzGerald and Deeth (1983) presented results that supported the view that the milk fat globule membrane offers little resistance to bacterial lipase, but it is not known if the enzyme preparations used in this work contained significant amounts of protease and/or phospholipase.

Shelley et al. (1986) concluded that the amount of lipase activity required to cause spoilage in a dairy product was dependent on the storage conditions and heat stability of the lipase originally present. Low activities may be significant in UHT milk as it is stored at ambient temperatures for many months. A much higher concentration of enzyme would be required for spoilage of short shelf life products stored under refrigeration. Bacterial lipolytic spoilage should not be a problem in refrigerated liquid milk products unless they have been grossly mishandled or the raw milk from which they were made was stored for long periods (Driessen, 1983; Shelley et al., 1986).

Assay methods for lipolytic enzymes have been reviewed by Lawrence (1967b) and Stead (1986) and will not be considered in this article.

7.2.1. Removal of Lipase Activity from Pasteurized Products

As with proteases, lipases from psychrotrophic bacteria are reputedly less stable in the temperature range 50–60°C (Christen and Marshall, 1984b; Driessen, 1983; FitzGerald et al., 1982; Fox and Stepaniak, 1983; Griffiths et al., 1981b). This biphasic inactivation has been attributed to the presence of two lipases of different thermostability (Cogan, 1977; Driessen and Stadhouders, 1974) or association with components of skim milk (Andersson et al., 1979). There may also be two different denatured states of lipase (Swaisgood and Bozoglu, 1984), one representing low-temperature inactivation (possibly involving active site histidine groups) (Christen and Marshall, 1985) and the other a more extensive high-temperature denaturation. Bucky et al. (1987a,b) have proposed the use of low-temperature heat treatments at 60°C in combination with UHT to deactivate bacterial lipases. There exists considerable between and within species variation in the susceptibility of bacterial lipases to low-temperature inactivation (FitzGerald et al., 1982; Griffiths et al., 1981b). Thus, the process may not have a widespread practical application. Furthermore, bacterial lipases show considerable ability to catalyze lipolysis of milk fat globules at temperatures around 55°C (Deeth and FitzGerald, 1983). Hence prolonged storage of whole milk or cream at this temperature would result in substantial lipolysis.

8. PROBLEMS ENCOUNTERED DURING THE DISTRIBUTION OF PASTEURIZED PRODUCTS

There is little doubt that adequate temperature control is the one overriding factor to be taken into account for maintenance of product quality during distribution from the processing site to the retailer (or consumer). The effect of temperature on the keeping quality of pasteurized products has been well documented through the years. The pronounced drop in expected shelf life with increasing temperature of storage is illustrated with data produced for good and bad quality pasteurized milk in our laboratory (Figure 3). Whereas the shelf lives of milks with high and low level contamination are significantly different at low temperatures, as the storage temperature increases, the shelf lives become similar.

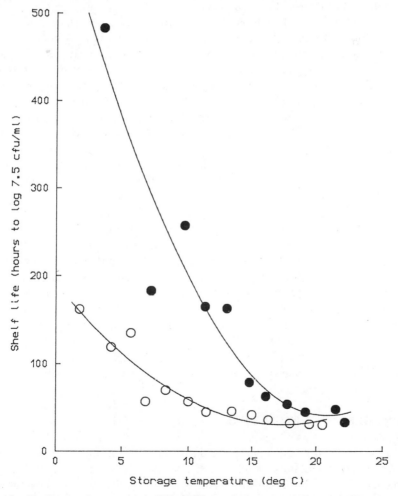

Figure 3.   The decrease in expected shelf life with increasing temperature for good (●) and poor (○) quality pasteurized milk.

Results achieved with pasteurized creams showed a 50% decrease in shelf life on raising the storage temperature from 4 to 8°C (Airey, 1978; Cox, 1975). Chandler and McMeekin (1985) have proposed a mathematical model for predicting the variation of growth rate of bacteria with temperature in pasteurized milk. These workers studied only five samples of milk that were of relatively poor quality. Their predictive model may not be suitable for use with good quality milks.

Ideally, short shelf life dairy products of good overall quality should be refrigerated as quickly as possible after the filling stage. Indeed, it has been recommended that the product should not be transported until its temperature is at or below 5°C since the first refrigeration step is probably the most important in the whole chain (Hamman, 1979). On leaving the factory, pasteurized products should be delivered to the retail outlet, and on to the consumer, as quickly as possible while maintaining as low a temperature as possible without prejudicing the physical character of the commodity.

A substantial (48.3%), but decreasing, volume of pasteurized milk is still delivered to the doorstep in the United Kingdom. This system provides many opportunities for temperature abuse of the product such as at the retail depot and, not least, on unrefrigerated delivery vehicles. The milk, however, tends to be consumed soon after delivery, and a long shelf life is not necessarily a prerequisite of this system. Blake (1979), in a review of the storage and transportation of pasteurized milk, considered low-temperature bottling, rapid bulk distribution, and overnight storage at not more than 7.2°C ample protection for milk distributed by doorstep delivery.

For distribution via large retail outlets, more stringent precautions should be taken to ensure adequate keeping quality for the consumer. The purchasing patterns for liquid milk are changing in Britain and are becoming more analogous to those seen in the United States. More and more people are buying milk in bulk during one or two shopping expeditions a week. Thus, an increased shelf life is essential to meet these new demands. Pasteurized products are transported greater distances and these will increase with the introduction of free trade within the European Economic Community. There are immense benefits to be gained in product quality from operating an efficient refrigerated storage and distribution network, and refrigerated transport plays a major role in this area (Nisbet, 1979). These vehicles are mostly designed to hold goods at a given temperature and not reduce the temperature of products in transit. Therefore, to ensure that the cold chain is effective, temperature checks on the product should be carried out at frequent intervals throughout the distribution chain.

9.   PROBLEMS ENCOUNTERED AT RETAIL OUTLETS

In a survey of the quality of pasteurized products at point of sale, 54% of retail packs of pasteurized milk purchased from small retailers had psychrotroph counts in excess of 1×10^3 cfu/mL at sale (Griffiths et al., 1985b). The figures for pasteurized milks from doorstep deliveries and large supermarkets were 31 and

19%, respectively. The poor performance of the small retailer was probably due to inadequate handling of the perishable commodity. For example, many small shops in the United Kingdom stock unrefrigerated pasteurized milk. Also, some samples of milk may be stored in warehouses at ambient temperatures before transfer to chilled cabinets. The effect of such breaks in the cold chain are not well quantified. The milk delivered to the doorstep undergoes a period of storage at ambient temperature, but the period of abuse is generally shorter than that observed in some of the small shops. Griffiths et al. (1985b) observed psychrotroph counts in excess of 1×10^5 cfu/mL in 4% of milks sampled from large supermarkets, which indicates that deficient handling is not confined to the small retailers. However, milk supplies in this survey varied in initial quality depending on hygiene standards within the manufacturing plant. The influence of this factor could not be easily assessed. Nevertheless, this work points out the pitfalls encountered when milk is badly handled and steps are being taken within the United Kingdom to ensure that milk displayed for sale in shops is held at chill temperatures.

The use of chilled selling units does not guarantee that milk will be held at low temperatures. Unless these are monitored regularly, faults can occur or the cabinets can be stocked incorrectly. As many as 36% of cabinets tested in a survey in the United States (Bodyfelt and Davidson, 1975) and 17% tested in the United Kingdom (Hamman, 1979) did not have an adequate performance. To safeguard against a breakdown in the operation of the cold chain, Hamman (1979) proposed at least four checks of product temperature and three of storage temperature. These should be (1) filling temperatures of the product, (2) after factory cooling, (3) on arrival at the distribution depot, (4) on arrival at the retail outlet, (5) temperature of the distribution cold store, (6) temperature of the retail cold store, and (7) temperature of retail selling units. It is strongly recommended that dairy products should be transported and displayed at temperatures not exceeding 5°C.

10.   THE ROLE OF THE CONSUMER

The final link in the cold chain involves the consumer, and, in some respects, may be the worst of all. The products may be stored at ambient temperatures, for example in cars, for several hours before being transported home. Domestic refrigerators may have temperatures set above those recommended for dairy products. At this stage no control can be exerted on the way the product is handled apart from an attempt to educate the customer by explicit instructions on the retail pack. In the United Kingdom this usually involves a sell-by code with a customer instruction (e.g., "Keep refrigerated—eat within two days of purchase"). However, this is not always a safeguard of product quality. Griffiths et al. (1985b) showed that 3% of cartons of pasteurized milks purchased from retail outlets had a residual sell-by life of more than 7 days, whereas over a quarter of all the samples purchased had acceptable shelf lives in excess of 7 days. In contrast, the results with pasteurized cream were much poorer. Of the cream samples tested, 80% had a manufacturers' anticipated shelf life of 7 days or more, but only 60% actually fulfilled this criterion. Consumers would, therefore, be disappointed with the keeping quality of the creams purchased in one out of every four purchases.

11.   ENVIRONMENTAL HEALTH ASPECTS OF PASTEURIZED DAIRY PRODUCTS

As well as expecting a product that has satisfactory organoleptic and keeping quality characteristics, the consumer expects pasteurized dairy products to be safe to eat. The incidence of milk-borne disease in the United Kingdom and the United States has recently been reviewed (Bryan, 1983; Galbraith and Pusey, 1984; Sharp et al., 1985; Sharp, 1987). The implementation of eradication schemes means that tuberculosis and brucellosis no longer feature as milk-borne diseases in Britain. Since the introduction of compulsory pasteurization of milk in Scotland in 1983, there have been no outbreaks of milk-borne infection in the general population. Some reports of infection due to milk have been reported in farming communities that still have access to raw milk. However, sporadic outbreaks of milk-borne disease associated with ingestion of pasteurized products have been recorded in other parts of the United Kingdom and in the United States. These are invariably linked with postpasteurization contamination. Todd (1987) stated that 1.6% of the incidents of food poisoning outbreaks in Canada in 1980 were due to milk, but it is not clear whether these were caused by the consumption of raw or pasteurized milk.

The ability of pathogens associated with milk and milk products to survive pasteurization and grow at refrigeration temperatures is outlined in Table 11. The changing nature of the diseases attributed to milk consumption can be seen from Table 12. The incidence of salmonellosis and campylobacteriosis has increased and other causative organisms such as *Yersinia enterocolitica* and *Listeria monocytogenes* have been identified (Bryan, 1983; Galbraith and Pusey, 1984). The food-borne pathogens of recent concern have been reviewed by Doyle (1985). Organisms that have been linked to recent outbreaks of food poisoning following ingestion of pasteurized milk products are discussed below.

11.1.   Salmonellosis

The growth of *Salmonella* spp. is arrested between 5 and 10°C (Riemann, 1969), although the organism can remain viable in milk for periods up to 4 days (Velimirovic, 1984). *Salmonella* spp. have been isolated from about 5% of raw milk supplies in the United States (McManus and Lanier, 1987). Of 20 outbreaks of salmonellosis reported in England and Wales during 1983–1984, in only one, involving five people, was pasteurized milk implicated (Sharp et al., 1985). This milk was contaminated after heat treatment as proper pasteurization is known to

TABLE 11 Pathogens Associated with Pasteurized Milk

Organism	Survival of Pasteurization	Growth at <6°C in Milk	References
Staphylococcus aureus	−	−	Evans et al. (1970); Bergey (1986)
Salmonella typhi	−	?	Evans et al. (1970)
Salmonella typhimurium	−	?	Bradshaw et al. (1987)
Escherichia coli	−	+	Evans et al. (1970); Palumbo (1986)
Yersinia enterocolitica	−	+	Lovett et al. (1982); Francis et al. (1980)
Campylobacter jejuni	−	−	Waterman (1982); Doyle and Roman (1985)
Clostridium spp.	+	−	Bryan (1983)
Bacillus cereus	+	+	Griffiths et al. (1986c)
Listeria monocytogenes	−	+	Bradshaw et al. (1985); Bergey (1986)

TABLE 12 Incidence of Bacterial Infections Associated with Milk and Milk Products[a]

Bacterial Infection	1940–1949 E&W[b]	USA[b]	1950–1959 E&W	USA	1960–1969 E&W	USA	1970–1979 E&W	USA	1980–1982 E&W	USA	1983–1985 E&W	USA
Disappearing and Declining Infections												
Brucellosis	1.35	8	12.0	4	15.7	9	2.81	1	0.7	0	0	0
Diphtheria	0.06[c]	1	0	0	0	0	0.13[d]	0	0.1[d]	0	0	0
Dysentery	4.57	4	2.4	3	0	0	0	0	0	0	0	0
Paratyphoid B	0.68	n.k.[b]	0.7	n.k.	15.3	n.k.	0	n.k.	0	n.k.	0	0
Scarlet fever/ streptococcal sore throat,	0.68	27	0.1	8	0	0	0	0	0	0	1.2	0.09
Streptococcus zooepidemicus											0.2	4.96
Staphylococcal intoxication	0.35	26	40.9	50	6.9	30	0	5	0	0	0	0
Tuberculosis (bovine)	88.77	n.k.	4.8	n.k.	1.0	n.k.	0.22	n.k.	0	n.k.	0	0
Typhoid fever	0.16	17	0	3	0	0	0	0	0	0	0	0
Unknown aetiology	0.48	26	10.5	17	0	31	0	41	0	0	0	0
Rare Milk-Borne Infections												
Escherichia coli enteritis	0	0	0	0	0	0	0.04	1	0	0	0	0
Clostridium welchii	0	n.k.	0	n.k.	0	n.k.	0	n.k.	5.2	n.k.	0	0
Q fever	0	0	0.6	1	2.4	0	0.26	0	0	0	0	0
Possibly Increasing Milk-Borne Infections												
Salmonellosis	2.90	7	28.0	21	58.7	28	13.94	41	22.9	50	63.5	92.52
Campylobacteriosis	n.k.	<1	n.k.	0	n.k.	0	82.60	3	71.1	40	34.6	0.58
Yersiniosis	n.k.	n.k.	n.k.	n.k.	n.k.	n.k.	n.k.	1	n.k.	n.k.	0.5	0.99
Listerosis	n.k.	n.k.	n.k.	n.k.	n.k.	n.k.	n.k.	n.k.	n.k.	n.k.	0	0.86
Totals	37177	n.k.	4157	n.k.	1017	n.k.	4585	n.k.	1480	n.k.	980	17324

[a]Percentage of reported milk-borne infections per decade.
[b]E & W, England and Wales (Galbraith and Pusey, 1984); USA, United States of America (Bryan, 1983); n.k., not known.
[c]Corynebacterium diphtheria.
[d]C. ulcerans.

destroy these pathogens. Most of the outbreaks of salmonellosis associated with pasteurized milk have been shown to be the result of inadequate pasteurization or contamination after heat treatment (Marth, 1969).

In the space of less than 1 month in 1985 more than 16,000 confirmed cases of salmonellosis occurred in Illinois and surrounding states, the largest outbreak of its kind ever recorded in the United States (Lecos, 1986). Low-fat pasteurized milk (2% butterfat) was implicated as the source of this infection, which was caused by a rare strain of *S. typhimurium* (Margolis, 1985). A comprehensive survey of the processing plant showed that a design fault could have allowed unpasteurized milk to enter pasteurized skim milk lines and this may have been responsible for the contamination with salmonella (Lecos, 1986). Subsequent experimentation on the strain involved confirmed that isolates would not survive proper pasteurization (Bradshaw et al., 1987).

11.2. *Yersinia enterocolitica*

The frequent association of *Yersinia enterocolitica* with raw milk (Hughes, 1979; Lee, 1977; Schiemann and Toma, 1978) coupled with its ability to grow at refrigeration temperatures (Binet, 1983; Francis et al., 1980; Stern et al., 1980), makes this bacterium a potential hazard for processors. Although *Y. enterocolitica* coule be isolated from 48.1% of raw milk samples tested, all the isolates were environmental and nonvirulent (McManus and Lanier, 1987). In addition this organism is unable to grow competitively in milk (Stern et al., 1980).

Yersinia enterocolitica has been isolated from pasteurized milk (Black et al., 1978; Hughes, 1980; Moustafa et al., 1983; Schiemann, 1978). Chocolate milk was incriminated in the first major outbreak (Black et al., 1978). The milk apparently became contaminated after pasteurization. Contaminated pasteurized milk has also been implicated in food poisoning caused by *Yersinia* spp. in the United Kingdom (Sharp et al., 1985). Routine samples of milk from retail outlets were positive for *Yersinia*, and a bulk holding tank at the processing site was suggested as a source of contamination. Samples of milk were negative for the organism after the tank had been properly cleaned. No other cases of infection were then reported. Pasteurized milk has been implicated as the vehicle for yersiniosis in the United States (Bryan, 1983; Tacket et al., 1984), but again the organism was introduced as a contaminant, probably from milk crates, after heat treatment (Aulisio et al., 1982; Stanfield et al., 1985). The organism was able to survive more than 21 days on the outside of milk cartons stored at 4°C.

Hughes (1979) suggested that certain strains of *Y. enterocolitica* could survive pasteurization. When detailed work was carried out using the supposedly heat-resistant strains, pasteurization was shown to destroy these bacteria (Lovett et al., 1982). Temperatures up to 62.8°C were used in this work, but extrapolation of the results indicates a d-value of 0.2 sec at 74.4°C for the most heat-resistant species studied. Similar results (d-value = 0.1 sec at 73.6°C) for the heat resistance of *Y. enterocolitica* in skim milk were achieved by Hanna et al. (1977).

11.3. Campylobacter

In the period between 1978 and 1982, 25 outbreaks of enteritis were recorded in England and Wales due to *Campylobacter* spp. where milk was strongly implicated as the vector (Galbraith and Pusey, 1984). Of these, three were associated with ingestion of pasteurized milk. All three could be explained on the basis of inadequate pasteurization, contamination from unsterile containers, or addition of raw milk to pasteurized material. No cases of *Campylobacter* enteritis associated with pasteurized milk had been reported in the United States up to 1982 (Bryan, 1983).

Campylobacter jejuni does not grow at temperatures below 30°C (Doyle and Roman, 1981). Neither does it survive well in milk at 4°C (Boer et al., 1984; Doyle and Roman, 1982; Koidis and Doyle, 1984), although there is an isolated report that the organism survives in milk for up to 112 days at this temperature (Sandstedt, 1982). This latter work also suggested that an infective dose of only 500 organisms was required to cause illness. *Campylobacter jejuni* is isolated only from a small percentage of raw milks (Doyle and Roman, 1982; Lovett et al., 1983; McManus and Lanier, 1987). The bacterium is heat sensitive and proper pasteurization gives complete protection against the spread of *Campylobacter* by milk (Christopher et al., 1982; Doyle and Roman, 1981; Gill et al., 1981; Waterman, 1982). All these facts suggest that pasteurized dairy products are not a major source of disease caused by this organism.

11.4. Listeria monocytogenes

Listeria monocytogenes is a small, Gram-positive, non-spore-forming, aerobic, rod-shaped bacterium. The organism is the causative agent of listeriosis, which usually results in septicemia, meningitis, or abortion and occurs most commonly in the newborn or immunosuppressed patients. There is mounting, though inconclusive, evidence that zoonotic transmission (i.e., transmissable from other vertebrates to man) of *Listeria* is not an infrequent event (Hird, 1987). Ruminant animals seem to be at greatest risk and the occurrence of the disease is strongly associated with silage feeding (Fenlon, 1986). Considerable recent attention has been focused on this bacterium because of reports that pasteurized milk may contain *L. monocytogenes* and can act as a vehicle for spread of infection (Fernandez Garayzabal et al., 1986; Fleming et al., 1985).

In an outbreak of listeriosis thought to be due to ingestion of contaminated pasteurized milk, the organism was isolated from 12% of raw milks sampled at the processing plant as well as from milk filters (Fleming et al., 1985). However, the organism was of a different phage type than the epidemic isolate, and attempts to isolate the organism from the pasteurized milk were unsuccessful. Development of the illness was strongly linked to the consumption of pasteurized milk. Postpasteurization contamination was ruled out as multiple inspections of the plant did not reveal a potential source. It was also difficult to explain why only whole milk and 2% fat milk were affected and not skim milk as all were processed on the same day with the same equipment. The milks were supplied by different farm groups, which led to the suggestion that *L. monocytogenes* was heat resistant. The heat resistance, it was proposed, may be conferred by the ability of this bacterium to exist as an intracellular parasite within monocytes.

Listeria cells were found, on average, in 4.2% of raw milks from farm bulk tanks in areas of the United States, but the concentration was probably less than 1/mL (Lovett et al., 1987). Most of the strains isolated were pathogenic in adult mice. *Listeria* spp. can grow at temperatures as low as 1°C and have been shown to grow in milk and cream at 4°C (Donnelly and Briggs, 1986; Rosenow and Marth, 1987). Thus, if cells are resistant to pasteurization, then the organism poses a threat to the safety of pasteurized products. Work that supports the view that *Listeria* can survive pasteurization has been produced (Bearns and Girard, 1958). *Listeria monocytogenes* suspended in concentrate and pasteurized milk at initial levels approaching 1×10^5 cfu/mL was able to endure heating during the manufacture of nonfat dry milk (Doyle et al., 1985) and cottage cheese (Ryser et al., 1985). Doyle et al. (1987) suggest that the organism can survive the minimum HTST treatment (71.7°C, 15 sec) required by the U.S. Food and Drug Administration for pasteurizing milk. These workers used milk from cows inoculated with *L. monocytogenes*. The milks were heated in a plate-heat exchanger at temperatures from 71.7 to 73.9°C for 16.4 sec and from 76.4 to 77.8°C for 15.4 sec. A range of 1.5 to 9.2 *Listeria* cells was noted in each milk polymorphonuclear leukocyte before heat treatment in 11 of the 12 trials. There was no evidence that the organism grew within the leukocytes during refrigerated storage. *Listeria monocytogenes* was isolated from milk in 6 of the 9 pasteurization trials conducted at the lower temperature range, but in none of 3 experiments done at the higher temperatures. Raw milk used in one of the latter trials contained few *Listeria* cells, however. The authors conclude that heat resistance was conferred on the bacterial cells due to their presence in the leukocytes. Holding raw milk at 4°C for 4 days resulted in complete disintegration of the leukocytes, but experiments were not conducted on the stored milks to measure the heat resistance of free *Listeria* cells.

A series of recent papers casts doubt on the ability of *L. monocytogenes* to survive pasteurization. Cells of the organism freely dispersed in milk (Bradshaw et al., 1985; Bunning et al., 1986; Donnelly et al., 1987; Twedt, 1986), cream (Bradshaw et al., 1987), and internalized by mouse phagocytes (Bunning et al., 1986) could not endure pasteurization. Free and internalized cells had similar *d*-values at temperatures up to 68.9°C. The heat resistance of these bacteria was dependent on the method used for its determination (Donnelly et al., 1987). Other experiments using artificially and naturally contaminated milk containing up to 1×10^4 *Listeria* cells/mL showed that the organism failed to grow after pasteurization at 72°C for 16 sec (Farber et al., 1987). The majority of studies conducted on the thermal resistance of *L. monocytogenes* indicate that it can be destroyed by effective pasteurization procedures (Table 13), and its occurrence in pasteurized milk is due to recontamination. Pasteurized milks that contained *Listeria* spp. (Fernandez Garayzabal et al., 1986) also had large numbers

TABLE 13   Ability of *Listeria monocytogenes* to Survive Pasteurization

Expected Survival after Pasteurization	Heat Treatment Applied	Reference
Yes	61.7°C/35 min ($d_{61.7} = 10.9$ min)	Bearns and Girard (1958)
No	73°C/15 sec	Donker-Voet (1963)
No	70°C/20 sec	Ikonomov (1969)
No	72°C/20 sec	Obiger (1976)
No	71°C/42 sec	Stajner et al. (1979)
No	71.7°C/16 sec ($d_{71.7} = 0.8$–1.1 sec)	Bradshaw et al. (1985, 1987)
No	71.7°C/16 sec ($d_{71.7} = 0.9$ sec)	Twedt (1986)
No	71.7°C/16 sec ($d_{71.7} = 1.6$ sec free cells) ($d_{71.7} = 1.9$ sec internalized cells)	Bunning et al. (1986)
No	71.7°C/16 sec ($d_{71.7} = {<}0.6$ sec)	Donnelly and Briggs (1986); Donnelly et al. (1987)
Yes	72.4°C/16.4 sec	Doyle et al. (1987)
No	72°C/16 sec	Farber et al. (1987)

($> 1 \times 10^3$ cfu/mL) of Enterobacteriaceae present, a sure sign that these milks were subject to postprocess contamination. The primary source of *Listeria* spp. in processing plants is probably floors and floor drains and effective measures for control of this organism have been described (Coleman, 1986).

11.5.   *Aeromonas hydrophila*

Aeromonas hydrophila is one of a group of pathogens that is emerging as a food borne organism of increasing concern (Buchanan and Palumbo, 1985). The organism is capable of growth at 1°C (Eddy, 1960b), and clinical isolates can grow between 4 and 40°C (Palumbo et al., 1985). The organism can be isolated from pasteurized products (Phillips et al., 1981a), but it is doubtful whether the strains are of clinical significance. The organism is heat sensitive so its presence in heat-treated products is indicative of postpasteurization contamination.

11.6.   *Clostridium perfringens*

Spores of *Clostridium perfringens* can survive pasteurization but refrigerated storage prevents the outgrowth of spores and multiplication of vegetative cells (Bryan, 1983). An outbreak of food poisoning due to this organism was described in the United Kingdom in 1980 but was attributed to milk powder (Galbraith and Pusey, 1984).

11.7. *Bacillus cereus*

Spores of *Bacillus cereus* can survive pasteurization, germinate, and then grow in refrigerated dairy products (Griffiths et al., 1986c). Milk and ice cream have been implicated as vehicles in outbreaks of gastroenteritis caused by *B. cereus* (Gilbert, 1979). Emetic strains of the organism have been isolated from raw milk, albeit in low numbers and on a limited number of occasions (Phillips and Griffiths, 1986b). These toxigenic strains are not usually psychrotrophic and need to be ingested in large numbers for adverse affects to become apparent. Galbraith and Pusey (1984) make no reference to this organism in their article on milk-borne infectious diseases, which indicates that pasteurized milk is probably of little consequence in transmission of food poisoning by *B. cereus*.

Other strains of *Bacillus* spp. are now considered to cause human illness, but there have been no reports of pasteurized milk as a vector for their transmission.

Several other pathogenic bacteria can grow in milk and milk products, but this article has focused on those of current interest. As a result of outbreaks and problems associated with pasteurized milk over the last few years (Table 14), particularly the salmonellosis incident at Illinois, the U.S. Food and Drug Administration implemented a Dairy Safety Initiative Program (Kozak, 1987). Results of this have revealed a significant number of microbiological problems with finished dairy products, indicating postpasteurization contamination to be a widespread problem. *Listeria* and *Yersinia* spp. have been recovered from numerous plants leading to guidelines being established for the prevention of environmental contamination of dairy plants. In the United Kingdom processor awareness of the problems connected with post-heat treatment contamination has led to the setting-up of the Scottish Pasteurized Milk Testing Scheme. This will serve as a useful monitor of the incidence of recontamination, and help to maintain the high standards of safety related to pasteurized products in the United Kingdom.

12. CONCLUSIONS

Bacteria are introduced into milk from the environment at all points along the production/processing sequence from when the cow is milked to the point when the product is filled, and even, on rare occasions, after this point. Contamination can never be eliminated completely so it is important to note the critical stages. Marth (1981) has highlighted the need for concerted effort by all people involved in the production and processing of milk to maintain a high quality product.

The introduction into milk of spore-forming bacteria, particularly those able to grow at refrigeration temperatures, should be reduced to a minimum. It appears that the bulk of these organisms are introduced into milk due to soiling of the udder. Is this really the critical stage or are inadequately cleaned farm bulk tanks a more important source? Can psychrotrophic *Bacillus* spp. be more effectively controlled by modifications to udder washing procedures or adoption of novel processing techniques? These are questions that demand an answer as

TABLE 14 Some Microorganisms (and Sources) Linked with Food Poisoning in Pasteurized Products

Date	Location	Number of Persons Affected	Probable Source of Infection	Product	Microorganisms Involved	Reference
1965–66	USA (nationwide)	29+	N.K.[a]	Powdered milk	Salmonella newbrunswick	Collins et al. (1968)
1972	Rumania	270	N.K.	Milk	Bacillus cereus	Vlad and Vlad (1972)
1975	UK (England and Wales)	2	N.K.	Cream	Bacillus cereus	McSwiggan et al. (1975)
1975	USA (Louisiana)	43+	N.K.	Milk	Salmonella newport	Blouse et al. (1975)
1975	USA (Colorado)	339+	Inadequate pasteurization	Cheese	Salmonella heidelberg	Fontaine et al. (1980)
1976	USA (Onieda County, NY)	36	Chocolate syrup	Milk + added chocolate syrup	Yersinia enterocolitica	Black et al. (1978)
1968	Canada (British Columbia)			Powdered milk	Bacillus cereus	Schmidt et al. (1976)
1978	USA (Arizona)	23+	N.K.	Milk	Salmonella typhimurium	Dominguez et al. (1979)
1979	USA (Oregon)	1+	N.K.	Powdered milk	Salmonella agona, typhimurium	Furlong et al. (1979)
1979	UK (Luton, Beds)	3500	N.K.	Milk	Campylobacter sp.	Jones et al. (1981)
1980	UK (England and Wales)	30	Unsterilized containers?	Milk	Campylobacter sp.	Galbraith and Pusey, (1984)
1981	UK (Wessex)	25	Food handler	Cream	Staphylococcus aureus	Anon (1982a)
1982	UK (Horsham, Sussex)	400	Raw milk mixed in	Milk	Campylobacter sp.	Anon (1983)
1982	USA (Tennessee, Arkansas, Mississippi)	172	Mud contaminating milk crates	Milk	Yersinia enterocolitica	Aulisio et al. (1982), Tacket et al. (1984)
1982	UK (Scotland)	179	Failure to pasteurize	Milk	Campylobacter sp.	Sharp (1987), Wood et al. (1984)
1983	Canada		Inadequately heat-treated milk	Cheese	Salmonella muenster	Anon (1982b)
1983	USA (Massachusetts)	49	Postpasteurization contamination	Milk	Listeria monocytogenes	Fleming et al. (1985)
1983	UK (England and Wales)	5	Postpasteurization contamination	Milk	Salmonella spp.	Sharp (1987)
1984	Canada (Ontario)	1500+	Inadequately heat-treated milk	Cheese	Salmonella typhimurium phage type 10	D'Aoust et al. (1985)
1984	UK (England and Wales)	2	Postpasteurization contamination	Milk	Yersinia spp.	Sharp (1987)
1984	USA (Kentucky)	16	Inadequate pasteurization	Milk	Salmonella typhimurium	Adams et al. (1984)
1985	USA (Los Angeles, Orange Counties, California)	86	Inadequate pasteurization?	Cheese	Listeria monocytogenes	US PHS (1985)
1985	USA (Illinois)	14000	Postpasteurization contamination?	Milk	Salmonella typhimurium	Sun (1985)
1985	UK (England and Wales)	3	N.K.	Milk	Yersinia spp.	Sharp (1987)
1985	UK (Guernsey, Channel Isles)	50	Inadequate pasteurization	Milk	Salmonella spp.	Sharp (1987)
1985	Sweden	153	Postpasteurization contamination	Milk	Salmonella spp.	Sharp (1987)
1985	UK (England, Wales and Scotland)	48	N.K.	Powdered milk	Salmonella ealing	Anon (1986)

[a]N.K. not known.

there is evidence that the incidence of spoilage of pasteurized products by heat-resistant, psychrotrophic bacteria is increasing and, as longer shelf lives are sought, they will assume greater importance.

It is generally recognized that heat-resistant proteases, lipases, and other enzymes produced by psychrotrophs are secreted when bacterial numbers reach approximately 5×10^6 cfu/mL and above. To introduce a margin for error milk should be processed before the count exceeds 1×10^6 cfu/mL. These enzymes, if present, produce off-flavors in the raw milk that will be retained in the final product. Enzymic activity may also remain in the pasteurized milk or cream and continue to degrade the substrate. It is essential, therefore, that the quality of the raw milk be monitored throughout the period from collection at the farm to immediately before processing to ensure that the psychrotroph count remains as low as possible. Rapid bacterial enumeration techniques have a role to play in accomplishing effective monitoring. Particular attention should be paid to storage temperatures, both on the farm and at the factory. The temperature should preferably not exceed 4°C and recent work suggests that 2°C has advantages in terms of storage life. If the initial quality of the raw milk cannot be guaranteed, or if prolonged storage at the creamery becomes necessary, consideration should be given to the introduction of a thermization step. Milk producers and their representative organizations within Britain have done much in the last 5 years to improve the quality of raw milk, with the result that in the majority of cases, raw milk quality does not compromise product quality.

Arguably the crucial step in the production of pasteurized products takes place after heat treatment. If the pasteurization process is properly carried out, the product emerging from the heat exchanger will be essentially free of Gram-negative psychrotrophs and the major pathogens associated with raw milk. The only organisms to remain are heat-resistant bacteria, and only a relatively small proportion of these are able to grow at low temperatures. Pasteurized products fulfilling these criteria should have shelf lives in excess of 14 days at 6°C. However, in 80% or more of samples of pasteurized milks and creams, appreciable numbers of Gram-negative psychrotrophs are present. These severely limit the keeping quality and are introduced solely as a result of inadequate plant hygiene and bad process management. Other organisms can also be introduced at this stage, including pathogens, with potentially disastrous consequences as evidenced by the sporadic outbreaks of food poisoning caused by the consumption of pasteurized milk. Fortunately these incidents are rare, but highlight the need for continued diligence by processing plant workers in operating good manufacturing practices. The cleanliness of the plant and the environment should be monitored using appropriate microbiological tests coupled with effective action when evidence of contamination is found.

Even with the minimum amount of post-heat treatment contamination, pasteurized products will still contain appreciable numbers of bacteria. To discourage the growth of these organisms it is essential that the temperature of the product be rapidly cooled to at least 4°C, and that this temperature is maintained throughout distribution as well as the period of storage and display at the retail outlet. In this respect, the performance of many retailers in the United Kingdom could be improved substantially. Monitoring of the temperature of product and storage and display areas at critical points in the cold chain should be a matter of routine. Action should be taken when any unacceptably high readings are found. Consumer education through detailed information on packages regarding the effect of temperature on shelf life is widely used in the United States and should be introduced on a wider scale in the United Kingdom.

The maintenance of high standards of hygiene by milk producers and processors coupled with strict control of the temperature during distribution and retailing will ensure that the public can continue to enjoy pasteurized dairy products that are safe to eat over an acceptable period of time.

REFERENCES

Adams, D. M., and Brawley, T. G. (1981). Heat resistant bacterial lipases and ultra-high temperature sterilization of dairy products. *J. Dairy Sci.* **64**, 1951–1957.

Adams, D. M., and Brawley, T. G. (1983). Factors influencing the heat-resistance lipase of *Pseudomonas. J. Food Sci.* **46**, 673–676.

Adams, D. M., Barach, J. T., and Speck, M. L. (1975). Heat resistant proteases produced in milk by psychrotrophic bacteria of dairy origin. *J. Dairy Sci.* **58**, 828–834.

Adams, D., Well, S., Brown, B. F., Gregorio, S. Townsend, L. Skaggs, J. W., and Hinds, M. W. (1984). Salmonellosis from inadequately pasteurized milk–Kentucky. *U.S. Morbid. Mortal. Weekly Rep.* **33**(36), 505–506.

Airey, F. K. (1978). Market cream processing—factors affecting locations. *J. Soc. Dairy Technol.* **31**, 148–155.

Alford, J. A., and Elliott, L. E. (1960). Lipolytic activity of micro-organisms at low and intermediate temperatures. 1. Action of *Pseudomonas fluorescens* on lard. *Food Res.* **25**, 296–303.

Alford, J. A., and Pierce, D. A. (1961). Lipolytic activity of microorganisms at low and intermediate temperatures. 3. Activity of microbial lipases at temperatures below 0°C *J. Food Sci.* **26**, 518–524.

Alichandis, E., and Andrews, A. T. (1977). Some properties of the extracellular protease produced by the psychrotrophic bacterium *Pseudomonas fluorescens* strain AR-11. *Biochim. Biophys. Acta* **485**, 424–433.

Alkanhal, H. A., Frank, J. F., and Christen, G. L. (1985). Microbial protease and phospholipase C stimulate lipolysis of washed cream. *J. Dairy Sci.* **68**, 3162–3170.

Anas, K. E. (1977). Bacterial content of milk after passage through different sections of a plate pasteurizer. *Nord. Mejerindust.* **4**, 324–326.

Anderson, E. B., and Meanwell, L. J. (1933). Studies in the bacteriology of low temperature pasteurization. 1. The significance of heat-resistant organisms in raw milk supplies. *J. Dairy Res.* **4**, 213–225.

Anderson, M. (1983). Milk lipase and off-flavour development. *J. Soc. Dairy Technol.* **36**, 3–7.

Andersson, R. E. (1980). Microbial lipolysis at low temperatures. *Appl. Environ. Microbiol.* **39**, 36–40.

Andersson, R. E., Hedlund, C. B., and Jonsson, U. (1979). Thermal inactivation of a heat-resistant lipase produced by the psychrotrophic bacterium *Pseudomonas fluorescens. J. Dairy Sci.* **62**, 361–367.

Anon (1982a). Disease attributed to dairy products. *Br. Med. J.* **285**, 1664.

Anon (1982b). Presence of *Salmonella muenster* in Ontario cheese. *Ontario Dis. Surveill. Report* **13**, 143.

Anon (1983). Communicable disease associated with milk and dairy products, 1982. *Br. Med. J.* **288**, 466–467.

Anon (1984). General code of hygiene practice for the dairy industry. *Int. Dairy Fed. Bull.* **178**, 5–12.

Anon (1986). Salmonellosis due to dried baby milk. *Commun. Med.* **8**, 246.

Auclair, J. E. (1954). The inhibition of microorganisms by raw milk. III Distribution and properties of 2 inhibitory substances, Lactenin 1 and Lactenin 2. *J. Dairy Res.* **21**, 323–336.

Auclair, J. (1986). Processes avoiding recontamination of pasteurized milk. *Int. Dairy Fed. Bull.* **200**, 15–16.

Aulisio, C. C. G., Lanier, T. M., and Chappel, M. A. (1982). *Yersinia enterocolitia* 0: 13 associated with an outbreak in three southern states. *J. Food Protection* **45**, 1263.

Bandler, D. K., Gravani, R. B., Kinsella, J. K., Ledford, R. A., Scnyk, G. F., Shipe, W. F., Wolff, E. T., and Zall, R. R. (1981). The milk quality situation in New York State. *J. Dairy Sci.* **64** (Suppl. 1), 56.

Banks, J. G., Board, R. G., and Sparks, N. H. C. (1986). Natural antimicrobial systems and their potential in food preservation of the future. *Biotechnol. Appl. Biochem.* **8**, 103–147.

Banks, J. M., Griffiths, M. W., Phillips, J. D., and Muir, D. D. (1986). The yield and quality of Cheddar cheese produced from thermised milk. *Daily Ind. Int.* **51**(7), 31–35.

Banks, J. M., Griffiths, M. W., Phillips, J. D., and Muir, D. D. (1987). A comparison of the effects of storage of raw milk at 2°C and 6°C on the yield and quality of Cheddar cheese. *Food Microbiol.* **5**, 9–16.

Barach, J. T., Adams, D. M., and Speck, M. L. (1976). Low temperature inactivation in milk of heat-resistant proteases from psychrotrophic bacteria. *J. Dairy Sci.* **59**, 391–395.

Barach, J. T., Adams, D. M., and Speck, M. L. (1978). Mechanism of low temperature inactivation of a heat-resistant bacterial protease in milk. *J. Dairy Sci.* **61**, 523–528.

Barron, W. (1984). A practical look at CIP. *Dairy Ind. Int.* **49**(6), 34–39.

Bearns, R. E., and Girard, K. E. (1958). The effect of pasteurization on *Listeria monocytogenes. Can. J. Microbiol.* **4**, 55–61.

Bengtsson, K., Gardhage, L., and Isaksson, B. (1973). Gelation in UHT treated milk, whey and casein solution. The effect of heat resistant proteases. *Milchwissenschaft* **28**, 495–499.

Berg, M. G. van den (1986). Pasteurization of milk. Design and operation. *Int. Dairy Fed. Bull.* **200**, 35–47.

Bergey's Manual of Determinative Bacteriology (1974). 8th ed. R. E. Buchanan, ed. Williams & Wilkins, Baltimore.

Bergey's Manual of Systematic Bacteriology 1. (1984). 2. (1986). J. G. Holt, ed. Williams & Wilkins, Baltimore.

Bhadsavle, C. J., Shehata, T. E., and Collings, E. B. (1972). Isolation and identification of psychrophilic species of *Clostridium* from milk. *Appl. Microbiol.* **24**, 699–702.

Billing, E., and Cuthbert, W. A. (1958). Bitty cream: The occurrence and significance of *Bacillus cereus* spores in raw milk supplies. *J. Appl. Bacteriol.* **21**, 65–78.

Binet, F. (1983). Survival of *Yersinia enterocolitica* in milk. *Tech. Laitière* **977**, 43–45, 47–51.

Bishop, J. R., and White, C. H. (1986). Assessment of dairy product quality and potential shelf-life—a review. *J. Food Protection* **49**, 739–753.

Bjorgum, I., Oterholm, B., and Solberg, P. (1978). Thermization of milk *Proc. XXth Int. Dairy Cong.* E, 614–615.

Bjork, L. (1980). Enzymatic stabilization of milk—utilization of the milk peroxidase for the preservation of raw milk. *Int. Dairy Fed. Bull. Document* **126**, 5–7.

Bjork, L., Rosen, C. G., Marshall, V., and Reiter, B. (1975). Antibacterial activity of the lactoperoxidase system in milk against pseudomonads and other gram-negative bacteria. *Appl. Microbiol.* **30**, 199–204.

440   Pasteurized Dairy Products

Black, R. E., Jackson, R. J., Tsai, T., Medvesky, M., Shayegani, M., Feeley, J. C., MacLeod, K. I. E. and Wakelee, A. M. (1978). Epidemic *Yersinia enterocolitica* infection due to contaminated chocolate milk. *N. Engl. J. Med.* **298**, 76–79.

Blake, F. G. B. (1979). Storage and transport of pasteurized milk. *J. Soc. Dairy Technol.* **32**, 72–77.

Blankenagel, G. (1976). Relationship between the bacteriological quality of raw milk and that of pasteurized milk and cream. *Kieler Milchwirtschaftliche Forsch.* **34**, 158–162.

Blouse, L. E., Jr., Gengler, R. E., Lathrop, G. D., Hodder, R. A., Nowosiwsky, T., and Caraway, C. T. (1975). A common-source outbreak of *Salmonella newport*—Louisiana. *U.S. Morbid. Mortal. Weekly Rep.* **24**, 413–414.

Bockelmann, I. von (1970). The composition of the thermoresistant flora in cold-stored milk. *XVIII Int. Dairy Cong.* **1E**, 105.

Bockelmann, I. von (1974). Wachstum von Bakterien in Rohmilch während einer verlängerten Kaltlagerung. *FIL-IDF 19th Int. Milchw. Kong.* **ID**, 439–440.

Bodyfelt, F. (1980). Heat resistant psychrotrophs affect quality of fluid milk. *Dairy Rec.* **81**(3), 96–98.

Bodyfelt, F. W. (1986). A survey and analysis of Oregon grade A raw milk for flavor quality and shelf-life potential. Presented at the 75th Oregon Dairy Industries Conference, Springfield, Oregon. February 12, 1986.

Bodyfelt, F. W., and Davidson, W. D. (1975). Temperature control. I: A procedure for profiling temperatures of dairy products in stores. *J. Milk Food Technol.* **38**, 734–737.

Boer, E. de, Hartzog, B. J.., and Borst, G. H. A. (1984). Milk as a source of *Campylobacter jejuni*. *Netherlands Milk Dairy J.* **38**, 183–194.

Boer, E. de, Hartzog, B. J., Driessen, F. M., and Schmidt, D. G. (1982). Growth of thermoresistant streptococci and deposition of milk constituents on plates of heat exchangers during long operation times. *J. Food Protection* **45**, 806–812.

Bradshaw, J. G., Peeler, J. R., Corwin, J. J., Hunt, J. M., Tierney, J. T., Larkin, E. P., and Twedt, R. M. (1985). Thermal resistance of *Listeria monocytogenes* in milk. *J. Food Protection* **48**, 743–745.

Bradshaw, J. G., Peeler, J. T., Corwin, J. J., Barnett, J. E., and Twedt, R. M. (1987a). Thermal resistance of disease-associated *Salmonella typhimurium*. *J. Food Protection* **50**, 95–96.

Bradshaw, J. G. Peeler, J. T., Corwin, J. J., Hunt, J. M., and Twedt, R. M. (1987b). Thermal resistance of *Listeria monocytogenes* in dairy products. *J. Food Protection* **50**, 543–544.

Bramley, A. J., and Dodd, F. H. (1984). Reviews of the progress of dairy science: Mastitis control—progress and prospects. *J. Dairy Res.* **51**, 481–512.

Bramley, A. J., and Neave, F. K. (1975). Studies on the control of coliform mastitis in diary cows. *Br. Vet. J.* **131**, 160–169.

Bramley, A. J., McKinnon, C. H., Staker, R. T., and Simpkin, D. C. (1984). The effect of udder infection on the bacterial flora of the bulk milk of ten dairy herds. *J. Appl. Bacteriol.* **57**, 317–323.

Breuil, C., and Kushner, D. J. (1975). Partial purification and characterization of the lipase of a facultatively psychrophilic bacteria (*Acinetobacter* 016). *Can. J. Microbiol.* **21**, 434–441.

British Standards Institution (1982). Code of practice for equipment and procedures for cleaning and disinfecting of milking machine installations. B S No 5226, London 1982.

Brown, J. V., Wiles, R., and Prentice, G. A. (1979). The effect of a modified Tyndallization process upon the sporeforming bacteria of milk and cream. *J. Soc. Dairy Technol.* **32**, 109–112.

Brown, J. V., Wiles, R., and Prentice, G. A. (1980). The effect of different time–temperature pasteurization conditions upon the shelf life of single cream. *J. Soc. Dairy Technol.* **33**, 78–79.

Bryan, F. L. (1983). Epidemiology of milk-borne diseases. *J. Food Protection.* **46**, 637–649.

Buchanan, R. L., and Palumbo, S. A. (1985). *Aeromonas hydrophila* and *Aeromonas sobria* as potential food poisoning species: A review. *J. Food Safety* **7**, 15–29.

Bucky, A. R., Hayes, P. R., and Robinson, D. S. (1986). Lipase production by a strain of *Pseudomonas fluorescens* in whole milk and skimmed milk. *Food Microbiol.* **3**, 37–44.

Bucky, A. R., Robinson, D. S., and Hayes, P. R. (1987a). Enhanced deactivation of bacterial lipases by a modified UHT treatment. *Int. J. Food Sci. Technol.* **22**, 35–40.

Bucky, A. R., Hayes, P. R., and Robinson, D. S. (1987b). A modified ultra high temperature treatment for reducing microbial lipolysis in stored milk. *J. Dairy Res.* **54**, 275–282.

Bunning, V. K., Crawford, R. G., Bradshaw, J. G., Peeler, J. T., Tierney, J. T., and Twedt, R. M. (1986). Thermal resistance of intracellular *Listeria monocytogenes* cells suspended in raw bovine milk. *Appl. Environ. Microbiol.* **52**, 1398–1402.

Burton, H. (1986). Microbiological aspects. *Int. Dairy Fed. Bull.* **200**, 9–14.

Campbell, J. R., and Marshall, R. T. (1975). *The Science of Providing Milk for Man.* McGraw-Hill, New York.

Cannon, R. Y. (1970). Types and populations of microorganisms in the air of fluid milk plants. *J. Milk Food Technol.* **33**, 19–21.

Cannon, R. Y. (1972). Contamination of raw milk with psychrotrophic sporeformers. *J. Dairy Sci.* **55**, 669.

Carini, S., Mucchetti, G., and Neviani, E. (1985). Lysozyme: Activity against clostridia and use in cheese production—a review. *Microbiol. Aliments Nutr.* **3**, 299–320.

Chandan, R. C., Shahani, K. M., and Holly, R. G. (1964). Lysozyme content of human milk. *Nature* **204**, 76–77.

Chandler, R. E., and McMeekin, T. A. (1985). Temperature function integration and its relationship to the spoilage of pasteurized homogenized milk. *Austr. J. Dairy Technol.* **40**, 37–41.

Chaudhary, R. A., Tuckey, S. L., and Witter, L. D. (1960). Heat resistance of three stains of psychrophilic organisms added to skim milk for cottage cheese manufacture. *J. Dairy Sci.* **43**, 1774–1782.

Chilliard, Y., and Lamberet, G. (1984). [Milk lipolysis: Various types, mechanisms, factors of variation, practical significance.] *Lait* **64**, 544–578.

Chrisope, G. L., and Marshall, R. T. (1976). Combined action of lipase and microbial phospholipase C on a model fat globule emulsion and raw milk. *J. Dairy Sci.* **59**, 2024–2030.

Christen, G. L., and Marshall, R. T. (1984a). Selected properties of lipase and protease of *Pseudomonas fluorenscens* 27 produced in four media. *J. Dairy Sci.* **67**, 1680–1687.

Christen, G. L., and Marshall, R. T. (1984b). Thermostability of lipase and protease of *Pseudomonas fluorescens* 27 produced in various broths. *J. Dairy Sci.* **67**, 1688–1693.

Christen, G. L., and Marshall, R. T. (1985). Effect of histidine on thermostability of lipase and protease of *Pseudomonas fluorescens* 27. *J. Dairy Sci.* **68**, 594–604.

Christopher, F. M., Smith, G. C., and Vandergant, C. (1982). Effect of temperature and pH on the survival of *Campylobacter fetus. J. Food Protection* **45**, 253–259.

Chung, B. H., and Cannon, R. Y. (1971). Psychrotrophic sporeforming bacteria in raw milk supplies. *J. Dairy Sci.* **54**, 448.

Cogan, T. M. (1977). A review of heat resistant lipases and proteinases and the quality of dairy products. *Irish J. Food Sci. Technol.* **1**, 95–105.

Coghill, D. (1982). Studies on thermoduric psychrotrophic bacteria in south east Queensland dairy products. *Aust. J. Dairy Technol.* **37**, 147–148.

Coghill, D., and Juffs, H. S. (1979a). Comparison of the Moseley keeping quality test for pasteurized milk and cream products with other tests of shorter duration. *Aust. J. Dairy Technol.* **34**, 118–120.

Coghill, D., and Juffs, H. S. (1979b). Incidence of psychrotrophic sporeforming bacteria in pasteurized milk and cream products and effect of temperature on their growth. *Aust. J. Dairy Technol.* **34**, 150–153.

Coghill, D. M., Mutzelburg, I. D., and Birch, S. J. (1982). Effect of thermization on the bacteriological and chemical quality of milk. *Aust. J. Dairy Technol.* **37**, 48–50.

Coleman, W. W. (1986). Controlling *Listeria hysteria* in your plant. *Dairy Food Sanitation* **6**, 555–557.

Collins, E. B. (1961). Resistance of certain bacteria to cottage cheese cooking procedures. *J. Dairy Sci.* **44**, 1989–1996.

Collins, E. B. (1981). Heat resistant psychrotrophic microorganisms. *J. Dairy Sci.* **64**, 157–160.

Collins, R. N., Treger, M. D., Goldsby, J. B., Boring, J. R., III, Coohon, D. B., and Barr, R. N. (1968). Interstate outbreak of *Salmonella newbrunswick* infection traced to powdered milk. *J. Am. Med. Assoc.* **203**, 838–844.

Cousin, M. A. (1982). Presence and activity of psychrotrophic microorganisms in milk and dairy products. A review. *J. Food Protection* **45**, 172–207.

Cousins, C. M. (1977). Cleaning and disinfection in milk production. *J. Soc. Dairy Technol.* **30**, 101–105.

Cousins, C. M. (1978). Milking techniques and the microbial flora of milk. *XXth Int. Dairy Cong.*, Paris: Congress lecture.

Cousins, C. M., and Bramley, A. J. (1981). The microbiology of raw milk. In R. K., Robinson, ed., *Dairy Microbiology.* Vol. 1. *The Microbiology of Milk.* Applied Science, Barking, pp. 119–163.

Cousins, C. M., and McKinnon, C. H. (1979). Cleaning and disinfection in milk production. In C. C. Thiel and F. H. Dodd, eds., *Machine Milking* NIRD-HRi Technical Bulletin No. 1, Reading, pp. 286–329.

Cousins, C. M., Sharpe, M. E., and Law, B. A. (1977). The bacteriological quality of milk for Cheddar cheesemaking. *Dairy Ind. Int.* **42**(7), 12–17.

Cox, W. A. (1975). Problems associated with bacterial spores in heat-treated milk and dairy products. *J. Soc. Dairy Technol.* **28**, 59–68.

Crawford, R. J. M. (1967). Bulk milk collection and milk quality. *J. Soc. Dairy Technol.* **20**, 114–129.

Credit, C., Hedeman, R., Heywood, P., and Westhoff, D. (1972). Identification of bacteria isolated from pasteurized milk following refrigerated storage. *J. Milk Food Technol.* **35**, 708–709.

Crossley, E. L. (1948). Studies on the bacteriological flora and keeping quality of pasteurized liquid cream. *J. Dairy Res.* **15**, 261–276.

Cullen, G. A., and Herbert, C. N. (1967). Some ecological observations on microorganisms inhabiting bovine skin, teat canals and milk. *Br. Vet. J.* **123**, 14–25.

Daemen, A. L. H. (1981). The destruction of enzymes and bacteria during the spray-drying of milk and whey. 1. The thermoresistance of zone enzymes and bacteria in milk and whey with various total solids content. *Netherlands Milk Dairy J.* **35**, 133–144.

Dalahay, B. K., and Abbo, A. (1982). Extracellular proteases from *Pseudomonas* sp. isolated from raw milk. *21st Int. Dairy Cong., Moscow* **1**(2), 487–488.

D'Aoust, J.-Y., Warburton, D. W., and Sewell, A. M. (1985). *Salmonella typhimurium* phage-type 10 from Cheddar cheese implicated in a major Canadian foodborne outbreak. *J. Food Protection* **48**, 1062–1066.

Davies, F. L. (1975a). Heat resistance of *Bacillus* species. *J. Soc. Dairy Technol.* **28**, 69–78.

Davies, F. L. (1975b). Discussion of papers presented at symposium on bitty cream and related problems. *J. Soc. Dairy Technol.* **28**, 85–90.

Davies, F. L. (1977). The role of various milk fractions and the importance of somatic cells in the formation of germinant(s) for *Bacillus cereus* when milk is pasteurized. *J. Dairy Res.* **44**, 555–568.

Davies, F. L., and Wilkinson, G. (1977). A germinant for *Bacillus cereus* derived from pasteurized milk. In A. N. Barker, J. Wolf, D. J. Ellar, G. J. Dring, and G. W. Gould, eds. *Spore Research 1976.* Academic Press, London, pp. 699–709.

Davis, J. G. (1981). Microbiology of cream and dairy desserts. In R. K. Robinson, ed. *Dairy Microbiology.* Vol. 2. *The Microbiology of Milk Products.* Applied Science, London, pp. 31–89.

Deeth, H. C., and FitzGerald, C. H. (1983). Lipolytic enzymes and hydrolytic rancidity in milk and milk products. In P. F. Fox, ed. *Developments in Dairy Chemistry 2. Lipids.* Applied Science, London, pp. 195–239.

Dempster, J. F. (1968). Distribution of psychrophilic microorganisms in different dairy environments. *J. Appl. Bacteriol.* **31**, 290–301.

Dommett, T. W., Baseby, L. J., and Swain, A. J. (1986). Effects of storage conditions in a final factory on raw milk microbiological quality. *Aust. J. Dairy Technol.* **41**, 23–27.

Dominguez, L. B., Kelter, A., Marks, F. J., Press, W. B., and Starko, K. M. (1979). Salmonella gastroenteritis associated with milk—Arizona. *U.S. Morbid. Mortal. Weekly Rep.* **28**, 117, 119–120.

Donker-Voet, J. (1963). My view on the epidemiology of *Listeria* infections. In M. L. Gray, ed. *Second Symposium on Listeria Infection*. Montana State College, Bozeman, pp. 133–139.

Donnelly, C. W., and Briggs, E. H. (1986). Psychrotrophic growth and thermal inactivation of *Listeria moncytogenes* as a function of milk composition. *J. Food Protection* **49**, 994–998.

Donnelly, C. W., Briggs, E. H., and Donnelly, L. S. (1987). Comparison of heat resistance of *Listeria monocytogenes* in milk as determined by two methods. *J. Food Protection* **50**, 14–17, 20.

Donnelly, L. S., and Busta, F. F. (1981). Anaerobic sporeforming microorganisms in dairy products. *J. Dairy Sci.* **64**, 161–166.

Downey, W. K. (1980). Review of the progress of dairy science: Flavour impairment from pre- and post manufacture lipolysis in milk and dairy products. *J. Dairy Res.* **47**, 237–252.

Doyle, M. P. (1985). Food-borne pathogens of recent concern. *Annu. Rev. Nutr.* **5**, 25–41.

Doyle, M. P., and Roman, D. J. (1981). Growth and survival of *Campylobacter fetus* subsp. *jejuni* as a function of temperature and pH. *J. Food Protection* **44**, 596–601.

Doyle, M. P., and Roman, D. J. (1982). Prevalence and survival of *Campylobacter jejuni* in unpasteurized milk. *Appl. Environ. Microbiol.* **44**, 1154–1158.

Doyle, M. P., Meske, L. M., and Marth, E. H. (1985). Survival of *Listeria monocytogenes* during the manufacture and storage of non fat dry milk. *J. Food Protection* **48**, 740–742.

Doyle, M. P., Glass, K. A., Beery, J. T., Garcia, G. A., Pollard, D. J., and Schulty, R. D. (1987). Survival of *Listeria monocytogenes* in milk during high temperature, short-time pasteurization. *Appl. Environ. Microbiol.* **53**, 1433–1438.

Driessen, F. M. (1983). Lipases and proteinases in milk. Occurrence, heat inactivation, and their importance for the keeping quality of milk products. Nederlands Instituut voor Zuivelonderzoek Verslag V236.

Driessen, F. M., and Stadhouders, J. (1971). Heat stability of lipase of *Alcaligenes viscolactis* 23al. *Netherlands Milk Dairy J.* **25**, 141–144.

Driessen, F. M., and Stadhouders, J. (1974). Thermal activation and inactivation of exocellular lipases of some Gram-negative bacteria common in milk. *Netherlands Milk Dairy J.* **28**, 10–22.

Dring, R., and Fox, P. F. (1983). Purification and characterization of a heat-stable lipase from *Pseudomonas fluorescens* AFT 29. *Irish J. Food Sci. Technol.* **7**, 157–171.

Druce, R. G., and Thomas, S. B. (1972). Bacteriological studies on bulk milk collection. Pipeline milking plants and bulk milk tanks as sources of bacterial contamination of milk. *J. Appl. Bacteriol.* **35**, 253–270.

Duthie, C., Shipe, W., and Hotchkiss, J. (1985). Effect of low-level carbonation on the keeping quality of processed milk. *J. Dairy Sci.* **68** (Suppl. 1), 69.

Eddy, B. P. (1960a). The use and meaning of the term psychrophilic. *J. Appl. Bacteriol.* **23**, 189–190.

Eddy, B. P. (1960b). Cephalotrichous, fermentative Gram-negative bacteria: the genus *Aeromonas*. *J. Appl. Bacteriol.* **23**, 216–249.

Elliker, P. R., Sing, E. L., Christensens, L. J., and Sandine, W. F. (1964). Psychrophilic bacteria and keeping quality of pasteurized products. *J. Milk Food Technol.* **27**, 69–75.

Evans, D. A., Hankinson, D. J., and Litsky, W. (1970). Heat resistance of certain pathogenic bacteria in milk using a commercial plate heat exchanger. *J. Dairy Sci.* **53**, 1659–1665.

Fairbairn, D. J., and Law, B. A. (1986). Proteinases of psychrotrophic bacteria: their production, properties, effects and control. *J. Dairy Res.* **53**, 139–177.

Farber, J. M., Sanders, G. W., Emmons, D. B., and McKellar, R. C. (1987). Heat resistance of *Listeria monocytogenes* in artificially-inoculated and naturally-contaminated raw milk. *Dairy Food Sanitation* **7**, 520.

Felmingham, D., and Juffs, H. S. (1977). Comparison and evaluation of keeping quality tests for pasteurized milk and cream products. *Aus. J. Dairy Technol.* **32**, 158–162.

Fenlon, D. R. (1986). Growth of naturally occurring *Listeria* spp. in silage: A comparative study of laboratory and farm ensiled grass. *Grass Forage Sci.* **41**, 375–378.

Fernandez Garayzabal, J. F., Dominguez Rodriguez, L., Vazquez Boland, J. A. Blanco Cancelo, J. L., and Suarez Fernandez, G. (1986). *Listeria monocytogenes* in pasteurized milk. *Can. Microbiol.* **32**, 149–150.

FitzGerald, C. H., and Deeth, H. C. (1983). Factors influencing lipolysis by skim milk cultures of some psychrotrophic microorganisms. *Aus. J. Dairy technol.* **38**, 97–103.

FitzGerald, C. H. Deeth, H. C., and Coghill, D. M. (1982). Low temperature inactivation of lipase from psychrotrophic bacteria. *Aus. J. Dairy Technol.* **37**, 51–54.

Flake, J. C., Parker, A. C. Smathers, J. B., Saunders, A. K., and marth, E. H. (1972). *Methods for production of high-quality raw milk.* Int. Assoc. Milk, Food Environ. Sanit., Shelbyville, IN.

Fleming, D. W., Cochi, S. L., MacDonald, K. L., Brondum, J., Hayes, P. S. Plikaytis, B. D., Holmes, M. B., Audurier, A., Broome, C. V., and Reingold, A. L. (1985). Pasteurized milk as a vehicle of infection in an outbreak of listeriosis. *N. Engl. J. Med.* **312**, 404–407.

Foley, J., and Buckley, J. (1978). Pasteurization and thermization of milk and blanching of fruit and vegetables. In W. K. Downey, ed. *Food Quality and Nutrition Research. Priorities for Thermal Processing.* Applied Science, London, pp. 191–216.

Fontaine, R. E., Cohen, M. L., Martin, W. T., an Vernon, T. M. (1980. Epidemic salmonellosis from Cheddar cheese: surveillance and prevention. *Am. J. Epidemiol.* **111**, 247–253.

Fowler, G. G. (1979). The potential of Nisin. *Food Manufacture* **54**(2), 57–59.

Fox, P. F., and Stepaniak, L. (1983). Isolation and some properties of extracellular heat-stable lipases from *Pseudomonas fluorescens* strain AFT 36. *J. Dairy Res.* **50**, 77–89.

Francis, D. W., Spaulding, P. L., and Lovett, J. (1980). Enterotoxin production and thermal resistance of *Yersinia enterocolitica* in milk. *Appl. Environ. Microbiol.* **40**, 174–176.

Franklin, J. G. (1967). The incidence and significance of *Bacillus cereus* in milk—2. *Milk. Ind.* **61**(5), 34–37.

Franklin, J. G. (1969). Some bacteriological problems in the market milk industry in the UK. *J. Soc. Dairy Technol.* **22**, 100–112.

Franklin, J. G., Williams, D. J., and Clegg, L. F. L. (1956). A survey of the number and types of aerobic mesophilic spores in milk before and after commercial sterilization. *J. Appl. Bacteriol.* **19**, 46–53.

Furlong, J. D., Lee, W., Foster, L. R., and Williams, L. P. (1979). Salmonellosis associated with consumption of nonfat powdered milk—Oregon. *U.S. Morbid. Mortal. Weekly Rep.* **28**, 129–130.

Galbraith, N. S., and Pusey, J. J. (1984). Milkborne infectious disease in England and Wales 1938–1982. In D. L. J. Freed, ed., *Health Hazards of Milk*. Baillière Tindall, London, pp. 27–59.

Galesloot, Th.E. (1953). Some aspects of the bacteriology of pasteurized milk IV. The deterioration of laboratory pasteurized milk. *Netherlands Milk Dairy J.* **7**, 1–14.

Galesloot, Th.E. (1955). The effect of the heat treatment on the keeping quality of pasteurized milk. *Netherlands Milk Dairy J.* **9**, 237–248.

Gebre-Egziabher, A., Humbert, E. S., and Blankenagel, G. (1980). Heat-stable proteases from psychrotrophs in milk. *J. Food Protection* **43**, 197–200.

Gehriger, G. (1980). Multiplication of bacteria in milk during farm storage. *Int. Dairy Fed. Bull.* 120, 22–24.

Gilbert, R. (1979). *Bacillus cereus* Gastroenteritis. In H. Riemann and F. L. Bryan, eds., *Foodborne Infections and Intoxications.* Academic Press, New York, pp. 495–518.

Gill, C. O., and Tan, K. H. (1979). Effect of carbon dioxide on growth of *Pseudomonas fluorescens. Appl. Environ. Microbiol.* **38**, 237–240.

Gill, K. P. W., Bates, P. G., and Lander, K. P. (1981). The effect of pasteurization on the survival of *Campylobacter* species in milk. *Br. Vet. J.* **137**, 578–584.

Gillies, A. J. (1971). Significance of thermoduric organisms in Queenland Cheddar cheese. *Aust. J. Dairy Technol.* **26**, 145–149.

Gilmour, A., and Rowe, M. T. (1981). Microorganisms associated with milk. In R. K. Robinson, ed., *Dairy Microbiology.* Vol. 1. *The Microbiology of milk.* Applied Science, Barking, pp. 35–75.

Gilmour, A., MacElhinney, R. S., Johnston, D. E., and Murphy, R. J. (1981). Thermisation of milk. Some microbiological aspects. *Milchwissenschaft* **36**, 457–461.

Goudkov, A. V., and Sharpe, M. E. (1965). Clostridia in dairying. *J. Appl. Bacteriol.* **28**, 63–73.

Greene, V. W., and Jezeski, J. J. (1954). Influence of temperature on the development of several psychophilic bacteria of dairy origin. *Appl. Microbiol.* **2**, 110–117.

Griffiths, M. W. (1983). Synergistic effects of various lipases and phospholipase C on milk fat. *J. Food Technol.* **18**, 495–505.

Griffiths, M. W. (1986). Use of milk enzymes as indices of heat-treatment. *J. Food Protection* **49**, 696–705.

Griffiths, M. W., and Phillips, J. D. (1984a). Effect of aeration on extracellular enzyme synthesis by psychrotrophs growing in milk during refrigerated storage. *J. Food Protection* **47**, 697–702.

Griffiths, M. W., and Phillips, J. D. (1984b). Detection of post-pasteurization contamination of cream by impedimetric methods. *J. Appl. Bacteriol.* **57**, 107–114.

Griffiths, M. W., and Phillips, J. D. (1986). The application of the preincubation test in commercial dairies. *Aust. J. Dairy Technol.* **41**, 71–79.

Griffiths, M. W., Phillips, J. D., and Muir, D. D. (1981a). Development of flavour defects in pasteurized double cream during storage at 6 and 10°C. *J. Soc. Dairy Technol.* **34**, 142–146.

Griffiths, M. W., Phillips, J. D., and Muir, D. D. (1981b). Thermostability of proteases and lipases from a number of species of psychrotrophic bacteria of dairy origin. *J. Appl. Bacteriol.* **50**, 289–303.

Griffiths, M. W., Phillips, J. D., and Muir, D. D. (1984a). Methods for rapid detection of post-pasteurization contamination in cream. *J. Soc. Dairy Technol.* **37**, 22–26.

Griffiths, M. W., Phillips, J. D., and Muir, D. D. (1984b). Pre-incubation test to rapidly identify post-pasteurization contamination in milk and single cream. *J. Food Protection* **47**, 391–393.

Griffiths, M. W., Phillips, J. D., and Muir, D. D. (1985a). Post-pasteurization contamination—the major cause of failure of fresh dairy products. *Hannah Res. 1984*, 77–87.

Griffiths, M. W., Phillips, J. D., and Muir, D. D. (1985b). The quality of pasteurized milk and cream at point of sale. *Dairy Ind. Int.* **50**, 25–31.

Griffiths, M. W., Phillips, J. D., and Muir, D. D. (1986a). The effect of subpasteurization heat treatments on the shelf-life of milk. *Dairy Ind. Int.* **51**(5), 31–35.

Griffiths, M. W., Hurvois, Y., Phillips, J. D., and Muir, D. D. (1986b). Elimination of spore-forming bacteria from double cream using sub-UHT temperatures. I. Processing conditions. *Milchwissenschaft* **41**, 403–405.

Griffiths, M. W., Hurvois, Y., Phillips, J. D., and Muir, D. D. (1986c). Elimination of spore-forming bacteria from double cream using sub-UHT temperatures. II. Effect of processing conditions on spores. *Milchwissenschaft* **41**, 474–478.

Griffiths, M. W., Phillips, J. D., and Muir, D. D. (1987). Effect extended of low-temperature storage on the bacteriological quality of raw milk. *Food Microbiol.*, **5**, 75–87.

Grosskopf, J. C., and Harper, W. J. (1969). Role of psychrophilic sporeformers in long life milk. *J. Dairy Sci.* **52**, 897.

Grosskopf, J. C., and Harper, W. J. (1974). Isolation and identification of psychrotrophic sporeformers in milk. *Milchwissenschaft* **29**, 467–470.

Guirguis, A. H., Griffiths, M. W., and Muir, D. D. (1983). Sporeforming bacteria in milk I Optimization of heat treatment for activation of spores of *Bacillus* species. *Milchwissenschaft* **38**, 641–644.

Guirguis, A. H., Griffiths, M. W., and Muir, D. D. (1984). Sporeforming bacteria in milk. II. Effect of carbon dioxide addition on heat activation of spores of *Bacillus* species. *Milchwissenschaft* **39**, 144–146.

Hamman, J. H. (1979). Transport and storage of short shelf-life dairy products. *J. Soc. Dairy Technol.* **32**, 125–128.

Hanna, M. O., Stewart, J. C., Carpenter, Z. L., and Vanderzant, C. (1977). Heat resistance of *Yersinia enterocolitica* in skim milk. *J. Food Sci.* **42**, 1134–1136.

Harding, F. (1987). The impact of central testing on milk quality. *Dairy Ind. Int.* **52**(1), 17–19.

Heldman, D. R., Hedrick, T. I., and Hall, C. W. (1964). Air-borne microorganism populations in food packaging areas. *J. Milk Food Technol.* **27**, 245–251.

Hill, A. W. (1986). Mastitis, the non-antibiotic approach to control. In M. Bateson, C. L. Benham, and F. A. Skinner, eds., *The Society for Applied Bacteriology Symposium Series No. 15. Microorganisms in Agriculture.* Blackwell Scientific Publications, Oxford, pp. 935–1035.

Hird, D. W. (1987). Review of evidence for zoonotic listeriosis. *J. Food Protection* **50**, 429–433.

Honer, C. (1981). Raw milk psychrotrophs and the importance of their control. *Dairy Rec.* **82**(8), 104-D, 106, 108.

Hughes, D. (1979). Isolation of *Yersinia enterocolitica* from milk and a dairy farm in Australia. *J. Appl. Bacteriol.* **46**, 125–130.

Hughes, D. (1980). Repeated isolation of *Yersinia enterocolitica* from pasteurized milk in a holding vat at a dairy factory. *J. Appl. Bacteriol.* **48**, 383–385.

Hussong, R. V., and Hammer, B. W. (1931). The pasteurizing efficiencies secured with milk from individual farms. *Bull. Agric. Exp. Station Iowa* No. 286.

Hutchinson, E. M. S. (1974). A study of *Bacillus cereus* biotypes with special reference to liquid milk. Ph.D. thesis. Queens University, Belfast, Northern Ireland.

Hutchinson, E. M. S. (1975). An approach to the biotyping of *Bacillus cereus* strains with special reference to milk. *J. Soc. Dairy Technol.* **28**, 79.

Ienistea, C., Chitu, M. and Roman, A. (1970). Heat resistance in milk of some strains of Group D streptococci from pasteurized milk and the influence exerted on their growth by selective media. *Zbl. Bakt.* I. *Abt. Orig.* **215**, 173–181.

Ikonomov, L. (1969). Microbiological studies on the pasteurization of ewes milk. VI. Comparison of effects of HTST pasteurization at different temperatures. *Vet. Nauki Sofia* **6**(6), 71–76.

Jackman, D. M., Bartlett, F. M., and Patel, T. R. (1983). Heat-stable proteases from psychrotrophic pseudomonads: Comparison of immunological properties. *Appl. Environ. Microbiol.* **46**, 6–12.

Janzen, J. J., Bishop, J. R., and Bodine, A. B. (1982). Relationship of protease activity to shelf-life of skim and whole milk. *J. Dairy Sci.* **65**, 2237–2240.

Jayne-Williams, D. J., and Franklin, J. G. (1960). *Bacillus* spores in milk—Part I. *Dairy Sci. Abstr.* **22**, 215–221.

Jenness, R., and Koops, J. (1962). Preparation and properties of a salt solution which stimulates milk ultrafiltrate. *Netherlands Milk Dairy J.* **16**, 153–164.

Joergensen, K. (1980). Bacteriological contamination from the surface of the teat and udder. *Int. Dairy Fed. Bull.* 120, 11–15.

Johns, C. K. (1962). The coliform count of raw milk as an index of udder cleanliness. *XVIth Int. Dairy Cong.* C365.

Johnston, D. E., Murphy, R. J., Gilmour, A., McGuiggen, J. T. M., Rowe, M. T., and Mullan, W. M. A. (1987). Manufacture of Cheddar cheese from thermized cold-stored milk. *Milchwissenschaft* 42, 226–230.

Johnston, D. W., and Bruce, J. (1982). Incidence of thermoduric psychrotrophs in milk produced in the west of Scotland. *J. Appl. Bacteriol.* 52, 333–337.

Jones, P. H., Willis, A. T., Robinson, D. A., Skirrow, M. B., and Josephs, D. S. (1981). *Campylobacter* enteritis associated with the consumption of free school milk. *J. Hygiene Cambridge* 87, 155–162.

Juffs, H. S. (1975). Proteolysis detection in milk. III. Relationships between bacterial populations, tyrosine value and organoleptic quality during cold storage of milk and cream. *J. Dairy Res.* 42, 31–41.

Juffs, H. S., and Babel, F. J. (1975). Inhibition of psychrotrophic bacteria by lactic cultures in milk stored at low temperatures. *J. Dairy Sci.* 58, 1612–1619.

Juffs, H. S., and Doelle, H. W. (1968). Some properties of the extracellular proteolytic enzymes of the milk-spoiling organism *Pseudomonas aeruginosa* ATCC 10145. *J. Dairy Res.* 35, 395–398.

Juffs, H. S., Hayward, A. C., and Doelle, H. W. (1968). Growth and proteinase production in *Pseudomonas* spp. cultivated under various conditions of temperature and nutrition. *J. Dairy Res.* 35, 385–393.

Kaufmann, O. W., and Andrews, R. H. (1954). The destruction rate of psychrophilic bacteria in skim milk. *J. Dairy Sci.* 37, 317–327.

Keen, N. T., and Williams, P. H. (1967). Effect of nutritional factors on extracellular protease production by *Pseudomonas lachrymans*. *Can. J. Microbiol.* 13, 863–871.

Kessler, H. G., and Hovak, F. P. (1984). Effect of heat treatment and storage conditions on the keeping quality of pasteurized milk. *Milchwissenschaft* 39, 451–454.

Keynan, A., and Evenchik, Z. (1969). Activation. In G. W. Gould and A. Hurst, eds., *The Bacterial Spore*. Academic Press, London, pp. 359–396.

Khan, I. M., Dill, C. W., Chandan, R. C., and Shahani, K. M. (1967). Production and properties of the extracellular lipase of *Achromobacter lipolyticum*. *Biochim. Biophys. Acta* 132, 68–77.

King, J. S., and Mabbit, L. A. (1982). Preservation of raw milk by the addition of carbon dioxide. *J. Dairy Res.* 49, 439–447.

Kishonti, E. (1975). Influence of heat resistant lipases and proteases in psychrotrophic bacteria on product quality. *Int. Dairy Fed. Annu. Bull. Document* 86, 121–124.

Kishonti, E., and Sjöström, G. (1970). Influence of heat resistant lipases and proteases in psychrotrophic bacteria on product quality. *18th Int. Dairy Cong. Sydney* 1E, 501.

Korhonen, H. (1980). A new method for preserving raw milk. *World Anim. Rev.* 35, 23–29.

Kocak, H. R., and Zadow, J. G. (1985). The effect of low-temperature-inactivation treatment on age gelation of UHT whole milk. *Aust. J. Dairy Technol.* 40, 53–58.

Koidis, P., and Doyle, M. P. (1984). Procedure for increased recovery of *Campylobacter jejuni* from inoculated unpasteurized milk. *Appl. Environ. Microbiol.* 47, 455–460.

Kozak, J. J. (1987). Regulatory response to the problem of pathogenic bacteria in the dairy industry. *J. Dairy Sci.* 70 Suppl. 1, 57.

Kurzweil, R., and Busse, M. (1973). Total count and microflora of freshly drawn milk. *Milchwissenschaft* 28, 427–431.

Labots, H., and Hup, G. (1964). *Bacillus cereus* in raw and pasteurized milk. II. The occurrence of slow and fast germinating *Bacillus cereus* in milk and their significance in the enumeration of *Bacillus cereus* spores. *Netherlands Milk Dairy J.* 18, 167–176.

Labots, H., Hup, G., and Galesloot, En.Th.E. (1965). *Bacillus cereus* in raw and pasteurized milk. III. The contamination of raw milk with *Bacillus cereus* spores during its production. *Netherlands Milk Dairy J.* 19, 191–221.

Lakshminarasim, V., and Iya, K. K. (1955). Studies on the microoccoci in milk. I. Incidence and Distribution. *Indian J. Dairy Sci.* **8**, 67–77.

Langeveld, L. P. M., Cuperus, F., and Stadhouders, J. (1973). Bacteriological aspects of the keeping quality at 5°C of reinfected and non-reinfected pasteurized milk. *Netherlands Milk Dairy J.* **27**, 54–65.

Law, B. A. (1979). Reviews of the progress of dairy science: Enzymes of psychrotrophic bacteria and their effects on milk and milk products. *J. Dairy Res.* **46**, 573–588.

Law, B. A., and Mabbitt, L. A. (1983). New methods for controlling the spoilage of milk and milk products. In T. A. Roberts and F. A. Skinner, eds., *Food Microbiology: Advances and Prospects.* Academic Press, London, pp. 131–150 (Soc. Appl. Bact. Symp. Series No. 11).

Law, B. A., Andrews, A. T., and Sharpe, M. E. (1977). Gelation of ultra-high-temperature-sterilized milk by proteases from a strain of *Pseudomonas fluorescens* isolated from raw milk. *J. Dairy Res.* **44**, 145–148.

Law, B. A., Andrews, A. T., Cliffe, A. J., Sharpe, E. M., and Chapman, H. R. (1979). Effect of proteolytic raw milk psychrotrophs on Cheddar cheese-making with stored milks. *J. Dairy Res.* **46**, 497–509.

Lawrence, R. C. (1967a). Microbial lipases and related esterases. Part I. Detection, distribution and production of microbial lipases. *Dairy Sci. Abstr.* **29**, 1–8.

Lawrence, R. C. (1967b). Microbial lipases and related esterases. Part II. Estimation of lipase activity, characterization of lipases, recent work concerning their effect on dairy products. *Dairy Sci. Abstr.* **29**, 59–70.

Lawrence, R. C., Fryer, T. F., and Reiter, B. (1967). The production and characterization of lipases from a micrococcus and a pseudomonad. *J. Gen. Microbiol.* **48**, 401–418.

Lawton, W. C., and Nelson, F. E. (1955). Influence of sublethal treatment with heat or chlorine on the growth of psychrophilic bacteria. *J. Dairy Sci.* **38**, 380–386.

Leali, L. (1965). General survey of bulk milk collection systems in various countries (Discussion on section A). *Int. Dairy Fed. Ann. Bull.* Part II, 142.

Lecos, C. (1986). Of microbes and milk: probing America's worst salmonella outbreak. *FDA Consumer* **20**, 18–21.

Lee, W. J. (1977). An assessment of *Yersinia enterocolitica* and its presence in foods. *J. Food Protection* **40**, 486–489.

Lewis, S. J., and Gilmour, A. (1987). Microflora associated with the internal surfaces of rubber and stainless steel milk transfer pipeline. *J. Appl. Bacteriol.* **62**, 327–333.

Longstaff, G. W. (1985). Central testing in England and Wales and its impact on milk quality. *J. Soc. Dairy Technol.* **38**, 10–13.

Lovett, J., Bradshaw, J. G., and Peeler, J. T. (1982). Thermal inactivation of *Yersinia enterocolitica* in milk. *Appl. Environ. Microbiol.* **44**, 517–519.

Lovett, J., Francis, D. W., and Hunt, J. M. (1983). Isolation of *Campylobacter jejuni* from raw milk. *Appl. Environ. Microbiol.* **46**, 459–462.

Lovett, J., Francis, D. W., and Hunt, J. M. (1987). *Listeria monocytogenes* in raw milk: Detection incidence and pathogenicity. *J. Food Protection* **50**, 188–192.

Lu, J. Y., and Liska, B. J. (1969). Lipase from *Pseudomonas fragi*. I. Purification of the enzyme. *Appl. Microbiol.* **18**, 104–107.

Lück, H. (1981). Quality control in the dairy industry. In R. K. Robinson, ed., *Dairy Microbiology.* Vol. 2. *The Microbiology of Milk Products.* Applied Science, London, pp. 279–324.

Lück, H., Dunkeld, M., and van den Berg, M. (1980). Shelf-life tests on pasteurized milk. *South Afr. J. Dairy Technol.* **12**, 107–112.

Luedecke, L. O., and Harmon, L. G. (1966). Thermal resistance of *Pseudomonas fragi* in milk containing various amounts of fat. *Appl. Microbiol.* **14**, 716–719.

McKellar, R. C. (1981). Development of off-flavors in ultra-high temperature and pasteurized milk as a function of proteolysis. *J. Dairy Sci.* **64**, 2138–2145.

McKellar, R. C. (1982). Factors influencing the production of extracellular proteinases by *Pseudomonas fluorescens. J. Appl. Bacteriol.* **53**, 305–316.

McKellar, R. C., and Cholette, H. (1983). Purification and physical properties of the extracellular proteinases from three strains of *Pseudomonas fluorescens. Abstr. Annu. Meet. Am. Soc. Microbiol.* P27, p. 258.

McKenzie, D. A. (1962). Milk testing—a forward look. *J. Soc. Dairy Technol.* **15**, 207–212.

McKenzie, D. A. (1965). Thermoduric Bacteria. In J. G. Davis, ed., *The Supplement to a Dictionary of Dairying.* Leonard Hill, London, p. 1682.

McKinnon, C. H., and Pettipher, G. L. (1983). A survey of sources of heat-resistant bacteria in milk with particular reference to psychrothrophic spore-forming bacteria. *J. Dairy Res.* **50**, 163–170.

McLarty, R. M., and Robb, J. (1968). Farm tanker milk supplies cause concern. *Dairy Ind.* **33**, 536–539.

McManus, C., and Lanier, J. M. (1987). *Salmonella, Campylobacter jejuni* and *Yesinia enterocolitica* in raw milk. *J. Food Protection* **50**, 51–55.

McSwiggan, D. A., Gilbert, R. J., and Fowler, F. W. T. (1975). Food poisoning associated with pasteurized cream. Public Health Laboratory Service Communicable Disease Report No. 43.

Macaulay, D. M., Hawisko, R. Z., and James, N. (1963). Effect of pasteurization on survival of certain psychrophilic bacteria. *Appl. Microbiol.* **11**, 90–92.

Magdoub, M. N. I., Shehata, A. E., El-Sanragy, Y. A., and Hassan, A. A. (1983). Growth parameters of psychrotrophic *Bacillus* spp. isolated from raw milk. *Asian J. Dairy Res.* **2**, 97–105.

Malik, R. K., and Mathur, D. K. (1984). Purification and characterization of a heat-stable protease from *Pseudomonas* sp. B-25. *J. Dairy Sci.* **67**, 522–530.

Margolis, J. D. (1985). Salmonellosis outbreak—Hillfarm Dairy: Melrose Park Illinois, pp. 24–29. Final Task Force Report, Sept. 13, 1985.

Marin, A., Mawhinney, T. P., and Marshall, R. T. (1984). Glycosidic activities of *Pseudomonas fluorescens* on fat-extracted skim milk, buttermilk and milk fat globule membranes. *J. Dairy Sci.* **67**, 52–59.

Marshall, J. H. (1985). Hygiene on the farm. *J. Soc. Dairy Technol.* **38**, 3–6.

Marshall, R. T. (1982). Relationship between the bacteriological quality of raw milk and the final products. A review of basic information and practical aspects. *Kieler Milchwirtschaft. Forsch.* **34**, 149–157.

Marshall, R. T., and Appel, R. (1973). Sanitary conditions in twelve fluid milk processing plants as determined by use of the rinse filter method. *J. Milk Food Technol.* **36**, 237–241.

Marshall, R. T., and Marstiller, J. K. (1981). Unique response to heat of extracellular protease of *Pseudomonas fluorescens* M5. *J. Dairy Sci.* **64**, 1545–1550.

Marth, E. H. (1969). Salmonellae and salmonellosis associated with milk and milk products. A review. *J. Dairy Sci.* **52**, 283–315.

Marth, E. H. (1981). Assessing the quality of milk. *J. Dairy Sci.* **64**, 1017–1022.

Martin, J. H. (1981). Heat resistant mesophilic microorganisms. *J. Dairy Sci.* **64**, 149–156.

Martin, J. H., Harper, W. J., and Gould, I. A. (1966). Ultra-high temperature effects on selected *Bacillus* species. *J. Dairy Sci.* **49**, 1367–1370.

Martins, S. B., Hodapp, S., Dufour, S. W., and Kraeger, S. J. (1982). Evaluation of a rapid impedimetric method for determining the keeping quality of milk. *J. Food Protection* **45**, 1221–1226.

Mayerhofer, H. J., Marshall, R. T., White, C. H., and Lu, M. (1973). Characterization of a heat-stable protease of *Pseudomonas* P26. *Appl. Microbiol.* **25**, 44–48.

Mead, G. C. (1985). Enumeration of pseudomonads using cephaloridine–fucidin–cetrimide agar (CFC). *Int. J. Food Microbiol.* **2**, 21–26.

Mikolajcik, E. M. (1970). Thermodestruction of *Bacillus* spores in milk. *J. Milk Food Technol.* **33**, 61–63.

Mikolajcik, E. M. (1978). Psychrotrophic sporeformers: A possible keeping-quality problem in market milk. *Am. Dairy Rev.* **40**(4), 34A, 34D.

Mikolajcik, E. M. (1979). Psychrotrophic bacteria and dairy product quality I. Major organisms involved and defects produced. *Cultured Dairy Prod. J.* **14**(4) 6–10.

Mikolajcik, E. M., and Koka, M. (1968). Bacilli in milk. 1. Spore germination and growth. *J. Dairy Sci.* **51**, 1579–1582.

Mikolajcik, E. M., and Simon, N. T. (1978). Heat resistant psychrotrophic bacteria in raw milk and their growth at 7℃. *J. Food Protection* **41**, 93–95.

Mitchell, G. E., Ewings, K. N., and Bartley, J. P. (1986). Physicochemical properties of proteinases from selected psychrotrophic bacteria. *J. Dairy Res.* **53**, 97–115.

Moseley, W. K. (1958). What is being learned from total counts on cottage cheese? *Proc. 51st Annu. Conv. Milk Ind. Found.* (Laboratory Section), p. 27.

Moseley, W. K. (1980). Pinpointing post-pasteurization contamination. *J. Food Protection* **43**, 414.

Møller-Madsen, A. (1965). Bacteriological investigations of raw milk. *Int. Dairy Fed. Annu. Bull.* Part II, 202–207.

Mottar, J. (1981). Heat resistant enzymes in UHT milk and their influence on sensoric changes during uncoded storage. *Milchwissenschaft* **36**, 87–91.

Mottar, J., and Driessen, F. M. (1987). Methods for the determination of the activity of indigenous and bacterial proteinases in milk and milk products. A review. *Int. Dairy Fed. Bull. Document* 209, 41–44.

Mourgues, R., and Auclair, J. (1973). Durée de conservation à 4 et 8°C du lait pasteurisé conditioné aseptiquement. *Le Lait* **53**, 481–490.

Mourgues, R., Deschamps, N., and Auclair, J. (1983). Influence de la flore thermo-résistance du lait cru sur la qualité de conservation du lait pasteurisé exempt de recontaminations post-pasteurisation. *Le Lait* **63**, 391–404.

Moustafa, M. K., Ahmed, A. A-H., and Marth, E. H. (1983). Occurrence of *Yersinia enterocolitica* in raw and pasteurized milk. *J. Food Protection* **46**, 276–278.

Mrozek, H. (1970). Hygiene difficulties in food manufacture. *Arch. Hyg.* **154**, 240–246.

Muir, D. D., and Phillips, J. D. (1984). Prediction of the shelf-life of raw milk during refrigerated storage. *Milchwissenschaft* **39**, 7–11.

Muir, D. D. Kelly, M. E., Phillips, J. D., and Wilson, A. G. (1978a). The quality of blended raw milk in creameries in south-west Scotland. *J. Soc. Dairy Technol.* **31**, 137–144.

Muir, D. D., Kelly, M. E., and Phillips, J. D. (1978b). The effect of storage temperature on bacterial growth and lipolysis in raw milk. *J. Soc. Dairy Technol.* **31**, 203–208.

Muir, D. D., Phillips, J. D., and Dalgleish, D. G. (1979). The lipolytic and proteolytic activity of bacteria isolated from blended raw milk. *J. Soc. Dairy Technol.* **32**, 19–23.

Mull, L. E., Pilkhane, S. V., Richter, R. L., and Smith, K. L. (1974). Shelf-life of pasteurized milk stored at 4.5 and 7.0C. *J. Milk Food Technol.* **37**, 530–531.

Murray S. K., Kwan, K. K. H., Skura, B. J., McKellar, R. C. (1983). Effect of nitrogen flushing on the production of proteinase by psychrotrophic bacteria in raw milk. *J. Food Sci.* **48**, 1166–1169.

Nadkarni, S. R. (1971). Bacterial lipase. I. Nutritional requirements of *Pseudomonas aeruginosa* for production of lipase. *Enzymologia* **40**, 286–301.

Nashif, S. A., and Nelson, F. E. (1953a). The lipase of *Pseudomonas fragi*. II. Factors affecting lipase production. *J. Dairy Sci.* **36**, 471–480.

Nashif, S. A., and Nelson, F. (1953b). The lipase of *Pseudomonas fragi*. III. Enzyme action in cream and butter. *J. Dairy Sci.* **36**, 481–488.

Nisbet, J. (1979). Refrigerated transport. *J. Society Dairy Technology* **32**, 77–81.

Obiger, G. (1976). Studies on heat resistance of important pathogens during milk pasteurization. *Arch. Lebensmittelhyg.* **27**, 137–144.

O'Donnell, E. T. (1975). A study of lipase enzymes of psychrophilic bacteria. Ph.D. Thesis, University of Strathclyde, Glasgow, Scotland.

Olson, H. C. (1964). A study of the factors affecting shelf-life. *Milk Ind.* **55**(6), 36–40.

Orr, M. J., McLarty, R. M., and Baines, S. (1960). Further studies on bulk milk collection. *Dairy Ind.* **25**, 360–364.

Orr, M. J., McLarty, R. M., and Baines, S. (1965). Further investigations of alternate-day collection of bulk milk. *Dairy Ind.* **30**, 278–284.

Orr, M. J., McLarty, R. M., McCance, M. E., and Baines, S. (1964). Alternate day collection of bulk milk. *Dairy Ind.* **29**, 169–173.

Overcast, W. W., and Adams, G. A. (1966). Growth of certain psychrophilic bacteria in pasteurized milk as influenced by previous excessive psychrophilic growth in the raw milk. *J. Milk Food Technol.* **29**, 14–18.

Overcast, W. W., and Atmaram, K. (1974). The role of *Bacillus cereus* in sweet curdling of fluid milk. *J. Milk Food Technol.* **37**, 233–236.

Overcast, W. W., and Skean, J. D. (1959). Growth of certain lipolytic microorganisms at 4°C and their influence on free fat acidity and flavor of pasteurized milk. *J. Dairy Sci.* **42**, 1479–1485.

Owens, J. J. (1978). Lecithinase positive bacteria in milk. *Process Biochem.* **13**(1), 13–14, 30.

Palmer, J. (1980). Contamination of milk from the milking environment. *Int. Dairy Fed. Bull.* 120, 16–21.

Palumbo, S. A. (1986). Is refrigeration enough to restrain food-borne pathogens? *J. Food Protection* **49**, 1003–1009.

Palumbo, S. A., Morgan, D. R., and Buchanan, R. L. (1985). The influence of temperature, NaCl and pH on the growth of *Aeromonas hydrophila*. *J. Food Sci.* **50**, 1417–1421.

Panes, J. J., Parry, D. R., and Leech, F. B. (1979). Report of a survey of the quality of farm milk in England and Wales in relation to EEC proposals. Ministry of Agriculture, Fisheries and Food, London.

Pâquet, D., Driou, A., Bracquent, P., and Linden, G. (1987). Effect of refridgerated storage of milk on proteolysis. Relationship to the proteose-peptone content. *Netherlands Milk Dairy J.* **41**, 81–92.

Pastore, M. (1968). [Some considerations on the hygienic collection of refridgerated milk in bulk tanker vehicles and an evaluation of the methods of analysis]. *Ind. Agrarie* **6**, 88–91.

Patel, G. B., and Blankenagel, G. (1972). Bacterial counts of raw milk and flavor of the milk after pasteurization and storage. *J. Milk Food Technol.* **5**, 145–148.

Patel, T. R., Bartlett, F. M., and Hamid, J. (1983). Extracellular heat-resistant proteases of psychrotrophic pseudomonads. *J. Food Protection* **46**, 90–94.

Peterson, A. C., and Gunderson, M. F. (1960). Some characteristics of proteolytic enzymes from *Pseudomonas fluorescens*. *Appl. Microbiol.* **8**, 98–104.

Phillips, J. D., and Griffiths, M. W. (1985). Bioluminescence and impedimetric methods for assessing shelf-life of pasteurized milk and cream. *Food Microbiol.* **2**, 39–51.

Phillips, J. D., and Griffiths, M. W. (1986a). Estimation of Gram-negative bacteria in milk: A comparison of inhibitor systems for preventing Gram-positive bacterial growth. *J. Appl. Bacteriol.* **60**, 491–500.

Phillips, J. D., and Griffiths, M. W. (1986b). Factors contributing to the seasonal variation of *Bacillus* spp. in pasteurized dairy products. *J. Appl. Bacteriol.* **61**, 275–285.

Phillips, J. D., and Griffiths, M. W. (1987a). A note on the use of the Catalasemetre in assessing the quality of milk. *J. Appl. Bacteriol.* **62**, 223–226.

Phillips, J. D., and Griffiths, M. W. (1987b). The relation between temperature and growth of bacteria in dairy products. *Food Microbiol.* **4**, 173–185.

Phillips, J. D., Griffiths, M. W., and Muir, D. D. (1981a). Factors affecting the shelf-life of pasteurized double cream. *J. Soc. Dairy Technol.* **34**, 109–113.

Phillips, J. D., Griffiths, M. W., and Muir, D. D. (1981b). Growth and associated enzymic activity of spoilage bacteria in pasteurized double cream. *J. Soc. Dairy Technol.* **34**, 113–118.

Phillips, J. D., Griffiths, M. W., and Muir, D. D. (1983a). Effect of nisin on the shelf-life of pasteurized double cream. *J. Soc. Dairy Technol.* **36**, 17–21.

Phillips, J. D., Griffiths, M. W., and Muir, D. D. (1983b). Accelerated detection of post-heat-treatment contamination in pasteurized double cream. *J. Soc. Dairy Technol.* **36**, 41–43.

Prentice, G. A., and Neaves, P. (1986). The role of microorganisms in the dairy industry. In M. Bateson, C. L. Benham, and F. A. Skinner, eds., *Society for Applied Bacteriology Symposium Series No. 15. Micro-organisms in Agriculture.* Blackwell Scientific Publications, Oxford, pp 43s–57s.

Randolph, H. E., Freeman, T. R., and Peterson, R. W. (1965). Keeping quality of market milk obtained at retail outlets and at processing plants. *J. Milk Food Technol.* **28**, 92–96.

Read, R. B. Jr., Bradshaw, J. G. Dickerson, R. W., Jr., and Peeler, J. T. (1968). Thermal resistance of salmonellae isolated from dry milk. *Appl. Microbiol.* **16**, 998–1001.

Reimerdes, E. H. (1982). Changes in the proteins of raw milk during storage. In P. F. Fox, ed., *Developments in Dairy Chemistry—I. Proteins.* Applied Science, London, pp. 271–288.

Reiter, B. (1981). The impact of the lactoperoxidase system on the psychrotrophic microflora in milk. In T. A. Roberts, G. Hobbs, J. H. B. Christian, and N. Skovgaard, eds., *Psychrotrophic Microorganisms in Spoilage and Pathogenicity.* Academic Press, London, pp. 73–85 (Soc. Appl. bact. Tech. Ser. No. 13).

Reiter, B., and Marshall, V. M. (1979). Bactericidal activity of the lactoperoxidase system against psychrotrophic *Pseudomonas* spp. in raw milk. In A. D. Russell, and R. Fuller, eds., *Cold Tolerant Microbes in Spoilage and the Environment.* Academic Press, New York, pp. 153–164.

Richard, J. (1981). Taxomony and ecology of raw milk psychrotrophic *Pseudomonas*. In T. A. Roberts, G. Hobbs, J. H. B. Christian, and N. Skovgaard, eds., *Psychrotrophic Microorganisms in Spoilage and Pathogenicity.* Academic Press, London, pp. 117–125.

Richardson, B. C. (1981). The purifications and characterization of a heat-stable protease from *Pseudomonas fluorescens* B52. *N.Z. J. Dairy Sci. Technol.* **16**, 195–207.

Ridgway, J. D. (1954). A note on the seasonal variations of the keeping quality of commercial sterilized milk. *J. Appl. Bacteriol.* **17**, 1–5.

Riemann, H. (1969). Food processing and preservation effects. In H. Riemann, ed., *Food-Borne Infections and Intoxications.* Academic Press, New York, pp. 489–541.

Rosenow, E. M., and Marth, E. H. (1987). Growth of *Listeria monocytogenes* in skim, whole and chocolate milk and in whipping cream during incubation at 4, 8, 13, 21 and 35°C. *J. Food Protection* **50**, 452–459.

Rowe, M. T., and Gilmour, A. (1982). Growth, enzyme production and changes in oxygen tension occurring during batch cultivation of psychrotrophic *Pseudomonas fluorescens* strains. *Milchwissenschaft* **37**, 597–600.

Ryser, E. T., Marth, E. H., and Doyle, M. P. (1985). Survival of *Listeria monocytogenes* during manufacture and storage of cottage cheese. *J. Food Protection* **48**, 746–750, 753.

Sandstedt, K. (1982). Campylobacter transmission via food, especially milk. *XIV Nord. Veterinaerkong., København*, 6–9 July 1982, rapporter 225–226.

Schiemann, D. A. (1978). Association of *Yersinia enterocolitica* with the manufacture of cheese and occurrence in pasteurized milk. *Appl. Environ. Microbiol.* **36**, 274–277.

Schiemann, D. A., and Toma, S. (1978). Isolation of *Yersinia enterocolitica* from raw milk. *Appl. Environ. Microbiol.* **35**, 54–58.

Schmitt, N., Bowner, E. J., and Willoughby, B. A. (1976). Food poisoning outbreak attributed to *Bacillus cereus*. *Can. J. Public Health* **67**, 418–422.

Schröder, M. J. A. (1984). Origins and levels of post-pasteurization contamination of milk in the dairy and their effects on keeping quality. *J. Dairy Res.* **51**, 59–67.

Schröder, M. J. A., and Bland, M. A. (1983). Post-pasteurization contamination and shelf-life of HTST-pasteurized milk when filled in a liqui-Pak conventional or Model 820A cartoning machine. *J. Soc. Dairy Technol.* **36**, 43–49.

Schröder, M. J. A., and Bland, M. A. (1984). Effect of pasteurization temperature on the keeping quality of whole milk. *J. Dairy Res.* **51**, 569–578.

Schröder, M. J. A., Cousins, C. M., and McKinnon, C. H. (1982). Effect of psychrotrophic post-pasteurization contamination on the keeping quality at 11 and 5°C of HTST-pasteurized milk in the UK. *J. Dairy Res.* **49**, 619–630.

Schultze, W. D., and Olson, J. C. (1960). Studies on psychrophilic bacteria. I. Distribution in stored commercial dairy products. *J. Dairy Sci.* **43**, 346–350.

Seiler, H., Stör, S., and Busse, M. (1984). Identification of coryneform bacteria isolated from milk immediately after heating and following refrigerated storage. *Milchwissenschaft* **39**, 346–348.

Seminega, T., Hunbert, G., Le Déaut, J.-Y., and Linden, G. (1985). Determination de l'activité proleolytique dans le lait et les produit laitiers. *La Technique Laitière* No. 1001, 47–54.

Senhaji, A. F., and Loncin, M. (1975). Protection des microorganismes par les matieres grasses au cours des traitements thermique. *Ind. Alimentaires Agric.* **92**, 611–617.

Sharma, J. K., Malik, R. K., and Mathur, D. K. (1984). Isolation and identification of proteolytic psychrotrophic sporeforming bacteria from raw milk supplies at an experimental dairy in India. *J. Soc. Dairy Technol.* **37**, 96–98.

Sharp, J. C. M. (1987). Infections associated with milk and dairy products in Europe and North America 1980–1985. *Bull. W. H. O.* **65**(3), 397–406.

Sharp, J. C. M., Paterson, G. M., and Barrett, N. J. (1985). Pasteurization and the control of milkborne infection in Britain. *Br. Med. J.* **291**, 463–464.

Shehata, T. E., and Collins, E. B. (1971). Isolation and identification of psychrophilic species of *Bacillus* from milk. *Appl. Microbiol.* **21**, 466–469.

Shehata, T. E., Duran, A., and Collins, E. B. (1971). Influence of temperature on the growth of psychrophilic strains of *Bacillus*. *J. Dairy Sci.* **54**, 1579–1582.

Shelley, A. W., Deeth, H. C., and MacRae, I. C. (1986). Growth of lipolytic psychrotrophic pseudomonads in raw and ulta-heat-treated milk. *J. Appl. Bacteriol.* **61**, 395–400.

Shipe, W. F., and Senyk, G. F. (1981). Effects of processing conditions on lipolysis in milk. *J. Dairy Sci.* **64**, 2146–2149.

Shipe, W. F., Senyk, G. F., Adler, E. J., and Ledford, R. A. (1982). Effect of infusion of carbon dioxide on the bacterial growth in fluid milk. *J. Dairy Sci.* **65** (Suppl. 1), 77.

Sillén, G. (1987). Modern bactofuges in the dairy industry. *Dairy Ind. Int.* **52**(6), 27–29.

Sing, E. L. (1972). Program for quality assurance of finished dairy products. *J. Milk Food Technol.* **35**, 207–212.

Skura, B. J., Craig, C., and McKellar, R. C. (1986). Effect of disruption of an N_2-overlay on growth and proteinase production in milk by *Pseudomonas fluorescens*. *Can. Inst. Food Sci. Technol.* **19**, 104–106.

Solberg, P., Mabbitt, L. A., Naudts, M., and Schipper, C. J. (1974). Methods for assessing the bacteriological quality of cooled bulk milk from the farm. *Int. Dairy Fed. Bull.* **83**, 1–20.

Stadhouders, J. (1982). Cooling and thermization as a means to extend the keeping quality of raw milk. *Kieler Milchwirtschaft. Forsch.* **34**, 19–28.

Stadhouders, J., Hup, G., and Langeveld, C. P. M. (1980). Some observations on the germination, heat resistance and outgrowth of fast-germinating and slow germinating spores of *Bacillus cereus* in pasteurized milk. *Netherlands Milk Dairy J.* **34**, 215–228.

Stadhouders, J., Hup, G., and Nieuwenhof, F. F. J. (1985). Silage and cheese quality. Netherlands Institute Zuivelonderzoek, Mededeling M19A.

Stajer, B., Zakula, S., Kovincic, I., and Gallic, M. (1979). Heat resistance of *Listeria monocytogenes* and its survival in raw milk products. *Vet. Glasnik* **33**, 109–112.

Stanfield, J. T., Jackson, G. J., and Aulisio, C. C. G. (1985). *Yersinia enterocolitica*: Survival of a pathogenic strain on milk containers. *J. Food Protection* **48**, 947–948.

Stanley, S. O., and Rose, A. H. (1967). On the clumping of *Corynebacterium xerosis* as affected by temperature. *J. Gen. Microbiol.* **48**, 9–23.

Stead, D. (1983). A fluorimetric method for the determination of *Pseudomonas fluorescens* AR11 lipase in milk. *J. Dairy Res.* **50**, 491–502.

Stead, D. (1986). Microbial lipases: Their characteristics, role in food spoilage and industrial uses. *J. Dairy Res.* **53**, 481–505.

Stepaniak, L., and Fox, P. F. (1983). Thermal stability of an extracellular proteinase from *Pseudomonas fluorescens* AFT36. *J. Dairy Res.* **50**, 171–184.

Stepaniak, L., and Fox, P. F. (1985). Isolation and characterization of heat stable proteinases from *Pseudomonas* isolate AFT21. *J. Dairy Res.* **52**, 77–89.

Stepaniak, L., Birkeland, S. E., Sørhaug, T., and Vagias G. (1987a). Isolation and partial characterization of heat stable proteinase, lipase and phospholipase C from *Pseudomonas fluorescens* P1. *Milchwissenschaft* **42**, 75–79.

Stepaniak, L., Birkeland, S. E., Vagias, G., and Sørhaug, T. (1987b). Enzyme-linked immunosorbent assay (ELISA) for monitoring the production of heat stable proteinases and lipases from *Pseudomonas*. *Milchwissenschaft* **42**, 168–172.

Stepaniak, L., Fox, P. F., and Daly, C. (1982a). Influence of the growth of *Pseudomonas fluorescens* AFT36 on some technologically important characteristics of milk. *Irish J. Food Sci. Technol.* **6**, 135–146.

Stepaniak, L., Fox, P. F., and Daly, C. (1982b). Isolation and general characterization of heat-stable proteinase from *Pseudomonas fluorescens* AFT36. *Biochim. Biophys. Acta* **717**, 376–383.

Stern, N. J., Pierson, M. D., and Kotuln, A. W. (1980). Growth and competitive nature of *Yersinia enterocolitica* in whole milk. *J. Food Sci.* **45**, 972–974.

Stewart, D. B. (1975). Factors influencing the incidence of *B. cereus* spores in milk. *J. Soc. Dairy Technol.* **28**, 80–85.

Stone, M. J., and Rowlands, A. (1952). 'Broken' or 'bitty' cream in raw and pasteurized milk. *J. Dairy Res.* **19**, 51–62.

Stone, L. S., and Zottola, E. A. (1985). Effect of cleaning and sanitizing on the attachment of *Pseudomonas fragi* to stainless steel. *J. Food Sci.* **50**, 951–956.

Suhren, G. (1983). Occurrence and levels of heat-resistant proteinases and their effects on UHT-treated dairy products. *Int. Dairy Fed. Bull. Document* 157, 17–25.

Suhren, G., Heeschen, W., and Tolle, A. (1975). Free fatty acids in milk and bacterial activity. *Annu. Bull. Int. Dairy Fed. Document* 86, 51–57.

Sun, M. (1985). Desperately seeking *Salmonella* in Illinois. *Science* **228**, 829–830.

Swaisgood, H. E., and Bozoglu, F. (1984). Heat inactivation of the extracellular lipase from *Pseudomonas fluorescens* MC50. *J. Agric. Food Chem.* **32**, 7–10.

Symons, M. H., Clements, R. S., Mitchell, G. E., and Ewings, K. N. (1985). The immunological relationship of bacterial proteases from some psychrotrophic bacteria. *N.Z.J. Dairy Sci. Technol.* **20**, 173–178.

Tacket, C. O., Narain, J. P., Sattin, R., Lofgren, J. P., Konigsberg, C., Rendtorff, R. C., Rausa, A., Davis, B. R., and Cohen, M. L. (1984). A multistate outbreak of infections caused by *Yersinia enterocolitica* transmitted by pasteurized milk. *J. Am. Med. Assoc.* **251**, 483–486.

Tekinsen, O. C., and Rothwell, J. (1974). A study of the effect of storage at 5°C on the microbial flora of heat-treated market cream. *J. Soc. Dairy Technol.* **27**, 57–62.

Te Whaiti, I. E., and Fryer, T. F. (1978). Production and heat stability in milk of proteinases and lipases of psychrotrophic pseudomonads. *Proc. XXth Int. Dairy Congr.* **E**, 303–304.

Thomas, S. B. (1966). Sources, incidence and significance of psychrotrophic bacteria in milk. *Milchwissenschaft* **21**, 270–275.

Thomas, S. B. (1969). Methods of assessing the psychrotrophic bacterial content of milk. *J. Appl. Bacteriol.* **32**, 269–296.

Thomas, S. B. (1974). The influence of the refrigerated farm bulk milk tank on the quality of the milk at the processing dairy. *J. Soc. Dairy Technol.* **27**, 180–187.

Thomas, S. B., and Druce, R. G. (1969). Psychrotrophic bacteria in refrigerated pasteurized milk: A review. Part 2. *Dairy Ind.* **34**, 430–433.

Thomas, S. B., and Druce, R. G. (1971). Bacteriological quality of alternate day collected farm bulk tanks milk. *Dairy Sci. Abstr.* **33**, 339–342.

Thomas, S. B., and Thomas, B. F. (1977a). The bacterial content of milking machines and pipeline milking plants. A review. *Dairy Ind. Int.* **42**(4), 7–12.

Thomas, S. B., and Thomas, B. F. (1977b). The bacterial content of milking machines and pipeline milking plants. Part II of a review. *Dairy Ind. Int.* **42**(5), 16–23.

Thomas, S. B., and Thomas, B. F. (1977c). The bacterial content of milking machines and pipeline milking plants. Part III of a review. *Dairy Ind. Int.* **42**(7), 19–25.

Thomas, S. B., and Thomas, B. F. (1977d). The bacterial content of milking machines and pipeline milking plants. Part IV. Coli-aerogenes bacteria. *Dairy Ind. Int.* **42**(11), 25–33.

Thomas, S. B., and Thomas, B. F. (1978a). The bacterial content of milking machines and piepline milking plants. Part V. Thermoduric organisms. *Dairy Ind. Int.* **43**(5), 17–25.

Thomas, S. B., and Thomas, B. F. (1978b). The bacterial content of milking machines and pipeline milking plants. Part VI. Psychrotrophic bacteria. *Dairy Ind. Int.* **43**(10), 5–10.

Thomas, S. B., Hobson, P. M., Bird, E. R., King, K. P., Druce, R. G., and Cox, D. R. (1962). The microflora of raw milk as determined by plating on Yeastrel milk agar incubated at 30°C. *J. Appl. Bacteriol.* **25**, 107–115.

Thomas, S. B., Jones, M., Hobson, P. M., Williams, G., and Druce, R. G. (1963). Microflora of raw milk and farm dairy equipment. *Dairy Ind.* **28**, 212–219.

Thomas, S. B., Druce, R. G., Davies, A., and Bear, J. S. (1966a). Bacteriological aspects of bulk milk collection. *J. Soc. Dairy Technol.* **19**, 161–169.

Thomas, S. B., Druce, R. G., and Owen-Jones, E. (1966b). Bulk milk collection. Bacteriological quality of milk on arrival at creameries in road tankers. *J. Soc. Dairy Technol.* **19**, 222–224.

Thomas, S. B., Druce, R. G., Peters, G. J., and Griffiths, D. G. (1967). Incidence and significance of thermoduric bacteria in farm milk supplies: A reappraisal and review. *J. Appl. Bacteriol.* **30**, 265–298.

Thomas, S. B., Druce, R. G., and Jones, M. (1971). Influence of production conditions on the bacteriological quality of refrigerated farm bulk tank milk—a review. *J. Appl. Bacteriol.* **34**, 659–677.

Tinuoye, O. L., and Harnon, L. G. (1975). Growth of thermoduric psychrotrophic bacteria in refrigerated milk. *Am. Dairy Rev.* **37**(9), 26–30.

Todd, E. C. D. (1987). Foodborne and waterborne disease in Canada—1980 Annual summary. *J. Food Protection* **50**, 420–428.

Tolle, A. (1980). The microflora of the udder. *Int. Dairy Fed. Bull.* **120**, 4–10.

Torrie, J. P., Cholette, H., Froehlich, D. A., and McKellar, R. C. (1983). Growth of an extracellular proteinase-deficient strain of *Pseudomonas fluorescens* on milk and milk protein. *J. Dairy Res.* **50**, 365–374.

Turbutt, P. A., Seaman, A., and Woodbine, M. (1960). Microbiological aspects of the thermoduric *Microbacterium* genus. *Dairy Sci. Abstr.* **22**, 543–548.

Twedt, R. M. (1986). Thermal resistance characteristics of *Listeria monocytogenes. J. Food Protection* **49**, 849.

Underwood, H. M., McKinnon, C. H., Davies, F. L., and Cousins, C. M. (1974). Sources of *Bacillus* spores in raw milk. *19th Int. Dairy Congr. New Delhi* **1E**, 373–374.

United States Public Health Service, Centers for Disease Control. (1985). Listeriosis outbreak associated with Mexican-style cheeses—California. *U.S. Morbid. Mortal. Weekly Rep.* **34**, 357–359.

Velimirovic, B. (1984). Epidemiology and control of foodborne diseases in Europe. In B. Velimirovic, ed., *Infectious Diseases in Europe. A Fresh Look*. World Health Organization, Regional office for Europe, Copenhagen, pp. 207–253.

Visser, S., Slangen, K. J., and Hup, G. (1975). Some bitter peptides from rennet-treated casein. A method for their purification, utilizing chromatographic separation on silica gel. *Netherlands Milk Dairy J.* **29**, 319–334.

Vlad, A., and Vlad, A. (1972). Alimentary toxic-infection due to *Bacillus cereus*. *Microbiol. Parazitol. Epidemiol.* **17**, 513–536.

Waes, G. (1976). Aerobic mesophilic spores in raw milk. *Milchwissenschaft* **31**, 521–524.

Wainess, H. (1981). Shelf-life and environmental aspects. *Int. Dairy Fed. Bull.* **130**, 71–74.

Wainess, H. (1982). Hygienic design and construction of equipment used in dairy plants. *Int. Dairy Fed. Bull.* **153**, 3–10.

Washam, C. J., Olson, H. C., and Vedamuthu, E. P. (1977). Heat-resistant psychrotrophic bacteria isolated from pasteurized milk. *J. Food Protection* **40**, 101–108.

Waterman, S. C. (1982). The heat sensitivity of *Campylobacter jejuni* in milk. *J. Hyg. Cambridge* **88**, 529–533.

Watrous, G. H., Jr., Barnard, S. E., and Coleman II, W. W. (1971). Bacterial concentrations in raw milk, immediately after laboratory pasteurization and following 10 days storage at 7.2°C. *J. Milk Food Technol.* **34**, 282–284.

Weckbach, L. S., and Langlois, B. E. (1977). Effect of heat treatments on survival and growth of a psychrotroph and on nitrogen fractions in milk. *J. Food Protection* **40**, 857–862.

West, F. B., Adams, D. M., and Speck, M. L. (1978). Inactivation of heat resistant proteases in normal ultra-high temperature sterilized skim milk by a low temperature treatment. *J. Dairy Sci.* **61**, 1078–1084.

West, I. G., Griffiths, M. W., Phillips, J. D., Sweetsur, A. W. M., and Muir, D. D. (1986). Production of dried skim milk from thermised milk. *Dairy Ind. Int.* **51**(6), 33–34.

White, C. H., and Marshall, R. T. (1973). Reduction of shelf-life of dairy products by a heat-stable protease from *Pseudomonas fluorescens* P26. *J. Dairy Sci.* **56**, 849–853.

Wilkinson, G., and Davies, F. L. (1973). Germination of spores of *Bacillus cereus* in milk and milk dialysates. Effect of heat treatment. *J. Appl. Bacteriol.* **36**, 485–496.

Wilkinson, G., and Davies, F. L. (1974). Some aspects of the germination of *Bacillus cereus* in milk. In A. N. Barker, G. W. Gould, and J. Wolf, eds., *Spore Research 1973*. Academic Press, London, pp. 153–159.

Williams, D. J. (1956). The rates of growth of some thermoduric bacteria in pure culture and their effects on tests for the keeping quality of milk. *J. Appl. Bacteriol.* **19**, 80–94.

Wilson, C. D., and Richards, M. S. (1980). A survey of mastitis in the British dairy heard. *Vet. Record* **106**, 431–435.

Witter, L. C. (1961). Psychrophilic bacteria—a review. *J. Dairy Sci.* **44**, 983–1015.

Wood, D. S., Collins-Thompson, D. L., Irvine, D. M., and Myhr, A. N. (1984). Source and persistance of *Salmonella muenster* in naturally contaminated Cheddar cheese. *J. Food Protection* **47**, 20–22.

Zall, R. R. (1980). Can cheesemaking be improved by heat treating milk on a farm? *Dairy Ind. Int.* **45**(2), 25–31, 48.

14

THE FORMATION AND DETERMINATION OF ETHYL CARBAMATE IN ALCOHOLIC BEVERAGES

James F. Lawrence, B. Denis Page, and Henry B. S. Conacher

Food Research Division, Bureau of Chemical Safety, Food Directorate
Health and Welfare Canada
Banting Research Centre
Ottawa, Ontario, Canada

1. INTRODUCTION

Ethyl carbamate (urethane, structure in Figure 1) has been shown to be a potent carcinogen (Mirvish, 1968; Allen et al., 1982; IARC monograph, 1974) and thus any human exposure to this substance is of concern. It has been known that ethyl carbamate can be naturally present in fermented beverages in low μg/L

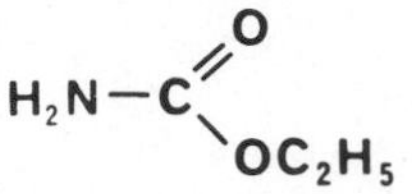

$$H_2N-C\overset{\displaystyle O}{\underset{\displaystyle OC_2H_5}{\diagup}}$$

Figure 1. Structure of ethyl carbamate, urethane: mp 48–50°C; bp 182–184°C. Sublimes readily at 103°C and 54 mm pressure; 1 g dissolves in 0.5 mL water, 0.8 mL alcohol, 0.9 mL chloroform, 1.5 mL ether, and 32 mL olive oil.

concentrations as a result of the fermentation process itself (Ough, 1976a). It also can form as a result of reaction of ethanol with urea or the antimicrobial agent, diethyl pyrocarbonate (Ough, 1976b; Löfroth and Gejvall, 1971). However, the latter compound was banned in the United States in 1972 (Federal Register, 1972) and the Joint FAO/WHO Expert Committee on Food Additives (World Health Organization, 1972) concluded that ethyl carbamate was permissible in soft drinks at 10 μg/L. It was not until 1985 that interest in ethyl carbamate in alcoholic beverages greatly increased. The concern resulted from findings of the Liquor Control Board of Ontario in Canada, which reported elevated levels of the substance in certain types of beverages. The levels in some cases were far higher than would be expected through natural production from the fermentation process alone. Since then, ethyl carbamate has been found in many alcoholic beverages including both distilled and nondistilled products of several countries (Andrey, 1987; Baumann and Zimmerli, 1986; Bertrand and Triquet-Pissard, 1986; Dennis et al., 1986; Mildau et al., 1987). The Canadian government recently established maximum permissible levels of ethyl carbamate in alcoholic beverages. These are shown in Table 1. The United States is considering imposing an interim action level for ethyl carbamate particularly in whiskey (Food Chemical News, 1987). As a result, there has been a great deal of activity recently in the development of improved and simplified methods as well as investigations by the alcoholic beverage industry to determine the source of high levels of ethyl carbamate.

This chapter presents an overview of the ethyl carbamate problem including a brief summary of its toxicology and investigations into its formation, and to a greater extent discusses the various methods developed to accurately determine it at low μg/L levels in alcoholic beverages.

TABLE 1   Regulatory Limits for Ethyl Carbamate in Alcoholic Beverages in Canada

Product	Limit (μg/kg)
Table wine	30
Fortified wines (sherries and ports)	100
Distilled spirits	150
Fruit brandies and liqueurs	400

2. TOXICOLOGY

The greatest concern about the toxicology of ethyl carbamate is its strong carcinogenicity. Nettleship et al. (1943) first reported that the compound induced pulmonary tumors in mice. It was later shown to exert a high level of mouse strain specificity for lung adenoma induction (Cowan, 1950; Shapiro and Kirschbaum, 1951). Other studies confirmed the genotoxicity of ethyl carbamate in nonrodent species (Mirvish, 1968; IARC, 1974) including nonhuman primates (Adamson and Sieber, 1983).

Additional studies on ethyl carbamate and vinyl carbamate have indicated that the latter compound is a more potent carcinogen and is suspected as being the main proximate carcinogenic metabolite of ethyl carbamate (Allen et al., 1986; Dahl et al., 1978, 1980). WiKman-Coffelt and Berg (1976) in their failed attempt to develop a radioimmunoassay method for ethyl carbamate discovered that both goats and humans were naturally exposed to enough ethyl carbamate or other material that caused them to build up natural antibodies against ethyl carbamate-like bonds. This indicated that perhaps the human body may be capable of handling low levels of the compound. In general, however, the toxicology studies done to date clearly support the potential of ethyl carbamate for being a human carcinogen.

3. FORMATION IN ALOCOHOLIC BEVERAGES

The initial interest in ethyl carbamate was stimulated by the findings of Löfroth and Gejvall (1971) through their studies on diethyl pyrocarbonate, an antibacterial compound that was used in alcoholic and nonalcoholic beverages. They showed that white wine and beer could react with diethyl pyrocarbonate to produce elevated levels of ethyl carbamate. Figure 2 illustrates the reaction. Diethyl pyrophosphate reacts with ammonia present in the beverages at neutral or basic pH to yield ethyl carbamate and CO_2. Although the rate of reaction is reduced at lower pH, the reaction can still occur. Ethyl carbamate was produced when diethyl pyrocarbonate was added to freshly squeezed orange juice (Joe et al., 1977). Ough (1976b) carried out detailed studies on the formation of ethyl carbamate from diethyl pyrocarbonate addition to wine and found that actual production was less than predicted, likely due to other competitive reactions.

The production of ethyl carbamate through natural means during the fermentation process has been studied by several groups. Ough (1976a) found

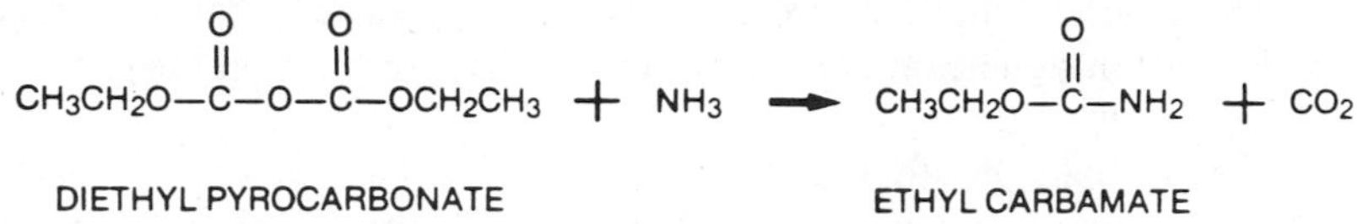

$$CH_3CH_2O-\overset{\overset{O}{\|}}{C}-O-\overset{\overset{O}{\|}}{C}-OCH_2CH_3 \; + \; NH_3 \;\longrightarrow\; CH_3CH_2O-\overset{\overset{O}{\|}}{C}-NH_2 \; + \; CO_2$$

DIETHYL PYROCARBONATE   ETHYL CARBAMATE

Figure 2.   Formation of ethyl carbamate by reaction of diethyl pyrocarbonate with ammonia.

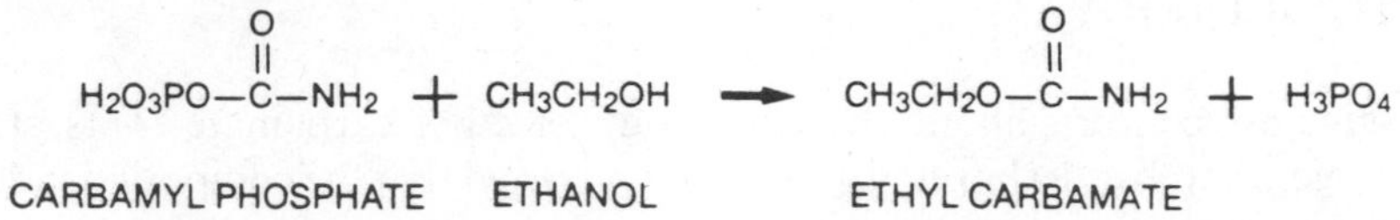

$$H_2O_3PO-\overset{\displaystyle O}{\overset{\|}{C}}-NH_2 \;+\; CH_3CH_2OH \;\longrightarrow\; CH_3CH_2O-\overset{\displaystyle O}{\overset{\|}{C}}-NH_2 \;+\; H_3PO_4$$

CARBAMYL PHOSPHATE   ETHANOL   ETHYL CARBAMATE

Figure 3. Ethanolysis of carbamyl phosphate to yield ethyl carbamate.

that one of the most likely sources of ethyl carbamate was ethanolysis of carbamyl phosphate as shown in Figure 3. His studies showed that the concentration of ethyl carbamate did indeed increase with increased levels of carbamyl phosphate in a model system.

A number of other constituents that may be present in fermentation mixtures have been examined for their potential to produce ethyl carbamate. Ough (1986) recently reported on results of incubation studies with urea, (which is known to react directly with ethanol to form ethyl carbamate, Merck Index, 1983), citrulline, arginine, and ornithine. Citrulline appeared to have an influence on ethyl carbamate formation. In later studies (Ough, unpublished results) arginine was clearly implicated in the formation of ethyl carbamate. However, Zimmerli et al. (1986) reported in their model studies that arginine produced no ethyl carbamate after incubation with ethanol at either room temperature or with heat. Their conclusions on work with stone-fruit brandies implicated the reaction of cyanhydric acid with 1,2-dicarbonyl compounds in the presence of ethanol. Certainly from work done in our laboratories and reported by Zimmerli et al. (1986), Andrey (1987), and Mildau et al. (1987), stone-fruit brandies have been found to have substantially higher levels of ethyl carbamate than wines or other distilled beverages, sometimes being greater than 1 mg/L.

Heating of alcoholic beverages has also been found to produce increased levels of ethyl carbamate. We have found that heating wine and sherry samples containing $< 20\,\mu g/L$ ethyl carbamate increased the levels to $200{-}300\,\mu g/L$. Bertrand and Triquet-Pissard (1986) generated ethyl carbamate levels in wine sometimes greater than 1 mg/L simply by heating the sample at 90°C for 6 hr. Ough (1986) also reported a positive influence of heat on ethyl carbamate formation. This effect may be the cause of some formation of ethyl carbamate in distilled spirits where fermentation mixtures are heated.

The distillation process has also been studied and ways in which the levels of ethyl carbamate could be reduced have been discussed (Bertrand and Triquet-Pissard, 1986; Tanner, 1986). The main problems arise from the difficulty in altering the brewing or distillation process to remove the substance while not affecting taste or appearance of the final product. Even if ethyl carbamate is removed from the fermentation mixture, the resulting clean distillate may re-form ethyl carbamate in significant amounts after storage, especially in the presence of light (Tanner, 1986).

The effect of light on ethyl carbamate formation has been also studied by Andrey (1987) and Mildau et al. (1987). Both reported increases in ethyl carbamate levels in fruit brandies when the samples were exposed to either

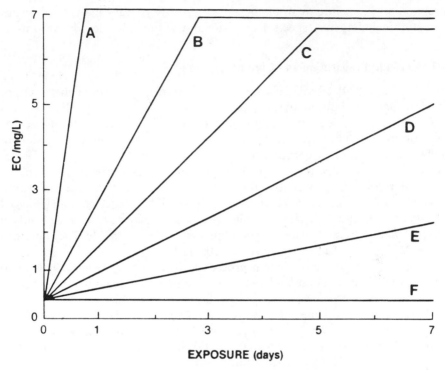

Figure 4.   Formation of ethyl carbamate in the presence of light. (Samples stored in colorless glass bottles.) (A) Daylight; (B) UV light 366 nm; (C) daylight (overcast); (D) incandescent light; (E) daylight (brown bottle); (F) stored in darkness. Samples contained 78% ethanol.

daylight or artificial light in clear glass bottles. It appeared that light between 330 and 340 nm had the greatest influence on ethyl carbamate formation. Figure 4 shows results obtained by Mildau et al. (1987) on the formation of ethyl carbamate due to light.

4.   DETERMINATION OF ETHYL CARBAMATE IN ALCOHOLIC BEVERAGES

Löfroth and Gejvall (1971) first determined ethyl carbamate at trace levels using isotopic dilution analysis with tritium-labeled diethyl pyrocarbonate to demonstrate the formation of ethyl carbamate in wine, beer, and orange juice. A radioimmunoassay technique for wine was attempted without success (Wikman-Coffelt and Berg, 1976). Gas chromatography (GC) has been the method of choice for determination of ethyl carbamate at μg/L or mg/L concentrations. The remainder of this section will discuss the various approaches taken for GC determination of ethyl carbamate. These include mainly a variety of sample preparation procedures, chromatographic separations, and the use of various detectors including mass spectrometry with and without chemical derivatization.

4.1. Sample Preparation and Chromatography

Walker et al. (1974) described the first GC procedure for ethyl carbamate at trace levels. Their method involved saturating wine with sodium chloride and extracting the resulting mixture with three successive portions of chloroform. The chloroform extracts were combined and filtered through anhydrous sodium sulfate and reduced to 10 mL using a Snyder column. The extract was further cleaned up by passage through a Florisil column. Ethyl carbamate was eluted with chloroform then concentrated to a small volume and then exchanged with ethyl acetate. The concentrated extract was analyzed by packed column GC (10% OV-17/5% carbowax 1540) and the compound detected with a Coulson electrolytic conductivity detector operating in the nitrogen mode. They obtained 80–100% recovery from wine spiked at 200 μg/L.

Essentially the same extraction procedure was used by Ough (1976a) who applied the method to foods and fruit juices in addition to alcoholic beverages. He obtained a mean recovery of 66.5% at spiking levels of 10–75 μg/L.

Joe et al. (1977) modified the procedure of Walker et al. (1974) by adjusting the pH of the wine to 7.5 before extraction and using methylene chloride for extraction of the salt-saturated wine. In addition, they employed two cleanup columns, one consisting of silica gel-deactivated alumina and the other of acid-Celite. The changes enabled the authors to detect as low as 1 μg/L ethyl carbamate in wine with recoveries at 10 μg/L averaging 71%. Detection was by either flame ionization (FID) or alkali flame ionization (AFID) although the latter was found to be more selective, being less susceptible to interferences.

In recent years improvements in GC resolution using capillary columns coupled with improved detectors such as the Hall electrolytic conductivity (HECD) or the nitrogen–phosphorus thermionic (NPD) detectors has enabled some simplification in the sample preparation process for alcoholic beverages. The use of mass selective detectors or complete mass spectrometers as routine GC detectors instead of only for confirmation has also enabled simplified sample preparation techniques for ethyl carbamate while still maintaining selectivity with excellent sensitivity.

A method developed by the Liquor Control Board of Ontario in Canada and later evaluated by the Canadian Federal Department of Health and Welfare (Conacher et al., 1987) involved essentially diluting between 15 and 50 g of the alcoholic beverage to about a 10% ethanol content then saturating the sample with sodium chloride and extracting with three successive volumes of methylene chloride. The extractions were carried out in large centrifuge bottles and the mixtures centrifuged to produce a rapid separation of the phases. The extracts were combined and passed through anhydrous sodium sulfate according to Walker et al. (1974) then ethyl acetate added as a keeper. The extract was reduced to about 2 mL by rotary evaporation then diluted exactly to 5 mL with ethyl

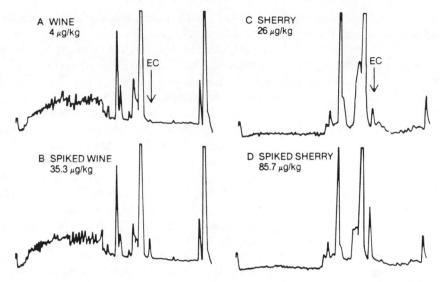

Figure 5. Chromatograms of ethyl carbamate in spiked and unspiked samples of wine and sherry. EC, ethyl carbamate. From Conacher et al. (1987), with permission of the Association of Official Analytical Chemists.

acetate for analysis by capillary GC with temperature programming on a 30-m × 0.33-mm-i.d. DB WAX (0.5-μm film) fused silica column using an HECD for detection. Mean recovery of ethyl carbamate from alcoholic beverages fortified at the Canadian guideline levels (see Table 1) was 81% with a coefficient of variation of 7%. Figures 5 and 6 show typical results obtained for wine, sherry,

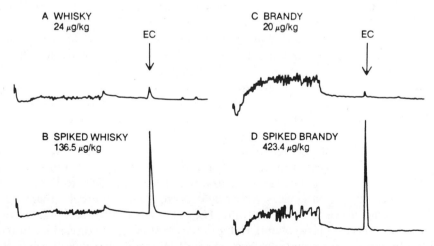

Figure 6. Chromatograms of ethyl carbamate in spiked and unspiked samples of whiskey and brandy. EC, ethyl carbamate. From Conacher et al. (1987), with permission of the Association of Official Analytical Chemists.

whiskey, and brandy. It can be seen that ethyl carbamate can be easily detected and quantitated at the regulatory guidelines.

Of all the methods available to date, the Canadian procedure has been evaluated most thoroughly. The method was subjected to a five-laboratory study, four using the Hall detector and one using a mass spectrometer. The overall between-laboratory coefficients of variation ranged from 12.5 to 21% depending on type of alcoholic beverage and the level of ethyl carbamate determined.

Other workers have reported similar sample preparation methods for ethyl carbamate. These involved methylene chloride extraction for GC–mass spectrometry determination of wines (Gaetanno and Matta, 1987) or distilled spirits (Aylott et al., 1987); the use of the same solvent for extraction from KCl-saturated alcoholic beverages and employing GC capillary column-switching and FID detection (van Ingen et al., 1987); and chloroform extraction of potassium carbonate saturated alcoholic beverages for capillary GC-mass spectrometry determination (Bebiolka and Dunke, 1987).

Cairns et al. (1987) described a procedure involving acetone extraction, followed by a water–dichloromethane partition and a Florisil column cleanup that yielded a determination limit of about 20 μg/L using an HECD. Bertrand and Triquet-Pissard (1986) employed ether to extract ethyl carbamate from the salt-saturated samples of brandy or wine.

Several workers have evaluated solid phase extraction as a means for isolating ethyl carbamate from alcoholic beverages. Dennis et al. (1986) used a Chemtube or Extrelute tube to adsorb samples of wine or whiskey that had been diluted to a final alcoholic content of less than 5%. The ethyl carbamate was removed from the cartridge with methylene chloride, which was then dried, concentrated, and passed through a Florisil Sep-Pak cartridge for further cleanup. Ethyl carbamate was eluted with 7% methanol in methylene chloride. The eluate was concentrated to about 1 mL for analysis by gas chromatography employing a CP Wax 52 CB column with either a thermal energy analyzer (TEA), a Hall, or a mass spectrometer as a detector. Wasserfallen and Georges (1987) used a similar approach. However, instead of diluting the sample they distilled off the ethanol before adding the liquid to a Chem-Elute tube. Baumann and Zimmerli (1986) used an Extrelute tube to adsorb 20 mL of beverage before removing nonpolar constituents with pentane. The ethyl carbamate was eluted with a mixture of methylene chloride and ethyl acetate (2:1) after the excess pentane was blown from the tube by a stream of nitrogen.

In certain types of alcoholic beverages where ethyl carbamate levels may be fairly high, direct injection of the sample into a gas chromatograph has been attempted using either mass spectrometry or nitrogen specific detection. Shultz and Renner (1986) used a mass selective detector to quantitate ethyl carbamate in fruit brandies down to about 100 ppb. They injected 1 μL of beverage onto a Carbowax 20M capillary column. Heisz et al. (1987) employed direct injection of plum brandy samples onto a Carbowax 20M capillary column using the split/splitless technique. The compound was detected selectively using an NPD. The range of interest was in the ppm range. While direct injections of samples, particularly distilled products, without cleanup are possible with selective detectors when the ethyl carbamate concentrations are in the ppm range, application to quantitation at low ppb levels may be more difficult, especially if done on a routine basis. Also, it is unlikely that nondistilled alcoholic beverages could be analyzed as easily at these concentrations.

4.2. Mass Spectrometry

Mass spectrometry has been used extensively for either the primary detection of ethyl carbamate or for confirmation of results obtained using other detection systems. Figure 7 shows a mass spectrum of ethyl carbamate using electron impact (EI) ionization from the work of Lau et al. (1987). The ion at m/z 74 corresponds to $[M—CH_3]^+$ $(CH_2={}^+O—CO—NH_2)$ and the base peak at m/z 62 derives from the "McLafferty $+1$" rearrangement reaction as shown in Figure 8. The ions at m/z 44 and 45, tentatively assigned as $O{=}C{=}^+NH_2$ and $[H(CO)NH_2]^+$, respectively, were not useful for diagnostic purposes due to interferences from chemical background at low resolution.

The m/z 62 ion appears to be most suitable as the quantitation ion not only because it is the strongest ion, but because it is susceptible to fewer interferences. For example, the m/z 74 ion is common to all alkyl methyl esters and thus is a less unique fragment. Using the m/z 62 fragment as little as 0.5 μg/L of ethyl carbamate in wine can be detected at a resolution of 4000.

Chemical ionization (CI) techniques using isobutane or methane have also been evaluated (Lau et al., 1987). Isobutane produces almost exclusively the $[M + H]^+$ ion at m/z 90 whereas methane CI yields m/z 62 and 90 ions of similar intensity. As low as 1 μg/L ethyl carbamate can be detected using methane CI.

Cairns et al. (1987) evaluated mass spectrometry/mass spectrometry for ethyl carbamate confirmation. Daughter ions were produced by collision-activated dissociation (CAD) with argon gas and were structrually related to the parent ion. However, comparison of the CAD results with simple CI mass spectrometry were

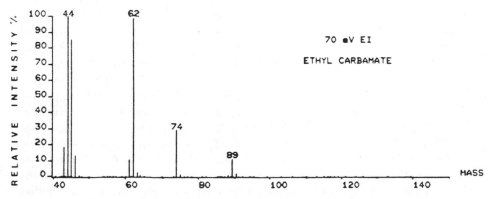

Figure 7.   Electron impact (70 eV) mass spectrum of ethyl carbamate.

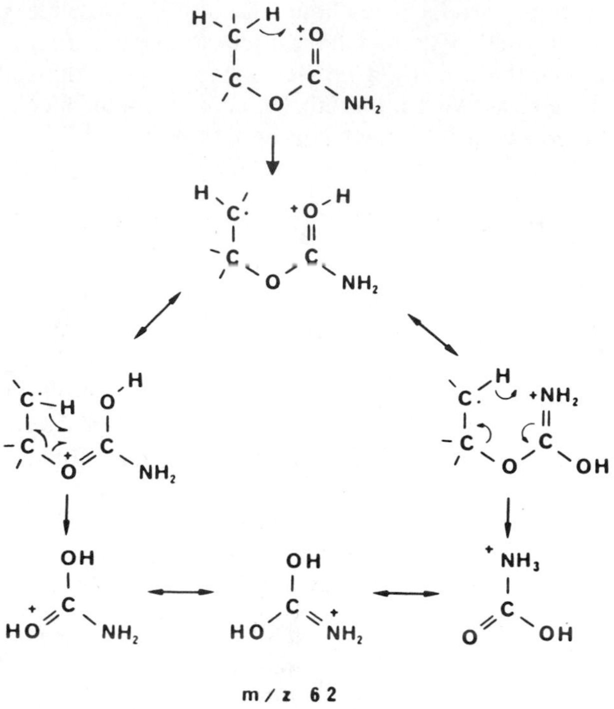

m / z   6 2

Figure 8.   "McLafferty + 1" rearrangement for ethyl carbamate.

not as good as anticipated. It appeared that optimization of the CAD parameters is very important in obtaining accurate results.

4.3.   Comparison of Detection Systems

A number of detectors have been used for determination of ethyl carbamate by gas chromatography. The Hall electrolytic conductivity detector has been used by a number of workers and because of its selectivity toward nitrogen containing compounds, it has enabled the detection of ethyl carbamate down to low μg/L levels with minimum sample preparation (Conacher et al., 1987). The nitrogen thermionic detector was found to be useful for the detection of ethyl carbamate in a variety of alcoholic beverages (Baumann and Zimmerli, 1986). Detection limits were about 20 μg/L. The flame ionization detector was found to be adequate for ethyl carbamate when two-dimensional capillary GC was employed (van Ingen et al., 1987). Those workers initially injected the sample extract onto a nonpolar (CP-SIL 5 CB) column and at the proper time switched columns so that the ethyl carbamate eluting from the first column was transferred to a second more polar (CP-WAX 52) column for separation and quantitation by flame ionization. As low as 1 μg/L ethyl carbamate could be detected although the quantitation limit was about 10 μg/L.

TABLE 2   Comparison of Ethyl Carbamate Results (μg/L) Obtained with different Detectors

	Detector		
Sample	Mass Spectrometry	TEA	Hall
Bourbon whiskey	(a) 216	204	176
	(b) 212	208	184
Scotch whiskey	(a) 75	80	72
	(b) 77	99	84
Red wine	(a) 22	16	13
	(b) NA[a]	13	10

[a]NA, not analyzed.

In a comparison of GC detectors used on identical extracts, Conacher et al. (1987) found that the Hall detector performed equally as well as mass spectrometric detection above 20 μg/L ethyl carbamate for both distilled and nondistilled alcoholic beverages. However, the mass spectrometric detection limits were about 10 times better than the Hall, being about 0.5 μg/L.

Dennis et al. (1986) compared results obtained by mass spectrometry with those obtained using either a thermal energy analyser or a Hall detector. Table 2 shows results they obtained for two whiskey and a red wine sample. The results agreed very well. The thermal energy analyzer, however, was preferred to the Hall detector for routine use because of better reliability and slightly better detection limits (1 μg/L, compared to the Hall, 2–5 μg/L). The detection limit with mass spectrometric detection was 1 μg/L with the sample cleanup they employed.

4.4. Chemical Derivatization

Several attempts at determining ethyl carbamate by gas chromatography after chemical derivatization have been reported. Walker et al. (1974) formed the trifluoroacetate derivative as shown in Figure 9 and confirmed the structure of the product by mass spectrometry. Although the TFA derivative was significantly more sensitive with the electron capture detector than with the alkali-flame ionization detector, the former was not useful for actual sample analysis due to sample interferences. The authors used the derivatization technique to confirm results obtained by direct determination using the Coulson electrolytic conductivity detector. Detection limits using packed column GC were about 100 μg/L ethyl carbamate.

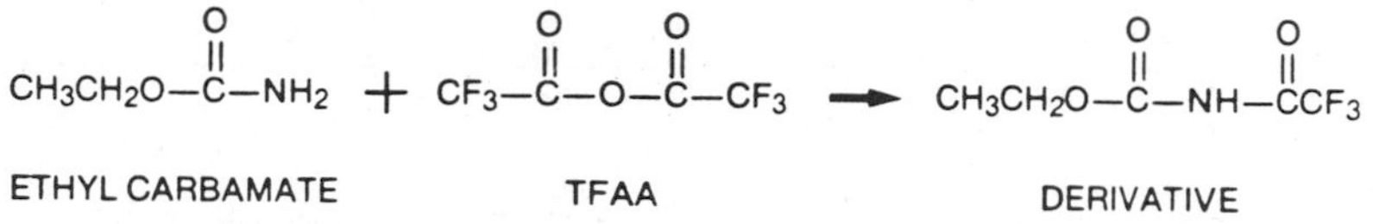

$$CH_3CH_2O-\overset{\overset{\text{O}}{\|}}{C}-NH_2 \ + \ CF_3-\overset{\overset{\text{O}}{\|}}{C}-O-\overset{\overset{\text{O}}{\|}}{C}-CF_3 \ \longrightarrow \ CH_3CH_2O-\overset{\overset{\text{O}}{\|}}{C}-NH-\overset{\overset{\text{O}}{\|}}{C}CF_3$$

ETHYL CARBAMATE   TFAA   DERIVATIVE

Figure 9.   Reaction of trifluoroacetic anhydride (TFAA) with ethyl carbamate.

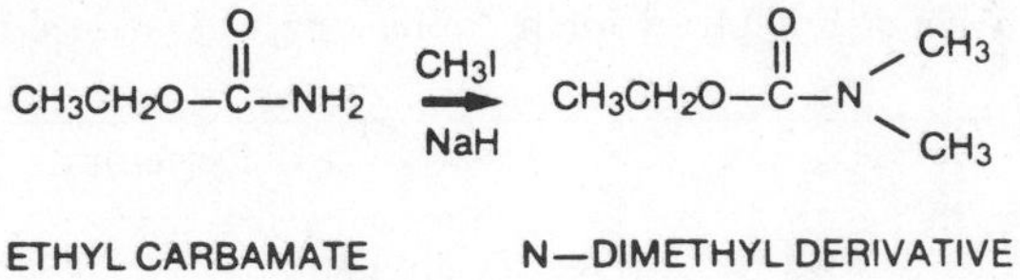

$$CH_3CH_2O-\overset{\overset{O}{\|}}{C}-NH_2 \quad \xrightarrow[\text{NaH}]{CH_3I} \quad CH_3CH_2O-\overset{\overset{O}{\|}}{C}-N\overset{CH_3}{\underset{CH_3}{<}}$$

ETHYL CARBAMATE   N—DIMETHYL DERIVATIVE

Figure 10. Reaction of ethyl carbamate with methyl iodide to form the N-dimethyl derivative.

Bailey et al. (1986) studied the response of ethyl carbamate and its *N*-dimethyl derivative to Hall electrolytic conductivity and nitrogen thermionic detection and found substantial differences. The product was prepared using methyl iodide in the presence of sodium hydride as shown in Figure 10.

These authors observed a 10-fold increase in sensitivity by thermionic detection for the derivative as compared to the parent compound while with the Hall detector the dimethyl product was about 10 times less sensitive. They attributed these differences to the detection mechanisms involved. For actual sample analysis the thermionic detector proved fully satisfactory for the derivative. Figure 11 shows typical results for a wine sample. Detection limits were comparable to those obtained for direct ethyl carbamate analysis using the Hall detector.

Kobayashi et al. (1987) prepared an alkylated derivative of ethyl carbamate for application to its detection in sake. The reaction is shown in Figure 12. The

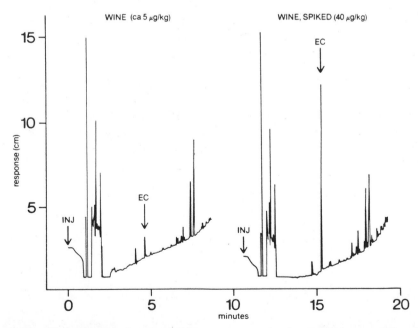

Figure 11. Chromatogram of a wine sample containing ∼ 5 μg/L ethyl carbamate (EC) and the same sample spiked with 40 μg/L. Sample extracts derivatized with methyl iodide before analysis.

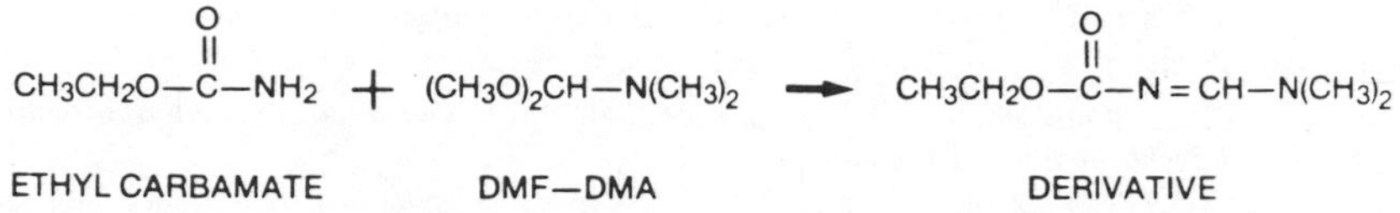

Figure 12. Reaction of ethyl carbamate with N,N-dimethylformamide–dimethylacetal (DMF–DMA).

derivative was detected by GC using FID. Because the nonselective FID was employed sample purification required extraction from the sake followed by Florisil column cleanup, then derivatization and isolation of the product using solid phase extraction with an Extrelute column. The results were confirmed by GC–mass spectrometry.

4.5 CONCLUSION

The majority of toxicological evidence indicates that ethyl carbamate is likely a human carcinogen and thus efforts to minimize its presence in alcoholic beverages is strongly justified. Although recent research indicates that the compound can form as the result of the reaction of ethanol with various components in the beverages and as the result of certain treatments (e.g., heat and light), the full understanding of its production is not yet complete. The problems associated with its removal stem from the difficulty in maintaing the integrity of the beverages in terms of palatibility after any special treatment to remove the ethyl carbamate. Research in this area is continuing at a good pace.

Good analytical methodology is essential for monitoring the presence of ethyl carbamate in alcoholic beverages. Many methods have appeared in the last few years. The most popular seem to be solvent extraction (or solid phase extraction) of the samples followed by direct analysis of the extracts by capillary GC using a polar column (Carbowax type) with Hall electrolytic conductivity or mass spectrometric detection. In the cases of distilled spirits direct sample injection into capillary GC systems with mass spectrometric detection appears to be adequate even for routine analysis.

REFERENCES

Adamson, R. H., and Sieber, S. M. (1983). Chemical carcinogenesis studies on non-human primates. In R. Langenbach, S. Nesnow, and J. M. Rice, eds., *Organ and Species Specificity in Chemical Carcinogenesis*. Plenum Press, New York, pp. 129–156.

Allen, J. W., Sharief, Y., and Langenbach, R. (1982). An overview of ethyl carbamate (urethane) and its genotoxic activity. In R. R. Tice, D. L. Costa, and K. M. Schaich, eds., *Gentoxic Effects of Airborne Agents*. Plenum Press, New York, pp. 443–460.

Allen, J. W., Stoner, G. D., Pereira, M. A., Backer, L. C., Sharief, Y., Hatch, G. G., Campbell, J. A., Stead, A. G., and Nesnow, S. (1986). Tumorigenesis and genotoxicity of ethyl carbamate and vinyl carbamate in rodent cells. *Cancer Res.* **46**, 4911–4915.

Alyott, R. I., McNeish, A. S., and Walker, D. A. (1987). Determination of ethyl carbamate in distilled spirits using nitrogen specific and mass spectrometric detection. *J. Inst. Brew.* **93**, 382–386.

Andrey, D. (1987). A simple gas chromatography method for the determination of ethyl carbamate in spirits. *Z. Lebensum. Unters. Forsch.* **185**, 21–23.

Bailey, R., North, D., Myatt, D., and Lawrence, J. F. (1986). Determination of ethyl carbamate in alcoholic beverages by methylation and gas chromatography with nitrogen-phosphorus thermionic detection. *J. Chromatogr.* **369**, 199–202.

Baumann, U., and Zimmerli, B. (1986). Gas chromatographic determination of urethane (ethyl carbamate) in alcoholic drinks. *Mitt. Gebiete. Lebensm. Hyg.* **77**, 327–332.

Bebiolka, H., and Dunkel, K. (1987). Determination of ethyl carbamate in alcoholic beverages with capillary gas chromatography/mass spectrometry. *Deutsche Lebens. Rundsch.* **83**, 75–76.

Bertrand, A., and Triquet-Pissard, R. (1986). Ethyl carbamate in wine brandies: observations on its origin. *Cannaissance Vigne Vin* **20**, 131–136.

Cairns, T., Siegmund, E. G., Luke, M. A., and Doose, G. M. (1987). Residue levels of ethyl carbamate in wines and spirits by gas chromatography and mass spectrometry/mass spectrometry. *Anal. Chem.* **59**, 2055–2059.

Conacher, H. B. S., Page, B. D., Lau, B. P., Lawrence, J. F., Bailey, R., Calway, P., Hanchay, J. P., and Mori, B. (1987). Capillary column gas chromatographic determination of ethyl carbamate in alcoholic beverages with confirmation by gas chromatography-mass spectrometry. *J. Assoc. Off. Anal. Chem.* **70**, 749–751.

Cowan, P. N. (1950). Strain differences in mine to the carcinogenic action of urethane and its noncarcirogenicity in chicks and guinea pigs. *Br. J. Cancer* **5**, 245–253.

Dahl, G. A., Miller, J. A., and Miller, E. C. (1978). Vinyl carbamate as a promutagen and a more carcinogenic analogue of ethyl carbamate. *Cancer Res.* **38**, 3793–3804.

Dahl, G. A., Miller, E. C., and Miller, J. A. (1980). Comparative carcinogenicities and mutagenicities of vinyl carbamate, ethyl carbamate and ethyl N-hydroxycarbamate. *Cancer Res.* **40**, 1194–1203.

Dennis, M. J., Howarth, N., Massey, R. C., Parker, I., Scotter, M., and Startin, J. R. (1986). Method for the analysis of ethyl carbamate in alcoholic beverages by capillary gas chromatography. *J. Chromatogr.* **369**, 193–198.

Federal Register, (1972). August 2, **37** (149), 15426.

Food Chemical News. (1987). DISCUS proposes 125 ppb ethyl carbamate action level in whiskey. *Food Chem. News* June 15, 17–20.

Gaetano, G., and Matta, M. (1987). Determination of urethane in wines and spirits by gas chromatography/mass spectrometry. *Bull. O. I. V.* **60**, 671–672.

Heisz, O., Fritsch, H., and McCreadie, S. W. S. (1987). Ethyl carbamate in slivowitz-development of the method. *Labor Praxis* **11**, 306–310.

IARC Monograph. (1974). 'Urethane' In *IARC Monograph on the Evaluation of Carcinogenic Risk of Chemicals to Man—Some Antithyroid and Related Substances, Nitrofurans and Industrial Chemicals, Vol.* 7, Lyon, pp. 111–140.

Joe, F. L., Jr., Kline, D. A., Miletta, M., Roach, J. A. C., Roseboro, E. L., and Fazio, T. (1977). Determination of urethane in wines by gas-liquid chromatography and its confirmation by mass spectrometry. *J. Assoc. Off. Anal. Chem.* **60**, 509–516.

Kobayashi, K., Toyoda, M., and Saito, Y. (1987). Determination of ethyl carbamate in sake by alkylation and gas chromatography with a flame ionization detector. *J. Food Hyg. Soc. Jpn.* **28**, 330–335.

Lau, B. P. Y., Weber, D., and Page, B. D. (1987). Gas chromatography-mass spectrometric determination of ethyl carbamate alcoholic beverages. *J. Chromatogr.* **402**, 233–241.

Löfroth, G., and Gejvall, T. (1971). Diethyl pyrocarbonate: Formation of urethane in treated beverages. *Science* **174**, 1248–1250.

Merck Index. (1983). Urethan. *Merck Index*, 10th ed. Merck & Co., Rahway, p. 1411.

Mildau, G., Preuss, A., Frank, W., and Heering, W. (1987). Ethyl carbamate in alcoholic drinks: Improved analysis and formation due to light. *Deutsche Lebensm. Rundsch.* **83**, 69–74.

Mirvish, S. S. (1968). The carcinogenic action and metabolism of urethane and N-hydroxyurethane. *Adv. Cancer Res.* **11**, 1–42.

Nettleship, A., Henshaw, P. S., and Meyer, H. L. (1943). Induction of pulmonary tumors in mice with ethyl carbamate (urethane). *J. Natl. Cancer Inst.* **4**, 309–319.

Ough, C. S. (1976a). Ethyl carbamate in fermented beverages and foods. I. Naturally occurring ethyl carbamate. *J. Agric. Food Chem.* **24**, 323–328.

Ough, C. S. (1976b). Ethyl carbamate in fermented beverages and foods. II. Possible formation of ethyl carbamate from diethyl pyrocarbonate addition to wine. *J. Agric. Food Chem.* **24**, 328–331.

Ough, C. S. (1986). Formation of ethyl carbamate during food and beverage processing. *Proc. Toxicol. Forum, Aspen*, July 14–18, 1986, pp. 387–392.

Shapiro, J. R., and Kirschbaum, A. J. (1951). Intrinsic tissue response to induction of pulmonary tumors. *Cancer. Res.* **11**, 644–647.

Shultz, J., and Renner, R. (1986). Ethyl carbamate (urethane) in fruit brandy-fast identification and quantitative determination with the mass selective detector (MSD). *GIT Fachz. Lab.* **30**, 1215–1220.

Tanner, H. (1986). Experiments to avoid high urethane concentrations in distilled beverages. *Proc. Euro Food Tox II, Zurich*, October 15–18, 1986, p. 255.

van Ingen, R. H. M., Nijssen, L. M., van den Berg, F., and Maarse, H. (1987). Determination of ethyl carbamate in alcoholic beverages by two-dimensional gas chromatography. *J. H. Res. Chrom.* **10**, 151–152.

Walker, G., Winterlin, W., Fouda, H., and Seiber, J. (1974). Gas chromatographic analysis of urethan (ethyl carbamate) in wine. *J. Agric. Food Chem.* **22**, 944–947.

Wasserfallen, K., and Georges, P. (1987). Gas chromatographic determination of urethane in spirits and mashes. *Z. Lebensm. Unters Forsch.* **184**, 392–395.

Wikman-Coffelt, J., and Berg, H. W. (1976). Radioimmunoassay method for analysis of ethyl carbamate in wine. *Am. J. Enol. Vitic.* **27**, 115–117.

World Health Organization. (1972). Joint FAO/WHO Committee on Food Additives, Series No. 4, pp. 67–74.

Zimmerli, B., Baumann, U., Nageli, P., and Battaglia, R. (1986). Occurrence and formation of ethyl carbamate (urethane) in fermented foods. Some preliminary results. *Proc. Euro Food Tox II, Zurich*, October 15–18, 1986, pp. 243–248.

15

FOOD RESIDUES FROM PESTICIDES AND ENVIRONMENTAL POLLUTANTS IN ONTARIO

Richard Frank and Brian D. Ripley

Agricultural Laboratory Services Branch, Ontario Ministry of Agriculture and Food, Guelph, Ontario, Canada

7.4.   Soil Residues of Persistent Organochlorine Insecticides in Food and
Tobacco

8. Conclusions
References

This chapter addresses topics on the issue of pesticide and environmental
pollutant residues in Canadian produced food and feed, with special emphasis
on Ontario.

1. PESTICIDE USE IN CANADIAN AND ONTARIO AGRICULTURE

In 1982 total cash receipts of $19.5 billion were generated by 313,490 commercial
farms across Canada. Costs of production accounted for $13.6 billion, leaving a
net of $5.9 billion. Statistics on feed and food production for 1985 appear in
Table 1. Wheat production amounted to 24 million tonnes, of which 58% was
exported. Other cereal crops exported were rye 34%, barley 2%, and grain corn
6.4%. Domestically, 76% of the oats, 55% of the barley, 63% of the grain corn, and
23% of the rye were fed to 12 million cattle and calves, 11 million hogs, 88 million
poultry, and 61 thousand sheep raised on Canadian farms for meats, eggs, and
dairy products.

In 1985 Canadian farmers bought pesticides worth a retail value of $671
million (Table 2). The greatest volume of the sales was for herbicides in the
production of cereal grains in the Prairie Provinces; in fact, almost 74% of the
total Canadian pesticide sales or 79% of the total herbicide sales were for use on
the Prairies and 2, 4-D and MCPA were the major herbicides. On a nation-wide

TABLE 1   Food and Feed Production in Canada during 1985

Commodity	Production (metric tonnes)	Commodity	Production (metric tonnes)
Wheat	23,899,000	Corn grain	7,293,000
Oats	2,997,000	Corn fodder	10,127,000
Barley	12,247,000	Buckwheat	28,000
Rye	598,000	Peas, dry	181,000
Mixed grains	1,446,000	Soybeans	1,063,000
Flaxseed	920,000	Beans (dry) and lentils	119,000
Canola/rapeseed	3,463,000	Mustard seed	133,000
Sunflower seed	82,000	Hay	23,788,000
Fruit	725,967	Vegetables	4,971,000
Maple syrup	14,000	Meats	2,693,000
Honey and wax	78,000	Milk	2,682,000

TABLE 2   Retail Sales of Agricultural Chemicals in Canada, 1984[a]

Sales in Millions of Dollars

| Pesticides | Western Canada (Prairie Provinces and British Columbia) | Eastern Canada | | Total | % |
		Ontario	Quebec and Maritimes		
Herbicides	453	92	28	573	85.4
Seed treatment	21	3	1	25	5.7
Insecticides	14	20	11	45	6.7
Fungicides	7	10	11	28	4.2
Total	495	125	51	671	
%	73.8	18.6	7.6		100

[a]Elliott (1987).

basis 85% of the pesticide sales were herbicides and only 6.7% were insecticides and 4.2% fungicides.

The details on pesticide use in Canada are not known; neither is the detailed picture on residues. A much more complete picture is available for the Province of Ontario and hence Ontario has been chosen for an in-depth study of the issue.

The Province of Ontario covers a land area of 91.74 million hectares, of which 6.04 million hectares or 6.6% is included in census farms. Improved land for the economic production of food, feed, and fiber occupies 4.16 million hectares or 4.5% of the Province. The overall inventory of land utilization among these agricultural holdings appears in Table 3 and is based on a 1985 economic survey. The amounts of pesticides used on these lands were based on a 1983 farm survey (Table 3).

Every 5 years the Ontario Ministry of Agriculture and Food conducts farm surveys into pesticide use in agriculture. Surveys were carried out in 1973 (Roller, 1975), 1978 (Roller, 1979), and 1983 (Magee, 1984). In 1983, 8719 metric tonnes of pesticides was used on 2.142 million hectares or 51.5% of the area in agricultural production (Table 3). Figure 1 shows the total current amount of pesticides applied to the various crops. The mean application rate of all the various types of pesticides on the different crops is shown in Figure 2. In 1983, herbicides represented 63.3% of the total volume of pesticides used, most of which were applied around planting time and hence left little or no residues in crops or soil at harvest time. Most of those pesticides leaving persistent residues in soils were used to produce animal feeds, especially corn silage (214,480 ha) and grain corn (902,436 ha), where 3653 tonnnes of herbicides or 66.2% of the total volume of herbicides was used.

Nematocides represented 18.5% of the total volume of pesticides used. Most of these were applied 2 weeks prior to transplanting tobacco and no detectable residues have been identified on the cured tobacco leaf.

TABLE 3 **Areas of Ontario Devoted to Food and Feed Production in Ontario[a] and the Volume of Pesticide Use from the 1983 Farm Survey[b]**

Crops	Total Area (× 10³ ha)	Treated Area (%)	Amounts Applied in Metric Tonnes (%)					
			Fungicides	Growth Regulators	Herbicides	Insecticides	Nematocides	All Pesticides
Cereals	852	62.8	0.1(<0.1)	—	376.1(6.8	<0.1(<0.1)	—	376.3(4.3)
Corn	1052	98.8	—	—	3653.3(66.2)	145.2(33.7)	—	3798.5(43.6)
Field beans	32	98.4	—	—	57.4(1.0)	1.3(0.3)	—	58.7(0.7)
Fruit	28.4	100	411(72.5)	—	7.89(0.1)	143.8(33.4)	—	562.6(6.4)
Hay and Pasture	1720	1.9	—	—	24.7(0.4)	0.24(<0.1)	—	24.9(0.3)
Soybean	364	99.2	—	—	1281.9(23.2)	—	—	1281.9(23.2)
Tobacco	40.5	100.0	1.8(0.2)	590.8(100)	17.8(0.3)	39.9(9.3)	1610.6(100)	2260.2(25.9)
Vegetables	72.9	100.0	154.7(27.3)	—	100.9(1.8)	100.5(23.3)	—	356.1(6.4)
Total	4161.8	51.5	566.9	590.8	5520	430.9	1610.6	8719.2
Percent	—	—	6.5	6.8	63.3	4.9	18.5	

[a]OMAF (1986).
[b]Magee (1984).

476

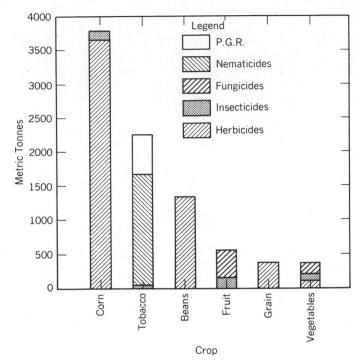

Figure 1. Histogram of the total pesticide volumes (metric tonnes) used by type of crop with a breakdown of the major groups of pesticides. P.G.R., plant growth regulators.

Growth regulators represented 6.8% of the total volume of pesticides used and these were applied mainly to tobacco. This did not include maleic hydrazide (MH), which is not registered and therefore not permitted for use on tobacco. In fact since 1968 the use of MH has been actively discouraged by a policing action that has involved sampling, analyzing, and identifying the sprayed crop on the sales floor. Buyers penalized growers by offering lower prices and this has maintained a crop free of residues (Table 4). The growth regulators include the C_{10} alcohols that leave only trace residues on cured tobacco leaf.

Fungicides represented 6.5% of the total volume of pesticides used in Ontario and 99.7% of this volume was applied to fruit and vegetables. A portion of these fungicides is used directly on ripening fruit and maturing vegetables and hence residues are frequently present on harvested food commodities.

Insecticides represented 4.9% of the total volume of pesticides used in Ontario. The major portion, 56.7%, was applied to fruit and vegetables, much of this being close to harvest and hence with a potential to leave terminal residues. A further 33.7% was used on corn at planting time and hence leaving little or no terminal residues in food products.

An expansion of the data found in Table 3 appears in Tables 5 and 6 and gives the extent of use of fungicides, insecticides, and herbicides on specific fruits and

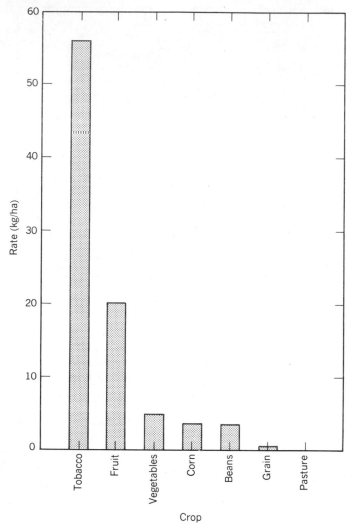

Figure 2. Histogram of the rate of total pesticide applied per hectare and by type of crop.

vegetables. It can be observed that almost 100% of fruit crops are treated with fungicides and insecticides, while use on vegetables varies widely from commodity to commodity. The actual volume of total active ingredient used varies from 1.11 kg/ha for peas to 32.9 kg/ha for plums.

2. LEGISLATIVE CONTROLS ON PESTICIDE SALES AND USE IN CANADA

Before a farmer can purchase and use a pesticide, the data packages on a compound must satisfy several acts and regulations at both the federal and

TABLE 4   Ontario Grown Flue Cured Tobacco[a] Analyzed for Maleic Hydrazide 1968–1985[b]

Year	Samples Analyzed	Samples with MH	Estimated Use on Violations (kg/ha)
1968–1970	14,263	17	0.6–3.0
1970–1985	16,001	4	0.6–1.9
1975–1980	9,383	17	0.5–3.0
1980–1985	8,824	9	0.5
Total	48,471	47 (0.1%)	

[a]Production 77 to 100 million kg/annum.
[b]Braun et al. (1983).

provincial levels. Federal legislation controls pesticide registration, the sale, transportation, and residues in feed and food. Provincial legislation controls storage, handling, and field application.

There is considerable legislation in Canada that deals with pests and pesticides. Most of it is written not only to protect human and animal health, but to safeguard environmental quality, by ensuring that pesticides are used in a proper and knowledgeable manner. In this section only the more important pieces of the legislation are veviewed as they pertain to foods residues. While there is both federal and provincial legislation, provincial acts and regulation are largely a refinement of federal acts and hence provincial legislation will be only briefly discussed.

2.1. Pest Control Products Act

This is a federal statute under which every pesticide product must be registered by Agriculture Canada before it can be sold in Canada. Before registration is granted, the manufacturer must provide scientific evidence that the pesticide product is effective for the claims made on the label and that it is safe when used as directed. Registration is product specific meaning each pesticide product must be supported with a datum package.

When applying to Agriculture Canada for registration of a new active ingredient, a company must submit the following types of data:

1. Physical and chemical properties of all ingredients in the pesticide product.
2. Toxicology studies for both the active ingredients and the formulated products:
   a. Acute studies from oral, dermal, inhalation, and ocular irritation exposure.
   b. Long- and short-term chronic toxicity feeding studies on rodents and nonrodents.
   c. Special studies on reproduction, teratology, mutagenicity, and carcinogenicity.

TABLE 5 Major Fruit Crops Produced in Ontario, 1985[a] and Pesticide Use on These Crops from the 1983 Farm Survey[b]

Commodity	Area Grown (ha)	Area Treated (%)			Rate on Total Area (kg/ha)			
		Fungicides	Insecticides	Herbicides	Fungicides	Insecticides	Herbicides	Total
Apples	10,484	99.7	100	2.8	22.58	6.13	0.06	28.77
Cherries	1,301	100	100	14.8	14.20	4.42	0.12	18.75
Grapes	9,752	100	100	37.3	21.53	3.28	0.33	25.13
Peaches	3,028	100	100	38.0	12.38	6.59	0.46	19.43
Pears	1,342	96.0	100	28.5	13.06	5.38	0.17	18.61
Plums	758	97	100	37.9	26.18	6.16	0.55	32.89
Raspberries	343	86.3	20.0	39.8	4.19	2.18	1.89	8.26
Strawberries	1,665	82.5	90.5	97.7	5.74	1.48	1.07	8.29

[a]OMAF (1986).
[b]Magee (1984).

TABLE 6 Major Vegetable Crops Produced in Ontario, 1985[a] and Pesticide Use on These Crops from the 1983 Farm Survey[b]

Commodity	Area Grown (ha)	Area Treated (%)			Rate on Total Area (kg/ha)			
		Fungicides	Insecticides	Herbicides	Fungicides	Insecticides	Herbicides	Total
Asparagus	1,541	0	99.7	38.9	0	1.55	1.61	3.16
Beans	2,763	74.1	0.4	25.9	0.74	0.01	0.69	1.44
Beets	253	0	0	100	0	0	4.73	4.73
Cole crops	4,137	38.8	100	33.9	1.17	1.91	0.13	3.21
Carrots	2,148	59.0	79.8	58.4	4.60	2.64	1.09	7.72
Corn, sweet	18,130	0	99.7	38.9	0	1.32	1.45	2.77
Cucumbers	1,931	84.2	89.8	84.5	2.09	1.61	2.14	5.85
Onions	2,534	81.1	97.7	67.5	5.52	2.48	8.52	16.52
Peas	7,227	0	3.8	56.4	0	0.43	0.68	1.11
Peppers	1,369	56.6	94.9	87.5	1.35	3.16	0.80	5.33
Potatoes	15,188	50.8	97.8	71.0	2.96	2.19	1.23	6.38
Pumpkin and squash	439	16.4	52.2	62.5	0.58	0.76	1.37	2.71
Tomatoes	12,781	100	82.6	72.9	6.10	0.95	1.02	8.07

[a]OMAF (1986).
[b]Magee (1984).

3. Terminal residues in or on foods with analytical methodology and metabolic studies on plants and animals.
4. Environmental studies on degradation, metabolism, and disposal including toxic effects on birds, mammals, aquatic organisms, predators, parasites, and honeybees.
5. Efficacy studies.

These data packages are reviewed by specialists in four federal departments:

1. Agriculture Canada—pass judgment on the acceptability of analytical methods and pest control effectiveness.
2. Health and Welfare Canada—pass judgment on the acceptability of the toxicology and residue data relative to preharvest intervals and food residue tolerances.
   a. Assesses occupational, environmental, and public health risks and effectiveness of protective clothing.
   b. Assesses health effects of residues in potable water.
3. Environment Canada—assesses disposal of pesticide containers and unwanted products, decontamination of spills.
   a. Evaluates environmental persistence and fate.
4. Fisheries and Oceans Canada—evaluates toxicity to aquatic organisms.

Pesticide products which were registered prior to the present stringent evaluation are currently being reviewed and where important data gaps on health or environmental effects occur, manufacturers are being asked to submit the missing data. Pesticide products that after retesting pose an imminent health problem have been removed from the market by cancellation or suspension of the registration by Agriculture Canada or have been voluntarily withdrawn by the manufacturer.

Full registrations are granted for a 5-year period and renewals depend on the product meeting the latest prescribed standards. Once registration has been approved, each product is assigned a Pest Control Products Registration Number that must appear on the label. It is an offence for vendors to sell or for applicators to apply pesticides for any purposes other than those outlined on the label.

Products are granted one of the following registration statuses:

1. Full Registration—the data packages are complete and no further questions are outstanding with the Registrant.
2. Temporary Registration—the data packages are not complete and there are studies to be completed or questions to be answered; however, these are of a nonpivotal nature.
3. Temporary (Restricted Class) Registration—the data packages are incomplete and the registration is granted because of an emergency need. Pivotal studies are lacking and major questions are unanswered.

2.2. Food and Drugs Act

Through the authority of this act, the Health Protection Branch of Health and Welfare Canada has the regulatory power to prohibit the sale of any food that is adulterated or has in or on it any poisonous or harmful substances. Hence, the sale of a food may be prohibited if residues of a pesticide are found in or on it. In certain cases, however, where it has been determined by appropriate toxicity studies and based on acceptable daily intake that small amounts of a pesticide may be safely ingested, provision is made under the Food and Drug Regulations to exempt these pesticides from the act providing the amount of residue does not exceed the maximum residue limit (MRL) listed in the regulations. Where no MRL has been established for a pesticide or a commodity, a residue must be at the "negligible level," normally below 0.1 mg/kg but in some cases below 0.05 mg/kg. The numbers of pesticides that have an MRL are listed in Table 7.

Crops that carry chemical residues in excess of the established MRL are subject to seizure by the Health Protection Branch. A total of 182 active ingredients appear on the list for enforcement in human food. The list includes pesticides registered in Canada and also some that can appear on imported commodities (Table 7). This list does not include tolerances for residues in feeds or tobacco; no MRL exist for tobacco and allowable levels in feed fall under the Feeds Act.

2.3. Pesticide Residue Compensation Act

Farmers whose crops have been seized by the Health Protection Branch under authority of the Food and Drugs Act, may be compensated under the Pesticide Residue Compensation Act, provided the pesticide was applied strictly according to label directions or recommendations in federal government publications.

TABLE 7   Number of Pesticides Listed under Canadian Food and Drugs Act and Regulations with and without Maximum Residue Limits (MRL)[a]

	MRL		No MRL		
Pesticide	Canadian Registration	No Canadian Registration	Canadian Registration	No Canadian Registration	Total
Growth regulators	6	0	4	0	10
Herbicides	9	0	47	10	66
Fungicides	21	2	9	0	32
Insecticides	41	10 (6)[b]	23	0	74
Total	77	12	83	10	182

[a]Health and Welfare (1986).
[b]Six insecticides (persistent organochlorines) used in the past, registrations now cancelled. MRL retained to allow for environmental residues in food.

2.4. Plant Quarantine Act

This statute is administered by Agriculture Canada and its purpose is to prevent
the introduction or spread of insects, nematodes, plant diseases, and other pests
that may be destructive to Canadian agricultural or forestry crops. The agency
can require the treatment of a serious pest outbreak with a specific pesticide. If
this should result in the contamination of food with a pesticide concentration
above the MRLs, then the food must be destroyed, and compensation paid.

2.5. Feeds Act

Agricultural Canada administers this Act, which provides for the regulation and
control of livestock feeds offered for sale. Feeds containing pesticides are
registered under the Feeds Act; however, the requirements of the Pest Control
Products Act apply with respect to efficacy and safety. No feed can be sold
containing a material that could leave residue levels of a poisonous substance in
the tissues of an animal that may be offered for sale and would violate the Food
and Drug regulations.

   The regulations made pursuant to the Feeds Act include (1) a list of deleterious
substances, (2) standards for registration of feed/pesticide combinations, and (3)
procedures for cancellation and suspension of feed/pesticide combinations.

2.6. Other Federal Acts

There are several other acts that affect the sale and movement of pesticides and
directly or indirectly affect food residues.

1. The Fertilizer Act regulates the registration of fertilizer–pesticide
    combinations.
2. The Seed Act regulates the requirement that grain treated with a poisonous
    material shall be thoroughly stained with a conspicuous color or the
    container must be labeled as prescribed under the Pest Control Products
    Act. This is to prevent such treated food entering the food chain, either as
    human food or animal feeds.
3. The Meat Inspection Act regulates germicides, insecticides, and roden-
    ticides that can be used in a registered establishment if they meet the MRL
    laid out in the Food and Drugs Act and Regulations.
4. The Fruit, Vegetable and Honey Act regulates sanitation and processing of
    products as to meet food residue standards of pesticides defined in the Food
    and Drugs Act and Regulations.
5. The Canada Grain Act regulates the handling and grading of grain for
    domestic and export markets. It prohibits selling or storing grain treated
    with mercurials or other poisonous substances at licensed elevators. The act
    requires that infested grain shall be treated according to the directives from
    the Canadian Grain Commission, Winnipeg, Manitoba and makes other

provisions with respect to grain fumigation. Grain must thereafter meet the provisions of the Food and Drugs Act and Regulations before sale.

2.7. Provincial Legislation

Most provinces have legislation covering the use and abuse of pesticides that allows for (1) licensing of vendors and applicators, (2) scheduling of products for special handling by restricting certain uses considered unduly hazardous, (3) restricting some uses to special permits, (4) certifying growers wanting to buy and handle highly toxic pesticides, and (5) regulating the disposal of empty containers.

After a registered label is granted, manufacturers normally seek provincial approval either from a legal or/and a promotional point of view. Some provinces require that a registered product must first be scheduled before sale. Most Provincial Governments have annual crop production guides for food producers detailing recommended pesticide products and most manufacturers seek approval in these publications.

3. OCCURRENCE AND FATE OF PESTICIDE RESIDUES

A detailed discussion of the large number of factors that affect the deposition, dissipation, persistence, and removal of pesticides on treated crops and the resultant food products is beyond the scope of this article and the reader is referred to other reviews. However, some brief comments are necessary to put later sections in perspective.

The placement, timing, and type of pesticide application suggest the likelihood of a terminal residue being present in a particular foodstuff. Table 8 illustrates the different types of applications and the potential for terminal residues to occur. The high risk of residues occurs with those applications that are directly applied to the crop late in the season and close to harvest, or with those applied immediately after the crop is harvested as post harvest protective measures. The low risk from an early season soil application can increase when systemic pesticides are used.

The dissipation of residues after foliar application often follows a bi- or triphasic curve. Typically, these curves are characterized by a rapid initial decline followed by a slower, more persistent phase. Often, first-order kinetics is adequate to explain the overall behavior. Figure 3 illustrates a typical dissipation curve. An increase in rate of application will increase the residue deposit and offset the dissipation curve accordingly. On the other hand, the vapor pressure of the pesticide and application technique will influence the amount of material reaching the plant canopy. In general, the decreasing volatility of a pesticide is reflected in greater deposits and persistence.

Figure 3 also shows the MRL allowed for the crop. In order for food residues to meet the MRL established by the Food and Drugs Act and Regulations, the pesticide label and local crop production recommendations

TABLE 8 Placement, Timing, and Techniques of Applying Pesticide and Potential Risk to Leave Food Residues

Placement	Timing	Type	Application Technique[a]	Potential to Leave Food Residue	Pesticide use (metric tonnes)
1. Soil	Precrop	Nematocides, Fumigants	Injection	Very low	
2. Soil	Immediate preplanting	Herbicides	Incorporation	Very low	
3. Soil	Planting	Herbicides	LP boom sprayer	Very low	6817
4. Soil	Planting	Insectides, Fungicides	Planting water in furrow	Very low	
5. Seed	Planting	Insecticides, Fungicides	Mixed with seed	Very low	
6. Weeds	Seedling	Herbicides	LP boom sprayer	Very low	
7. Crop	Seedling and established	Herbicides	LP boom sprayer, aerial sprayers	Low	731
8. Crop	Growing crop	Insecticides, Fungicides, Growth regulators	HP boom sprayers, air blast	Low	902
9. Crop	Maturing crop	Insecticides, Fungicides, Growth regulators	HP boom sprayers, air blast	Medium	
10. Crop	Immediately prior to harvest	Insecticides, Fungicides, Growth regulators	HP boom sprayers, air blast	High	270
11. Crop	Postharvest	Insectides, Fungicides, Fumigants	Dip, spray, injection	High	
12. Livestock	All ages	Insecticides	HP boom sprayers	High	

[a]LP, low pressure; HP, high pressure.

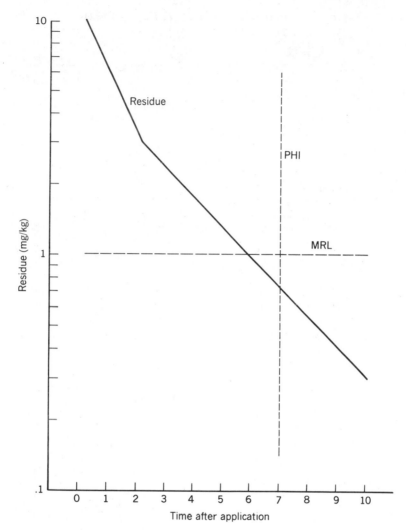

Figure 3. Typical dissipation of a pesticide with the maximum residue limit (MRL) and preharvest interval (PHI) indicated.

specify the pre-harvest interval (PHI) in days required between the last application of a pesticide and the harvesting of the commodity. Different PHIs are required for each pesticide and each commodity in order to meet the MRL. Table 9 shows a few examples.

Individual pesticides exhibit characteristic degradation patterns that differ from others and this behavior can be modified on different crops. Figure 4 gives the terminal residues of acephate on eight different crops. As might be expected, a leafy crop like lettuce had the highest residues whereas a buried crop like potatoes had the lowest. The surface area to mass ratio of a particular crop may affect the

TABLE 9   Examples of Preharvest Intervals (PHI) on Labels

	Pesticide	Pest	PHI (days)	MRL (mg/kg)
		Vegetables		
Celery	Carbaryl	Leafhoppers	3	5.0
Sweet corn	Carbaryl	European corn borer	1	1.0
Lettuce	Dimethoate	Aphids	7	2.0
Pepper	Dimethoate	Aphids	30	0.5
		Fruits		
Apples	Captan	Apple scab	7	5.0
Sweet cherries	Captan	Brown rot	2	5.0
Pears	Mancozeb	Pear scab	30	7.0
Peaches	Dimethoate	Tarnished plant bug	40	<0.1

level of residues found for example the solid tomato versus hallow pepper, while the systemic behavior of some chemical may lead to an accumulation of translocated chemical in certain portions of the crop, underground carrots. Figure 4 also illustrates the differences in dissipation of the same chemical on a variety of substrates.

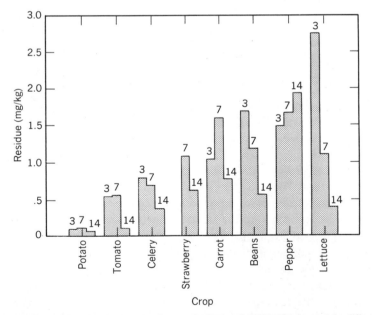

Figure 4.   Residues of acephate (including its metabolite, methamidophos) on eight different crops 3, 7, and 14 days after application (mean residues from 0.55 and 1.1 kg ai/ha for 2 years) (Frank et al., 1984).

TABLE 10   Captan Disappearance from Greenhouse and Field Grown Tomatoes[a]

PHI (days)	Captan Residue (mg/kg)[b]	
	Field Tomatoes (1983)	Greenhouse Tomatoes (1983)
0	0.73 a	0.88 ab
1	0.79 a	1.45 a
2	0.31 b	1.32 ab
5	0.59 a	1.04 ab
10	0.19 b	0.47 b
15	0.08 b	0.77 ab
Application rate	1.8 kg ai/ha	1.9 kg ai/ha
Number applications	2	2

[a]Frank et al. (1987c).
[b]Values followed by the same letter in each treatment were not significantly different at the $p = 0.05$ level according to Duncan's multiple range test.

The terminal residues of a pesticide remaining in a specific food depend on many factors including the amount applied, the number of applications, climatic conditions, soil type, crop growth, and the interval between last application and harvest. An illustration of this is shown in the comparison of the disappearance of captan from tomatoes grown indoors and outdoors (Table 10) where rainfall and other climatic conditions have affected the persistence or breakdown of the fungicide between indoors and crops.

Climatic conditions vary widely across Canada from the Atlantic to the Pacific Oceans. Seasonal temperature conditions can vary from very hot (30°C) to cool (15°C) and from very wet to dry. These multiple climatic combinations lead to very different rates of dissipation and/or degradation of pesticide residues. Hence terminal residues may vary between Provinces, creating problems in some years. Willis and McDowell (1987) presented comparative half-life disappearance on foliage for some pesticides under North American climatic conditions and for different crops. Because of these variations, application and dissipation rates must be ascertained under the local growing conditions and appropriate changes made to ensure that the PHI is sufficiently long to allow the residue on foodstuffs to meet acceptable MRLs.

Despite the seasonal and yearly changes in weather, long-term trends indicate that local recommendations are adequate and the PHIs are sufficient to allow foods to meet the MRL. Figure 5 shows the dissipation curves for a metalaxyl–mancozeb mixture on three varieties of lettuce over 3 years. Examination of this graph illustrates numerous points regarding the behavior of pesticides. Mancozeb has a higher residue deposit because it was applied at six to eight times the rate of metalaxyl. Metalaxyl, because of its higher vapor pressure, shows a pronounced early loss of volatile residues whereas mancozeb is more persistent. However,

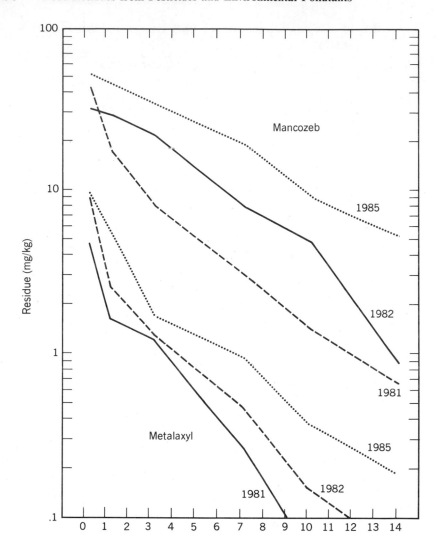

Figure 5. Dissipation of a metalaxyl–mancozeb mixture on lettuce for 3 years (mean residues for cos, leaf, and endive lettuce after application of 1:8 or 1:6 mixture at 1.8 kg ai/ha).

since metalaxyl is absorbed to become systemic while mancozeb is only slightly absorbed to act surficially it is therefore more prone to removal by rainfall.

A greater variation in residue occurs due to the morphology of the crop plant treated. Figure 6 shows the residue dissipation of a metalaxyl-mancozeb mixture on cos, leaf, endive, and head lettuce. The greater surface area and smaller mass of the leafy cultivars produced higher residues than with the denser head variety. The outside leaves of lettuce, cabbage, and celery contain high residues whereas the protected inner head has very low residues (Table 11). Normal trimming of

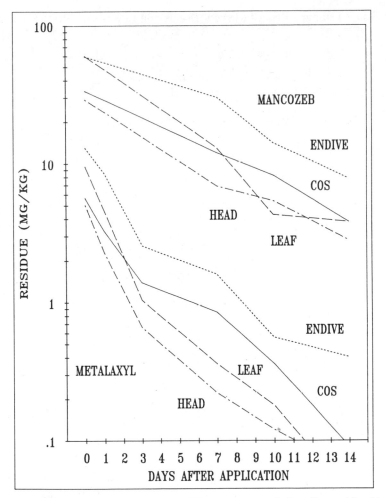

Figure 6. Dissipation of a metalaxyl–mancozeb mixture on cos, leaf, endive, and head lettuce in 1985 (mean residue from 1:8 mixture at 1.8 kg ai/ha).

celery for market purposes removed 60% of the residue and further trimming prior to eating removed an additional 20 to 30%.

The plant canopy can also intercept pesticide spray and this results in lower but variable residues on fruit. Table 12 shows that the residues of captan found on sweet and sour cherries and peaches depend on their position on the tree. Similarly, Table 13 shows the residues of parathion, captan, and cypermethrin measured on the foliage, calyx, and fruit of treated strawberry plants.

Pesticide residues on various foodstuffs following sale may decrease during storage and transportation to the consumer, as well as by following various culinary practices prior to consumption. Table 14 shows the fate of captan on five fruit crops after normal kitchen practices. Rinsing fruit under tap water

TABLE 11 Culinary Procedures and Processing to Remove Pesticide Residues

1. Removing Outer Leaves

Residues of EBDC (mg/kg)

	Outer Leaves	Exposed Inner Leaves
Cabbage	33	1.2
Lettuce	25	0.8

2. Trimming

Residues on Celery (mg/kg) (% Removal)

	No Trimming	Normal Trimming	Further Kitchen Trimming
EBDC	9.1 (0)	3.1 (66)	1.0 (90)
Diazinon	2.2(0)	0.94(57)	0.30(86)

3. Home Jam Making

Strawberries	Endosulfan	Captan
Raw Fruits no Calyx (mg/kg)	0.069	0.18
Jam (mg/kg)	0.055	0.065
Reduction (%)	20	64

4. Processing Company

Residues (mg/kg)

Tomato Product	EBDC	ETU	EBDC Removal (%)
Raw fruit	2.34	0.02	0
Juice	0.47	0.07	80
Whole pack (no lye wash)	0.10	0.03	96
Whole pack (lye wash)	<0.10	0.01	96+

5. Processed Tomatoes

Residues (mg/kg)

Tomato Product	EBDC	ETU
Whole pack	<0.10	0.02
Paste	0.11	0.01
Juice	0.12	<0.01
Soup	0.10	0.01
Ketchup	<0.10	<0.01

TABLE 12   Captan Residue on Fruit Sampled from Different Areas of Trees Treated with an Airblast Sprayer[a]

Part of Tree	Sweet Cherry 1982	Sweet Cherry 1983	Sour Cherry 1983	Peach 1983
Upper inside	1.7a[b]	3.5a	4.8a	6.2a
Upper outside	2.6ab	8.8b	9.5a	6.7a
Lower inside	1.7a	2.7a	9.6a	15.5a
Lower outside	4.8b	12.0b	18.8b	14.0a
Mean values				
Upper part	2.2	6.2	7.2	6.5
Lower part	3.3	7.4	14.2	14.8
Whole tree	2.7	6.9	10.7	10.5

[a]Northover et al. (1986).
[b]Means in the same column followed by different letters differ significantly ($p = 0.05$) by Duncan's multiple range test.

TABLE 13   Residue Deposit on Different Parts of a Strawberry Plant after Application at 1 kg ai/ha

Pesticide	Residue on Strawberry Plants (mg/kg) Foliage	Calyx	Fruit
Parathion	58	5.6	0.19
Captan	21	1.5	0.015
Cypermethrin	56	15	0.28

removed up to 93% of the residue from cherries, 58% from peaches, and 14% from strawberries. Other culinary practices and cooking removed 99.7% from apples and 96% from strawberries. Making wine reduced residues below the detection limit. Similarly, processing tomatoes removed residues (Table 11). In a study with a commerical food processor, residues of EBDC were reduced 80% by juicing and 96–100% by packing as whole fruit. In a case where homemade jam was prepared, little reduction in residue was observed with endosulfan but more than one-half of the captan was degraded or removed.

4.   MONITORING FOOD AND FEED FOR PESTICIDE RESIDUES IN CANADA

Most monitoring programs are designed to check food and feed commodities for those pesticides used in their production (Table 5) and that can lead to a potential

TABLE 14   Culinary and Processing Activities That Reduced Captan Residues on Food before Consumption[a]

Culinary Treatment	Residue of Captan (mg/kg)	Captan Removal (%)
Strawberry		
With calyx	2.37	0
Raw frozen 3 months	2.16	0
Rinsed and dried	2.05	14
Calyx removed	1.52	36
Raw rinsed cool water	1.35	43
Rinsed warm water	0.86	63
Cooked	0.11	95
Rinsed and cooked	0.10	96
Cherries		
Raw Fruit	2.7	0
15 sec rinse	0.8	70
30 sec rinse	0.6	78
60 sec rinse	0.3	89
Apples		
Raw	9.9	0
Rinsed and wiped	1.2	88
Wiped and cooked	0.5	95
Washed and peeled	1.2	88
Peeled and cooked	0.3	97
Washed, peeled and cooked	0.03	99.7
Grapes		
Raw fruit	2.2	0
Crushed fruit	1.9	24
Pressed fermented juice	<0.01	99.5+
Wine	<0.01	99.5+
Peaches		
Raw fruit	11.0	0
5 sec rinse	4.9	55
10 sec rinse	4.6	58
10 sec brush and rinse	3.3	70

[a]Frank et al. (1983a), Northover et al. (1986), and Ritcey et al. (1987).

for residues to be present, some of which may exceed the MRLs outlined in legislation. Most often these involve fruit and vegetables but also tobacco and occasionally some of the major field crops.

While the Federal Government has the responsibility of ensuring that imported commodities meet the standards in the Food and Drugs Act and Regulations, domestic food supplies in Canada are monitored by both federal and provincial agencies. The following major agencies at the federal level were involved in monitoring food and feed in 1985:

1. Health Protection Branch, Health and Welfare Canada, whose main mandate is to protect the consumer from residues of foreign substances on their food supply as prescribed under the Food and Drugs Act; this includes monitoring both imported and domestic foods.
2. Plant Health and Plant Products Directorate, Agriculture Canada whose main interest is to ensure livestock feeds are free of harmful residues and hence domestically produced animal products are safe for consumption by the general public. This is carried out under the Feed and Fertilizer Acts.
3. Food Inspection Directorate, Agriculture Canada, whose main interests lie in ensuring that meat products are free of undesirable or harmful residues of substances used to treat poultry and domestic livestock. This is carried out under the authority of Meat Inspection Act.
4. Canadian Grain Commission whose main interest is to ensure Canadian grain meets world health standards under their mandate to market Canadian grain.

Provincial agricultural departments monitor food supplies produced within their borders primarily to ensure that primary producers are applying pesticides in accordance with the label and according to provincial recommendations and that neither results in food residues in violation of health standards.

The following provincial agencies were monitoring foods in 1985:

1. British Columbia—Ministry of Agriculture and Fisheries
2. Alberta—Alberta Agriculture
3. New Brunswick—Department of Agriculture
4. Ontario—Ministry of Agriculture and Food

In order to provide the reader with a sense of the extent of monitoring programs in Canada data for the 1985 program are presented as a typical year. The program summary presented in Table 15 reveals that 35 major food commodities were monitored for the residues of a combined total of 136 pesticides. The total number of samples analyzed was 4879 and 1564 or 32% contained detectable residues. Fifteen or 0.31% were in violation of MRL allowed in the Food and Drugs Act and Regulations.

This monitoring program should not be confused with sampling and analysis

TABLE 15 Summary of the Residue Analyses Programs for Canadian Produced Foods and Feeds, 1985[a]

Commodity	Type	Number of Pesticides Analyzed	Number of Samples	Number of Samples with Detectable Residues	Percentage Violation of MRL
Fruit	10	109	910	121	0.1
Vegetables	9	115	447	123	1.3
Grain (cereal)	2	35	324	162	0.4[b]
Meats	9	32	2325	647	0.0
Eggs	1	22	163	4	0.0
Milk	1	21	449	394	0.0
Milk products	2	16	7	7	0.0
Feeds	1	47	254	15	—
Total	35	136	4879	1564	0.31

[a]Agriculture Canada (1986).
[b]One sample of 233 analyzed for malathion contained 35.0 mg/kg when MRL is 8.0 mg/kg; 99 (42%) of 233 had detectable residues of malathion averaging 0.46 mg/kg.

TABLE 16 Violations of MRL in 1985 Based on Analyses of 910 Samples of Fruit and 446 Samples of Vegetables Domestically Produced in Canada[a]

Commodity (Violations)	Number Analyzed (Total)	Number (Violation/ Analyzed)	Identity	Mean Residue (mg/kg)	MRL (mg/kg)
Apricot	20	8/20	Iprodione	0.56	0.10
Carrot	125	1/57	Linuron	0.19	0.10
Cauliflower	11	1/11	Endosulfan	2.2	1.0
Celery	14	2/11	Endosulfan	3.8	1.0
Lettuce	11	1/11	Endosulfan	2.0	2.0
Potato	261	1/209	Aldicarb	0.6	0.5

Commodity (Nonviolations)	Number Analyzed	Commodity (Nonviolations)	Number Analyzed	Commodity (Nonviolations)	Number Analyzed
Apples	589	Peaches	20	Broccoli	10
Blueberries	5	Pears	10	Cucumbers	5
Cherries	19	Raspberries	20	Tomatoes	5
Grapes	15	Strawberries	211	Turnips	5

[a]Agriculture Canada (1986).

of foods and feeds for regulatory action. Monitoring programs are designed to develop a general picture of food residues and subsequently look into the reasons that lead to excessive residues. Regulatory action is normally a follow-up on a specific commodity and specific pesticide where a violation of the Food and Drug Regulations has occurred or is suspected and legal actions are required.

In 1985 one (0.4%) of 233 cereal grain samples analyzed was in violation of the 8 mg/kg MRL for the insecticide malathion (Table 15). Six violations (1.3%) were identified among 447 vegetable samples analyzed. These involved the herbicide linuron on carrots, the insecticides endosulfan on cauliflower, celery, and lettuce, and aldicarb on potato (Table 16). Eight violations (0.1%) occurred among 910 fruit samples analyzed. All eight involved iprodione on apricots, where no MRL had been granted under the Food and Drug Regulations for this fungicide on apricot.

In 1985, 589 samples of apples were analyzed for 28 herbicides and growth regulators, 60 insecticides and miticides, and 13 fungicides and no violations were found (Table 17). One growth regulator, 11 insecticides and miticides, and 5 fungicides were detected, but none was above the MRLs. The incidence of

TABLE 17   Monitoring for Pesticide Residues in Apples Collected from Across Canada in 1985[a]

Pesticide[b]	Number Samples Analyzed	Number with Detectable Residues (%)	Range of Residues (mg/kg)	MRL (mg/kg)
Benomyl (F)	99	14 (14)	0.010–0.056	5.0
Captan (F)	137	4 (3)	0.004–0.34	5.0
Carbaryl (I)	137	4 (3)	0.018–0.12	5.0
Carbofuran (I)	137	1 (< 1)	0.01	0.1
Daminozide (GR)	4	4 (100)	0.6–13.0	30.0
Diazinon (I)	326	4 (1)	0.02–0.07	0.75
Dicofol (A)	57	26 (46)	0.002 0.68	3.0
Dimethoate (I)	326	4 (1)	0.02–0.07	2.0
Diphenylamine (F)	26	26 (100)	0.02–1.7	5.0
Endosulfan (I)	67	12 (18)	0.005–0.14	2.0
Ethion (I)	326	4 (1)	0.01–0.06	2.0
Methomyl (I)	54	7 (13)	0.012–0.10	0.5
Pentachlorophenol (F)	8	8 (100)	0.0001	0.1
Phosalone (I)	67	20 (30)	0.022–0.38	5.0
Phosmet (I)	67	1 (2)	0.47	10.0
Pirimicarb (A)	37	2 (5)	0.07–0.08	0.5
Tetrachlorphenol (F)	8	2 (25)	0.0001	0.1
Total	589			

[a]Agriculture Canada (1986).
[b]A, acaricide; F, fungicide; H, herbicide; GR, growth regulator; I, insecticide.

TABLE 18   Monitoring for Pesticide Residues in Strawberries Collected from Across Canada in 1985[a]

Pesticide[b]	Number Samples	Residues Detected Number (%)	Range of Residues (mg/kg)	MRL (mg/kg)
Captan (F)	22	20(91)	0.01–2.2	5.0
Carbaryl (I)	53	1(2)	0.032	0.1
Carbofuran (I)	53	3(6)	0.022–0.056	0.1
Dicofol (A)	22	1(5)	0.25	3.0
Dimethoate (I)	92	3(3)	0.02–0.75	1.0
Endosulfan (I)	22	4(18)	0.08–0.41	1.0
Iprodione (F)	22	1(5)	0.3	5.0
Malathion (I)	101	1(1)	0.01	0.1
Total	211			

[a]Agriculture Canada (1986).
[b]A, acaricide; F, fungicide; I, insecticide.

detectable residues varied from less than 1% for carbofuran to 100% for daminozide, diphenylamine, and pentachlorophenol. A similar review of 211 strawberry samples showed that the analyses covered 56 insecticides, 23 herbicides and growth regulators, and eight fungicides (Table 18) Residues of six insecticides and two fungicides were detected; all levels were below the MRLs. The incidence of a positive residue varied from 1% for malathion to 91% for captan.

5.   MONITORING FOODS FOR PESTICIDE RESIDUES IN ONTARIO

The Ontario Ministry of Agriculture and Food (OMAF) monitors food commodities grown in the Province for pesticide residues. This combined with use pattern data illustrate a more complete picture on food residues than can be obtained at the federal level.

Between 1980 and 1986, 1088 fruit and vegetable samples were collected and analyzed for a wide range of pesticides. The analytical procedures were designed to include between 80 and 95% of the insecticides and fungicides used by growers to produce their fruit and vegetable crops. The analytical methodologies included the following pesticide types:

1. Insecticides—organochlorine, organophosphorus, synthetic pyrethroid, and methylcarbamate.
2. Fungicides—organochlorine, phthalimide, dicarboximide, acylalanine, and dithiocarbamates.

3. Herbicides—s-triazines, phenylureas, chloroacetamides, bipyrilium, and others, especially those used on root crops like potatoes, carrots, and onions.

During the 1980–1986 period, 21 fruit and vegetable samples (1.9%) of the 1088 analyzed contained residues of a pesticide above the MRL. The violations appeared in 5 of 21 fruit and vegetables collected and details on commodity, pesticide, year, and residue appear in Table 19.

As outlined in Section 1, farm surveys on pesticide use have been carried out in Ontario, hence it was possible to correlate the farm use of pesticides with residues found on raw product offered for sale. To illustrate this point, several fruit and vegetable commodities are selected for in-depth study in the rest of this section. Pesticide use data were compiled from data collected from farm questionnaires. The residue data were generated annually from samples collected at the farm gate or from farmer's markets and in particular the Food Terminal in Toronto where OMAF inspectors sampled produce on delivery.

TABLE 19   Violation of the Food and Drugs Act and Regulations in Ontario Grown Fruit and Vegetables, 1980–1986[a]

Commodity	Number Analyzed	Number Violations	Pesticide	Residue (mg/kg)	Year	MRL (mg/kg)
Apples	305	1	Diphenylamine	6.7	1982	5.0
		2	Phosalone	5.9, 6.2	1981	5.0
Peaches	46	2	Dicofol	3.3, 6.3	1981	3.0
		1	Azinphosmethyl	2.2	1981	2.0
Pears	10	1	Dicofol	11.	1986	3.0
Strawberries	150	1	Azinphosmethyl	2.1	1986	1.0
		1	Carbofuran	4.5	1984	0.4
		2	Captan	7.1, 8.9	1983, 86	5.0
Potatoes	110	8	Aldicarb	0.67 ± 0.15	1983–85	0.5
		2	Linuron	0.55, 3.8	1984	0.1

Commodity (Non-violations)	Number Analyzed	Commodity (Nonviolations)	Number Analyzed	Commodity (Nonviolations)	Number Analyzed
Asparagus	34	Cole crops	48	Onions	10
Bean, green	50	Corn, sweet	28	Peppers	10
Blueberries	10	Cucumbers	72	Raspberries	19
Carrot	28	Grapes	10	Rutabaga	10
Celery	10	Lettuce	10	Tomatoes	72
Cherries	46				

Total samples   1088
Total violations   21, 1.9%

[a]Frank et al. (1987a, b).

TABLE 20 Compilation of Pesticide Use in Ontario Apple Production and Terminal Residues on the Product Offered for Sale, 1980–1986[a]

Pesticide	Crop Treated (%)	Rate (kg ai/ha)	Detection Limit (mg/kg)	Detectable Residues (%)	Residues (mg/kg) Mean ± SD	Maximum	MRL (mg/kg)
Azinphosmethyl	91.2	1.05	0.1	1.3	0.26 ± 0.10	0.45	2.0
Carbaryl	22.1	1.67	0.01	3.6	0.03 ± 0.01	0.04	5.0
Deltamethrin	14.1	0.01	0.005	0.0	—	—	0.1
Demeton	1.9	0.81	0.005	0.0	—	—	0.75
Diazinon	16.1	1.62	0.005	0.3	0.04	0.04	0.75
Dicofol	20.0	1.57	0.005	24.2	0.086 ± 0.129	0.76	3.0
Dimethoate	4.1	1.20	0.005	0.0	—	—	2.0
Endosulfan	0.4	1.70	0.005	5.2	0.014 ± 0.003	0.05	2.0
Ethion	0.2	1.00	0.005	0.0	0.13	0.13	2.0
Fenvalerate	46.4	0.13	0.005	0.0	—	—	0.1
Malathion	10.3	1.68	0.005	0.0	—	—	2.0
Methidathion	1.2	1.65	0.01	0.0	—	—	0.5
Methoxychlor	0.4	0.20	0.005	0.0	—	—	14.0
Parathion	1.9	0.26	0.005	0.7	0.023 ± 0.015	0.038	1.0
Permethrin	5.8	0.20	0.005	0.0	—	—	1.0
Phosalone	97.7	1.27	0.01	40.6	0.42 ± 0.45	6.2	5.0
Phosmet	83.9	2.12	0.02	14.8	0.193 ± 0.110	1.1	10.0
Pirimicarb	0.2	0.85	0.05	0.0	—	—	0.5
Captan	94.3	3.78	0.005	45.9	0.089 ± 0.119	1.0	5.0
Captafol	0.4	21.6	0.025	0.0	—	—	0.1
Diphenylamine		Harvest dip	0.01	69.5	1.29 ± 0.38	6.7	5.0
Mancozeb	99.4	7.10 ⎫	0.1	27.1	0.30 ± 0.17	0.76	7.0
Metiram	95.7	10.94 ⎬	0.01				
Daminozide	85.0	3.30		72.4	1.24 ± 0.81	3.9	30.0

[a]Frank et al. (1989).

TABLE 21 Compilation of Pesticide Use in Ontario Cherry Production and Terminal Residues on the Raw Product Offered for Sale, 1980–1986[a]

Pesticide	Crop Treated (%)	Rate (kg ai/ha)	Detection Limit (mg/kg)	Detectable Residues (%)	Residues (mg/kg)		MRL (mg/kg)
					Mean ± SD	Maximum	
Azinphosmethyl	83.5	1.56	0.5	0.0	—	—	1.0
Carbaryl	18.9	2.50	0.01	6.7	0.33 ± 0.10	0.50	10.0
Diazinon	9.2	1.50	0.005	0.0	—	—	0.75
Dicofol	19.7	2.06	0.002	47.8	0.08 ± 0.10	0.30	3.0
Endosulfan	0.3	1.00	0.002	41.3	0.02 ± 0.10	0.43	2.0
Parathion	1.4	0.22	0.01	0.0	—	—	1.0
Permethrin	20.0	0.02	0.005	10.0	0.005	0.005	0.1
Phosalone	95.9	1.57	0.05	58.7	0.17 ± 0.22	1.20	6.0
Phosmet	25.1	2.10	0.02	15.2	0.04 ± 0.05	0.14	7.0
Captan	100	9.89	0.002	76.1	0.38 ± 0.29	4.90	5.0
Captafol	15.7	1.59	0.002	0.0	—	—	10,2[b]
Chlorothalonil	0	0	0.01	10.0	0.02 ± 0.01	0.03	0.1
Ferbam	19.7	1.91	0.01	0.0	—	—	7.0
Folpet	19.1	2.20	0.002	0.0	—	—	25.0
Iprodione	4.3	0.75	0.01	20.0	0.29 ± 0.05	0.43	5.0

[a]Frank et al. (1987a).
[b]MRL 10 mg/kg sour cherries, 2 mg/kg sweet cherries.

TABLE 22 Compilation of Pesticide Use in Ontario Grape Production and Terminal Residues on the Raw Product Offered for Sale, 1980–1986

Pesticide	Crop Treated (%)	Rate (kg ai/ha)	Detection Limit (mg/kg)	Detectable Residues (%)	Residues (mg/kg)		MRL (mg/kg)
					Mean ± SD	Maximum	
Azinphosmethyl	99.6	0.66	0.5	50.0	0.26 ± 0.18	0.57	5.0
Carbaryl	83.7	1.79	0.01	0.0	—	—	5.0
Diazinon	7.9	1.25	0.005	0.0	—	—	0.75
Dicofol	6.7	1.35	0.0	10.0	0.063	0.063	3.0
Endosulfan	5.4	2.25	0.001	10.0	0.037	0.037	1.0
Parathion	83.7	0.27	0.01	0.0	—	—	1.0
Phosalone	1.4	0.90	0.05	10.0	0.014	0.014	5.0
Phosmet	33.9	1.25	0.02	10.0	0.99	0.99	10.0
Captan	84.1	2.08	0.002	50.0	0.29 ± 0.41	0.95	5.0
Folpet	21.0	1.80	0.002	0.0	—	—	25.0
Iprodione	11.7	0.70		0.0	—	—	10.0
Ferbam	0.8	2.02 ⎫					
Mancozeb	100	3.57 ⎭	0.1	100	1.36 ± 0.83	2.5	7.0
Vinclozolin	Not registered		0.01	10.0	0.09	0.09	0.1
Daminozide	20	2.20	0.01	30.0	0.54 ± 0.40	0.90	10.0

TABLE 23 Compilation of Pesticide Use in Ontario Peach Production and Terminal Residues on the Raw Product Offered for Sale, 1980–1986[a]

Pesticide	Crop Treated (%)	Rate (kg ai/ha)	Detection Limit (mg/kg)	Detectable Residues (%)	Residues (mg/kg)		MRL (mg/kg)
					Mean ± SD	Maximum	
Azinphosmethyl	98.2	1.24	0.5	26.1	0.83 ± 0.22	2.2	2.0
Carbaryl	14.3	3.12	0.01	26.7	0.22 ± 0.30	0.67	10.0
Dicofol	16.2	1.50	0.002	58.7	0.47 ± 0.33	6.4	3.0
Dimethoate	6.0	1.25	0.005	0.0	—	—	0.1
Endosulfan	57.0	0.54	0.002	58.7	0.074 ± 0.060	0.32	2.0
Malathion	0.2	0.80	0.005	0.0	—	—	6.0
Parathion	50.5	0.23	0.01	4.3	0.051 ± 0.031	0.084	1.0
Permethrin	10.5	0.02	0.005	5.0	0.051	0.051	1.0
Phosalone	4.9	1.12	0.05	15.2	0.17 ± 0.12	0.41	4.0
Captan	100	6.33	0.002	89.1	1.25 ± 1.51	9.2, 3.4[b]	25, 5[b]
Captafol	0.8	2.50	0.002	0.0	—	—	15.0
Dichloran	Postharvest Dip		0.002	6.5	0.011 ± 0.015	0.03	15.0
Ferbam	29.5	4.25					
Mancozeb	0.3	2.88	0.1	0.0	—	—	7.0
Metiram	0.2	3.20					
Iprodione	42.5	2.20	0.01	30.0	2.43 ± 1.54	3.60	5.0

[a]Frank et al. (1987a).

[b]MRL until 1983 was 25 mg/kg and highest residue was 9.2; after 1983 it was 5.0 mg/kg and highest has been 3.4 mg/kg.

TABLE 24 Compilation of Pesticide Use in Ontario Pear Production and Terminal Residues on the Raw Product Offered for Sale, 1980–1986

Pesticide	Crop Treated (%)	Rate (kg ai/ha)	Detection Limit (mg/kg)	Detectable Residues (%)	Residues (mg/kg)		MRL (mg/kg)
					Mean ± SD	Maximum	
Azinphosmethyl	93.7	1.01	0.5	40.0	0.38 ± 0.49	1.1	2.0
Carbaryl	2.3	2.95	0.01	0.0	—	—	5.0
Deltamethrin	79.9	0.005	0.005	0.0	—	—	0.1
Diazinon	0.7	1.40	0.005	0.0	—	—	0.75
Dicofol	39.1	1.42	0.002	80.0	1.80 ± 3.76	11.0	3.0
Endosulfan	95.8	1.80	0.001	60.0	0.54 ± 0.79	1.6	2.0
Fenvalerate	97.4	0.08	0.005	0.0	—	—	0.1
Parathion	19.3	0.40	0.01	0.0	—	—	1.0
Permethrin	21.2	0.19	0.005	0.0	—	—	1.0
Phosalone	23.9	0.90	0.05	20.0	0.31 ± 0.10	0.42	2.0
Phosmet	49.3	1.95	0.2	20.0	0.55 ± 0.35	0.92	10.0
Captan	85.8	5.59	0.002	70.0	0.79 ± 0.97	2.2	5.0
Ferbam	14.3	2.05	0.1				
Mancozeb	85.4	4.35	0.1	20.0	0.26 ± 0.10	0.36	7.0
Metiram	19.1	3.98	0.1				
Zineb	9.8	2.82	0.1				
Iprodione	12.1	0.80	0.01	10.0	0.59	0.59	0.1

TABLE 25 Compilation of Pesticide Use in Ontario Raspberry and Strawberry Production and Terminal Residues on the Product Offered for Sale, 1980–1986[a]

Pesticide	Crop Treated (%)	Rate (kg ai/ha)	Detection Limit (mg/kg)	Detectable Residues (%)	Residues (mg/kg)		MRL (mg/kg)
					Mean ± SD	Maximum	
Raspberry							
Azinphosmethyl	38.4	1.52	0.5	21.1	0.087 ± 0.019	0.11	2.0
Carbaryl	15.5	2.60	0.01	20.0	2.72 ± 1.59	4.6	10.0
Demeton	5.5	0.60	0.005	0.0	—	—	0.1
Dicofol	7.5	0.85	0.002	10.5	0.24 ± 0.21	0.47	3.0
Dimethoate	11.0	3.00	0.005	0.0	—	—	0.1
Malathion	27.4	4.19	0.005	0.0	—	—	8.0
Mevinphos	15.1	0.61	0.005	0.0	—	—	0.25
Permethrin	4.5	0.01	0.005	5.3	0.077	0.077	0.1
Captan	27.4	2.00	0.002	21.0	0.085 ± 0.084	0.21	5.0
Ferbam	71.2	3.33	0.1	0.0	—	—	7.0
Strawberry							
Azinphosmethyl	40.4	0.70	0.5	3.3	0.45 ± 0.85	2.1	1.0
Carbaryl	2.7	1.10	0.01	0.0	—	—	7.0
Carbofuran	39.8	0.53	0.01	8.7	0.33 ± 0.63	4.5	0.4
Dicofol	6.3	0.45	0.002	0.0	—	—	3.0
Dimethoate	32.4	1.14	0.005	0.7	0.04 ± 0.01	0.05	1.0
Endosulfan	71.4	0.59	0.002	56.0	0.13 ± 0.11	0.58	1.0
Malathion	3.1	1.11	0.005	0.0	—	—	8.0
Methoxychlor	1.3	1.35	0.01	2.0	0.008 ± 0.003	0.014	14.0
Mevinphos	7.5	0.50	0.005	0.0	—	—	0.25
Permethrin	1.3	0.2	0.005	0.0	—	—	0.1
Captan	86.4	6.17	0.002	90.0	2.00 ± 1.51	14, 8.9[b]	25, 5[b]
Chlorothalonil	5.6	1.02	0.01	0.0	—	—	7.0
2,4-D	23.5	0.50	0.01	33.3	0.02 ± 0.01	0.03	0.1

[a]Frank et al. (1987a).
[b]MRL 25 mg/kg up to 1983, highest residue 14 mg/kg; after 1983 MRL 5.0 mg/kg and highest residue 8.9 mg/kg.

The apple is the most important fruit grown in Ontario and is selected for study. The analytical monitoring program covered 19 of the 26 insecticides and acaricides found to be used in the 1983 farm survey. These 19 pesticides represented 82% of the volume used (excluding mineral oil). The fungicides analyzed involved four of the nine used on the farm; however, these four represented 96% of the amount applied. Table 20 gives the percentage of the apple hectarage treated with an average rate of application of insecticides, acaricides, and fungicides. The detection limit of the analytical procedure, the percentage of samples with detectable residues, the residues measured, and the maximum residue limits (MRL) appear for comparison with the use pattern data. Over the 7-year period three samples were in violation of the Food and Drugs Act and Regulations. Close correlation was observed between the percentage of crop treated and percentage of samples with detectable residues. For example, dicofol was applied to 20% of the crop hectarage and 25% of the samples had detectable residues. Many of the pesticides applied had disappeared by harvest time and were not detected in any of the apple samples. Endosulfan on the other hand was detected in more samples than would be predicted based on the farm survey.

Tables 21 to 25 provide similar data on the farm use of pesticides included in the monitoring of cherries, grapes, peaches, pears, raspberries and strawberries, respectively, along with the terminal residues found in the raw commodities offered for sale. The percentage of the crops treated with each of these pesticides and the amount applied were taken from a 1983 farm survey. For example, all cherries were treated with captan and almost all received phosalone. While good correlation was observed between the percentage of cherries treated and percentage of the fruit with detectable residues for most pesticides a notable exception was observed with endosulfan, where the survey showed only 0.3% of the cherries were treated, yet detectable residues were found on 41% of raw product. For peaches the correlation was excellent with 57% of the produce treated with endosulfan and 59% with detectable residues. With some pesticides, no terminal residues were found although the crop had been treated. Examples were carbaryl on cherries, parathion and folpet on grapes, and ferbam on raspberries and cherries. The explanation may have been that the pesticide treatments were made either at blossom time or on the immature fruit and had disappeared by harvest time. Other compounds like parathion, mevinphos, and malathion applied to grapes and peaches break down rapidly under Ontario conditions and leave little or no residue within 7–10 days. Therefore, unless used closer to harvest, residues would rarely appear on the raw product. Others, like permethrin, are used in such low amounts that the analytical detection limit of 0.005 mg/kg would be inadequate to find a residue. Yet others, like azinphosmethyl, where the methodology is relatively insensitive and the limits of detection are high (0.5 mg/kg), are difficult to measure unless a special procedure is adopted.

In Table 26 an estimate is made of the total pesticides used on four fruit crops. This is compared with the estimated amounts remaining as terminal residues on the harvested raw commodity. For apples, the single highest residual was

TABLE 26 The Use of Six Pesticides on Four Fruit Commodities and the Amount in the Final Raw Product

Commodity	Pesticide	Total (ha)	Treated (%)	Application (kg ai/ha)	Total Use (kg)	Production (tonnes)	Detectable Residue (%)	Mean Residue (mg/kg)	Estimated Residue (kg)	Remaining on Fruit (%)
Apples	Carbaryl	10,484	22.1	1.67	3,869	182,697	4.0	0.03	0.22	0.006
	Dicofol		20.0	1.57	3,292		24.7	0.082	3.70	0.11
	Endosulfan		0.4	1.70	713		7.4	0.014	0.19	0.027
	Phosalone		97.7	1.27	13,008		40.4	0.45	33.2	0.26
	Phosmet		83.9	2.12	18,648		13.7	0.22	5.51	0.03
	Captan		94.3	3.78	37,370		48.2	0.082	7.22	0.02
Cherries	Carbaryl	1,301	18.9	2.50	615	7,783	6.1	0.33	0.17	0.028
	Dicofol		19.6	2.06	528		47.8	0.08	0.30	0.056
	Endosulfan		0.3	1.00	3.9		41.3	0.02	0.06	1.6
	Phosalone		95.9	1.57	1,959		58.7	0.17	0.78	0.04
	Phosmet		25.1	2.10	686		15.2	0.04	0.05	0.007
	Captan		100.0	9.89	12,867		76.1	0.38	2.25	0.017
Peaches	Carbaryl	3,028	14.3	3.12	1,351	62,901	80.0	0.22	11.07	0.82
	Dicofol		16.2	1.50	736		58.7	0.47	17.35	2.4
	Endosulfan		57.0	0.54	932		58.7	0.074	2.73	0.29
	Phosalone		4.9	1.12	166		15.2	0.17	1.63	0.98
	Phosmet		98.2	3.28	9,753		65.2	1.53	62.75	0.64
	Captan		100.0	6.33	19,167		89.1	1.25	70.06	0.37
Strawberries	Carbaryl	1,665	2.7	1.10	49.4	12,854	0.0	<0.01	<3.5	0.007
	Dicofol		6.3	6.95	100		0.0	<0.002	<1.6	0.0016
	Endosulfan		71.4	0.59	701		56.0	0.13	0.94	0.13
	Captan		86.4	6.17	8,886		90.0	2.00	23.14	0.26

TABLE 27 Compilation of Pesticide Use in the Production of Asparagus, Green Beans, and Cole Crops in Ontario and the Terminal Residues in Raw Product Offered for Sale, 1980–1986[a]

Commodity Number of (Samples)	Pesticide	Crop Treated (%)	Rate (kg ai/ha)	Detection Limit (mg/kg)	Detectable Residues (%)	Residue (mg/kg) Mean ± SD	Residue (mg/kg) Maximum	MRL (mg/kg)
Asparagus (34)	Carbaryl	51.5	1.37	0.01	26.7	0.37 ± 0.68	1.20	10.0
	Endosulfan	5.5	1.40	0.001	3.1	0.011	0.011	0.1
	Malathion	18.0	1.10	0.005	0.0	—	—	6.0
	Methoxychlor	36.6	1.50	0.01	0.0	—	—	14.0
	Mevinphos	19.5	0.45	0.005	0.0	—	—	0.25
Green beans (50)	Carbofuran	0.8	0.36	0.01	0.0	—	—	0.1
	Mevinphos	0.8	1.31	0.005	0.0	—	—	0.1
Cole crops (48)	Azinphosmethyl	15.1	0.42	0.5	0.0	—	—	1.0,[b,c] 0.5[d]
	Carbaryl	3.9	1.00	0.01	7.4	0.70	0.70	5.0
	Carbofuran	1.1	0.31	0.01	0.0	—	—	0.1
	Chlordane	1.9	3.10	0.002	0.0	—	—	0.1
	Endosulfan	42.0	0.75	0.001	12.5	0.021 ± 0.025	0.058	1.0
	Fensulfothion	0.6	4.32	0.05	0.0	—	—	0.1
	Fenvalerate	26.2	0.085	0.005	0.0	—	—	0.1
	Malathion	0.6	1.00	0.005	0.0	—	—	6.0,[c] 0.5[b,d]
	Methamidophos	44.5	1.43	0.02	0.0	—	—	1.0,[b] 0.5[c,d]
	Permethrin	59.4	0.21	0.005	0.0	—	—	0.5[b,c], 0.1[d]
	Captan	0.6	1.14	0.002	0.0	—	—	0.1
	Chlorothalonil	35.4	1.71	0.01	13.8	0.42 ± 0.52	1.2	5.0
	Mancozeb	4.2	0.98 ⎱						5.0
	Zineb	1.0	1.80 ⎰	0.1	12.5	0.85 ± 0.80	1.6	7.0[e]

[a]Frank et al. (1987b).
[b]MRL for broccoli.
[c]MRL for cabbage.
[d]MRL for cauliflower.
[e]Zineb equivalent for EBDC.

TABLE 28 Compilation of Pesticide Use in the Production of Sweet Corn, Cucumber, and Peppers in Ontario and the Terminal Residues in Raw Product Offered for Sale, 1980–1986[a]

Commodity Number of (Samples)	Pesticide	Crop Treated (%)	Rate (kg ai/ha)	Detection Limit (mg/kg)	Detectable Residues (%)	Residue (mg/kg)		MRL (mg/kg)
						Mean ± SD	Maximum	
Corn, sweet (28)	Carbaryl	9.0	1.50	0.01	0.0	—	—	0.1
	Carbofuran	97.6	0.83	0.01	0.0	—	—	0.1
	Fonofos	33.9	1.11	0.005	0.0	—	—	0.1
	Permethrin	0.6	0.1	0.005	0.0	—	—	0.1
Cucumbers (72)	Azinphosmethyl	16.6	0.52	0.5	0.0	—	—	0.5
	Carbaryl	1.9	0.54	0.01	0.0	—	—	3.0
	Carbofuran	0.6	0.50	0.01	0.0	—	—	0.1
	Dicofol	Greenhouse crop		0.002	9.7	0.11 ± 0.26	0.70	3.0
	Endosulfan	87.9	0.64	0.002	45.8	0.044 ± 0.036	0.19	1.0
	Methoxychlor	62.7	1.51	0.01	2.8	0.027 ± 0.031	0.049	14.0
	Mevinphos	1.8	0.40	0.005	0.0	—	—	0.1
	Permethrin	0.9	0.10	0.005	0.0	—	—	0.5
	Captafol	1.9	1.44	0.002	0.0	—	—	2.0
	Chlorothalonil	70.6	2.22	0.01	6.3	0.014 ± 0.008	0.028	5.0
	Zineb	1.9	1.08	0.1	12.5	0.10 ± 0.01	0.14	4.0
Pepper (10)	Carbaryl	87.3	1.31	0.01	0.0	—	—	5.0
	Carbofuran	66.7	1.05	0.01	50.0	0.05 ± 0.04	0.11	0.1
	Deltamethrin	49.5	0.01	0.005	0.0	—	—	0.75
	Diazinon	2.1	1.00	0.005	0.0	—	—	0.75
	Endosulfan	64.5	0.91	0.001	0.0	—	—	1.0
	Malathion	4.8	1.2	0.005	0.0	—	—	0.5
	Methoxychlor	6.4	1.10	0.01	0.0	—	—	14.0
	Pirimicarb	3.9	0.41	0.01	0.0	—	—	0.1
	Chlorothalonil	28.2	0.9	0.01	0.0	—	—	0.1
	Mancozeb	30.7	1.5	0.1	0.0	—	—	7.0

[a]Frank et al. (1987b).

TABLE 29 Compilation of Pesticide Use in Ontario Production of Potatoes and the Terminal Residues in Raw Product Offered for Sale, 1981–1985[a]

Pesticide	Crop Treated (%)	Rate (kg ai/ha)	Detection Limit (mg/kg)	Detectable Residue (%)	Residue (mg/kg) Mean ± SD	Residue (mg/kg) Maximum	MRL (mg/kg)
Aldicarb	15.8	1.80	0.1	81.6	0.26 ± 0.13	0.88	0.5
Azinphosmethyl	42.4	3.28	0.5	0.0	—	—	0.1
Carbaryl	3.1	1.45	0.01	0.0	—	—	0.2
Carbofuran	66.8	1.68	0.01	0.0	—	—	0.1
Chlorpyrifos	13.5	3.47	0.005	0.0	—	—	0.1
Cypermethrin	44.4	0.06	0.005	0.0	—	—	0.1
Deltamethrin	18.4	0.005	0.005	0.0	—	—	0.1
Endosulfan	1.2	1.08	0.001	0.0	—	—	0.1
Methidathion	5.2	0.25	0.05	0.0	—	—	0.1
Methamidophos	2.4	0.89	0.002	0.0	—	—	0.1
Phorate	4.3	0.70	0.005	0.0	—	—	0.1
Phosmet	2.0	1.12	0.2	0.0	—	—	0.1
Captafol	14.7	0.83	0.00	0.0	—	—	0.1
Chlorothalonil	34.1	3.43	0.01	0.0	—	—	0.1
Mancozeb	11.4	3.00 }	0.1	0.0	—	—	0.1
Metiram	25.3	4.84 }					
Chlorpropham	Harvest time	0.55	0.02	6.1	0.18 ± 0.10	0.46	15.0
Diquat	3.0	3.02	0.01	5.0	0.02 ± 0.01	0.03	0.1
EPTC	7.9	0.95	0.01	0.0	—	—	0.1
Linuron	0.6	0.90	0.01	2.0	1.61 ± 1.87	3.8	0.1
Metobromuron	4.6	1.40	0.01	0.0	—	—	0.1
Metolachlor	50.2	0.55	0.01	0.0	—	—	0.1
Metribuzin	25.0	1.70	0.01	0.0	—	—	0.1
Monolinuron	1.0		0.01	0.0	—	—	0.1

[a]Frank et al. (1987b).

510

TABLE 30 Compilation of Pesticide Use in Ontario Production of Tomatoes and the Terminal Residues on Raw Product Offered for Sale, 1981–1986[a]

Pesticide	Crop Treated (%)	Rate (kg ai/ha)	Detection Limit (mg/kg)	Detectable Residue (%)	Residue (mg/kg) Mean ± SD	Residue (mg/kg) Maximum	MRL (mg/kg)
Azinphosmethyl	29.7	0.60	0.5	0.0	—	—	1.0
Carbaryl	32.5	1.10	0.01	5.2	0.06 ± 0.01	0.08	5.0
Carbofuran	6.3	4.70	0.01	5.2	0.01 ± 0.01	0.02	0.1
Chlorpyrifos	5.2		0.005	1.4	0.02 ± 0.01	0.03	0.1
Dimethoate	2.4	0.30	0.005	0.0	—	—	0.5
Endosulfan	12.6	0.50	0.001	41.7	0.056 ± 0.116	0.18	1.0
Malathion	0.9	0.60	0.005	1.4	0.01 ± 0.01	0.02	3.0
Permethrin	1.8	0.20	0.005	10.0	0.10	0.10	0.5
Captan	3.5	1.75	0.002	5.6	0.03 ± 0.04	0.10	5.0
Captafol	91.3	1.1	0.002	0.0	—	—	5.0
Chlorothalonil	92.7	2.8	0.01	21.7	0.07 ± 0.13	0.31	5.0
Mancozeb	50.3	2.7	0.1 ⎫	18.8	0.15 ± 0.09	0.30	4.0
Maneb	26.2	2.6	0.1 ⎭				

[a]Frank et al. (1987c).

phosalone at 33.2 kg remaining following the use of 13,008 kg; this represented a residue of 0.03%. This was a relatively low percentage residue when compared to dicofol where 1.1% of the original application remained on the raw product. For cherries, captan left the greatest residue at 2.25 kg following the use of 12,867 kg, a 0.017% remainder. For peaches, captan and phosmet where present at 70 and 62.7 kg, respectively, representing a 0.37 and 0.64% residue. These were not as great as dicofol, where a residue of 2.4% remained. With strawberries 23 kg of captan remained following the use of 8886 kg, a residual of 0.26%. The closeness to harvest in the application of these six pesticides influenced the percentage remaining on the fruit product, which ranged from relatively high to relatively low. As an example, captan residues varied from 0.02 to 0.37%, carbaryl varied from 0.006 to 0.82%, dicofol 0.56 to 2.4%, endosulfan 0.016 to 1.6%, phosalone 0.04 to 0.98%, and phosmet 0.007 to 0.04%.

A similar farm survey and monitoring program was conducted on vegetables. Fewer pesticides were found to be used on vegetables and those sought in the monitoring survey are listed in the Tables 27 to 30. No residues were found in or on green beans and sweet corn to the limits of the procedures. Carbaryl and endosulfan were the only residues found on asparagus. These two along with chlorothalonil and the EBDC fungicides were found on cole crops. With cucumber, 5 of the 10 pesticides sought in the monitoring procedure were measurable; these included dicofol that was found on the greenhouse crop and not on the field crop, along with endosulfan, methoxychlor, chlorothalonil, and zineb on both crops (Table 28). With green peppers, only carbofuran was identified on the raw product. On potatoes, aldicarb was found on 82% of the produce checked, yet the survey showed only 16% of growers using the product. However, these figures are somewhat biased since in 1983, the year of the survey, an independent survey on aldicarb use indicated that at least 60% of potato farmers were using this insecticide. Moreover, monitoring data were biased by the fact that samples were included from known users of this insecticide. The high residues of linuron on 2.0% of the samples remains unexplained. Among the vegetables, tomatoes has one of the longest lists of recommended pesticides, many of which left terminal residues; however these residues were very low and rarely exceeded 0.1 mg/kg.

6. LEGISLATION CONTROLLING ENVIRONMENTAL CONTAMINANT RESIDUES ON FOOD FROM INDUSTRIAL CHEMICALS AND DISCONTINUED PERSISTENT PESTICIDES

Some of the pertinent legislation controlling both the sale and allowable food residues for discontinued pesticides, as well as some of the persistent industrial chemicals, occurs in the Pest Control Products Act and/or the Food and Drugs Act. Since these pieces of legislation have been discussed previously readers are referred to Section 2.

Several additional acts and regulations now control the movement and sale of persistent hazardous substances in commerce. The following are the important new acts intended to protect the environment and avoid contamination of food and feed.

6.1. Environmental Contaminants Act

This act, administered by Environment Canada, provides for the protection of human health and the environment from substances that have or potentially can contaminate the environment. It gives the federal government the power to demand information from manufacturers about substances that may be harmful to the environment or to human health, and enables action to be taken in consultation with the provinces and other federal government departments to prevent or control their use or release into the environment. Substances affected include compounds such as PCB, mirex, PBB, and HCB.

6.2. Hazardous Products Act

This act, administered by Environment Canada, provides for the classification of those hazardous products likely to endanger the health and safety of the public. It further allows for the restriction of sale, importation, and advertisement of such products. This legislation will control new and old organochlorine industrial chemicals that have appeared in food, for example, PCB, PBB, mirex, and HCB.

6.3. Environmental Protection Act

This legislation, also administered by Environment Canada, is intended to extend the Hazardous Products Act by strengthening the classification process and the requirement for toxicological data while protecting water quality and the environment. This act has considerable authority to prosecute offenders.

6.4. Transportation of Dangerous Goods

This act, administered by Transport Canada, applies to handling, offering for transport, and transporting dangerous goods, by any means, whether or not for hire or reward and whether or not the goods originated from or are destined for any place in Canada. It also applies to transportation of dangerous goods by ships, vessels, and aircraft registered in Canada whether the movement is in or outside Canada. Dangerous goods are products, substances, or organisms scheduled in the act and include poisonous and infectious substances.

Provisions exist for inspection and/or seizure of dangerous goods in contravention of the act where there exists a serious and imminent danger to life, health, property, or the environment.

6.5. Other Federal Acts

1. The **Ocean Dumping Control Act** regulates the dumping of waste and other hazardous chemicals, including pesticides, in the ocean. It prohibits the dumping of such substances in the marine environment unless specific permission by permit is issued by Fisheries and Oceans Canada.
2. The **Fish Inspection Act** prohibits the trafficking of any fish, intended for human consumption, that is tainted, decomposed, or unwholesome and provides for the inspection thereof.

7. RESIDUES OF PERSISTENT ORGANOCHLORINE PESTICIDES AND INDUSTRIAL CHEMICALS IN FOODS AND TOBACCO

Current legislation is designed to (1) regulate the introduction of new persistent toxic compounds, (2) control the losses of toxic substances to the environment, and (3) decontaminate soils or areas where high levels of contamination have occurred. Meanwhile this section addresses the following current issues, that pertain to persistent toxic compounds:

1. Residues of persistent organochlorines on foods arising from industrial activities, for example, PCB, PBB, HCB, polychloronaphthalenes, mirex, PCDD, and PCDF.
2. Residues of persistent organochlorine pesticides in foods arising from current nonagricultural uses, for example, PCP.
3. Residues of persistent organochlorine pesticides in food arising from environmental contamination from past uses, for example, DDT, dieldrin, heptachlor expoxide.
4. Residues of persistent organochlorine pesticides in food and tobacco from persistent residues in soils, for example, DDT and dieldrin.

7.1. Residues of Industrial Chemicals in Food

Not all residues found in food occur from agricultural use. Residues of HCB and PCB are two substances that have appeared in food and were derived from urban-industrial activities. Other substances have caused concern as potential contaminants. These include (1) polybrominated biphenyl (PBB), a serious contaminant in feed and food in Michigan that did not appear in Ontario grown produce, (2) mirex, which was found in fish from Lake Ontario but did not appear in terrestrially produced food in Ontario, and (3) polychloro-naphthalenes, which appeared in wildlife tissues and were suspected in food commodities at trace levels.

During the 1970s PCB residues became ubiquitous in meats, eggs, and cows' milk (Tables 31 and 32). Recent analyses have shown a marked decline in these

TABLE 31 Residues of Organochlorine Insecticides and Industrial Chemicals in Meats, Milk, and Eggs[a]

Animal Product	Years	Samples (Carcasses)	Mean Residue in Extractable Fat of All Samples (μg/kg)							
			Chlordane	DDT	Dieldrin	Endosulfan	HCB[b]	HCH[b]	HE[b]	PCB
Beef	1969–70	137(835)	NM[d]	123	33	<0.1	10	16	1	239
	1975–76	21(53)	1	44	23	<0.1	NM	32	3	145
	1981[c]	197(990)	1	4	6	0.4	<2	3	4	5
	1985–86	20	<1	<5	<2	<0.1	<2	<2	<1	<5
Pork	1969–70	35(170)	NM	187	12	<0.1	NM	30	<1	330
	1973–74	5(18)	<1	42	2	<0.1	2	NM	<1	24
	1981	38(190)	<1	5	<2	<0.1	NM	<2	<1	3
	1985–86	20	<1	<5	<2	<0.1	<2	<2	<1	<5
Poultry	1969–70	19(95)	NM	391	28	NM	NM	9	<1	946
	1975–76	9(30)	22	139	20	10	9	<1	6	391
	1981–82	50(180)	5	2.3	<2	1.5	<0.1	1	5	5
	1985–86	45		<5	3	<0.1	<0.1	3	<1	<5
Cows' milk	1970–71	337	<1	125	35	<0.1	5	<1	<1	85
	1973	350	<1	51	15	<0.1	7	<1	<1	115
	1977	380	<1	15	11	1	0.2	NM	4	35
	1983	359	0.3	12	6	0.4	0.7	9.3	4	24
Hen eggs	1969–70	30(120)	NM	188	8	<0.1	NM	5	<1	405
	1973–74	22(64)	<1	137	11	<0.1	108	<1	<1	255
	1981–82	51(101)	<1	4	<0.5	<0.1	0.6	<1	<1	10

[a]Frank et al. (1983b, 1985a, b).
[b]HCB, benzene hexachloride; HCH, hexachlorocyclohexane; HE, heptachlor epoxide.
[c]Fenthion identified in 7/197 samples, 171 ± 256ug/kg.
[d]NM, not measured.

515

TABLE 32 Declining Organochlorine Residues as Observed by the Increasing Frequency of Residue Levels below 10 μg/kg in the Fat of Animal Tissue[a]

Animal Product	Years	Percentage Samples below 10 μg/kg							
		Chlordane	DDT	Dieldrin	Endosulfan	HCB[b]	HCH[b]	HE[b]	PCB
Beef	1969–70	100	0.8	10.2	100	50	73.0	96.9	0.0
	1975–76	100	9.6	66.7	100	NM[c]	85.7	100	9.5
	1981	100	73.1	92.9	100	100	97.5	97.4	78.2
	1985–86	100	100	100	100	100	100	100	100
Pork	1969–70	100	0.0	82.9	100	NM	85.7	100	0.0
	1973–74	100	18.7	100	100	100	100	100	18.8
	1981	100	92.1	100	100	NM	81.1	100	92.1
	1985–86	100	100	100	100	100	100	100	100
Poultry	1969–70	NM	0.0	21	NM	NM	90	95	0.0
	1975–76	100	22.0	33	94	80	100	89	0.0
	1981–82	100	100	100	100	100	100	96	84
	1985–86	100	100	100	100	100	95.6	100	100
Cows' milk	1970–71	100	0.0	0.0	92.4	100	98.2	97.0	0.0
	1973	100	3.5	9.0	100	81.2	100	97.5	0.0
	1977	98.4	49.0	59.1	99.4	97.5	100	99.0	13.6
	1983	99.7	69.4	95.3	99.4	100	93.3	99.7	95.6
Hen eggs	1969–70	NM	40.0	67.0	100	NM	82	100	0.0
	1973–74	100	44.0	73.0	100	30	100	100	3.0
	1981–82	100	100	100	100	100	100	100	50

[a]Frank et al. (1983b, 1985a, b).
[b]HCB, benzene hexachloride; HCH, hexachlorocyclohexane; HE, heptachlor epoxide.
[c]NM, not measured.

TABLE 33 PCB Residues above 0.1 mg/kg in Fats of Meats and Milk[a]

Animal Products	Year	Percentage above 0.1 mg/kg	Animal Products	Year	Percentage above 0.1 mg/kg
Beef	1969–70	75.0	Poultry meat	1969–70	30.0
	1971–72	38.4		1971–72	66.0
	1973–74	51.5		1973–74	25.0
	1975–76	61.9		1975–76	64.0
	1977–78	20.0		1979–80	0.0
	1979–80	20.5		1981–82	0.0
	1981	1.0		1985–86	0.0
	1985–86	0.0			
Pork	1969–70	66.7	Hen eggs	1969–70	80.0
	1971–72	29.2		1971–72	49.0
	1973–74	37.5		1973–74	87.0
	1975–76	0.0		1978–80	25.0
	1979	0.0		1981–82	0.0
	1981	0.0			
	1985–86	0.0	Cows' milk	1970–71	36.4
				1973	54.5
				1977	3.9
				1983	1.1

[a]Frank et al. (1983b, 1985a, b)

residues. The highest mean residues were observed in poultry meat at 946 μg/kg in 1969–1970, and these levels had declined to less than 5 μg/kg by 1985–1986. At the same time, the percentage of samples exceeding 10 μg/kg decreased from 100 to 0%. Actionable residues, under the Food and Drugs Act, were set at 0.1 mg/kg for the fats of meat and eggs and 0.2 mg/kg for milk fat. The results of monitoring programs appear in Table 33 and reveal that a high percentage of samples exceeded these levels in 1969–1970 but by 1985–1986 all meat fats samples were below this level. The above findings are in keeping with the actions carried out since 1971.

A survey of fruit and vegetables in 1986 found no detectable residues of industrial contaminants.

In 1971 a voluntary action was agreed on whereby PCB in all open systems uses would be phased out of use by the manufacturers. By 1976 legislation was enacted to control the disposal of PCB waste and to search for alternatives to replace their use in closed systems. Up to 1976, wastes of PCBs were disposed of openly in the environment and resulted in residues reported earlier in agricultural animal products.

During the PBB (polybrominated biphenyl) incident in Michigan, when this compound was inadvertently fed to livestock and resulted in environmental and human exposure, the Ontario border was watched carefully to prevent feed, meat,

TABLE 34   Results of Testing Ontario Produce for PCDD and PCDF, 1987[a]

Commodity	Chlorodibenzodioxin (ng/kg)					Chlorodibenzofurans (ng/kg)				
	Tetra	Penta	Hexa	Hepta	Octa	Tetra	Penta	Hexa	Hepta	Octa
Pork	<6	<6	<6	<2	<5	<4	<4	<4	<3	<2
Chicken	<4	<3	<3	<2	17	<3	<3	<2	<2	<2
Hamburger	<2	<1	<1	<2	3	<1	<1	<1	<1	<1
Prime beef	<2	<1	<1	<1	12	<2	<1	<1	<1	<1
Eggs	<4	<4	<4	<5	<6	<9	<4	<5	<6	<6
Milk	<2	<2	<4	<5	<7	<3	<3	<4	<5	<5
Apples	<0.8	<0.6	<0.7	<0.4	<2	<2	<0.7	<0.4	<0.5	<0.3
Peaches	<0.5	<0.3	<0.2	<0.3	<2	<1	<0.3	<0.2	<0.3	<0.3
Potatoes	<0.2	<0.1	<0.3	<0.2	1	<2	<1	<0.5	<0.2	<0.3
Tomatoes	<0.3	<0.1	<0.4	<0.2	<0.6	<3	<2	<0.2	<0.3	<0.2
Wheat	<0.7	<0.7	<0.3	<0.2	0.6	<4	<0.7	<0.3	<0.3	<0.3

[a]OMAF/OME (1988).

and livestock from crossing into Canada. Ontario monitored this situation carefully to minimize its introduction.

Mirex is another persistent toxic industrial product that was used as a fire retardant in household building materials and that could potentially accumulate in food. Quantities used in Ontario were small and confined. Residues that appeared in Lake Ontario fish as a result of manufacturing losses, however, did not appear in agricultural produce or animal feed.

Some PCB formulations are known to contain polychlorodibenzofurans (PCDF) as contaminants. These substances along with the polychlorodibenzodioxins (PCDD) appear in the environment in both feed and water and have been reported in some foods. Both PCDF and PCDD occur in fly ash of incinerators and in the polychlorinated phenolic wood preservatives (see Section 7.2) and the two make up the most potent sources of these two persistent toxic group of compounds. Residues of these contaminants have been sought in Ontario food commodities and results appear in Table 34. It should be noted that few residues were found above the detection limits, which were in the low ng/kg range.

7.2.   Nonagricultural Use of Persistent Pesticides Appearing in Food

The timber and lumber industries use the wood preservative fungicide polychlorinated phenol, which contains tri-, tetra-, and pentachlorophenol together with trace contaminants of polychlorodibenzodioxins (PCDD) and polychlorodibenzofurans (PCDF) dealt with in Section 7.1. Data are presented in Table 35 showing that detectable residues of tetra- and pentachlorphenol have appeared in most animal tissues analyzed between 1982 and 1986. The highest levels were present in beef, eggs, and milk.

TABLE 35   Polychlorinated Phenolic Residues in Tissues of Domestic Livestock and Polutry, 1982–1986[a]

Food or Species	Tissue	Year	Nondetected Residues (No.)	Detected Residues (No.)	T3CP (μg/kg)	T4CP (μg/kg)	P5CP (μg/kg)
Forage crops	Animal feed	1982–84	0	14	<5	10	17
Dairy	Milkfat	1982	3	39	<10	<10	29
Beef	Tallow	1982	0	1	<1	<1	7
	Abdominal fat	1985	7	25	<1	<1	6
	Abdominal fat	1986	16	4	<1	4	37
	Back muscle	1986	4	0	<1	<1	<1
Goat	Abdominal fat	1985–86	1	2	<1	<1	5
Rabbit	Abdominal fat	1985	4	5	<1	<1	5
Pork	Diaphragm muscle	1984	8	17	<1	2	5
	Kidney	1984	0	25	<1	4	7
	Liver	1984	0	25	<1	4	25
	Abdominal fat	1985	4	30	<1	4	25
	Abdominal fat	1986	5	15	<1	1	5
	Back muscle	1986	3	0	<1	<1	<1
Mutton	Abdominal fat	1985	5	9	<1	<1	4
	Abdominal fat	1986	7	2	<1	<1	5
Hen	Eggs	1986	3	18	<1	4	35
Broiler	Abdominal fat	1985	0	11	<1	4	7
	Abdominal fat	1986	2	8	<1	2	19
	Back muscle	1986	3	0	<1	<1	<1
Turkey	Abdominal fat	1985	0	6	<1	<1	5
	Abdominal fat	1986	5	0	<1	<1	<1

[a]Frank et al. (1988).

7.3.   Persistent Organochlorine Insecticides Used in Agriculture But Not on the Commodities Analyzed.

Throughout the 1960s data were accumulated that indicated persistent organochlorine pesticides were increasingly being stored in fatty tissues of domestic farm animals. These residues were frequently shown to be in violation of health standards for meats, eggs, milk, and other animal products. Monitoring surveys were carried out between 1967 and 1969 that involved 1651 bulk tankers delivering milk to 286 dairies in the province. Twenty-two bulk tankers or 1.3% were found hauling milk that had residues in violation of Health and Welfare actionable levels. Analytical results indicated that 17 tankers had dieldrin residues and one tanker had heptachlor epoxide residues in milkfat above the allowable 0.1 mg/kg, and three had t-DDT residues and one had lindane residues in milkfat above the permitted 1 mg/kg. At the same time a further 212 or 12.5% of the tankers contained milk with residues below the actionable levels but of concern to health officials. In all cases, except for lindane, the insecticides had been used on food or fodder crops or on timber and were introduced to the farm unknowingly to the owners of the dairy herds.

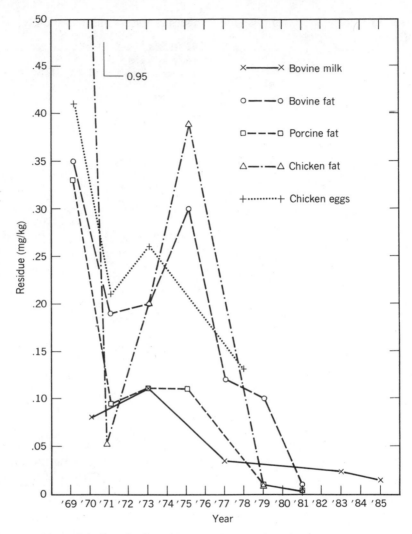

Figure 7.   Rise and decline of PCB residues in animal products from 1969 to 1985 in Ontario.

In 1969 Ontario banned all agricultural uses of aldrin, dieldrin, and heptachlor. In 1970 the use of DDT was restricted to permit use and all outdoor uses were banned in late 1971. Other persistent organochlorine insecticides (OCI) were removed at the federal and/or provincial levels during the early 1970.

Monitoring programs were set up to follow the decline in terminal residues in meat, eggs, and milk and these studies are continued to the present day. A summary of the findings appear in Tables 31 and 32 and Figure 7. Monitoring 2012 milk samples since 1970 has not revealed further violations of OCI. In fact the data have shown a marked decline in the average OCI residues and a

TABLE 36   Violation of Pesticides, Meat, and Milk, 1969–1986[a]

Animal Product	Total Samples Analyzed	Violations	Year	Pesticide	Residues (mg/kg)	MRL (mg/kg)
Beef	505	1	1970	DDT	11.3	7.0
		1	1981	Fenthion	0.6, 0.4	0.1
Pork	122	1	1971	DDT	7.7	7.0
Poultry	137	2	1970, 71	Dieldrin	0.7, 0.4	0.25
Hen eggs	118	0				
Cows' milk	2012	0				

[a]Frank et al. (1983b, 1985a, b).

substantial increase in the percentage of samples that have fallen below 0.01 mg/kg.

Between 1969 and 1986, 764 samples of meat were analyzed for both organochlorine and organophosphorus insecticides and only six samples have been found in violation. Two each involved DDT, dieldrin, and fenthion (Table 36).

None of the persistent organochlorine insecticides listed as having been detected, with the exception of HCH, was used on domestic animals. Even the HCH residue that was made up of three isomers did not all come from topical applications; only lindane, the γ-isomer, is still recommended and used on livestock, although a 30-day waiting period is required before slaughter.

7.4.   Soil Residues of Persistent Organochlorine Insecticides in Food and Tobacco

The persistent organochlorine insecticides were used in the production of tobacco up to 1969 when the registration of aldrin, dieldrin, and heptachlor was cancelled. In 1970 DDT was restricted and in 1971 the use of DDT and endrin was cancelled. In 1978 chlordane and endosulfan were deregistered. The decline in the OCI residues for the period 1970 to 1985 is found in Table 37. Residues of replacement pesticide compounds are given. Many of the newer insecticides are now being used and leave no detectable residues.

In spite of no agricultural use of DDT in 15 years, persistent residues that remain in tobacco soils have resulted in DDT residue levels in cured tobacco leaves. These have changed little in 10 years (1975–1986).

From 1980 to 1986, the 1088 fruit and vegetable samples analyzed were also examined for persistent organochlorine residues and no residues were found in these foods with the exception of cucumbers, where dieldrin was found in 1981, 1982, and 1983, in 30, 27, and 14% of the produce, respectively. The mean levels of dieldrin in cucumbers where residues were detected for the 3 years were 0.009, 0.006, and 0.031 mg/kg, respectively, and are the result of persistent residues in the soil.

TABLE 37   Pesticide Residues in Flue Cured Tobacco from 34 Farms between 1970 and 1985[a]

Pesticide[b]	Residues (mg/kg) in Dried Cured Tobacco Leaf				Last Year of Use
	1970	1975	1980	1985	
t-DDT (I)	1.83	0.89	0.62	0.57	1971
Dieldrin (I)	0.080	0.035	0.081	0.027	1969
Endrin (I)	0.11	0.06	0.018	0.067	1970
Heptachlor epoxide (I)	0.023	0.018	0.017	0.003	1969
Chlordane (I)	<0.001	0.076	0.008	0.004	1978
t-Endosulfan (I)	3.67	3.19	0.081	0.005	1978
Lindane (I)	<0.001	0.010	0.005	0.008	—
Permethrin (I)	—	—	0.39	0.012	—
Chlorpyriphos (I)	<0.005	<0.005	<0.005	<0.005	—
Acephate and methamidophos (I)	—	—	0.105	0.010	—
Metalaxyl (F)	—	—	0.36	<0.010	—
Diphenamid (H)	—	—	0.68	0.19	—

[a]Frank et al. (1987d).
[b]I, insecticide; F, fungicides; H, herbicide.

8. CONCLUSIONS

Canada has a wide range of legislative acts to regulate pesticides and industrial chemicals to prevent the contamination of food. In the 1960s and early 1970 a potential problem of organochlorine insecticide (OCI) residues in food was rectified by the cancellation of all persistent OCI registrations and/or their removal from agricultural use. Thus further build-up of the persistent organochlorines was prevented in both the environment and in food. A slow decline in food residues of these substances was observed and this has continued over the past 15 years.

Federal and provincial agencies, as well as many of the food processors are actively working together to use replacement and newer compounds to ensure that Canadian standards are not violated. Currently between 0.5 and 2% of samples analyzed annually are above the MELs; however, these are confined to specific commodities and attempts are being made annually to reduce this number to zero by changing use patterns for these chemicals.

Some pesticides leave detectable residues in raw food and many of these can be significantly reduced by culinary practices, cooking, and processing so that the quantities actually consumed are a fraction of those found in the raw product. Attempts are being made to reduce agricultural dependence on pesticides through the use of integrated pest management programs and the use of alternative control methods. Studies are continuing to help produce food with nondetectable residues through reducing pesticide rates and increasing intervals to harvest. The aim is to give the consumer the quality of food being demanded.

REFERENCES

Agriculture Canada. (1986). Food Residue Monitoring Consolidation Program. Pesticide Directorate, Agriculture Canada, Ottawa.

Braun, H. E., Frank, R., and Hoffman, I. (1983). Estimation of MH in Ontario tobacco between 1968–1982 using a rapid screening procedure. *Tobacco Sci.* **27**, 99–100.

Elliott, J. H. (1987). Crop Protection Institute of Canada, 6 Lansing Square, Willowdale, Ontario M2J 1T5.

Frank, R., Braun, H. E., and Stranek, J. (1983a). Removal of captan from treated apples. *Arch. Environ. Contam. Toxcial.* **12**, 265–269.

Frank, R., Braun, H. E., Fleming, G. (1983b). Organochlorine and organophosphorus residues in fat of bovine and porcine carcasses marketed in Ontario, Canada from 1969 to 1981. *J. Food Protection* **46**, 893–900.

Frank, R., Ritcey, G., Braun, H. E., and McEwen, F. L. (1984). Disappearance of acephate residues from beans, carrots, celery, lettuce, peppers, potatoes strawberries and tomatoes, *J. Econ. Entomol.* **77**, 1110–1115.

Frank, R., Braun, H. E., Sirons, G. H., Rasper, J., and Ward, G. G. (1985a). Organochlorine and organophosphorus insecticides and industrial pollutants in the milk supplies of Ontario—1983. *J. Food Protection* **48**, 499–504.

Frank, R., Rasper, J., Barun, H. E., and Ashton, G. (1985b). Disappearance of organochlorine residues from abdominal and egg fats of chickens, Ontario, Canada, 1969–1982. *J Assoc. Off Anal. Chem.* **68**, 124–129.

Frank, R., Braun, H. E., and Ripley, B. D. (1987a). Residues of insecticides, fungicides, and herbicides in fruit produced in Ontario, Canada, 1980–1984. *Bull. Environ. Contam. Toxicol.* **39**, 272–279.

Frank, R., Braun, H. E., and Ripley, B. D. (1987b). Residues of insecticides, fungicides, and herbicides on Ontario grown vegetables, 1980–85. *J. Assoc. Off. Anal. chem.* **70**, 1081–1086.

Frank, R., Braun, H. E., and Ritcey, G. (1987c). Disappearance of captan from field- and greenhouse-grown tomato fruit in relationship to time of harvest and amount of rainfall. *Can J. Plant. Sci.* **67**, 355–357.

Frank, R., Braun, H. E., Suda, P., Ripley, B. D., Clegg, B. S., Beyaert, R. P., and Zilkey, B. F. (1987d). Pesticide residues and metal contents in flue-cured tobacco and tobacco soils of southern Ontario, Canada 1980–85. *Tobacco Sci.* **31**, 40–45.

Frank, R., Stonefield, K. I., and Lukens, H. (1988). Monitoring wood shaving litter and animal products for polychlorophenol residues, Ontario, Canada, 1978–1986. *Bull. Environ. Contam. Toxicol.* **40**, 468–474.

Frank, R., Braun, H. E., and Ripley, B. D. (1989). Monitoring Ontario-grown apples for agricultural chemicals used in their production, 1978–86. *Food Addit. Contam.* **6**, 227–234.

Health and Welfare Canada. (1986). The Food and Drugs Act and Regulations with Amendments to December 1986. Queens Printers and Controller of Stationary, Ministry of Supply and Services, Ottawa.

Magee, W. (1984). Survey of pesticide use in Ontario 1983. Economics Information Report 84–05, ISBN 0-7743-99459-7 Ontario Ministry of Agriculture and Food, Legislative Buildings, Queen's Park, Toronto, Ontario, M7A 1B6.

Northover, J., Frank, R., and Braun, H. E. (1986). Dissipation of captan residues from cherry and peach fruits. *J. Agric. Food Chem.* **34**, 525–529.

OMAF. (1986). Agricultural Statistics for Ontario 1985, Publication 20, Ontario Ministry of Agriculture and Food, Legislative Buildings, Queen's Park, Toronto, Ontario M7A 1B6.

OMAF/OME. (1988). Polychlorinated Dibenzo-p-dioxins and Polychlorinated Dibenzo furnas and other Organochlorine Contaminants in Food. ISBN 0-7729-4608-6. Queen's Printer for Ontario, Toronto.

Ritcey, G., Frank, R., McEwen, F. L., and Braun, H. E. (1987). Captan residues on strawberries and estimates of exposure to pickers. *Bull. Environ. Contam. Toxicol.* **38**, 840–846.

Roller, N. F. (1975). Survey of pesticide use in Ontario, 1973. Economics Information, Ontario Ministry of Agriculture and Food, Toronto, Ontario, M7A 1B6.

Roller, N. F. (1979). Survey of pesticide use in Ontario, 1978. Economics Information, Ontario Ministry of Agriculture and Food, Toronto, Ontario, M7A 1B6.

Willis, G H., and McDowell, L. L. (1987). Pesticide persistence on foliage. *Rev. Environ. Contam. Toxicol.* **100**, 23–73.

16

HUMAN EXPOSURE PATHWAYS TO SELECTED ORGANOCHLORINES AND PCBs IN TORONTO AND SOUTHERN ONTARIO

Katherine Davies

Department of Public Health, Toronto, Ontario, Canada

1. INTRODUCTION

Humans are exposed to organochlorines and PCBs in the environment and several studies have reported residues of these chemicals in samples of adipose tissue (Mes et al., 1982) and human milk (Mes et al., 1986) from Canadians. Human exposure pathways have not, however, been well characterized and the relative importance of different pathways has not yet been established for specific chemicals or for groups of chemicals. This preliminary study is intended to provide an overview of human exposure to selected organochlorines and PCBs through three nonoccupational exposure pathways: food, air, and drinking water. This was done by using background concentrations of these chemicals and estimates of the average amounts of air inhaled, drinking water ingested, and food consumed. Previous research suggests that food is likely to be an important exposure pathway for some organochlorines (Campbell et al., 1965).

2. METHOD

Average annual human exposure to selected organochlorines and PCBs in Toronto and Southern Ontario was estimated by multiplying previously reported background concentrations in air, drinking water, and food by estimates of the average amounts of air inhaled, drinking water ingested, and food consumed, respectively. Previously reported background concentrations of selected organochlorines and PCBs in air, drinking water, and food water used to estimate annual human exposure, although it has been necessary to compare data from different laboratories using different analytical methods with different minimum detection levels. Estimates of the average amounts of air inhaled, drinking water ingested, and food consumed were also taken from previously reported studies. The average amount of air inhaled was assumed to be 15 breaths a minute with a tidal volume of half a liter (Menzel and McClennan, 1980). This is

TABLE 1   Estimated Annual Purchase of Fresh Food by Ontario Residents (1982) and Relative Proportions in Composite Samples

	Estimated Annual Purchase[a] (kg/person/year)	Composite (%)
Fresh Meat and Eggs		
Beef muscle	9.70	16.17
Hamburger	7.14	11.90
Pork muscle		
(including cured)	12.10	20.17
Veal	1.02	1.70
Lamb and mutton	0.76	1.27
Liver	0.65	1.08
Chicken	15.13	25.22
Eggs	13.50	22.50
Total	60.00	100.00

TABLE 1 (*Continued*)

	Estimated Annual Purchase[a] (kg/person/year)	Composite (%)
Root Vegetables Including Potatoes		
Potatoes	24.12	81.05
Carrots	3.93	13.21
Radishes	0.34	1.14
Turnips/rutabagas	0.72	2.42
Others[b]	0.64	2.15
Total	29.75	100.00
Fresh Fruit		
Apples	11.69	68.60
Peaches/nectarines	2.56	15.02
Pears	1.88	11.03
Plums	0.91	5.34
Total	17.04	100.00
Leafy and Other above Ground Vegetables		
Green/wax beans	0.66	2.16
Broccoli	1.12	3.66
Cabbage	1.77	5.78
Cauliflower	1.88	6.14
Celery	1.76	5.75
Corn	0.74	2.42
Cucumber	2.58	8.43
Lettuce	6.63	21.67
Mushroom	0.74	2.42
Onion	3.99	13.04
Pepper	1.08	3.53
Tomato	5.52	18.04
Other seed/gourd[c]	1.16	3.79
Other leaf/stalk[d]	0.97	3.17
Total	30.60	100.00

[a] Calculated from Statistics Canada (1982).
[b] Other root vegetables were beets.
[c] Other seed/gourd were zucchini, squash, and vegetable marrow.
[d] Other leaf/stalk were green onion stalks.

equivalent to 3941 m³ a year. The average amount of drinking water ingested was assumed to be 1.5 L/day (Gillman, 1987). This is equivalent to 547.5 L/year.

The average amounts of different food items consumed were estimated from food purchase data for Ontario residents (Statistics Canada, 1982). This is likely to overestimate actual consumption because of the amount of food purchased and then discarded. These data are the only recent reliable information relevant to average dietary patterns for Ontario residents, although it should be recognized that individual dietary habits are likely to be diverse. Dietary surveys of Canadians have been undertaken (Smith, 1971; Smith et al., 1972, 1973, 1975), but they were not used for this study because they were done several years ago, and may not reflect existing food consumption patterns. The data on consumption of meat and dairy products in a more recent study (McLeod et al., 1980) of pesticide residues in the Canadian diet data compare moderately well with the food purchase data.

The food purchase data report the amounts of all foods purchased weekly by an average family in each province for 1982 (Statistics Canada, 1982). Amounts shown in the report were divided by the average number of individuals in a family (2.74) and multiplied to provide estimates of the amount of different foods purchased per person per year (see Table 1).

3. RESULTS

3.1. Human Exposure to Selected Organochlorines and PCBs in Air

Most air monitoring has been conducted in relation to specific point source discharges, so there are few data available on ambient atmospheric concentrations on organochlorines and PCBs in Toronto and Southern Ontario. Ambient atmospheric concentrations of some organochlorine pesticides and PCBs in the Great Lakes basin were measured in 1981, and are shown in Table 2 (Eisenreich et al., 1981). The estimated human exposures were calculated and are listed in the table.

Studies of PCB concentrations in ambient air in 1979 and 1980 in Ontario have been conducted (Singer et al., 1983). These suggest that concentrations of PCBs are higher in industrial and urban areas than in rural and suburban areas. The concentrations reported for 1979 are signficantly higher than those shown in Table 2 although the 1980 concentrations are much lower.

Most data on ambient atmospheric concentrations are for outdoor air, although most people are outside for limited periods of time, depending on prevailing weather conditions. Although outdoor air is the source of all indoor air, indoor concentrations can be higher because chemicals are less able to disperse. No studies have been reported on concentrations of these chemicals in indoor air in Ontario. Some studies have been conducted elsewhere although most of them report atmospheric concentrations of pesticides following treatments for pests, such as termites. Such treatments frequently give rise to detectable concentrations of pesticides, and although these cannot be construed

TABLE 2 1981 Ambient Atmospheric Concentrations and Estimated Human Exposure to Organic Chemicals in the Great Lakes Basin

Chemical	Range[a] (ng/m³)	Mean[a] (ng/m³)	Estimated Mean Human Exposure[b] (μg/year)
Total PCBs	0.4–3.0	1.0	3.942
Total DDT	0.01–0.05	0.03	0.118
α-HCH	0.25–0.4	0.3	1.183
γ-HCH	1.0–4.0	2.0	7.884
Dieldrin	0.01–0.1	0.05	0.197
HCB	0.1–0.3	0.2	0.788
Total PAH	10–30	20	78.840
Anthracene	0.1–1.0	0.6	2.365
Phenanthrene	0.1–1.0	0.6	2.365
Pyrene	0.1–4.0	1.1	4.336
Benzo[a]anthracene	0.1–1.0	0.5	1.971
Perylene	0.1–2.0	0.6	2.365
Benzo[a]pyrene	0.1–2.0	1.0	3.942
Dibutyl phthalate	0.5–5.0	2.0	7.884
Diethylhexyl phthalate	0.5–5.0	2.0	7.884

[a]See Eisenreich et al. (1981).
[b]Calculated by the authors assuming a tidal volume of half a liter and a frequency of 15 breaths a minute (Menzel and McClennan, 1980).

as being background concentrations, several surveys have shown that the majority of urban residents use pesticides (Frankie and Levenson, 1978; Levenson and Frankie, 1983).

3.2. Human Exposure to Organochlorines and PCBs in Drinking Water

Toronto's drinking water system is the largest in Ontario and uses water from Lake Ontario. Water samples are analyzed 2 or 3 times a year for the presence of approximately 170 chemicals, of which about 135 are organic and 35 are inorganic. Between 1978 and 1984, 51 chemicals were identified during routine monitoring, although most were detected infrequently and at concentrations well below the provincial drinking water objectives. Average concentrations have been calculated using 1978–1981 data on treated drinking water from all four Metropolitan Toronto filtration plants. Assuming an average consumption of 1.5 L/day, human exposure estimates were calculated from these average concentrations (Davies, 1987). Some of these average concentrations and the estimated human exposures are shown in Table 3.

Drinking water is also used for many purposes other than direct ingestion that could result in human exposure. The most obvious of these are bathing and showering. Recent evidence suggests that organic chemicals can be absorbed through skin (Wester et al., 1983; Wester and Maibach, 1985). The extent of

TABLE 3   Mean Concentrations and Estimated Mean Human Exposure to Selected Persistent Toxic Chemicals Detected in Toronto's Drinking Water 1978–1984

Chemical	Average Concentration in Drinking Water[a] (μg/L)	Estimated Mean Human Exposure[b] (mg/year)
Organics		
α-HCH	0.034	0.0186
β-HCH	0.0005	0.0003
γ-HCH	0.0004	0.0002
Aldrin	0.002	0.0011
Dieldrin	0.001	0.0005
DDE	0.0003	0.0002
Dibutyl phthalate	2.547	1.3945
Dioctyl phthalate	0.133	0.0728
Bisethylhexyl phthalate	63.54[c]	34.7881
Diethyl phthalate	0.0107	0.0058
Atrazine	0.03	0.0164
Hexachlorobenzene	0.002	0.0011
Inorganics		
Arsenic	0.15	0.0821
Boron	25.8	14.1255
Cadmium	1.2	0.6570
Chromium	1.8	0.9855
Cyanide	0.81	0.4434
Mercury	0.06	0.0328
Selenium	0.16	0.0876

[a]See Davies (1987).
[b]Calculated by the author, assuming an average consumption of 1.5 L/day.
[c]Sample contamination suspected.

absorption depends on the nature of the chemical, its concentration, the solvent, and the site and duration of dermal exposure. One study of volatile organics estimated that dermal absorption contributed between 29 and 91% of the total dose from drinking water (Brown et al., 1984). It is, therefore, likely that dermal exposure to volatile organochlorines in drinking water contributes to overall human exposure. Inhalation during showering and bathing could also be a significant exposure pathway.

3.3.   Human Exposure to Organochlorines and PCBs in Food

Foods grown in Ontario have been analyzed individually and in composites for the presence of organochlorines and PCBs, although dietary intake of these chemicals from different types of food has not been estimated.

TABLE 4 Average 1980 Concentrations and Estimated Mean Human Exposure to Persistent Toxic Chemicals in the Edible Portion of Fish from the Toronto Harbor

	Rainbow Trout		Northern Pike		White Sucker	
Chemical	Average Concentration[a] (mg/kg wet wt.)	Estimated Mean Human Exposure[b] (mg/year)	Average Concentration (mg/kg wet wt.)	Estimated Mean Human Exposure (mg/year)	Average Concentration (mg/kg wet wt.)	Estimated Mean Human Exposure (mg/year)
PCBs	0.395	0.209	0.636	0.337	0.280	0.148
Total DDT	0.099	0.052	0.173	0.092	0.02	0.011
Chlordane	0.048	0.025	0.061	0.032	0.018	0.009
HCB	0.004	0.002	0.003	0.002	0.003	0.002
Mirex	0.01	0.005	0.033	0.017	0.001	0.0005
Heptachlor	ND	—[c]	ND[d]	—	ND	—
α- and β-HCH	0.029	0.015	0.015	0.008	0.003	0.002
γ-HCH	0.002	0.001	Trace	—	ND	—
Aldrin	ND	—	ND	—	ND	—
Mercury	0.07	0.037	0.34	0.180	0.160	0.085
Cadmium	0.04	0.021	0.04	0.021	0.043	0.023
Lead	0.91	0.482	1.14	0.604	2.40	1.272
Arsenic	0.10	0.053	0.157	0.083	0.085	0.045
Selenium	0.27	0.143	0.245	0.130	0.237	0.126

[a]See Dredging Subcommittee (1983).
[b]Calculated by the author, assuming an estimated average consumption of 0.53 kg/year (calculated from Statistics Canada, 1982).
[c]—, information not available.
[d]ND, not detected.

3.3.1. Fish

Concentrations of organochlorines and PCBs in fish in the Great Lakes basin have been extensively studied and concentrations of many organochlorines in fish in Lake Ontario have decreased since the mid-1970s. Toronto and Southern Ontario residents, however, also consume marine fish and freshwater fish from other lakes, as well as commercially grown fish. Data on organochlorines and PCBs in marine fish and fish from other lakes are limited and commercially grown fish are unlikely to contain significant amounts of these chemicals.

1980 data from the OMOE for fish in Toronto Harbour contain results for the edible portion of rainbow trout, northern pike, and white sucker. These are shown in Table 4 (Dredging Subcommittee, 1983). The estimated human exposures have been calculated from food purchase data for Ontario residents published by Statistics Canada (Statistics Canda, 1982), using an estimated average annual consumption of freshwater fish of 0.53 kg/year. This estimate is somewhat lower than those used by provincial and federal agencies to quantify ingestion of toxic chemicals from fish. These exposure estimates are likely to overestimate actual exposure because of the reasons given above.

3.3.2. Dairy Products

Concentrations of organochlorines and PCBs in cows' milk samples from Ontario have been reported by Frank et al. (1970, 1979, 1985a). These show that the average concentrations of organochlorines decreased between 1967 and 1983. The most recent published report examines data from 1983 (Frank et al., 1985a). They are shown in Table 5. This shows that dieldrin, DDE, heptachlor, heptachlor epoxide, and α-HCH were found in over 90% of the samples. Other organochlorines and PCBs were detected less frequently.

TABLE 5   Average Concentrations of Organochlorine Residues in Cows' Milk in Ontario—1983 and Estimated Mean Human Exposure

Chemical	Average Concentration		Estimated Mean Human Exposure[c] (mg/year)
	μg/kg butterfat[a]	μg/kg whole milk[b]	
Total DDT	12	0.46	0.011
α-HCH	6.2	0.24	0.006
γ-HCH	4.8	0.18	0.004
Heptachlor epoxide	3.4	0.13	0.003
Dieldrin	6.2	0.24	0.006
HCB	0.67	0.02	0.0004
Total PCBs	24	0.91	0.021
PCP	29	1.10	0.026

[a]See Frank et al. (1985a)
[b]Calculated assuming a butterfat content of 3.8%.
[c]Calculated assuming an estimated annual average consumption of 23.62 L/year (calculated from Statistics Canada, 1982).

TABLE 6 Summary of Average Concentrations and Estimated Mean Human Exposure to Organochlorines in Beef and Pork Fat from Ontario (1981) and Estimated Human Exposures

	Beef			Pork		
Chemical	Mean Concentration[a] (μg/kg fat)	SD[a]	Estimated Mean Human Exposure[b] (mg/year)	Mean Concentration[a] (μg/kg fat)	SD[a]	Estimated Mean Human Exposure[c] (mg/year)
Total DDT	12	52	0.040	5	7	0.0113
Dieldrin	5.7	7.2	0.019	<1	—	—
Heptachlor epoxide	4.2	8.0	0.014	<1	—	—
Chlordane	1.1	2.5	0.003	<1	—	—
α-HCH	6.6	6.7	0.022	1.1	1.8	0.002
γ-HCH	2.9	23.6	0.010	13	39	0.0294
PCB	9.5	22.5	0.032	3	5	0.007

[a]See Frank et al. (1983).

[b]Calculated by the author, assuming an average fat content of 20% and an estimated average consumption of 16.81 kg/year (calculated from Statistics Canada, 1982).

[c]Calculated by the author, assuming an average fat content of 30% and an estimated average consumption of 7.55 kg/year (calculated from Statistics Canada, 1982).

3.3.3. Meat and Eggs

Residues of organochlorines and PCBs have been determined in fat from beef and port carcasses marketed in Ontario between 1969 and 1981 (Frank et al., 1983), and in chicken and egg fat between 1969 and 1982 (Frank et al., 1985b). The results showed that mean concentrations of total PCBs, DDT residues, and dieldrin have decreased over time. Chlordane and heptachlor epoxide were rarely detected in the early 1970s; however, the incidence in beef and chicken fat increased in 1973 and remained detectable through 1981. γ-HCH residues in both beef and pork fat fluctuated from year to year and appeared to vary with the amount of γ-HCH used to control insect pests, however, it was rarely detected in chicken and egg fat. A summary of the most recent (1981) data is shown in Tables 6 and 7. Estimates of human exposure were calculated using food purchase data from Statistics Canada (Statistics Canada, 1982).

Data on residues of organochlorines in meat have been reported by Agriculture Canada (Agriculture Canada, 1985). The estimated human exposure calculated from these 1984–1985 data compare well with the 1981 data for Ontario beef and pork in Table 6. Most of the estimated exposure are within an order of magnitude, and some are in closer agreement. The estimated human exposures calculated from the 1984–1985 data can also be compared to the 1981–1982 exposure estimates for chicken shown in Table 7. The 1984–1985 exposure estimates for DDT are less than the 1981–1982 estimates and α- and β-HCH, chlordane, endosulfan, endrin, heptachlor, HCB, and methoxychlor were not detected in any of the 1984–1985 samples. This is likely to reflect the decreased use of these organochlorine pesticides.

TABLE 7   Summary of Organochlorine Residues in Chicken and Egg Fat (1981–1982) and Estimated Mean Human Exposure

	Chicken		Eggs	
Chemical	Mean Concentration[a] (μg/kg fat)	Estimated Mean Human Exposure[b] (μg/year)	Mean Concentration[a] (μg/kg fat)	Estimated Mean Human Exposure[c] (μg/year)
Total DDT	2.3	6.18	3.9	8.95
Dieldrin	0.74	1.99	ND	—
Heptachlor epoxide	4.6	12.36	ND	—
Chlordane	4.6	12.36	ND	—
γ-HCH	0.4	1.07	ND	—
PCB	4.8	12.90	ND	—

[a]See Frank et al. (1985b).
[b]Calculated by the author, assuming an average fat content of 17% and an estimated average consumption of 15.81 kg/year (calculated from Statistics Canada, 1982).
[c]Calculated by the author, assuming an average fat content of 17%, and an estimated average consumption of 13.50 kg/year (calculated from Statistics Canada, 1982).

3.3.4.   Fruit and Vegetables

Organochlorine pesticide residues have also been reported in fruit and vegetables grown in Ontario (Frank et al., 1982; Braun et al., 1982). These studies analyzed individual samples of different fruits and vegetables for the presence of pesticides not discussed in this paper.

3.3.5.   Food Composites

One preliminary study has analyzed fresh food composites of Ontario grown produce and estimated the dietary intake of organochlorines, PCBs, polychlorinated dibenzodioxins, and dibenzofurans from these composites (Davies, 1988). Five composite samples were prepared and analyzed (fruit, root vegetables including potatoes, meat and eggs, cows' milk, and leafy vegetables). The results of this study suggest that most exposure to these chemicals would come from meat and eggs and the least from leafy vegetables. However, this study examined fresh food grown only in Ontario and assumed that Ontario residents consume only Ontario-grown food. It did not include fresh food grown elsewhere or processed food that could be consumed by Ontario residents.

3.4.   Overview of Human Exposure

Two overviews of exposure to PCBs and selected organochlorines have been developed using the average exposure estimates calculated from the air, drinking water and food data discussed in this paper. These are shown in Tables 8 and 9. Both of these overviews include data on fresh food grown in Ontario only.

The first overview uses the data on air, drinking water, and individual food items discussed above. The data of Davies (1988) on fruit and vegetables have also been included. The results suggest that representative Ontario adults would receive the majority of their nonoccupational exposure to the organochlorines and PCBs studied from fresh food. Drinking water was the second largest source of nonoccupational exposure and air was the smallest source.

The second overview of exposure also uses the air and drinking water data discussed above and Davies' estimates of the annual intake of selected organochlorines and PCBs. This is shown in Table 9 and suggests that representative Ontario adults would receive the majority of their nonoccupational exposure to the organochlorines and PCBs studied from fresh foods. This estimate is almost the same as that shown in Table 8, although different data were used to calculate the exposure estimates. Most notably, freshwater fish were included in Table 8, but were excluded from Table 9 for reasons discussed in Davies (1988). Despite this inclusion, and the relatively high proportions of PCB and HCB exposure from freshwater fish, the overall proportion of exposures received from food is similar in Tables 8 and 9. This is because some of the exposure estimates developed for cows' milk, beef/pork, and chicken/eggs are lower in Table 9 than those calculated for these food items in Davies (1988), from which the data in Table 9 were taken.

TABLE 8 Overview of Human Exposure

| Food | Chemical | | | | | | | Average % by Food Type | Average % by Exposure Route |
	PCB	Total DDT	Aldrin/ Dieldrin	Heptachlor	HCB	α,β-HCH	γ-HCH		
Fish[a]									
mg/year	0.209	0.052	ND	ND	0.002	0.015	0.001		
% fresh food Exposure	70.8	2.3	0	0	47.6	33.1	2.0	25.6	—
Cows' milk[b]									
mg/year	0.021	0.011	0.006	0.003	0.0004	0.006	0.004		
% fresh food Exposure	7.1	0.5	0.7	10.3	9.5	13.2	8.0	7.9	—
Beef/Pork[c]									
mg/year	0.04	0.053	0.019	0.014	—	0.024	0.0394		
% fresh food Exposure	13.5	2.4	2.3	48.2	—	52.9	77.2	30.3	—
Chicken/Eggs[d]									
mg/year	0.013	0.015	—	0.012	—	—	0.001		
% fresh food Exposure	4.4	0.7	—	41.3	—	—	1.9	7.1	—

Fruit/Vegetables[e]								
mg/year	0.012	2.075	0.781	ND	0.0018	0.0003	0.00534	—
% fresh food Exposure	4.1	94.1	96.8	0	42.8	0.6	10.4	29.0
Total Fresh Food								
mg/year	0.295	2.206	0.806	0.029	0.0042	0.0453	0.051	—
% total exposure	98.6	99.9	99.7	100	68.8	69.5	83.8	88.6
Drinking Water[f]								
mg/year	ND	0.0002	0.0016	ND	0.0011	0.0189	0.0002	—
% total exposure	0	0.02	0.2	0	18.0	28.9	0.3	6.8
Air[g]								
mg/year	0.0039	0.0001	0.0002	—	0.0008	0.001	0.0078	—
% total exposure	1.3	0.01	0.02	—	13.1	1.5	12.8	4.1
Total Exposure								
mg/year	0.2989	2.2063	0.8078	0.029	0.0061	0.0652	0.060	

[a]See Table 4 for rainbow trout.
[b]See Table 5.
[c]See Table 6 (estimated total average exposure from beef and pork).
[d]See Table 7 (estimated total average exposure from chicken and egg).
[e]See Davies (1988).
[f]See Table 3.
[g]See Table 2.

TABLE 9   Overview of Human Exposure

| | Estimated Average Human Exposure | | | | | | |
| Chemical | Fresh Food[a] | | Drinking Water[b] | | Air[c] | | Total |
	mg/year	%	mg/year	%	mg/year	%	mg/year
α-HCH	0.011	36.7	0.018	60.0	0.001	3.3	0.03
γ-HCH	0.1313	94.2	0.0002	0.14	0.0078	5.6	0.139
Aldrin/dieldrin	1.01	99.8	0.0016	0.16	0.0002	0.02	2.329
Total DDT	2.328	88.9	0.0002	0.008	0.0001	0.004	0.018
HCB	0.016	88.9	0.0011	6.1	0.0008	4.44	0.036
PCB	0.032	88.9	ND	—	0.0039	10.8	0.036
Average % by all exposure routes		84.73		11.1		4.03	

[a]See Davies (1988).
[b]See Table 3.
[c]See Table 2.

4. DISCUSSION

This study suggests that fresh food could be the largest exposure pathway for the chemicals investigated, and that drinking water and air are likely to be minor contributors. This preliminary study, however, includes several assumptions and simplifications. For example, processed foods and foods grown outside the Southern Ontario region were not included and the lack of total diet data for Ontario residents made generalizations necessary. Despite this, it is still likely that representative Toronto and Southern Ontario adults receive the majority of their exposure to the organochlorines and PCBs investigated in this study through food.

ACKNOWLEDGMENTS

The author acknowledges the contribution of Lisa Richman for organizing the manuscript. Doug Hallett is thanked for proposing the study. This research was partially funded by the International Joint Commission.

REFERENCES

Agriculture Canada (1985). Food Residue Monitoring Consolidation Program. Agriculture Canada, Ottawa.

Braun, H. E., Ritcey, G.M., Ripley, B. D., McEwen, F. L., and Frank, R. (1982). Studies of the disappearance of nine pesticides on celery and lettuce grown on muck soils in Ontario, 1977–1980. Pest. Sci. **13**, 119–128.

Brown, H. S., Bishop, D., and Rowan, C. (1984). The role of skin absorption as a route of exposure for volatile organic compounds in drinking water. *Am. J. Public Health* **74**, 479–484.

Campbell, J. E., Richardson, L. A., and Schafer, M. L. (1965). Insecticide residues in the human diet. *Arch. Environ. Health* **10**, 831–836.

Davies, K. (1987). Use of water quality data and the Ontario drinking water objectives to assess human exposure to chemicals in Toronto's drinking water 1978–1984. In L. Grima, and D. Fowle, eds., *Information Needs for Environmental Risk Management in Canada*. University of Toronto Press, Toronto.

Davies, K. (1988). Concentrations and dietary intake of selected organochlorines, including PCBs, PCDDs and PCDFs in fresh food composites grown in Ontario, Canada. *Chemosphere* **17** (2), 263–276.

Dredging Subcommittee. (1983). Evaluation of Dredged Material Disposal Options for Two Great Lakes Harbours Using the Water Quality Board Dredging Subcommittee Guidelines. International Joint Commission, Windsor, Ontario.

Eisenreich, S. J., Looney, B. B., and Thornton, I. D. (1981). Airborne organic contaminants in the Great Lakes ecosystem. *Environ. Sci. Technol.* 15(1), 30–38.

Frank, R., Braun, H. E., and McWade, J. W. (1970). Chlorinated hydrocarbon residues in the milk supply of Ontario, Canada. *Pest. Monit. J.* **4**, 31–41.

Frank, R., Braun, H. E., Holdrinet, M., Sirons, G. J., Smith, E. H., and Dixon, D. W. (1979). Organochlorine insecticides and industrial pollutants in the milk supply of southern Ontario— 1977. *J. Food Protection* **42**(1), 31–37.

Frank, R., Braun, H. E., Ritcey, G., McEwen, F. L., and Sirons, G. J. (1982). Pesticide residues in onions and carrots grown on organic soils, Ontario 1975 to 1980. *J. Econ. Entomol.* **75**, 560–565.

Frank, R. Braun, H. E., and Fleming, G. (1983). Organochlorine and organophosphorus residues in fat of bovine and porcine carcasses marketed in Ontario, Canada from 1969 to 1981. *J. Food Protection* **46**, 893–900.

Frank, R. Braun, H. E., Sirons, G. J., Rasper, J., and Ward, G. G. (1985a). Organochlorine and organophosphorus insecticides and industrial pollutants in the milk supply of Ontario—1983. *J. Food Protection* **48**(6), 499–504.

Frank, R., Rasper, J., Braun, H. E., and Ashton, G. (1985b). Disappearance of organochlorine residues from abdominal and egg fats of chickens, Ontario, Canada, 1969–1982. *J. Assoc. Off. Anal. Chem.* **68**, 124–129.

Frankie, G. W., and Levenson, H. (1978). Insect problems and insecticide use: Public opinion, information and behaviour. In G. W. Frankie and C. S. Koehler, eds., Perspectives in Urban Entomology. Academic Press, New York.

Gillman, A. (1987). Department of National Health and Welfare. Personal communication.

Levenson, H., and Frankie, G. W. (1983). A study of homeowner attitudes towards arthropod pests and practices in three U.S. metropolitan areas. In, C. W. Frankie, and C. S. Koehler, eds., *Urban Entomology: Interdisciplinary Perspective*. Praeger, New York.

McLeod, H. A., Smith, D. C., and Bluman, N. (1980). Pesticide residues in the total diet in Canada— V—1976–1978. *Food Safety* **2**, 141–164.

Menzel, D. B., and McClellan, R. O. (1980). Toxic responses of the respiratory system. In, J. Doull, C. D. Klaassen, and M. O. Amdur, eds., *Casarett and Doull's Toxicology*. Macmillan, New York.

Mes, J., Davies, D. J., and Turton, D. (1982). Polychlorinated biphenyl and other chlorinated hydrocarbon residues in adipose tissue of Canadians. *Bull. Environ. Contam. Toxicol.* **28**, 97–104.

Mes, J., Davies, D. J., Turton, D., and Sun, W. F. (1986). Levels and trends of chlorinated hydrocarbon contaminants in the breast milk of Canadian women. *Food Add. Contam.* **3**, 313–322.

Singer, E., Jarv, T., and Sage, M. (1983). Survey of polychlorinated biphenyls in ambient air across the province of Ontario. In D. Mackay, S. Paterson, S. J. Eisenreich, and M. S. Simmons, eds., *Physical Behaviour of PCBs in the Great Lakes*. Butterworth, New York.

Smith, D. C. (1971). Pesticide residues in the total diet in Canada. *Pest. Sci.* **2**, 92–95.

Smith, D. C., Sandi, E., and Leduc, R. (1972). Pesticide residues in the total diet in Canada—II—1970. *Pest Sci.* **3**, 207–210.

Smith, D. C., Leduc, R., and Charbonneau, C. (1973). Pesticide residues in the total diet in Canada—III—1971. *Pest. Sci.* **4**, 211–214.

Smith, D. C., Leduc, R., and Tremblay, L. (1975). Pesticide residues in the total diet in Canada—IV 1972 and 1973. *Pest. Sci.* **6**, 75–82.

Statistics Canada. (1982). Family Food Expenditure in Canada. Statistics Canada, Ottawa.

Wester, R C., Ducks, D. A. W., Maibach, H. I., and Anderson, J. (1983). Polychlorinated biphenyls: Dermal absorption, systemic elimination, and dermal wash efficiency. *J. Toxicol. Environ. Health* **12**, 511–519.

Wester, R. C., and Maibach, H. I. (1985). In vivo percutaneous absorption and decontamination of pesticides in humans. *J. Toxicol. Environ. Health* **16**, 25–37.

17

ENVIRONMENTAL CONTAMINANTS IN MEXICAN FOOD

Lilia A. Albert

División de Estudios sobre Contaminación Ambiental
Instituto Nacional de Investigaciones sobre Recursos Bióticos
Xalapa, Ver., México

1. BACKGROUND

Until recently, it was considered that the main public health problems in Mexico were those resulting from the presence of microorganisms and parasites in water and food (Fernández de Castro et al., 1982). Because of this, the presence of chemical contaminants in food has not been considered a priority by the responsible government agencies; as a result, the presence of environmental chemical contaminants in food in Mexico has not been properly studied.

Prior to 1945, this contamination was not important in the country since, up to then, agricultural practices were traditional, with no use or very low use of chemicals, and there was relatively little need to preserve food before its comercialization. However, at that time, intensive use of pesticides was started in Mexcio; initially, this use was only for agricultural production; however, after a few years, the same products were used in the campaign against malaria. This campaign was part of the worldwide effort to erradicate malaria and, in Mexico, it was started in 1956 with the support of the World Health Organization; for this reason, large quantities of pesticides, mainly DDT, were used in the country. After this, the "green revolution"—with its emphasis on monocultures and its requirements for high concentrations of chemicals to obtain better yields in agriculture—was also introduced.

At the same time, a change in the model of agricultural development was initiated in Mexico (Warman, 1983). As a result of the new model the best lands in the country are devoted, at present, to industrial crops (cotton, tobacco, coffee) or to vegetables for export (tomato, cucumber, mango, pepper, and so on).

In consequence, marginal lands are mainly devoted to the production of staples, such as corn and beans. These crops are grown either for self-consumption or for sale in the local markets and, for them, traditional agricultural practices are mainly used. After some years, this has resulted in the parallel development of two types of agriculture in Mexico, which are progressively more distant: agriculture for export or industry and agriculture for the internal market.

An additional factor that has contributed to the growing contamination of Mexican food is industrial development; this development started after the Second World War and has been carried out without any environmental controls, causing extensive pollution of water, soil, and air in some areas, which results, eventually, in the ontamination of food.

At present, some 50% of all pesticides used in the country are insecticides; of these, close to 25% are organochlorine compounds and 25% organophosphorous, carbamic, or pyrethroid pesticides. The remaining 50% are fungicides, herbicides, and other pesticides. Most pesticides used in the country (approximately 75%) are devoted to industrial or export crops (Mena, 1987).

Toward 1973, due to pressures from importing countries—mainly the United States—a change in the pattern of use of pesticides was started in Mexico; in this way, in the regions devoted to export crops, the use of organophosphorous or carbamic pesticides rapidly replaced organochlorine compounds. This was due to the frequent violations of tolerances for the presence of persistent pesticide residues in the export crops detected when they were analyzed at the border within the U.S. Food and Drug Administration (US-FDA) sampling program.

The US-FDA recognizes that after the strengthening of its surveillance program, 5 years ago, the number of pesticide-related violations in Mexican produce has decreased greatly; however, these violations still occur with enough frequency to justify the fact that the US-FDA carries out as much analyses of Mexican food as those carried out for all the other countries exporting food to the United States. In this way, annually the US-FDA carries out 2500 analyses of food produced in the United States, 2500 of Mexican produce, and 2500 of produce from all the other countries exporting food to that country (McMahon and Burke, 1986). Through these data, an approximate idea of the magnitude of the problem of the presence of pesticide residues in Mexican crops can be obtained.

Despite these surveillance efforts and official U.S. pressure on the Mexican government, recent reports from that country still focus on the problem of the safety of Mexican food exported to the United States. It is mentioned, for example, that between 1979 and 1985, fruits and vegetables from Mexico contained nearly twice the levels of illegal residues as American-grown food (Kistner, 1987).

However, the pressure from importing countries has not been enough to persuade the Mexican government that the problem of the contamination of Mexican food is real and severe and that this problem, besides its adverse impact on international trade, must also have adverse effects on the health of the Mexican population.

This low level of awareness is shown, for example, in the fact that in 1987— nearly 15 years after laboratories to certify the quality of crops for export were established—there does not exist in Mexico one single laboratory devoted to certify—on a permanent basis—that the residues of contaminants in food for the internal market are within the internationally accepted tolerances.

The consequences of this lack of interest in the safety of food sold in the internal market are evident in several common and easily verifiable facts: the high levels of residues—very often in excess of international tolerances—in food of animal origin; the presence of high concentrations of organophosphorous and carbamic pesticides in fruits and vegetables that are consumed as such, without cooking; and the presence of aflatoxins (and quite probably other mycotoxins) in staples like corn and beans.

2. LEGAL ASPECTS

The main legal instruments for the control of chemical contamination of food in Mexico are, at present, the Ley de Sanidad Agropecuaria (Animal and Vegetable Health Law) and the Ley General de Salud (General Health Law) (SARH, 1974; SSA, 1984). Both include sections on the chemical contaminants of food—

especially pesticide residues; however, in general, these sections lack the corresponding regulations and technical norms that in Mexican law are essential to define the procedure for analysis and the details of particular food items and residues. Because of these deficiencies, these laws cannot be enforced in practice.

Other characteristics of the Mexican legal situation in this regard are the following:

1 The legal control is fragmented among several agencies from different sectors (health, agriculture, commerce, etc.). The lack of a single responsible agency for all aspects of the problem makes it impossible to enforce even the few existing regulations and justifies the lack of an explicit interest of the sectors to issue the missing regulations. Is also has caused the surge of new agencies whose fields of action overlap with each other; this excess of agencies makes it even more difficult to exert efficient controls.

2. It is also evident that the legislation in this area is not complete, since essential norms and regulations are missing. Besides, some limits and tolerances are established for contaminants that usually are not found in Mexican food and there exist no limits or tolerances for those that do appear. Although there are some tolerances for pesticide residues in vegetables, the norms required to complement them have never been established and no practical limits for pesticide residues in food of animal origin have been issued.

3. Besides these deficiencies in the legal framework for pesticide residues in food, there is no mechanism to ensure that the crops rejected by the importing countries are not sold in the national market. The lack of this mechanism and of a national certification mechanism for food to be sold in the country promotes a practice in which food items that would not be acceptable, for safety and health reasons, in other countries are sold in Mexico without problems.

It should also be mentioned that the agencies that control and supervise food production do not take into account the presence of a particular contaminant in a given food until the importing countries demand it is controlled.

4. Another characteristic of this legislation is its obsolescence, since it does not include the majority of the food contaminants that are of present interest and the few control mechanisms included could have been useful 20 years ago but are mainly useless at present.

In view of this, it can be concluded that the legal mechanisms, rules, and regulations in force in Mexico for the control of food contaminants suffer from several deficiencies and cannot be enforced in practice.

Besides, the control of the sources of food contaminants has been limited, fragmentary, and discontinuous and has occurred, generally, in response to acute, isolated, and well-defined problems, for example, the poisoning and death of many individuals in Tijuana, as a result of the contamination of bread ingredients with parathion that had been transported in the same vehicle with them (Márquez-Mayaudón et al., 1968).

3. PRESENT SITUATION

Although the problem of environmental contamination of food in Mexico is of national dimensions, it has special characteristics in each region of the country. In this way, food items from regions devoted to export agriculture are characteristically contaminated with nonpersistent pesticides of high acute toxicity, while food from other regions, in particular, those where cotton is or has been the main crop, frequently have high levels of organochlorine pesticide residues; this is true, in particular, of food of animal origin.

Therefore, according to the data available at present, in Mexico it is safer to predict that a particular food will be simultaneously contaminated with several types of contaminants than to predict that it will not be contaminated. For example, on the basis of the experience obtained to date, it can be said that practically any contaminant can be found in Mexican food, provided that proper analytical methodology is applied. In this way, the presence of polychlorobiphenyls and phthalates in several food items, in particular, those industrialized before consumption, has been shown (Albert and Aldana, 1982; Albert et al., 1987a) and the presence of excessive amounts of lead in some canned foods (Parada et al., 1975) and in fish (Albert and Badillo, 1986) has also been shown.

An additional problem is the presence, in some regions of the country, of very high quantities of arsenic in drinking water; this causes a severe regional environmental health problem, although this chapter is not the proper place to discuss it.

Even though it is not generally accepted, it can be said that the environmental contamination of food in Mexico is, at least, as severe as the contamination of water and, possibly, worse than that of soil and air—considering the country as a whole and not isolated regions of it—since, as mentioned earlier, environmental food contamination is partially a consequence of pollution in other substrates and, therefore, all the existing deficiencies in the control of environmental pollution in the country contribute markedly to the contamination of Mexican-produced food.

As already stated, up to this time, this problem has not been a priority for the responsible agencies; therefore, an effective control of the sources of these contaminants has not been implemented. In consequence, it can be said that, at present, the Mexican population is continuously exposed to several of these contaminants. For some of them, for example, pesticides, nitrites, nitrates, and some food additives, this exposure can be expected to have occurred during the last 40 years, starting with the industrialization of the country and the introduction of new technologies in agriculture; however, for other contaminants, such as lead from the deficient glazing of earthenware pots and pans, the problem has been recognized since the last century (Ruiz-Sandoval, 1878) and it has not been solved to date.

In terms of public health and in the context of a population whose life expectancy has increased considerably, it is to be expected that the problems caused be food contamination will appear, eventually, as chronic and degenerative diseases whose magnitude and social costs cannot be predicted at present.

There have been some isolated efforts in Mexico to establish control programs for food contaminants; these programs have had the support of several international agencies; however, due to several reasons—some of them already mentioned—the only program with some results has been the certification that Mexican food devoted to export fulfills the requirements of the importing countries.

In fact, the results of these controls do not meet the requirements of even the most superficial cost–benefit analysis, since the investment has been extremely high and, as a result, not all the laboratories established for this purpose are working, most of the equipment is now useless, and the few laboratories that are working are not doing so at full capacity or providing completely reliable results. Under these conditions, the benefits are minimal in comparison with the needs of the country in this area, the investments have not had the expected returns, and the needs are evident everywhere, mainly in legislation, manpower, methodology, and total diet studies.

Therefore, a profound reorganization of all the official mechanisms to meet these needs in the short term is essential in order to guarantee that the food of the Mexican population is free from residues of environmental contaminants that might endanger its health now and in the future.

4.   MAIN CONTAMINANTS

According to the available information, chlorinated hydrocarbons are the main environmental contaminants of Mexican food. The predominance of these compounds surely results from the widespread use—past and present—of organochlorine pesticides in the country. In this way, despite official declarations to the contrary, in 1986 these products still constituted close to 25% of the total pesticides used (Mena, 1987). In spite of this situation, this problem has not been studied with the depth it deserves.

Possibly, the next largest group of contaminants of Mexican food is organophosphorous pesticides. Its presence in food is due to several deficiencies in the prevailing agricultural practices: collecting crops before the safety period has expired, use of pesticides not allowed in some crops, use of pesticides in larger quantities than those allowed, an so on. The problem posed by the residues of these pesticides is potentially very important since they are, in general, of high acute toxicity and they are being used in food that is consumed raw, for example, fruits and vegetables such as lettuce, tomato, and cucumber. However, there are still fewer studies in this area than in the former case and many more are needed.

Another problem of great potential severity is the presence of mycotoxins in food, in particular in grains and their derivatives. This problem is caused, in part, by the particular climate of large parts of the country, which fosters the reproduction of mycotoxin-producing fungi, and it is compounded by deficient storing conditions for grains, since, very often, they are simply piled on the fields under a plastic cover and, in this way, they are easily contaminated with mycotoxin-producing fungi. In this area many more studies are needed, because mycotoxins mainly affect food items devoted to the national market and, therefore, there has not been any pressure from importing countries to solve the problem.

In addition, there are no studies on the presence in food of heavy metals of environmental origin, since the few existing studies are mainly on the presence in food of lead of industrial origin or on the presence of lead in locally made earthenware. In the same way, there are no studies on other contaminants, for example, hydrocarbon residues in seafood. The few existing studies in this area are focused on the problem of environmental quality and not on food safety.

In a final analysis, it can also be stated that the presence of some contaminants in food, for example, polychlorobiphenyls, is due to a problem of environmental pollution since, apparently, these contaminants are transferred to food from the packing materials, which are generally made from recycled paperboard and therefore have accumulated these contaminants along the way.

For these reasons, the following discussion is centered on pesticides, in particular, organochlorine pesticides, since more data are available on them.

5. ORGANOCHLORINE PESTICIDES

In the case of Mexico, the importance of these contaminants is due to their widespread use—both past and present—for agricultural and public health purposes.

For example, in 1986 (Mena, 1987) the following organochlorine pesticides were registered and used in the country, while those indicated were also produced in Mexico: DDT (approximately 4000 tons/year), BHC (approximately 2000 tons/year), toxaphene (approximately 2000 tons/year), endrin (approximately 400 tons/year), heptachlor, chlordane, chlorobenzilate, and endosulfan. Until very recently, aldrin, dieldrin, chlordimeform (Galecron), and other minor products were also in use.

Most of these pesticides have been of importance in the cotton-growing regions; however, some of them, like aldrin and dieldrin, have also been used to control soil pests in potato, corn, and other crops.

In most of the cotton-growing regions, due to pest resistance arising from the excessive use of pesticides, cotton crops were no longer economically feasible and had to be replaced, mainly by cultivating animal feeds such as sorghum, soybeans, and alfalfa. It is evident that this increases the rate of entrance to the trophic chain of the chlorinated pesticide residues that remain in the soil in these regions. Because of this, although at present smaller quantities of these pesticides are used in those regions, there is an adverse effect of the prevailing soil contamination on food of animal origin, not only in those areas but, also, in other regions of the country, since the animal feeds produced in the former cotton-growing regions are not only for local use.

It must be mentioned that no total diet studies have been carried out in Mexico; this type of study would be extremely useful, both to establish the real exposure of the population to these and other contaminants and to allow the authorities to implement adequate control measures for those contaminants that are presently in excess of acceptable daily intakes.

5.1.   Alfalfa

In Mexico there have been few studies on the presence of organochlorine pesticide resides in vegetables. In relation to feeds, there is one study on alfalfa (Albert et al., 1975) and another on mixed feeds (Saval, 1976). Given the widespread use of alfalfa the results of the first of these studies are summarized.

TABLE 1   Chlorinated Hydrocarbon Residues in Alfalfa: Qualitative Analysis[a]

Compound	Positive Samples/Total	Frequency (%)	Samples with Residues above Traces
α-HCH	14/14	100	5/14
β-HCH	7/14	50	3/14
γ-HCH	14/14	100	4/14
Aldrin	2/14	14	—
Dieldrin	4/14	14	—
p,p'-DDE	12/14	86	4/14
p,p'-DDD	9/14	64	2/14
p,p'-DDT	14/14	100	13/14

[a]Albert et al. (1975).

TABLE 2   Chlorinated Hydrocarbon Residues in Alfalfa: Quantitative Analysis[a]

Compound	Concentrations[b]		
	Minimum	Maximum	Average
α-HCH	T[c]	0.57	0.06
β-HCH	ND[d]	0.02	Insuf.[e]
γ-HCH	T	0.44	0.04
p,p'-DDE	T	0.29	0.03
p,p'-DDD	ND	0.05	Insuf.
p,p'-DDT	T	0.70	0.08

[a]Albert et al. (1975).
[b]$\mu g/g$, whole weight basis.
[c]T; traces ($0.007\ \mu g/g >$ traces $> 0.001\ \mu g/g$).
[d]ND, not detectable.
[e]Insufficient values for a significant mean.

In Table 1 in qualitative results of this study are presented; the predominance of the HCH isomers and that of DDT and its derivatives as residues in this substrate can be easily seen. It is also evident that in the region of origin of these samples, DDT was in common use at the time since it was found in all the samples.

The concentrations of these contaminants are included in Table 2. The importance of DDT as the major contaminant in this feed is confirmed, since both its maximum and average concentrations exceed by far those of the other residues present. It can also be noticed that the proportion of DDE to DDT is below 50%, which confirms the hypothesis that either this crop was grown in an area with very recent use of DDT, or that DDT was used as an insecticide on it.

5.2. Milk

There are more available studies on milk than on other products and some of them afford interesting data. The more indicative are summarized here.

5.2.1. Fluid Milk

There are three studies available on fluid milk. Two of them were carried out some years ago, to screen for pesticide residues in food in Mexico (Albert and Reyes, 1975; Capella et al., 1975). The first was on fluid milk available in Mexico City and its results are summarized in Tables 3 and 4.

It can be seen in Table 3 that residues of seven different compounds were found in all the samples of milk obtained in Mexico City whose origin was the region known as Comarca Lagunera, while this was true only for p, p'-DDE and p, p'-DDT for milk samples from other regions.

In Table 4 it can be observed that the average concentrations of residues were

TABLE 3   Chlorinated Hydrocarbon Residues in Fluid Milk Available in Mexico City[a]: Comparison of Qualitative Results and Origin of the Samples

	Positive Samples/Total	
Compound	Samples from Comarca Lagunera	Other Samples
α-HCH	5/5	7/9
β-HCH	5/5	1/9
γ-HCH	3/5	4/9
Dieldrin	5/5	5/9
p, p'-DDE	5/5	9/9
p, p'-DDD	5/5	5/9
p, p'-DDT	5/5	9/9
Compounds/sample	6–7	2–7

[a]Albert and Reyes (1975).

TABLE 4   Chlorinated Hydrocarbon Residues in Fluid Milk Available in Mexico City[a]: Comparison of Quantitative Results and Origin of the Samples

| | Average Concentrations[b] | |
| | Samples from | Other |
Compound	Comarca Lagunera	Samples
α-HCH	0.04	0.03
β-HCH	0.32	Insuf.[c]
γ-HCH	0.04	0.03
Dieldrin	0.04	Insuf.
p,p'-DDE	0.82	0.05
p,p'-DDD	0.06	Insuf.
p,p'-DDT	0.09	0.02

[a]Albert and Reyes (1975).
[b]μg/g, fat basis.
[c]Insufficient values for a significant mean.

higher in the samples from Comarca Lagunera than in the other samples. Although this was true for all compounds, it was specially so for β-HCH (0.32 μg/g vs. insufficient), p,p'-DDE (0.82 vs. 0.05 μg/g), and p,p'-DDT (0.09 vs. 0.02 μg/g).

From these data it was concluded that the milk samples from Comarca Lagunera were more contaminated with these residues than other samples of milk available in Mexico City. Comarca Lagunera is a region located in the central northern part of Mexico (see Figure 1) and, for a long time, was the main

Figure 1.   Areas of study. (A) Mexico City; (B) Comarca Lagunera; (C) Soconusco.

cotton-growing region in the country; some data (Bordas, 1973) indicate that up to 1% of the world production of DDT was used some years ago in this region.

Because of these facts, this region was selected for a follow-up study (Albert et al., 1987); its results are presented in Tables 5 and 6.

In order to further clarify the meaning of these data, Figures 2 to 5 show the following: a comparison of the number of different residues per sample for both

TABLE 5   Chlorinated Hydrocarbon Residues in Fluid Milk From Comarca Lagunera, Mexico[a]: Qualitative Analysis

Compound	Positive Samples[b]/Total	Frequency (%)
α-HCH	15/15	100
β-HCH	15/15	100
γ-HCH	—	—
HCB[c]	7/15	47
Dieldrin	12/15	80
H.E.[d]	5/15	33
p,p'-DDE	15/15	100
p,p'-DDD	15/15	100
p,p'-DDT	15/15	100

[a] Albert et al. (1987b).
[b] With concentrations above traces.
[c] Hexachlorobenzene.
[d] Heptachlor epoxide.

TABLE 6   Chlorinated Hydrocarbon Residues in Fluid Milk from Comarca Lagunera, Mexico[a]: Quantitative Analysis

| Compound | Concentrations[b] | | |
	Minimum	Maximum	Average
α-HCH	0.04	0.07	0.06
β-HCH	0.11	0.30	0.2
γ-HCH	T[c]	T	—
HCB[d]	T	0.04	0.014
Dieldrin	T	0.05	0.02
H.E.[e]	T	0.015	0.007
p,p'-DDE	1.3	3.7	2.56
p,p'-DDD	0.03	0.08	0.05
p,p'-DDT	0.06	0.15	0.10

[a] Albert et al. (1987b).
[b] μg/g, extractable lipid basis.
[c] Traces (0.007 μg/g > traces > 0.001 μg/g).
[d] Hexachlorobenzene.
[e] Heptachlor epoxide.

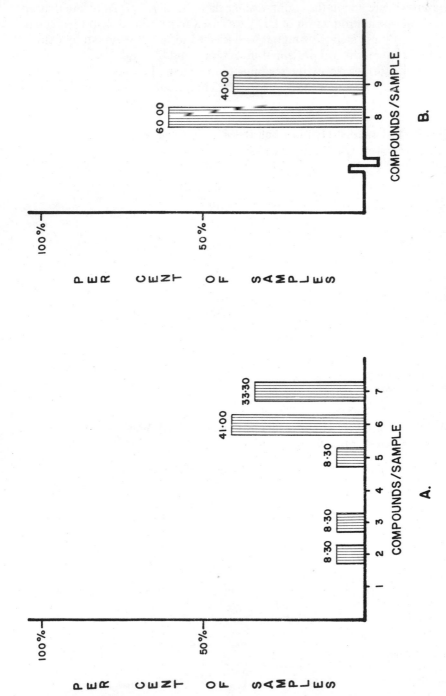

Figure 2. Comparison of the number of different residues per sample. (A) Mexico City (Albert and Reyes, 1975). (B) Comarca Lagunera (Albert et al., 1987b).

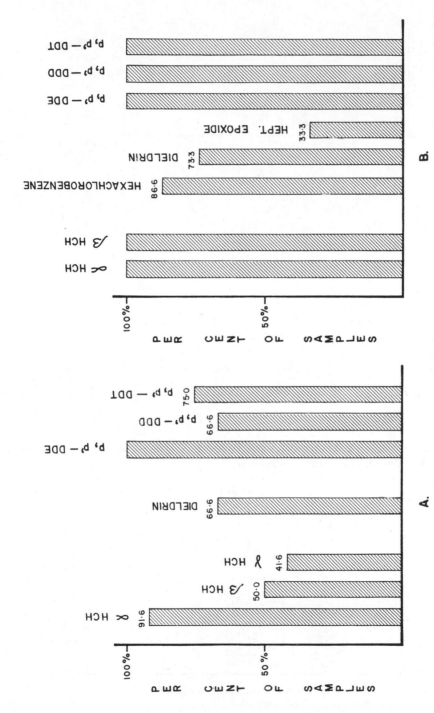

Figure 3. Frequency of some residues in the samples studies. (A) Mexico City (Albert and Reyes, 1975). (B) Comarca Lagunera (Albert et al., 1987b).

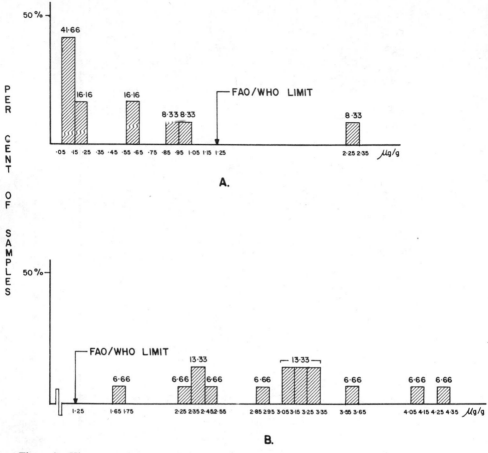

Figure 4. Histogram of total DDT concentrations in milk samples. (A) Mexico City (Albert and Reyes, 1975). (B) Comarca Lagunera (Albert et al., 1987b).

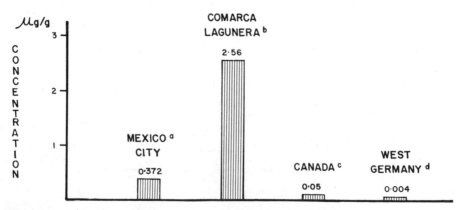

Figure 5. Total equivalent DDT for milk samples analyzed in both Mexican studies compared to data from studies in Canada and West Germany.

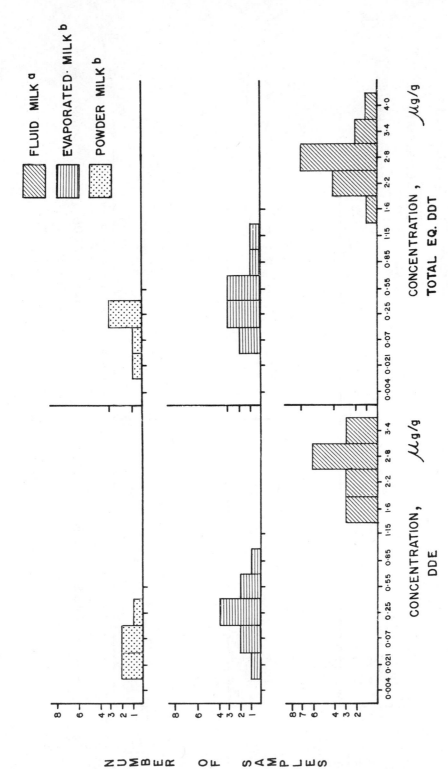

Figure 6. Analysis of concentration of DDT in evaporated and powdered milk.

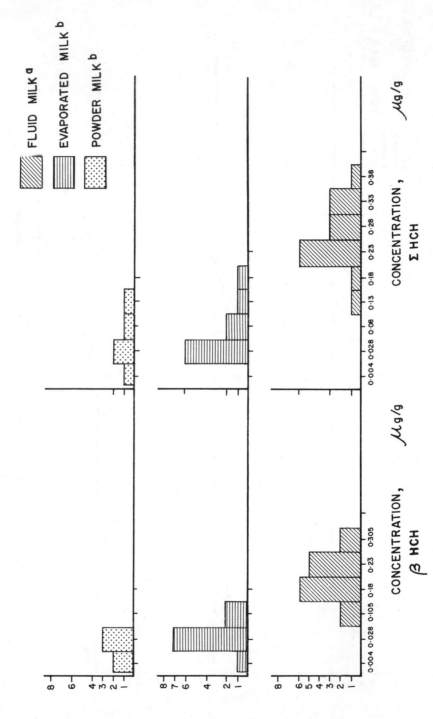

Figure 7. Analysis of concentration of HCH in evaporated and powdered milk.

studies, the frequency of some particular residues in the samples studied, the histogram of total DDT concentrations in milk samples from the first and the follow-up studies, and, finally, a comparison of total equivalent DDT for the milk samples analyzed in both Mexican studies in relation to data from studies carried out in Canada (Frank et al., 1975) and West Germany (Washüttl, 1974).

5.2.2. Evaporated and Powdered Milk

Results of the studies on the quantitative analyses of evaporated and powdered milk—both standard and baby formula type—are presented in Figures 6 and 7.

It can be observed that in comparison with the results from fluid milk described in Section 5.2.1, the samples of processed milk (Albert et al., 1987c) had lower frequencies and lower concentrations of these contaminants. These concentrations are compared with those from a study carried out in Canada (Ritcey et al., 1972) in Table 7.

In this table it can be seen that the major difference between Canadian and Mexican products was the higher number of residues per sample in the Mexican products and the presence of HCH isomers in both types of Mexican milk and of endrin in one of them. It should also be noted that the concentrations of DDT derivatives were higher for the Mexican evaporated milk. Although these concentrations were not in excess of those recommended by FAO/WHO (1986) for this type of product, it should be remembered that in Mexico these types of milk are used mainly for the nourishment of children; therefore these residues constitute an important hazard for this sector of the population.

TABLE 7   Chlorinated Hydrocarbon Residues in Evaporated Milk and Powdered Milk in Mexico: Comparison with Results from Canada[a]

Compound	Evaporated Milk		Powdered Milk	
	Mexico[b]	Canada[c]	Mexico[d]	Canada[c]
α-HCH	0.02	—	0.04	—
β-HCH	0.05	—	0.02	—
γ-HCH	0.015	0.01	0.03	—
Dieldrin	0.02	0.019	0.03	0.026
Endrin	—	—	0.06	—
H.E.[e]	0.015	0.005	0.02	0.018
p,p'-DDE	0.37	0.052	0.05	0.127
p,p'-DDD	0.05	0.036	0.06	0.025
p,p'-DDT	0.05	0.051	0.13	0.110

[a] Average concentrations, μg/g extractable lipid basis.
[b] Albert et al. (1982).
[c] Ritcey et al. (1972).
[d] Albert et al. (1988).
[e] Heptachlor epoxide.

5.3.   Dairy Products

There have been several studies carried out on dairy products. Here their main results are presented.

5.3.1.   Cheese

The results of three studies on Mexican cheese are available. In Figure 1 the areas selected to obtain the samples for these studies are shown; these were, for the first study, Mexico City and, for the follow-up studies, the regions known as Comarca Lagunera and Soconusco (Albert and Alpuche, 1987a; Albert and Bárcenas, 1987).

Mexico City was selected since it is the major urban center in the country and products from the whole country are sold in the local market. For this reason, it was expected that the analysis of cheese samples available there would provide important information on the general state of this problem in the country.

The other two regions were selected for the follow-up studies since, in the first study, the samples from these places had consistently more contaminants per sample and higher concentrations of them. For a long time, both regions were major cotton-growing areas; for this reason it was also expected that through the analysis of cheese samples, some indications of the effect that this crop has had on the environmental contamination by chlorinated hydrocarbons in dairy products in Mexico could be obtained.

In Table 8 the results of the qualitative analysis of chlorinated hydrocarbon residues in cheese from these studies are summarized.

From these results it can be concluded that, as was found for fluid milk and alfalfa, the major organochlorine residues to be found in cheese were p,p'-DDT and its transformation products p,p'-DDE and p,p'-DDD; it is also evident that the residues of p,p'-DDT, as such, were less frequent in the follow-up studies, since this product was found in only 18.2 and 30% of the samples, in contrast to the situation found 10 years ago, when it was much more frequent. This is consistent with the assertions of government officials regarding the restrictions in the use of DDT, however, it is not consistent with their stating that this product is no longer used in Mexico.

Since both follow up studies were carried out at the same time, it is also evident that the products from Comarca Lagunera are more contaminated than those from Soconusco, since except for heptachlor epoxide, the residues of chlorinated hydrocarbons were far more frequent in the samples from Comarca Lagunera where 12 different compounds were identified, against only 9 for Soconusco. It must also be noticed that p,p'-DDE was found in all the samples studied— regardless of their origin.

Of importance, because of the possibility of adverse effects to the consumer, is the presence of hexachlorobenzene and heptachlor epoxide residues in the samples of the follow-up studies; these products had not been found in the earlier study. In Table 9 the quantitative results of these studies are compared.

It can be seen that the levels for p,p'-DDE and p,p'-DDT in Mexican cheese are approximately the same through time in Mexico City and Soconusco and that

TABLE 8   Chlorinated Hydrocarbon Residues in Cheese in Mexico:
Percentage of Positive Samples

	Cheese Obtained in		
Compound	Mexico City[a]	Comarca Lagunera[b]	Soconusco[c]
HCB[d]	—	100	20
α-HCH	100	72.7	50
β-HCH	20	81.8	60
γ-HCH	63	36.4	70
Aldrin	6.6	9.1	—
Dieldrin	33.3	9.1	—
Endrin	10	27.2	—
Heptacloro	—	9.1	30
H.E.[e]	—	36.4	80
p, p'-DDE	86.7	100	100
p, p'-DDD	66.7	45.5	30
p, p'-DDT	86.7	18	30

[a] Albert and Reyes (1978).
[b] Albert and Alpuche (1987).
[c] Albert and Bárcenas (1988).
[d] Hexachlorobenzene.
[e] Heptachlor epoxide.

these levels are about eight times more for p, p'-DDE and close to three times more for p, p'-DDT that the levels found in 1967 in the United States (Duggan, 1967). However p, p'-DDE concentrations are much higher in the samples from Comarca Lagunera. In relation to heptachlor epoxide, its average level in Mexico is approximately four times more than the levels found in that country.

In Table 10 it can be observed that the frequency of the major contaminants found in these studies was definitively higher in the Mexican samples than in those from the United States.

From this it was concluded that in the cotton-growing regions of Mexico, the levels of persistent contaminants in soil—and probably in water—should be very high and that these contaminants are continuously entering the local trophic chains, bioaccumulating in each level, until they finally reach the general population, through food, in very high concentrations.

Some calculations indicate that the ingestion of 100 g of this type of cheese, with the average concentrations found in those studies, would provide the consumer with quantities of residues between 60 and 100% of the acceptable daily ingest (FAO/WHO, 1986). It should also be stressed that the consumer is exposed to at least nine different contaminants from this type of food; this should be a serious cause of concern, in view of the possibility of synergistic effects from these substances.

TABLE 9 Chlorinated Hydrocarbon Residues in Mexican Cheese: Results of the Quantitative Analysis

Origin of the Samples	Average Concentrations[a]						References
	HCB[b]	β-HCH	γ-HCH	H.E.[c]	p,p'-DDE	p,p'-DDT	
Mexico City	—	0.27	0.05	—	0.49	0.15	Albert and Reyes (1978)
Comarca Lagunera	Insuf.[d]	0.12	Insuf.	Insuf.	1.39	Insuf.	Albert and Alpuche (1987)
Soconusco	Insuf.	0.04	0.10	0.16	0.47	0.15	Albert and Bárcenas (1988)
United States	—	—	—	0.036	0.066	0.042	Duggan (1967)
West Germany	—	—	—	0.04	—	0.08	Heeschen (1972)

[a] Only the major contaminants are included; μg/g, extractable lipid basis.
[b] Hexachlorbenzene.
[c] Heptachlor epoxide.
[d] Insuficient values for a significant mean.

TABLE 10 Percentage Frequency of Chlorinated Hydrocarbon Residues in Mexico Cheese: Comparison with Results from the United States

Origin of the Samples	Compound[a]						Reference
	HCB	β-HCH	γ-HCH	H.E.	p,p'-DDE	p,p'-DDT	
Mexico City	—	20	63	—	86.7	86.7	Albert and Reyes (1978)
Comarca Lagunera	100	81.8	36.4	36.4	100	18.2	Albert and Alpuche (1987)
Soconusco	20	60	70	80	100	30	Albert and Bárcenas (1988)
United States	—	—	7.1	23.2	41.8	25.3	Duggan (1967)

[a] Only the major contaminants are included.

5.3.2. Cream from Milk and Butter

There have been two studies reported on cream in milk and only one on butter. The first of the studies on cream was only presented in congress (Laborín et al., 1983). The second study (Albert et al., 1987d) was carried out with samples obtained in Mexico City but originating in several parts of the country. In Table 11 the results of the qualitative analyses are presented and Table 12 includes a summary of the quantitative data.

The number of residues of different compounds fluctuated from four to nine.

TABLE 11   Chlorinated Hydrocarbon Residues in Mexican Cream from Milk and Butter[a]

	Cream from milk		Butter	
Compound	Positive Samples/Total	Frequency (%)	Positive Samples/Total	Frequency (%)
α-HCH	12/12	100	7/7	100
β-HCH	3/12	25	4/7	57.1
γ-HCH	10/12	83	6/7	85.7
Aldrin	4/12	33	—	—
Dieldrin	10/12	83	6/7	85.7
H.E.[b]	3/12	25	1/7	14.3
p,p'-DDE	12/12	100	7/7	100
p,p'-DDD	8/12	66.6	6/7	85.7
p,p'-DDT	9/12	75	5/7	71.4

[a] Albert et al. (1986).
[b] Heptachlor epoxide.

TABLE 12   Chlorinated Hydrocarbon Residues in Mexican Cream from Milk and Butter[a]: Quantitative Analysis[b]

	Range		Average Concentrations	
Compound	Cream	Butter	Cream	Butter
α-HCH	T[c]–0.04	0.02–0.12	0.02	0.06
β-HCH	ND[d]–0.42	ND–1.33	0.22	0.49
γ-HCH	ND–0.03	T–0.06	0.02	0.03
Dieldrin	ND–0.04	T–0.03	0.02	0.02
p,p'-DDE	0.02–0.31	0.04–4.31	0.07	0.67
p,p'-DDD	ND–0.05	0.02–0.19	0.03	0.07
p,p'-DDT	ND–0.07	ND–0.67	0.04	0.18

[a] Albert et al. (1986).
[b] $\mu g/g$, extractable lipid basis.
[c] Traces ($0.007\,\mu g/g > 0.001\,\mu g/g$).
[d] Not detectable.

TABLE 13 Chlorinated Hydrocarbon Residues in Hens' Eggs from Four Mexican Regions: Qualitative Analysis

Region	Compound[a]								Reference
	β-HCH	γ-HCH	Dieldrin	Endrin	H.E.[b]	p,p'-DDE	p,p'-DDD	p,p'-DDT	
Mexico City	—	1/15	6/15	4/15	1/15	14/15	2/15	12/15	Albert et al. (1981a)
Comarca Lagunera	—	5/13	5/13	4/13	—	13/13	7/13	7/13	Albert et al. (1981a)
Monterrey	15/15	—	6/15	6/15	15/15	15/15	15/15	15/15	Albert and Loera (1983)
Comarca Lagunera	—	5/13	1/13	1/13	12/13	13/13	12/13	12/13	Albert and Alpuche (1988)
Soconusco	6/18	2/18	4/18	—	—	16/18	10/18	12/18	Albert and Reyes (1986)

[a]Only the major contaminants are included.
[b]Heptachlor epoxide.

TABLE 14 Chlorinated Hydrocarbon Residues in Hens' Eggs from Four Mexican Regions: Average Concentrations[a]

	Region				
Compound	Mexico City[b]	Comarca Lagunera[b]	Monterrey[c]	Comarca Lagunera[d]	Soconusco[e]
HCB	—	—	—	0.024	—
β-HCH	—	—	0.003	—	0.103
γ-HCH	—	0.01	—	0.036	Insuf.
Dieldrin	0.10	0.01	0.002	—	0.042
Endrin	0.11	0.01	0.004	—	—
Heptachlor	—	—	—	0.006	—
H.E.[f]	—	—	0.002	0.003	—
p,p'-DDE	0.10	0.07	0.040	0.236	0.318
p,p'-DDD	—	0.02	0.002	—	0.044
p,p'-DDT	0.18	0.07	0.004	0.018	0.210

[a] μg/g, whole egg basis.
[b] Albert et al. (1981a).
[c] Albert and Loera (1983).
[d] Albert and Alpuche (1988)
[e] Albert and Reyes (1986).
[f] Heptachlor epoxide.

563

The presence of aldrin in some of the cream samples was interpreted as an indication that these samples were not only cream from milk but had some vegetable milk substitutes besides.

It can be observed that regarding class and number of contaminants, the same pattern already described was found in these samples, that is, DDT derivatives and HCH isomers were the major contaminants of these products.

In relation to the results summarized in Table 12, it can be observed that residue concentrations were higher for butter than for cream from milk, in particular for DDT derivatives. There can be many reasons for these differences, among them the different regions of origin of the samples; however, another possibility might be the presence in these cream samples of some vegetable milk substitutes. As already described for cheese and fluid milk, the higher concentrations of residues were found in the samples from Comarca Lagunera.

5.4. Eggs

In Mexico there have been five studies on the presence of chlorinated hydrocarbon residues in hens' eggs; of them, the first two were carried out in 1975 and its results were published jointly (Albert et al., 1981a); the third was carried out in 1978 (Albert and Loera, 1983). Later, two follow-up studies, both on regions devoted mainly—at present or in the recent past—to cotton crops, were carried out (Albert and Alpuche, 1987b; Albert and Reyes, 1987). These were, as has already been described for cheese, the Comarca Lagunera and Soconusco regions (see Figure 1). The qualitative results of the five studies are summarized in Table 13 and, in Table 14, those from the quantitative analyses are shown.

It can be seen that a tendency toward the disappearance of endrin and dieldrin residues from these products is evident in recent years but, instead, residues of heptachlor epoxide tend to appear; also, p,p'-DDT and its transformation products continue to be the major contaminants, both in frequency and in concentration.

The increase with time of these concentrations in the products from Comarca Lagunera should also be noticed. Besides, it is important that although the concentrations of p,p'-DDT in hens' eggs are lower at present, this substance in still found; this shows that its use has not been discontinued despite official restrictions; therefore, it can be stated that these restrictions are not being enforced in practice. These results are compared in Figure 8.

5.5.   Chicken Meat

There is only one report on chicken meat (Saval, 1976); although it is known that government laboratories have finished similar studies, their results have never been published and are not available.

The qualitative and quantitative results of this study are presented in Table 15 and 16.

The presence of high concentrations of dieldrin, endrin, and p,p-DDT in this

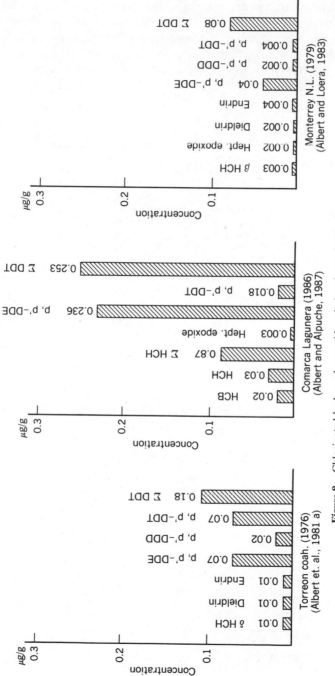

Figure 8. Chlorinated hydrocarbon residues in hens' eggs in Mexico.

substrate should be stressed. These results, together with those of the studies on hens' eggs, point to a severe problem of contamination of fowl in Mexico, which might occur through mixed feeds, since their main ingredients are grown mainly on the highly polluted lands that used to be devoted to cotton crops.

It should be mentioned that no data on the presence of these residues in beef or pork are available, although it is possible that these types of food have been studied by some government agencies.

TABLE 15   Chlorinated Hydrocarbon Residues in Chicken Meat from Mexico[a]: Quantitative Analysis

Compound	Positive Samples/Total	Frequency (%)
α-HCH	7/16	43.7
γ-HCH	9/16	56.2
Aldrin	8/16	50
Dieldrin	12/16	75
Endrin	11/16	68.7
H.E.[b]	4/16	25
p, p'-DDE	16/16	100
p, p'-DDD	12/16	75
p, p'-DDT	15/16	93.7

[a]Saval (1976).
[b]Heptachlor epoxide.

TABLE 16   Chlorinated Hydrocarbon Residues in Chicken Meat from Mexico[a]: Quantitative Analysis

Compound	Concentrations[b] Minimum	Maximum	Average
α-HCH	ND[c]	0.06	0.02
γ-HCH	ND	0.05	0.03
Aldrin	ND	0.12	0.05
Dieldrin	T[d]	0.78	0.14
Endrin	ND	0.60	0.12
H.E.[e]	ND	0.07	0.03
p, p'-DDE	0.1	1.0	0.4
p, p'-DDD	ND	0.24	0.15
p, p'-DDT	T	1.2	0.3

[a]Saval (1976).
[b]μg/g, extractable lipid basis.
[c]Not detectable.
[d]Traces (0.007 μg/g > traces > 0.001 μg/g).
[e]Heptachlor epoxide.

5.6.   Freshwater Fish

Two studies have been done on the presence of chlorinated hydrocarbon residues in freshwater fish; for the first (Albert et al., 1980b), some edible fish generally available in Mexico City ("carpa" and "charal") were selected. The results showed that these species had excessive amounts of chlorinated hydrocarbon residues.

For the second study (Albert and Figueroa, 1985), freshwater species from the Blanco river in Veracruz State were chosen. Its results confirmed those of the first study; besides, some of the samples also had residues of contaminants of industrial origin: dimethyl, dibutyl, and di-2-ethyl-*n*-hexyl phthalates (DMP, DBP, and DEHP, respectively).

Both studies can be taken as an indication of the impairment of superficial waters in Mexico by these types of contaminants; the results—although not

TABLE 17   Chlorinated Hydrocarbon Residues in Freshwater Fish Obtained in Mexico City[a]: Qualitative Analysis

Compound	Positive Samples/Total	Frequency (%)
α-HCH	7/12	58.3
β-HCH	5/12	41.6
γ-HCH	7/12	58.3
Dieldrin	5/12	41.6
Endrin	4/12	33.3
Chlordane α	3/12	25
Chlordane γ	3/12	25
p, p'-DDE	12/12	100
p, p'-DDD	10/12	83
p, p'-DDT	10/12	83

[a] Albert et al. (1980b).

TABLE 18   Chlorinated Hydrocarbon Residues in Freshwater Fish in Mexico[a]: Quantitative Analysis

Compound	Concentrations[b]		
	Minumum	Maximum	Average
α-HCH	ND[c]	0.78	0.11
β-HCH	ND	0.32	0.06
γ-HCH	ND	0.34	0.03
p, p'-DDE	0.05	27.30	8.77
p, p'-DDD + p, p'-DDT	T[d]	33.85	7.51

[a] Albert et al. (1980b).
[b] μg/g, extractable lipid basis.
[c] Not detectable.
[d] Traces (0.007 μg/g > traces > 0.001 μg/g).

TABLE 19 Chlorinated Hydrocarbon and Phthalate Residues in Aquatic Organisms from the Blanco River, Veracruz, Mexico[a]

Organism (Number)	Compound[b]						Pollutants/Sample
	β-HCH	p,p'-DDE	p,p'-DDD	p,p'-DDT	DMP[c]	DEHP[d]	
Poecilia mexicana (1)	+	+	+	−	−	−	3
Dorosema anale (1)	−	−	+	+	−	−	2
Machrobranchium sp. (1)	−	+	−	−	+	+	3
Centropomus paralletus (1)	−	+	−	+	−	+	3
Ophiascion imiceps (1)	−	−	−	+	−	+	2
Cynoscium arenarius (1)	−	−	−	−	+	−	2
Oreochromis ailoticus (2)	+	+	+	+	−	+	5
Penaeus sp. (1)	+	+	−	+	−	−	3
Callenectus sp. (1)	−	−	−	+	+	−	2
Gobiomorus dormitor (1)	−	+	+	+	−	−	3
Rhamdia guatemalensis (1)	−	+	+	+	−	−	3

[a] Albert and Figueroa (1985).
[b] Only the major contaminants are shown.
[c] Dimethyl phthalate.
[d] Di-2-ethyl-*n*-hexyl phthalate.

TABLE 20 Chlorinated Hydrocarbon and Phthalate Residues in Organisms from the Blanco River, Veracruz, Mexico[a]

Organism	Compound[b]					
	β-HCH	p,p'-DDE	p,p'-DDD	p,p'-DDT	DMP	DEHP
P. mexicana	0.08	0.66	0.78	—	—	—
D. anale	T	—	0.3	0.09	—	—
Machrobranchium sp.	T	0.36	—	—	382.0	791.2
C. parallelus	—	1.35	—	2.08	—	451.0
O. imiceps	—	—	—	2.12	—	9876.1
C. arenarius	—	—	—	—	4.03	432.6
O. niloticus	15.12	135.8	394.6	1947.17	—	8061.5
O. niloticus	T	3.27	0.28	—	—	167.6
Penaeus sp.	0.67	205.51	—	1383.3	—	1410.6
Callenectus sp.	T	—	—	732.2	488.77	—
G. dormitor	T	0.11	0.09	0.37	—	—
R. guatemalensis	—	0.24	0.11	0.01	—	—

[a] Albert and Figueroa (1985).
[b] Concentrations in μg/g, extractable lipid basis.

enough—should alert the authorities to the potential dimensions of this problem.

In Table 17 the qualitative results of the first study are shown: Table 18 includes the quantitative data and, in Tables 19 and 20, the qualitative and quantitative results of the second study are presented.

In the first study, the contaminants found fluctuated from three to eight different compounds per sample; it should be mentioned that, for some samples, the p, p'-DDE peak was extremely wide and could have masked at least four others with similar retention times. The samples with the higher number of contaminants were those of the fish known as "charal" and came from the Lake of Chapala in Jalisco State in the west of Mexico.

As for the second study, it can be seen in Table 19 that the organism with the higher number of contaminants was tilapia (*Oreochromis niloticus*) while the lower number of contaminants was found in white trout (*Cynoscium arenarius*); this had residues of only two contaminants, both phthalates.

It can be added to the data presented in Table 20 that the samples with more contaminants were obtained in the Alvarado Lagoon, where the Blanco river flows into the sea. Those with the lower number of contaminants were from the beginning of the river; however, these particular samples had the highest concentrations of phthalates. It was also found that the organisms with higher values of residues were omnivorous and benthic (*Oreochromis, Penaeus, Callenectus*) and they generally eat deteritus.

It should be stressed that p, p'-DDT was identified in 8 of the 12 samples of the second study. Doubtlessly, this high frequency points to the continuing use of this insecticide in the country. It must be remembered that to date, up to 4000 tons of it are produced annually in Mexico. At the same time, the results of this study indirectly point toward an important pollution problem of water and sediments by these residues; although these results are for the Blanco river only, there is no reason to predict that the situation in the rest of the country could be better.

Both studies show that the contamination of superficial waters with these residues in Mexico must be high and that it is already affecting freshwater organisms adversely. Besides, this contamination affects food items that are consumed by large sectors of the Mexican population.

It should also be evident that the corresponding authorities—in this case those of fisheries, health, and environment—should take, as soon as possible, whatever steps are needed to analyze this problem in depth and to control it efficiently in the short term, in view of the adverse repercussions that it might have, not only for the economy of the affected regions, but also on the conservation of these environments and on the health of their populations.

6. ORGANOPHOSPHOROUS PESTICIDES

As mentioned earlier (Section 1) a series of relatively recent changes in Mexican agriculture has altered the pattern of use of pesticides in the country. As a result, there are at present two different types of Mexican agriculture: one with easy access to modern technology and good support on irrigation, roads, and chemical inputs—including pesticides and fertilizers—and another, marginal, located in low quality lands, without irrigation and depending on rains; this has access only to "tied" loans provided by a para-state bank, which consist mainly of fertilizers and the pesticides produced by a para-state company (FERTIMEX), especially organochlorine pesticides, parathion, and methylparathion.

The first type of agriculture is located in the best lands of the country and is devoted to industrial monocultures (cotton, coffee, tobacco) or to crops that will be exported, such as fruits and vegetables; in these lands, organophosphorous and carbamic pesticides are mainly used, with a growing participation of pyrethroids.

These pesticides are expensive in Mexico and can be afforded only by some farmers. In contrast, the marginal lands are devoted to corn, beans, and other staples that will be sent to the local markets or used for self-consumption. On them, organochlorine and organophosphorous pesticides are mainly used.

The first organophosphorous insecticides to be used in Mexico were ethyl- and methylparathion; later, malathion was added to them (Narro-Reyes, 1979). These products are still widely used; for example, malathion is still the preferred product for public health campaigns against malaria and dengue.

These products are also widely used in agriculture, besides other organophosphorous pesticides such as monocrotophos, diazinon, methamidophos, dimethoate, phorate, acephate and, ethion, and residues of these compounds have been found in the surveillance programs carried out on Mexican produce by the U.S. Food and Drug Administration.

As previously mentioned (Section 1), as a result of pressures exerted by importing countries, starting in 1973, Mexico established several laboratories to certify the quality of its export crops. The results of the analyses peformed in these laboratories have not been published and are not available to the general public; however, data published in the United States (Kistner, 1987) show that more than 15% of the beans and more than 13% of the peppers exported by Mexico to the United States had residues above the tolerances or residues of pesticides not allowed in a particular food item.

Only two studies have been published on this in Mexcio. The first (Albert et al., 1979) included vegetables such as rice, avocado, tomato, strawberry, and lettuce. All the samples analyzed had residues; these fluctuated from two to five different compounds per sample. In all cases, one or more samples with residues in excess of the tolerances recommended by FAO/WHO (1986) were found. These results are shown in Table 21.

It can be seen from this Table 21 that the most frequent residues were those of parathion, malathion, and guthion. In Table 22 a summary of the quantitative results is presented. It can be seen that the main contaminant in excess of the tolerances recommended by FAO/WHO was methylparathion.

The second study (Loera and Hernández, 1987) was a screening to determine which of several food items consumed without cooking had residues of these

TABLE 21 Organophosphorous Pesticide Residues in Mexican Vegetables[a]: Qualitative Analysis

Vegetable	Percentage Frequency					
	Guthion	Malathion	Diazinon	Methylparathion	Parathion	Phorate
Rice	40	70	—	40	30	70
Avocado	50	80	20	60	80	—
Strawberry	50	—	—	30	70	—
Tomato	10	70	—	50	70	60
Lettuce	70	70	30	—	20	—

[a]Albert et al. (1979).

TABLE 22 Organophosphorous Pesticide Residues in Mexican Vegetables: Quantitative Results[a]

Compound[b]	Vegetable					Samples in excess of FAO/WHO tolerances (%)
	Rice	Avocado	Strawberry	Tomato	Lettuce	
Parathion	ND[c]–1.5	ND–1	ND–0.8	ND–0.8	ND–0.2	10
Methylparathion	ND–2	ND–0.4	ND–0.8	ND–1	—	26
Guthion	ND–1.5	ND–0.5	ND–0.4	—	ND–0.8	2

[a]Albert et al. (1979).
[b]Only the major contaminants are included.
[c]Not detectable.

TABLE 23   Organophosphorous Pesticides in Vegetables[a]: Qualitative Analysis

Pesticide	Sample[b]										
	1	2	3	4	5	6	7	8	9	10	11
Ethylparathion	+	−	−	−	−	−	−	−	−	−	−
Mevinhos	−	−	+	+	−	−	−	−	−	−	−
Methamidophos	−	+	−	−	−	−	−	+	−	+	+
Phosvel	−	−	+	+	−	+	+	−	+	−	−

[a]Loera and Hernández (1987).
[b]1. Tomato; 2. lettuce; 3. hot pepper; 4. onion; 5. carrot; 6. cucumber; 7. banana; 8. papaw; 9. pineapple; 10. lemon; 11. orange.

compounds. The study was carried out in a medium size city (Xalapa, Ver.) and, for it, the following fruits or vegetables were selected: tomato, hot pepper, lettuce, onion, cucumber, carrot, banana, papaw, pineapple, lemon, and orange. The qualitative results of this screening are sumarized in Table 23.

From these few available results it can be concluded that this problem has not been thoroughly studied although it evidently poses a important toxicological hazard for the Mexican population, since these compounds are characterized by a high acute toxicity and in Mexico the exposure to them through food is also high.

It must be also evident that—as has already been mentioned for other products—the existing control mechanisms for the use of pesticides do not work efficiently in practice, since it is not otherwise possible to explain the presence of these nonpersistent residues in Mexican produce.

7. AFLATOXINS

Contamination of food with aflatoxins is a serious problem in Mexico. This is due, partially, to the humid and tropical climate prevalent in large parts of the country and, partially, to the primitive storing methods used for most grains. However, these contaminants have received only occasional attention by government agencies and research groups. Further, most of the existing studies are available only as congress abstracts (Albert and Flores, 1979; Albert and Farfán, 1979) or have not been published at all (Rosiles, 1979, personal communication; Carbajal, 1987) and there is only one that has been published (Fernández de Castro et al., 1982).

However, from the existing data it is known that from 10 to 20% of the corn tortillas sold in Mexico City are contaminated with aflatoxins. In the same way, it was found that a large part of corn imported by Mexico to supplement the national production—which is insufficient to feed the growing population—is also contaminated with aflatoxins. In addition, since this corn is not of first

**TABLE 24   Aflatoxins in Mexican Food:
Survey Results 1979–1981**[a]

Food	Frequency (%)
Natural peanuts	47.32
Processed peanuts	6.57–31.42
Beans	5.33
Sorghum	2.9/
Corn	0.94

[a]Fernández de Castro et al. (1982).

quality, it is also more easily contaminated with aflatoxins producing fungi than if the grain was intact. It has been found that corn and beans of local production are also affected by this problem and that mixed feeds are affected as well. The data published by authorities are given in Table 24.

It can be concluded that this problem is of particular importance for the Mexican population, since the severe adverse effects, both short and the long term, of aflatoxins are well known.

It is evident that further studies in this area are needed at least as urgently as in the other cases discussed earlier. This is so, not only because of the severe adverse effects of these toxins, but also because it is not known whether other mycotoxins could also be present and which food items could be affected.

8. CONCLUSIONS

From the information reviewed here, it is evident that the Mexican population is exposed to a high number of environmental contaminants through food and that the concentrations of these contaminants often exceed those recommended by the international agencies.

However, with the information available at present, it is not possible to establish the magnitude of this risk; also, it is not possible to select a single type of food as the one that poses the greatest or the least risk for the population; in this way, if a vegetable has residues of a single pesticide, but this pesticide is phosvel, it cannot be evaluated if the risk of ingestion will be higher than if the vegetable was contaminated with aflatoxin B_1 or if it was a staple contaminated with DDT and its derivatives.

From the data presented here it can be concluded that this problem originates from a deficient perception by responsible authorities of the significance, for the health of the Mexican population, of the environmental contamination of food — especially in the long term — that results in a lack of urgency to establish food surveillance systems as one of the major priorities of the Mexican government.

It is also evident that the existing control mechanisms for the use of pesticides in Mexico are not working properly.

In addition, from the data on the presence of organochlorine pesticide residues in food of animal origin from the cotton-growing regions of the country, it can be concluded that, in them, the change from cotton to feed crops (sorghum, alfalfa, soybeans) was not evaluated adequately, since it has facilitated the entrance to the trophic web of high quantities of persistent pesticide residues that were formerly accumulated in soil but now are present in the animals fed with those feeds and in their products such as meat, milk, and eggs.

The high concentrations of these residues that have been found in Mexican human milk (Albert et al., 1978, 1981b; Slorach and Vaz, 1983) and human adipose tissue (Albert et al., 1980a) can be traced back to the continual ingestion of food contaminated with these residues.

In view of this, it should be evident that the problem posed by the presence of several of these contaminants in Mexican food is potentially very serious and that more data are urgently needed to evaluate it properly. Also, it is essential that the Mexican government put into practice, as soon as possible, effective control mechanisms in order to guarantee that the Mexican population have access to food that is within the internationally accepted guidelines, particularly in regard to its levels of environmental contaminants.

ACKNOWLEDGMENTS

The author acknowledges the invaluable continuous support of Dr. Josef E. Herz and Guillermo Massieu, which made possible most of the studies reviewed here, and that of Dr. Arturo Gómez-Pompa, which made possible their continuation.

REFERENCES

Albert, L., and Aldana, P. (1982). Polychlorinated biphenyls in Mexican cereals and their packings. *J. Environ. Sci. Health, part B*, **17**(5), 515–525.

Albert, L. A., and Alpuche, L. (1987). Determinación de plaguicidas organoclorados en quesos de la Comarca Lagunera, México. *Toxicologia* (*Uruguay*), **2**(1), 61–85.

Albert, L. A., and Alpuche, L. (1988). Plaguicidas organoclorados en huevo de gallina procedente de la Comarca Lagunera. *Rev. Soc. Quim. Méx.*, **32**(6), 195–203.

Albert, L. A., and Badillo, F. (1986). Plomo y zinc en organismos del río Blanco, Veracruz. *Rev. Soc. Quím. Méx.* **30**(5), 281.

Albert, L. A., and Bárcenas, C. (1988). Contaminación por plaguicidas organoclorados en muestras de queso procedentes del Soconusco, Chiapas, México. *Rev. Soc. Quim. Méx.*, **32**(3), 78–85.

Albert, L., and Farfán, N. (1979). Contaminación por aflatoxinas en maíz y frijol del estado de Puebla. *Rev. Soc. Quím. Méx.* **23**(5), 285.

Albert, L. A., and Figueroa, A. (1985). Evaluación preliminar de la contaminación por organoclorados y ftalatos en organismos del río Blanco, Veracruz. *Rev. Soc. Quím. Méx.* **29**(4), 198.

Albert, L., and Flores, A. (1979). Contaminación por aflatoxinas en alimentos balanceados para aves en el estado de Puebla. *Rev. Soc. Quím. Méx.* **23**(5), 281.

Albert, L., and Locra, R. (1983). Plaguicidas organoclorados VII. Residuos de plaguicidas organoclorados en huevo de gallina procedente de Monterrey. *Rev. Soc. Quím. Méx.* **27**(1), 12–17.

Albert, L. A., and Reyes, R. (1975). Organochlorine pesticide residues in milk and milk products in México. *Rev. Soc. Quím. Méx.* **19**(5), 215.

Albert, L., and Reyes, R. (1978). Plaguicidas organoclorados II. Contaminación de algunos quesos mexicanos por plaguicidas organoclorados. *Rev. Soc. Quím. Méx.* **22**(2), 65–72.

Albert, L. A., and Reyes, M. J. (1986). Plaguicidas organoclorados en huevo de gallina procedente del Soconusco, Chiapas, México. *Rev. Soc. Quím. Méx.* **30**(5), 280.

Albert, L., Reyes, R., and Saval, S. (1975). Pesticide residue problems in Mexico. *Rev. Soc. Quím. Méx.* **19**(5), 217.

Albert, L., Cebrián, M., Vega, P., Ruiz, I., and Hornández-Román, P. (1978). Organochlorine pesticides in human milk in Mexico. Abstract Volume VI-801, IVth International Congress of Pesticide Chemistry, Zurich, Switzerland.

Albert, L., Martínez-Dewane, M. G., and González, M. E. (1979). Plaguicidas organofosforados I. Residuos de insecticidas organofosforados en algunos alimentos mexicanos. *Rev. Soc. Quím. Méx.*, **23**(4), 189–196.

Albert, L., Méndez, F., Cebrián, M. E., and Portales, A. (1980a). Organochlorine pesticide residues in human adipose tissue in Mexico: Results of a preliminary study in three Mexican cities. *Arch. Environ. Health* **35**(5), 262–269.

Albert, L., Reyes, R., and Saval, S. (1980b). Plaguicidas Organoclorados en Algunos Peces Comestibles de Agua Dulce. II Simposium sobre Contaminación de Alimentos, Associación Mexicana de Mujeres Médicos Veterinarios Zooctenistas, México, D.F.

Albert, L., Loera, R., and Saval, S. (1981a). Plaguicidas organoclorados IV. Residuos de plaguicidas organoclorados en huevo de gallina. Estudio preliminar en dos ciudades mexicanas. *Biótica* **6**(3), 325–338.

Albert, L., Vega, P., and Portales, A. (1981b). Organochlorine pesticide residues in Mexican human milk. *Pest. Mon. J.* **15**(3), 135–138.

Albert, L., Vega, P., and Nava, E. (1982). Plaguicidas organoclorados VI. Residuos de plaguicidas organoclorados en leches evaporadas mexicanas. *Biótica* **7**(3), 473–482.

Albert, L., Loera, R., and Tamez, G. (1987a). Ftalatos en avenas industrializadas mexicanas y en sus empaques. *Toxicología (Uruguay)*, in press.

Albert, L. A., Loera, R., Reyes, R., and Vega, P. (1987b). Residuos de plaguicidas organoclorados en leches pasteurizadas mexicanas. *Ciencia Interam.*, in press.

Albert, L. A., Vega, P., Aguirre-Beltrán, G., and Aldana-Torres, P. (1988). Residuos de plaguicidas organoclorados en leches deshidratadas y leches del tipo "maternizada." *Biótica*, **13**, 59–67.

Albert, L. A., Flores, J., and Reyes, R. (1986). Determinación de residuos de plaguicidas organoclorados en cremas y mantequillas mexicans. *Rev. Toxicol.*, **3**, 51–63.

Bordas, E. (1973). El empleo de los insecticidas agrícolas y la contaminación en el embiente rural mexicano. In *Memoria, la. Reunión Nacional sobre Problemas de Contaminación Ambiental.* Subsecretaría de Mejoramiento del Ambiente, México, D.F., 1222 pp.

Capella, S., Manjarrez, A., Labastida, C., and Santarriaga, L. (1975). Nivel de la contaminación por p,p'-DDT y p,p'-DDE en la leche que se consume en la ciudad de México. *Rev. Soc. Quím. Méx.* **19**(5), 217.

Carbajal, M. M. (1987). Micotoxinas de la Tortilla. Su efecto cancerígeno y teratogénico, 2a. Reunión de Investigaciones Universitarios en Contaminación Ambiental, México, D.F.

Duggan, R. E. (1967). Chlorinated pesticide residues in fluid milk and other dairy products in the United States. *Pest. Mon. J.* **1**(3), 2–8.

FAO/WHO. (1986). *Guide to Codex Recommendations Concerning Pesticide Residues.* FAO/WHO, Rome, 122 pp.

Fernández de Castro, D., Hernández, H. A., and Suárez, R. (1982). Contaminación de alimentos. In M. López-Portillo, Comp., *El Medio Ambiente en México: Temas, Problems y Alternativas.* Fondo de Cultura Económica, México, D.F., pp. 211–232.

Frank, R., Smith, E. H., Braun, H. E., Holdrinet, M., and McWade, J. W. (1975). Organochlorine insecticides and industrial pollutants in the milk of the southern region of Ontario, Canada. *J. Milk Food Technol.* **38**(2), 65–72.

Heeschen, W. (1972). Analysis for residues in milk and milk products. In F. Korte and F. Coulston, eds., *Environmental Quality and Safety,* Vol. 1. Academic Press, New York, pp. 229–234.

Kistner, W. (1987). Deadly harvest. *This World,* February 8, pp. 8–9.

Laborín, R., Ruiz, J. A., Muñiz, J. J., and Ogura, T. (1983). Determinación de insecticidas clorados en crema láctea. *Rev. Soc. Quím. Méx.* **27**(5), 293.

Loera, R., and Hernández, M. R. (1987). Determinación de plaguicidas organofosforados en vegetales que se expenden en Xalapa, Ver. *Rev. Soc. Quím. Méx.* **31**(4), 189.

McMahon, B. D., and Burke, J. A. (1986). Expanding and tracking the capabilities of pesticide multiresidue methodology used in the U.S. Food and Drug Administration's pesticide monitoring programs. Abstract 5D-01, VIth International Congress of Pesticide Chemistry, Ottawa.

Márquez-Mayaudón, E., Fujigaki, A., Moguel, C., and Aranda, B. (1968). Problemas de contaminación con pesticidas, caso Tijuana (1967). *Salud Púb. Méx.* **10**(3), 293–300.

Mena, J. (1987). El Uso de Plaguicidas en México, lr. Taller Nacional de Prevención de Riesgos en el Uso de Plaguicidas, Xalapa, Ver., México.

Narro-Reyes, J. G. (1979). El uso de plaguicidas en la agricultura mexicana. In R. N. Ondarza, ed., *Los Reguladores de las Plantas y los Insectos.* CONACYT, México, D.F., pp. 27–40.

Parada, E., Velasco, O., and Avila, M. (1975). Determinación del contenido de plomo en alimentos enlatados. *Rev. Tecnol. Aliment. (Méx.)* **10**, 170–173.

Ritcey, W. R., Savary, G., and McCully, K. A. (1972). Organochlorine insecticide residues in human milk, evaporated milk and some milk substitutes. *Can. J. Public Health* **63**, 125–132.

Ruiz-Sandoval, G. (1878). Envenenamiento lento por el plomo en habitantes de Oaxaca. *Gaceta Méd. Méx.* **13**, 393.

Saval, S. (1976). *Determinación de Residuos de Plaguicidas Organoclorados en Alimento Balanceado para Aves, Carne de Pollo y Huevo.* Thesis, Escuela Nacional de Ciencias Biológicas, Instituto Politécnico Nacional, México, D.F., 83 pp.

SARH. (1974). *Ley de Sanidad Fitopecuaria de los Estados Unidos Mexicanos.* Secretaría de Recursos Hidraúlicos, México, D.F., 29 pp.

SARH. (1979). *Tolerancias de Plaguicidas.* Secretaría de Agricultura y Recursos Hidraúlicos, México, D.F., 31 pp.

Slorach, S. A., and Vaz, R. (1983). *Assessment of Human Exposure to Selected Organochlorine Compounds Through Biological Monitoring.* United Nations Environment Programme, World Health Organization and Swedish National Food Administration, Uppsala, Sweden, 134 pp.

SSA. (1984). *Ley General de Salud.* Ed. Libros Económicos, México, D.F., 167 pp.

Washüttl, V. J. (1974). Pesticide in Milch and Milch Produkten. *Wein Tieräztl. Mschr.* **61**(2), 44–51.

Warman, A. (1983). El problema del campo. In P. González-Casanova and E. Florescano, Coords., *México, Hoy,* Siglo XXI editores, México, D.F., 7th ed., pp. 108–120.

18

MONITORING OF PESTICIDE RESIDUES IN FOOD IN HUNGARY

Ferenc Hargitai*

Department of Plant Protection and Agrochemistry
Ministry of Agriculture and Food
Budapest, Hungary

*Present address: Plant Protection Chemicals and Intermediates, div. 82, Chemolimpex, P.O. Box 12, H-1805 Budapest, Hungary.

1. INTRODUCTION

In Hungary, prior to the Second World War, the manufacture, marketing, and usage of pesticides were already under governmental license (official approval) in recognition of the hazards of the predominantly used arsenic and mercuric compounds. The continuity in the strictness of regulations can also be followed in the orders issued after World War II. Regulations pertaining to pesticides are traditionally issued by the Minister of Agriculture and Food with inputs from the authorities of health and environmental agencies for the past two decades. The reasons for the banning of aldrin, dieldrin, DDT, and HCH in Hungary in 1968, the first country in the world do so, is hardly emphasized in the technical literature. At that time, this measure gave rise to great debate, especially in Europe, but our views have proved correct.

In order to provide high proficiency in plant protection activities as well as ensure the observance of regulations, a national plant protection network was established in 1954 that consisted of 20 plant protection stations (one in each of the 19 counties and one in the capital) and a center for the coordination of the work of stations from a professional point of view. From the early 1970s a laboratory for pesticide residue analysis was established in all plant protection stations. Since 1974, standardized methods have been applied in the pesticide residue analytical laboratories.

It can be seen from this brief historical statment that the Hungarian Government attaches special importance to plant protection activities. This stems from the fact that agricultural production plays and always has played a very significant role in the economy of the country. This tendency is expected to continue as some other countries are faced with the problem of overproduction.

2. BRIEF OVERVIEW OF HUNGARIAN AGRICULTURE

The total land area of Hungary is only 93000 km^2 (9.3 million ha) but 70% of this area is suitable for agricultural production, an extraordinary high percentage by international standards. Approximately half of the agricultural produce comes from plant production, while animal husbandary accounts for the other portion. About 20% of the Gross National Product and 25% of the total export is of agricultural origin. Practically one-third of the agricultural area is devoted to production for export purposes. About 16% of the employed population work in the agricultural sector.

The dominance of large-scale farming systems is a characteristic feature of Hungarian agriculture. About 77% of the total agricultural area (6.2 million ha) is used by the agricultural cooperatives while state farms account for only 14%. Table 1 shows that the average size of the farms is hardly less than 7000 and 4000 ha, respectively (Hargitai, 1984).

Though the total area of private farms is only 9% of the total arable land, it represents a significant production capacity, especially in hand labor intensive cropping. The government provides both economic and political support for the development of this sector. Regarding the success of the Hungarian agriculture and its high production level, there is a need for special and close cooperation between small holdings and large-scale farms in the field of production.

Because of an advantageous geographical situation and favorable conditions available for plant production, almost all but tropical crops can be cultivated in Hungary. Table 2 shows the growing area of major crops as well as the corresponding yield results for the last 10 years.

Apart from the significant expansion of the sunflower growing area there have been no special changes in the structure of crop production. Any modification and fluctuation regarding the growing area is primarily influenced by the profitability of the given crop and its hand labor demand. The relatively

TABLE 1   Number and Average Size of Farms in Hungary

Share of Land Owners	Area (ha)	Number of Farms	Average Size of Farms (ha)
State farms	850,400	127	6,695
Cooperatives	4,805,800	1,262	3,805
Private (farms and gardens)	535,400	1.5 million	0.35

TABLE 2   Growing Area and Production of Large-scale Farms

Crops	Growing Area (thousands of ha)		Production (thousands of metric tons)	
	1976–1980	1981–1985	1976–1980	1981–1985
Wheat	1274	1307	5180	6048
Barley	237	274	769	1004
Corn	1297	1111	6292	6791
Sunflower	185	309	298	612
Sugarbeet	118	115	3975	4461
Peas and beans	85	77	302	351
Rape	53	56	80	86
Potato	84	54	1194	980
Tomato	15	9	384	252
Green pepper	21	14	177	143
Cabbage	6	4	115	89
Cucumber	5	2	47	26
Onion	7	6	114	129
Apple	67	65	968	1139
Pear	12	11	87	105
Peach	20	15	90	86
Grape	183	157	836	784

stable structure of plant production underscores the tasks of plant protection and, within this, the scope of official work of pesticide residue control.

3. PRESENT SITUATION OF CHEMICAL PLANT PROTECTION

3.1. Trends in Pesticide Usage

Hungarian agriculture emphasizes the combined application of different plant protection methods with complex pest management. Nevertheless, intensive crop production, as in other countries with developed agriculture, can be ensured mainly by chemical control based on pesticide use. The usage of pesticides was significantly increased through the 1960s and the beginning of the 1970s while between 1975 and 1985 it was practically stable and just recently has shown a decreasing tendency (see Table 3) (Hargitai and Baranyai, 1987).

The stability and even the slight reduction of pesticide usage are the result of a conscious plant protection policy. In 1985, the per unit usage still exceeded the 5 kg/ha (in 1987 it was reduced) which ranked Hungary among the most intensive chemical consumer countries. The state subsidy provided for the purchase of pesticides has ceased, the scope of environmental requirements has become stricter, the values of maximum residue limits (MRL) of pesticides generally have decreased, the reliability of pest forcasting has been strengthened, and the possibilities for the widespread introduction of integrated pest management methods into the practice have improved. All these factors make possible the stabilization of pesticide usage at a lower level without imposing risk to the crop production (Hargitai, 1986).

Presently in Hungary, the application of 550 chemical products are registered. This seems to be a low number compared to that of the developed countries in Europe or the United States. Considering that these products represent 300 different active ingredients, of which 250 are regularly used in large quantity, it is understandable that the structure of pesticide usage, with reference to the assortment, is quite modern. More than half of the pesticides used are produced

TABLE 3   Pesticide Use in Hungary

	Usage in 1000 Metric Tonnes of Active Ingredients		
	1975	1980	1985
Insecticides	3.1	3.2	2.8
Fungicides	13.7	15.0	13.6
Herbicides	12.5	13.7	14.7
Others	1.5	1.8	1.4
Total	30.8	33.7	32.5

locally while about 80% of imported products are in the form of active ingredients.

3.2. Education of Plant Protection Specialists

In Hungary, pesticide application, with the exception of areal sprays, is carried out by the members and employees of the cooperatives and the state farms, respectively. Thus the efficacy of pesticide application and the decrease of risks of occupational exposure depend to a large extent on the professional knowledge of the workers.

Since 1970 it is stipulated by Ministerial Decree that large-scale farms must employ graduate plant protection specialists with the application itself executed only by specially trained, skilled workers. The Hungarian training system in this field has been presented in details elsewhere (Hargitai, 1979). The distribution of specialists working in plant protection in 1987 is shown in Table 4.

Pesticides can be ranked into two main categories: the first is for those products that can be exclusively used in large-scale farms while the second group can be accessed freely in the trade. Products of the former category are more toxic and special knowledge is needed in their application. All the large-scale farms and private growers who sell their produce in the market must keep a record of all chemicals they use.

3.3. The Network of Plant Protection Stations

The organization of the Hungarian plant protection network can be regarded as special since plant protection, plant quarantine, plant nutrient supply, soil reclamation, and agricultural irrigation are under one organization. The official name for this institutional unit is the plant Protection and Agrochemistry Station (the proposed future name is Phytosanitary and Soil Protection Station). There are altogether 19 stations—one in each of the 19 counties plus one more in the capital. Official control of plant protection activities on the total arable land of more than 6 million ha is carried out by these 20 stations, which are also responsible for the execution of the biological, physical, chemical, environmental, and residue analytical tests required for the registration of pesticides. They are also involved in pest forecasting and the development of certain plant protection

TABLE 4   Plant Protection Specialists Working in Large-Scale Farms, 1987

Specialist	Number
Plant protection engineers	1,997
Plant protection technicians	594
Skilled plant protection workers	35,000

techniques; they also take part in common and postgraduate training of specialists.

4. PURPOSES AND SYSTEM OF PESTICIDE RESIDUE ANALYSIS

Samples taken from agricultural products at the market are analyzed to ensure the safety of the consumer. This market control apparently does not exclude the selling of contaminated goods, especially in the case of rapidly perishable ones. Even if MRL values found were unacceptably high by the time results were available the commodities would have been sold. Consequently, market control, that is, the control of the "market basket," serves primarily as an exercise in human hygiene.

However, for the producers and consumers, it is more important that possible problems be discovered well before the foods get to the market and the goods should meet both export standards as well as the national guidelines. Thus, the basic interest of all concerned parties should be prevention.

The marketing of goods that meet the above criteria can be ensured only if (a) the safety margins for the pesticide are clearly defined during the time of official evaluation prior to registration, and (b) the control of pesticide residues is carried out during or just before the harvest, before the transportation of the goods to the market (selective survey of field control).

In accordance with Hungarian standards, prior to registration, pesticide residue level must be determined under local conditions. Results derived from foreign laboratories are useful but can never replace the local measurements that take into account the climatic and soil condition, the plant types, and the "good agricultural practice" prevailing in Hungary.

In order to offer assistance to growers and to protect the health of consumers, a residue analytical network was established at the beginning of the 1970s by the Hungarian administration with the aim of preventing the trade of plant products already at the farm gate if pesticide residues measured exceed the acceptable MRL. In conformity with this aim, laboratories were placed at the plant protection and agrochemistry stations. If food products contain pesticide residue higher than the level set in the mutual Order of the Minister of Agriculture and Food and the Minister of Public Health the marketing of the product will be temporarily banned by the plant protection authorities. Afterward the final decision will be made on the fate of such products by the health authorities in terms of elimination, "dilution," or special treatment of the product.

Pesticide residues are also measured at random by institutes controlling food quality and the laboratories of health authorities. However, the total number of such random samples is far less than that evaluated by the plant protection network and thus is less suitable for drawing sound conclusions. For this reason I prefer to use the data of the Plant Protection Agrochemistry Center and laboratories of the stations.

5. RESULTS OF THE PESTICIDE RESIDUE ANALYSIS

5.1. Early Studies

The regular measurement of pesticide residues was introduced in the plant protection network in 1968. In accordance with the techniques available at that time, agar diffusion and thin-layer chromatography were used for detection, which was later followed by spectrophotometric and gas chromatographic methods. In the early years, the residues of chlorinated hydrocarbons and organic phosphorous compounds were mainly measured. Between 1974 and 1977 the capacity of the laboratories was gradually increased and the reliability of the analyses was very good as the collaborative tests showed (Ambrus et al., 1978). During this period almost 90,000 analyses were carried out. Analyses classified by the origin of the samples are represented in Table 5.

No residues could be detected at all (i.e., residue level was less than limit of determination) for organic phosphorous compounds, dithiocarbamates, and benomyl in 90.8, 72.4, and 52.0% of the samples screened, respectively. However, the number of samples in which the dinocap residue exceeded the acceptable level in apples was unexpectedly high. This could be due mainly to the fact that an unreasonably low MRL value had been set in Hungary compared to that used internationally.

In order to facilitate handling and evaluation of the data base, a computerized data processing system was introduced. This made possible the grouping of data from different points of view.

5.2. The Number of Investigations since 1978

Since 1978 the network of pesticide residue laboratories has been working at full capacity. Tests pertain to the control of crops and food commodities produced in Hungary and imported, and exported, foodstuffs as well. The total number of the pesticide residue analyzed is shown in Table 6.

It can be seen that the annual number of samples stabilized around 15,000. The

TABLE 5   Pesticide Residue Analysis from 1974 to 1977

Type of Sampling	Number of Samples	Number of Investigations
Selective survey (field control)	8,235	19,503
Market control	9,080	22,622
Export control	2,934	10,427
Import control	7,449	34,171
Total	27,698	86,724

TABLE 6   Total Number of Pesticide Residue Analyses

Year	Number of Samples Analyzed	Number of Active Ingredients	Total Number of investigations
1978	15,055	216	47,769
1979	14,670	219	59,332
1980	12,709	306	36,616
1981	15,062	296	33,166
1982	14,508	230	26,440
1983	17,393	195	32,272
1984	16,301	203	48,224
1985	15,637	189	45,264
1986	14,620	195	40,144

actual number of the investigations is approximately three times higher. The number of active ingredients evaluated is quite high. In practice a much less active ingredient is to be expected, the residues of which reach or exceed the tolerance level. But it is a common practice that residues of newly introduced chemicals should be regularly checked for the first years. In addition, the evaluations for registration purposes are also included in the data of Table 6.

5.3.   Control of Imported Products

It has been mentioned in Section 2 that a wide variety of crops can be grown in Hungary. As a result, only primarily tropical fruits, especially spices, coffee, and cacao, are imported. Since plant protection practices, problems, and climatic conditions of the countries from which the products originate are significantly different from those in Hungary, control of pesticide residue levels always becomes more rigorous if a new foreign trading company, a new exporting country, or a new crop commodity appears on the scene. It is standard practice that a preliminary sample of goods is required, together with the scope of information concerning the plant protection practice of the producing country, prior to delivery. The regular exchange of information is well promoted by bilateral plant protection agreements. On the basis of analytical test results obtained from the preliminary samples, residue analysis can be rapidly carried out at the time of the arrival of the consignment.

In Hungary, the plant quarantine service might refuse import consignments not only if they are infected by quarantine pests but also if the plant products were treated with a pesticide not registered in Hungary or the pesticide residue level is higher than the acceptable one.

The entrance of imported plant products might happen in two ways: (a) with final permit—in this case randomized control is carried out, and (b) with temporary permit—the consignment may enter but cannot be marketed unless a favorable pesticide residue level is determined.

Between 1978 and 1980 the number of control tests of imported goods that

TABLE 7 Analysis of Samples from Imported Consignments

Year	Number of samples	Number of analysed plant sorts	Number of active ingredients	Number of investigations	R > T cases in % of the investigations	R < MD cases in % of the investigations
1978	3503	64	62	18658	0,30	76,4
1979	3818	44	81	31754	0,17	73,5
1980	2152	32	76	13576	0,28	72,0
1981	1456	40	59	7629	0,35	91,3
1982	1251	30	67	5892	0,31	89,8
1983	1345	32	70	7532	0,12	91,6
1984	1172	20	66	7595	0,08	89,4
1985	1282	29	68	8301	0,13	89,3
1986	1295	37	70	7804	0,13	90,8

failed was high. On the basis of our experience since then, a reasonable selection criterion has become possible that resulted in the decrease of the failure rate of controlled import samples (see Table 7).

In the table R > T means that the residue level found in the sample is higher than the tolerance (MRL) value while R < MD indicates that the residue level is below the limit of detection.

Since 1983 the number of R > T cases has been significantly reduced. This reduction cannot be explained by the fact that agricultural practice has changed in the exporting country or that better partners have been found by foreign trade agencies, though the latter was likely to have contributed to it. The reduction can be attributed to the fact that Hungary, in accepting the recommendations of the Codex Committee on Pesticide Residues, accepted MRL values for some of the imported products that are higher than the national standards. The recommended standards for some food products and active ingredients are presented in Table 8.

Some active ingredients or groups of active ingredients can be regularly detected in several food products at a level very close to or just above of MRL values. The most characteristic pairs are as follows:

Coffee—DDT and its metabolites, dieldrin, diazinon
Cacao bean—DDT, lindane, aldrin, endrin
Groundnut—endrin
Tea—DDT
Orange—parathion, methylparathion, quinalphos
Lemon—parathion, methylparathion, quinalphos.

TABLE 8   Accepted Pesticide Tolerance levels in Some Imported Products

Active Ingredient	Plant Product	Hungarian MRL (ppm)	Accepted Codex MRL (ppm)
Inorganic bromide	Citruses, dried fruits	30	30–250
Methylazinphos	Fruits, vegetables	0.5	2.0–4.0
Bromophos	Leek	0.5	2.0
Chlorpyrifos	Citruses	0.1	0.3
DDT plus metabolits	Tropical fruits, spices	0.1	0.2–0.3
Diazinon	Fruits, vegetables	0.1–0.5	0.7
Dichlorvos	Bean, lentil, coffee bean	0.2–0.5	2.0
Lindane	Huckleberry	1.0	3.0
Parathion	Citruses	—	1.0
Methylparathion	Citruses	0.2	1.0
Quinalphos	Citruses	0.1	1.0
Trichlorphon	Vegetables	0.1	0.2

TABLE 9 Pesticide Residue Survey of Local Food Production (Selective Survey, Market and Export Control)

Year	Number of samples	Number of analysed plant sorts	Number of active ingredients	Number of investigations	R > T cases in % of the investigations	R < MD cases in % of the investigations
1978	4494	88	128	11361	0,41	89,7
1979	5157	88	127	14052	0,38	91,0
1980	5415	79	146	14370	0,52	87,2
1981	5216	87	132	11369	0,54	74,8
1982	4408	88	125	10855	0,56	81,1
1983	5309	90	125	11976	0,42	83,2
1984	4658	98	137	15546	0,36	89,3
1985	4504	71	125	16122	0,18	91,0
1986	4844	69	134	16317	0,11	90,6

589

Sometimes parathion, DDT, chloropyriphos, metidathion, and α-HCH can be detected at a higher concentration than is acceptable in grape fruit, lentils, kiwi fruit, mandarins, and ginger.

5.4. Control of Pesticide Residue of National Plant Products

About one-third of total manpower of pesticide residue laboratories is utilized for preharvest and market control of plant products in Hungary. It can be seen from

TABLE 10   Specific Production, Consumption, and the Export of Agricultural Products

Product	Production Per Capita (kg) 1980	1985	Consumption Per Capita (kg) 1980	1985	Export (thousand metric tons) 1980	1985
Wheat	567	618	115.1	110.0	813.6	2002.5
Corn	623	641	feedstuff		82.9	227.8
Potato	130	129	61.2	54.1	—	—
Vegetables	198	203	79.6	75.6	307	398
Fruits	154	143	74.9	71.0	535	463
Wine (in liter)	53	27	34.8	24.8	209 million	270 million
Meat	193	217	73.8	79.1	480	585
Milk, milk products	231	234	166.1	182	—	—

TABLE 11   Distribution of Crop Sampling

Crop	Percentage of the Total Sampling 1980	1982	1984	1986
Apple	16	19.4	16.4	22
Pear	7	4.4	2.8	3.1
Grape	6.1	7.9	5.1	4.5
Potato	5.6	4.6	4.0	2.7
Corn	4.4	0.8	0.8	3.6
Green pepper	4.4	5.7	6.4	7.0
Tomato	4.2	4.1	4.4	4.7
Wheat	3.6	4.1	6.0	4.8
Peach	3.5	3.9	2.6	3.3
Green peas	2.7	3.5	2.9	5.2
Sunflower	2.4	3.9	3.1	3.5
Sour cherry	2.3	3.5	3.1	2.5
Cucumber	1.6	1.7	3.1	5.8
Cabbage	1.6	2.4	2.4	2.5
Cherry	1.5	2.6	2.4	2.3
Lettuce	0.4	0.9	1.7	2.6

data of Table 9 that the number of samples processed annually totals 4500–5500 while that of active ingredients amounts to 120–140. Recently the investigations have intensified but the assortment of agricultural products sampled has been reduced as a consequence of the experience from earlier years.

Since sampling takes place deliberately to search for misuse of pesticides rather than at random, and we refer mainly to field control instead of market control, data obtained during the past years are quite encouraging, with R > T and R < MD cases being 0.3–0.6 and 90% of samples, respectively. The quantity

TABLE 12   Structure of Sampling in the Field and from the Market in 1985

Crop	Number of Samples	Number of Laboratories Sampling
Apple	1028	19
Green pepper	339	19
Tomato	268	19
Grape	245	18
Potato	239	18
Wheat	186	14
Cucumber	162	19
Sour cherry	149	19
Sunflower	128	15
Peach	116	16
Pear	114	19
Watermelon	106	13
Cherry	87	16
Peas	85	12
Cabbage	82	16
Strawberry	71	13
Lettuce	70	13
Radish	69	13
Raspberry	67	9
Green peas	64	12
Apricot	59	10
Currant	57	10
Rape	55	8
Plum	53	13
Onion	46	10
Carrot	35	13
Green beans	35	11
Sweet corn	34	5
Cauliflower	30	10
Rice	29	3
Beans	28	4
Kohlrabbi	27	10
Chives	24	6
Savoy	24	8
Chinese cabbage	23	2

of a given plant in crop production, local consumption, and the export structure all influence the practice of sampling. Data reflecting this are shown in Table 10.

Though cereal dominates the crop production, its plant protection is far less intensive than that of vegetables and fruits that, in most cases, are likely to go directly to the consumer. For this reason residue control of the later crops must be deemed more important. This is reflected in Table 11, showing the rate of sampling of individual products.

Although the total number of crops monitored generally exceeds 70 on a yearly basis, about 35 crops account for 90–95% of the samples in accordance with their relative importance in agricultural production. The structure of sampling in 1985 and the number of plant protection stations taking part in the control of the given crop variety are shown in Table 12. Number 19 means that all the plant protection stations surveyed the same crop, while low numbers refer to some special crops (e.g., rice, chinese cabbage).

In order to assess the number of R > T cases, it should be noted that in 1980–1982 some national MRL values were decreased, which in turn called for the alteration of existing agricultural practice. Tables 13–16 show some important MRL values for apple, strawberry, green pepper, and lettuce in Hungary, the Federal Republic of Germany, the Netherlands, Austria, and Yugoslavia. The comparison is feasible because of the close geographical situation of these countries (Hargitainé and Ambrus, 1989).

For some commodities produced in Hungary, among R > T cases, dithiocarbamates and diquat are widely identified during the period; triazophos and endosulfan were also detected but only occasionally. Pesticides that occur frequently in domestic crops are shown in Table 17.

Apart from pesticides listed in Table 17, others have also been detected less frequently as shown below:

1981—methylparathion, quinalphos
1982—diazinon, dimethoate, folpet, mevinphos, fenitrothion, methylparathion
1983—phosphamidon, carbofuran, quinalphos, metalaxyl, methydathion, methylparathion, mevinphos, methylpyrimiphos, trichlorphos, copper compounds
1984—diazinon, methylparathion, sulfotep, dichlorvos, phorate, methidathion, quinalphos, phosalone, amitraze, deltametrin, fenitrothion, fenarimol, metalaxyl.
1985—benomyl, captan, carboxin, methylparathion, chlorpropylate, sulfotep, DNBP acetate
1986—amitraz, deltametrin, diazinon, phosmetilan, carbofuran, chlorpropham, mevinphos.

5.5. Trends in Pesticide Exposure of Consumers in Hungary

The trend in the exposure of Hungarian consumers to pesticides can be derived from the available data from the import-field-market surveys. Owing to the

TABLE 13 Changes in Some MRL Levels in Apple (ppm)

Active Ingredient	Hungary 1980	Hungary 1985	Federal Republic of Germany 1980	Federal Republic of Germany 1985	Netherlands 1980	Netherlands 1985	Austria 1980	Austria 1985	Yugoslavia 1980	Yugoslavia 1985
Benomyl	1.0	2.0	2.0	2.0	2.0	3.0	2.0	2.0	10.0	2.0
Captan	10.0	5.0	15.0	15.0	15.0	15.0	15.0	15.0	15.0	15.0
Copper	10.0	10.0	20.0	20.0	20.0	20.0	15.0	15.0	15.0	15.0
Dimethoate	1.0	2.0	1.5	1.5	0.6	0.6	1.5	1.5		
Dithiocarbamates	3.0	2.0	2.0	2.0	2.0	0.5	2.0	2.0	2.0	2.0
Folpet	10.0	5.0	15.0	15.0	15.0	10.0	15.0	15.0	15.0	15.0
Lindane	1.0	0.5	1.5	1.0	2.0	2.0	1.5	1.5	1.0	1.0
Methidathion	0.2	0.5	0.3	0.3			0.3	0.3	0.1	0.1
Methylparathion	0.5	0.2	0.15	0.2	0.5	0.2	0.15	0.15	0.5	0.5
Methylthiophanate	1.0	2.0	2.0	2.0	2.0	2.0	2.0	2.0		

TABLE 14 Changes in Some MRL Levels in Strawberry (ppm)

Active Ingredient	Hungary 1980	Hungary 1985	Federal Republic of Germeny 1980	Federal Republic of Germeny 1985	Netherlands 1980	Netherlands 1985	Austria 1980	Austria 1985	Yugoslavia 1980	Yugoslavia 1985
Benomyl	1.0	2.0	1.5	1.5	2.0	3.0	1.5	1.5	10.0	2.0
Dimethoate	1.0	1.0	1.5	1.5	0.6	0.6	1.5	1.5	0.5	0.5
Dithiocarbamates	3.0	2.0	2.0	2.0	2.0	3.0	2.0	2.0	3.0	2.0
Folpet	10.0	5.0	15.0	15.0	20.0	20.0	15.0	15.0	15.0	15.0
Lindane	1.0	0.5	1.5	1.0	2.0	2.0	1.5	1.5	1.0	1.0
Methidathion	0.2	0.2	0.2	0.2			0.2	0.2	0.1	0.1
Methylparathion	0.5	0.2	0.15	0.2	0.5	0.2	0.15	0.15	0.5	0.5
Methylthiophanate	1.0	2.0	1.5	1.5	2.0	2.0	1.5	1.5		
Trichlorphon	1.0	1.0	0.5	0.5	0.5	0.5	0.5	0.5	0.5	0.5

TABLE 15 Changes in Some MRL Levels in Green Pepper (ppm)

Active Ingredient	Hungary		Federal Republic of Germany		Netherlands		Austria		Yugoslavia	
	1980	1985	1980	1985	1980	1985	1980	1985	1980	1985
Benomyl	1.0	2.0	1.0	1.0	2.0	3.0	1.0	1.0	5.0	1.0
Captan	10.0	5.0	15.0	15.0	15.0	15.0	15.0	15.0	15.0	15.0
Copper	10.0	10.0	20.0	20.0			15.0	15.0	15.0	15.0
Dimethoate	1.0	1.0	1.5	1.5	0.6	0.6	1.5	1.5	0.5	0.5
Dithiocarbamates	3.0	3.0	2.0	2.0	2.0	0.5	2.0	2.0	3.0	2.0
Lindane	1.0	1.0	1.0	1.0	2.0	2.0	1.0	1.0	1.0	1.0
Malathion	2.0	2.0	3.0	3.0	0.5	0.5	3.0	3.0	0.5	0.5
Tetradifon	0.1	0.1	1.5	1.5	1.5	1.5	1.5	1.5	1.0	1.0
Trifluralin	0.1	0.1	0.05	0.05			0.05	0.05		

TABLE 16 Changes in Some MRL Levels in Lettuce (ppm)

Active Ingredient	Hungary		Federal Republic of Germany		Netherlands		Austria		Yugoslavia	
	1980	1985	1980	1985	1980	1985	1980	1985	1980	1985
Benomyl	1.0	2.0	1.0	1.0	2.0	3.0	1.0	1.0	5.0	1.0
Dithiocarbamates	3.0	2.0	2.0	2.0	2.0	2.0	2.0	2.0	3.0	2.0
Folpet	10.0	5.0	15.0	15.0	15.0	15.0	15.0	15.0	2.0	15.0
Lindane	1.0	1.0	2.0	1.0	2.0	2.0	1.5	1.5	1.0	1.0
Malathion	2.0	2.0	3.0	3.0	0.5	0.5	3.0	3.0	0.5	0.5
Mevinphos	0.1	0.2	0.5	0.5	0.1	0.1			0.1	0.1
Pirimicarb	0.5	0.5			1.0	1.0				
TMTD	3.0	3.0	2.0	2.0	3.0	3.0	2.0	2.0	3.0	3.0

TABLE 17 Number of Cases Exceeding the MRL in Samples of Domestic Origin

Year	Dithiocarbamatesn	Endosulfan	Triazophos	Diquat	Other
		Active Ingredient			
1978	14			8	
1979	18			5	
1980	21			3	3 quinalphos
1981	26			3	
1982	20			6	4 metalaxyl
1983	8	6	7	5	
1984	11	4	4	4	
1985	10	2		6	3 fenarimol
1986	7		3		3 phos-phamidon

TABLE 18 Exposure of the Domestic Consumers Pesticide Residue (Samples include Imported and Domestic Products)

Year	Number of Investigations	R > T cases[a] in % of the Investigations	R < MD cases[a] in the % of the Investigations
1978	30,019	0.34	81.3
1979	45,806	0.23	78.8
1980	27,946	0.41	79.6
1981	18,998	0.46	81.3
1982	16,747	0.47	83.6
1983	19,508	0.20	86.3
1984	23,141	0.14	89.4
1985	24,423	0.15	90.4
1986	24,121	0.12	90.7

[a]See text for the definition of R > T and R < MD.

measures taken following the conclusion of analyses and the successful modification of plant protection technologies, the rate of R > T cases decreased from 0.46 to 0.12% while that of R < MD increased from 81 to 91% between 1981 and 1986 (Table 18).

6. SPECIAL CONTROL PROGRAMS

6.1. Control of Dithiocarbamate Residues

Table 17 show that in Hungary , as in other countries, the residues of active ingredient of dithiocarbamate-type pesticides can be regularly detected. The

TABLE 19   Results of the Special Program on
Dithiocarbamates Residues

Year	Number of Target Sampling	R > T[a] Cases in % of the Samples
1979	1083	1.6
1980	1468	1.4
1981	1692	1.5
1982	1175	1.7
1983	1246	0.6
1984	1320	0.8
1985	1381	0.7
1986	1385	0.5

[a]See text for definition of R > T.

number of R > T cases has particularly increased when, following international recommendations, MRL values were decreased. In the effort to reduce dithiocarbamate residue levels, widespread investigations have been initiated to evaluate the effects of dosages, time and frequency of application, spraying volume, and nozzles (Hargitai, 1983). As a result of the recommended new use pattern developed on the basis of these studies the number of R > T cases has decreased sharply (Table 19).

No significant difference could be found between active ingredients of different dithiocarbamate-type pesticides.

6.2.   Control of the Production in Private Greenhouses

Foodstuffs consumed fresh must always be controlled more carefully. This refers especially to the products grown in sheet enclosures or greenhouses as well as to the out of season vegetables grown in the field. Since the harvest is frequently continuous, the observance of preharvest monitoring is very essential. For the control of the pesticide residues private growers have been selected who are not specially trained in plant protection and are carrying out this activity less consciously than specialists working at large-scale farms. Sampling took place mainly as a field survey rather than as market control; consequently levels determined show a less favorable picture as compared to that of market control. In 1982, 8 foodproducts and 144 samples were checked while in 1986, 26 food products and 796 samples were checked (Table 20).

With the conclusion that the number of R > T cases is not expected to increase at first harvest, this particular monitoring activity has since been discontinued.

6.3.   Survey Relating to Environmental Protection

The monitoring of pesticide residue level in surface water and sediments has been going on since 1975 on a regular basis, although the number of samples and the

TABLE 20 Checking of Pesticide Residues of First Fruits Produced at Small Private farms

Crop	Number of Samples		Number of Investigations	
	1982	1986	1982	1986
Green pepper	41	160	119	416
Lettuce	12	124	35	325
Tomato	14	101	41	260
Cucumber	19	89	53	253
Radish	21	58	62	146
Chives		33		82
Savoy	19	29	49	78
Cabbage	8	23	24	62
Chinese cabbage	10	19	26	44
Potato		19		46
Cauliflower		13		34
Green peas		11		29
Total	144	796	418	2150
R > T cases (%)	0.6	0.42	0.2	0.14

TABLE 21 Monitoring of Pesticide Residues in Surface Water in Hungary

Year	Number of Sampling Spot	Number of Samples	Number of Positive Samples	Percentage of Positive Samples
1976	44	526	121	23.0
1977	68	566	163	28.8
1978	62	510	63	12.3
1979	59	240	21	8.7

TABLE 22 Level of Pesticides in Surface Water in Hungary

Active Ingredient	Maximum Residues Found (mg/L)	Toxic Level to Fish (LC_{50})
Atrazine	0.08	95
2,4-D	0.30	250
DDT plus metabolits	0.0021	0.3
Lindane	0.0011	0.2

TABLE 23 Contamination of Surface Water and Sediment with Pesticides

	1975		1980		1985	
	Water	Sediment	Water	Sediment	Water	Sediment
Number of sampling spots	68	45	60	52	45	17
Number of samples	467	78	325	102	395	42
Positive samples (%)	35	31	19	30	11	26
Average level of residues (ppm)						
Atrazine	0.004	0.009	0.006	0.002	0.0009	0.004
2,4-D	0.0003	—	0.0001	—	0.0002	—
DDT plus lindane	0.003	0.0008	0.002	0.0003	0.0002	0.0003

range of investigations have varied. Although aquaculture is an important branch of agriculture in Hungary this monitoring program is aimed primarily at gaining some understanding of the fate and distribution of pesticides in the environment. The number of studies made between 1976 and 1979 is given in Table 21.

The column headed "positive" refers to samples containing detectable quantities of pesticide residue. In spite of the banning of chlorinated hydrocarbons, DDT and its metabolites as well as HCH isomers have been frequently detected. Following springtime weed control, atrazine and 2, 4-D derivates can be detected at the beginning of summer months but they do not seen to represent any real risk to the water ecosystem. The detected maximum values of these herbicides are generally a thousand times lower than the LC_{50} values for fish (Table 22).

Values obtained in surface water and sediments between 1975 and 1985 are given in Table 23. It is interesting to note that in spite of regular use of atrazine and 2, 4-D, its level of environmental contamination is not rising and the data even suggest a significant decrease.

In the last years carbofuran and EPTC have been detected in water in some cases but at low concentration and without persistency.

REFERENCES

Ambrus, Á., Hargitainé, A. É., and Györfi, L. (1978). Experience of pesticide residue examinations between 1974–1977. *Növényvédelem* **XIV**(7), 289–299.

Center of Plant Protection and Agrochemistry. (1978–1986). Annual Report, Budapest, published for internal use.

Hargitai, F. (1979). Training and post graduate training of plant protection specialists in Hungary. In Abstracts of papers, IX. International Congress of Plant Protection, Washington D.C., 370 p.

Hargitai, F. (1983). Prospects for improving pesticide application in Hungary. *EPPO Bull.* **13**(3), 345–349.

Hargitai, F. (1984). Chemicals in plant protection. *New Hung. Exp.* **34**(4), 3–6.

Hargitai, F. (1986). Possibilities of a harmony between the chemical plant protection and environment-conscious farm management. *Növényvédelem* **XXII**(7), 289–291.

Hargitai, F., and Baranyai, F. (1987). Pest management under large-scale farming systems. In Abstracts of the 11th international Congress on Plant Protection, Manila, pp. 149–150.

Hargitainé A. É., and Ambrus, Á. (1989). The influence of the export requirements on the plant protection technologies. *Magyar Mezögazdaság*, in press.

19

USE OF ORGANOCHLORINE PESTICIDES FOR PEST CONTROL IN BUILDINGS: THE EFFECT ON OCCUPANTS

Conway Ivan Stacey

School of Applied Chemistry
Curtin University of Technology
Bentley, Western Australia

---

1. INTRODUCTION

After World War II the organochlorine pesticides such as DDT were being hailed as the champions of mankind in his continuing battle against insect pests that affected his agriculture and health. The advent of gas chromatographic techniques, in particular the development of the electron capture detector, enabled the analysis of samples containing ppm levels of these compounds and the world was soon alerted to their persistent nature in the ecosystem. Racheal

601

Carson in her book *Silent Spring* was one of the first to sound a note of warning concerning their possible adverse effects in nature.

Since then a numer of well-documented reports have been published showing the harmful effects caused by the introduction of the organochlorine pesticides into the food chains of a number of species of animals, birds, and fish. These reports highlighted the particular vulnerability of species at the top of food chains. Man being such a species, but having the ability to influence his environment, it was not surprising that governments in many countries introduced legislation banning or imposing controls on their use on food crops and pastures. The result has been a decline in the level of the organochlorine pesticide residues in man's food chain and in most cases in man himself.

2. MONITORING PESTICIDE LEVELS

Monitoring the level of organochlorine pesticides in the human food chain has proved relatively simple, being performed through market basket surveys based on the main diet of the surveyed country. By and large these surveys have been carried out by government instrumentalities of the country, as a means of checking the efficiency of their imposed restrictions. These surveys have shown that within a decade of legislation being introduced by a government, banning or restricting the use of organochlorine pesticides for use in agriculture related to food production, the levels in the food chain decreased markedly.

This does not imply that such surveys are good guides to general environmental contamination. A much better guide is a survey of the residue levels in humans and to this end a number of body tissues and fluids have been used as monitoring media. Thus, adipose tissue was used in the monitoring of farm workers in Costa Rica (Barquero and Constenla, 1986). Urine samples were used as an indication of occupational exposure in Yugoslavia (Vasilic et al., 1987) and in the monitoring of citrus workers in Florida (Duncan and Griffith, 1985). Blood appears to be a more useful medium for monitoring and has been used in Japan for testing the residue levels in pest control operators (Naguchi, 1985) and in a survey of the organochlorine contaminants in individuals in Nigeria (Atuma and Okor, 1986).

Over the years human milk has frequently been used, with success, as a medium for checking the effect of legislation on food chain contamination. In a Norwegian survey (Skaare, 1981) the observation was made that human milk is a good indicator substance in monitoring for organochlorine pesticide contamination. A similar observation was made in a Western Australian survey (Stacey et al., 1985); however it was noted that values for the same donor could vary widely from day to day, from feed to feed, and for samples taken at various stages of an individual feed. Thus if human milk is to be used to monitor the general levels of contamination by these compounds, the practice used in Sweden (Noren, 1983) would appear to have advantages. Here the analysis was performed on pooled samples from the Mothers Milk Centre in Stockholm. This would appear to nullify the often large variation in individual samples that necessitates a large number of samples being analyzed to overcome the problem.

The Western Australian survey also highlights the fact that while analysis of pooled milk provides a good guide to overall environmental contamination, it does not allow for the detection and identification of unsuspected occupational, regional, or specific problems.

3. NONFOOD CHAIN SURVEYS

Analysis of pesticide residue levels in human milk does not usually offer a viable alternative to adipose tissue, urine, or blood in the monitoring for occupational exposure. The much higher levels of contact in many occupations has, in recent years, resulted in increased monitoring of workers with programs including pest control operators in Japan (Naguchi, 1985), the United States (Reinant et al., 1986), and Sweden (Wiklund et al., 1986), agricultural workers in the United States (Duncan and Griffith, 1985; Duncan et al., 1986; Coye et al., 1986), Sweden (Wiklund et al., 1986), and Costa Rica (Barquero and Constenla, 1986), manufacturers and formulators in the United States (Levine et al., 1986) and India (Kashyap, 1986), and female green house workers in Russia (Kundiev et al., 1986).

The concern for occupational exposure to high levels of contamination for long periods is understandable and justified, however, very little appears to have been done concerning the effects on individuals who work or live in buildings that have been treated with pesticides or alternatively work or live near other sources of contamination.

There have been a few studies in which vapor and surface area concentrations within building have been determinated. Some residential buildings in the United States that had been treated for insect pest control with chlordane and heptachlor gave both air and surface area concentrations that exceeded the limits recommended by the United States National Academy of Science (Jurinski, 1984). It has also been reported (Lewis et al., 1986) that from monitoring for nonoccupational exposure to pesticides in indoor air the results were generally in agreement with stratification of householders by the degree of pesticide use.

A group that could fall into the category of individuals who undergo long periods of exposure to pesticides not directly related to their occupation are nursing mothers whose residences have been treated for pest control. Although it is not common in these times to use organochlorine pesticides inside the house, it is still common practice in some countries, including Australia, for the pest control operators to apply them around and under the building for the control of termites.

4. PESTICIDES IN HUMAN MILK

A source of concern over the years has been the presence of organochlorine pesticide residues in human milk and through this their injestion by infants. However, studies have shown that as the levels in the food chain decrease, so the levels of their residues or those of their metabolites in human milk also decrease (Skaare, 1981; Acker, 1981; Noren, 1983; Stacey et al., 1985).

Although a continuing decrease was being noted during the last 20 years, the popular press continued to report, often in an irresponsible manner, on new surveys that showed the presence of pesticides in human milk. These reports, often taken out of context, frequently engendered concern in nursing mothers, which on occasion led to reactions that produced adverse results.

An example of the adverse effect that can result was observed in some of the Third World countries where, due to grave concern caused by publicity and advertising which pointed to the fact that breast milk had a higher level of organochlorine pesticide contamination than did formula milk, mothers gave up the traditional breast-feeding of their infants in favor of formula feeding. Consequently the infants no longer received the benefits of breast-feeding, including the important aquisition of the mother's immunity to disease and sickness. Moreover, the high cost of formula milk accompanied by the low family income frequently resulted in the formula milk being made up below strength, in order to eke out the supply, with the ultimate result being malnutrition of the infant. Combined with this the lack of hygienic conditions for its preparation and the poor quality of the water supply often resulted in disease or sickness that could prove fatal to the malnurtured infant.

It was not only Third World countries that suffered from this type of reporting and advertising. During the late 1960s and early 1970s many articles in the Australian press highlighted the fact that levels of organochlorine residues in human milk were higher than those in cows' milk and suggestions were made that breast-fed babies were being poisoned. As a result, many mothers opted for bottle feeding rather than breast-feeding. A survey carried out in Western Australia (Stacey and Thomas, 1975) showed, in fact, that infants in Western Australia were receiving only slightly more (0.012 mg/kg day) than the current FAO/WHO acceptable daily intake for humans (0.010 mg/kg day).

In 1979 a press report in Australia suggested that, because of the residue levels present in their milk, mothers should not breast-feed their babies beyond 6 months. Based on the Western Australian survey this did not appear logical since, although the actual intake of pesticide rose with the intake of breast milk, when related to the body weight of the infant the results showed that by the 6 month the level was down to the FAO/WHO figure and after that it decreased even further (Table 1).

Soon after the age of 3 months the infant would be receiving the total milk produced by the mother, which would be insufficient for its daily requirement and would therefore need a further supplement. The DDT from this source has not been taken into account in the calculations.

A survey carried out in Sydney (Siyali, 1973) showed the levels of the organochlorines in breast milk to be in the range of the Western Australian and overseas values while one carried out in Queensland (Miller and Fox, 1973) showed the values to be somewhat higher and, in fact, greater than the FAO/WHO recommended values.

TABLE 1   Daily Intake of DDT in Western Australian Infants (1970–1971)

Age	Weight (kg)[a]	Daily Breast Milk Intake[b] (mL)	Daily DDT Intake[d] mg	mg/kg day
At birth	3.3	495	0.039	0.012
3 months	6.2	945	0.074	0.012
6 months	8.0	1000[c]	0.078	0.010
9 months	9.5	1000[c]	0.078	0.008

[a] Based on 50 percentile weight of Western Australian (W.A.) infants.
[b] Based on W.A. value for milk required (150 ml/kg day).
[c] W.A. mothers average daily milk production (1 L).
[d] Based on the survey average of 0.078 ppm.

4.1. Effect on Infants

Although the level of organochlorine residues in human milk and hence the daily intake by the infants has been well surveyed for over a decade, there appears to be no clear picture as to the effect of these low levels on the well-being of the child. In fact, there appear to be differing opinions in the literature.

One of the worst scenarios would appear to be given in a 1955–1961 survey in Turkey (Cripps et al., 1984) where many breast-fed children under the age of 1 year died from a disease known as pembe yara or "pink sore," which was attributed to their mother's ingestion of hexachlorobenzene, used as a fungicide on wheat seedlings.

In North Carolina a survey (Rogen and Gladen, 1985) that studied the levels of PCBs and DDE in milk and other fluids and noted growth, morbidity, and development in children concluded that certain kinds of morbidity that occur in breast-fed children might represent the result of chemical contamination of milk.

In West Germany a survey (Acker, 1981) found that although levels of PCB, HCB, and β-BHC had decreased over the previous decade, the levels were still so high that the amounts consumed by breast-fed infants exceeded the acceptable daily intake. But in the opinion of the experts the advantages connected with alimentation with human milk outweigh the eventual risk by the contaminent. A similar conclusion was reached regarding breast-feeding in Sweden despite levels of DDT complex in some samples of human milk exceeding the FAO/WHO proposed acceptable daily intake (Hofvander et al., 1981).

Levels of HCB and PCB were shown to be characteristic of highly industrialized countries (Niessen et al., 1984) and thus the levels in children of German mothers were in general higher than in children of Turkish mothers. The report identified two areas requiring urgent attention as the development of further ways of reducing environmental sources and studies of the possible pahtogenetic effect on the health of children.

Since polyhalogenated compounds will be present in the environment and the body for many decades and reach the infant at an early stage of development, through placental transfer and breast-feeding, it has been suggested (Weisenberg, 1986) that further studies on their toxicological effects should be undertaken.

The levels absorbed by breast-fed infants in France have been reported (Klein et al., 1986) to not exceed the acceptable daily doses and to be lower than those reported in both Canada and Italy but higher than in formulated milk. Similarly a Norwegian study (Skaare, 1981) looked at the levels of HCH, DDT, DDE, and PCB and evaluated the results toxicologically by comparison with the maximum residue limits and acceptable daily intakes and concluded that the levels, at that time, did not represent any threat to the infants health.

The general concensus appears to be that although the levels of some organochlorine pesticides may exceed the acceptable daily intake the advantages of breast-feeding, nutritional, immunological, pyschological, etc., outweigh the disadvantages, and it should therefore be encouraged. However, the Turkish reports should sound a note of warning since they show that specific forms of contamination, which may be of a regional nature, can cause levels of a particular pesticide to be far above the normal values and thus may pose a threat to the infants' health.

5. DIELDRIN LEVELS IN WESTERN AUSTRALIA

In 1979–1980 a survey (Stacey et al., 1985) was carried out in Western Australia to determine whether restrictions on the use of organochlorine pesticides imposed in the early 1970s, by the Australian Government, had been effective. A previous survey (Stacey and Thomas, 1975), in which samples from 1970–1971 had been analyzed, acted as a benchmark since the restrictions had not been introduced at that time.

Where the first survey had used only 23 donors from the urban areas of Perth, the second survey used 140 donors from both urban (45) and rural (95) districts and showed no significant difference between the two groups. As expected the levels of DDT and HCB showed a decrease, while there was a slight increase in BHC in the urban area. A significant increase in the level of dieldrin was unexpected since it was one of the organochlorine pesticides that had been banned for use on food crops and pastures. The rise was more unexpected since market basket surveys had indicated its decrease in the food chain. Table 2 compares the results of the two surveys.

TABLE 2   Comparison of Two Western Australian Surveys

Pesticide	1970–1971 Survey (ng/g)	1979–1980 Survey (ng/g)
Total DDT	78	46
HCB	25	8
BHC	Trace	1
Dieldrin	5	9

TABLE 3   Dieldrin Levels Related to Suburb and Treatment

| Suburb | Treatment | Number of Donors | Dieldrin Levels in Milk (ng/g) | |
			Range	Mean
New	Yes	6	7–24	16
Old	Yes	18	3–24	11
New	No	6	5–13	9
Old	No	13	2–8	5

As dieldrin is a metabolite of aldrin the uses of both were investigated in trying to determine the reason for the increase of dieldrin in the milk. As the values for both urban and rural sample were similar it was more convenient to concentrate on the urban donors. It had been noted that one of the permitted uses of both aldrin and dieldrin at the time was for the protection of buildings against termites. This gave a lead that, when followed up with a questionnaire to the donors, suggested that the use of aldrin (dieldrin although allowed was no longer in common use) in antitermite treatment was partly responsible for the high values in the milk of mothers who occupied houses treated in this way. A statistical analysis of the results further strengthened this view (Table 3).

It should be noted that chlordane, heptachlor, and metabolites were not included in either of these two surveys.

In an attempt to prove the above supposition a further study was undertaken in 1980–1981 (Stacey and Tatum, 1985) in which a biased sample of donors was used. Fourteen donors each supplied samples at regular intervals for a number of months in an attempt to determine whether the levels of dieldrin in their milk changed to any great extent in the period after treatment of their homes.

Two problems became apparent very early in the study. One was the lack of appropriate donors. Following the 1979–1980 survey reports in the local press

TABLE 4   Sumary of Pesticide Residues: 1980–1981 Survey

| Pesticide | Levels in Whole Milk (ng/g) | |
	Range	Mean
HCB	1–33	9
BHC	0–4	1
Chlordane	0–66	6
Heptachlor	0–13	1
Heptachlor epoxide	0–29	4
Dieldrin	2–35	13
Total DDT	3–159	42

were such that nursing mothers were becoming increasingly wary about having their homes treated during the period of lactation or even during their pregnancy. The second was that many of the pest control agencies were no longer using aldrin in their treatments, having changed to chlordane or heptachlor or a mixture of both. Because of this chlordane, heptachlor, and heptachlor epoxide were included in the survey. The results are sumarized in Table 4.

The only significant difference between these results and those of the 1979–1980 survey was in the mean of the dieldrin. In only 3 of the 14 cases studied was aldrin the most recently used pesticide and the results of these suggested that the levels continued to rise until the seventh or eighth month after treatment, at which time they level off and start to decline.

Even though the most recent treatment may not have been with aldrin, the treatment prior to this survey would, in most cases, undoubtedly have been so, since it had previously been the most commonly used pesticide for this type of treatment. The high values shown by donors whose homes had been treated annually (Table 5) showed there was a link between the use of aldrin for antitermite treatment of homes and the level of dieldrin residues in the milk of nursing mothers inhabiting them.

Two cases in this study were of particular interest since it had been possible to obtain samples of milk before and after the home was treated. The levels of the pesticide used or its metabolite showed a sharp increase in the milk of both donors soon after the treatment of their home, with the peak levels being reached

TABLE 5   Comparison of Dieldrin Levels with Treatment Records

Donor	Dieldrin Level (ng/g)	Treated Yearly	Comments
1	14	Yes	
2	14	Yes	
3	10	Yes	Most recent treatment with chlordane
4	9	No	Aldrin used 2.5 years before
5	26	Yes	
6	21	Yes	
7	10	Yes	
8	19	Yes	
9	19	Yes	No details available
10	13	Yes	
11	7	No	3 years since previous treatment
12	9	Yes	Nonorganochlorines used in last two treatments
13	8	No	
14	16	Yes	

between 3 and 5 weeks. In one case the pesticide used was chlordane, while heptachlor was used in the other. In both cases the levels returned to pretreatment values by about the fifteenth week. Thus, for 3–4 months after a home is treated with the organochlorine pesticides, the nursing mother absorbs the vapors and transmits them to the infant by way of the milk. It would be reasonable to assume that much of this uptake is due to inhalation of the vapors and, if this is the case, then it would be reasonable to assume that other residents in the house would also have an increased uptake without the nursing mothers' mechanism for reducing the level through lactation. Thus it would appear that the infant is at a greater risk than other inhabitants since it would be receiving levels from inhalation as well as from the mother's milk. No survey, testing the levels in the infant under these conditions, appears to have been carried out.

A recent survey was undertaken by the author in an attempt to detect any correlation that may exist between the levels of the organochlorine vapor in the air of the dwelling after treatment and the levels in the mother's milk. Unfortunately only one donor has so far been available who fitted the criteria for the survey. This required the donor to be lactating at the time the house was to be treated, thus enabling a pretreatment sample of both milk and household air to be obtained. After treatment air and milk samples where to be taken for at least 2 months. Although only one donor has been found the results (Table 6) are very promising and indicate a clear correlation between heptachlor in the air samples and heptachlor epoxide in the milk samples. Further donors are being sought. However, in Western Australia, few families are allowing their houses to be treated with the organochlorine pesticides while there is a pregnant or lactating female in residence.

While this evidence demonstrates an increase in the level of organochlorines in human milk for nursing mothers living in treated buildings it makes no attempt to determine the effect this exposure has on the residents, in particular the infants. The press has reported cases where the inhabitants have complained of headaches and nausea, particularly in the days immediately following the

TABLE 6   Relationship between Air and Milk Sample

Day after Treatment	Air ($\mu g/m^3$) heptachlor	Whole Milk (ng/g)	
		Heptachlor	Heptachlor Epoxide
−1	0.09	Tr[a]	4
+3	1.2	6	13
+7	1.3	3	30
+14	0.62	Tr	42
+21	0.38	Tr	36
+35	0.42	2	47
+49	0.35	4	27
+63	0.26	1	15

[a]Tr indicate value less than 1 ng/g.

treatment. The infant, unfortunately, is not in the position to complain. However, if, as suggested it is receiving a dose from both inhalation and from the mother's milk it would be reasonably to assume that it also is being affected.

6. STREAKY BAY SCHOOL EXPERIENCE

The previous surveys dealt only with residential buildings and the mother's milk offered a very good gauge of the contamination. Buildings other than residences are also treated with organochlorines for pest control and these are often occupied for several hours a day and could therefore be expected to produce some problems.

During the latter part of 1986 the Streaky Bay Area School in South Australia was treated with aldrin for the control of termites. The teachers and parents of the children expressed concern that aldrin had been misapplied and that the school had become contaminated. Concern was heightened when some of the students displayed symptoms similar to those attributed to aldrin poisoning.

In February 1987 the Public Health Service of the South Australian Health Commission commenced investigating the situation. By March the South Australian Government decided, because of aldrin contamination, that the school should not be reoccupied until the South Australian Health Commission advised that it was safe to do so. In June a report was published (Calder, 1987) that discussed the problem and considered that the decontamination procedures, which included replacement of carpets, had reduced the levels of chronic contamination to a point where the school could be reoccupied. Although the report suggested that the symptoms shown by the students were probably due to a viral infection it did recommend that further blood tests be offered to those individuals with blood dieldrin levels of above 4 ng/mL to ensure that the expected decline in blood levels following the decotamination is, in fact, taking place.

This report offers another example of how the nonoccupational exposure to organochlorine pesticides can influence the levels of their residues in the human body and serves to further emphasize the need for very strict controls and supervision on their use in situations where the public can become exposed.

7. CONCLUSION

Very strict regulations have been imposed by many governments of the world controlling the use of organochlorine pesticides in agriculture in an attempt to reduce the intake by humans through their food chain. Where these restrictions, have been applied and policed the levels in the food chain, as measured by market basket surveys, have been seen to decrease. Regardless of this instances of their entry into the human body via other routes still occur. One route has been shown to be through their use in the control of pests such as termites. While instances such as the Streaky Bay School attract public attention and receive publicity through the press and hence government action, their continuing use and often misuse by pest control opperators in household treatment still go on.

In Western Australia most bank and home financing institutions insist on pretreatment for new houses or a certificate of inspection or treatment for existing houses as a condition of granting a housing loan. This means that in newly developing housing locations large amounts of organochlorine pesticides are injected into the soil to form a barrier before the laying of the concrete rafts on which the house is built. The presence of high levels of residues in the breast milk of nursing mothers living in such areas suggests that not only are they receiving doses from their own homes, if recently treated, but also from other treatments in the area. This is not unrealistic in light of results from Sweden (Bidleman et al., 1987) that suggest that the levels of PCBs in the atmosphere there may, in part, have come from Eastern Eupope.

Thus, although the levels of the organochlorines in the food chain are being well monitored and as a result are decreasing, there is need for continual vigilance in order to identify other sources of contamination that may cause isolated problems. This cannot be left to governments or individual industries as these may have vested interests. Continued monitoring of humans by independent groups or individuals using all possible media, including milk, is therefore essential.

REFERENCES

Acker, L. (1981). The contamination of human milk with chlororganic pesticides. (Authors translation.). *Geburtshilfe Fraunheilkd.* **41**(12), 882–886.

Atuma, S. S., and Okor, D. I. (1986). A preliminary survey of organochlorine contaminants in Nigerian environment. *Int. J. Environ. Stud.* **26**(4), 321–327.

Barquero, M., and Constenla, M. A. (1986). Organochlorine pesticide residues in the adipose tissue of Costa Rican farmers. *Turrialba* **36**(2), 191–196.

Bidleman, T. F., Wideqvist, U., Jansson, B., and Soederund, R. (1987). Organochlorine pesticides and PCB's in the atmosphere of southern Sweden. *Atoms. Environ.* **21**(3), 641–654.

Calder, I. (Chairman). (1987). Review of the toxicity of aldrin: Report of the Ministerial Committee. Committee of Review of the Toxicity of Aldrin. Public Health Service, South Australian Health Commission. 119 p.

Coye, M. J., Lowe, J. A., and Maddy, K. J. (1986). Biological monitoring of agricultural workers exposed to pesticides. II. Monitoring of intact pesticides and their metabolites. *J. Occup. Med.* **28**(8), 628–636.

Cripps, D. J., Peters, H. A., Gocmen, A., and Dogramici, I. (1984). Porphyria turcica due to hexachlorobenzene: A 20 to 30 year follow-up study on 204 patients. *Br. J. Dermatol.* **111**(4), 413–422.

Duncan, R. C., and Griffith, J. (1985). Monitoring study of urinary metabolites and selected symtomatology among Florida citrus workers. *J. Toxicol. Environ. Health* **16**(3–4), 509–512.

Duncan, R. C., Griffith, J., and Konefal, J. (1986). Comparison of plasma chlorinesterase depression among workers occupationally exposed to organophosphorus pesticides as reported by various studies. *J. Toxicol. Environ. Health* **18**(1), 1–11.

Hofvander, Y., Hagman, U., Linder, C. E., Vaz, R., and Slorach, S. A. (1981). WHO collaborative breast feeding study. I. Organochlorine contaminants in individual samples of Swedish human milk, 1978–1979. *Acta Paediatr. Scand.* **70**(1), 3–8.

Jurinski, N. B. (1984). The evaluation of chlordane and heptachlor vapour concentrations within buildings treated for insect pest control. *Indoor Air, Proc. Int. Conf. Indoor Air Qual. Clim. 3rd.* 4(*PB85–104214*), 51–56.

Kashyap, S. K. (1986). Health surveillance and biological monitoring of pesticide formulators in India. *Toxicol. Lett.* 33(1–3), 107–114.

Kundiev, Y. I., Krasnyuk, E. P., and Viter, V. P. (1986). Specific features of the changes in the health status of workers exposed to pesticides in greenhouses. *Toxicol. Lett.* 33(1–3), 85–89.

Klein, D., Dillon, J. C., Jirou-Najou, N. I., Gagey, M. J., and Debry, G. (1986). Elimination kinetics of organochlorine compounds in the 1st week of breast feeding. *Food Chem. Toxicol.* 24(8), 869–873.

Levine, M. S., Fox, N. L., Thompson, B., Taylor, W., Darlington, A. C., Van der Hoeden, J., Emmett, E. A., and Rutten, W. (1986). Inhibition of esterase activity and an undercounting of circulating monocytes in a population of production workers. *J. Occup. Med.* 28(3), 207–211.

Lewis, R. G., Bond, A. E., Fitz-Simons, T. R., Johnson, D. E., and Hsu, J. P. (1986). Monitoring for non-occupational exposure to pesticides in indoor and personal respiratory air. *Proc.-APCA Annu. Meet.* 79, (vol. 2). 86/37.4, 15 pp.

Miller, G. J., and Fox, J. A. (1973). Chlorinated hydrocarbon pesticide residues in Queensland human milks. *Med. J. Aust.* 2(6), 261–264.

Naguchi, N. (1985). Epidemiological studies of chlordane for termite prevention. I. Chlordane residue of houses treated for termites and results of blood tests of pest control operators. *Okayama Iqakkai Zasshi.* 97(3/4), 315–326.

Niessen, K. H., Ramolla, J., Binder, M., Brugmann, G., and Hofmann, U. (1984). Chlorinated hydrocarbons in adipose tissue of infants and toddlers: Inventory and studies on their association with intake of mother's milk. *Eur. J. Petiatr.* 142(4), 238–244.

Noren, K. (1983). Organochlorine contaminants in Swedish human milk from the Stockholm region. *Acta Paediatr. Scand.* 72(2), 259–264.

Reinent, J. C., Nielsen, A. P., Lunchick, C., Hermandez, O., and Mazzetta, D. H. (1986). The United States Environmental Protection Agency's guidelines for applicator exposure monitoring. *Toxicol. Lett.* 33(1–3), 183–191.

Rogen, W. J., and Gladen, B. C. (1985). Study of human lactation for effects of environmental contaminants: The North Carolina Breast Milk and Formula Project and some other ideas. *Environ. Health Perspect.* 60, 215–221.

Siyali, D. S. (1973). Polychlorinated biphenyls, hexachlorobenzene and other organochlorine pesticides in human milk. *Med. J. Aust.* 2(17), 815–818.

Skaare, J. U. (1981). Persistent organochlorinated compounds in Norwegian human milk in 1979. *Acta Pharmacol. Toxicol. (Copenh.)* 49(5), 384–389.

Stacey, C. I., and Thomas, B. W. (1975). Organochlorine pesticide residues in human milk, Western Australia—1970–1971. *Pest. Monit. J.* 9(2), 64–66.

Stacey, C. I., Perriman, W. S., and Whitney, S. (1985). Organochlorine pesticide residue levels in human milk: Western Australia, 1979–1980. *Arch. Environ. Health* 40(2), 102–108.

Stacey, C. I., and Tatum, T. (1985). House treatment with organochlorine pesticides and their levels in human milk—Perth, Western Australia. *Bull. Environ. Contam. Toxicol.* 35(2), 202–208.

Vasilic, Z., Drevenkar, V., Frobe, Z., Stengl, B., and Tkalcevic, B. (1987). The metabolites of organophosphate pesticides in urine as an indicator of occupational exposure. *Toxicol. Environ. Chem.* 14(1–2), 111–127.

Weisenberg, E. (1986). Hexachlorobenzine in human milk: A polyhalogenated risk. *IARC Sci. Publ.* 77, 193–200.

Wiklund, K., Dich, J., and Holm, L. E. (1986). Testicular cancer among agricultural workers and licensed pesticide applicators in Sweden. *Scand. J. Work Environ. Health* 12(6), 630–631.

20

NATURALLY OCCURRING FUNGAL TOXINS

James J. Pestka and William L. Casale
Department of Food Science and Human Nutrition, Michigan State University, East Lansing, Michigan

1. INTRODUCTION

Certain fungi produce low-molecular-weight compounds that are toxic to other organisms. In general, these toxins are not essential constituents of primary metabolic pathways. These "secondary metabolites" are often strain specific and are not essential for survival of the producing fungus (Bu'Lock, 1975). They may, however, confer selective advantages under some conditions, such as enhancing the ability to compete with other microorganisms for substrate, or increasing the capacity of parasites to infect and colonize their hosts. The effects of toxic fungal metabolites have been observed for centuries, and in the last several decades hundreds of these compounds have been described and characterized. Antibiotics (pencillin, etc.) are examples of fungal products, toxic to other fungi and to prokaryotes, that are used clinically. Some plant pathogens produce low-molecular-weight compounds that are involved in plant disease development (Scheffer, 1983; Yoder, 1980) (such a toxin was a key factor in the severe Southern

Corn Leaf Blight epidemic of 1970). This review focuses specifically on those low-molecular-weight fungal metabolites that are toxic to humans or animals. These compounds have been termed "mycotoxins."

Historically, ergotism was the first mycotoxicosis known to occur in humans. It occurred following ingestion of flours of rye and other cereals that were infected with *Claviceps purpea* and *C. paspali*. Pistils of flowering grains can become contaminated by *Claviceps* ascospores resulting in invasion of the grain kernel and elaboration of pharmacologically active ergot alkaloids. The fungus is readily visible as dark sclerotial masses or 'ergot bodies." One form of ergotism, first documented in Europe during the early middle ages, is characterized by vasoconstriction, muscle pain, and hot/cold sensations in the extremities. In severe cases, necrosis, gangrene, and limb loss occur and hence the disease was given the name "St. Anthony's fire." It was said that a pilgrimage to St. Anthony's shrine often conferred a cure for this disease—an event probably brought about by the physical removal of the victims from their contaminated food supply. A convulsive form of ergotism involving the nervous system has also been described (van Rensburg and Altenirk, 1974). The Salem witchcraft trials of the late 1600s have been circumstantially associated with this form of ergotism (Matossian, 1982). By the seventeenth century the association between sclerotia-contaminated rye and ergotism was made and the disease has been essentially eliminated as a human health concern. Effective control measures include use of *Claviceps*-resistant cereal grains, physical removal of ergot bodies by screening, application of fungicides, and crop rotation. The history (Bove, 1970), chemistry

TABLE 1   Summary of Naturally Occurring Fungal Toxins

Group	Reported Toxic Manifestations	Occurrence
Aflatoxins	Hepatoxic, hepatocarcinogenic, teratogenic	Corn, peanuts, cottonseed, rice, treenuts, sorghum, dairy products.
Citrinin	Nephrotoxic	Barley, wheat, oats
Cyclopiazonic acid	Necrosis of GI tract, hepatotoxic	Corn, peanuts, cheese
Ergot alkaloids	Convulsions; vasconstriction, necrosis of extremities	Rye, cereal grains
Ochratoxins	Nephrotoxic, teratogenic, carcinogenic, immunotoxic	Barley, corn, oats, rye
Patulin	Stomach lesions	Apple products
Penicillic acid	Acute toxicity, carcinogenic	Corn, beans
Sterigmatocystin	Hepatotoxic, hepatocarcinogenic	Wheat, barley, rice, green coffee bean
Trichothecenes	Feed refusal, emesis, reduced weight grain, necrosis of GI tract, immunotoxicity, immunoglobulin A nephropathy	Corn, wheat
Zearalenone	Hyperestrogenism, infertility abortion, carcinogenic	Corn, wheat, moldy hay

(van Rensburg and Altenirk, 1974), occurrence (Busby and Wogan, 1979), and pharmacology (Berde and Schild, 1978) of ergotism have been reviewed extensively.

As summarized in Table 1, a number of mycotoxins have been identified and these elicit a wide range of toxic effects such as immediate vomiting (trichothecenes), impaired reproduction (zearalenone), liver cancer (aflatoxins), and immunosuppression (trichothecenes, zearalenone, aflatoxins). Figure 1 reveals the multiplicity of structures found for some of the major mycotoxin groups found in the United States and worldwide. The principal means of human and animal exposure to mycotoxins is through consumption of contaminated food. Commodities affected typically include peanuts, corn, wheat, barley, treenuts, cassava, sorghum, and various animal feeds and animal products such as milk and meats. The UN Food and Agriculture Organization has estimated that 25%

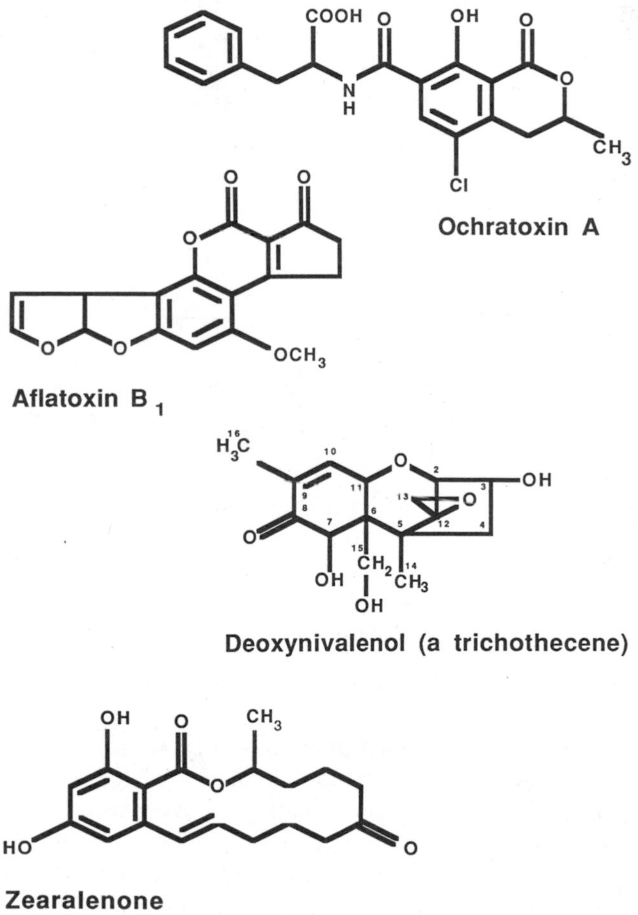

Ochratoxin A

Aflatoxin B₁

Deoxynivalenol (a trichothecene)

Zearalenone

Figure 1. Chemical structures of commonly occurring mycotoxins.

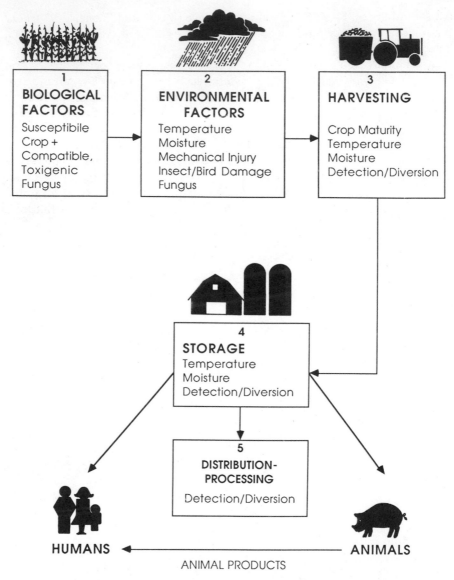

Figure 2. Factors affecting occurrence of mycotoxins in the food chain.

of the world's food crops are ruined by mycotoxins (Hesseltine, 1986). Figure 2 illustrates the genetic and environmental factors which contribute to the occurrence of mycotoxins in the human food chain. As exemplified in the United States (Figure 3), climatic patterns largely dictate the regions where specific mycotoxin groups occur. Thus, *Aspergillus* and *Penicillium* toxins (aflatoxins, ochratoxins) occur in the southern half, whereas the *Fusarium* toxins (trichothecenes, zearalenone) occur in the northern half of the country.

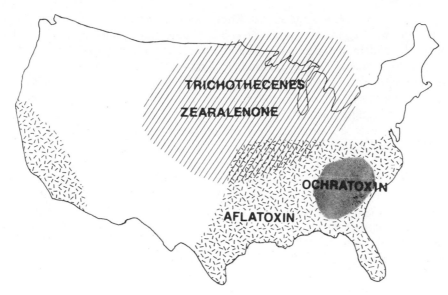

Figure 3. Regional distribution of the four most common mycotoxins during a typical year in the United States.

Although much is still to be discovered about the chemistry, biosynthesis, and molecular mechanisms of toxicity of mycotoxins, a considerable amount is known and is well represented in numerous reviews (Cole and Cox, 1981; Kadis et al., 1971; Mirocha and Pathre, 1973; Purchase, 1974; Rodricks and Hesseltine, 1977; Wyllie and Morehouse, 1978). In this chapter we provide a brief overview of the four major groups of mycotoxins focusing primarily on those historical, clinical, and toxicological aspects, that apply directly to management of human and animal health.

2. AFLATOXINS

Of the mycotoxins, the aflatoxins, a family of difuranocoumarins that is produced by toxigenic strains of *Aspergillus flavus* and *A. parasiticus*, have received the greatest notoriety because of their ability to act as hepatotoxins and hepatocarcinogens. These toxins were first identified during the early 1960s in association with an epidemic known as "Turkey X disease" in which over 100,000 turkey poults died after ingesting toxic peanut meal (Austwick, 1978; Goldblatt, 1969). This disease was characterized by anorexia, lethargy, and weakness of the wings. Histopathological evaluation indicated damage to liver parenchyma and proliferation of bile duct epithelium. Because investigators were unable to identify bacterial pathogens or viruses as a cause of the disease, toxicity was suspected. Most of the cases occurred within 100 miles of London and all were associated with feeds associated with a single mill. Brazilian peanut meal was identified as a common constituent of suspect feed. After toxic extracts of the peanut meal were subjected to successive chromatography, a blue fluorescent compound was purified and demonstrated experimentally to cause the symptoms of turkey "X" disease. Subsequently it was determined that *A. flavus* produced the compound and the toxin was named aflatoxin, after the fungus. Nearly concurrent with this was the discovery of trout hepatoma epidemics in hatcheries in the western United States (Sinnhuber et al., 1968). It was ascertained that trout reared on dry rations for 2 or more years developed massive liver tumors. Ultimately aflatoxin-contaminated cotton seed meal was identified as the etiological agent of trout hepatoma (Goldblatt, 1969).

Given the proper humidity and temperatures, *A. flavus* and *A. parasiticus* are capable of growing and elaborating aflatoxins on almost any organic substrate. Agricultural commodities that can be contaminated by the aflatoxins include corn, peanuts, cottonseed, cassava, copra, and many types of treenuts (Davis and Deiner 1978). Four naturally occurring aflatoxins, aflatoxin B_1 (AFB_1), aflatoxin B_2 (AFB_2), aflatoxin G_1 (AFG_1), and aflatoxin G_2 (AFG_2), have been identified. AFB_1 (Figure 1) is encountered most often in nature.

Aflatoxins can be acutely toxic, carcinogenic, mutagenic, and teratogenic (Busby and Wogan, 1979). The rank order for toxicity, carcinogenicity, and mutagenicity among the four naturally occurring aflatoxins is $AFB_1 > AFG_1 > AFB_1 > AFG_2$, indicating that the unsaturated terminal furan of AFB_1 is critical for determining the degree of biological activity of this mycotoxin family. Extensive studies on the metabolic fate of the aflatoxins and their toxic mode of action have been conducted. These compounds, like other nonpolar xenobiotics, are metabolized primarily in the liver by microsomal mixed-function oxidases and cytosolic enzymes. Most of the primary metabolites are further detoxified by conjugation with glucuronic acid or sulfate and eliminated via urine and feces. AFM_1, which results from hydroxylation of AFB_1, has about 3% of the mutagenic potency of AFB_1 in the *Salmonella* mutagenesis assay (Wong and Hsieh, 1976) and is also less toxic and carcinogenic. AFM_1 has been shown, however, to be excreted in the milk of animals that have ingested AFB_1. Stoloff (1980), in a summary of several studies, estimated that cows ingesting AFB_1 in their diet at a level of 300 ng/g will produce milk containing 1 ng/mL AFM_1 24 hr later. Detectable AFM_1 disappears 4–5 days after withdrawl from AFB_1-contaminated feed.

The carcinogenicity and mutagenicity of AFB_1 most likely result from the formation of a reactive epoxide at the 8, 9 position of the terminal furan and its subsequent covalent binding to nucleic acid (Essigmann et al., 1982). These nucleic acid adducts can last for weeks after initial formation. Covalent binding of the 8, 9-epoxide to nucleophilic moieties in proteins is also likely, but specific characterization of these adducts has not yet been carried out. An important means of detoxifying the active AFB_1 epoxide is via formation of a glutathione conjugate (Moss et al., 1983). Another major metabolite, 8, 9-dihydro-8, 9-dihydroxyaflatoxin B_1 (AFB_1-diol) appears after spontaneous or enzymatic reaction of the 8, 9-epoxide with water (Lin et al., 1977; Neal and Colley, 1979) and after the degradation of AFB_1-modified DNA (Wang and Cerutti, 1980). A functional role for AFB_1-diol in aflatoxicosis has been postulated on the basis that, under neutral and alkaline conditions, the metabolite becomes a dialdehydic phenolate ion that is capable of forming Schiff base adducts with amino groups of cellular proteins (Neal et al., 1981).

At chronic levels, aflatoxins cause marked decreases in productivity and the immune response of domestic animals (CAST, 1979). Reduced growth rate in livestock and poultry is the most common effect of aflatoxin in feed. AFB_1 is acutely hepatotoxic with LD_{50}s ranging from 0.3 to 40 mg/kg depending on the species. At high levels, toxicity occurs in 3–6 hr after ingestion and is evidenced by necrosis of the hepatocytes, derangement of the normal clotting mechanisms, and capillary fragility. This can lead to widespread hemorrhaging and eventual death (Carlton and Szczech, 1978; Cysewski et al., 1968). In 1974, acute aflatoxin toxicity occurred in India during which over 400 persons were poisoned with 106 dying following consumption of moldy, aflatoxin-contaminated corn (Krishnamachari et al., 1975; van Rensburg, 1977). The rural population that was affected was dependent on a single food crop (corn) they produced themselves. Warm moist conditions prevailed, undoubtedly leading to *Aspergillus* colonization and aflatoxin elaboration at levels ranging from 0.25 to 15.6 ppm. A similar incident occurred in Kenya with a mortality rate of 60% being observed (Ngindu et al., 1982).

Species susceptible to AFB_1-induced hepatocarcinogenesis include the rat, monkey, marmoset, ferret, trout, and salmon. The rat and rainbow trout are among the most sensitive of the species with ppb levels of AFB_1 in the diet being adequate to cause liver tumors in test animals (Hsieh, 1985). Epidemiological studies have demonstrated that a positive correlation exists between current exposure to aflatoxins in foods and the incidence of human hepatic cancer (PHC) in areas of Africa, the Phillipines, Southeast Asia, and China (Bullatao-Jayme et al., 1982; Peers and Linsell, 1973; Peers et al., 1976; Shank et al., 1972; Shank, 1979; Wang et al., 1983). This hypothesis is supported by the observation that the AFB_1 guanine adduct, a repair product of AFB_1 covalent binding to DNA is detectable in urine following ingestion of dietary AFB_1 (Autrip et al., 1983).

Epidemiological studies on the carcinogenic risk of AFB_1 to humans have been criticized for several reasons (Hsieh, 1986; Stoloff, 1986). First, current exposure was correlated to the current liver cancer rate even though the latency period for PHC may be as long as 30 years. Second, lifetime risk estimates did not take into account the age distribution of the populations studied. Third, estimates of the average AFB_1 intake were based on inaccurate assumptions relative to the average AFB_1 concentration in food, average food intake, and types of food ingested by a population.

Stoloff (1983) designed a U.S. study that eliminated these flaws and included aflatoxin exposure levels in the same range as the highest exposure levels observed in the African or Asian epidemiological studies. This regional population had the same lifetime risk of PHC as the related population in the rest of the United States that had been exposed to low levels of aflatoxin.

However, it was determined that individuals of Oriental descent residing in the United States were at significantly higher risk than the Caucasian population. Chinese males in the 30- to 49-year-old group were at 33 times greater risk. Whereas increased AFB_1 consumption was not a factor in this increased risk, Stoloff (1983) proposed that increased exposure to hepatitis viral infections in Asian as well as African populations was perhaps a critical factor in risk of PHC.

In response to the controversy over the above-described hypotheses, a multifactorial etiology theory that includes AFB_1 and hepatitis B virus, and that is based on recent investigations on the molecular genetic basis of hepatocarcinogenesis, is now gaining support (Hsieh, 1986; Harris and Sun, 1984). This theory suggests that AFB_1 is a significant initiator and promoter of human PHC whereas hepatitis B virus is an effective promoter enhancing PHC development untreated by AFB_1 and other environmental carcinogens. Thus the two agents synergistically contribute to human hepatocarcinogenesis.

If one then accepts the premises that AFB_1 can be hepatotoxic and contribute to human liver cancer, what dietary levels can be considered as safe? The issue is complicated by the fact that appearance of aflatoxins in a commodity in the United States is largely dictated by complex biological factors and weather conditions during a given year (Figure 2). Aflatoxins are relatively stable to processing and no FDA-approved methods exist for its detoxification in foods. Initial regulatory levels for aflatoxins in food during the 1960s and 1970s were based on limits of detection of available analytical methods and limited knowledge of their toxic and carcinogenic potential. In the United States the current action levels is 20 ppb for total aflatoxins in food and feed with comparable regulatory levels existing in most developed countries. Efforts to lower the allowable aflatoxin level to 15 ppb in the United States, now easily detectable with improved analytical methods, were unsuccessful because of the absence of evidence for improved health benefits relative to high costs of implementation due to increased surveillance and decreased product availability. An action level of 0.5 ppb exists in the United States for AFM_1 in milk because of the potential risks for growing children who consume large quantities of milk. This regulatory level may be questionable because AFM_1 exhibits a much lower level of toxic and carcinogenic potency than AFB_1. Stoloff (1986) has proposed that raising the allowable aflatoxin level in foods is feasible and economical based on available experimental and epidemiological data. However, the inevitable rejection of such a premise by an increasingly health-conscious public and the negative political ramifications of implementing an increased action level for a known carcinogen make such a possibility very unlikely.

3. TRICHOTHECENES

Strains of saprophytic and plant parasitic fungi in the genera *Fusarium*, *Myrothecium*, *Trichothecium*, *Stachybotrys*, *Cephalosporium*, and *Verticimonos-*

porium (Marasas et al., 1984; Ueno, 1983) elaborate a diverse group of mycotoxins that is known collectively as trichothecenes. The trichothecenes are esters of sesquiterpenoid alcohols containing the trichothecane tricyclic ring system that include some of the most potent protein synthesis inhibitors known (Cole and Cox, 1981; Godtfredsen et al., 1967). More than 45 different trichothecenes have been isolated and characterized (Ueno, 1983). All naturally occurring tricho-thecene mycotoxins possess a double bond at C9–C10 and an epoxide at C12–C13, and are therefore designated 12, 13-epoxytrichothecenes (Tamm, 1977) (Figure 1). Symptoms that, in the light of current knowledge, may be suspected as caused by trichothecene mycotoxicosis have been reported since the nineteenth century. These outbreaks were generally associated with consumption of contaminated grains. Presentation of the more significant epidemics is useful in demonstrating the complexity of contributory factors and clinical signs of trichothecene mycotoxicoses. This will be followed by a discussion of some specific biological effects of these compounds.

"Taumelgetreide" (staggering grains) toxicosis was described by the mycolo-gist, Woronin (1891), in the Ussuri district of eastern Siberia. Humans consuming suspect millet and barley suffered from headache, vertigo, chills, nausea, vomiting, and visual disturbances. Farm animals showed similar symptoms, in addition to feed refusal. Woronin found the grain contaminated with several fungi, mostly notably, *Fusarium roseum* and *Gibberella saubinetti*.

Alimentary toxic aleukia (ATA), or septic angina, is a condition described in the Soviet Union since the nineteenth century (Joffe, 1978). ATA was reported in Eastern Siberia in 1913 (Yefremov, 1944), and outbreaks occurred in a gradually widening area. A major epidemic of ATA occurred in the district of Orenberg during WWII (Forgacs and Carll, 1962; Joffe, 1971, 1978; Mayer, 1953). As a consequence of the war, food was scarce and many families were forced to collect grain (millet, wheat, barley) in the spring from fields that had been covered with snow. ATA occurred primarily in these families and in people who purchased the toxic overwintered grain or grain products. ATA is characterized by inflamm-ation of the mouth and gastrointestinal mucosa, resulting in vomiting, diarrhea, and abdominal pain, which may appear after a single meal of contaminated food (Joffe, 1978). These initial symptoms often disappear after several days and the patient feels well, but there is a progressive leukopenia, agranulocytosis, and eventually multiple hemorrhaging, nectrotic angina, sepsis, and exhaustion of the bone marrow. *Fusarium poae* and *F. sporotrichioides* were toxigenic fungi consistently isolated from overwintered grain associated with ATA. These original isolates, and other isolates of these species, have subsequently been shown to produce the trichothecenes (Joffe, 1974; Ueno et al., 1972b, 1973; Yagen and Joffe, 1976). Most of the toxin accumulated in the spring, and growth of toxigenic fungi and toxin production were enhanced by relatively high tempera-tures, dense snow cover that prevented the soil from freezing to its usual depth, and alternate thawing and freezing. Conditions for mycotoxin production were particularly favorable in the spring of 1944, the year that the severity of the ATA epidemic peaked. More than 10% of the population of the Orenberg district was affected in 1944, with greater than 60% mortality. The disease declined during the next several years, and was absent in 1948 and 1949.

Stachybotryotoxicosis affects principally domestic animals that consume contaminated hay and straw, or contact these toxic materials used for bedding. The mycotoxins responsible for this disease, the satratoxins and stachybotryotoxins (Eppley, 1977), are produced by the saprophyte *Stachybotrys atra* (*S. alternans*, *S. chartrum*). There was widespread fatal intoxication of horses, calves, swine, and poultry in the Ukraine and central Europe during the 1930s (Hintikka, 1977, 1978b). The disease continued to occur, though less severely, throughout the 1940s and 1950s, but since then, stachybotryotoxicosis has been less frequent. Ruminants are less sensitive to *Stachybotrys* toxins, apparently because of the alkalinity of their digestive system, satratoxins being most active at low pH conditions (Noskov and Ogryskov, 1967; Spesivtseva, 1964). Horses are especially sensitive (1 mg of pure toxin can be fatal; Forgacs, 1972). Typically, stachybotryotoxicosis resulting from ingestion of contaminated material starts with irritation of areas that first contact toxin-contaminated feed (Hintikka, 1978b). Continued consumption of toxic feed leads to progressive leukopenia and thrombocytopenia. Reduced blood coagulation and septicemia is common. There is a disruption of digestion and appetite decreases or is absent; the animal has difficulty swallowing and loses weight. When the mycotoxicosis has progressed this far, the result is usually death of the animal. Contact with contaminated bedding may cause some of the symptoms described above, or mainly dermal lesions, that are common in swine (Hintikka, 1978c). Human stachybotryotoxicosis occurs primarily in farm workers that handle contaminated hay or straw used for animal feed or bedding (Forgacs, 1972; Hintikka, 1978a; Sarkisov et al. 1971). Stachybotryotoxicosis has also been reported in humans that use hay or straw for bedding or as fuel to heat their homes. Symptoms in humans generally include dermatitis, catarrhal angina, rhinitis, pharyngitis, and conjunctivitis.

Trichothecene mycotoxins are also believed to cause dendrochiotoxicosis, a disease first described in the Soviet Union in 1937 (Ueno, 1983). Dendrochiotoxicosis was first observed in horses with dermal lesions around the mouth and nose, cyanosis, hemorrhaging, accelerated pulse, leukopenia, and sudden death (Bilai, 1970). *Dendrochium toxicum* (= *Myrothecium roridum*) was isolated from feed and cotton associated with these symptoms. Although mycotoxins from these samples were not identified, *M. roridum* is now know to produce macrocyclic trichothecenes: roridins and verrucarins (Tamm, 1977).

Horses in the Hokkaido province of Japan have been affected with a disease of unknown etiology that has developed in the last several decades. Symptoms include convulsion, cyclic movements, disturbed respiration, decreased heart rate, retarded reflexes, and in 10–15% of affected horses, death (Ueno, 1983). The disease, occurring primarily in the winter and early spring when horses are fed dried bean hulls, has been called "been hull poisoning." Bean hulls implicated in the disease have been found to be contaminated with *F. solani* (Ueno et al., 1972a), a producer of the trichothecenes, T-2 toxin and neosolaniol.

Corn and small grains may become infected by plant pathogenic *Fusarium* species in the field, leading to significant crop losses. If toxigenic fusaria are involved, mycotoxicoses can occur in humans and animals consuming contaminated grain. Wheat and barley in Japan have been severely damaged by "akakabi-byo" (red-mold disease) when cool, wet conditions prevailed during bloom, maturation, and harvesting of the crop (Ueno et al., 1971). Red-mold disease and associated toxicoses in humans and animals have been reported sporadically since 1890. Vomiting, food and feed refusal, diarrhea, hemorrhaging, and death resulted from ingestion of contaminated grain (Ueno et al., 1971). Severe damage to the wheat and barley crop occurred in Southern Japan after a long, rainy season in 1963. *F. nivale* was isolated from infected grain and in culture this strain produced the trichothecenes, nivalenol and fusarenone-X (Tatsuno et al., 1968). Red-mold disease was again prevalent in 1970 when the wheat harvest was only 15% and barley 60% that of a normal year (Yoshizawa, 1983). Samples of infected barley from this epidemic contained *F. roseum* (*F. graminearum*); chemical analysis detected nivalenol and deoxynivalenol at 5–7 mg/kg grain (Morooka et al., 1972).

Mycotoxicoses associated with *Fusarium*-infected grains are a significant problem to the animal production industries in the United States and Canada. Moldy-corn toxicosis occurs in swine and cattle primarily in the north–midwestern United States. Cool, wet conditions favor fungal colonization of grain and mycotoxin accumulation (Bamburg, 1983), especially when corn has not dried sufficiently prior to harvest. Such conditions accompanied an extremely severe outbreak in the U.S. corn belt in 1972, which was exacerbated by storage of corn at high moisture content due to energy shortages for operating corn dryers (Vesonder, 1983). Animals ingesting toxic, moldy corn may exhibit vomiting, feed refusal, diarrhea, reduced milk production in dairy cattle, and hemorrhaging in the liver, stomach, heart, lungs, bladder, kidney, and intestines (Bamburg, 1983). The hemorrhagic syndrome is characteristic of mold-corn toxicosis, and may lead to death due to massive blood loss from intestinal hemorrhaging into the abdominal cavity. Hsu et al. (1972) detected 2 ppm T-2 toxin in moldy corn implicated in hemorrhagic syndrome resulting in the death of 20% of a Wisconsin dairy herd. This was the first trichothecene identified in association with moldy-corn toxicosis. Vesonder et al. (1973) isolated a new trichothecene from corn associated with emesis and feed refusal, and named this compound "vomitoxin" (= deoxynivalenol) (Figure 1). Although moldy-corn toxicosis may involve a variety of trichothecene mycotoxins, deoxynivalenol occurs most frequently. This toxin has also been found in wheat (Hart and Braselton, 1983; Neish and Cohen, 1981; Scott et al., 1981), barley (Neish and Cohen, 1981; Scott et al., 1981; Yoshizawa and Morroka, 1973), and mixed feed (Mirocha et al., 1976). Several fungal species may be present, however, *F. graminearum* (asexual stage of *Gibberella zeae*) is the trichothecene-producing fungus most often isolated from samples associated with moldy-corn toxicosis, and in this context some strains of *F. graminearum* also elaborate the estrogenic, nontrichothecene mycotoxin, zearelenone (Hart et al., 1982; Neish and Cohen,

1981), which is discussed later. Since trichothecenes are recalcitrant to inactivation during milling and processing and practical methods of detoxification do not exist, humans are at risk from food produced from contaminated grain. In a recent survey, 60% of breakfast cereals tested contained deoxynivalenol (Trucksess et al., 1986). Currently, the U.S. Food and Drug Administration does not have an enforcable action level for deoxynivalenol or other trichothecenes.

The 12, 13-epoxytrichothecenes are among the most potent low-molecular-weight inhibitors of protein synthesis in eukaryotic cells (McLaughlin et al., 1977). Trichothecenes bind to a single site on 60 S eukaryotic ribosomes and block peptidyltransferase activity (Barbacid and Vazquez, 1974; Cannon et al., 1976a, b; Carrosco et al., 1973; Carter and Cannon, 1978; Wei et al., 1974). Ueno et al. (1968) first demonstrated the ability of a trichothecene, nivalenol, to inhibit protein synthesis in both intact rabbit reticulocytes and a cell-free system extracted from rabbit reticulocytes. There has since been considerable interest in trichothecenes as inhibitors of eukaryotic protein synthesis, and they are being considered as candidates for cancer chemotherapy both directly and as antibody conjugates (Oen et al. 1985; Ueno, 1983).

Of most concern to human health is the possibility that vomitoxin and other trichothecenes may modulate immune function because low and moderate levels are likely to escape detection and enter the human diet through grain-based processed foods. Immunotoxicity is a frequently observed effect in field cases of subacute mycotoxin exposure in livestock (Pier et al., 1980). Experimentally, repeated exposure of animals to trichothecenes results in markedly increased susceptibility to bacterial, fungal, and viral infection (Boonchuvit et al., 1975; Corrier and Ziprin, 1987; Friend et al., 1983a, b; Fromentin et al., 1981; Kanai and Kondo, 1984; Pestka et al., 1987; Tai and Pestka, 1988; Tryphonas et al., 1986). In one case study, trichothecenes were isolated from household air samples and molded ductwork of a house where occupants suffered from unexplained asthma and flu-like symptoms for 5 years (Croft et al., 1986); it was noted that replacement of the contaminated ductwork ultimately eliminated these problems although workers carrying out the work exhibited the same clinical signs as described previously for the occupants.

Other experimental effects of trichothecenes include decreased humoral response to T-dependent antigens, increased response to T-independent antigens, increased skin graft rejection times, and depressed acute phase response (Forsell et al., 1986; Mann et al., 1983; Masuko et al., 1977; Rosenstein et al., 1979; Tryphonas et al., 1986). Lymphocytes from animals treated with low levels of trichothecene exhibit enhanced mitogenic responses, whereas higher doses are inhibitory (Friend et al., 1983a; Lafarge-Frayssinet et al., 1979); similar effects are observable *in vitro* (Cooray and Lindahl-Kiessling, 1987; Forsell et al., 1985; Forsell and Pestka, 1985; Miller and Atkinson, 1986; Pestka and Forsell, 1988). T-2 toxin is more effective at inhibiting protein synthesis *in vivo* and *in vitro* in lymphoid tissue and cell culture, respectively, than in nonlymphoid types suggesting that trichothecenes may be selectively toxic to lymphocytes (Lafarge-Frayssinet et al., 1979). We have demonstrated that dietary exposure of mice to deoxynivalenol alters regulation of immunoglobulin production and results in elevated serum levels of IgA (Forsell et al., 1986). Recently we have determined that deoxynivalenol-exposed mice accumulate glomerular IgA in a manner similar to that found in human IgA nephropathy—the commonest form of glomerulonephritis worldwide (Pestka and Moorman, 1988). Gyongyossy-Issa and Khachatourians (1984, 1985) demonstrated specific accumulation of T-2 toxin by lymphocytes and estimated approximately 10^5 putative toxin receptors per lymphocyte, although it was not clear from these reports whether accumulation was actually the result of interaction with ribosomes. It should be noted that trichothecenes can also alter macrophage function (Miller and Atkinson, 1986; Sorenson et al., 1986) suggesting that these may also be a target for trichothecene immunotoxicity.

The ability of trichothecenes to suppress immune surveillance might be an unknown factor in human cancer. Appearance of vomitoxin in corn and corn beer has been correlated to areas of high esophageal cancer incidence in South Africa (Marasas et al., 1977, 1979; Thiel et al., 1982). Both leukemia and tumors have been demonstrated in the gastrointestinal tract and other organs of rats exposed to T-2 toxin (Schoental, 1977, 1979; Schoental et al., 1979). T-2 toxin is negative in the *Salmonella typhimurium* mutagenesis assay, but it induces single strand in DNA breakage of rat spleen and thymus tissue *in vitro* and *in vivo* (Larfarge-Frayssinet et al., 1981) and is a weak promoter in 7, 12-dimethyl-benz[*a*]anthracene initiated mouse skin cells (Lindenfelser et al., 1974).

4. ZEARALENONE

Another important mycotoxin elaborated by strains of *Fusarium* is the β-resorcyclic acid lactone, zearalenone. Although less toxic than some of the other mycotoxins, zearalenone is unusual in that its principal observed effects are estrogenic. Zearalenone has been detected in corn and small grains, and implicated in hyperestrogenism in swine, cattle, and poultry (Allen et al., 1981; Meronuck et al., 1970; Miller et al., 1973). Swine are the most sensitive of domestic animals, and livestock may be affected by feed containing zearalenone at 0.5–1.0 mg/kg (Kuiper-Goodman et al., 1987). Several species of *Fusarium* have been reported to produce zearalenone (Caldwell and Tuite, 1970; Mirocha et al., 1977), however, *F. graminearum* (= *F. roseum* "Graminearum"; asexual stage of *Gibberella zeae*) is most significant because it so frequently infects grains used for feed and food, and is capable of producing large quantities of the mycotoxin [Eugenio et al. [1970] report up to 1900 ppm based on dry weight of corn substrate).

The first report of estrogenic mycotoxicosis in the United States was from Buxton (1927), who associated tumefaction of the vulva with consumption of moldy corn by swine in Iowa. Stob et al. (1962) demonstrated estrogenic syndrome in swine fed corn experimentally inoculated with *G. zeae* and isolated a compound that had a uterotropic effect in ovariectomized mice. Christensen et al.

(1965) isolated from *Fusarium* cultures a compound that they called "F-2" having the same absorption spectrum as the compound that Stob et al. (1962) described, with uterotropic activity in rats. The structure of zearalenone (= F-2) was elucidated by Urry et al. (1966). At least 15 naturally produced derivatives of zearalenone have since been identified (Richardson et al., 1985); although there are varying degrees of toxicity among the derivatives (Hurd, 1977), the effects of interactions between different zearalenone derivatives are not known.

Symptoms of hyperestrogenism may develop in 4 days to 1 week after ingestion of zearalenone-contaminated feed, and disappear within 3–4 weeks after consumption of toxic feed ceases (Eriksen, 1968; Koen and Smith, 1945). Prepuberal swine show more intense hyperestrogenic symptoms than mature animals, and tumefaction of the vulva in these young gilts is most characteristic of zearalenone-induced hyperestrogenism (Kurtz et al., 1969). The clinical symptoms (described in greater detail by Kurtz and Mirocha, 1978) include hyperemia and edematous swelling of the vulva with slightly turbid vaginal discharge (Bristol and Djurikovic, 1971), mammary gland enlargement and hypertrophy of the nipples, and, in severe cases, vaginal and rectal prolapse (Kurtz and Mirocha, 1978). Zearalenone has also been considered as playing a role in infertility and abortions in animals (Miller et al., 1973; Voluntir et al., 1971). Although zearalenone intoxication is not usually fatal, vaginal and rectal prolapse may lead to secondary bacterial infections that can result in death of the animal (Buxton, 1927; Kurtz and Mirocha, 1978). It should be noted that zearalenone and trichothecenes can occur together in *Fusarium*-infected grain. Therefore, symptoms due to trichothecene toxicosis (described in the previous section) may accompany symptoms of zearalenone-induced hyperestrogenism after consumption of contaminated feed.

Evidence for the carcinogenicity of zearalenone is, at present, limited. Neoplastic effects were not associated with zearalenone intake in 2-year studies with F344/N (NTP, 1982) or Wistar (Bescci et al., 1982) rats (zearalenone provided in diet at up to 2–3 mg/kg body weight). There was, however, positive trends in the incidence of pituitary adenomas in male and female B6C3F$_1$ mice and hepatocellular adenomas in female mice (NTP, 1982) when zearalenone was administered in the diet at up to 100 ppm (equivalent to 15.8 mg/kg body weight for males, and 18.5 mg/kg body weight for females). Although the nonneoplastic effects of zearalenone are similar to other exogenous estrogens, increases in mammary gland tumors, uterine horn adenocarcinomas, cervical adenocarcinomas, osteosarcomas, and ovarian tumors related to intake of DES or estradiol-17β by various strains of mice (Highman et al., 1980; IARC, 1979) did not occur in B6C3F$_1$ mice exposed to zearalenone. Based on the positive results in mice, zearalenone has been placed in the category of "limited evidence" of carcinogenicity by the International Agency for Research in Cancer (1983). The possible role of zearalenone in carcinogenesis is not clear since most evidence indicates that it does not induce point mutations in somatic cells. Zearalenone was negative in the *Salmonella typhimurium* assay (Bartholomew and Ryan, 1980; Boutibonnes, 1979;

Ingerowski et al., 1981; Kuczuk et al., 1978; Wehner et al., 1978), point-mutation assay with *Saccharomyces cerevisiae* (Kuczuk et al., 1978), and *in vitro* point-mutation assay using mouse lymphoma cells (Truhaut et al., 1985). However, recombination tests with *Bacillus subtilis* (Ueno and Kubota, 1976) and *B. thurigensis* (Boutibonnes and Loquet, 1979) showed zearalenone to have a DNA-damaging effect. Sister chromatid exchange was slightly enhanced by zearalenone (3 μg/mL) in human peripheral blood lymphocyte cultures, and at 30 μg/mL, zearalenone completely inhibited DNA synthesis (Cooray, 1984).

Several reviews on the occurrence of zearalenone in agricultural commodities are available (Bennett and Shotwell, 1979; Kuiper-Goodman ct al., 1987, Scott, 1978; Senti, 1979; Shotwell, 1977). Zearalenone is quite stable in stored grain and its concentration may increase in moist, *Fusarium*-contaminated grain (Kuiper-Goodman et al., 1987). Milling of corn does not destroy zearalenone, although it may be concentrated in certain fractions (Bennett and Anderson, 1978; Bennett et al., 1976, 1978). Zearalenone can survive processing of wheat flour (including baking) (Matsuura et al., 1981) and fermentation (Bennett et al., 1981; Lovelace and Nyathi, 1977; Martin and Keen, 1978). Several treatment have been used successfully to detoxify zearalenone-contaminated grain, including alkali (Bennett et al., 1980; Lasztity et al., 1979), chlorine (Sarudi et al., 1979), preservatives (Kallela and Saastamoinen, 1982), food additives (Matsuura et al., 1979, 1981), and physical methods such as density segregation (Huff and Hagler, 1985). Not only is zearalenone found in animal feed, but it has been detected in products intended for human consumption, for example, corn meal in the United States (Ware and Thorpe, 1978; Warner and Pestka, 1986), breakfast cereals in the United Kingdom (Norton et al., 1982), and beer and fermented products in Africa (Lovelace and Nyathi, 1977; Martin and Keen, 1978). There may also be considerable carryover of zearalenone into milk from dairy cows consuming contaminated feed (Mirocha et al. 1981), and this may present a very significant exposure risk, particularly to children. Zearanol (zeranol), a zearalenone derivative, is used as a growth-promoting agent for beef cattle. Although no residual zearalanol (detection limit of assay = 20 ng/g) is permitted in uncooked edible tissues of cattle and sheep in the United States (Code of Federal Regulations, 1985), recent analyses with increased sensitivity indicate that actual residues of zearalanol may be 0.2–1.0 ng/g animal tissue (Dixon and Russell, 1986). Kuiper-Goodman et al. (1987) have calculated the daily intake of zearalenone and metabolites for the highest relative consumption group, 1- to 4-year-old males and females, to be 50–100 μg/kg body weight day from corn products and 50–70 μg/kg body weight day from milk (mean—90th percentile of consumers, respectively). In a study to assess the health risks to humans, ovariectomized mature female cynomolgous monkeys were used, and a no hormonal effect level of less than 50 μg/kg body weight day was determined for α-zearalanol, a derivative more actively estrogenic than zearalenone (Griffin et al., 1984). It is important to note that zearalenone may contribute only a fraction of the overall intake of exogenous estrogens.

5. OCHRATOXINS

The ochratoxins are a group of seven related isocoumarin derivatives linked to phenylalanine (Figure 1) that are produced by species of *Aspergillus* and *Penicillium* (Busby and Wogan, 1979). Ochratoxin A (OA), the most commonly encountered of the group, is highly nephrotoxic to monogastric animals and has also been shown to be immunotoxic and teratogenic. OA is frequently detected in Scandanavian and Balkan countries and occasionally in the United States in commodities such as barley, corn, wheat, oats, rye, peanuts, hay, and green coffee beans and as residues in pork (CAST, 1979). The toxin has been related to endemic kidney disease in swine and poultry in Denmark and Sweden. Pigs fed diets containing OA at 200 ng/g developed pale swollen kidneys characterized by atrophy of the proximal tubules and interstitial cortical fibrosis (Krogh, 1977). These symptoms are similar to those occurring in humans during the course of endemic Balkan kidney disease in Yugoslavia, Bulgaria, and Romania. The incidence of this human disease can be correlated regionally to OA content of foods in the Balkan countries. OA has been found in 6.5% of blood samples collected from inhabitants living in an edemic area in Yugoslavia (Hult et al., 1982). In rodents, ingestion of OA results in DNA single-strand breakage in kidney and liver (kane et al., 1986) and renal and hepatic tumors (Kanisawa et al., 1978; Kanisawa, 1983; Bendele et al., 1983). Austwick (1981) reported a high incidence of urinary tract tumours in patients suffering endemic nephropathy. Because of the possibility of potential effects in humans, OA occurrence in food is subject to regulation in some European countries; similar regulations for OA do not currently exist in the United States.

6. CONCLUSION

Mycotoxin contamination of human food and animal feed will remain a problem of global dimensions throughout the observable future. From a present day perspective, the aflatoxin, trichothecene, zearalenone, and ochratoxin groups pose the greatest challenge. Critical research needs in mycotoxicology exist in the areas of toxicology, control, economic, and risk management (Hesseltine et al. 1985).

In relation to toxicology, although much is known about the acute toxic and *in vitro* effects of mycotoxins, relatively little information is available concerning the long-term chronic effects of ingesting these compounds. Perhaps the most crucial question is whether these compounds subtly alter immune function. The potential for synergistic effects among known mycotoxins as well as the existence of undescribed mycotoxins is also of considerable interest.

What means are available for controlling entry of mycotoxins into the food chain? As shown in Figure 2, a multitude of factors contribute to mycotoxin occurrence. Besides moisture control during storage, the most effective means of preventing human and animal exposure to mycotoxins is via detection and diversion. In general, conventional chromatographic detection of mycotoxins is time consuming, tedious, and costly, and thus precludes routine screening. As an alternative, our laboratory and others have developed monoclonal antibody-based immunoassays for the aflatoxin, trichothecene, and zearalenone groups that facilitate the rapid, specific, and sensitive assay of these compounds in food (Pestka, 1988). Another possible means of mycotoxin control is via detoxification. Ammoniation of aflatoxin-contaminated feed is one example of such an approach (CAST, 1979). Over the long term, a better understanding of biochemistry and genetics of mycotoxigenesis, fungal ecology, and plant–fungus interactions is needed. This information is critical in developing appropriate biotechnological strategies for intervention such as using resistant crops or blocking toxin elaboration by the fungus.

A better understanding of the economic effects of mycotoxins is needed at the global level. For example, how great are the effects of reduced feed efficiency, reproductive interference, and immunosuppression on the animal production industry? Information is needed on the dollar losses due to crops deemed unusable because of mycotoxin contamination and how these commodities can be alternatively processed and utilized.

Finally, toxicological control and economic information must be integrated into an effective worldwide risk management scheme. Central to such a scheme would be establishment of "acceptable" and "unacceptable" levels of mycotoxin contamination such as already exist for the aflatoxins. Such levels would be based on reliable toxicological data and weighed against current control technologies, economic considerations, and nutritional/health benefits of the affected commodities. Such a risk management scheme must by its very nature be dynamic and responsive to new information and technologies.

REFERENCES

Allen, N. K., Mirocha, C. J., Aakus-Allen, S., Bitgood, J. J., Weaver, G., and Bates, F. (1981). Effect of dietary zearalenone on reproduction of chickens, *Poult. Sci.* **60**, 1165–1174.

Austwick, P. C. (1981). Balkan nephropathy. *The Practitioner* **225**, 1031–1038.

Austwick, P. K. C. (1978). Aflatoxicosis in poultry. In T. D. Wyllie and L. G. Morehouse, eds., *Mycotoxic Fungi, Mycotoxins, Mycotoxicoses, An Encyclopedic Handbook*. Dekker, New York, Vol. 2, pp. 279–301.

Autrup, H., Bradley, A., Shamsuddin, A. K. M., Wakhisi, J., and Wasunna, A. (1983). Detection of putative adduct with fluorescence characteristics identical to 2,3-dihydro-2-(7'-guanyl)-3-hydroxy aflatoxin B_1 in human urine collected in Murang'e district, Kenya. *Carcinogen* **4**, 1193–1195.

Bamburg, J. R. (1983). Biological and biochemical actions of trichothecene mycotoxins. In F. E. Hahn, ed., *Progress in Molecular and Subcellular Biology*, Springer-Verlag, Berlin, pp. 41–110.

Barbacid, M., and Vazquez, D. (1974). Binding of [acetyl-^{14}C]trichodermin to the peptidyl transferase centre of eukaryotic ribosomes. *Eur. J. Biochem.* **44**, 437–444.

Bartholomew, R. M., and Ryan, D. S. (1980). Lack of mutagenicity of some phytoestrogens in the *Salmonella*/mammalian microsome assay. *Mutat. Res.* **78**, 317–321.

Becci, P. J., Voss, K. A., Hess, F. G., Gallo, M. A., Parent, R. A., and Stevens, K. R. (1982). Long-term carcinogenicity and toxicity study of zearalenone in the rat. *J. Appl. Toxicol.* **2**, 247–254.

Bendele, S. A., Carlton, W. W., Krogh, P., and Lillehoj, E. B. (1983). Ochratoxin A carcinogenesis in the mouse. *Abstr. 3rd Mycol. Cong., Tokyo*, p. 21.

Bennett, G. A., and Anderson, R. A. (1978). Distribution of aflatoxin and/or zearalenone in wet-milled corn products: A review. *J. Agric. Food Chem.* **26**, 1055–1060.

Bennett, G. A., and Shotwell, O. L. (1979). Zearalenone in cereal grains. *J. Am. Oil Chem. Soc.* **56**, 812–819.

Bennett, G. A., Peplinski, A. J., Brekke, O. L., and Jackson, L. K. (1976). Zearalenone: Distribution in dry-milled fractions of contaminated corn. *Cereal Chem.* **53**, 299–307.

Bennett, G. A., Vandegraft, E. E., Shotwell, O. L., Watson, S. A., and Bocan, B. J. (1978). Zearalenone: Distribution in wet-milling fractions from contaminated corn. *Cereal Chem.* **55**, 455–461.

Bennett, G. A., Shotwell, O. L., and Hesseltine, C. W. (1980). Destruction of zearalenone in contaminated corn. *J. Am. Oil Chem. Soc.* **57**, 245–247.

Bennett, G. A., Lagoda, A. A., Shotwell, O. A., and Hesseltine, C. W. (1981). Utilization of zearalenone-contaminated corn for ethanol production. *J. Am. Oil Chem. Soc.* **58**, 974–976.

Berde, B., and Schild, H. O. (eds.) (1978). *Ergot Alkaloids and Related Compounds*. Springer-Verlag, New York.

Bilai, V. I., and Pidoplisko, N. M. (1970). *Toxigenic Microscopical Fungi*. Naukova Dumka, Kiev.

Boonchuvi, B., Hamilton, P. B., and Burmeister, R. (1975). Interaction of T-2 toxin with *Salmonella* infections in chickens. *Poult. Sci.* **54**, 1693–1696.

Bove, F. J. (1970). *The Story of Ergot*. S. Karger, New York.

Boutibonnes, P. (1979). Antibacterial activity of zearalenone. *Can. J. Microbiol.* **25**, 421–423.

Boutibonnes, P., and Loquet, C. (1979). Antibacterial activity, DNA-attacking ability and mutagenic ability of the mycotoxin zearalenone. *IRCS Med. Sci.* **7**, 204.

Bristol, F. M., and Djurikovic, S. (1971). Hyperestogenism in female swine as the result of feeding mouldy corn. *Can. Vet. J.* **12**, 132.

Bulatao-Jayme, J., Almero, E. M., Castro, M. C., Jardeleza, M. T., and Salamat, L. A. (1982). A case-control dietary study of primary liver cancer risk from aflatoxin exposure. *Int. J. Epidemiol.* **11**, 1–5.

Bu'Lock, J. D. (1975). Secondary metabolism in fungi and its relationship to growth and development. In J. E. Smith and D. R. Berry, eds., *The Filamentous Fungi*. Edwards Arnold, London, Vol. 1, pp. 33–58.

Busby, W. F. Jr., and Wogan, G. F. (1979). Mycotoxins and mycotoxicoses. in H. Riemann and F. L. Bryan, eds., *Food-Borne Infections and Intoxications*. Academic Press, New York, 2nd ed., pp. 519–610

Buxton, E. A. (1927). Mycotic vaginitis in gilts. *Vet. Med.* **22**, 736–742.

Caldwell, R. W., and Tuite, J. (1970). Zearalenone production in field corn in Indiana. *Phytopathology* **60**, 1696–1697.

Cannon, M., Jimenez, A., and Vazquez, D. (1976a). Competition between trichodermin and several other sesquiterpene antibiotics for binding to their receptor site(s) on eukaryotic ribosomes. *Biochem. J.* **160**, 137–145.

Cannon, M., Smith, K. E., and Carter, C. J. (1976b). Prevention, by ribosome-bound nasceut polyphenylalanine chains, of the functional interaction of T-2 toxin with its receptor site. *Biochem. J.* **156**, 289–294.

Carlton, W. W., and Szczech, G. M. (1978). Mycotoxicoses of laboratory animals. In T. D. Syllie and L. G. Morehouse, eds., *Mycotoxic Fungi, Mycotoxins, Mycotoxicoses, An Encyclopedic Handbook*. Dekker, New York, Vol. 2, pp. 333–338.

Carrasco, L., Barbacid, M., and Vazquez, (1973). The trichodermin group of antibiotics, inhibitors of peptide bond formation by eukaryotic ribosomes. *Biochim. Biophys. Acta* **312**, 368–376.

Carter, C. J., and Cannon, M. (1978). Inhibition of eukaryotic ribosomal function by the sesquiterpenoid antibiotic fusarenon-X. *Eur. J. Biochem.* **84**, 103–111.

CAST. (1979). Aflatoxin and other mycotoxins: An agricultural perspective. Report No. 80. Council for Agricultural Science and Technology, Ames. Iowa.

Christensen, C. M., Nelson, G. H., and Mirocha, C. J. (1965). Effect on the white rat uterus of a toxic substance isolated from *Fusarium. Appl. Microbiol.* **13**, 653–659.

Code of Federal Regulations, 21 CFR. (1985). Chapter 1, Zeranol. 556.760, pp. 477–481. U.S. Govt. Printing Office, Washington, D.C.

Cole, R. J., and Cox, R. H. (1981). *Handbook of Toxic Fungal Metabolites.* Academic Press, New York.

Cooray, R. (1984). Effects of some mycotoxins on mitogen-induced blastogenesis and SCE frequency in human lymphocytes. *Food Chem. Toxicol.* **22**, 529–534.

Cooray, R., and Lindahl-Kiessling, K. (1987). Effect of T2 toxin on the spontaneous antibody-secreting cells and other non-lymphoid cells in the murine spleen. *Food Chem. Toxicol.* **25**, 25–29.

Corrier, D. E., and Ziprin, R. L. (1987). Immunotoxic effects of T-2 mycotoxin on cell-mediated resistance to *Listeria monocytogenes* infection. *Vet. Immunol. Immunopathol.* **4**, 11–21.

Croft, W. A., Jarvis, B. B., and Yatawara, C. S. (1986). Airborne outbreak of trichothecene toxicosis. *Atmos. Environ.* **20**, 519–522.

Cysewski, S. J., Pier, S. J., Engstrom, G. W., Richard, J. L., Dougherty, R. W., and Thurston, J. R. (1968). Clinical features of acute aflatoxicosis in swine. *Am. J. Vet. Res.* **29**, 1577–1582.

Davis, N. D., and Diener, U. L. (1978). Mycotoxin, In L. R. Beuchat, ed., *Food and Beverage Mycology.* AVI, Westport, Ct, pp. 397–444.

Dixon, S. N., and Russell, K. L. (1986). Radioimmunoassay of the anabolic agent zeranol. IV. The determination of zeranol concentrations in the edible tissues of cattle implanted will zernol (Ralgro). *J. Vet. Pharmacol. Ther.* **9**, 94–100.

Eppley, R. M. (1977). Chemistry of stachybotryotoxicosis. In J. V. Rodricks, C. W. Hesseltine, and M. A. Mehlman, eds., *Mycotoxins in Human and Animal Health.* Pathotox Publishers, Park Forest South, IL, pp. 285–293.

Eriksen, E. (1968). Oestrogene factorer i muggest korn. Vulvoreginitis hos svin. (Estrogen factors in moldy grain. Vulvovaginitis in hogs.) *Nord. Vet. Med.* **20**, 396–401.

Essigmann, J. M., Croy, R. G., Bennett, R. A., and Wogan, G. N. (1982). Metabolic activation of aflatoxin B_1: Patterns of DNA adduct formation, removal, and excretion in relation to carcinogenesis. *Drug Metab. Rev.* **13**, 581–602.

Eugenio, C. P., Christensen, C. M., and Mirocha, C. J. (1970). Factors affecting production of the mycotoxin F-2 by Fusarium roseum. *Phytopathology* **60**, 1055–1057.

Forgacs, J. (1972). Stachybotryotoxicosis. In S. Kadis, A. Ciegler, and S. J. Ajl, eds., *Microbial Toxins.* Academic Press, New York, Vol. 8, pp. 95–128.

Forgacs, J., and Carll, W. T. (1962). Mycotoxicoses. *Adv. Vet. Sci.* **7**, 273–282.

Forsell, J. H., and J. J. Pestka. (1985). Relation of 8-ketotrichothecene and zearalenone analog structure to inhibition of mitogen-induced human lymphocyte blastogenesis. *Appl. Env. Microbiol.* **50**, 1304–1307.

Forsell, J. H., Witt, M. F., Tai, J.-H., Jensen, R., and Pestka, J. J. (1986). Effects of chronic exposure to dietary deoxynivalenol (vomitoxin) and zearalenone on the growing B6C3F1 mouse. *Food Chem. Toxicol.* **24**, 213–219.

Friend, S. C. E., Babiuk, L. A., and Schiefer, H. B. (1983a). The effects of dietary T-2 toxin on the immunological function and herpes simplex reactivation in Swiss mice. *Toxicol. Appl. Pharmacol.* **69**, 234–244.

Friend, S. C. E., Schaefer, H. B., and Babiuk, (1983b). The effect of dietary T-2 toxin on acute herpes simplex virus type 1 infection in mice. *Vet. Pathol.* **20**, 737–760.

Fromentin, H., Salazar-Mejicanos, S., and Mariat, F. (1981). Experimental cryptococcosis in mice treated with diacetoxyscirpenol, a mycotoxin of fusarium. *Sabourandia* **19**, 311–313.

Goldblatt, L. A. (1969). Introduction. In L. A. Goldblatt, ed., *Aflatoxin*, Academic Press. New York, pp. 1–11.

Godtfredsen, W. O., Grove, J. F., and Tamm, C. (1967). Zur Nomenklatur einer neuren Klasse von Sesquiterpenen. [In German, English summary]. *Helv. Chim. Acta* **50**, 1666–1668.

Griffin, T. B. Singh, A. R., and Coulson, F. (1984). *No Hormonal Effect in Non-human Primates of Oral Zeranol.* Report to the International minerals and Chemical Company, Terre Haute, Indiana.

Gyongyossy-Issa, M. I. C., and Khachatourians, G. G. (1984). Interaction of T-2 toxin with murine lymphocytes. *Biochim. Biophys. Acta* **803**, 197–202.

Gyongyossy-Issa, M. I. C., and Khachatourians, G. G. (1985). Interaction of T-2 toxin and murine lymphocytes and the demonstration of a threshold effect on macromolecular synthesis. *Biochim. Biophys. Acta* **844**, 167–173.

Harris, C. C., and Sun, T. (1984). Multifactoral etiology of human liver cancer. *Carcinogenesis* **5**, 697–701.

Hart, L. P., and Braselton, Jr., W. E. (1983). Distribution of vomitoxin in dry milled fractions of wheat infected with *Gibberella zeae*. *J. Agric. Food Chem.* **31**, 657–659.

Hart, L. P., Braselton, W. E., Jr., and Stebbins, T. C. (1982). Production of zearalenone and deoxynivalenol incommercial sweet corn. *Plant Dis.* **66**, 1133–1135.

Hesseltine, C. W. (1986). Global significance of mycotoxins. In P. S. Steyn and R. Vleggar, eds., *Mycotoxins and Phycotoxins*. Elsevier, Amsterdam, pp. 1–18.

Highman, B., Greenman, D. L., Norwell, M. J., Former, J., and Shellenberger, T. E. (1980). Neoplastic and preneoplastic lesions induced in female C_3H mice by diets containing diethylstilbestrol or 17-beta-estradiol. *Environ. Pathol. Toxicol.* **4**, 81–95.

Hintikka, E. L. (1977). Stachybotryotoxicosis is a veterinary problem. In J. V. Rodricks, C. W. Hesseltine, and M. A. Mehlman, eds., *Mycotoxins in Human and Animal Health*. Pathotox Publishers, Park Forest South, IL, pp. 277–284.

Hintikka, E.-L. (1978a). Human stachybotryotoxicosis. In T. D. Wyllie and L. G. Morehouse, eds., *Mytotoxic Fungi, Mycotoxins, Mycotoxicoses*. Dekker, New York, Vol. 3, pp. 87–89.

Hintikka, E.-L. (1978b). Stachybotryotoxicosis in horses. In T. D. Wyllie and L. G. Morehouse, eds., *Mycotoxic Fungi, Mycotoxins, Mycotoxicoses*. Dekker, New York, Vol. 2, pp. 182–185.

Hintikka, E. L. (1978c). Stachybotryotoxicosis in swine. In T. D. Wyllie and L. G. Morehouse, eds., *Mycotoxic Fungi, Mycotoxins, Mycotoxicoses*. Dekker, New York, Vol. 2, pp. 268–273.

Hsieh, D. P. H. (1986). The role of aflatoxin in human cancer. In P. S. Steyn and R. Vleggaar, eds., *Mycotoxins and Phycotoxins*. Elsevier, Amsterdam, pp. 447–456.

Hsu, I.-C., Smalley, E. B., Strong, F. M., and Ribelin, W. E. (1972). Identification of T-2 toxin in moldy corn associated with lethal toxicoses in dairy cattle. *Appl. Microbiol.* **24**, 684–690.

Huff, W. E., and Hagler, W. M., Jr. (1985). Density segregation of corn and wheat naturally contaminated with aflatoxin, deoxynivalenol and zearalenone. *J. Food Protection* **48**, 416–420.

Hult, K., Plestina, R., Ceovic, S., Habazin-Novak, V., and Radic, B. (1982). Ochratoxin A in human blood: Analytical results and confirmation tests from a study in connection with Balkan endemic nephropathy. In: Technical University Vienna, ed., *Mycotoxins Phycotoxins*, p. 338.

Hurd, R. N. (1977). Structure activity relationships in zearalenones. In J. V. Rodricks, C. W. Hesseltine, and M. A. Mehlman, eds., *Mycotoxins in Human and Animal Health*. Pathotox Publishers, Park Forest South, IL, pp. 379–391.

Ingerowski, G. H., Scheutwinkel-Reich, M., and Stan, H. J. (1981). Mutagenicity studies on veterinary anabolic drugs with the *Salmonella*/microsome test. *Mutat. Res.* **91**, 93–98.

International Agency for Research on Cancer (IARC). (1979). IARC monographs on the *Evaluation of the Carcinogenic Risk of Chemicals to Humans: Sex Hormones (II)*. IARC, Lyon, France, Vol. 21, pp. 173–213.

International Agency for Research on Cancer (IARC). (1983). IARC monographs on the *Evaluation of the Carcinogenic Risk of Chemicals to Humans: Some Food Additives, Feed Additives and Naturally Occurring Substances.* IARC, Lyon, France, Vol. 31, pp. 279–291.

Joffe, A. Z. (1971). Alimentary toxic aleukia. In S. Kadis, A. Ciegler, and S. J. Ajl, eds., *Microbial Toxins.* Academic Press, New York, Vol. 7, pp. 139–189.

Joffe, A. Z. (1974). Toxicity of *Fusarium poae* and *F. sporotrichiodes* and its relation to alimentary toxic aleukia. In I. F. H. Purchase, ed., *Mycotoxins.* Elsevier, Amsterdam, pp. 229–262.

Joffe, A. Z. (1978). *Fusarium poae* and *F. sporotrichioides* as principal causal agents of alimentary toxic aleukia. In T. D. Wyllie and L. G. Morehouse, *Mycotoxic Fungi, Mycotoxins, Mycotoxicoses: An Encyclopedic Handbook.* Dekker, New York, Vol. 3, pp. 21–86.

Kadis, S., Ciegler, A., and Ajl, S. J., eds. (1971). *Microbial Toxins.* Academic Press, New York, pp. 207–292.

Kallela, K., and Saastamoinen, I. (1982). The effects of "Gasol" grain preservative dosages on the growth of *Fusarium graminearum* and the quantity of the toxin zearalenone. *Nord. Vet. Med.* **34,** 124–129.

Kanai, K., and Kondo, E. (1984). Decreased resistance to mycobacterial infection in mice fed a trichothecene compound (T-2 toxin). *Jpn. J. Med. Sci. Biol.* **37,** 97–104.

Kane, A., Creppy, E. E., Roth, A., Roschenthaler, R., and Dirheimer, G. (1986). Distribution of the [³H]-label from low doses of radioactive ochratoxin A ingested by rats, and evidence for DNA single-strand breaks caused in liver and kidneys. *Arch. Toxicol.* **58,** 219–224.

Kanisawa, M. (1983). Carcinogenicity of ochratoxin A and citrinin. *Abstr. 3rd Mycol. Cong. Tokyo,* p. 136.

Kanisawa, M., and Susuki, S. (1978). Induction of renal and hepatic tumors in mice by ochratoxin A, a mycotoxin. *Gann* **69,** 599–600.

Koen, J. S., and Smith, H. C. (1945). An unusual case of genital involvement in swine associated with eating moldy corn, *Vet. Med.* **40,** 131–133.

Krishnamachari, K. A. V. R., Bhat, R. V., Nagarajan, V., and Telak, T. B. G. (1975). Hepatitis due to aflatoxicosis. An outbreak in Western India. *Lancet* **I**: 1061–1063.

Krogh, P. (1977). Ochratoxins. IN J. V. Rodricks, C. W. Hesseltine, and M. A. Mehlman, eds., *Mycotoxins in Human and Animal Health.* Pathotox Publishers, Park Forest South, IL, pp. 489–498.

Kuczuk, M., Benson, P., Heath, H., and Hayes, A. (1978). Evaluation of the mutagenic potential of mycotoxins using *Salmonella typhimurium* and *Saccharomyces cerevisiae. Mut. Res.* **53,** 11–20.

Kuiper-Goodman, T., Scott, P. M., and Watanabe, H. (1987). Risk assessment of the mycotox in zearalenone. *Reg. Toxicol. Pharmacol.* **7,** 253–306.

Kurtz, H. J., and C. J. Mirocha. (1978). Zearalenone (F-2) induced estrogenic syndrome in swine. In T. D. Wyllie and L. G. Morehouse, eds., *Mycotoxic Fungi, Mycotoxins, Mycotoxicoses: an Encyclopedia Handbook.* Dekker, New York, Vol. 2, pp. 256–268.

Kurtz, H. J., Nairn, M. E., Nelson, G. H., Christensen, C. M., and Mirocha, C. J. (1969). Histologic changes in the genital tracts of swine fed estrogenic mycotoxin. *Am. J. Vet. Res.* **30,** 551–556.

Lafarge-Frayssinet, C., Lespinats, G., Lafont, P., Loisillier, F., Mousset, S., Rosenstein, Y., and Frayssinet, C. (1979). Immunosuppressive effects of fusarium extracts and trichothecenes: Blastogenic response of murine splenic and thymic cells to mitogens. *Proc. Soc. Exp. Biol. Med.* **106,** 302–311.

Lafarge-Frayssinet, C., Declortre, F., Mousset, S., Martin, M., and Frayssinet, C. (1981). Induction of DNA single-strand breaks by T-2 toxin, a trichothecene metabolites of fusarium: Effect on lymphoid organs and liver. *Mut. Res.* **88,** 115–123.

Lasztity, R., Bekes, F., Torley, D., Villanyi, M., Gyorey-Vadon, E., Kovacs, I., Nedelkovits, J., Salgo, A., Zaigmond, A., et al. (1979). Detoxification of agricultural products contaminated with *Fusarium* toxin. *Hung. Teljes* **16,** 518 (in Hungarian).

Lin, J., Miller, J. A., and Miller, E. C. (1977). 2,3-Dihydro-2(guan-7-yl)-3-hydroxy-aflatoxin B_1, a major acid hydrolysis product of aflatoxin B_1-DNA or ribosomal RNA adducts formed in hepatic microsome-mediated reactions and rat liver *in vivo*. *Cancer Res.* **34**, 4430–4438.

Lindenfelser, L. A., Lilehoj, E. B., and Burmesiter, A. P. (1974). Aflatoxin in trichothecene toxins: Skin tumour induction and synergistic toxicity in white mice. *J. Natl. Cancer Inst.* **74**, 113–116.

Lovelace, C. E. A., and Nyathi, C. D. (1977). Estimation of the fungal toxins, zearalenone and aflatoxin, contaminating opaque maize beer in Zambia. *J. Sci. Food. Agric.* **28**, 288–292.

Mann, D. D., Buening, G. M., Osweiler, G. D., and Hook, B. S. (1983). Effect of subclinical levels of T-2 toxin on the bovine cellular immune system. *Can. J. Comp. Med.* **308**–312.

Marasas, W. F. O., Krick, N. P. J., van Rensburg, S. J., Steyn, M., and Schalkwyk, G. C. (1977). Occurrence of zearalenone and deoxynivalenol, mycotoxins produced by *Fusarium graminearum* Schwabe, in maize in Southern Africa. *S. Afr. J. Sci.* **73**, 346–349.

Marasas, W. F. O., van Rensburg, S. J., and Mirocha, C. J. (1979). Incidence of *Fusarium* sp. and the mycotoxins deoxynivalenol and zearalenol, in corn produced in esophageal cancer areas of Transkei. *J. Agric. Food Chem.* **27**, 1108–1112.

Marasas, W. F. O., Nelson, P. E., and Toussoun, T. A. (1984). *Toxigenic Fusarium Species: Identity and Mycotoxicology*. The Pennsylvania State University Press, University Park, 328 pp.

Martin, P. M. D., and Keen, P. (1978). The occurrence of zearalenone in raw and fermented products from Swaziland and Lesotho. *Sabrouaudia* **16**, 15–22.

Masuko, H., Ueno, Y., Otokawa, M., Yaginuma, K. (1977). The enhancing effect of T-2 toxin on delayed hypersensitivity in mice. *Jpn. J. Med. Sci. Biol.* **30**, 159–163.

Matossian, M. K. (1982). Ergot and the Salem witchcraft affair. *Am. Sci.* **70**, 355–357.

Matsuura, Y., Yoshizawa, T., and Morooka, N. (1979). Stability of zearalenone in aqueous solutions of some food additives. *J. Food Hyg. Soc. Jpn.* **20**, 385–390. (in Japanese).

Matsuura, Y., Yoshizawa, T., and Morooka, N. (1981). Effect of food additives and heating on the decomposition of zearalenone in wheat flour. *J. Food Hyg. Soc. Jpn.* **22**, 293–298 (in Japanese).

Mayer, C. F. (1953). Endemic panmyelotoxicosis in the Russian grain belt. Part I. The clinical aspect of alimentary toxic aleukia (ATA); a comprehensive review. *Military Surg.* **113**, 173–189.

McLaughlin, C. S., Vaughn, M. H., Campbell, I. M., Wei, C. M., Stafford, M. E., and Hansen, B. S. (1977). Inhibition of protein synthesis by trichothecenes. In J. V. Rodricks, C. W. Hesseltine, and M. A. Mehlman, Eds., *Mycotoxins in Human and Animal Health*. Pathotox Publishers, Park Forest South, IL, pp. 263–273.

Meronuck, R. A., Garren, K. H., Christensen, C. M., Nelson, G. H., and Bates, F. (1970). Effects of turkey poults and chicks of rations containing corn invaded by *Pencillium* and *Fusarium* species. *Am. J. Vet. Res.* **31**, 551–555.

Miller, K. and Atkinson, H. A. C. (1986). The *in vitro* effects of trichothecenes on the immune system. *Food Chem. Toxicol.* **24**, 545–549.

Miller, J. K., Hacking, A., Harrison, J., and Gross, V. J. (1973). Stillbirths, neonatal mortality, and small litters in pigs associated with the ingestion of *Fusarium* toxin by pregnant sows. *Vet. Rec.* **93**, 555–559.

Mirocha, C. J., and Pathre, S. V. (1973). Identification of the toxic principle in a sample of poaefusarin. *Appl. Microbiol.* **26**. 719–724.

Mirocha, C. J., Pathre, S. V., Schauerhamer, B., and Christensen, C. M. (1976). Natural occurrence of Fusarium toxins in feedstuff. *Appl. Environ. Microbiol.* **32**, 553–556.

Mirocha, C. J., Pathre, S. V., and Christensen, C. M. (1977). Zearalenone. In J. V. Rodricks, C. W. Hesseltine, and M. A. Mehlman, Eds., *Mycotoxins in Human and Animal Health*. Pathotox Publishers, Park Forest South, IL, pp. 345–364.

Mirocha, C. J., Pathre, S. V., and Robison, T. S. (1981). Comparative metabolism of zearalenone and transmission into bovine milk. *Food Cosmet. Toxicol.* **19**, 25–30.

Morooka, N., Uratsuchi, N., Yoshizawa, T., and Yamamota, H. (1972). Studies on the toxic substances in barley infected with *Fusarium* spp. *J. Food Hyg. Soc. Jpn.* **13**, 368–375.

Moss, E. J., Judah, D. J., Przybylski, M., and Neal, G. (1983) Some mass-spectral and NMR analytical studies of a glutathione conjugate of aflatoxin B_1. *Biochem. J.* **210**, 227–232.

National Toxicology Program (NTP). (1982). *Carcinogenesis Bioassay of Zearalenone in F344/N Rats and B6C3F1 Mice.* NTP, Technical Report Series No. 235, Dept. of Health and Human Services, Research Triangle Park, NC.

Neal, G. E., and Colley, P. J. (1979). The formation of 2, 3-dihydro-2, 3-dihydroxy aflatoxin B_1 *in vitro* by rat liver microsomes. *FEBS Lett.* **101**, 382–386.

Neal, G. E., Judah, D. J., Stripe, F. and Patterson, D. S. (1981). The formation of 2, 3-dihydroxy-2, 3-dihydro-aflatoxin B_1 by the metabolism of aflatoxin B_1 by liver microsomes isolated from certain avain and mammalian species and the possible role of this metabolite in the acute toxicity of aflatoxin B_1. *Toxicol. Appl. Pharmacol.* **58**, 431–438.

Neish, G. A. and Cohen, H. (1981). Vomitoxin and zearalenone production by *Fusarium graminearum* from winter wheat and barley in Ontario. *Can. J. Plant. Sci.* **61**, 811–815.

Ngindu, A., Kenya, P. R., Ocheng, D. M., Omondi, T. N., Ngure, W., Gatei, D., Johnson, B. K., Ngira, J. A., Nandwa, H., Jansen, A. J., Kaviti, J. N., and Siongok, T. A. (1982). Outbreak of acute hepatitis caused by aflatoxin poisoning in Kenya. *Lancet* 1346–1348.

Norton, D. M., Toule, G. M., Cooper, S. J., Partington, S. R., and Chapman, W. B. (1982). In G. A. Pepin, D. S. P. Paterson, and D. E. Gray, eds., *Proceedings, Fourth Meeting on Mycotoxins in Animal Disease.* Min. Agr. Fisheries Food, Alnwick, Northumberland, UK, pp. 77–81.

Noskov, A. I., and Ogryskov, S. E. (1967). Zur diagnostik der Stachybotriotoxicose der rinder. *Turkey vses. Inst. Vet. Sanit.* **28**, 21–27.

Oen, K., Schroeder, M. L., and Krzekotowska, D. (1985). Pokeweed mitogen and *Staphylococcus aureus* Cowan I induced immunoglobul in A synthesis by lymphocytes of IgA deficient blood donors. *Clin. Exp. Immunol.* **62**, 387–396.

Peers, F. G., and Linsell, C. A. (1973). Dietary aflatoxins and liver cancer-a population based study in Kenya. *Br. J. Cancer* **27**, 473–484.

Peers, F. G., Gilman, G. A., and Linsell, C. A. (1976). Dietary aflatoxins and human liver cancer. A study in Swaziland. *Int. J. Cancer* **17**, 67–176.

Pestka, J. J. (1988). Immunochemical assay of mycotoxins: a model for hazardous residue screening in foods. *J. Assoc. Off. Anal. Chem.* **71**, 1075–1081.

Pestka, J. J., and Forsell, J. H. (1988). Inhibition of human lymphocyte transformation by the macrocycllic trichothecenes roridin A and verrucarin A. *Toxicol. Lett.* **41**, 215–222.

Pestka, J. J. and Moorman, M. A. (1988). IgA nephropathy in mice exposed to the trichothecene deoxynivalenol. *FASEB J.* **2**, A1112.

Pestka, J. J., Tai, J. H., Witt, M. F., Dixon, D. E., and Forsell, J. H. (1987). Suppression of immune response in the B6C3F1 mouse after dietary exposure to the Fusarium toxins deoxynivalenol (vomitoxin) and zearalenone. *Food Chem. Toxicol.* **25**, 297–304.

Pier, A. C., Pier, T. L., and Cysewski, S. J. (1980). Implications of mycotoxins in animal disease. *J. Am. Vet. Med. Assoc.* **176**, 719–724.

Purchase, I. F. H., ed. (1974). *Mycotoxins.* Elsevier, Amsterdam.

Richardson, K. E., Hagler, W. M., Jr., and Mirocha, C. J. (1985). Production of zearalenone, alpha- and beta-zearalenol, and alpha- and beta-zearalanol by *Fusarium* spp. in rice culture. *J. Agric. Food Chem.* **33**, 862–866.

Rodricks, J. V., Hesseltine, C. W., and Mehlman, M. A., eds. (1977). *Mycotoxins in Human and Animal Health.* Pathotox Publishers, Park Forest South, IL, 807 pp.

Rosenstein, Y., Lafarge-Frayssinet, C., Lespinats, G., Loisillier, F., Lafont, P., and Frayssinet, C. (1979). Immunosuppressive activity of fusarium toxins. Effects on antibody synthesis and skin grafts of crude extracts, T-2 toxin and diacetoxyscirpenol. *Immunology* **36**, 111–117.

Sarkisov, A. H., Koroleva, V. P., Kvasnina, E. S., and Grezin, V. F. (1971). *Diagnosis of the Fungal Diseases in Animals. Mycosis and Mycotoxicosis*. Kolos, Moscow, pp. 84–91.

Sarudi, I., Kupai, J., Maleczki, J., Galambos, A., Csukas, B., Ormos, Z., Pataki, K., and Horvath, L. (1979). Treatment of fodders contaminated with zearalenone F_2 toxin. *Hung. Teljes* **16**, 38 (in Hungarian).

Scheffer, R. P. (1983). Toxins as chemical determinants of plant disease. in J. M. Daly and B. J. Deverall, eds., *Toxins and Plant Pathogenesis*. Academic Press, New York, pp. 1–40.

Schoental, R. (1977). The role of nicotinamide and of certain other modifying factors in diethylnitrosanine carcinogenesis: Fusaria mycotoxins and spontaneous tumours in mice. *Cancer* **40**, 1833–1840.

Schoental, D. (1979). The role of fusarium mycotoxins in the etiology of tumours of the digestive tract and of certain other organs in man and animals. *Frot. Gastrointest. Res.*. **4**, 17–24.

Schoental, R., Joffe, A. Z., and Yagen, B. (1979). Cardiovascular lesions and various tumours found in rats given T-2 toxin, a trichothecene metabolite of *Fusarium. Cancer Res.* **39**, 2179–2189.

Scott, P. M. (1978). Mycotoxins in feeds and ingredients and their origin. *J. Food. Protection* **41**, 385–398.

Scott, P. M., Lau, P., and Kanhere, S. R. (1981). Gas chromatography with electron capture and mass spectrometric detection of deoxynivalenol in wheat and other grains. *J. Assoc. Off. Anal. Chem.* **64**, 1364–1371.

Senti, F. R. (1979). Global perspective on mycotoxins. In *Perspective on Mycotoxins*. FAO, Rome, pp. 15–120.

Shank, R. C. (1978). Mycotoxicoses of man: Dietary and epidemiological conditions. In T. D. Wyllie and L. G. Morehouse, eds., *Mycotoxic Fungi, Mycotoxins, Mycotoxicoses, An Encyclopedic Handbook*. Dekker, New York, Vol. 3, pp. 1–19.

Shank, R. C., Bhamarapravati, N., Gordon, J. E., and Wogan, G. N. (1972). Dietary aflatoxins and human liver cancer. IV. Incidence of primary liver cancer in two municipal populations in Thailand. *Food Cosmet. Toxicol.* **10**, 171–179.

Shotwell, O. L. (1977). Assay methods for zearalenone and its natural occurrence. In J. V. Rodricks, C. W. Hesseltine, and M. A. Mehlman, eds., *Mycotoxins in Human and Animal Health*. Pathotox, Park Forest South, IL, pp. 403–413.

Sinnhuber, R. O., Wales, J. H., Ayres, J. L., Engebrecht, R. H., and Amend, D. L. (1968). Dietary factors and hepatoma in rainbow trout. *J. Natl. Cancer Inst.* **41**, 711–718.

Sorenson, W. G., Gerberick, G. F., Lewis, D. M., and Castranova, V. (1986). Toxicity of mycotoxins for the rat pulmonary macrophage *in vitro. Environ. Health Perspect.* **66**, 45–53.

Spesivtseva, N. A. (1964). Mycoses and mycotoxicoses. In A. H. Sarkisov, ed., *Mycotoxicosis*. Kolos, Moscow, pp. 357–386.

Stob, M., Baldwin, R. S., Tuite, J., Andrews, F. N., and Gillette, K. G. (1962). Isolation of an anabolic uterotropic compound from corn infested with *Gibberella zeae. Nature (London)* **196**, 1318–1320.

Stoloff, L. (1980). Aflatoxin M_1 in perspective, *J. Food Protection* **43**, 226–230.

Stoloff, L. (1983). Aflatoxin as a cause of primary liver-cell cancer in the United States: A probability study. *Nutr. Cancer* **5**, 165–186.

Stoloff, L. (1986). A rationale for the control of aflatoxin in human foods. In P. S. Steyn and R. Uleggar, eds., *Mycotoxins and Phycotoxins*. Elsevier, Amsterdam, pp. 457–471.

Tai, J. H. and Pestka, J. J. (1988). Impaired murine resistance to *Salmonella typhimurium* following oral exposure to the trichothecene T-2 toxin. *Food Chem. Toxicol.* **126**, 619–698.

Tamm, C. (1977). Chemistry and biosynthesis of trichothecenes. In. J. V. Rodricks, C. W. Hesseltine, and M. A. Mehlman, eds., *Mycotoxins in Human and Animal Health*. Pathotox Publishers, Park Forest South, IL, pp. 209–228.

Tatsuno, T., Saito, M., Enomoto, M., and Tsunoda. (1968). *Chem. Pharm. Bull.* **16**, 2519–2520.

Thiel, P. G., Meyer, C. J., and Marasas, W. F. O. (1982). Natural occurrence of moniliformin together with deoxynivalenol and zearalenone in Transkeian corn. *J. Agric. Food Chem.* **30**, 308–312.

Trucksess, M., Flood, M. T., and Page, S. N. (1986). Thin layer chromatographic determination of deoxynivalenol in processed grain products. *J. Assoc. Off. Anal. Chem.* **69**, 35–36.

Truhaut, R., Shubik, P., and Tuchmann-Duplessis, H. (1985). Zeranol and 17-beta-estradiol: A critical review of the toxicological properties when used as anabolic agents. *Regul. Toxicol. Pharmacol.* **5**, 276–283.

Tryphonas, H., Iverson, F., So, Y., Nera, E. A., McGuire, P. F., O'Grady, L., Clayson, D. B., and Scott, P. M. (1986). Effects of deoxynivalenol (vomitoxin) on the humoral and cellular immunity of mice. *Toxicol. Lett.* **30**, 137–150.

Ueno, Y. (1983). Historical background of trichothecene problems. In Y. Ueno, ed., *Trichothecenes: Chemical, Biological and Toxicological Aspects.* Elsevier, Amsterdam, pp. 1–6.

Ueno, Y., Hosoya, M., Morita, Y., Ueno, I., and Tatsuno, T. (1968). Inhibition of the protein synthesis in rabbit reticulocyte by nivalenol, a toxic principle isolated from *Fusarium nivale*-growing rice. *J. Biochem.* **64**, 479–485.

Ueno, Y., Ishikawa, Y., Nakajima, M., Sakai, K., Ishii, K., Tsunoda, H., Saito, M., Enomoto, M., Ohtsubo, K., and Umeda, M. (1971). Toxicological approaches to the metabolites of *Fusaria*. I. Screening of toxic strains. *Jpn, J. Exp. Med.* **41**, 257–272.

Ueno, Y., Ishii, K., Sakai, K., Kanaeda, S., Tsunoda, H., Tanaka, T., and Enomoto, M. (1972a). Toxicological approaches to the metabolites of *Fusaria*. IV. Microbial survey on "bean-hull poisoning of horses' with the isolation of toxic trichothecenes neosolaniol and T-2 toxin of *Fusarium solani* M-1-1. *Jpn. J. Exp. Med.* **42**, 187–203.

Ueno, Y., Sato, N., Ishii, K., Sakai, K., and Enomoto, M. (1972b). Toxicological approaches to the metabolites of *Fusaria*. V. Neosolaniol; T-2 toxin and butenolide, toxic metabolites of *Fusarium sporotrichiodes* NRRL 3510 and *Fusarium poae* 3287. *Jpn. J. Exp. Med.* **42**, 461–472.

Ueno, Y., Sato, N., Ishii, K., Sakai, K., Tsunoda, H., and Enomoto, M. (1973). Biological and chemical detection of trichothecene mycotoxins of *Fusarium* species. *Appl. Microbiol.* **25**, 699–704.

Urry, W. H., Wehrmeister, H. L., Hodge, E. D., and Hily, P. H. (1966). The structure of zearalenone. *Tetrahedron Lett.* **27**, 3109–3114.

van Rensburg, S. J. (1977). Role of epidemiology in the elucidation of mycotoxin health risks. In J. V. Rodricks, C. W. Hesseltine, and M. A. Mehlman, eds., *Mycotoxins in Human and Animal Health.* Pathotox Publishers, Park Forest South, IL, pp. 699–711.

van Rensburg, S. J., and Altenkirk, B. (1974). *Claviceps purpea*—Ergotism. In I. H. F. Purchase, ed., *Mycotoxins*, Elsevier, New York, pp. 67–96.

Vesonder, R. F. (1983). Toxicoses and natural occurrence in North America. In Y. Ueno, ed., *Trichothecenes, Chemical, Biological and Toxicological Aspects.* Elsevier, Amsterdam, pp. 210–217.

Vesonder, R. F., Cleigler, A., and Jensen, A. H. (1973) isolation of the emetic principle from *Fusarium*-infected corn. *Appl. Microbiol.* **26**, 1008–1010.

Voluntir, V., Popescu, I., Jivanescu, I., Maga Minzat, R., Purcel Vlah, M., Constantinescu, S., and Filip, M. (1971). Aspecte ale stachibotriotoxic ozei si fusariotoxicozei la pore (Aspects of stachybotryotoxicosis and fusariotoxicosis in swine). *Rev. Zooteh. Med. Vet.* **21**, 68–00.

Wang, T. V., and Cerutti, P. A. (1980). Spontaneous reactions of aflatoxin B_1 modified deoxyribonucleic acid *in vitro. Biochemistry* **19**, 1692–1698.

Wang, Y., Yeh, P., Li, W., and Liu, Y. (1983). Correlation between geographical distribution of liver cancer and aflatoxin B_1 climate conditions. *Acta Sinica, Ser. B* 431–437.

Ware, G. M., and Thorpe, C. W. (1978). Determination of zearalenone in corn by high pressure liquid chromatography and fluorescence detection. *J. Assoc. Off. Anal. Chem.* **61**, 1058–1062.

Warner, R., and Pestka, J. J. (1987). ELISA survey of retail grain-based products for zearalenone and aflatoxin B_1. *J. Food. protection* **50**, 502–506.

Wehner, F., Marasas, W., and Thiel, P. (1978). Lack of mutagenicity to *Salmonella typhimurium* of some *Fusarium* mycotoxins. *Appl. Environ. Microbiol.* **35**, 659–662.

Wei, C.-M., Campbell, I. M., McLaughlin, C. S., and Vaughan, M. H. (1974). Binding of trichodermin to mammalian ribosomes and its inhibition by other 12, 13-epoxy-trichothecenes. *Mol. Cell. Biochem.* **3**, 215–219.

Wong, J. J., and Hsieh, D. D. (1976). Mutagenicity of aflatoxin related to their metabolism and carcinogenic potential. *Proc. Natl. Acad. Sci. U.S.A.* **73**, 2241–2244.

Woronin, M. (1891). Uber das Taumel-Getriede in Sud-Ussurien. *Bot. Z.* **49**, 81–93.

Wyllie, T. D., and Morehouse, L. G. (1978). *Mycotoxic Fungi, Mycotoxins, Mycotoxicoses: An Encyclopedic handbook.* Dekker, New York, 3 vols.

Yagen, B., and Joffe, A. Z. (1976). Screening of toxic isolates of *Fusarium poae* and *Fusarium sporotrichioides* involved in causing alimentary toxic aleukia. *Appl. Environ. Microbiol.* **32**, 423–427.

Yefremov, V. V. (1944). Alimentary toxic aleukia (septic angina). *Hyg. Sanit.* **7–8**, 18–45.

Yoder, O. C. (1980). Toxins in pathogenesis. *Annu. Rev. Phytopathol.* **18**, 103–129.

Yoshizawa, T. (1983). Toxicoses and natural occurrence in Japan. In Y. Ueno, ed., *Trichothecenes, Chemical, Biological and Toxicological Aspects.* Elsevier, Amsterdam, pp. 195–209.

Yoshizawa, T., and Morooka, N. (1973). Deoxynivalenol and its monoacetate: New mycotoxins from *Fusarium roseum* and moldy barley. *Agric. Biol. Chem.* **37**, 2933–2934.

21

VIRAL DISEASE TRANSMISSION BY SEAFOOD

Ricardo De Leon and Charles P. Gerba

Department of Microbiology and Immunology and Department of
Nutrition and Food Science, University of Arizona, Tucson, Arizona

1. **Epidemiology of Viral Transmission by Shellfish**
    1.1. Introduction
    1.2. Hepatitis
        1.2.1. Hepatitis A
        1.2.2. Non-A, Non-B Hepatitis
        1.2.3. Hepatitis B
    1.3. Gastroenteritis
        1.3.1. Norwalk Virus
        1.3.2. Small Round Viruses
        1.3.3. Astrovirus
        1.3.4. Human Calicivirus
    1.4. Other Diseases Associated with Shellfish Consumption
2. **Virus Occurrence in Shellfish**
    2.1. Virus Contamination, Transport, and Survival
        2.1.1. Sources of Virus Contamination
        2.1.2. Virus Transport in Water
        2.1.3. Virus Survival in the Marine Environment
    2.2. Occurrence of Enteric Viruses in Marine Food
    2.3. Virus Persistence in Shellfish
3. **Virus Removal from Shellfish**
    3.1. Depuration and Relaying
    3.2. Fate of Viruses during Cooking and Processing of Shellfish
4. **Risk Assessment**
    4.1. Minimal Infectious Dose of Enteric Viruses
    4.2. Risk of Secondary Spread of Disease
    4.3. Estimated Morbidity and Mortality for Enteric Pathogens
    4.4. Risk Assessment for Shellfish Transmitted Viral Infections: Research
        Needs

---

1. EPIDEMIOLOGY OF VIRAL TRANSMISSION BY SHELLFISH

1.1. Introduction

Transmission of human viral disease via consumption of seafood was initially recognized in the 1950s when the first documented outbreaks of shellfish-associated hepatitis were reported (Roos, 1956; Lindberg-Braman, 1956). Since then more than 100 outbreaks of hepatitis and viral gastroenteritis have been associated with consumption of sewage-contaminated shellfish in the United States (Richards, 1985). Shellfish-associated viral diseases appear to be on the rise in the United States (Richards, 1985) (Figure 1). In England and Wales 98 outbreaks of illness associated with shellfish have been documented between 1965 and 1986 (Appleton, 1987). Only two of these outbreaks were due to bacterial pathogens, 57 to viral gastroenteritis, and 11 to hepatitis A virus. Several of the outbreaks involved novel enteric viruses like small round viruses, astroviruses, and caliciviruses. Twenty-six of those outbreaks did not have a defined cause but a viral aetiology was suspected (Appleton, 1987). It is now widely recognized that shellfish can serve as effective vehicles in the transmission of viral hepatitis and gastroenteritis.

Bacterial standards for shellfish quality have been in effect since 1925 when the National Shellfish Sanitation Program was developed (Frost, 1925). In the United States, bacteriological standards for shellfish growing waters are 70

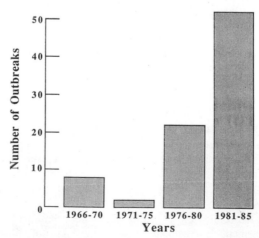

FIGURE 1. Reported   outbreaks   of   shellfishassociated gastroenteritis (United States).

coliforms/100 mL maximum and shellfish meat should not contain more than 230 fecal coliforms/100 g (Wood, 1976). The standards have proved to be very effective in preventing outbreaks of disease due to enteric bacterial pathogens. For example, outbreaks of typhoid associated with shellfish consumption have not occurred in the United States since 1955 (Verber, 1984). None the less, outbreaks of viral-associated illness have continued to occur, suggesting that currently used bacterial indicators are not adequate to prevent viral disease by shellfish consumption. Virus standards for shellfish have not been established primarily due to the lack of adequate detection methodology.

Shellfish are edible bivalve mollusks of the class *Pelecypoda*, that is, oysters, mussels, and clams. They provide approximately 8% of the total world catch of aquatic animals and are commonly eaten raw (WHO, 1974). Consumption of shellfish supports a 173 million dollar a year industry in the United States (Gottfried et al., 1987). Their filter-feeding mechanism consists of sieving food particles suspended in their environment by siphoning organelles and mucous membranes. If their surrounding water is contaminated by viruses and bacteria, the mucous membranes may entrap the pathogens and transfer them to their digestive tract. Since shellfish are usually eaten whole and raw, they may act then as passive carriers of the human pathogens. Viral disease outbreaks have been associated with all major types of edible bivalve mollusks and human enteric

TABLE 1   Enteric Viruses of Man[a]

Virus (Number of Types)	Disease Caused
Enteroviruses	
Poliovirus (3)	Meningitis, paralysis, fever
Echovirus (31)	Meningitis, diarrohea, rash, fever, respiratory disease
Coxsackievirus A (23)	Meningitis, herpangina, fever, respiratory disease
Coxsackievirus B (6)	Myocarditis, congenital heart anomalies, pleurodynia, respiratory disease, fever, rash, meningitis
New enteroviruses (4)	Meningitis, encephalitis, acute hemorrhagic conjunctivities, fever, respiratory disease
Hepatitis type A (1)	Infectious hepatitis
Norwalk virus (1?)	Diarrhea, vomiting, fever
Calicivirus (1)	Gastroenteritis
Astrovirus (1)	Gastroenteritis
Reovirus (3)	Not clearly established
Rotavirus (4 or more)	Diarrhea, vomiting
Adenovirus (2)	Respiratory disease, eye infections, gastroenteritis
Coronavirus (1)	Gastroenteritis
Snow-Mountain agent (?)	Gastroenteritis
Epidemic non-A, non-B hepatitis(?)	Hepatitis
Small round viruses (?)	Gastroenteritis?

[a] From Melnick and Gerba (1980).

TABLE 3   Enteric Viruses Documented Epidemiologically as Causes of Shellfish Associated Illness

Hepatitis
Non-A non-B hepatitis
Norwalk
Snow Mountain agent
Small round viruses (SRV)
Astroviruses
Calicivirus

pathogens have been isolated from shellfish obtained from both open and closed harvesting areas (Gerba and Goyal, 1978; Vaughn et al., 1980).

Human enteric viruses appear to be the major cause of shellfish-associated viral disease. There are over 110 known enteric viruses that are excreted in human feces and ultimately find their way into domestic sewage (WHO, 1979; Melnick and Gerba, 1980). Enteric viruses are divided into several groups based on morphologic, physical, chemical, and antigenic differences (Table 1). These viruses cause a wide variety of illnesses such as hepatitis, fever, diarrhea, paralysis, meningitis, and myocarditis. The list of known enteric viruses has grown rapidly in the last two decades as better methods have become available for their detection (Melnick and Gerba, 1980). Enteric viruses are parasites of animal and man believed to be principally transmitted by the fecal–oral route. They are excreted in large numbers in the feces of infected individuals and can almost always be detected in domestic sewage effluents (Slade and Ford, 1983). Also, they are more resistant to common sewage treatment processes including chlorination as commonly practiced than bacterial pathogens (Bitton, 1980).

Enteric viruses may enter the ocean through direct discharge of domestic sewage, sewage contaminated rivers and streams, ocean disposal of domestic sewage sludge, and through boat wastes (Goyal, 1984). Although there are over 110 known enteric viruses, only a few have been shown epidemiologically to be transmitted by shellfish (Table 2). The lack of methods for virus detection and the difficulty of recognizing viral disease outbreaks have probably precluded the list from growing longer. Thus far, there is epidemiological evidence for hepatitis A, non-A, non-B hepatitis, Norwalk, Snow Mountain agent, astroviruses, caliciviruses, and small round viruses as causes of shellfish-associated illness.

1.2.   Hepatitis

1.2.1.   Hepatitis A

The hepatitis A (HAV) virus is a member of the enterovirus group and shares with them basic characteristics,— single-stranded RNA genome, 28-nm diameter, and facal–oral transmission. Hepatitis A causes a higher incidence of symptomatic

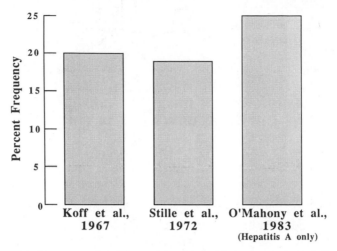

FIGURE 2. Frequency of shellfish-associated hepatitis as a percentage of all cases.

infections than other enteroviruses, up to 95% during outbreaks (Lednar et al., 1985). Common symptoms of HAV infection are dark urine, nausea, vomiting, malaise, fever, chills, and jaundice (Benenson et al., 1980). Infections of HAV have been reported to be asymptomatic in 84% of children under 2 years of age, 50% in ages 3–4, and 20% in children older than 5 years (Hadler et al., 1980), 19% in adolescents (Tabor et al., 1979), and 11% in adults (Benenson et al., 1980; Decker et al., 1979). Fulminant hepatitis A, although rare, has a very high mortality rate, 65% (7/11) (Tabor, 1984; Gerety, 1984). The high rate of HAV asymptomatic infections in children may contribute greatly to nonepidemic prevalence of the disease in other ages (Tabor, 1984).

Outbreaks of shellfish-associated infectious hepatitis were not documented until the early 1960s in the United States. Since then nearly 1500 cases of shellfish-associated HAV have been reported (Richards, 1985). Although the number of sporadic cases of HAV infections is difficult to ascertain, it is considered to be more significant than outbreak-related cases (O'Mahony et al., 1983). The significance of shellfish in the transmission of hepatitis A has been determined by several epidemiological studies (Figure 2). A recent case controlled study reported that 25% of the hepatitis A cases in southeast England could be attributed to the consumption of shellfish (O'Mahony et al., 1983). Consumption of contaminated mollusks accounted for an estimated 19% of the infectious hepatitis cases in Frankfurt, Germany (Stille et al., 1972). In the United States, Koff et al. (1967) conducted a prospective study to determine the modes of transmission of nonepidemic HAV cases among 10 Boston hospitals. Consumption of raw shellfish was found to be more frequent in infectious hepatitis patients (34/185) than in controls (10/185). Ingestion of steamed clams was more common in patients (13/104) than in matched controls. In the study of Koff et al. ingestion of raw shellfish or steamed clams was considered as frequent a potential exposure to hepatitis A as was contact with jaundiced persons.

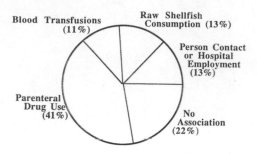

FIGURE 3. Risk factors associated with acquisition of non-A, non-B hepatitis.

1.2.2. Non-A, Non-B Hepatitis

Four recent reports have implicated shellfish as a source of non-A, non-B hepatitis (Caredda et al., 1981, 1986; Alter et al., 1982; Bamber et al., 1983). In a study of hepatitis patients in Baltimore, Maryland, Alter et al. (1982) found that 42% of patients with acute viral hepatitis were diagnosed as non-A, non-B hepatitis. The majority of patients in the study were not hospitalized; the figure is a good approximation of non-A, non-B hepatitis in that community. Raw shellfish consumption was associated with 12.5% of the cases and the risk factor was the third most common after parenteral drug use and history of blood transfusion (Figure 3). Tabor (1985) has suggested three possible non-A, non-B viruses, one of them, a fecal–orally transmitted virus, may be the cause of the shellfish-associated illness and waterborne outbreaks.

1.2.3. Hepatitis B

Hepatitis B viruses have been demonstrated in oysters near a hospital sewage outfall (Mahoney et al., 1974). The Australia (Au) antigen was used as a marker for hepatitis B in the study. However, shellfish have never been shown to be a cause of infection.

1.3. Gastroenteritis

1.3.1. Norwalk Virus

The Norwalk agent was discovered in 1972 in Norwalk, Ohio by immune electron microscopy of an infectious stool filtrate (Kapikian et al., 1972). Norwalk virus is considered the first recognized human gastroenteritis virus of medical importance. The infectivity and pathogenicity of the virus have been studied by means of adult human volunteers (Cukor and Blacklow, 1984). The attack rate to adult human volunteers is 50% where typical gastroenteritis symptoms and reinfection occur in the presence of serum antibody (Blacklow et al., 1972; Wyatt et al., 1974). A short-term resistance of 4–14 weeks has been reported (Kapikian et al., 1978; Blacklow et al., 1979). Forty-two percent of the outbreaks of nonbacterial gastroenteritis investigated by the Centers for Disease Control in 1976–1980 were attributed to Norwalk virus (Kaplan et al., 1982).

The first documented outbreak of shellfish-associated Norwalk gastroenteritis involved over 2000 persons throughout Australia (Murphy et al., 1979). In the United States, the first cases of shellfish-associated gastroenteritis attributed to Norwalk virus, and documented as such, occurred in 1980 after individuals consumed oysters from Florida (Gunn et al., 1982). In 1982, in New York state, there were 103 well-documented outbreaks in which over 1000 persons became ill with gastroenteritis from eating clams and oysters. Norwalk virus was identified in both clam and oyster specimens as well as serologically in five of seven of the outbreaks (Morse et al., 1986). Importation of depurated English clams has led to enteric illness due to Norwalk virus (Richards, 1985). A recent review by Guzewich and Morse (1986) indicates that Norwalk virus illness associated with shellfish is a continuing problem in the United States.

1.3.2.   Small Round Viruses

A group of small round viruses (SRV) has been reported as the cause of numerous outbreaks of shellfish-associated gastroenteritis (Appleton and Pereira, 1977; Gill et al., 1983). These viruses do not appear to be serologically related to the Norwalk or hepatitis A viruses. Diagnosis of SRVs is done by electron microscopy and information on the characteristics of the viruses are virtually nonexistent. The small round virus group most likely represents more than one virus type and has received several names such as Snow Mountain agent (Dolin et al., 1982), Hawaii agent (Thornhill et al., 1977), W-Ditchling agent (Appleton et al., 1977), and Cockle agent (Appleton and Pereira, 1977). Gill et al. (1983) investigated an outbreak involving small round virus illness from Pacific oysters. An attack rate of 79% was determined for individuals consuming oysters. The number of cases was 181 and 11/16 people who consumed only one oyster experienced illness.

More recently, the Snow Mountain agent has been reported as the cause of several clam-associated outbreaks of gastroenteritis (Truman et al., 1987). Forty-three percent of individuals interviewed had gastroenteritis. The attack rate for people who ate only baked clams was 18%, for those who ate only raw clams the rate was 60%, and 84% for those who ate both raw and baked clams. The Snow Mountain agent was identified in one stool and seroconversion was observed in 66% of cases tested. A few individuals were found to seroconvert to both Norwalk and Snow Mountain agent and one individual seroconverted only to Norwalk, suggesting that both viruses may have been involved in the outbreak.

Some of the small round virus members are suspected of belonging to the Parvovirus family. Currently, it is not clear whether parvoviruses are involved in gastroenteritis associated with shellfish consumption. Involvement of some parvovirus-like particles in dual infections with Norwalk has been suspected of originating from enhanced replication of persistent parvoviruses during the Norwalk infection (Caul, 1987b). More research is still needed to differentiate the parvoviruses as a separate group of shellfish-associated gastroenteritis viruses. However, the small round parvovirus-like particles have been found in the shellfish incriminated in the outbreaks (Appleton, 1987).

1.3.3. Astrovirus

The virus was initially detected in feces of children with gastroenteritis (Appleton and Higgins, 1975). The name of astrovirus was given by Madeley and Cosgrove (1975) due to a characteristic 5–6 pointed star-like form seen in electron microscopy. Symptomatic infection has been found in 80% of babies infected with astroviruses (Madeley, 1979). Infection of volunteers who have detectable serum antibody does not result in diarrhea (Kurtz et al., 1979).

Recently, an outbreak of astrovirus gastroenteritis has been associated with the consumption of oysters (Caul, 1987a). The outbreak, which occurred at an officers' dinner in a naval base, had two phases. The first phase was caused by a small round virus and the second phase occurred 4 days later after recovery from the SRV illness. The second phase consisted of another bout of diarrhea, this time shedding large numbers of astroviruses.

1.3.4. Human Calicivirus

Human caliciviruses have been isolated from feces and are 28–34 nm in diameter. These viruses have a characteristic morphology of 32 "capped" depressions in icosahedral symmetry. Some human calicivirus strains more commonly affect children, causing vomiting and diarrhea, whereas other strains affect all age groups with more flu-like symptoms (i.e., fever, malaise, and nausea) (Cubitt, 1987).

A human calicivirus has recently been shown to be transmitted by shellfish (Cubitt, 1987). The outbreak, due to raw oysters, was characterized by high attack rate in all age groups. Cubitt (1987) regards Norwalk virus as a calicivirus although this has yet to be confirmed. If Norwalk is indeed a calicivirus, then this family would contain two members that have been documented to be transmitted by shellfish.

1.4. Other Diseases Associated with Shellfish Consumption

In addition to hepatitis and gastroenteritis, two epidemiological studies have suggested a possible link between Creutzfeldt–Jakob disease and consumption of raw shellfish. However, these studies were of a limited scope and the link has not been conclusively established (Bobowick et al., 1973; Davanipour et al., 1985).

2. VIRUS OCCURRENCE IN SHELLFISH

Enteric viruses are shed in large numbers in the feces of infected individuals. In feces a range of 10^6-10^{10} virus particles/g can be found depending on the virus type and the stage of the infection (Rodgers, 1981; Slade and Ford, 1983). More than 110 different enteric viruses have been detected in sewage (Melnick and Gerba, 1980) although only a few have been associated with shellfish consumption at this point. The sources of virus contamination of shellfish harvesting beds are thus related to feces or sewage pollution.

2.1. Virus Contamination, Transport, and Survival

2.1.1. Sources of Virus Contamination

The direct sources of viruses are raw sewage or partially treated effluents, sludge, and boat dumping. The indirect sources involve pathways of contaminated water through land application and sludge burial with subsequent runoff via rivers and perhaps groundwater to estuaries and the coastline. Resuspension of contaminated sediment by means of dredging operations may find its way to nearby estuaries. More information on the sources of virus contamination of oceans can be found in Bishop (1983), Goyal (1984), and Grimes (1986). In the past ocean disposal of raw sewage and sludge was a common practice but at least 43 countries utilize the 1975 guidelines stipulated in an international convention on "Prevention of Marine Pollution by Dumping of Wastes and other Matter" (Park and O'Connor, 1981; Grimes, 1986).

2.1.1.1. Sewage Discharge. In 1979, 5500 million gallons per day (MGD) of domestic sewage was discharged directly into the coastal waters adjacent to large U.S. cities, small coastal islands, and small coastal communities (Anon., 1979). Of these discharges, 2000 MGD were by means of ocean outfalls and 3500 from small coastal communities (Anon., 1979). In the United States, many coastal towns and cities discharge wastewater to the ocean (Grimes, 1986). At present, 500 municipal sewage treatment plants discharge effluents into estuaries and 70 municipal plants into coastal waters. There are very few pipelines discharging into the open ocean (OTA, 1987).

Sewage, specially raw, countains very high numbers of enteric viruses. Estimated ranges of viruses in raw sewage are 6,000 to 490,000/L in water saving countries like Israel and in developing countries (Buras, 1976; Rao et al., 1978). In developed countries concentrations of viruses in raw sewage tend to be lower; 100–9000/L have been reported (Ruiter and Fujioka, 1978; Slade and Ford, 1983). A raw wastewater median virus concentration of 10,000/PFU/L was calculated by Leong (1983) from a review of the literature. Secondary treatment by activated sludge reduces virus concentration by an overall 80%. Virus concentrations in secondary effluent average 60–120 PFU/L (Slade and Ford, 1983). The discharge of wastewater into the open ocean in not considered as harmfull since mixing and dilution are more efficient and the level of biological activity is low (Bishop, 1983). The discharge into coastal waters and estuaries has the potential of impacting shellfish beds. In 1982, the U.S. Environmental Protection Agency required all communities to treat to a minimum of secondary treatment before discharging but high costs of treatment have forced many communities to apply for waivers for primary treatment only (Bishop, 1983).

2.1.1.2. Sludge Discharge. Due to the tendency of enteric viruses to become solid associated, sludges may contain high levels of them. Also, solid association contributes to greater virus persistence (Slade and Ford,1983). Primary sludges may contain up to 1 million viruses/L. Treated sludge may contain less viruses, from 0 to 100/L, depending on the treatment (Slade and Ford, 1983). In 1977,

5,134,000 tons of sludge was dumped in the New York and mid-Atlantic bights alone (Bishop, 1983). Federal law required all sludge dumping to cease in 1981 but deadline extensions have been granted due to high costs of alternate disposal systems. The requirement of increased wastewater treatment before disposal increases the need for sludge discharge into the ocean by coastal communities (Bishop, 1983). The Office of Technology Assessment (1987) has suggested that the dumping of sludge be done in the deep ocean.

2.1.1.3.   Boat Dumping. The combined wastes of commercial and recreational vessels in the United States have been estimated equivalent to a community of 500,000 (Hopper and Myrick, 1971). Due to the concentration of boating activity in small harbor and coastal areas the potential local impact could be high (Bishop, 1983). To diminish the impact of boat dumping on shellfish harvesting beds the U.S. Environmental Protection Agency has designated these areas as zero-discharge. The discharge in coastal areas is regulated by the use of sanitation devices in vessels authorized for discharge at designated nautical miles from the coast (Anon., 1978). However, the difficulty of enforcing boat discharges gives this source of contamination high potential impact on estuaries (Bishop, 1983).

2.1.2.   Virus Transport in Water

The transport of viruses over long distances in surface and marine water is well documented. Metcalf et al. (1974) detected enteroviruses 13 km from their source in the Houston ship channel. Dahling and Safferman (1979) detected at least 30% of the enteric viruses found at the source 300 km upstream and 7.1 days of transport distance. In the marine environment Hugues et al. (1981) detected viruses in the same concentration 200 m at the coast and at the beaches.

2.1.3.   Virus Survival in the Marine Environment

Numerous studies have been conducted on the survival of viruses in marine water and reviews on the subject have been published by Gerba and Goyal (1978) and more recently by Goyal (1984). Although the survival of enteric viruses in marine waters appears to be less than that observed in freshwater environments (Kutz and Gerba, 1988) they are still capable of prolonged survival. Enteric viruses have been reported to survive from 2 to 130 days in seawater in laboratory studies and generally survive longer in such environments than coliform bacteria (Melnick and Gerba, 1980). Goyal et al. (1979) were able to isolate enteroviruses from sediment and blue crabs at the Philadelphia dump site 18 months after sludge disposal had stopped. Almost all previous studies on virus survival have been concerned only with enteroviruses and coliphages. However recent studies with hepatitis A virus suggest it is capable of longer survival than other enteric viruses (Bosch and Shields, 1987). Information on the survival of Norwalk virus, small round viruses, caliciviruses, and astroviruses in water is nonexistent.

Environmental factors such as temperature, salinity, pH, solar radiation, association of viruses to solids, and microbial antagonism have been found to affect virus survival in water (Kapuscinski and Mitchel, 1980). Block (1983) has divided the factors affecting virus survival into physical, chemical, and biological nature. Physical factors include light, temperature, hydrostatic pressure, adsorption, and aggregation. Chemical factors include pH, ionicity, cations, heavy metals, organic chemicals, and disolved oxygen. Biological factors include virus types, bacterial and algal activity, and predation by protozoa (Block, 1983).

Thus far, the physical factors of temperature and virus association to solids (adsorption and aggregation) have been found to play major roles in virus survival in water (Block, 1983). At temperatures below 10°C enteric viruses could be expected to survive for several months (Gerba and Goyal, 1986). Inactivation of 99.9% of poliovirus 3 in Lee River water was obtained only after 9 weeks at 5–6°C (Poynter, 1968) Echovirus 6 decreased 1 log in titer at 3–5°C in seawater and 5 logs at 22°C (Won and Ross, 1973). The inactivation rate of poliovirus 2 increases 15 times when the temperature of autoclaved seawater is increased from 4 to 22°C (Denis at al., 1977). Virus association to sewage solids or marine sediments may protect viruses against inactivation, perhaps by reducing the rate of thermoinactivation (Liew and Gerba, 1980). Generally when sediment is present inactivation rates of viruses in seawater-moistened sand tended to be 4.5-fold slower than in seawater alone (Gerba and Goyal, 1986). This probably explains why Goyal et al. (1984) were able to detect enteric viruses in marine sediments at the Philadelphia sludge dumpsite 18 months after disposal had stopped. Solid association may also protect viruses against common wastewater treatment disinfection practices (Hejkal et al., 1979).

2.2. Occurrence of Enteric Viruses in Marine Food

Fish, shellfish, and crabs that live in water contaminated by domestic wastes are frequently found to contain enteric bacteria and viruses. In addition, it has been demonstrated in laboratory studies that lobsters, sandworms, detrital feeding fish, conch, and aplysia can accumulate enteric viruses (Sigel et al., 1976; Gerba and Goyal, 1978). Field studies have demonstrated the presence of enteric viruses in another marine organism. Goyal et al. (1984) reported the isolation of enteroviruses from blue crabs from the city of Philadelphia sludge dump site in the North Atlantic. However, only shellfish have been implicated in the transmission of enteric viral illness, probably because they are eaten raw or may not be as thoroughly cooked as other seafoods. In addition, shellfish as filter feeders tend to concentrate viruses from the water in which they are growing. The occurrence of enteroviruses in shellfish taken from the coastal waters of the United States has been documented in numerous studies (Gerba and Goyal, 1978; Goyal et al., 1979; Vaughn et al., 1980; Ellender et al., 1980; Wait et al., 1983). In these studies, enteroviruses were isolated from areas open to shellfish harvesting (Figure 4). The percentage of virus-positive shellfish range from 9–40% for open waters to 13–40% for closed waters (Goyal, 1984).

Landry et al. (1982), in a study designed to mimic shellfish virus uptake in low to moderately contaminated waters by either feces-associated or monodispersed virus, found an uptake-depuration equilibrium at virus exposure of 0.01 PFU/mL. They suggest that at water column contamination levels of 0.01

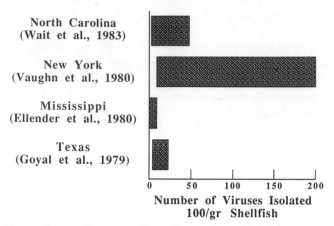

North Carolina
(Wait et al., 1983)

New York
(Vaughn et al., 1980)

Mississippi
(Ellender et al., 1980)

Texas
(Goyal et al., 1979)

0   50   100   150   200

Number of Viruses Isolated
100/gr Shellfish

FIGURE 4. Range of reported concentrations of enteroviruses in shellfish collected from areas open to harvesting.

PFU/mL or less virus accumulation by shellfish may not be efficient. Bioaccumulation was most efficient with high concentrations of solid-associated virus. However, the same group of researchers (Vaughn et al., 1980) had previously reported the sporadic isolation of enteroviruses in shellfish from a natural environment with contamination levels of 0.001 PFU/mL. The mechanisms and conditions of bioconcentration and depuration need to be better defined before tolerable water contaminant levels could be widely established. Risk assessment may help determine what contaminant levels could be acceptable (see Section 4).

No statistically significant relationship has been shown between concentrations of enteric viruses and coliform or fecal coliform bacteria in shellfish or shellfish-growing waters (Goyal et al., 1979; Wait et al., 1983). In addition, viruses have been isolated on numerous ocasions from areas open to harvesting as previously discussed. Grohmann et al. (1981) in a study with human volunteers and oysters meeting bacteriological standards reported 52 cases of illness with Norwalk causing 37%. Bacterial standards are thus not adequate for the control of virus-contaminated shellfish and Richards (1985) has proposed the use of poliovirus testing as an interim standard. Gene probe techniques may soon allow the implementation of virus testing for the regular monitoring of shellfish quality (see Section 5).

2.3. Virus Persistence in Shellfish

Although no multiplication of human enteric viruses in shellfish has been demonstrated their uptake by the shellfish and other marine animals may act to prolong their survival (Hedstrom and Lyke, 1964). Enteric viruses may survive for weeks or months in refrigerated marine foods (Gerba and Goyal, 1978). In one outbreak of gastroenteritis caused by the Norwalk virus, the shellfish had been frozen 15 weeks before consumption (Linco and Grohmann, 1980). Poliovirus has been reported to survive from 30 to 90 days in refrigerated oysters (DiGirolamo et al., 1970). Poliovirus has been found to survive up to 300 days on peeled prawns at $-20°C$ (Eyles, 1983) and prolonged survival of virus in crabs has also been reported (DiGirolamo and Daley, 1973).

3. VIRUS REMOVAL FROM SHELLFISH

3.1. Depuration and Relaying

Depuration and relaying are methods used for the self-purification of con-taminated shellfish. Both methods rely on the ability of the shellfish to eliminate contaminating microorganisms from the digestive tract through normal feeding, digestion, and excretion activities. Artificial purification or depuration of shellfish takes place in tanks or basins that are filled with disinfected water and are maintained in stable environmental conditions. The more common means of water disinfection for use in shellfish depuration tanks is ultraviolet light or ozonation. Physiological processes of shellfish are very sensitive to chlorine and thus it is not used in depuration basins (Gerba and Goyal, 1978). Depuration at a temperature of 20°C and a salinity of at least 75% of the original source seem to be efficient (Liu et al., 1967). Free flowing systems are more efficient for depuration than static systems (DiGirolamo et al., 1975). Conventional depuration time is 48 hr. When shellfish are relayed they are moved to areas in the natural environment that are unpolluted. In this case, initial microbial load and temperature have been found to determine the efficiency of the decontamination process (Cook and Ellender, 1986). Also Cook and Ellender determined that oysters under physiological stress decontaminated at slower rates than healthy oysters. Healthy oysters required 7 days at a temperature above 10°C for appropriate cleaning. Depuration of shellfish for commercial purposes is expens-ive and is practiced only on a limited scale in the United States (Cook and Ellender, 1986). Relaying is the more common practice.

The rates of elimination for bacterial pathogens and fecal indicator bacteria are very similar but fecal coliform levels do not correlate with virus elimination (Cook and Ellender, 1986). Recent research also suggests that hepatitis A virus may not be eliminated as readily as other enteroviruses during depuration (Sobsey, 1987b). This probably explains the outbreaks of hepatitis A associated with depurated shellfish (Richards, 1985). A study by Grohmann et al. (1981) with human volunteers consuming depurated oysters indicated that conventional depuration (48 hr) is not sufficient for the complete elimination of Norwalk (37% of illnesses in study) and other enteric viruses. All the depurated oysters in the human volunteer study met bacteriological standards. It appears, therefore, that fecal coliforms are not useful as end-point indicators of depuration or relaying. However, proper depuration of shellfish will reduce the risk of viral infection, but it is no guarantee of a virus-free product (Metcalf et al., 1980). The depuration and relaying of shellfish has been recently reviewed (Richards, 1988).

3.2. Fate of Viruses during Cooking and Processing of Shellfish

Steamed and undercooked shellfish have been implicated in outbreaks of Norwalk virus and hepatitis A virus (Feingold, 1973; Morse et al., 1986). Shells of soft shelled clams opened 1 min after steaming and that is not sufficient to kill all contaminating viruses (Koff and Sear, 1967). Up to 7–13% of added poliovirus survived in oysters even after four commonly used processing methods: steaming, frying, baking, and stewing (DiGirolamo et al., 1970). Heat penetration studies showed that the internal temperature of shellfish was not sufficient to inactivate all viruses present. In animal feeding studies, hepatitis A virus failed to be totally inactivated after thermal treatment of oysters at 140°F for 19 min (Peterson et al., 1978). A boiling temperature for at least 20 min has been advocated (Mosley and Galainbos, 1975). By contrast, a recent study by Millard et al. (1987) on heat inactivation of poliovirus and hepatitis A indicates that bioaccumulated virus can be inactivated after 2 min of immersion in boiling water or 2.5 min of steaming. In this study, cockles' shells opened after only 30 sec of immersion in boiling water with an internal temperature of 65°C. Variability of shellfish species, size, time after harvest, contamination level, temperature prior to experiment, and other factors most likely account for the discrepancy between the studies mentioned. An even greater degree of variability in cooking conditions is to be expected at the consumer level and thus the difficulty in establishing the minimum cooking time required for complete inactivation of viruses.

Radiation has also been tested as a means to inactivate viruses in shellfish. Poliovirus 1 in oysters is able to survive relatively high doses of γ radiation (DiGirolamo et al., 1972). In this study, to inactivate more than 90% of the virus present, a dose of 400 krad is required, which results in undesirable organoleptic changes.

4.   RISK ASSESSMENT

In order to assess the potential health impact of virus-contaminated shellfish and of virus contamination of the overlaying water in shellfish harvesting beds it is necessary to assess the risks associated with the microorganisms known to be transmitted by shellfish. Shellfish have been known to transmit viruses capable of causing serious illness and mortality in man (Gerba and Goyal, 1978). The viruses known to be transmitted by shellfish are also responsible for waterborne disease (Craun, 1986). Enteric pathogens are the most common cause of water and foodborne illness today (NRC, 1985; Craun, 1986). Between 1971 and 1980 man-made chemicals were responsible for less than 5% of all reported waterborne illness in the United States (Craun, 1986). Thus, it is essential to develop methodologies to properly assess the risks of viruses ingested during consumption of shellfish.

4.1.   Minimal Infectious Dose of Enteric Viruses

Crucial to risk assessment is empirical knowledge of the level or concentration of a contaminant necessary to cause a health effect. Ideally, a maximum contaminant level for potentially harmful substances should be established on firm epidemiological evidence where cause and effect can be clearly quantified to determine a minimum or no-risk level. Unfortunately, epidemiology cannot quantitatively demonstrate cause and effect for pathogens (CST, 1983). The minimum concentration or level of pathogenic organisms that is necessary to cause a health effect is called minimum infectious dose (MID). Exact data on MID for humans is generally not possible because of the extreme cost, unethical nature of human experimentation, and uncertainty in extrapolating dose–response curves to low exposure level. In spite of the many difficulties in estimating MIDs of viruses pathogenic to man some data have been obtained by several investigators and Ward and Akin (1984) have recently reviewed the literature. The results indicate that relatively low numbers of enteric viruses, perhaps one or two tissue culture plaque-forming units (PFU), are capable of causing infection. In several studies, small numbers of viruses, primarily vaccine strains, produced infection in human subjects. Kaprowski et al. (1956) fed poliovirus 1 in gelatin capsules to adult volunteers and infected 2 of 3 subjects with 2 PFU of the virus. Katz and Plotkin (1967) administered attenuated poliovirus 3 (Fox) by nasogastric tube to infants and infected 2 of 3 with 10 $TCID_{50}$ and 3 of 10 with 1 $TCID_{50}$ of the virus. Minor et al. (1981) administered attenuated poliovirus 1 vaccine orally and infected 3 of 62-month-old infants with 50 $TCID_{50}$ of the virus. The most extensive studies to date on dose–response by enteric viruses have been conducted by Schiff et al. (1984). Over 100 healthy adult volunteers were fed various doses of echovirus 12, a mild pathogen, in drinking water. Probit analysis gave an average MID of 17 PFU. To date, the minimal infectious dose of enteric viruses that will cause infection in a human consuming shellfish has not been determined. The minimum infectious dose of viruses when consumed in shellfish may be greater than those calculated in water or by direct ingestion in gelatin capsules. Mossel and Oei (1975) have suggested that substances clear the stomach much more quickly between meals than when consumed with food. However, there is evidence in the literature that consumption of a single contaminated oyster is sufficient for disease (Gill et al., 1983).

4.2.   Risk of Secondary Spread of Disease

Unlike risks associated with toxic chemicals in shellfish, individuals who do not actually consume shellfish are also at risk. This indirect risk is due to spread of the pathogen by person-to-person contact or by subsequent fecal contamination of other materials with which noninfected individuals may come in contact. This risk, called secondary and tertiary spread of microorganisms, has been well documented during waterborne outbreaks caused by the Norwalk virus (Gerba et al., 1985). The secondary attack rate of Norwalk during waterborne outbreaks has been deternimed to be about 30%. Appleton et al. (1981) identified secondary cases of infection during foodborne outbreaks of suspected viral aetiology.

4.3. Estimated Morbidity and Mortality for Enteric Pathogens

Not all enteric virus or parasite-infected individuals will develop clinical illness. Asymptomatic infections are very common among some of the enteroviruses. Numerous factors determine whether clinical illness will develop or not, such as immune status of the host, type of microorganism, virulence of the particular strain, and the route of infection. For hepatitis A virus the percentage of individuals with clinically observed illness is low for children (usually <5%) (Evans, 1982). The frequency of clinical hepatitis A in adults is estimated at 75%; however, during waterborne outbreaks it has been observed as high as 97% (Lednar et al., 1985). For rotavirus, the frequency of clinical symptoms is greatest in children and lowest in adults (Gerba et al., 1985). The observed frequencies of symptomatic infections for various enteroviruses may range from 1% for poliovirus to more than 75% for some of the coxsackie B viruses (Cherry, 1981).

Mortality rates are likewise affected by multiple factors that determine likelihood of clinical illness development. The risk of mortality from hepatitis A virus infection is 0.6% (CDC, 1985). Mortality due to infection with other enteroviruses in North America and Europe has been reported from <0.1 to 1.8% (Assaad and Borecka, 1977). Mortality rates available for hepatitis A and selected enteroviruses are summarized in Table 3. These values most likely represent only hospitalized cases of enterovirus infection.

4.4. Risk Assessment for Shellfish Transmitted Viral Infections: Research Needs

Three major components are needed for a risk assessment of infection by consumption of contaminated shellfish: an extrapolation model, an MID of

TABLE 3   Mortality Rates for Enteroviruses[a]

Enterovirus	Mortality Rate (%)
Hepatitis A	0.6
Coxsackie A2	0.5
Coxsackie A4	0.5
Coxsackie A9	0.26
Coxsackie A16	0.12
Coxsackie B	0.59–0.94
Echo 6	0.29
Echo 9	0.27
Polio 1	0.9

[a]From ODC (1985); Assaad and Borecka (1977).

enteric pathogens when consumed in shellfish, and a distribution model for pathogens in shellfish.

Several extrapolation models are available and Haas (1983) compared the simple exponential model, a modified exponential model (beta), and the log-normal (or log-probit) model with the experimental dose–response data available for enteroviruses. Using Haas' improved dose–response model Gerba and Haas (1986) arrived at estimated annual risks of infection from one enterovirus in 1000 L of drinking water (assuming ingestion of 2 L/day) using infectivity data from several studies. They found that a significant risk (i.e., greater than 1:10,000) of infection may result from very low numbers of viruses in drinking water. To date, no information is available on the actual MID of enteric viruses when consumed in shellfish and the type of distribution that viruses may have in contaminated shellfish. If data on the minimal infectious dose of enteric viruses when bioconcentrated by shellfish were available some basic risk assessment calculations could be made assuming the distribution of viruses to be Poisson. Data on average consumption of shellfish will also be needed. Even if this information is available the actual risks would be greater since secondary and tertiary spread of virus has not been included in the models for risk assessment thus far used for drinking water. In contrast, existing immunity would lower the risk for some pathogens like polio and hepatitis A since lifelong protection has been demonstrated (Evans, 1982). Immunity would not play a role in Norwalk or echovirus 12 infection where existing antibodies are not protective and multiple infection and disease can occur (Blacklow et al., 1979; Schiff et al., 1984).

5. METHODS FOR VIRUS DETECTION IN SHELLFISH

Various methods have been developed for the recovery of enteric viruses from shellfish and other seafoods. These methods have been summarized by Sobsey (1982, 1985, 1987a). Briefly, the methods consist of shucking the shellfish and homogenizing a part or whole in buffer. In earlier methods the homogenate was centrifugued to eliminate large particulate matter and the supernatant assayed directly on cells. Due to the large volumes of supernatant and to cytotoxic components in shellfish that interfered with the assay, concentration methods were developed to extract viruses from shellfish tissue. The concentration methods involve the adsorption of viruses to shellfish meat by lowering the pH of the homogenate followed by elution in a reduced volume of a high pH buffer. The adsorption–elution method is easily adapted to the different species of shellfish and has even been utilized for freshwater clams by the authors (De Leon et al., 1986). The eluates can be further concentrated to a small volume (<30 mL) by floculation, hydroextraction, ultracentrifugation, or ultrafiltration.

In the adsorption–elution procedure the pH, the salinity, and the solid/volume ratio need to be optimized for each shellfish species for maximum efficiency of recovery. Also, since shellfish concentrates are assayed for viruses in animal cell culture, it is prudent to test for toxicity and plaquing inhibitors and select a method that will reduce their interference. Enteric viruses in shellfish are assayed by cell culture and thus the limitation on the types of viruses for which there is information. Most studies on enteric viruses in shellfish have been limited to enteroviruses because of their ease of detection in cell culture. Richards (1985) has suggested to use the enterovirus, polio, as an interim indicator of virus contamination of shellfish since it is easily assayed. Cell culture techniques have been developed recently for growth and detection for hepatitis A, which can be used for environmental samples (Provost and Hilleman, 1979; Cromeans et al., 1987). However, long assay time and cost seriously limit monitoring efforts for enteric viruses in shellfish and their growing waters. Cell culture techniques are presently not available for Norwalk, small round viruses, or astroviruses. Recently, Margolin et al. (1986) described the use of gene probes to detect low numbers of infectious units of poliovirus or hepatitis A virus in environmental samples in much shorter periods of time than cell culture. The same investigators have used gene probes to detect small numbers of polioviruses in shellfish concentrates (Margolin et al., 1987). The use of gene probe technology to detect viruses in shellfish would reduce the cost and time of assay sufficiently to allow for routine monitoring of virus contamination in shellfish.

REFERENCES

Alter, M. J., Gerety, R. J., Smallwood, L. A., Sampliner, R. E., Tabor, E., Deinhardt, F., Frosner, G., and Matanoski, G. M. (1982). Sporadic non-A, non-B hepatitis: Frequency and epidemiology in an Urban U.S. Population. *J. Infect. Dis.* **145**, 886–893.

Anon. (1978). New regulations tighten U.S. grip on vessel sewage disposal. *Marine Eng./Log* **83**, 37–39.

Anon. (1979). Proceedings of the workshop on national needs and priorities for ocean pollution, research and development and monitoring. 600/8–79, U.S. Environmental Protection Agency.

Appleton, H. (1987). Small round viruses: Classification and role in food-borne infections. In *Novel Diarrhoea Viruses.* Ciba Foundation Symposium. Wiley, Chichester, U.K., pp. 108–125.

Appleton, H., and Higgins, P. G. (1975). Viruses and gastroenteritis in infants. *Lancet* **1**, 1297.

Appleton, H., and Pereira, M. S. (1977). A possible virus aetiology in outbreaks of food-poisoning from cockeles. *Lancet* **2**, 780–781.

Appleton, H., Buckley, M., Thorn, B. T., Cotton, J. L., and Henderson, S. (1977). Virus-like particles in winter vomiting disease. *Lancet* **1**, 409–411.

Appleton, H., Palmes, S. R., and Gilbert, R. J. (1981). Foodborne gastroenteritis of unknown aetiology: A virus infection? *Br. Med. J.* **282**, 1801–1802.

Assaad, F., and Borecka, J. (1977). Nine-year study of WHO virus reports on fatal virus infections. *Bull. W.H.O.* **55**, 445–453.

Bamber, M., Thomas, H. C., Bannister, B., and Sherlock, S. (1983). Acute type A, B, and non-A, non-B hepatitis in a hospital population in London: Clinical and epidemiological features. *Gut* **24**, 561–564.

Benenson, M. W., Takafuji, E. T., Bancroft, W. H., Lemon, S. M., Callahan, M. C., and Leach, D. A. (1980). A military community outbreak of hepatitis type A related to transmission in a child care facility. *Am. J. Epidemiol.* **112**, 471–481.

Bishop, P. L. (1983). *Marine Pollution and Its Control.* McGraw-Hill, New York.

Bitton, G. (1980). *Introduction to Environmental Virology.* Wiley, New York.

Blacklow, N. R., Dolin, R., Fedson, D. S., DuPont, H., Northrop, R. S., Hornick, R. B., and Chanock, R. M. (1972). Acute infectious nonbacterial gastroenteritis: Aetiology and pathogenesis. *Ann. Intern. Med.* **76**, 993–1008.

Blacklow, N. R., Cukor, G., Bedigian, M. K., Echeveria, P., Greenberg, H. B., Schreiber, D. S., and Trier, J. S. (1979). Immune response and prevalence of antibody to Norwalk enteritis virus as determined by radioimmunoassay. *J. Clin. Microbiol.* **10**, 903–909.

Block, J. C. (1983). Viruses in environmental waters. In G. Berg, ed., *Viral Pollution of the Environment.* CRC Press, Boca Raton, FL.

Bobowick, A. R., Brody, J. A., Mathews, M. R., Roos, R., and Gajdusek, D. C. (1973). Creutzfeld-Jakob disease: A case-control study. *Am. J. Epidemiol.* **98**, 381–394.

Bosch, A., and Shields, P. A. (1987). Survival of hepatitis A virus and poliovirus in seawater and marine sediments. *Abstr. Annu. Mtg. Am. Soc. Microbiol.*, p. 295.

Buras, N. (1976). Concentration of enteric viruses in wastewater and effluent: A two year study. *Wat. Res.* **10**, 295.

Caredda, F., d'Arminio Monforte, A., Rossi, E., Lopez, S., and Moroni, M. (1981). Non-A, non-B hepatitis in Milan. *Lancet* **2**, 48.

Caredda, F., Antinori, S., Re, T., Pastecchia, C., and Moroni, M. (1986). Acute non-A, non-B hepatitis after typhoid fever. *Br. Med. J.* **292**, 1429.

Caul, E. O. (1987a). Discussion. Astroviruses: Human and animal. In *Novel Diarrhoea Viruses.* Ciba Foundation Seminar. Wiley, Chichester, U.K., pp. 102–107.

Caul, E. O. (1987b). Discussion. Small round viruses: Classification and role in food-borne infections. In *Novel Diarrhoea Viruses.* Ciba Foundation Seminar. Wiley, Chichester, U.K., pp. 120–125.

CDC (Centers for Disease Control). (1985). Hepatitis Surveillance. Report No. 40, CDC Atlanta, GA.

Cherry, J. D. (1981). Nonpolio enteroviruses: Coxsachieviruses, Fchoviruses, and Enteroviruses. In, R. D. Feigin and J. D. Cherry, eds., *Textbook of Pediatric Infectious Diseases.* W. B. Saunders, Philadelphia, pp. 1316–1365.

Cook, D. W., and Ellender, R. D. (1986). Relaying to decrease the concentration of oyster-associated pathogens. *J. Food Protection* **49**, 196–202.

Craun, G. F. (1986). *Waterborne Diseases in the United States.* CRC Press, Boca Raton, FL.

Cromeans, T., Sobsey, M. D., and Fields, H. A. (1987). Development of a plaque assay for a cytopathic, rapidly replicating isolate of hepatitis A. *J. Med. Virol.* **22**, 45–56.

CST (Committee on Science and Technology). (1983). A review of risk assessment methodologies. U.S. House of Representatives, Washington, D.C.

Cubitt, W. D. (1987). The candidate caliciviruses. In *Novel Diarrhoea Viruses.* Ciba Foundation Symposium. Wiley, Chichester, U.K., pp. 126–143.

Cukor, G., and Blacklow, N. R. (1984). Human viral gastroenteritis. *Microbiol. Rev.* **48**, 157–179.

Dahling, D. R., and Safferman, R. S. (1979). Survival of enteric viruses under natural conditions in a subartic river. *Appl. Environ. Microbiol.* **38**, 1103.

Davanipour, Z., Alter, M., Sobel, E., Asher, O. M., and Gajdvsek, D. C. (1985). A case-control study of Creutzfeldt-Jacob disease. *Am. J. Epidemiol.* **122**, 443–451.

Decker, R. H., Overby, L. R., Ling, C. M., Frosner, G., Deinhardt, F., and Boggs, J. (1979). Serologic studies of transmission of hepatitis A in humans. *J. Infect. Dis.* **139**, 74–82.

De Leon, R., Payne, H. A., and Gerba, C. P. (1986). Development of a method for poliovirus detection in freshwater clams. *Food Microbiol.* **3**, 345–349.

Denis, F., Brisou, J. F., and Dupuis, T. (1977). Survie dans l'eau de mer de 20 souches de virus a ADN er ARN. *J. Fr. Hydrol.* **8**, 25.

DiGirolamo, R., and Daley, M. (1973). Recovery of bacteriophage from contaminated chilled and frozen samples of edible west coast crabs. *Appl. Microbiol.* **25**, 1020–1022.

DiGirolamo, R., Liston, J., and Matches, J. R. (1970). Survival of virus in chilled, frozen and processed oysters. *Appl. Microbiol.* **20**, 58–63.

DiGirolamo, R., Liston, J., and Matches, J. R. (1972). Effects of irradiation on the survival of virus in west coast oysters. *Appl. Microbiol.* **24**, 1005–1006.

DiGirolamo, R., Liston, J., and Matches, J. (1975). Uptake and elimination of poliovirus by west coast oysters. *Appl. Microbiol.* **29**, 260–264.

Dolin, R., Reichman, C., Roessner, K. D., Tralka, T. S., Scooley, R. T., Gary, W., and Morens, D. (1982). Detection by immune electron microscopy of the Snow Mountain agent of acute viral gastroenteritis. *J. Infect. Dis.* **146**, 184–189.

Ellender, R. D., Mapp, J. B., Middlebrooks, B. L., Cook, D. W., and Cake, E. W. (1980). Natural enterovirus and fecal coliform contamination of Gulf Coast oysters. *J. Food Protection* **43**, 105–110.

Evans, A. S. (1982). Epidemiological concept and methods. In A. S. Evans, ed., *Viral Infection of Humans.* Plenum, New York, pp. 1–32.

Eyles, M. J. (1983). Assessment of cooked prawns as a vehicle for transmission of viral disease. *J. Food Protection* **46**, 426–428.

Feingold, A. (1973). Hepatitis from eating steamed clams. *J. Am. Med. Assoc.* **225**, 526–527.

Frost, H. W. (1925). Report of committee on the sanitary control of the shellfish industry in the United States. *Public Health Rep.* November 6, 1925. Suppl. **53**, 1–17.

Gerba, C. P., and Goyal, S. M. (1978). Detection and occurrence of enteric viruses in shellfish: A review. *J. Food Protection* **41**, 743–754.

Gerba, C. P., and Goyal, S. M. (1986). Development of a qualitative pathogen risk assessment methodology for ocean disposal of municipal sludge. U.S. Environmental Protection Agency. EC AO-CIN-493, Cincinnati, OH.

Gerba, C. P., and Haas, C. N. (1986). Assessment of risks associated with enteric viruses in contaminated drinking water. Symposium, Chemical and Biological Characterization of Sludges, Sediments, Dredge Spoils, and Drilling Muds. May 20–22, Cincinatti, OH.

Gerba, C. P., Singh, S. N., and Rose, J. B. (1985). Waterborne viral gastroenteritis and Hepatitis. *CRC Crit. Rev. Environ. Control* **15**, 213–236.

Gerety, R. J. (1984). Introduction. In R. J. Gerety, ed., *Hepatitis A.* Academic Press, New York, pp. 1–8.

Gill, O. N., Cubitt, W. D., McSwiffan, O. A., Watney, B. M., and Batlett, C. L. R. (1983). Epidemic of gastroenteritis caused by oysters contaminated with small round structured viruses. *Br. Med. J.* **287**, 1532–1534.

Gottfried, M., Axtell, S. J., and Casey, M. B. (1987). Contaminated shellfish a problem in the stream of interstate commerce. *J. Environ. Health* **50**, 150–156.

Goyal, S. M. (1984). Viral pollution of the marine environment. *CRC Crit. Rev. Environ. Control* **14**, 1–32.

Goyal, S. M., Gerba, C. P., and Melnick, J. L. (1979). Human enteroviruses in oysters and their overlaying waters. *Appl. Environ. Microbiol.* **37**, 572–581.

Goyal, S. M., Adams, W. N., O'Malley, M. L., and Lear, D. W. (1984). Human pathogenic viruses at sewage sludge disposal sites in the middle Atlantic region. *Appl. Environ. Microbiol.* **48**, 758–763.

Grimes, D. J. (1986). Assessment of ocean waste disposal: Task 5. Human health impacts of waste constituents. Final Report. Congress of the United States. Office of Technology Assessment. Ocean Waste Disposal Project.

Grohmann, G. S., Murphy, A. M., Christopher, P. J., Auty, G., and Greenberg, H. B. (1981). Norwalk virus gastroenteritis in volunteers consuming depurated oysters. *Aust. J. Exp. Biol. Med. Sci.* **59**, 219–228.

Gunn, R. A., Janowski, H. T., Lieb, S., Prather, E. C., and Greenberg, H. (1982). Norwalk virus gastroenteritis following raw oyster consumption. *Am. J. Epidemiol.* **115**, 348–351.

Guzewich, J. J., and Morse, D. L. (1986). Sources of shellfish in outbreaks of probable viral gastroenteritis: Implications for control. *J. Food Protection* **49**, 389.

Haas, C. N. (1983). Estimation of risk due to low doses of microorganisms: A comparison of alternative methodologies. *Am. J. Epidemiol.* **118**, 573.

Halder, S. C., Webster, H. M., Erben, J. J., Swanson, J. E., and Maynard, J. E. (1980). Hepatitis A in day care centers. *N. Eng. J. Med.* **302**, 1222–1227.

Hedstrom, C. E., and Lyke, E. (1964). An experimental study on oysters as virus carriers. *Am. J. Hyg.* **79**, 134–142.

Hejkal, T. W., Wellings, F. M., LaRock, P. A., and Lewis, A. L. (1979). Survival of poliovirus within organic solids during chlorination. *App. Environ. Microbiol.* **38**, 114–118.

Hooper, M., and Myrick, H. (1971). Evaluation of solid and liquid waste discharge management techniques for commercial watercraft. *Proc. 3rd Annu. Offshore Tech. Conf. Am. Inst. Mining, Metallurg. Petrol. Eng.* **II**, 587–610.

Hugues, B., Lefevire, J. R., Plissier, M., and Cini, A. (1981). Distribution of viral and bacterial densities in sea water near a coastal discharge of treated domestic sewage. *Zbl. Bakt. Hyg. I. Abt. Orig.* **B173**, 509–516.

Kapikian, A. Z., Wyatt, R. G., Dolan, R., Thornhill, T., Kalica, A. R., and Chanock, R. M. (1972). Visualization by immune electron microscopy of a 27 nm particle associated with acute infectious nonbacterial gastroenteritis. *J. Virol.* **10**, 1075–1081.

Kapikian, A. Z., Greenberg, H. B., Cline, W. L., Kalica, A. R., Wyatt, R. G., James, H. J., Jr., Lloyd, N. L., Chanock, R. M., Ryder, R. W., and Kim, H. W. (1978). Prevalence of antibody to the Norwalk agent by a newly developed immune adherence haemagglutination assay. *J. Med. Virol.* **2**, 281–294.

Kaplan, J. E., Gary, G. W., Baron, R. C., Singh, N. Schonberger, L. B., Feldman, R., and Greenberg, H. B. (1982). Epidemiology of Norwalk gastroenteritis and the role of Norwalk virus in outbreaks of acute nonbacterial gastroenteritis. *Ann. Intern. Med.* **96**, 756–761.

Kaprowski, H., Norton, T. W., Jervis, G. A., Nelson, T. L., Chadwick, D., Nelsen, J. N., and Meyer, C. F. (1956). Clinical investigations of attenuated strains of poliomyelitis virus: Use as a method of immunization of children with living virus. *J. Am. Med. Assoc.* **160**, 954–966.

Kapuscinski, R. B., and Mitchell, R. (1980). Processes controlling virus inactivation in coastal waters. *Water Res.* **14**, 363.

Katz, M., and Plotkin, S. A. (1967). Minimal infective dose of attenuated poliovirus for man. *Am. J. Public Health* **57**, 1837–1840.

Koff, R. S., and Sear, H. S. (1967). Internal temperature of steamed clams. *N. Engl. J. Med.* **276**, 737–739.

Koff, R. S., Grady, G. F., Chalmers, T. C., Mosley, J. W., and Swartz, B. L. (1967). Viral hepatitis in a group of a Boston hospitals. III Importance of exposure to shellfish in a nonepidemic period. *N. Eng. J. Med.* **276**, 703–710.

Kurtz, J. B., Lee, T. W., Craig, J. W., and Reed, S. E. (1979). Astrovirus infection in volunteers. *J. Med. Virol.* **3**, 321–330.

Kutz, S. M., and Gerba, C. P. (1988). Comparison of virus survival in freshwater sources. *Wat. Sci. Tech.* **20**, 467–471.

Landry, E. F., Vaughn, J. M., Vicale, T. J., and Mann, R. (1982). Inefficient accumulation of low levels of monodispersed and feces-associated poliovirus in oysters. *Appl. Environ. Microbiol.* **44**, 1362–1369.

Lednar, W. M., Lemon, S. M., Kirkpatrick, J. W., Redfield, R. R., Fields, M. L., and Kelley, P. W. (1985). Frequency of illness associated with epidemic hepatitis A virus infections in adults. *Am. J. Epidemiol.* **122**, 226–233.

Leong, L. Y. C. (1983). Removal and inactivation of viruses by treatment processes for potable water and wastewater—a review. *Water Sci. Technol.* **15**, 91.

Liew, P., and Gerba, C. P. (1980). Thermostabilization of enteroviruses by estuarine sediment. *Appl. Environ. Microbiol.* **40**, 305–308.

Linco, S. J., and Grohman, G. S. (1980). The Darwin outbreak of oyster-associated viral gastroenteritis. *Med. J. Aust.* **1**, 211–213.

Lindberg-Braman, A. M. (1956). Clinical observations on the so-called oyster hepatitis. *Am. J. Publ. Health* **53**, 1003–1011.

Liu, O. C., Seraichekas, H. R., and Murphy, B. L. (1967). Viral pollution and self cleansing mechanism of hard clams. In G. Berg, ed., *Ttransmission of Viruses by the Water Route*. Interscience, New York.

Madeley, C. R. (1979). Viruses in the stools. *J. Clin. Pathol.* **32**, 1–10.

Madeley, C. R., and Cosgrove, B. P. (1975). 28 nm particles in faeces in infantile gastroenteritis. *Lancet* **2**, 451–452.

Mahoney, P., Fleichner, G., Millman, I., London, W. T., Blumberg, B. S., and Arias, I. M. (1974). Australia antigen: Detection and transmission in shellfish. *Science* **183**, 80–81.

Margolin, A. B., Hewlett, M. J., and Gerba, C. P. (1986). Use of a cDNA dot-blot hybridization technique for detection of enteroviruses in water. Water Quality Technology Conference Proceedings, American Water Works Association, Denver, CO, pp. 87–95.

Margolin, A. B., Bitrick, M. S., De Leon, R., and Gerba, C. P. (1987). Application of gene probes to poliovirus and hepatitis A virus detection in water and shellfish. *Proc. Oceans '87*, pp. 1746–1751.

Melnick, J. L., and Gerba, C. P. (1980). The ecology of enteroviruses in natural waters. *CRC Crit. Rev. Environ. Control.* **10**, 65.

Metcalf, T. G., Wallis, C., and Melnick, J. L. (1974). Virus enumeration and public health assessments in polluted surface water contributing to transmission of virus in nature. In J. F. Malina and B. P. Sagik, eds., Virus Survival in Water and Wastewater Systems. University of Texas, pp. 57–83.

Metcalf, T. G., Eckerson, D., Moulton, E., and Larkin, E. P. (1980). Uptake and depletion of particulate-associated polioviruses by the soft shell clam. *J. Food. Protection* **43**, 87–88.

Millard, J., Appleton, H., and Parry, J. V. (1987). Studies on heat inactivation of hepatitis A virus with special reference to shellfish. Part 2. Heat inactivation of hepatitis A virus in artificially contaminated cockles. *Epidem. Inf.* **98**, 406–414.

Minor, T. E., Allen, C. I., Tsiatis, A. A., Nelson, D. B., and D'Alissio, D. J. (1981). Human infective dose determination for oral poliovirus type 1 vaccine in infants. *J. Clin. Microbiol.* **13**, 388–389.

Morse, D. L., Guzewich, J. J., Hanrahan, J. P., Stricof, R., Shayegani, M., Deibel, R., Grabau, J. C., Nowak, N. A., Herrmann, J. E., Cukor, G., and Blacklow, N. R. (1986). Widespread outbreaks of clam- and oyster-associated gastroenteritis. *N. Engl. J. Med.* **314**, 678–681.

Mosley, J. W., and Galambos, J. T. (1975). Viral Hepatitis. In L. Schiff, ed., *Diseases of the Liver*. J. B. Lippincott, Philadelphia, pp. 500–593.

Mossel, D. A. A., and Oei, H. Y. (1975). Person-to-person transmission of enteric bacterial infection. *Lancet* **1**, 751.

Murphy, A. M., Grohmann, G. S., Christopher, P. J., Lopez, W. A., Davey, G. R., and Millsom, R. H. (1979). An Australian-wide outbreak of gastroenteritis from oysters caused by Norwalk virus. *Med. J. Aust.* **2**, 329–333.

NRC (Nation Research Council). (1985). An evaluation of the role of microbiological criteria for foods and food ingredients. National Academy Press, Washington, D.C.

O'Mahony, M. C., Gooch, C. D., Smyth, D. A., Thrussel, A. J., Bartlett, C. L. R., and Noah, N. D. (1983). Epidemic hepatitis A from cockles. *Lancet* **1**, 518.

OTA (Office of Technology Assessment). (1987). Wastes in Marine Environments. Congress of the United States, Washington, D.C. 20510.

Park, P. K., and O'Connor, T. P. (1981). Ocean dumping research: Historical and international development. In B. H. Ketchum, D. R. Kester, and P. K. Park, eds., *Ocean Dumping of Industrial Wastes*. Plenum, New York.

Peterson, D. A., Wolfe, L. G., Larkin, E. P., and Deinhardt, F. W. (1978). Thermal treatment and infectivity of hepatitis A virus in human feces. *J. Med. Virol.* **2**, 201–206.

Poynter, S. F. B. (1968). The problem of viruses in water. *Water. Treat. Exam.* **17**, 187.

Provost, P. J., and Hilleman, M. R. (1979). Propagation of human hepatitis A virus in cell culture in vitro. *Proc. Soc. Exp. Biol. Med.* **160**, 213–221.

Rao, V. C., Lakhe, S. B., and Waghmare, S. V. (1978). Developments in environmental virology in India. Indian Association for Water Pollution Control. *Tech. Annu.* **5**, 1–16.

Richards, G. P. (1988). Microbial purification of shellfish: A revier of depuration and relaying. *J. Food Microbiol.* **51**, 218–251.

Richards, G. P. (1985). Outbreaks of shellfish-associated enteric virus illness in the United States: Requisite for development of viral guidelines. *J. Food Protection* **48**, 815–823.

Rodgers, F. G. (1981). Concentration of viruses in faecal samples from patients with gastroenteritis. In M. Goddard and M. Buttler, eds., *Viruses and Wastewater Treatment.* Pergamon, New York, pp. 15–18.

Roos, R. (1956). Hepatitis epidemic conveyed by oysters. *Sven. Lakartidningen* **53**, 989–1003.

Ruiter, C. G., and Fujioka, R. S. (1978). Human enteric viruses in sewage and their discharge into the ocean. *Water, Air Soil Pollut.* **10**, 95–103.

Schiff, G. M., Stefanovic, G. M., Young, E. C., Sander, D. S., Pennekanp, J. K., and Ward, R. L. (1984). Studies of Echovirus-12 in volunteers: Determination of minimal infectious dose and the effect of previous infection on infectious dose. *J. Infect. Dis.* **150**, 858–866.

Sigel, M. M., Rippe, D. F., Beasley, A. R., and Dorsey, M. (1976). Systems for detecting viruses and viral activity. In G. Berg. H. L. Bodily, E. H. Lennette, J. L. Melnick, and T. G. Metcalf, eds., Viruses in water. American Public Health Association, Washington, D.C., p. 139.

Slade, J. S., and Ford, B. J. (1983). Discharge to the environment of viruses in wastewater, sludges and aerosols. In G. Berg, ed., *Viral Pollution of the Environment.* CRC Press, Boca Raton, FL, pp. 3–18.

Sobsey, M. D. (1982). Detection of viruses in shellfish. In C. P. Gerba and S. M. Goyal, eds., *Methods in Environmental Virology.* Decker, New York and Basel.

Sobsey, M. D. (1985). Procedures for the virological examination of seawater, shellfish and sediment. In A. E. Greenberg and D. A. Hunt, eds., *Laboratory Procedures for the Examination of Seawater and Shellfish.* American Public Health Association, Washington, D.C., pp. 81–118.

Sobsey, M. D. (1987a). Methods for recovering viruses from shellfish, seawater and sediments. In G. Berg, ed., *Methods for Recovering Viruses from the Environment.* CRC Press, Boca Raton, FL.

Sobsey, M. D. (1987b). Personal communication.

Stille, W., Kunkel, B., and Nerger, K. (1972). Oyster-transmitted hepatitis. *Dtsch. Med. Wschr.* **97**, 145–147.

Tabor, E. (1984). Clinical presentation of hepatitis A. In R. J. Gerety, ed., *Hepatitis A.* Academic Press, New York, pp. 47–53.

Tabor, E. (1985). The three viruses of non-A, non-B hepatitis. *Lancet* **1**, 743–745.

Tabor, E., Jones, R., Gerety, R. J., Drucker, J. A., and Colon, A. R. (1979). Asymptomatic viral hepatitis type A and B in an adolescent population. *Pediatrics* **62**, 1026–1030.

Thornhill, T. S., Wyatt, R. G., Kalica, A. R., Dolin, R., Chanock, R. M., and Kapikian, A. Z. (1977). Detection by immune electron microscopy of 26 to 27 nm virus-like particles associated with two family outbreaks of gastroenteritis. *J. Infect. Dis.* **135**, 20–27.

Truman, B. I., Madore, H. P., Menengus, M. A., Nitzkin, J. L., and Dolin, R. (1987). Snow Mountain agent gastroenteritis from clams. *Am. J. Epidemiol.* **126**, 516–525.

Uhnoo, I., Wadell, G., Svensson, L., and Johannson, M. E. (1984). Importance of enteric adenoviruses 40 and 41 in acute gastroenteritis in infants and young children. *J. Clin. Microbiol.* **20**, 365–372.

Vaughn, J. M., Landry, E. F., Thomas, M. Z., Vicale, T. J., and Panello, W. F. (1980). Isolation of naturally occurring enteroviruses from a variety of shellfish species residing in Long Island and New Jersey marine embayments. *J. Food Protection* **43**, 95–98.

Verber, J. L. (1984). Shellfish borne disease outbreaks. U.S. Public Health Service, Food and Drug Administration, Davisville, R.I., pp. 6–14.

Wait, D. A., Hackney, C. R., Carrick, R. J., Lovelack, G., and Sobsey, M. D. (1983). Enteric bacterial and viral pathogens and indicator bacteria in hard shell clams. *J. Food Protection* **46**, 493–496.

Ward, R. L., and Akin, E. W. (1984). Minimum infective dose of animal viruses. *CRC Crit. Rev. Environ. Control* **14**, 297–310.

Won, W. D., and Ross, M. (1973). Persistance of bacteria and virus in seawater. *J. Environ. Eng. Div. Am. Soc. Civ. Eng.* **99**, 205.

Wong, D. C., Purcell, R. H., Sreenivasan, M. A., Prasad, S. R., and Pavri, K. M. (1980). Epidemic and endemic hepatitis in India: Evidence for a non-A, non-B hepatitis virus aetiology. *Lancet* **2**, 876–878.

Wood, P. C. (1976). *Guide to Shellfish Hygiene.* World Health Organization, Geneva, Switzerland.

World Health Organization. (1974). *Fish and Shellfish Hygiene.* Tech. Rept. Ser. No. 550, p. 8. World Health Organization, Geneva, Switzerland.

World Health Organization. (1979). *Human Viruses in Water, Wastewater and Soil.* Technical Report Series. No. 639. World Health Organization, Geneva, Switzerland.

Wyatt, R. G., Dolin, R., Blacklow, N. R., DuPont, H., Buscho, R., Thornhill, T. S., Kapikian, A. Z., and Chanock, R. M. (1974). Comparison of three agents of acute infectious nonbacterial gastroenteritis by cross-challenge in voluteers. *J. Infect. Dis.* **129**, 709–714.

22

ACUTE ILLNESS FROM ENVIRONMENTAL FOOD CONTAMINANTS

Ewen C. D. Todd

Bureau of Microbial Hazards, Food Directorate, Health Protection Branch, Health and Welfare Canada, Ottawa, Ontario

1.   INTRODUCTION

Environmental contaminants that cause most illnesses in humans within a few hours or days after ingestion of a food are microbial in origin. Bacteria, viruses, parasites, and planktonic dinoflagellates that are etiologic agents in foodborne disease are found throughout the world in a variety of habitats. Some, such as *Listeria, Clostridium botulinum,* and *Bacillus cereus* are found in virgin as well as cultivated soil, plants, mud, and water and therefore seem to be native to these environments. Also, there are dinoflagellates and bacteria associated with fish and shellfish, that occur in nonpolluted seawater. However, there are others, such as *Salmonella* and *C. perfringens,* whose natural habitat is animal and human intestines, and may reside there in small numbers without causing illness; these, therefore, easily contaminate slaughterhouses, carcasses, and kitchen food-contact surfaces. *Clostridium* and *Bacillus* and most parasites produce spores or cysts that are resistant to heat and other forms of stress, and can easily reach the environment through soil particles, dust, water, and equipment to remain inactive until ingested or are transferred to a suitable nutrient capable of supporting growth. Spores can remain viable for years. Other bacteria like *Salmonella* and *Listeria*, even though they produce only vegetative cells, are also capable of surviving for long periods of time under stressed conditions. Man plays a role in the spread of these organisms through cultivation of fields, pollution of rivers and lakes, use of sewage as fertilizer, intense rearing of animals, and encouraging presence of birds, rodents, and lizards through improper storage of food. Many plants are poisonous but only a few of these come in

TABLE 1   Sources of Acute Bacterial Infection and Intoxication through Food

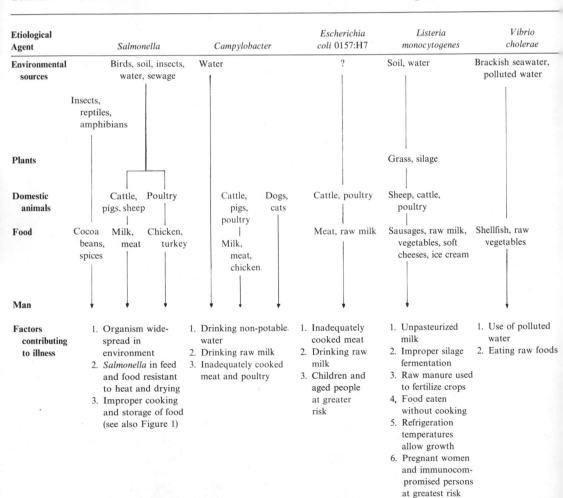

Etiological Agent	Salmonella	Campylobacter	Escherichia coli 0157:H7	Listeria monocytogenes	Vibrio cholerae
Environmental sources	Birds, soil, insects, water, sewage	Water	?	Soil, water	Brackish seawater, polluted water
	Insects, reptiles, amphibians				
Plants				Grass, silage	
Domestic animals	Cattle, Poultry, pigs, sheep	Cattle, pigs, poultry; Dogs, cats	Cattle, poultry	Sheep, cattle, poultry	
Food	Cocoa beans, spices; Milk, meat; Chicken, turkey	Milk, meat, chicken	Meat, raw milk	Sausages, raw milk, vegetables, soft cheeses, ice cream	Shellfish, raw vegetables
Man					
Factors contributing to illness	1. Organism widespread in environment 2. *Salmonella* in feed and food resistant to heat and drying 3. Improper cooking and storage of food (see also Figure 1)	1. Drinking non-potable water 2. Drinking raw milk 3. Inadequately cooked meat and poultry	1. Inadequately cooked meat 2. Drinking raw milk 3. Children and aged people at greater risk	1. Unpasteurized milk 2. Improper silage fermentation 3. Raw manure used to fertilize crops 4. Food eaten without cooking 5. Refrigeration temperatures allow growth 6. Pregnant women and immunocompromised persons at greatest risk	1. Use of polluted water 2. Eating raw foods

contact with commercial food supplies. Some toxic plant products, however, are entering the retail market without instructions to render them harmless. Chemicals such as heavy metals, solvents, and cleaning solutions can also cause acute illness, but these are normally introduced at the processing or postprocessing stages. Toxic chemicals in the soil may be accumulated by crop plant; for example, selenium in legumes *Astragalus* and *Neptunia* have caused toxicity in pasture animals, but human illness has not been documented (Reilly, 1980). Extraneous matter may come from environmental sources and cause damage, somtimes physical, in the alimentary system. Thus, there are a variery of ways that agents can come from the environment through the food supply to cause acute gastroenteritis and other syndromes. Broad outlines of these are shown in Tables 1 and 2. Control is difficult for most of them because of their widespread

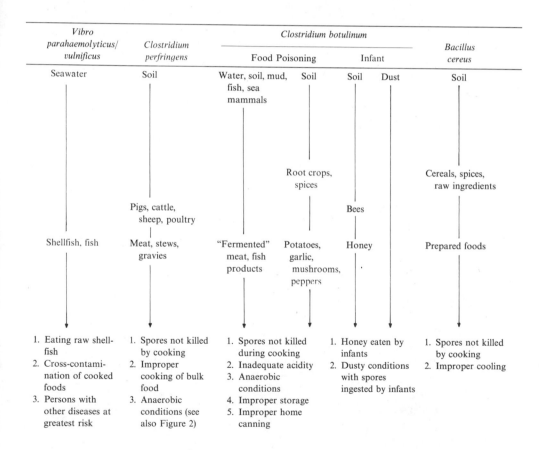

TABLE 2 Sources of Acute Viral and Parasitic Infections and Illnesses through Mycotoxins,

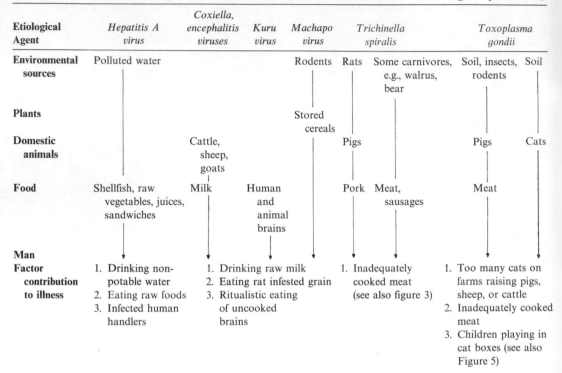

Etiological Agent	Hepatitis A virus	Coxiella, encephalitis viruses	Kuru virus	Machapo virus	Trichinella spiralis		Toxoplasma gondii	
Environmental sources	Polluted water			Rodents	Rats	Some carnivores, e.g., walrus, bear	Soil, insects, rodents	Soil
Plants				Stored cereals				
Domestic animals		Cattle, sheep, goats				Pigs	Pigs	Cats
Food	Shellfish, raw vegetables, juices, sandwiches	Milk	Human and animal brains		Pork	Meat, sausages	Meat	
Man								
Factor contribution to illness	1. Drinking non-potable water 2. Eating raw foods 3. Infected human handlers	1. Drinking raw milk 2. Eating rat infested grain 3. Ritualistic eating of uncooked brains			1. Inadequately cooked meat (see also figure 3)		1. Too many cats on farms raising pigs, sheep, or cattle 2. Inadequately cooked meat 3. Children playing in cat boxes (see also Figure 5)	

nature. For microorganisms, their destruction or at least inhibition is best achieved at the final product level by use of proper cooking and storage temperatures.

Environmental contaminants that cause acute illness will be listed by type, that is, bacteria, viruses, parasites, mycotoxins, seafood toxins, and toxic plants.

2. BACTERIA

2.1. *Salmonella*

2.1.1. *Introduction*

Salmonellosis is the most widespread and significant foodborne disease currently affecting both developed and developing nations. Salmonellae infect mammals,

Seafood Toxins, and Plants

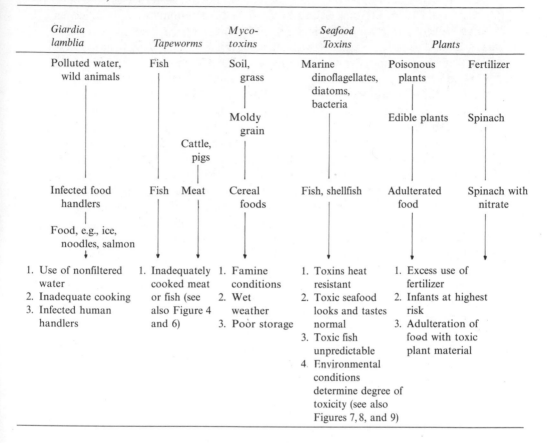

Giardia lamblia	Tapeworms	Myco-toxins	Seafood Toxins	Plants	
Polluted water, wild animals	Fish	Soil, grass	Marine dinoflagellates, diatoms, bacteria	Poisonous plants	Fertilizer
		Moldy grain		Edible plants	Spinach
	Cattle, pigs				
Infected food handlers	Fish Meat	Cereal foods	Fish, shellfish	Adulterated food	Spinach with nitrate
Food, e.g., ice, noodles, salmon					
1. Use of nonfiltered water 2. Inadequate cooking 3. Infected human handlers	1. Inadequately cooked meat or fish (see also Figure 4 and 6)	1. Famine conditions 2. Wet weather 3. Poor storage	1. Toxins heat resistant 2. Toxic seafood looks and tastes normal 3. Toxic fish unpredictable 4. Environmental conditions determine degree of toxicity (see also Figures 7, 8, and 9)	1. Excess use of fertilizer 2. Infants at highest risk 3. Adulteration of food with toxic plant material	

birds, reptiles, amphibians, fish, and insects but in most situations the hosts are not ill (Turnbull, 1979). They, however, can be effective transmitters of the organism to food animals, food ingredients, and processed food products. There are more than 2000 *Salmonella* serovars, some of which are host adapted, for example, *S. choleraesuis* in pigs, *S. dublin* in cattle, *S. pullorum* and *S. gallinarum* in poultry, and *S. typhi* and *S. paratyphi* in man. Many serovars such as *S. typhimurium* are capable of infecting a wide range of animals, including man, and are frequently involved in foodborne outbreaks.

2.1.2. Cocoa Beans

Reptiles and amphibians in warm countries are frequent carriers of servoars rare in temperate nations and may become a source of contamination for food processed in other countries (Chiodini and Sundberg, 1981). For instance, in 1973 and 1974 foil-covered chocolate balls, manufactured in the province of

Quebec, infected about 200 persons in Canada and the eastern United States (Craven et al., 1975; D'Aoust et al., 1975). The contaminant was identified as *S. eastbourne*, a strain that is typically found in west Africa. The infective dose was estimated to be 20–90 cells. It was surmised that the cocoa beans from west Africa were contaminated with excrement of lizards and insects when laid on the ground to dry in the sun. Processing in Quebec including dry-rosting of cocoa beans and heating of molten chocolate at 46–70°C for 10–96 hr. Despite this severe treatment *S. eastbourne* survived the thermal process and could be isolated from the incriminated chocolate up to 13 years after its manufacture (J.-Y. D'Aoust, Health Protection Branch, Ottawa, personal communication). The probable reason for the thermal resistance of *S. eastbourne* was the high level of cocoa fat in the chocolate. Another possible source of the organism was recontamination of the final product with contaminated cocoa bean dust. Unusual serovars have caused two other chocolate outbreaks—*S. napoli* in Italian chocolate bars exported to England (Gill et al., 1983) and *S. nima* from foil-wrapped Belgian chocolate coins imported into Canada (Jessop et al., 1986). In outbreaks associated with chocolate low infective doses have been documented. It is assumed the cocoa fat surrounding the cells prevented their destruction during the passage through the stomach. In addition, the stomach acidity would be less in the many children infected.

2.1.3. Spices

Spices, usually harvested with limited hygienic precautions in subtropical and tropical countries, are sometimes contaminated with *Salmonella*, perhaps through excretions of animals on the drying plant material. Most spices are incorporated into food to be cooked and any *Salmonella* present will probably be destroyed. But the risk is much greater where spices are added to food left standing for some time before consumption. In the Canadian Maritime provinces 15 cases of salmonellosis were linked to the use of black pepper contaminated with *S. weltevreden* (Handzel, 1974; Laidley et al., 1974). One of these cases was a truck driver who informed health authorities he had prepared meat sandwiches for his lunch, added pepper, and left them in the cab for several hours before they were eaten. Black pepper imported into Norway from Brazil has also been found to be contaminated, and 126 cases of *Salmonella oranienburg* arising from the pepper were confirmed in that country (Gustavsen and Breen, 1984).

2.1.4. Dairy Foods

Dairy products are at high risk because cows may be infected with the organism and excrete it into milk or the milk may be fecally contaminated. Cattle ingest *Salmonella* by grazing in the fields contaminated with farm animal feces, seagull excrement, polluted water, and other environmental sources (Reilly et al., 1981). Intense rearing of calves indoors has also led to infection (Peters, 1985). Animals may appear healthy but the carrier state can continue for months. Because dairy cattle may occasionally excrete *Salmonella* through the udder, cases of salmonellosis often occur in farming families drinking raw milk. How one cow can affect an industry is illustrated by the largest foodborne disease outbreak to occur in Canada when Cheddar cheese infected at least 2700 persons in the Maritime provinces Newfoundland and Ontario in 1983 and 1984 (Sharp, 1987). Worker error had allowed small amounts of raw milk to contaminate vats of pasteurized milk for several months. Analysis of milk from 327 producers was carried out and only one cow from one farm was found to be infected with the organism causing the outbreak, *S. typhimurium* page type 10. This cow was an intermittent shedder of this *Salmonella* from one teat only. Although low levels of *Salmonella* ($\leq 10/100$ g) were found in the cheese (D'Aoust et al., 1985), many people eating it were infected, probably because the fat content of the product protected the *Salmonella* aginst gastric acidity. Quantitative studies of incriminated cheese from the households of infected persons indicated that a single cell could have been infectious (D'Aoust, 1985). The cost to the company was at least \$10 million for recall and lost business. The largest North American outbreak occurred in Illinois in 1985 with 16,000 laboratory confirmed cases and an estimated 200,000 cases of illness resulting from consumption of improperly pasteurized milk (Lecos, 1986; Sharp, 1987). The source of contamination was never confirmed but a cross-connection between raw and pasteurized milk lines was postulated.

2.1.5. Poultry

Various foods including egg nog, shell and liquid eggs, ice cream, diet supplement, meringue pies, smoked fish, apple cider, raw and fermented beef and pork products, and poultry have been involved in foodborne outbreaks of salmonellosis (D'Aoust, 1989). Chicken and turkey are major contributors to the spread of *Salmonella* in foodservice establishments and homes. The organism frequently contaminates working surfaces in the kitchen from which it cross contaminates other foods, often leading to outbreaks. Carcasses of chickens, turkeys, and ducks have high levels of contamination (up to 80%), but the live birds are healthy carriers. Infection of poultry usually occurs relatively soon after hatching. Eggs may be infected through the oviduct of the hens or by penetration of the shell and membranes through fecal contamination. Although salmonellosis from egg sources is relatively rare today, 27 outbreaks between January 1985 and May 1987 in the northeastern United States were caused by Grade-A eggs infected with *S. enteritidis* (St-Louis et al., 1988). Transovarian contamination is suspected rather than cracked or fecally soiled eggs, and many cooking techniques do not totally destroy any salmonellae present. The two main serovars capable of causing human illness and infecting ovaries of layer hens are *S. enteritidis* and *S. typhimurium*. However, chicks are most likely to be infected through contaminated feed, drinking water, litter, and fecally soiled floors of poultry houses (Willinger et al., 1986). Animal feeds are usually made from by-products of animal processing or from inexpensive sources of protein, for example, fish meal (Williams, 1981). These may be contaminated during manufacture through inadequate heat processing and recontamination of the products by rodent, insect, and bird droppings (Lee, 1982; Morris et al., 1970).

The importance of contaminated animal feed is illustrated by the international spread of *Salmonella agona*. In the 1960s and 1970s increasing numbers of human infections of *S. agona* were reported in the United Kingdom and the United States (Clark et al., 1973). Epidemiologic studies traced the origin of these cases to fish meal processed in Peru. Poor sanitation contributed to widespread contamination in the fish-rendering plant. Although fish meal is not currently a significant source of *S. agona*, the serovar remains firmly entrenched in the United Kingdom and United States environments where it continues to cause foodborne illnesses. Infected poultry transfer the *Salmonella* to other birds during transportation to the slaughterhouse in poorly sanitized poultry crates, and also through defciencies in scalding, plucking, evisceration, and chilling at the processing plant (Simonsen et al., 1987). Various control measures have been implemented in different countries to reduce the importance of poultry as a source of human salmonellosis. Eradication of infected flocks, checking feed for contamination, chlorination of wash and chill water in processing plants, improved hygiene at the farm and slaughterhouse, competitive exclusion of pathogens by oral dosing of day-old chicks with gut flora from *Salmonella*-free adults, and irradiation of carcasses have been implemented in various countries with some degree of success but not on a scale large enough to substantially reduce the contamination level of birds at the retail level.

2.1.6.   Antibiotics in Feeds

Subtherapeutic levels of antibiotics have been used in domestic animals to protect against disease and to promote more rapid growth, but the practice may encourage increased bacterial resistance (O'Brien et al., 1982; Sharma et al., 1987). A recent study indicated that hamburger produced from beef cattle fed tetracycline-supplemented feed resulted in human salmonellosis (Holmberg et al., 1984). It is believed that resistant *Salmonella* in the hamburger entered the human gut but failed to multiple sufficiently to cause illness. However, when antibiotics were subsequently taken for other clinical conditions, the resistant *Salmonella* rapidly proliferated and produced symptoms in the human host. Similar problems related to chloramphenicol-treated dairy cows have been reported in California (Spika et al., 1987).

2.1.7.   Pets

Reptiles and amphibians kept as household pets are also important sources of salmonellosis, although it is usually close contact with the animals or tank water that leads to infection rather than cross-contamination of food. *S. java* and other serovars had such a significant impact on morbidity that the sale of turtles was banned in the United States and important into Canada forbidden in 1975 (Chiodini and Sundberg, 1981; D'Aoust and Lior, 1977). However, embryonated eggs are now being imported and hatchlings sold in pet shops, and numerous turtle-associated illnesses have been reported in recent years (Rogers and Johnstone, 1985). In fact, surveys have shown that many reptiles and amphibians in pet shops may carry a variety of *Salmonella* serovars (Chiodini and Sundberg, 1981; Todd and Styliadis, 1985). In addition, the use of subtherapeutic levels of antibiotics in shipping water for aquarium pets may select for antibiotic-resistant strains.

2.1.8. *Control*

The different routes of human *Salmonella* infection through the food chain are shown in Figure 1. Substantial reduction of *Salmonella* in the environment would be the most effective control measure but the most difficult to achieve. Some control may be possible in the foreseeable future at the farm and processing levels but not in wild life (Simonsen et al., 1987). At present, the ultimate responsibility for prevention of illness is in the food-service and home kitchen. If workers are properly educated they can destroy *Salmonella* by thorough cooking, avoid recontamination of food through proper disinfection of food-contact surfaces, and prevent bacterial growth by cooling prepared food rapidly (Bryan, 1979; Silliker, 1982).

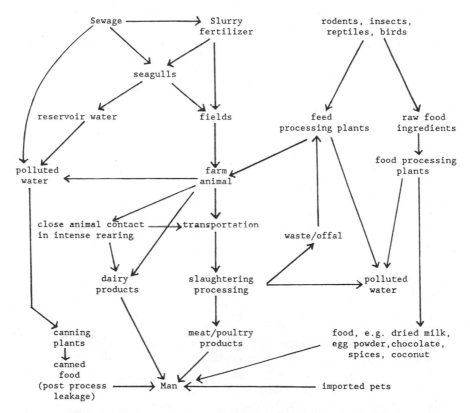

Figure 1.   Environmental sources of *Salmonella* in food and animal feed.

2.2. *Campylobacter*

2.2.1. The Disease

Campylobacter species cause a variety of animal infections but the one most likely to be transmitted through food to cause human illness is *C. jejuni*. This organism is isolated more often from stools in diarrheal cases, particularly children, than is *Salmonella* in most surveys carried out in developed countries (Stern and Kazmi, 1989). It is also one of the principal causes of acute diarrheal diseases in young children in developing countries (Glass et al., 1983). The minimum infective dose has yet to be determined but it may be low. One researcher developed campylobacteriosis after consuming 500 cells in 180 mL of milk (Robinson, 1981). Although the organism is sensitive to low acidity, certain foods may protect the organism from gastric juices, for example, high liquid content such as milk or water, high fat content such as hamburger, and foods with a good buffering capacity. The incubation period is normally 2 to 7 days after which symptoms of watery, bloody, or mucoid diarrhea, abdominal cramps, fever, and nausea develop. Recovery usually takes place spontaneously within a week but symptoms, particularly cramps, recur in 25% of cases. Jejenum, ileum, colon, and rectum may all be affected with hemorrhagic lesions and acute inflammation (Mandal et al., 1984). Although actual mechanisms of infection are not completely known, the organism is able to attach and colonize epithelial cells, then invade them and at the same time produce a cytotoxin and an enterotoxin. At present, invasiveness seems to be the most important factor in pathogenicity (Newell, 1984; Pange et al., 1987).

2.2.2. Foods Implicated

Campylobacter is frequently found in poultry, cattle, swine, sheep, dogs, cats (particularly puppies and kittens), and rodents. The organism, unlike *Salmonella*, does not occur in reptiles and amphibians because temperatures of $\geqslant 30°C$ are required for its growth. Poultry is probably the greatest source of *Campylobacter* for human infections. More than 50% of carcasses (up to 92% in one study) are contaminated and the organism can survive on chicken meat refrigerated for several days and frozen for 3 months (Yogasundram and Shane, 1986). Young chickens may be infected directly by adult birds or through water or feed contamination. The organism is also found in 38–59% of freshly slaughtered pig carcasses, but numbers of cells are less than in poultry. *Campylobacter* is rarely isolated from fresh beef, although beef, more than pork, has been implicated in outbreaks (Doyle, 1984). Cattle excrete the organism, particularly in the summer months, and cows have infected their calves but not other adult animals. Raw meat and poultry products are probably contaminated through feces and bile during slaughtering and processing. The organism does not grow in water but has been isolated from streams and the sea as well as sewage, effluent, and soil, although it does not appear to be long-lived in these environments (Blaser et al., 1984). It is assumed from circumstantial evidence that strains in the environment and animal intestines are pathogenic for humans, and the most likely sources for human infection are direct contact with domestic animals and pets and through ingestion of certain foods. The food association with frequently occurring sporadic cases is usually not well established, but outbreaks have mainly originated from contaminated water, raw milk, undercooked chicken and steak, raw hamburger, raw clams, and cake icing (Blaser and Reller, 1981; Park et al., 1982).

2.2.3. Control

Since *Campylobacter* is well established in animals, both domestic and wild, it will be impossible to prevent it from contaminating carcasses and milk. Thus raw foods of animal origin should be thoroughly cooked before consumption.

2.3. *Escherichia coli* 0157:H7

2.3.1. The Disease

Escherichia coli has been known to be responsible for travelers' and infant diarrhea for many years, but foodborne outbreaks are relatively rare in North America. However, in the last few years an apparently new variety of *E. coli* (serotype 0157:H7) has been reported to be a significant foodborne disease pathogen. The first outbreaks associated with *E. coli* 0157-H7 occurred in 1983 in the United States (Riley et al., 1983), and many others have occurred since in nursing homes, day-care centers, schools, and communities in North America and Europe (Hockin and Lior, 1987; Robaeys et al., 1987; Rowe et al., 1987; Smith et al., 1987). The main syndrome arising from an infection of this organism is hemorrhagic colitis, that is, abdominal cramps with bloody or sometimes watery diarrhea, and fever a few days after contaminated food has been eaten. Some victims may go into a coma and some, particularly children, may develop a hemolytic uremic syndrome (Karmali et al., 1983; Neill et al., 1985). In elderly persons, particularly residents of nursing homes, the fatality rate is higher than for most intestinal infectious diseases. The outbreak causing most deaths occurred in London, Ontario, in 1985 when 79 residents at a home for the aged were infected with the organism and 17 died (Carter et al., 1987; Pudden et al., 1985). It is believed that veal patties contained this strain of *E. coli*, which contaminated a cutting board used to prepare ham, turkey, and cheese sandwiches for another meal, without proper sanitation between the two operations. Also, the food handler preparing the sandwiches had recently recovered from diarrhea. These sandwiches were epidemiologically linked to the illness but none was left for testing.

Verocytotoxins, one identical to the shiga toxin produced by *Shigella dysenteriae* type 1, are probably responsible for the invasiveness of the organism and its ability to destroy the microvillae of the colon. However, attachment of the *E. coli* to the epithelial cells is essential before these toxins can be effective. A certain type of fimbrae produced on the cell structure achieves this, and is mediated by a 60-MDa plasmid (Levine, 1987). If one of these two factors is missing the disease does not occur.

2.3.2. Foods Implicated

Insufficiently cooked hamburger, chicken nuggets, raw milk, and possibly raw vegetables have been associated with outbreaks. The frequently occurring sporadic cases, like those from *Campylobacter*, have no known source of infection, but unlike this organism. *E. coli* 0157:H7 does not seem to be such a frequent contaminant of raw meat and poultry.

It has, however, been found in 3.7% of beef, 1.5% of pork, 1.5% of poultry, and 2.0% of lamb samples at the retail level (Doyle and Schoeni, 1987). Although it has rarely been isolated from raw milk and not at all, so far, from milk filters (Clark et al., 1987; Wells et al., 1987), there is some evidence that dairy cattle may be reservoirs. Contaminated beef and raw milk have been responsible for outbreaks (Borczyk et al., 1987; Ostroff et al., 1987), and, in addition, the organism has been isolated from a low percentage of healthy cows (mainly calves and heifers) both from general surveys (Wells et al., 1987) and from herds implicated in outbreaks (Borczyk et al., 1987; Ostroff et al., 1987). The *E. coli* was also found in a 1- to 3-week-old Argentinian calf suffering from coli bacillosis (Ørskov et al., 1987). The apparent rarity of the organism in the environment may be a true picture or it may reflect inadequate testing or sampling methods.

2.3.3. Significance and Control

It is not yet certain whether this organism has been present in the environment for many years although unrecognized, but there is some evidence it is relatively new and appears to be increasing (Whittam et al., 1987). Plasmid transfer within *E. coli* serotypes is possible and other varieties are known to be verocytotoxin producing, for example, 026:H11 (Levine, 1987). Therefore, there is a real probability that hemorrhagic colitis from *E. coli* infections will become a major foodborne disease in the future. Even today, its economic impact in foodborne disease is significant, costing over $46 million in Canada alone (Todd, 1989). Thorough cooking of meats, prevention of cross-contamination, and adequate temperature control of prepared foods should be carried out to avoid infection. Particular care should be taken in homes for the aged, schools, and day-care centers because the organism is most infectious to the elderly and the young. Consumption of raw milk also should be avoided.

2.4. Listeria monocytogenes

2.4.1. The Disease

Listeriosis is caused by *L. monocytogenes* with most isolates identified as serovar 4b. *L. innocua*, *L. ivanovii*, and *L. seeligeri* have been reported as rare animal disease agents and the last two have also caused a few human infections (McLauchlin, 1987). Listeriosis in humans is not recorded frequently, for example, 1 case in 333,000 adults and juveniles over 1 year old, and 1 perinatal case (pregnant mothers or mothers at birth) in 20,000 births in the United Kingdom. These figures are similar to those in the United States and the rest of

Europe, except for France where the incidence is four times higher (McLauchlin, 1987). Generally, the disease is not severe but serious consequences can occur. Adults can have bacteremia and infections of the central nervous system, such as meningitis. Immunocompromised persons are at greatest risk. In the perinatal cases mothers usually have a mild flu-like syndrome and fetuses can be infected (aborted or infected newborn) or young infants cross-infected in hospitals from babies already having listeriosis. The overall mortality rate is 46% (McLauchlin, 1987), although for the larger North American foodborne outbreaks it is somewhat less (29–44%). The infective mechanism is not entirely known and apparently healthy adults can be infected and die. Also, there may be many cases of subclinical infection that as yet cannot be measured.

2.4.2. The Environment

Although listeriosis in man and animals has been recognized for over 60 years, there has been an apparent increase in the disease since the mid-1970s. This is partly because identification of more isolates has been demanded through increased interest in *Listeria*, and also because of changes in agricultural practices that favor growth of the organism, such as use of big bale silage (Fenlon, 1985; Gouet, 1974) and untreated human sludge and animal slurry on land used for crops (Watkins and Sleath, 1981). *Listeria* can survive for 2 months in sewage sludge spread on land, buried carcasses for 4 months, and manure and foods for 1 year or more. The organism has also been found in water, uncultivated ground, leaf litter, silage, and farm and wild animals and birds (Russell et al., 1984). From these sources it is easy to understand how foodstuffs like raw meat and milk become contaminated. Even though it is ingested, however, disease does not necessarily occur since the transitory carrier state is not uncommon. In Copenhagen 1% of the normal population and 4–5% of abbatoir personnel are excreters of *Listeria* at any one time (Bojsen-Møller, 1972). An even higher percentage (8.8%) was recorded by Kampelmacher et al. (1976) for pregnant women in the Netherlands.

2.4.3. Foods Implicated

Sporadic cases are the ones most frequently encountered, although several common source outbreaks have been documented. Food was the suspected source in two of these (raw vegetables in Boston in 1979, and seafood in Auckland, New Zealand, in 1980; McLauchlin, 1987). However, it was the Nova Scotia outbreak in 1981 that confirmed the role of food in transmission of the disease (Schlech et al., 1983). A farmer fertilized his cabbage crop with sheep manure that had originated from infected animals. The cabbages were stored in a root cellar over the winter period and sold for production of coleslaw in the spring. Forty-one persons consuming the coleslaw were ill (7 adults and 34 perinatal cases), of which 18, including two adults, died. The organism was isolated from coleslaw left in the refrigerator of one of the victims, and also from the farm. The cabbages were assumed to be contaminated from the manure and the organism probably grew in the root cellar under the cool conditions (*Listeria* grows at 4°C), and became sufficiently numerous to infect the victims. In 1983 in the Boston area, 49 persons (42 immunosuppressed adults and 7 perinatal cases) drinking pasteurized milk were infected with *L. monocytogenes*, and 14 fatalities, including 12 adults occurred (Fleming et al., 1985). How the milk became contaminated is not known since the plant had apparently pasteurized the milk properly. The largest outbreak recorded took place in California 2 years later when 181 perinatal cases and 133 adults, mostly of hispanic origin, became infected through eating Mexican-style soft cheese (James et al., 1985). There were 65 perinatal and 40 adult deaths. The cheese plant was not operating properly and had mixed pasteurized milk with raw product to produce the cheese. The company went bankrupt and the vice-president was convicted on a criminal charge and sentenced to 60 days in jail and fined $9300 (Anon., 1986). This last outbreak, particularly, stimulated the dairy industry in North America to review its procedures and agree that *Listeria* should not be present in processed food products, such as pasteurized milk and cheese. The U.S. Food and Drug Administration is treating any *Listeria* organism as a potential pathogen and has recalled ice cream and soft cheeses from the retail market. In 1987 recalls of specific brands of Swiss soft cheese and Danish blue cheese were carried out internationally. Listeriosis was associated with consumption of the Swiss cheese (Anon., 1988).

2.4.4. Control

Control measures should include disinfection of farm premises where animal listeriosis is suspected, and also of infected animals including mastitic cows, proper silage fermentation to lower the pH, biothermal treatment of manure and other waste material, and pasteurization of food of animal origin (Russell et al., 1984). *Listeria* may not be completely eliminated, but its incidence should be reduced since it can have very serious consequences for some persons.

2.5.   *Vibrio* Infections

2.5.1.   Vibrio cholerae

Cholera, cased by *Vibrio cholerae* 01, has been a major cause of illness and mortality, especially where masses of people live or travel together and sanitation is poor. It is endemic in eastern India and Bangladesh and has spread in seven pandemics throughout the world along well-traveled trade routes from 1817 to 1971 (Sakazaki, 1979). It was this disease in London in 1854 that John Snow linked to contaminated water supplies, and polluted water is still the major way cholera is spread (Last, 1986). Inadequate environmental sanitation control and lack of potable water, especially if there is heavy demand on river or canal water for drinking, washing, or disposal of sewage, are important factors in the persistance of the disease. Explosive outbreaks may follow consumption of contaminated food. Climate influences the progression of the disease in the Indian subcontinent. In the Calcutta area of the river Ganges the peak occurs at the hot dry season, whereas in Dacca on the Bramhaputra delta the hot wet monsoon season fosters the most cases (Sakazaki, 1979). Persons of the lower socioeconomic stratum are most likely to be affected because they are often forced to use polluted water and also may be malnourished, making them more susceptible to infection. Healthy adults require over 10^9 cells to cause an infection (Gorbach, 1983). Typical cholera symptoms include sudden onset of vomiting followed by watery diarrhea without blood or mucus and rapid dehydration. If untreated, the loss of fluid from the gut is up to half of the body weight in 24 hr and the mortality rate is 50–75%. The cells multiply in the intestine after adhering to the epithelial cells and produce an enterotoxin (MW 84,000) that fixes to ganglioside receptors (Mims, 1982). Cells are not damaged, hence no blood or mucus in the stools, but electrolyte transport is affected. Adenyl cyclase is activated to increase the intracellular level of cyclic adenosine monophosphate, which, in turn, forces alkaline ions and water into the lumen. Dehydration can be prevented by ingestion or intravenous supply of an isotonic solution. Recovery is spontaneous. In Queensland, Australia, *V. cholerae* has recently been found in rivers and the organism appears capable of growing there and also in brackish water in the Gulf of Mexico (Anon., 1984; Blake, 1984; Bourke et al., 1987). However, in most water supplies the life of the *Vibrio* is a matter of days and it requires continual contamination from a polluted source to remain dagnerous. The organism, however, lives longer in seawater, and varieties of *V. cholerae* (non-01) and other *Vibro* species are natural residents of certain sea coasts, for example, in the Caribbean and southeast Asian seas. *V. cholerae* and these other species have caused foodborne outbreaks from contaminated fish and shellfish. For instance, mussels harvested from the Naples, Italy, harbor area in 1973 infected many persons and 22 died (Todd, 1987b). Public health officials were criticized for not recognizing the problem sooner and, in fact, some were given suspended prison sentences. The polluted mussel beds were eventually raked over and destroyed. Boiled crabs, raw oysters, and probably shrimp have been implicated in causing cholera in the United States (Blake, 1984). Apart from marine products, *V.* cholerae has contaminated raw vegetables including lettuce and cucumbers, cooked rice, and dates (Roberts, 1984) to cause outbreaks. Experimental inoculation studies confirm that the organism grows well on certain vegetables, for example, up to 10^8/g on rice, courgettes, and fennel, but did not do so on others, for example, bean sprouts or mushrooms (Kolvin and Roberts, 1982).

2.5.2. Vibrio parahaemolyticus

V. parahaemolyticus is the most important foodborne disease agent in Japan where much raw fish and shellfish are eaten (Sakazaki, 1979). It has also caused incidents in the United States, for instance, at clam bakes. Since this species is naturally present in the seawater in summer months, no polluted water supply is necessary for a source. Cooling cooked clams may be spattered with drops from raw shellfish above them awaiting cooking, and the organism will quickly grow under hot summer ambient temperatures. Cooked crabs, lobster, and shrimp have also been implicated. The sympotoms of nausea, vomiting, abdominal cramps, diarrhea, mild fever, and prostration usually take over 12 hr to develop.

The organism produces a heat-stable protein hemolysin (45,000 Da) that is lethal to mice, causes ileal loop dilation, and kills cultured intestinal epithelial and heart muscle cells. This hemolysin and probably other toxins are responsible for the syndrome (Glatz, 1987). The illness is not as severe as cholera and recovery, usually requiring little medical attention, occurs in 2–5 days. The more recently recognized *V. vulnificus*, however, can cause septicemia and death.

2.5.3. *Vibrio vulnificus*

Vibrio vulnificus is found only in warm waters and has caused gastroenteritis in persons eating raw oysters in the southern United States (Johnston et al., 1985). Wound infections from this and other marine *Vibrio* species have also been reported as far north as New Brunswick, Canada (Abbott, 1986; Tilton and Ryan, 1987), but *V. vulnificus* is the only one that is likely to cause death. Deep wounds, punctures, and skin abrasions can be penetrated by these organisms causing cellulitis and sometimes septicemia and those persons involved in harvesting, cleaning, and shucking shellfish are at high risk unless gloves are worn. Liver damage, for example, through infections or alcoholism, releases iron that favors the growth of the organism and makes it more likely the result will be fatal (Ratner, 1987; Tilton and Ryan, 1987).

2.5.4. *Significance and Control*

In summary, for foodborne disease in the United States, *Vibrio cholerae* 01 is associated with cooked seafood (mainly crabs) and does not appear to need much multiplication after contamination to cause illness; *V. cholerae* non-01 and *V. vulnificus* have caused illness only from consumption of raw shellfish (mainly oysters); and *V. parahaemolyticus* requires a time–temperature mishandling in order to reach enough numbers to infect people (Blake, 1984).

Control measures for infections from *Vibrio* species are limited, because these organisms can be expected in marine products when harvested from warm water. Harvesting and removing contents of shellfish should be done with care to avoid cuts or abrasions. Seafood should be cooked thoroughly (steamed for at least 5–6 min), not allowed to be contaminated with raw shellfish, and eaten quickly. Waders and bathers should ensure they have no open wounds for organisms to invade, and persons with liver conditions should be especially aware of the risks of eating seafood and contact with seawater.

2.6. *Clostridium perfringens*

2.6.1. *Diarrheal Infection*

There are five types of this bacterium determined by different combinations of toxins and enzymes produced. Types A and C are responsible for human intestinal infections. *C. perfringens* type A is widely distributed in soil and is a natural inhabitant of man and animals. However, stressful situations, as can occur before slaughter, lead to penetration of muscle tissue and organs (Hobbs, 1979). At abattoirs meat and poultry tend to be contaminated with spores present in feces. Flies, dust, and sewage sludge also contain the organism and are environmental sources for further spread. Therefore, it is not possible to prevent the organism from being present in food. Spores of *C. perfringens* germinate and grow rapidly under suitable conditions, for example, a cooked protein food that is slowly cooling down. Outbreaks are often associated with foods in bulk quantities, such as stews, gravies, soups, roasts, and meat pies. The organism can multiply every 10–12 min between 43 and 47°C (Gilbert et al., 1984) and can reach large numbers in only a few hours. For instance, a couple prepared chili and ate it for lunch. The remainder was left at room temperature for 6 hr before being reheated for supper. When it began frothing and giving off foul odors, they decided it was not suitable for the meal but decided to confirm this with a taste. They ate two teaspoonfuls each and agreed the product should not be served. However, they had not cooked the chili sufficiently to destroy the bacterial cells, and about 6 hr later they developed nausea, abdominal cramps, gas, and weakness. Subsequent analyses of the refrigerated chili showed that they had probably each consumed 10^9 cells of *C. perfringens* (Todd, 1982).

The cause of the illness is a protein enterotoxin produced in the intestine during sporulation. The enterotoxin (MW 34,362) whose activity increases 3-fold on trypsinization, binds to receptors on the brush border membrane of epithelial intestinal cells, and causes their disintegration through excess accumulation of calcium ions (Matsuda et al., 1986). Pain in the intestinal area and diarrhea are the usual result of this damage with an increased flow of water, sodium, and chloride ions into the lumen (McDonel, 1979). Millions of spores per gram of feces are excreted, which can easily contaminate the environment or food directly if personal hygiene is not meticulous. Type A illness is usually mild except for older persons who occasionally succumb to the disease.

2.6.2. *Enteritis–Necroticans and Pigbel*

A more severe form of poisoning is caused by type C strains, often present in animal feces. However, this occurs only under unusual conditions (Figure 2). In postwar Germany (1946–1949), when nutrition was poor, enteritis–necroticans (Darmbrand disease) affected many hundreds of people (Zeissler and Rassfeld-Steinberg, 1949). Symptoms were prolonged abdominal pain, diarrhea, and sloughing enteritis of the jejenum, ileum, and colon, sometimes with gangrene. The disease was often fatal. In healthy persons the strain is not one that causes

Humans/animal → Pigs →   Type C   → large number → production of β, δ, and
     feces   *C. perfringens*   of cells Φ toxins
                in cooked pork   ingested

Normal diet: Trypsin degrades toxins → no illness
Low protein diet: Low trypsin secretion/trypsin inhibitors → toxins remain →
               enteritis–necroticans

Figure 2. Action of *C. perfringens* type C to cause illness.

illness, because normal trypsin levels in the intestine would degrade the three toxins produced (β, δ, and Φ), but for persons on a low protein diet the amount of trypsin secreted was not enough (Jolivet-Reynaud et al., 1986). The same disease occurs regularly in the highlands of New Guinea and perhaps also in southeast Asia, the west Pacific, and South America (Murrell, 1982). The New Guinea syndrome is called pigbel because of its association with consumption of pork. The normal diet of the highland tribes is low in protein, but on special occasions quantities of pigs are ritually killed and cooked over fires with portions served to those present over several days and also taken some distances as gifts to persons not present. The custom of random human defecation in the bush, the coprophagous nature of the free-ranging pigs, the time–temperature storage abuse of the cooked pork, the feasters' relatively low trypsin secretion, and the consumption of sweet potatoes that contain a trypsin inhibitor, combine to make enteritis–necroticans likely (Figure 2). The mortality rate for untreated cases is between 44 and 85% (Smith and Smith, 1984), but this can be dramatically reduced to 15–25% after administration of a toxoid vaccine against type C *C. perfringens* (Murrell, 1982). Pigbel is probably going to continue to occur in the future as traditional feasting celebrations are hard to change, although rapid treatment can reduce the mortality and suffering.

2.6.3. Control

Although *C. perfringens* does not normally cuase serious infections, these can be prevented simply by limiting the number of cells ingested. Because the organism multiplies very rapidly in meat products, particularly under anaerobic conditions, any prolonged exposure of cooked meats at ambient temperatures should be avoided. In particular, ingestion of suspect foods by persons on a low protein diet, and, therefore, lacking adequate trypsin, should be prevented.

2.7. *Clostridium botulinum*

2.7.1. The Disease

Clostridium botulinum is an anaerobic spore-forming rod that produces a neurotoxin of seven different serological varieties (A–G), although the toxic effect is similar: prevention of release of acetylcholine from endings of cholinergic nerve cells resulting in paralysis of muscles controlled by these nerves. Symptoms include blurred vision, respiratory difficulty, nausea, vomiting, diarrhea and/or constipation, cramps, sore throat, muscle weakness, and paresthesia. The common types involving human foodborne illness are A, B and E. Types C and D are involved with animal diseases and types F and G are rarely found. Type F has caused two foodborne outbreaks, one in Scandinavia the other in the United States (Hauge, 1970; Midura et al., 1972) and G has not been known to cause human illness. Types A and some of B and F are proteolytic, that is, they digest meat to produce a characteristic putrefactive smell. Type E and some of types B, C, D, and F are nonproteolytic, which means their presence in a food is not betrayed by any strong odor. The toxins of E and nonproteolytic B and F require activation by gastric trypsin to give them the same toxicity as other botulinal toxins.

2.7.2. The Environment

Spores have been found in most countries where samples have been taken, although with varying frequency (Hauschild, 1989). Type A occurs predominantly in the western United States, Brazil, and Argentina, proteolytic type B in the eastern United States, and nonproteolytic type B in the United Kingdom, Denmark, and The Netherlands. Type E occurs more frequently in northern regions Scandinavia, Poland, Japan, USSR, Alaska, and Canada and also in water—Great Lakes, Baltic Sea, and Caspian Sea. Sediments between Denmark and Sweden and from the Baltic Sea shores contain up to 350,000 spores/kg, larger numbers than expected from land runoff, especially as type B is the predominant soil type in Denmark (Huss, 1980). Active growth of the organism probably occurs in this marine environment, in the St. Lawrence estuary, and in shallow and brackish lakes (Hauschild, 1989).

The organism is also present in healthy animals as small numbers of spores, but once the animals die growth and toxin production can take place in the carcasses. Gills and intestines of live fish are the most heavily contaminated (Huss and Pedersen, 1979), but dead fish contain many more cells. Large-scale outbreaks of botulism in salmon and trout reared in ponds have occurred in the western United States through the cannibalism of infected dead fish (Eklund et al., 1984).

2.7.3. Botulism in the Inuit

Fish-eating mammals also contain the organism and these have caused botulism in the Inuit people when food is prepared in traditional ways such as blubber and flippers "fermented" in skin pouches and jars (muktuk, utjak) at ambient room or outside temperatures. Sometimes the food is buried underground, where temperatures are probably cooler, but occasionally it has been hung up in the air and become "suntanned." Anaerobic conditions quickly develop under these storage procedures through the consumption of oxygen by spoilage organisms and any *C. botulinum* spores present can germinate and multiply. In Canada and Alaska most outbreaks are caused by abuse of seal or whale products, whereas in Japan and the USSR improperly fermented or home-processed fish are more likely to be responsible. Control measures are relatively limited where traditional foods are involved because the disease is relatively rare and there is little incentive to change or abolish dangerous practices. In Arctic Canada, however, there is a trend toward making large quantities of muktuk for sale at retail stores in the frozen state (Todd, 1988b). Any toxin present in frozen meat is preserved and can be destroyed only by thorough cooking, which is not often done. In the past, freezer loads of muktuk have had to be disposed of by health authorities because of contaminated product. Nevertheless, even if people contract botulism fewer of them die because of the presence of antitoxin in medical stations throughout the Arctic and also because of improved air evacuation procedures to fly patients to large hospitals where intensive care facilities are available. These medical care costs, however, are relatively high and the mean cost of six incidents examined in the Canadian Northwest Territories and northern British Columbia was $70,000 per incident ($7200 per case), with evacuation of patients being the most expensive component (31.2%). With an average of 8 cases and 1.5 deaths each year the economic and social cost of botulism in northern Canada is about $2 million, with $1.5 million being the value of lives lost to the illness (Todd, 1988b).

2.7.4.   *Botulism involving Home, Restaurant, and Processed Food*

Botulism has also occurred from consumption of a variety of other foods, some not typical of other foodborne disease outbreaks, for example, chicken pies, marinated kingfish, and home-bottled mushrooms from homes; sauteed onions, garlic in oil, baked potatoes, bottled jalapeno peppers, and potato salad from restaurants; and bottled mushrooms, canned hot peppers, canned tuna, and canned salmon through food processor mishandling (Hauschild, 1989). In the majority of these situations the heat-resistant spores of the organism grew and produced toxin during the subsequent lengthy storage period. In the United States canned tuna and salmon episodes, which cost over $100 million each, mainly in lost business to the industry (Todd, 1985c), improperly made seams or punctured metal probably allowed penetration of spores into the can contents from the hands and overalls of cannery workers. There was no evidence of improper retorting practices. However the organism enters cans, once there the conditions are ideal for growth—anaerobic atmosphere, lack of competitive bacteria, slow cooling, and long incubation before cans are opened. In addition, products like canned salmon, tuna, and vichyssoise soup are usually consumed without further cooking, which would destroy the toxin. These three products have caused illnesses and deaths in the last 25 years.

2.7.5.   *Infant Botulism*

It is not only through preformed toxin that *C. botulinum* can cause illness. In a few outbreaks, but more specifically in cases of infant botulism, the organism germinates in the gut and produces small amounts of toxins to debilitate persons over several weeks, and could be the cause of some sudden infant death syndrome cases (Arnon et al., 1981; MacDonald et al., 1986; Sugiyama, 1982). Although toxins and antitoxins are rarely found in infant sera, spores and toxin frequently occur in stools. Sources of *C. botulinum* for infants are not certain, since spores have been found in several foods and dust, but the link is strongest in honey. Honey that was fed to infants and caused botulism was subsequently found to contain spores of the same type. Because the infective dose may be quite low for some infants, the feeding of these foods to babies of 1 year or less is not recommended.

2.7.6.   *Control*

It is difficult to control the presence of spores of the organism in food generally because of its widespread distribution in soils, water, healthy animals, and root crops, and the consumption of small numbers of spores is not considered a hazard (except perhaps for certain infants). Prevention of spore germination is the key to stopping toxin production through destruction (retorting), inhibition by presence of oxygen, low acidity (pH < 4.6), low a_w (< 0.94], salt (> 7%), sugar (> 50%), nitrite, ascorbates or other additives, alcohol, refrigeration, or freezing. Often a combination of these is used, for example, for canned bacon 7% salt and 120 ppm nitric (Hauschild, 1989).

2.8. Bacillus cereus and Other Bacillus Species

2.8.1.   Bacillus cereus

Bacillus cereus is a spore-forming bacterium that is widely dispersed in soil, crops and dust, but had not been linked with foodborne disease until 1950 in Norway when vanilla sauce used in desserts poisoned 600 persons in 4 separate outbreaks (Gilbert, 1979). The sauce, which had counts of $10^7–10^8$ cells/mL, had been prepared the day before from cornstarch (up to 10^4 B. cereus/g) and left at room temperature in a large container. The symptoms were mild—nausea, abdominal cramps, and profuse watery diarrhea, and recovery was complete after 12 hr. There was little vomiting and fever. Since then many outbreaks have been documented, but rarely as large as the Norwegian ones. Although cereal-based products such as sauces, puddings, pasta, and baked foods have been implicated, so also have potatoes, green vegetables, pork, beef, lamb, and chicken (Gilbert, 1979; Johnston, 1984). The organism is reported to be a major cause of food poisoning from meat in Hungary. It is argued that the Hungarians like their meat dishes well spiced and spices often contain large numbers of B. cereus spores (Ormay and Novotny, 1969; Powers et al., 1976). Cooking is insufficient to destroy the spores that may germinate if the food is subsequently stored at room temperature.

2.8.2.   Diarrheal Illness

The agent causing the gastroenteritis is probably a protein toxin (diarrheal toxin) with a molecular weight of about 50,000 and is produced during the exponential growth phase of the organism. It is destroyed by heat (56°C for 5 min) and by trypsin and pronase. A large infective dose (normally $> 10^6$/g) seems necessary to produce an intestinal infection and the associated diarrheal toxin that stimulates the adenylate cyclic AMP system in the intestinal epithelium. This action is similar to that of cholera and Escherichia coli heat-stable enterotoxins (Johnston, 1984).

2.8.3.   Emetic Illness

In the 1970s a new type of B. cereus food poisoning was reported from the United Kingdom (Gilbert, 1979). Victims suffered from acute attacks of nausea and vomiting with some abdominal cramps but little diarrhea. The incubation period was shorter (1–5 hr) than for the diarrheal type (8–16 hr). The responsible agent appears to be a preformed toxin (emetic toxin) of < 10,000 Da and probably not protein. It is not affected by trypsin or pronase and is heat resistent. Its mode of action is not yet known (Johnston, 1984). The most frequently implicated food is boiled or fried rice from Chinese take-out restaurants. Rice is a major component of most Chinese take-out meals and the cereal is normally boiled in batches and kept for use in various portions, as required. Cooks are reluctant to refrigerate the cereal because it becomes sticky when cool and difficult to manipulate. Boiled rice is normally simply reheated, whereas fried rice is made from boiled rice mixed with beaten eggs and fried in a small amount of hot oil. The rice contains spores of *Bacillus* that germinate and cells multiply once the rice has been kept moist and warm for some time. The warmth is supplied by the proximity of the stove and the bulk-storing techniques used by the cooks. This problem is not restricted to the United Kingdom and has also occurred in other European countries, Canada, United States, Australia, and Japan. It is not certain how widespread the problem is in Asian countries where rice is a staple diet because different cooking practices may prevent undue growth of the bacterium. However, there is recent evidence that *B. cereus* is an important pathogen in India (Jauhari and Kulshreshtha, 1986) and China (Guangxian and Jingyeh, 1986).

2.8.4. *The Environment*

Bacillus cereus is frequently found in pasteurized milk, for example, 86.7% of bottled milk in Romania (Ionescu et al., 1966), and can grow to spoilage levels by producing flavor changes and proteinaceous fat droplets (bitty cream) through lecithinase activity. Reports of illness, however, are rare, probably because of a number of factors: *B. cereus* may be more prevalent in milk at one time of the year than another (Phillips and Griffiths, 1986), not all strains are capable of producing toxins, the refrigerator temperature may be too low to produce enough cells or toxin, and spoiled milk is not usually consumed. The following two incidents, nevertheless, show that both diarrheal and emetic forms of illness have been associated with such products. In the first, two girls eating cream with a dessert developed vomiting and diarrhea 8–10 hr later (McSwiggan et al., 1975). The cream contained about 10^6 cells/g, enough to cause symptoms but not quite enough to spoil the cream beyond palatability. In the second episode spoiled refrigerated milk was used to make pancakes in a school class because the teacher thought that "sour" milk was an appropriate ingredient (Todd et al., 1978). Nine children suffered vomiting and cramps 15–45 min after eating the pancakes. The remainder of the milk had about 10^9 *B. cereus*/mL.

In a similar way the organism is widespread in raw vegetables and has grown to large numbers in beans and lentils that have been soaked prior to cooking (Blakey and Priest, 1980), and also in bean and wheat seeds used in home-sprouting kits (Harmon et al., 1987). Illnesses are rarely reported, but an outbreak from contaminated bean sprouts did occur in Texas (Portnoy et al., 1973).

As seen from the above examples, it is not yet certain what determines the production of the diarrheal and emetic toxins. Foods implicated and different serotypes do not appear to be major determining factors. In fact, in some outbreaks both diarrheal and vomiting type symptoms seem to be present, or even intermediate symptoms. More work remains to be done to find out the significance of environmental conditions on toxicity.

2.8.5. Other Bacillus Species

Other species of *Bacillus*, usually *licheniformis* and *subtilis*, have rarely been implicated in foodborne illness in the United Kingdom and a few other countries. Meat, poultry, pasties, and baked foods seem to be the types of food involved. High counts of organisms in foods are associated with illnesses. Diarrhea is the main symptom caused by *B. licheniformis* and vomiting by *B. subtilis* (Kramer and Gilbert, 1989).

2.8.6. Control

To prevent *Bacillus* from causing illness food should be cooked in small batches and served immediately or cooled quickly. However, cooled boiled rice is not easy to prepare further as fried rice as it tends to clump. Refrigerated milk should not be kept for long periods of time, and soaking beans, sprouts, and so on, should be rinsed from time to time to prevent excessive bacterial build-up.

3. VIRUSES

3.1. Introduction

Most viruses that have been linked to human foodborne illness, for example, hepatitis A, Norwalk agent, and probably rotavirus, are transmitted through contamination by sewage, polluted water, or infected food workers. Since viruses do not grow in food, enough particles must be transferred onto the food to commence infection in the human gut. Foods implicated include salads, shellfish, juices, milk, and sandwiches (Cliver, 1979). Only a few rare diseases, at least to North Americans, involve transmission through other means.

3.2. Machupo Virus

Machupo virus, causing Bolivian hemorrhagic fever with malaise, headache, fever, and prostration, is carried by rodents and excreted in their urine (Simpson, 1984). Therefore, rodent contact with stored cereals is liable to result in contaminated grain, which, if not thoroughly cooked, could cause the fever. Relapses of the disease may occur and the mortality rate is up to 30%.

3.3. Tick-Borne Encephalitis virus

There are several viruses that are responsible for tick-borne encephalitis (Russian tick-borne virus complex, Russian spring–summer louping ill group) (Cliver, 1979; Simpson, 1984). The symptoms include headache, fever, vomiting, weakness, meningitis or encephalitis, and flaccid paralysis with recovery in 3 weeks, although mortality may be as high as 30%. Sheep and goats are infected from ticks and their milk can contain these viruses. Whereas raw milk is the usual medium for transmission early in the season, direct tick bites are more likely to be the cause of cases later in the season in the USSR.

3.4. *Coxiella*

Q fever is also spread through sheep, goats, and dairy cattle that are infested with ticks, contaminated placental and fetal tissues, and aerosols (Cliver, 1979; Marmion, 1984). In addition, the disease can be spread to farm and veterinary workers through direct skin contact with the infected animals. However, some cases occur through ingestion of raw milk. The causative organism, *Coxiella burnetii*, is a rickettsia, which is more resistant to heat and desiccation than most other viruses and can exist in the environment for long periods of time. The incubation period may take several weeks before there is an acute fever, pneumonia, and general malaise.

3.5. *Kuru Virus*

The most unusual of these foodborne viruses is kuru that causes a degenerative disease usually ending in death after 3–6 months (Kimberlin, 1984). Uncooked human brain tissue is consumed in ritual cannibalism in New Guinea and by this means the virus is spread from person to person. It seems women and children are more affected and a genetic factor may be important. The agent and action of kuru is similar to that of a better known viral disease, Creutzfeldt–Jacob disease, which causes progressive brain damage ending in dementia and death. Sheep and pig brains have been suggested as probable sources of virus and if these are consumed uncooked, the virus could start its infection. The incubation period is not known but probably is long.

4. PARASITES

4.1. *Trichinella spiralis*

4.1.1. *The Disease*

Trichinosis is one of the best known parasitic diseases in North America and people are well aware that pork must be well cooked. This, the use of home freezers for storing pork, and the fact that pigs are not fed on garbage to the same extent as in former years probably account for a progressive decrease in the number of cases in the last few decades, although 100,000 cases are still estimated to occur in the United States each year (Bennett et al., 1987). Encysted larvae in muscles of pigs (undercooked pork) develop into adult nematodes on ingestion. This stage is characterized by gastroenteritis in the host. Several days later sexual union between male and female nematodes produces larvae that migrate via the blood to striated muscle where they encyst and remain infective for many months

or years. Fever, edema of the eyelids, myalgia, headaches, and chills occur during
this period. Death may occur if there are massive infections, but mortality today is
generally less than 1% (Acha and Szyfres, 1980). Most sporadic cases of a mild
nature are probably not diagnosed, and the disease in a subclinical form may be
widespread.

4.1.2. The Environment

Trichinella occurs in pigs, wild carnivores, dogs, cats, mice, and rats, but some of
these are more important in spreading the parasite than others (Figure 3). Pigs
will eat feces of other pigs and also rodents if they can catch them. Man may be
infected through nonfermented sausages or rare pork. Human flesh rarely
transmits the organism because bodies are buried or cremated. However, some
east African tribes leave their dead or dying (the ones most likely to contain the
parasite because of a lifetime of exposure) to be scavenged by hyenas, which apart
from bush pigs are the only African animals to carry the parasite. In the northern
regions of North America and Europe, bears, foxes, lynxes, and so on eat carrion
and occasionally attack and eat their own species, such as bears. Although walrus
usually dig for clams they will also eat seals and other mammals if they can catch
them. Seals are also eaten by some whales and bears. Sled dogs in the Arctic are
widely infected through feeding on infected raw meat and any available carcasses
of wild animals. *Trichinella spiralis* cysts do not normally survive in a frozen
condition for long ($-15°C$ for $\geqslant 20$ days or $-30°C$ for $\geqslant 6$ days), but strains

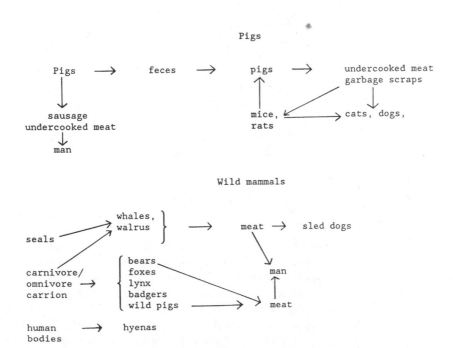

Figure 3.   Life cycle of *Trichinella spiralis* in pigs and wild mammals.

from northern regions are much more resistant. Large outbreaks from wild animal sources are rare, but in 1967 300 persons in Greenland were infected and 33 died (walrus being the probable source) (Acha and Szyfres, 1980) and in the Canadian Northwest Territories in 1974 29 Indians at a hunting camp were infected after eating raw or undercooked brown bear meat and 2 children died (Gullett et al., 1975).

4.1.3.   Control

Control measures, such as the following, may be partially successful in reducing the impact of the disease:

1. Heat-treat garbage and abattoir wastes before feeding to pigs.
2. Inspect carcasses for cysts (mainly for massive infections).
3. Cook pork and wild carnivore meat to $\geq 77°C$ or freeze to $-15°C$ for ≥ 20 days.
4. Heat or freeze nonfermented pork or game sausages properly before consumption.
5. Have adequate rodent control around pig farms.

4.2.   Taenia

4.2.1.   Taenia saginata

Tens of millions of cases of tapeworm infection are estimated to occur throughout the world. *T. saginata* infects man from consumption of inadequately cooked beef and may be increasing in Europe from the demand for more rare steaks. Infected persons excrete eggs (ova) and may defecate on ground where cattle graze, particularly if they are cattle herders. Sewage sludge on fields is also a source of eggs. In addition, birds and dung flies can also transport the parasite to areas where cattle feed. Eggs last for 3 weeks in hay (Healy and Juranek, 1979) and many months on pasture land (Acha and Szyfres, 1980). Humans suffer from gastroenteritis, weight loss, decreased appetite, and nervousness.

4.2.2.   Taenia solium

Man is also the definitive host for another tapeworm, *T. solium*, which is more associated with developing countries, especially where pigs are raised in primitive and unsanitary conditions. However, there are two forms of this disease (Figure 4). Intestinal adult tapeworms causes taeniasis and larval infection causes cysticercosis. The tapeworm may infect a person for up to 25 years with symptoms similar to those associated with *T. saginata*, but often the infection is subclinical. Ova (eggs) are continually excreted and contaminate the ground where pigs and occasionally other mammals feed, and a larval stage develops in various organs causing cysticercosis. When raw or undercooked flesh of infected animals are eaten by man the cycle is complete. However, man can also become the intermediate host by developing cystricerci through (1) ingestion of ova from

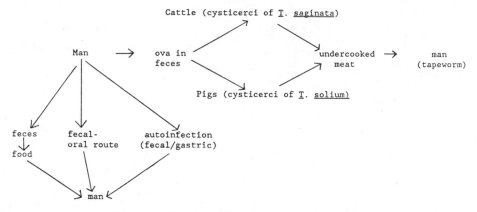

Figure 4. Life cycle of *Taenia saginata* and *T. solium.*

human feces that may contaminate food or water, (2) autoinfection by the fecal–oral route, and (3) reversed peristalsis or vomiting. The last route has not yet been proven. Under all these conditions any ova reaching the stomach would liberate the embryos that in turn would cause larval migration from the intestine into the body tissues. Cysticercosis of man is a serious illnesss and causes many deaths and neurologic disorders because the cysticerci reside in brain tissue, meninges, eye, heart, liver, and lungs.

4.2.3. Control

Control measures have been limited so far, especially in areas where poverty prevents proper separation of residents and their animals, lack of sanitation, and poor understanding of the problem. Also, slaughtered animals are usually not inspected. Abattoirs, properly operated, could freeze meat long enough to destroy the organism. In westernized countries the popularity of undercooked beef increases the chances of *T. saginata* infection. Good therapeutic treatment with proper drugs can eliminate the adult tapeworm but not the cysticerci.

4.3. *Toxoplasma gondii*

4.3.1. The Disease

This is one of the most widespread parasites in the world. One-third of the population is estimated to possess antibodies against these organisms (Acha and Szyfres, 1980). However, although infection is frequent, clinical disease is rare. It is estimated there are 2.4 million cases in the United States each year (Bennett er al., 1987). Most of them are mild with fever or no fever, a type of lymphadenopathy, and are often not properly diagnosed. More serious forms include severe visceral infections, myalgia, arthralgia, myocarditis, and pneumonia. Congenital infection mentally impairs 3300 infants in the United States alone with lifetime care costs of $430 million per annum (Roberts, 1985). The intracellular protozoan parasite

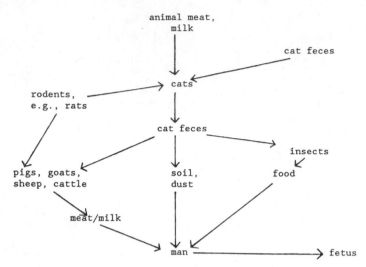

Figure 5. Life cycle of *Toxoplasma gondii.*

exists in three forms (Healy and Juranek, 1979): (1) a proliferative form that multiplies rapidly in the intestinal tissues and migrates to the lungs and other organs via the lymph vessels and blood stream; (2) in reaction to the body's immune response a cystic form develops in the tissues and persists in the host for its lifetime, particularly in the brain; ingestion of inadequately cooked meat of pigs, goats, sheep, and, to a lesser extent, cattle can start an infection of the proliferative form, as can (3) oocysts that form only in the intestines of the cat family. They have to remain in the soil for 2–4 days before sporulating to become infective. Sporulated oocysts are resistant in the environment and may last up to a year in soil. One-quarter to nearly one-half of all domestic cats are seropositive, and the older the animal the more likely it is to be positive. However, even infected cats excrete oocysts for only a short period at any one time (3–15 days). Maybe one-third of human infections originate from meat (pork, mutton, goat, beef) containing the cystic form and the rest from food contaminated by oocysts, and oocysts directly into the mouth (infants) through soil and dust containing cat feces. Unpasteurized goat and cow milk may also be an important source of the parasite for man and cats. Flies and cockroaches may help transfer oocysts from feces to food. The cat is the complete host and other animals including man the intermediate hosts only, where feces do not play any role in the spread of the disease. The complicated life cycle of this parasite is illustrated in Figure 5.

4.3.2. Control

Control measures are limited but the following can reduce the chances of infection:

1. Keep cats off farms where sheep, cattle, and pigs are reared.
2. Rear pigs intensively to reduce exposure of the animals to rats and cats.

3. Pasteurize goat and cow milk.
4. Do not keep cats as pets in homes with pregnant women or infants used to crawling outside.
5. Dispose of cat litter and feces before the oocysts can sporulate and treat sandboxes with boiling water.
6. Maintain fly and cockroach control.
7. Avoid eating raw or partially cooked meat.
8. Wash hands thoroughly after contact with cat faces or soiled litter.

4.4. *Giardia lamblia*

4.4.1. The Disease

Giardia lamblia is a parasite responsible for abdominal cramps and mucoid diarrhea, vomiting, and weight loss 1–4 weeks after it has been ingested. The normal medium for this organism is raw or inadequately treated water (Healy and Juranek, 1979; Kulda and Nohynkova, 1978; Meyer and Radulescu, 1979). *Giardia* apparently is much more widespread than it used to be and is a cause of major concern in small communities where it is too expensive to build a water filtering system. Normal chlorination does not destroy the resistant cysts. Campers have become infected through ingestion of river water containing sewage and also from apparently pure mountain streams; wild mammals are believed to be a continuing contaminating source. Ski resorts have had their giardiasis problems, probably because of a major influx of people into a small community without an adequate potable water supply. In Aspen, Colorado, in 1965/1966, 123 of 1096 skiers suffered from giardiasis. Well water was contaminated by leaking sewage pipes (Moore et al., 1969). In 1982 the resort town of Banff, Albert, had a major problem with the parasite and 121 cases including tourists were diagnosed between February and April (Wilson et al., 1982). The reservoir of the muncipal water supply high in the mountains was examined and beaver swimming in it were found to excrete *Giardia*. One of the victims was a bartender from Peach River, further north in the same province (Campsall et al., 1982). He had been on vacation in Banff in mid-February. At a wedding reception in March he was serving drinks with ice lifted from the ice bucket with his hands rather than tongs. As a result of this action 55 of the 100 guests developed the disease over the next 2 weeks. This parasite is now assumed to be native to the Rocky Mountains, and both Canadian and United States mountain parks notices alert visitors to the dangers of giardiasis from drinking untreated stream and lake water.

Giardia has also been responsible for two foodbrone outbreaks in the United States, one involving home-canned salmon (Osterholm et al., 1981) and the other noodle salad (Petersen et al., 1988). In each incident the preparer of the food in the home probably transmitted the organism during handling of the ingredients. Although *Giardia* does not multiply in food, as few as 10 cysts may be required to cause an infection, and, therefore, it is surprising more outbreaks are not reported in view of the extent of the organism in the environment (Petersen et al., 1988).

4.4.2. Control

Control measures include adequately filtered water, avoidance of raw water even in national parks, and good personal hygiene.

4.5. Fish Parasites

4.5.1. Pseudoterranova

Fish parasites have traditionally been considered nuisance organisms, more of a concern for the quality of the fish than the human diseases they may cause. For instance, *Pseudoterranova decipiens* (cod or seal worm) infect cod and related species and are present in the fillets of most cod caught in eastern Canadian waters, North Sea, and around Iceland (Hafsteinsson and Rizvi, 1987). Arctic stocks are not infected probably because the sea temperature is too cold for early development of the parasite. Worms found in fillets do not stimulate sales even though human infection is not likely if the fish is well cooked. Marine mammals, particularly the seal, are the definitive hosts that become infected through fish consumption. Their feces, in turn, serve to infect the fish. Marine copepods become the first intermediate host followed by invertebrates, including crustaceans, that act as the second intermediate host. Therefore, if raw fish is eaten there is a risk of illness, although relatively few cases of phocanemiasis have been reported (Hafsteinsson and Rizvi, 1987).

4.5.2. Anisakis

The most serious condition is caused by *Anisakis* worms that may penetrate the human intestinal wall to embed themselves in the submucosa and cause lesions. The symptomology may resemble an ulcer or gastric tumor and surgery is often performed. Raw, lightly salted, pickled, or smoked fish, particularly in The Netherlands and Japan, have caused hundreds of cases. Squid and octopus in Japan are also suspected of containing infectious organisms (Figure 1). Although the Dutch have been eating raw herring for centuries, the *Anisakis* problem was recognized only in 1938 when whole herring were kept cold at sea instead of being gutted soon after harvesting. The cold seems to stimulate the larvae to migrate from the gut into the fish muscle, and are, therefore, not detected on gutting at the processing stations (Healy and Juranek, 1979). The current interest in consumption of raw fish at sushi bars is a potential risk for increase in infections, although most fish eaten do not contain the parasite (Oshima, 1987).

4.5.3. Diphyllobothrium

Tapeworm *Diphyllobothrium* is most prevalent in northern temperate regions, and residents of the Baltic area and Inuit fisherman are those mostly infected. Pike, walleye, perch, eel, burbot, trout, and salmon are the main food fish involved. The adult tapeworm lives in man and other mammals, such as bear,

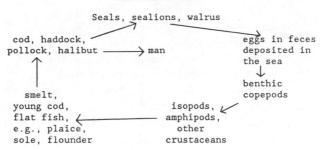

Pseudoterranova

Seals, sealions, walrus

cod, haddock,
pollock, halibut ⟶ man

eggs in feces
deposited in
the sea

benthic
copepods

smelt,
young cod,
flat fish,
e.g., plaice,
sole, flounder

isopods,
amphipods,
other
crustaceans

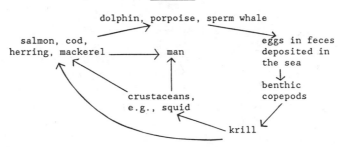

Anisakis

dolphin, porpoise, sperm whale

salmon, cod,
herring, mackerel ⟶ man

eggs in feces
deposited in
the sea

benthic
copepods

crustaceans,
e.g., squid

krill

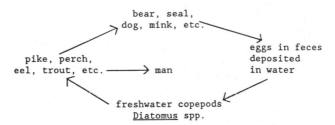

Diphyllobothrium

bear, seal,
dog, mink, etc.

eggs in feces
deposited
in water

pike, perch,
eel, trout, etc. ⟶ man

freshwater copepods
Diatomus spp.

Figure 6. Life cycle of marine and freshwater fish parasites affecting man.

seal, mink, pig, dog, and cat and may reach 3–10 feet in length. Eggs are released, which, when deposited into water, hatch into ciliated embryos. These may be consumed by copepods (intermediate host) that are, in turn, eaten by fish (Figure 6).

4.5.4. Control

The best control procedure is to make sure fish and crustaceans are well cooked, frozen, or properly salted and pickled before consumption.

5. MYCOTOXINS

5.1. Introduction

Aflatoxins and other mycotoxins can render grain, nut, and fruit crops unsuitable for human and animal feed. In addition, farm animals that have been fed moldy feed have often died. Human mycotoxicosis, however, has rarely been documented. Epidemiologic studies have correlated prevalence of chronic disease, particularly hepatoma in Africa and endemic nephropathy (fatal renal disease) in the Balkan countries in Europe with the degree of aflatoxin and ochratoxin contamination of foods, but acute toxicity has rarely been observed (Busby and Wogan, 1979).

5.2. Aflatoxins

Outbreaks of hepatitis have been recorded from Uganda (Serck-Hanssen, 1970), India (Krishnamachari et al., 1975), and Kenya (Ngindu et al., 1982) where moldy grain, mainly maize, was eaten. The best documented was in the Machakos district of Kenya in a mixed farming area. In 1980 the rains were sparse and there were food shortages. In contrast, the rains came early in 1981 and were heavy and prolonged. Stored foods including maize, sorghum, millet, beans, and peas were eaten once it was known that the 1981 crop would be successful. Unfortunately, some of these foods, mainly maize, contained high levels of aflatoxin and 20 persons were admitted to hospital with acute hepatitis between late March and early June. Twelve of these died with liver tissue at necropsy containing up to 89 ppb aflatoxin B_1. Maize from their homes had levels up to 12,000 ppb. The high levels of aflatoxin in food probably occurred because of two situations: (1) in 1980 even moldy produce, normally discarded, was kept because of the poor yield that year, and (2) in 1981 the moist conditions allowed growth of *Aspergillus* on stored foods. It is interesting to note that many grain-pecking local doves died in March after the rains had begun, and village dogs, which had consumed uneaten food from meals refused further food and died a few days before human illness occurred. In the Indian outbreak deaths of local dogs also preceded human hepatitis. Acute aflatoxicosis seems to be rare but chronic affects of mycotoxins may be more widespread, and if food shortages are noted in areas where mold growth could occur health authorities should counsel residents of the dangers of eating moldy food, and, if possible, provide replacement food, as occurred in Kenya once the outbreak had been recognized.

5.3. *Claviceps, Fusarium,* and *Cladosporium* Toxins

Historically, ergotism and alimentary toxic aleukia have been important human mycotoxic diseases. Recent cases, however, are rare. Wet springs and dry summers are particularly conducive to growth of *Claviceps purpurea* on rye and other cereal grain and epidemics were often associated with famine, which meant that the new crops with fresh, highly toxic sclerotia were likely to be eaten (Busby and Wogan, 1979). Ergotism is of two types, gangrenous and convulsive, both often fatal. The last major epidemics in the west occurred in the late nineteenth century. Problems still occur in other areas such as India where millet may be infected, but the illnesses are less severe. *Fusarium sporotrichioides, F. poae, Cladosporium,* and other fungi were probably responsible for epidemics of alimentary toxic aleukia in the USSR in the first half of the twentieth century, particularly from 1942 to 1947 (Joffe, 1983). Wheat, millet, and other grains that had overwintered often became infected with these fungi, but because of starvation of the population in rural areas these crops were eaten. Fatality rates were as high as 80% if enough toxic grain was eaten, and bone marrow destruction and pulmonary hemorrhages were frequent.

6. SEAFOOD TOXINS

6.1. Paralytic Shellfish Poison

6.1.1. The Disease

Paralytic shellfish poison (PSP) occurs primarily in temperature seas—the western Pacific coasts of the United States (including Alaska) and Canada, the Bay of Fundy and St. Lawrence estuary in eastern Canada, the North Sea, and the Chilean and Japanese coasts. However, outbreaks have been reported in recent years in warmer waters, for example, Philippines, Malaysia, Indonesia, Guatemala, and Venezuela (White et al., 1984; Yentsch, 1984). Dinoflagellates *Protogonyaulax* (*Alexandrium*) and *Pyrodinium* occur naturally in these and other coastal areas and are capable of producing a variety of related neurotoxins, substituted tetrahydropurine bases, called saxitoxins. Sodium conductance in nerve and muscle membranes is blocked by these causing paralysis of muscles. Symptoms of poisoning including tingling or burning sensation of lips, tongue, face, and extremities in limbs. Paresthesias may progress to numbness with difficulty in moving the body or even standing up. Other symptoms are lightheadedness, floating sensation, weakness, incoherent speech, thirst, headache, and temporary blindness. Respiratory difficulty may be encountered and if artificial respirators are not available, death may occur. Fatal levels of toxin have ranged from 5000 to > 30,000 mouse units. Even with adequate health care the case–fatality rate is about 8.5%. There are about 1600 cases with 300 deaths worldwide each year (Halstead and Schantz, 1984). However, poisoning does not result from ingestion of free dinoflagellates in the sea but only when they have been accumulated by marine life. The predominant food source that has caused most outbreaks in temperate seas is shellfish, particularly clams, oysters, mussels, cockles, scallops, and whelks. In tropical seas, however, chub mackerel and scabs feeding on *Pyrodinium bahamense* var *compressa* have caused PSP, and crabs of several genera have been implicated in a similar syndrome but the exact nature of the toxins is unknown (Halstead and Schantz, 1984; Maclean and White, 1985).

6.1.2. The Environment

For marine life to be toxic, dinoflagellate blooms are normally required (Figure 7). These tend to be seasonal, at least in temperate waters, and depend on water temperature, salinity, amount of sunshine (for photosynthesis), upwelling of cold, nutrient rich water, winds, tides, and elimination of toxic metal ions (these may be chelated by humic acid brought in from rivers). Areas of high tidal dissipation in the world may be important for dinoflagellate growth because these are the same as where PSP tends to occur, but a direct link has to be confirmed (Yentsch, 1985). In North America, the shellfish are most toxic in late summer and early fall. According to Yentsch (1985) red tides appear in one of four ways: (1) during development of the thermocline in spring and its breakup in the fall, (2) after a major storm to disrupt the thermocline, (3) gradual off-shore flow of surface waters replaced by on-shore subsurface waters at the thermocline, and (4) local microclimates in bays and estuaries. The last may be important for differences in shellfish toxicity levels that vary from one location to another in the same general area. As water temperatures fall, hypnocysts, or resting zygotes, form and remain in sediment until weather conditions in the spring bring cysts to the warmer surface where they germinate and grow rapidly. Unusual conditions, such as late summer hurricanes, may stir up more sediments than usual, and also break up the thermocline containing vegetative cells, and large blooms may follow. Hypnocysts are also toxic, and if water temperatures are warm enough for shellfish to feed, the cysts in shallow water, if disturbed by dredging or transplanting of shellfish, may be ingested by shellfish to render them toxic even in winter months. Shellfish vary in their ability to detoxify—some in weeks, others up to a year, for example, the Alaska butter clam occurring on northwest Pacific coasts. Thus, much of Alaskan and northern British Columbia coasts are permanently closed to the gathering of shellfish.

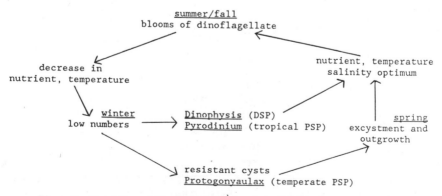

Figure 7. Paralytic and diarrhetic shellfish poison dinoflagellate annual cycles.

6.1.3. Control

Harvesting is carefully controlled in many countries because of the dangers of illness, and in Canada, the United States, and the United Kingdom areas with $> 80 \, \mu$g toxin/100 g shellfish are closed to commercial fishing and warning notices posted. Most illnesses occur because local harvesters or tourists do not see, or ignore, these notices. Samples of shellfish from specific areas of the coastline at risk are extracted and tested by injection into mice. Areas then may be closed or opened depending on the results. Since the toxin is not destroyed by heat or by acid preservation (pickling) there is no way of eliminating the toxin from the seafood once it has been harvested and no antitoxin is currently available. Therefore, in areas where testing is not done routinely and no posting of shorelines is carried out, caution should be taken in eating any shellfish if there is a history of PSP in the area. However, the following two examples show how blooms can occur in locations where problems have not been apparent before.

6.1.4. Recent problems

Four cases of PSP were recorded for the first time in Newfoundland after mussels gathered from one bay in the eastern part of the island were eaten in September 1982. Up to 1200 μg toxin/100 g was found in mussels sampled locally and also in two other Newfoundland sites separated by many hundreds of kilometers. In the preceding 30 or so years periodic testing had yielded no toxic shellfish and in the 2 years following the incident toxin levels gradually dropped to $< 80 \, \mu$g/100 g. Cysts of *Alexandrium tamarense* at 30–150/cm^2 were found in sediment in 15 samples near the outbreak location the following May. These data indicate that the organism has probably always been there but has never reached toxic levels before or since. The probable reason for blooms was an unusually long period (several weeks) of warm, sunny weather preceding the outbreak (White and White, 1985).

In August 1987 residents of five coastal villages in Guatemala gathered small clams from the beach and boiled them to make soup. During harvesting these clams became toxic over a 2-day period and 187 persons fell ill with PSP and 26 died (S. Hall, Food and Drug Administration, Washington, D.C., personal communication). The implicated clams contained 7500 μg PSP/100 g (Rodrigue et al., 1988). The clams are seasonal and are harvested during the summer months to supplement the villagers' diet. Aerial photographs showed than an off-shore bloom of *Pyrodinium bahamense* var *compressa* was carried onto the beach and rendered the clams toxic. Some villagers claimed to have seen a red tide but did not consider it significant because the previous year a similar tied had occurred without having any impact on health. Medical authorities had no experience with the disease and lack of proper treatment contributed to the deaths. It is possible that off-shore *Pyrodinum* blooms had been present during previous summers, but this was the first time PSP was associated with it.

To avoid similar problems periodic monitoring of shellfish in this region of Guatemala has been initiated (Rodrigue et al., 1988).

6.2. *Ciguatera*

6.2.1. *The Disease*

Ciguatera poisoning is another form of illness caused by dinoflagellates. Although the mortality rate is low ($< 1\%$) the number of cases worldwide may be over 50,000 each year (Ragelis, 1984). The food source of the illness is tropical and subtropical fish and, therefore, the disease affects people mainly in areas where these fish are caught, for example, Caribbean and Pacific island communities. Illnesses have also occurred when travelers have brought back toxic fish from the Caribbean to Canada (Todd, 1985b) and the United Kingdom (Tatnall et al., 1980). With an increasing demand for tropical fish by persons living in temperature lands some illnesses have been reported from persons eating in restaurants in Maryland, Massachusetts, Vermont and purchasing of fish from markets in Ontario and France (Baylet et al., 1978; Ragelis, 1984; Todd, 1985b; Vogt and Liang, 1986). Tourists are increasingly going for vacations to tropical seacoast areas and eating the local fish; they sometimes return home with persistent illness symptoms and visit their local hospitals for treatment. Physicians working in Tropical Disease Units in large hospitals in temperate countries seem to be the medical personnel most aware of ciguatera; so, the disease may be misdiagnosed in small communities. The symptoms are gastrointestinal occurring a few hours after consumption of a toxic fish, followed by neurological problems, such as paresthesias, numberness, loss of equilibrium, anxiety, blurred vision and hot–cold temperature reversal sensation. Muscle pains, pruritus, weakness, reduced blood pressure, and altered pulse may also occur (Anderson et al., 1983; Bagnis et al., 1979). There is evidence, at least in one incident, that the ciguatera toxin with its low molecular weight (1000–1500) passes across the placental barrier to affect fetuses (Pearn et al., 1982). A pregnant woman not only became ill herself from the toxin after eating coronation trout (Australian reef fish), but noticed tumultuous fetal movements and fetal shivering that lasted for 40 hr. A Cesarean section was performed 2 days later and the infant had a left-sided facial palsy. The baby was put into intensive care and gradually recovered, and seemed normal except he had not smiled at 6 weeks of age. As well as her typical ciguatera symptoms, the mother also experienced excrutiating pain in the nipples during attempted breast feeding.

Recovery from the gastrointestinal problems is normally rapid but much slower for the neurological symptoms. In fact, patients are often very weak for several weeks or even months after the event with recurrences possible if there has been excessive exercise or stress, or consumption of nontoxic fish or alcohol. Like PSP there is no apparent immunity to ciguatera poisoning and toxic fish eaten again could cause similar symptoms. In fact, in some island communities the condition is almost chronic, or, at least, is accepted as a part of life in a population that eats locally caught fish (Lewis, 1984).

6.2.2 *The Environment*

The ciguatoxin originates from dinoflagellates *Gambierdiscus toxicus* and *Prorocentrum* species, which associate with coral reefs (Bagnis et al., 1980; Tindall

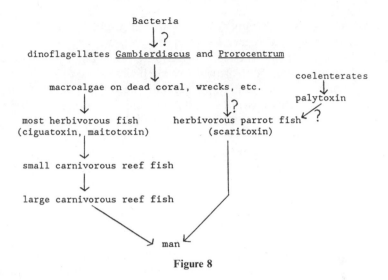

Figure 8

et al., 1984). It appears that these organisms attach themselves to macroalgae that grow mainly on dead coral surfaces (Figure 8). The presence of bacteria may also be important (Carlson et al., 1984). Herbivorous fish then browse on the algae and ingest the dinoflagellates. Ciguatoxin, a fat-soluble cyclic polyether related to okadaic acid, can accumulate in fish fatty tissues, and it appears to travel up the fish food chain as herbivores are eaten by carnivores (Ragelis, 1984). In general, the largest carnivores are the most toxic; these include barracuda, amberjack, red snapper, and grouper. As there is no way of eliminating ciguatera toxin from fish once they are caught, commercial fishing in ciguatera areas should cease. Unfortunately, the extent of a toxic area is not really known. Local knowledge will indicate that one reef is safe and another is not, or that fish caught at one season are more at risk than at another. Since the ultimate source of the toxicity is the macroalgae in association with the dinoflagellates, factors that determine their growth are critical but largely unknown except that the more exposed dead coral surfaces and coral detritus there are, the more likely algal growth will occur (Ragelis, 1984). Therefore, heavy storms, seismic shocks, reef blasting, dredging, and organic pollution are activities that might precede algal attachment and subsequent finding of toxic fish, (Grell, 1983; Lewis, 1984) but few ecological studies have been attempted to establish such relationships. Certainly, windward sides of islands tend to have more toxic fish than leeward sides because the effects of storms are more noticeable on the ecosystem (Lee, 1980). Underwater structures, such as bridges, wharfs, and sea walls, may also provide surfaces for algal attachment (Craig, 1980; Lewis, 1984) and some areas where wrecks occur have associations with ciguatoxic fish (Benchley, 1988).

6.2.3. Control

Some bans have been recommended by island governments, for example, no buying or selling of barracuda, amberjack, or blackjack in Puerto Rico (Ragelis,

1984). Red snapper and grouper, which are in prime demand for local consumption and export, are generally not controlled, although they have been responsible for many of the illnesses. Since fishing is an important industry in many of the tropical islands, control measures are likely to be costly in lost profits and unemployment if they are implemented effectively. Rapid testing of individual fish for ciguatoxin is a possibility, and has been done at the research level with a radioimmunoassay technique in Hawaii (Kimura et al., 1982; Ragelis, 1984). If this method can be refined enough to detect all toxic fish without too many false positive results, then large-scale monitoring would be possible to ensure the sale of safe fish for export and local fish markets.

6.3. Scombroid Poison

6.3.1. The Disease

Scombroid fish poisoning is better understood than some of the dinoflagellate-caused illnesses. It arises not from the fish consuming toxic plankton but from bacterial spoilage after the fish are harvested. Scombroid poisoning is associated with consumption of Scombridae fish, for example, tuna, mackerel, and other species, such as mahi-mahi, that have a high concentration of free histidine in their flesh. Unless these fish are kept cool after being caught, certain bacteria, for example, *Proteus morganii* and *Klebsiella pneumoniae*, which are in the skin, gills, and intestinal tract, start to multiply and penetrate the flesh to convert the histidine to histamine through histidine decarboxylase enzymes (Taylor, 1985). When the fish is consumed, the histamine is absorbed into the bloodstream, to cause symptoms of headache, nausea, a bloated feeling, rash, flushing of the skin, rapid heart beat, burning sensation in the mouth, numbness of face, and diarrhea. The action of the histamine is probably facilitated by other compounds in the fish, including inhibition of histamine-catabolizing enzymes diamine oxidase and histamine-N-methyltransferase in the victims' intestines (Arnold and Brown, 1978; Taylor et al., 1984). Cadaverine and aminoguanidine, amines found in spoiled fish, potentiate the uptake of histamine through apparent enzyme inhibition (Lyons et al., 1983). Detection of histamine in fish is usually by chemical means, for example, fluorometric assay, but is not routinely done in laboratories unless illness or spoilage problems are suspected. Incidents of scombroid poisoning have been reported from several countries, but published data are limited to a few, for example, Japan, United States, Canada, United Kingdom, and New Zealand (Taylor, 1985). Illnesses may be related to types of fish eaten. For instance, in Britain once herring stocks diminished through overfishing in the 1970s demand for mackerel went up. Landings rose from 8800 tons in 1972 to 320,900 tons in 1978, and even though only 10% was locally consumed this represents a considerable increase (Turnbull and Gilbert, 1982). Between 1976 and 1982 there were 65 incidents involving smoked mackerel with most (37) occurring in 1979 (Taylor, 1985). Once the problem became recognized there were fewer outbreaks in succeeding years through better hygienic practices. Many fermented foods contain high levels of histamine but generally do not cause illness, for example, sausages, wine, sauerkraut, cheese, and fermented fish. However, a few incidents involving Gruyere, Gouda, and Swiss cheese have been documented. A *Lactobacillus buchneri* strain capable of producing histamine was isolated from the incriminated Swiss cheese (Summer et al., 1985). In addition, patients on izoniazid therapy for tuberculosis treatment have become ill with typical scombroid type symptoms when eating Cheddar, Cheshire cheese, skipjack, and tuna. Izoniazid may potentiate histamine poisoning by inhibiting normal detoxification action, for example, the drug inhibits diamine oxidase (Taylor, 1985). The levels of histamine present in fish and other foods capable of causing illnesses, therefore, are going to vary. Amounts as low as a few mg/100 g to several hundred mg/100 g have been determined from foods causing illness.

It appears that scombroid poisoning is a worldwide phenomenon and is probably the most common type of fish poisoning (Russell and Maretic, 1986). However, it is not easy to predict which fish, or even parts of a fish, will contain histamine and which will not. There is some evidence that *Proteus marganii* and *Klebsiella pneumoniae* may contaminate fish after catching and are not present in all fish (only 1 in 10 frozen skipjack in one study) (Taylor, 1985). Also, spoilage will not proceed uniformly in stored fish. A combination of presence of histamine-producing bacteria, warm storage temperatures, and enough time for the decarboxylases to penetrate the muscles may not be available for all of a catch. In addition, spoilage can occur later in fish awaiting or during processing, for example, smoking or canning. These conditions may explain why there is often considerable variation in the amount of histamine found in samples taken from a lot suspected of causing illness (Murray et al., 1982).

6.3.2. Control

Because cooking or even retorting does not destroy histamine, the only way of limiting the level of histamine in fish is through good hygienic practices, for example prompt gutting and proper refrigeration or freezing of fish as soon as they are caught. Guidelines in different countries for regulatory action, however, range from 10 to 50 mg/100 g, and for fish are interpreted as evidence of decomposition or spoilage, irrespective of whether there is a risk for human illness or not. Bartholomew et al. (1987) recommend the following action based on presence of histamine in fish:

< 5 mg% histamine—normal and safe for consumption
5–20 mg% histamine—mishandled and possibly toxic
20–100 mg% histamine—unsatisfactory and probably toxic
> 100 mg% histamine—toxic and unsafe for consumption.

For cheese and other fermented foods not enough is known about the origin of the histamine, and, therefore, no control measures can be suggested. However, histamine or other amine-associated illnesses from cheese, wine, sausages, and so on appear either to be rare or mild.

6.4.   Other Fish and Shellfish Toxins

6.4.1.   Introduction

Many different kinds of fish and shellfish poisoning, as well as ones involving jellyfish, sea anemones, sea cucumbers, octopuses, crabs, and turtles, have been recorded, but limited information is available on the origin of the toxins and how they can be avoided. Most are heat resistant. Scaritoxin (from parrot fish) and maitotoxin (from tropical algal-browsing herbivorous fish) are closely related to ciguatoxin and probably originate from the same or similar dinoflagellates (Bagnis et al., 1974; Halstead, 1984; Ragelis, 1984), although recent evidence indicates that palytoxin from coelenterates may be the source of scaritoxin (Noguchi et al., 1987) (Figure 8).

6.4.2.   Clupeotoxin

Clupeotoxin is associated with herring, sprat, anchovies, and herring-like fish in the tropical Pacific Ocean and Caribbean Sea caught close to shore. Although dinoflagellate plankton are the suspected origin of the clupeotoxin, no organisms or toxins have been specifically linked to these fish (Halstead, 1984). A victim may experience a severe gastrointestinal upset followed by a drop in blood pressure and eventual collapse. The mortality rate is high and death can occur 15 min after the commencement of symptoms.

6.4.3.   Tetrodotoxin

Tetrodotoxin is found in certain puffer fish and many fatalities have occurred from people eating these (10 annually in Japan, Yotsu et al., 1987). In Japan the toxic varieties, including molas, blowfish, and globe fish, are especially sought after, and in restaurants these fish are carefully prepared by trained chefs to obtain the least toxic portions (exclusion of gonads, roe, intestines, and skin) (Fuhrman, 1983). Thin raw slices or cooked chunks have a delectable, delicate taste that makes these fish in high demand (Vietmeyer, 1984). It is also possible that occasional low levels of toxin in these prepared fish give a desirable euphoric effect to consumers. Intoxications are highest during the spawning season because the flavor is better. The origin of the toxin is just beginning to be elucidated. Fish raised in captivity are nontoxic but become toxic when kept in close contact with toxic fish, and bacteria such as *Pseudomonas* spp. are thought to play an important role (Figure 9). One unidentified species of *Pseudomonas* found on the skin of toxic fish produces tetrodotoxin in broth cultures. It is currently believed that this organism produces the toxin that is transmitted to the fish through the skin surface (Yotsu et al., 1987). Tetrodotoxin is also found in a California salamander, Central American frogs, highly toxic Australian blue-ringed octopus, and Japanese ivory shell and trumpet shell. The structure of the toxin, however, has unusual features and no readily identifiable precursor. At present, no bacteria have been identified as a possible source for these amphibians, cephalopods, and gastropods (Mosher and Fuhrman, 1984). Therefore, the environmental contamination of such animals may be the same source or differing sources with a similar metabolic pathway.

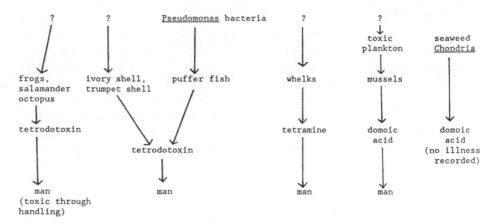

Figure 9. Origin of seafood toxins: tetrodotoxin, tetramine, and domoic acid.

6.4.4. Diarrhetic Shellfish Poison

Diarrhetic shellfish poison (DSP) is one recently recognized as being of dinoflagellate origin (Figure 8). *Dinophysis acuminata* and *fortii* have been linked to thousands of cases of gastroenteritis in Japan and Europe (Kat, 1985; Underdal et al., 1985; Yasumoto et al., 1984). The incubation period is short (about 4 hr) with the main symptoms of diarrhea, nausea, and vomiting lasting up to 3 days with no subsequent ill effects, and no fatalities have been recorded. This syndrome is similar to those caused by some bacteria and not typical of other dinoflagellate poisonings. However, one of the toxins isolated is a derivative of okadaic acid, a compound associated with some other dinoflagellates, for example, *Prorocentrum lima* (Kat, 1985).

6.4.5. Tetramine

Certain whelks (genera *Neptunea, Buccinum, Fusitriton*) are capable of causing severe headaches, dizziness, and abdominal cramps from tetramine found in the salivary glands (Halstead, 1984) (Figure 9). Even boiling the shells for 1 h will not destroy the toxin but removal of the glands will render the whelks edible. Illnesses have been reported from Japan (frequently) and Scotland (rarely) (Fleming, 1971; Millar and Dey, 1987). The red whelks have been eaten in Britain and other parts of Europe without problems being reported previous to the two Scottish incidents but tetramine has been found in the glands. Apparently, the Scottish fish merchants are aware of the problem and routinely remove the glands before exporting large shipments of red whelk (*Neptunea*) to Japan. It is the fishermen and buyers who are less familiar with the toxic nature of the red whelk and a government circular has been issued to explain the difference between the two types of whelks normally harvested, and to emphasize that red whelk should be sold only to merchants and not directly to the public (Millar and Dey, 1987). The origin of the tetramine is not known, but may depend on diet because of the varying degrees of toxicity found in European whelks.

6.4.6. Domoic Acid

In November and December 1987 a strange illness associated with mussels occurred in eastern Canada. Over 120 cases and two deaths were recorded from cultured mussels served in restaurants or bought from stores, mostly in the Montreal, Quebec, area (Perl et al., 1987). These had originated from a river estuary in eastern Prince Edward Island. An initial bout of gastroenteritis was followed in some cases, particularly elderly persons, by neurological problems such as disorientation, memory loss, and inability to communicate. Some of these were hospitalized for over 2 months. Mice injected intraperitoneally with extracts of mussels died after showing unusual neurological symptoms including a typical hind leg scratching syndrome. The source of the problem was sought through an intensive effort by many scientists and domoic acid was identified as the toxin (Wright et al., 1989). A method was quickly developed to quantify the compound and as much as 96 mg/100 g mussels was found in one supply eaten by ill persons. Because domoic acid has been used in Japan as an anthelminthic drug (oral ingestion) without any ill effects (Nisizawa, 1978), it has been questioned that the acid alone could be responsible for the disease, although domoic acid does cause neurological problems in animal studies (Shinozaki and Ishida, 1976; Coyle, 1983; Maeda et al., 1987). The domoic acid is obtained from Japaneses seaweed *Chondria armata* and a related species has been found in Prince Edward Island but not in the area where the toxic mussels were cultured. These were feeding on toxic plankton in a bloom that extended many miles offshore. The specific toxin-producing organism was identified as the diatom *Nitzschia pungens*, the dominant component of the plankton bloom (Bates et al., 1989). (Figure 9). In January 1988, the toxicity of the mussels decreased rapidly because of the die-off of the toxic plankton. This unique episode forced the closure of shellfish harvesting in the whole of eastern Canada until stocks could be cleared through intensive testing. A monitoring program was instituted in 1988 and domoic acid in amounts greater than 20 μg/s shellfish (the action level used to close an area) were found in several estuaries and bays of eastern Prince Edward Island and New bronswick in the fall and winter of 1988. No further illnesses, however, have been reported.

6.4.7. Conclusion

All the fish and shellfish toxins appear to come through the natural food chain in the sea and are not connected with human or animal pollution. At present, control measures are very limited, until adequate rapid testing methods become available and currently depend on local knowledge and experience. Most of the implicated foods are eaten locally within a limited area and do not substantially affect international trade. However, several countries currently do not allow importation of Japanese puffer fish even though there is a demand for it. The whole area of seafood toxins has yet to be explored thoroughly and probably many new ecological chains involving bacteria and dinoflagellates will be elucidated.

7. POISONOUS PLANTS

7.1. Introduction

Many plants are known to be poisonous and are normally avoided, or the toxic components removed or destroyed (Liener, 1980; Scott, 1969), but are occasionally eaten mistakenly as nontoxic species., for example, *Amanita* mushrooms (Rumack and Salzman, 1978). Sometimes these affect commercial foods such as herbal teas. For instance, an Ontario woman was found by neighbors in an acute hallucinating state, with a flushed appearance, dilated fixed pupils, dry mucous membranes, and a rapid heart beat (Todd, 1988a). She was taken to a hospital and diagnosed as a case of anticholinergic intoxication, and atropine was detected in her urine, blood, and gastric contents. All drugs in the home were tested, but these were not the source. However, she had recently begun drinking large quantities of mallow leaf herbal tea in order to lose weight, and this tea was found to contain atropine. Her family physician had treated a girl with similar symptoms earlier, but the case had not been toxicologically verified. Six weeks later a man in Alberta was admitted to a hospital with vomiting, dehydration, dry mouth, dizziness, double vision, difficulty in speaking, and delirium that lasted 12 hr. He had been drinking herb tea of the same brand that the Ontario woman had purchased, and the package, on analysis, contained a high atropine level. The dominant tea leaf present was not mallow (*Malva sylvestris*) but that of a plant of the Solanaceae, although it was not possible to determine the species. An attempt was made to germinate a seed from the tea but this did not succeed. A search of Edmonton retail stores showed that no more packages of this brand of tea were available. Therefore, the degree and source of contamination could not be ascertained.

7.2. *Spirulina*

Another environmentally determined plant illness is associated with *Spirulina*, which is a blue-green alga that is used as a protein supplement, especially for persons on diets (Scrimshaw, 1975). The product is harvested from freshwater or brackish ponds in the wild and the dried algal powder formed into tablets that are sold in health food stores. It is also produced commercially in Mexico, Thailand, and California. Two British Columbia women in separate incidents claimed that illness occurred from eating these tablets (Todd, 1987a). Symptoms of nausea, vomiting, retching, severe abdominal cramps, diarrhea, and weakness were noticed 0.5–2 hr after consuming the tablets on an empty stomach. The effect was milder if taken after a meal. The same symptoms appeared each time the tablets were taken. The tablets consumed by the two cases were from different lots made by the same manufacturer, probably in Mexico. Laboratory tests showed that the tablets were virtually sterile and did not contain bacterial toxins. Pathogenic bacteria do not survive well in the dried alga (Mahadevaswamy and Venkataraman, 1987). However, insect fragments and other vegetable matter were found.

Crushed tablets injected into mice intraperitoneally did not cause acute toxicity associated with some toxic algal strains, sometimes harvested with the *Spirulina*. For unknown reasons, some persons consuming single cell protein, including algae, have developed nausea, vomiting, and diarrhea, probably because of high protein, high nucleic acid, and high uric acid levels (Scrimshaw, 1975). Although the cause of illness in British Columbia is uncertain, it was noted that tablets consumed on an empty stomach had a more noticeable effect than those taken after a meal.

7.3. *Datura*

A third example is accidental contamination of a food by a toxic plant (Patton et al., 1984). A couple was preparing a hamburger meal in their home and the wife added what she thought was seasoning. Immediately afterward she recognized these were seed of Angels' Trumpets (*Datura suaveolens*) drying above the stove. She suggested to her husband that the food be thrown away but he recommended that the seeds be removed from the meat, which was attempted. After the meat had been cooked, they ate one hamburger each and an hour later they both had "high" feelings and began hallucinating. They also had severe diarrhea and a rapid heart beat before the husband collapsed and both were hospitalized. Gradual recovery took 2 days before they were discharged. Although they thought they had removed all the seeds from the meat clearly some had remained. Poisonings from the accidental mixing of seeds into food have been reported previously, but not in recent years (Scott, 1969).

7.4. Nitrate in Spinach

Nitrate poisoning from vegetables is rare but has been documented in infants over 20 years ago (Simon, 1966). In a more recent incident a 3-month-old baby developed vomiting, diarrhea, rapid heart beat, and a blue skin color, a condition known as methemoglobinemia normally associated with nitrite poisoning (Todd, 1985a). Various sources of nitrite were sought without success. The infant had been put on solid food including boiled vegetables the week before the illness, and when put back on infant formula had recovered rapidly. All of the vegetables were obtained from the retail store except for the spinach. Samples of this showed not nitrites but 996–1076 ppm nitrate. The father worked in a fertilizer plant and had heavily fertilized his garden soil. The spinach had been grown in the summer and frozen. The freezing had prevented any bacterial action from rendering the nitrate harmless.

7.5. Control

Acute plant poisonings are relatively rare, but care should be taken when wild plants or their parts, for example, mushrooms, fern fiddleheads, berries, and roots, are eaten. Cooking does not always destroy any toxins present. Poor harvesting techniques may cause a poisonous plant to be identified as edible or a mixture of plant parts to be gathered by mistake. Herbal preparations, often prepared under limited control conditions, are foods most at risk for this type of problem.

8. CONCLUSIONS

Many incidents of acute foodborne illness are documented throughout the world annually from pathogenic bacteria or their toxins. Most of these have been known about for many decades. Yet, control measures have not been very effective. Salmonellosis, campylobacteriosis, and infections from *E. coli* 0157:H7 are increasing each year in the developed countries. *Clostridium botulinum* is being implicated in foods previously unsuspected, such as onions, garlic, potatoes, and other vegetables. Recent contamination of soft cheeses, meat, and fish products by *Listeria* gives great concern to industries and governments alike. Parasitic infections and illnesses from seafood toxins are increasingly being identified on a worldwide scale. If the agent implicated is well entrenched in the environment, control measures appear to be limited to treatment at the final food process step. However, even here if toxins present are heat resistant or recontamination is likely, the product remains hazardous. Thus, man has, at least for the present, to live with the real possibility that these types of agents will be contaminants of food eaten every day and that prevention of illness depends on sufficient understanding of their nature and how to keep the risk to a minimum.

REFERENCES

Abbott, L. P. (1986). *Vibrio vulnificus* in New Brunswick. *Can. Dis. Wkly Rep. Health and Welfare Canada* **12**, 57–58.

Acha, P. N., and Szyfres, B. (1980). *Zoonoses and Communicable Diseases Common to Man and Animals.* Pan American Health Organization, Washington D.C., pp. 1–700.

Anderson, B. S., Sims, J. K., Wiebenga, N. II., and Sugi, M. (1983). The epidemiology of ciguatera fish poisoning in Hawaii, 1975–1981. *Hawaii Med. J.* **42**, 326–330, 332, 334.

Anon. (1984). Indigenous cholera—Queensland. *Commun. Dis. Intell., Dept. of Health, Australia* **84/8**, 1.

Anon. (1986). Mexican cheese maker receives 60-day jail sentence. *Food Chem. News* May 26, 51.

Anon. (1988). Foodborne listeriosis—Switzerland. *Can. Dis. Wkly Rep. Health and Welfare Canada* **14**(2), 7–8.

Arnold, S. H., and Brown, W. D. (1978). Histamine(?) toxicity from fish products. *Adv. Food Res.* **24**, 113–154.

Arnon, S. S., Damus, K., and Chin, J. (1981). Infant botulism: Epidemiology and relation to sudden death infant death syndrome. *Epidemiol. Rev.* **3**, 45–66.

Bagnis, R., Loussan, E., and Thevenin, S. (1974). Les intoxications par poissons perroquets aux Iles Gambier. *Med. Trop.* **34**, 523–527.

Bagnis, R., Kuberski, T., and Laugier, S. (1979). Clinical observation of 3,009 cases of ciguatera (fish poisoning) in the South Pacific. *Am. J. Trop. Med. Hyg.* **26**, 1067–1073.

Bagnis, R., Chanteau, S., Chungue, E., Hurtel, J. M., Yasumoto, T., and Inoue, A. (1980). Origins of ciguatera fish poisoning: A new dinoflagellate. *Gambierdiscus toxicus* Adachi and Fukuyo, definitely involved as a causal agent, *Toxicon* **18**, 199–208.

Bartholomew, B. A., Berry, P. R., Rodhouse, J. C., and Gilbert, R. J. (1987). Scombrotoxic fish poisoning in Britain: Features of over 250 suspected incidents from 1976 to 1986. *Epidem. Inf.* **99**, 775–782.

Bates, S. S., Bird, C. J., de Freitas, A. S. W., Foxall, R. A., Gilgan, M., Hanic, L. A., Johnson, G. A., McCulloch, A. W., Odense, P., Pocklington, R., Quilliam, M. A., Sim, P. G., Smith, J. C., Subba Rao, D. V., Todd, E. C. D., Walter, J. A., and Wright, J. L. C. (1989). Pennate diatom *Nitzchia pungens* as the primary source of domoic acid, a toxin in shellfish from eastern Prince Edward Island Canada. *Can. J. Fish. Aquat. Sci.* **47**, 1203–1215.

Baylet, R., Beccaria, C., Niausoat, P. M., Boyer, F., and Guillaud, M. (1978). Ichtyosarcotoxisme par ciguatoxine en France. *Pathol. Biol.* **26**, 95–97.

Benchley, P. (1988). Ghosts of war in the South Pacific. *Natl. Geograph.* **173**, 424–456.

Bennett. J. V., Holmberg, S. D., Rogers, M. F., and Solomon, S. L. (1987). Infectious and parasitic diseases. In R. W. Amler and H. B. Dull, eds. *Closing the Gap: The Burden of Unnecessary Illness.* Oxford University Press, New York, pp. 102–114.

Blake, P. A. (1984). Prevention of food-borne disease caused by *Vibrio* species. In R. R. Colwell, ed. *Vibrios in the Environment.* Wiley, New York, pp. 579–599.

Blakey, L. J., and Priest, F. G. (1980). The occurrence of *Bacillus cereus* in some dried foods including pulses and cereals. *J. Appl. Bacteriol.* **48**, 297–302.

Blaser, M. J., and Reller, L. B. (1981). *Campylobacter* enteritis. *N. Engl. J. Med.* **305**, 1444–1451.

Blaser, M. J., Taylor, D. N., and Feldman, R. A. (1984). Epidemiology of *Campylobacter* infections. In J. P. Butzler, ed., *Campylobacter Infection in Man and Animals.* CRC Press, Boca Raton, Fl, pp. 143–161.

Bojsen-Møller, J. (1972). Human listeriosis. Diagnostic, epidemiological and clinical studies. *Acta Pathol. Microbial. Scand. B.* Suppl. **229**, 1–157.

Borczyk, A. A., Karmali M. A., Lior, H., and Duncan, L. M. C. (1987). Bovine reservoir for verotoxin-producing *Escherichia coli* 0157:H7. *Lancet* **1**, 98.

Bourke, A. T. C., Ashdown, L. R., Cossins, Y. M., Hapgood, G. D., and Wilson, J. (1987). A cholera case—Queensland. *Commun. Dis. Intell., Dept. of Health, Australia* **87/12**, 7–9.

Bryan, F. L. (1979). Prevention of foodborne diseases in foodservice establishments. *J. Environ. Health* **41**: 198–206.

Busby, W. F., Jr., and Wogan G. N. (1979). Food-borne mycotoxins and alimentary mycotoxicoses. In H. Rieman and F. L. Bryan, eds. *Foodborne Infections and Intoxications.* Academic Press, New York, pp. 519–610.

Campsall, H., Koehler, U., and Masui, W. (1982). *Giardia lamblia* outbreak following a banquet. *Epidemiol. Notes Rep. Alberta Social Services and Community Health* **6**(9), 227–229.

Carlson, R. D., Morey-Gaines, G., Tindall, D. R., and Dickey, R. W. (1984). Ecology of toxic dinoflagellates from the Caribbean Sea. In E. P. Ragelis, ed., *Seafood Toxins.* ACS Symposium Series 262. American Chemical Society, Washington, D.C., pp. 271–287.

Carter, A. O., Borczyk, A. A., Carlson, J. A. K., Harvey, B., Hockin, J. C., Karmali, M. A., Krishnan, C., Korn, D. A., and Lior, H. (1987). A severe outbreak of *Escherichia coli* 0157:H7-associated hemorrhagic colitis in a nursing home. *N. Engl. J. Med.* **317**, 1496–1500.

Chiodini, R. J., and Sundberg, J. P. (1981). Salmonellosis in reptiles: A review. *Am. J. Epidemiol.* **113**, 494–499.

Clark, G. M., Kaufmann, A. F., and Gangarosa, E. J. (1973). Epidemiology of an international outbreak of *Salmonella agona. Lancet* **2**, 490–493.

Clarke, R. C., McEwen, S. A., Gannon, V. P., Valli, V. E. O., Lior, H., and Gyles, C. L. (1987). Isolation of verotoxin-producing *Escherichia coli* from milk filters and calves in Ontario. Abstr.

Int. Symp. Workshop on Verocytotoxin-Producing Infections, July 12–15, Toronto, Ontario, LFE-15

Cliver, D. O. (1979). Viral infections. In H. Riemann and F. L. Bryan, eds., *Food-Borne Infections and Intoxications*. Academic Press, New York, 2nd ed., pp. 299–342.

Coyle, J. T. (1983). Neurotoxic action of kainic acid. *J. Neurochem.* **41**, 1–11.

Craig, C. P. (1980). It's always the big ones that should get away. *J. Am. Med. Assoc.* **244**, 272–273.

Craven, P. C., Mackel, D. C., Baine, W. B., Barker, W. H., Gangarosa, E. J., Goldfield, M., Rosenfeld, H., Altman, R., Lachapelle, G., Davies, J. W., and Swanson, R. C. (1975). International outbreak of *Salmonella eastbourne* infections traced to contaminated chocolate. *Lancet* **1**, 788–793.

D'Aoust, J. Y. (1985). Infective dose of *Salmonella typhimurium* in Cheddar cheese. *Am. J. Epidemiol.* **122**, 717–720.

D'Aoust, J.-Y. (1989). *Salmonella*. In M. Doyle, ed., *Foodborne Bacterial Pathogens*. Dekker, New York, pp. 328–445.

D'Aoust, J. Y., and Lior, H. (1977). Pet turtle regulation and abatement of human salmonellosis. *Can. J. Publ. Health* **69**, 107–108.

D'Aoust, J. Y., Aris, B. J., Thisdele, P., Durante, A., Brisson, M., Dragon, D., Lachapelle, G., Johnston, M., and Laidley, R. (1975). *Salmonella eastbourne* outbreak associated with chocolate. *Can. Inst. Food Sci. Technol. J.* **8**, 181–184.

D'Aoust, J.-Y., Warburton, D. W., and Sewell, A. M. (1985). *Salmonella typhimurium* phage-type 10 from Cheddar cheese implicated in a major Canadian foodborne outbreak. *J. Food Protection* **48**, 1062–1066.

Doyle, M. P. (1984). *Campylobacter* in foods. In J.-P. Butzler, ed., *Campylobacter Infection in Man and Animals*. CRC Press, Boca Raton, Fl., pp. 143–161.

Doyle, M. P., and Schoeni, J. L. (1987). Isolation of *Escherichia coli* 0157:H7 from retail fresh meats and poultry. *Abstr. Intl. Symp. Workshop on Verocytotoxin–Producing Infections*, July 12–15, Toronto, Ontario, CEP-9.

Eklund, M. W., Poysky, F. T., Peterson, M. E., Peck, L. W., and Brunson, W. D. (1984). Type E botulism in salmonids and conditions contributing to outbreaks. *Aquaculture* **41**, 293–309.

Fenlon, D. R. (1985). Wild birds and silage as reservoirs in the agricultural environment. *J. Appl. Bacteriol.* **59**, 537–543.

Fleming, C. (1971). Case of poisoning from red whelk. *Br. Med. J.* **3**, 520–521.

Fleming, D. W., Cochi, S. L., MacDonald, K. L., Brondum, J., Hayes, P. S., Plikaytis, B. D., Holmes, M. B., Audurier, A., Broome, C. V., and Reingold, A. L. (1985). Pasteurized milk as a vehicle of infection in an outbreak of listeriosis. *N. Engl. J. Med.* **312**, 404–407.

Fuhrman, F. A. (1983). Toxic constituents of animal foodstuffs: eggs of fishes and amphibians. In M. Rechcigl, Jr., ed., *Handbook of Naturally Occurring Food Toxicants*. CRC, Boca Raton, Fl, pp. 301–311.

Gilbert, R. J. (1979). *Bacillus cereus* gastroenteritis. In H. Riemann and F. L. Bryan, eds., *Food-Borne Infections and Intoxications*. Academic Press, New York, 2nd ed., pp. 495–518.

Gilbert, R. J., Roberts, D., and Smith, G. (1984). Food-borne diseases and botulism. In G. R. Smith, ed., *Topley and Wilson's Principles of Bacteriology, Virology and Immunity, Vol. 3 Bacterial diseases*. Williams & Wilkins, Baltimore, 7th ed., pp. 477–514.

Gill, O. N., Sockett, P. H., Bartlett, C. L. R., Vaile, M. S. B., Rowe, B., Gilbert, R. J., Dulake, C., Murrell, H. C., and S. Salmaso (1983). Outbreak of *Salmonella napoli* infection caused by contaminated chocolate bars. *Lancet* **1**, 574–577.

Glass, R. I., Stoll, B. J., Huq, M. I., Struelens, M. J., Blaser, M., and Kibriya, A. K. M. G. (1983). Epidemiologic and clinical features of endemic *Campylobacter jejuni* infection in Bangladesh. *J. Inf. Dis.* **148**, 292–296.

Glatz, B. A. (1987). Genetic regulation of toxin production by foodborne microbes. In T. J. Montville, ed., *Food Microbiology Vol. 1, Concepts in Physiology and Metabolism.* CRC Press, Boca Raton, FL, pp. 103–130.

Gorbach, S. L. (1983). Infectious diarrhea. In M. H. Sleisenger and J. S. Fordtran eds., *Gastriontestinal Disease.* W. B. Saunders, Philadelphia, 3rd ed., pp. 925–965.

Gouet, P. (1974). Growth of *Listeria monocytogenes* in gnotoxenic silages of maize, rye grass, fescue and lucerne. In *Problems of Listeriosis.* 6th Intern. Meeting, Nottingham, England, pp. 161–163.

Grell, G. A. C. (1983). Ciguatera fish poisoning. *West Indian Med. J.* **32**, 63–65.

Guangxian, W., and Jingyeh, Y. (1986). Studies on condition of contamination, types and pathogenicity of *Bacillus cereus* in foods. Proc. 2nd World Congress Foodborne Infections and Intoxications. Inst. of Veter. Med., Berlin, pp. 203–210.

Gullett, J., Wilkinson, D., Blain, J. G., Boodran, C., and Tanner, C. (1975). Trichinosis outbreak— Northwest Territories. *Can. Dis. Wkly Rep. Health and Welfare Canada* **1**(28), 69–71.

Gustavsen, S., and Breen, O. (1984). Investigation of an outbreak of *Salmonella oranienburg* infections in Norway, caused by contaminated black pepper. *Am. J. Epidemiol.* **119**, 806–812.

Hafsteinsson, H., and Rizvi, S. S. H. (1987). A review of the sealworm problem: biology, implications and solutions. *J. Food Protection* **50**, 70–84.

Halstead, B. W. (1984). Miscellaneous seafood toxicants. In E. P. Ragelis, ed., *Seafood Toxins.* ACS Symposium Series 262. American Chemical Society, Washington, D. C., pp. 37–51.

Halstead, B. W., and Schantz, E. J. (1984). *Paralytic Shellfish Poisoning.* Offset Publication No. 79, WHO, Geneva.

Handzel, S. (1974). *Salmonella weltevreden* outbreak—Atlantic provinces. *Epidemiol. Bull., Health and Welfare Canada* **18**(5), 77.

Harmon, S. M., Kautter, D. A., and Solomon, H. M. (1987). *Bacillus cereus* contamination of seeds and vegetable sprouts grown in a home sprouting kit. *J. Food Protection* **50**, 62–65.

Hauge, S. (1970). Botulismetilfether i Norge i lopet avde siste 5–6 ar. *Norsk. Vet. (Tidskr.)* **82**, 259–261.

Hauschild, A. H. W. (1989). *Clostridium botulinum.* In M. Doyle, ed., *Foodborne Bacterial Pathogens.* Dekker, New York, pp. 111–189.

Healy, G. R., and Juranek, D. (1979). Parasitic infections. In H. Riemann and F. L. Bryan, eds., *Food-Borne Infections and Intoxications.* Academic Press, New York, 2nd ed., pp. 343–385.

Hobbs, B. C. (1979). *Clostridium perfringers* gastroenteritis. In H. Riemann and F. L. Bryan, eds., *Food-Borne Infections and Intoxications.* Academic Press, New York, 2nd ed., pp. 131–171.

Hockin, J., and Lior, H. (1987). Hemorrhagic colitis and haemolytic uremic syndrome caused by *Escherichia coli* 0157:H7 in Canada. *Can. Dis. Wkly Rep. Health and Welfare Canada* **13**(45), 203–204.

Holmberg, S. D., Osterholm, M. T., Senger, K. A., and Cohen, M. L. (1984). Drug-resistant *Salmonella* from animals fed antimicrobials. *N. Engl. J. Med.* **311**, 617–622.

Huss, H. H. (1980). Distribution of *Clostridium botulinum. Appl. Environ. Mirobiol.* **39**, 764–769.

Huss, H. H., and Pedersen, A. (1979). *Clostridium botulinum* in fish. *Nord. Vet. Med.* **31**, 214–221.

Ionescu, G., Ienistea, C., and Ionescu, C. (1966). Freeventa *B. cereus* in laptele crud si in laptele pasteurizat. *Mircobiol. Parazitol. Epidemiol.* **11**, 423–430.

James, S. M., Fannin, S. L., Agee, B. A., Hall, B., Parker, E. Vogt, J. Run, G., Williams, J., and Lieb, L. (1985). Listeriosis outbreak associated with Mexican-style cheese. *Morbid. Mortal. Wkly Rep.* **34**, 357–359.

Jauhari, J. S., and Kulshrestha, S. B. (1986). Food poisoning episode in a family due to *B. cereus.* Proc. 2nd World Congress Foodborne Infection and Intoxications. Inst. of Veter. Med., Berlin, pp. 203–210.

Jessop, J. H., Khanna, B., Black, W. A., Milling, M. E., Bowering, D., Hockin, J., and Lior, H. (1986). *Salmonella nima* in British Columbia. *Can. Dis. Wkly Rep. Health and Welfare Canada* **12**(41), 183–184.

Joffe, A. Z. (1983). Foodborne diseases: Alimentary toxic aleukia. In M. Rechcigl, Jr., ed., *CRC Handbook of Foodborne Disease of Biological Origin*. CRC Press, Boca Raton, FL, pp. 353–495.

Johnston, K. M. (1984). *Bacillus cereus* foodborne illness—an update. *J. Food Protection* **47**, 145–153.

Johnston, J. M., Becker, S. F., and McFarland, L. M. (1985). *Vibrio vulnificus*: Man and the sea. *J. Am. Med. Assoc.* **253**, 2850–2853.

Johnston, W. M., Lior, H., and Bezanson, G. S. (1983). Cytotoxic *Escherichia coli* 0157:H7 associated with haemorrhagic colitis in Canada. *Lancet* **1**, 76.

Jolivet-Reynaud, C., Popoff, M. R., Vinit, M.-A., Ravisse, P., Moreau, H., and Alouf, J. E. (1986). Enteropathogenicity of *Clostridium perfringens β* toxin and other clostridial toxins. *Zbl. Bakt. Microbiol. Hyg. Suppl.* **15**, 145–151.

Kampelmacher, E. H., Mass, D. E., and van Noorle Jansen, L. M. (1976). Occurrence of *Listeria monocytogenes* in feces of pregnant women with and without direct animal contact. *Zbl. Bakt. 1. Abt. Orig. A.* **234**, 238–242.

Karmali, M. A., Petric, M., Steele, B. T., and Lim, C. (1983). Sporadic cases of haemolytic-uraemic syndrome associated with faecal cytotoxin and cytotoxin-producing *Escherichia coli* in stools. *Lancet.* **1**, 619–620.

Kat, M. (1985). *Dinophysis acuminata* blooms. The distinct cause of Dutch mussel poisoning. In D. M. Anderson, A. W. White, and D. G. Baden, eds., *Toxic Dinoflagellates*. Elsevier, New York, pp. 73–77.

Kimberlin, R. (1984). Slow viruses: Conventional and unconventional. In F. Brown and G. Wilson, eds., *Topley and Wilson's Principles of Bacteriology, Virology and Immunity, Vol. 4 Virology*. Williams & Wilkens, Baltimore, 7th ed., pp. 487–510.

Kimura, L. H., Abad, M. A., and Hokama, Y. (1982). Evaluation of the radioimmunoassay (RIA) for detection of ciguatoxin (CTX) in fish tissues. *J. Fish. Biol.* **21**, 671–681.

Kolvin, J. L., and Roberts, D. (1982). Studies on the growth of *Vibrio cholerae* biotype eltor and biotype classical in foods. *J. Hyg. Cambridge* **89**, 243–252.

Karmer, J. M., and Gilbert, R. J. (1989). *Bacillus cereus* and other *Bacillus* species. In M. Doyle, ed. Foodborne Bacterial Pathogens. Dekker, New York, pp. 22–70.

Krishnamachari, K. A., Bhat, R. V., Nagaragan, V., and Tilak, T. B. (1975). Heptatitis due to aflatoxicosis. An outbreak in West India. *Lancet* **1**, 1061–1063.

Kulda, J., and Nohynkova, E. (1978). Flagellates of the human intestine and of intestines of other species. In J. P. Kreir, ed., *Parasitic Protozoa*, Vol. II, Academic Press, New York, pp. 1–130.

Laidley, R., Handzel, S., Severs, D., and Butler, R. (1974). *Salmonella weltevreden* outbreak associated with contaminated pepper. *Eprdemiol. Bull. Health and Welfare Canada* **18**(4), 62.

Last, J. M. (1986). Epidemiology and health information. In J. M. Last, ed., *Maxcy-Rosenau Public Health and Preventive Medicine*. Appleton-Century-Crofts, Norwalk, CT., 12th ed., pp. 9–74.

Lecos, C. (1986). Of microbes and milk: Probing America's worst *Salmonella* outbreak. *Dairy Food Sanit.* **61**, 136–140.

Lee. C. (1980). Fish poisoning with particular reference to ciguatera. *J. Trop. Med. Hyg.* **83**, 93–97.

Lee, J. A. (1982). The role of animal feeding stuffs in the causation of human salmonellosis in the United Kingdom. In H. Kurata and C. W. Hesseltine, eds., *Control of the Microbial Contamination of Foods and Feeds in International Trade: Microbial Standards and Specifications*. Saikon Publ., Tokyo, pp. 61–77.

Levine, M. M., Xu, J.-G., Kaper, J. B., Lior, H., Prado, V., Tall, B., Nataro, J., Karch, H., and Wachsmuth, K. (1987). A DNA probe to identify enterohemorrhagic *Escherichia coli* of 0157:H7 and other serotypes that cause hemorrhagic colitis and hemolytic uremic syndrome. *J. Infect. Dis.* **156**, 175–182.

Lewis, N. D. (1984). Ciguatera in the Pacific: Incidence and implications for marine resource development. In E. P. Ragelis, ed., *Seafood Toxins*. ACS Symposium Series 262. American Chemical Society, Washington, D.C., pp. 289–306.

Liener, I. E. (1980). *Toxic Constituents of Plant Foodstuffs*. Academic Press, New York, 2nd ed., pp. 1–502.

Lyons, D. E., Berry, J. T., Lyons, S. A., and Taylor, S. L. (1983). Cadaverine and aminoguanidine potentiate the uptake of histamine in vitro in perfused intestinal segments of rats. *Toxicol. Appl. Pharmacol.* **70**, 445–458.

MacDonald, K. L., Cohen, M. L., and Blake, P. A. (1986). The changing epidemiology of adult botulism in the United States. *Am. J. Epidemiol.* **124**, 794–799.

Maclean, J. L., and White, A. W. (1985). Toxic dinoflagellate blooms in Asia: A growing concern. In E. P. Ragelis, ed., *Seafood Toxins*. ACS Symposium Series 262. American Chemical Society, Washington, D. C., pp. 517–520.

Maeda, M., Kodama, T., Saito, M., Tanaka, T., Yoshizumi, H., Nomoto, K., and Fugita, T. (1987). Neuromuscular action of insecticidal domoic acid on the American cockroach. *Pest. Biochem. Physiol.* **28**, 85–92.

Mahadevaswamy, M., and Venkataraman, L. V. (1987). Bacterial contaminants in blue green alga Spirulina produced for use as biomass protein. *Arch. Hydrobiol.* **110**, 623–630.

Mandel, B. K., de Mol, P., and Butzler, J.-P. (1984). Clinical aspects of *Campylobacter* infections in humans. In J.-P. Butzler, ed., *Campylobacter Infection in Man and Animals*. CRC Press, Boca Raton, FL, pp. 21–31.

Marmion, B. P. (1984). Rickettsial diseases of man and animals. in G. R. Smith, ed., *Topley and Wilson's Principles of Bacteriology, Virology and Immunity, Vol. 3. Bacterial Diseases*. Williams & Wilkins, Bltimore, 7th ed., pp. 574–590.

Matsuda, M., Ozutsumi, K., Iwahashi H., and Sugimoto, N. (1986). Primary action of *Clostridium perfringens* type A enterotoxin on Hela and Vero cells in the absence of extracellular calcium: Rapid and characteristic changes in membrane permeability. *Biochem. Biophys. Res. Commun.* **141**, 704–710.

McDonel, J. L. (1979). The molecular mode of action of *Clostridium perfringens* enterotoxin. *Am. J. Clin. Nutr.* **32**, 210–218.

McLauchlin, J. (1987). *Listeria monocytogenes*, recent advances in the taxonomy and epidemiology of listeriosis in humans. *J. Appl. Bacteriol.* **63**, 1–11.

McSwiggan, D. A., Gilbert, R. J., and Fowler, F. W. T. (1975). Food poisoning associated with pasteurized cream. Commun. Dis. Surv. Public Health Lab. Service, London 75(43).

Meyer, E. A., and Radulescu, S. (1979). *Giardia* and giardiasis. *Adv. Parasitol.* **17**, 1–47.

Midura, T. F., Nygaard, G. S., Wood, R. M., and Bodily, H. L. (1972). *Clostridium botulinum* type F: Isolation from venison jerky. *Appl. Microbiol.* **24**, 165–167.

Millar, J. G., and Dey, A. (1987). Food poisoning due to the consumption of red whelks (*Neptunea antiqua*). *Commun. Dis. Scotland* **87**(38), 5–6.

Mims, C. A. (1982). *The Pathogenesis of Infectious disease*. Academic Press, New York, 2nd ed. pp. 1–297.

Moore, G. T., Cross, W. M., McGuire, D., Mollohan, C. S., Gleason, N. N., Healy, G. R., and Newton, L. H. (1969). Epidemic giardiasis at a ski resort. *N. Engl. J. Med.* **281**, 402–407.

Morris, G. K., Martin, W. T., Shelton, W. H., Wells, J. G., and Brachman, P. S. (1970). Salmonellae in fish meal plants: Relative amounts of contamination at various stages of processing and a method of control. *Appl. Microbiol.* **19**, 401–408.

Mosher, H. S., and Fuhrman, F. A. (1984). Occurrence and origin of tetrodotoxin. In E. P. Ragelis, ed., *Seafood Toxins*. ACS Symposium Series 262. American Chemical Society, Washington, D. C., pp. 333–344.

Murray, C. K., Hobbs, G., and Gilbert, R. J. (1982). Scombrotoxin and scombrotoxin-like poisoning from canned fish. *J. Hyg. Cambridge* **88**, 215–220.

Murrell, T. G. C. (1982). Enteritis necroticans (pigbel) an unrecognised preventable disease. *Clin. Med. J.* **95**, 843–848.

Neill, M. A., Agosti, J., and Rosen, H. (1985). Hemorrhagic colitis with *E. coli* O157:H7 preceding adult hemolytic uremic syndrome. *Arch. Int. Med.* **145**, 2215–2217.

Newell, D. G. (1984). Experimental studies of *Campylobacter enteritis.* In J. P. Butzler, ed., *Campylobacter Infections in Man and Animals.* CRC Press, Boca Raton, FL, pp. 113–131.

Ngindu, A., Johnson, B. K., Kenya, P. R., Ngira, J. A., Ocheng, D. M., Nandwa, H., Omondi, T. N., Jansen, A. J., Ngare, W., Kaviti, J. N., Gatei, D., and Siongok, T. (1982). Outbreak of acute hepatitis caused by aflatoxin poisoning in Kenya. *Lancet* **1**, 1346–1348.

Nisizawa, K. (1978). Marine algae from a viewpoint of pharmaceutical studies. *Jpn. J. Phycol.* **26**, 73–78.

Noguchi, T., Hwang, D. F., Arakawa, O., Daigo, K., Sato, S., Ozaki, H., Kawai, N., Ito, M., and Hashimoto, K. (1987). Palytoxin as the causative agent in the parrotfish poisoning. In P. Gopalakrishnakone and G. K. Tan, eds., *Progress in Venom and Toxin Research.* National University Singapore, Singapore, pp. 325–335.

O'Brien, T. F., Hopkins, J. D., Gilleece, E. S., Medeiros, A. A., Kent, R. L., Blackburn, B. D., Holmes, M. B., Reardon, J. P., Vergeront, J. M., Schell, W. L., Christenson, E., Bissett, M. L., and Morse, E. V. (1982). Molecular epidemiology of antibiotic resistance in *Salmonella* from animals and human beings in the United States. *N. Engl. J. Med.* **207**, 1–6.

Ormay, L., and T. Novotny (1969). The significance of *Bacillus cereus* food poisoning in Hungary. In E. H. Kampelmacher, M. Ingram, and D. A. A. Mossel, eds., *The Microbiology of Dried Foods.* Int. Assoc. Microbiol. Soc., Bilthoven, The Netherlands, pp. 279–285.

Ørskov, F., Ørskov, I., and Villar, J. A. (1987). Cattle as reservoir of verotoxin-producing *Escherichia coli* O157:H7. *Lancet* **2**, 276.

Oshima, T. (1987). Anisakiasis—is the sushi bar guilty? *Parasit. Today* **3**, 44–48.

Osterholm, M. T., Forfang, J. C., Ristenen, B. A., Dean, A. G., Washburn, J. W., Codes, J. R., Rude, R. A., and McGullough, J. G. (1981). An outbreak of foodborne giardiasis. *N. Engl. J. Med.* **304**, 24–78.

Ostroff, S. M., Griffin, P. M., Tauxe, R. V., Wells, J. G., Green, K. D., Lewis, J. H., Blake, P. A., and Kobayashi, J. M. (1987). Source tracing in an outbreak of *E. coli* O157:H7-induced illness. Abst. Int. Symp. Workshop on Verocytotoxin-Producing infections, July 12–15, Toronto, Ontario, CEP-8.

Pang, T., Wong, P. Y., Puthucheary, S. D., Sihotang, K., and Chang, W. K. (1987). In-vitro and in-vivo studies of a cytotoxin from *Campylobacter jejuni. J. Med. Microbiol.* **23**, 193–198.

Park, C. E., Lior, H., Pauker, P., and Purvis, U. (1982). *Campylobacter* food poisoning from steaks— Ontario. *Can. Dis. Wkly Rep. Health and Welfare Canada* **8**(36), 177–178.

Patton, R., Bergin, A. O., Todd, E., and Cole, T. (1984). *Datura* poisoning from hamburger—Ontario. *Can. Dis. Wkly Rep. Health and Welfare Canada* **10**(12), 45–46.

Pearn, J., Harvey, P., Lewis, R., and McKay, R. (1982). Ciguatera and pregnancy. *Med. J. Aus.* **1**, 57–58.

Perl, T., Bédard, L., Remis, R., Kosatsky, T., Hoey, J., Massé, R., Guimont, L., Labrecque, C., Desroches, F., Vézina, C., and Dionne, M. (1987). Intoxication following mussel ingestion in Montreal. *Can. Dis. Wkly. Rep. Health and Welfare Canada* **13**(49), 224–225.

Peters, A. R. (1985). An estimation of the economic impact of an outbreak of *Salmonella dublin* in a calf rearing unit. *Vet. Rec.* **117**, 667–668.

Petersen, L. R., Cartter, M. L., and Hadler, J. L. (1988). A food-borne outbreak of *Giardia lamblia. J. Inf. Dis.* **157**, 846–848.

Phillips, J. D., and Griffiths, M. W. (1986). Factors contributing to the seasonal variation of *Bacillus* spp. in pasteurized dairy products. *J. Appl. Bacteriol.* **61**, 275–285.

Portnoy, B. L., Geopfert, J. M., and Harmon, S. M. (1973). An outbreak of *Bacillus cereus* food poisoning resulting from contaminated vegetable sprouts. *Am. J. Epidemiol.* **103**, 589–594.

Powers, E. M. Latt, T. G., and Brown, T. (1976). Incidence and levels of *Bacillus cereus* in processed spices. *J. Milk Food Technol.* **39**, 668–670.

Pudden, D., Tuttle, N., Korn, D., Carlson, J., Carter, A., and Hockin, J. (1985). Hemorrhagic colitis in a nursing home—Ontario. *Can. Dis. Wkly Rep. Health and Welfare Canada* **11**, 169–170.

Ragelis, E. P. (1984). Ciguatera seafood poisoning—overview. In E. P. Ragelis, ed., *Seafood Toxins*. ACS Symposium Series 262. American Chemical Society, Washington, D. C., pp. 25–36.

Ratner, H. (1987). *Vibrio vulnificus. Infect. Control* **8**, 430–433.

Reilly, C. (1980). *Metal Contamination of Food*. Applied Science, Barking, England, pp. 1–235.

Reilly, W. J., Forbes, G. I., Patterson, G. M., and Sharp, J. C. M. (1981). Human and animal salmonellosis in Scotland associated with environmental contamination. *Vet. Rec.* **102**, 553–555.

Riley, L. W., Remis, R. S., Helgerson, D. S., McGee, H. B., Wells, J. G., Davis, B. R., Hebert, R. J., Olcott, E. S., Johnson, L. M., Hargett, N. T., Blake, P. A., and Cohen M. L. (1983). Hemorrhagic colitis associated with a rare *Escherichia coli* serotype. *N. Engl. J. Med.* **308**, 681–685.

Robaeys, G., Surmont, I., Lemmens, P., Coremans, G., Vantrappen, G., and Vandespitte, J. (1987). Haemorrhagic colitis and verotoxin-producing *Escherichia coli* 0157 in Belgium. *Lancet* **1**, 1493–1496.

Roberts, D. (1984). Bacteria pathogenic to man in foods of plant origin. In I. Kiss, T. Deak, and K. Incze, eds., *Microbial Associations and Interactions in Food*. D. Reidel, Boston, pp. 43–48.

Roberts, T. (1985). Microbial pathogens in raw pork, chicken and beef: benefit estimates for control using irradiation. *Am. J. Agric. Econ.* Dec. 957–965.

Robinson, D. A. (1981). Infective dose of *Campylobacter jejuni* in milk. *Br. Med. J.* **282**, 1584.

Rodrigue, D. C., Etzel, R. A., Kilbourne, E. M., and Blake, P. A. (1988). Epidemic paralytic shellfish poisoning, Gautemala. Abstr. Epidemic Intelligence Conference, April 18–22, Atlanta, p. 44.

Rogers, W. E., and Johnstone, T. (1985). *Salmonella poona* from pet turtles—British Columbia. *Can. Dis. Wkly Rep. Health and Welfare Canada* **11**(38), 117–119.

Rowe, B., Smith, H. R., Scotland, S. M., and Cross, R. J. (1987). Haemorrhagic colitis and haemolytic uraemic syndrome associated with vero cytotoxin-producing *Escherichia coli* (VTEC) in England and Wales. Abstr. Intl. Symp. and Workshop on Verocytotoxin-producing Infections, July 12–15, Toronto, Ontario, CEP-1.

Rumack, B. H., and Salzman, E. (1978). *Mushroom Poisoning: Diagnosis and Treatment*. CRC Press, Boca Raton, FL, pp. 1–263.

Russell, A. D., Yarnych, V. S., and Koulikovskii, A. V. (1984). Guidelines on disinfection in animal husbandry for prevention and control of zoonotic diseases. WHO document WHO/VPH/84.4.

Russell, F. E., and Maretic, Z. (1986). Scombroid poisoning: Mini-review with case histories. *Toxicon* **24**, 967–973.

Sakaguchi, G. (1979). Botulism. In H. Riemann and F. L. Bryan, eds., *Food-Borne Infections and Intoxications*. Academic Press, New York, 2nd ed., pp. 389–442.

Sakazaki, R. (1979). Vibrio infections. In H. Riemann and F. L. Bryan, eds., *Food-Borne Infections and Intoxications*. Academic Press, New York, 2nd ed., pp. 173–209.

Schlech, W. F., Lavigne, P. M., Bortolussi, R. A., Allen, A. C., Haldane, E. V., Wort, A. J., Hightower, A. W., Johnston, S. E., King, S. H., Nicholls, E. S., and Broome, C. V. (1983). Epidemic listeriosis—evidence for transmissions by food. *N. Engl. J. Med.* **308**, 203–206.

Scott, H. G. (1969). Poisonous plants and animals. In H. Riemann, ed., *Food-Borne Infections and Intoxications*. Academic Press, New York, 1st ed., pp. 543–604.

Scrimshaw, N. S. (1975). Single-cell protein for human consumption—an overview. In S. R. Tannenbaum and D. I. C. Wang, eds., *Single Cell Protein II*. The MIT Press, Cambridge, MA, pp. 24–25.

Serck-Hanssen, A. (1970). Aflatoxin induced fatal hepatitis? A case report from Uganda. *Arch. Environ. Health* **20**, 729–731.

Sharma, V. D., Singh, S. P., and Taku, A. (1987). *Salmonella* from commercial pork preparations: Isolation, drug-resistance and enterotoxigenicity. *Int. J. Food Microbiol.* **5**, 57–62.

Sharp, J. C. M. (1987). Infections associated with milk and dairy products in Europe and North America, 1980–85. *Bull. WHO* **65**(3), 397–406.

Shinozaki, H., and Ishida, M. (1976). Inhibition of quisqualate responses by domoic or kainic acid in crayfish opener muscle. *Brain Res.* **109**, 435–439.

Silliker, J. H. (1982). The *Salmonella* problem: Current status and future direction. *J. Food. Protection* **45**, 661–666.

Simon, C. (1966). Nitrite poisoning from spinach. *Lancet* **1**, 872.

Simonsen, B., Bryan, F. L., Christian, J. H. B., Roberts, T. A., Tompkin, R. B., and Silliker, J. H. (1987). Prevention and control of food-borne salmonellosis through application of hazard analysis critical control point (HACCP). *Int. J. Food Micro.* **4**, 227–247.

Simpson, D. I. H. (1984). Arenaviridae. In F. Brown and G. Wilson, eds., *Topley and Wilson's Principles of Bacteriology, Virology and Immunity, Vol 4 Virology*. Williams & Wilkins, Baltimore, 7th ed., pp. 255–265.

Smith, H. R., Rowe, B., Gross, R. J., Fry, N. K., and Scotland, S. M. (1987). Haemorrhagic colitis and verocytotoxin-producing *Escherichia coli* in England and Wales. *Lancet* **1**, 1062–1065.

Smith, J. W. G., and Smith, G. (1984). Gas gangrene and other clostridial infections of man and animals. In G. R. Smith, eds., *Topley and Wilson's Principles of Bacteriology, Virology and Immunology, Vol. 3. Bacterial diseases*. Williams & Wilkins, Baltimore, 7th ed., pp. 327–344.

Spika, J. S., Waterman, S. G., Soo Hoo, G. W., St-Louis, M. E., Pacer, R. E., James, S. M., Bissett, M. L., Mayer, L. W., Chiu, J. Y., Hall, B., Greene, K., Potter, M. E., Cohen, M. L., and Blake, P. A. (1987). Chloramphenicol-resistant *Salmonella newport* traced through hamburger to dairy farms. *N. Engl. J. Med.* **316**, 565–570.

St-Louis, M. E., Morse, D. L., Potter, M. E., Demelfi, T. M., Guzewich, J. J., Tauxe, R. V., and Blake, P. A. (1988). The emergence of Grade-A eggs as a major source of *Salmonella enteritidis* infections. *J. Am. Med. Assoc.* **259**, 2103–2107.

Stadhouders, J., Hup, G., and Langeveld, L. P. M. (1980). Some observations in the germination, heat resistance and outgrowth of fast-germinating and slow-germinating spores of *Bacillus cereus* in pasteurized milk. *Netherlands Milk Dairy J.* **34**, 215–228.

Stern, N. J., and Kazami, S. U. (1989). *Campylobacter jejuni*. In M. Doyle, ed., *Foodborne Bacterial Pathogens*. Dekker, New York, pp. 71–110.

Sugiyama, H. (1982). Infant botulism: Microbial ecological basis. In J. B. Robbins, J. C. Hill, and J. C. Sadoff, eds., *Seminars in Infectious Disease. Vol. 4. Bacterial Vaccines: International Symposium, Sept 15–18, 1980*. Thieme-Stratton, New York, pp. 42–47.

Summer, S. S., Speckhard, M. W., Somers, E. B., and Taylor, S. L. (1985). Isolation of histamine-producing *Lactobacillus buchneri* from Swiss cheese implicated in a food poisoning outbreak. *Appl. Environ. Microbiol.* **50**, 1094–1096.

Tatnall, F. M., Smith, H. G., Welsby, P. D., and Turnbull, P. C. B. (1980). Ciguatera poisoning. *Br. Med. J.* **281**, 948–949.

Taylor, S. L. (1985). *Histamine Poisoning Associated with Fish, Cheese and Other Foods*. WHO Monograph VPH/FOS/85.1, pp. 1–47.

Taylor, S. L., Hui, J. Y., and Lyons, D. E. (1984). Toxicology of scombroid poisoning. In E. F. Ragelis, ed., *Seafood Toxins*. ACS Symposium Series 262. American Chemical Society, Washington, D.C., pp. 417–430.

Terranova, W., and Blake, P. A. (1978). Current concepts: *Bacillus cereus* food poisoning. *N. Engl. J. Med.* **298**, 143–144.

Tilton, R. C., and Ryan, R. W. (1987). Clinical and ecological characteristics of *Vibro vulnificus* in the northeastern United States. *Diagn. Microbiol. Infect. Dis.* **6**, 109–117.

Tindall, D. R., Dickey, R. W., Carlson, R. D., and Morey-Gaines, G. (1984). Ciguatoxigenicdino-flagellates from the Caribbean Sea. In E. P. Ragelis, ed., *Seafood Toxins*. ACS Symposium Series 262. American Chemical Society. Washington, D.C., pp. 225–240.

Todd, E. C. D. (1985a). Foodborne and waterborne disease in Canada—1979 annual summary. *J. Food Protection* **48**, 1071–1078.

Todd, E. C. D. (1985b). Ciguatera poisoning in Canada. In D. M. Anderson, A. W. White, and D. G. Baden, eds., *Toxic Dinoflagellates*. Elsevier, New York, pp. 505–510.

Todd, E. C. D. (1985c). Economic loss from foodborne disease and non-illness related recalls because of mishandling by food processors. *J. Food Protection* **48**, 621–633.

Todd, E. C. D. (1987a). Foodborne and waterborne disease in Canada—1981 annual summary. *J. Food Protection* **50**, 982–991.

Todd, E. C. D. (1987b). Legal liability and its economic impact on the food industry. *J. Food Protection* **50**, 1048–1057.

Todd, E. C. D. (1988a). Foodborne and waterborne disease in Canada—1982 annual summary. *J. Food Protection* **51**, 56–65.

Todd, E. C. D. (1988b). Botulism in native peoples—an economic study. *J. Food Protection* **51**, 581–587.

Todd, E. C. D. (1989). Preliminary estimates of costs of foodborne disease in Canada and costs to reduce salmonellosis. *J. Food Protection* (in press).

Todd, E., and Styliadis, S. (1985). Comments. *Can. Dis. Wkly Rep. Health and Welfare Canada* **11**(38), 119–120.

Todd, E., Szabo, R., Graham, L. M., Edwards, G., Melling, J., Capel, B., and Parry, J. (1978). Foodborne illness from pancakes—Ontario. *Can. Dis. Wkly Rep. Health and Welfare Canada* **4**(43), 171–172.

Turnbull, P. C. B. (1979). Food poisoning with special reference to *Salmonella*—its epidemiology, pathogenesis and control. In H. P. Lambert, ed., *Clinics in Gastroenterology 8(3)*. W. B. Sanders, Philadelphia, pp. 663–714.

Turnbull, P. C. B., and Gilbert, R. J. (1982). Fish and shellfish poisoning in Britain. In E. F. Jelliffe and D. B. Jelliffe, eds., *Adverse Effects of Foods*. Plenum, New York, pp. 297–306.

Underdal, B., Yndestad, M., and Aune, T. (1985). DSP intoxication in Norway and Sweden, autumn 1984–Spring 1985. In D. M. Anderson, A. W. White, and D. G. Baden, eds., *Toxic Dinoflagellates*. Elsevier, New York, pp. 489–494.

Vietmeyer, N. D. (1984). The preposterous puffer. *Natl. Geograph.* **166**, 260–270.

Vogt, R. L., and Liang, A. P. (1986). Ciguatera fish poisoning. *Mortal. Morbid. Wkly Rep., Centers for Dis. Control., Atlanta* **35**, 263–264.

Watkins, J., and Sleath, K. P. (1981). Isolation and enumeration of *Listeria monocytogenes* from sewage, sewage study and river water. *J. Appl. Bacteriol.* **50**, 1–9.

Watson, W. A., and Kirby, F. D. (1984). The *Salmonella* problem and its control in Great Britain. In G. H. Snoeyenbos, ed., *Proceedings of the International Symposium on Salmonella*. American Association of Avian Pathology, University of Pennsylvania, Kennet Square, PA, pp. 35–47.

Wells, J. G., Shipman, L. D., Greene, K. D., Downes, F. P., Martin, M. L., Tauxe, R. V., and Wachsmuth, I. K. (1987). Isolation of *Escherichia coli* 0157:H7 and other Shiga-like vero toxin-producing *E. coli* from dairy cattle. Abstr. Int. Symp. Workshop on Verocytotoxin-producing Infections, July 12–15, Toronto, Ontario, LFE-4.

White, A. W., Anraku, M., and Hooi, K. (eds.). (1984). Proc. meeting Singapore, 11–14 Sept. *Toxic Red Tides and Shellfish Toxicity in South-East Asia*. Southeast Asian Fisheries Development Center, Bangkok, Thailand and Internatl. Development Res. Centre, Ottawa, Canada, pp. 1–133.

White, D. R. L., and White, A. W. (1985). First report of paralytic shellfish poisoning in Newfoundland. In D. M. Anderson, A. W. White, and D. G. Baden, eds., *Toxic Dinoflagellates*. Elsevier, New York, pp. 511–516.

Whittam, T. S., Wachsmuth, I. K., and Wilson, R. A. (1987). Clonal nature of 0157:H7 *Escherichia coli* associated with outbreaks of hemorrhagic colitis. Abstr. Int. Symp. Workshop on Verocytotoxin-producing Infections, July 12–15, Toronto, Ontario, GEN-1.

Williams, J. E. (1981). Salmonellas in poultry feeds—a worldwide review *World Poult. Sci. J.* **37**, 6–25.

Willinger, H., Flatscher, J., Dreier, F., and Wildner, T. (1986). Epidemiologische untersuchungen zum verkommen von Salmonellen in geflügelhaltungen. (Epidemiological investigations on the incidence of salmonellae in poultry in Austria.) *Wien. Tierärztl. Mschr.* **73**, 141–148.

Wilson, C. R., and Remington, J. S. (1980). What can be done to prevent congenital toxoplasmosis? *Am. J. Obstet. Gynecol.* **138**, 357–363.

Wilson, H. P. S., Stauffer, S. J., and Walker T. S. (1982). Waterborne giardiasis outbreak—Alberta. *Can. Dis. Wkly Rep. Health and Welfare Canada* **8**(20), 97–98.

Wright, J. L. C., Boyd, R. K., de Freitas, A. S. W., Falk, M., Foxall, R. A., Jamieson, W. D., Laycock, M. V., McCulloch, A. W., McInnes, A. G., Odense, P., Pathak, V. P., Quilliam, M. A. Ragan, M. A., Sim, P. G., Thibault, P., Walter, J. A., Gilgan, M., Richard, D. J. A., and Dewar, D. (1989). Identification of domoic acid, a neuroexcitatory amino acid, in toxic mussels from Eastern Prince Edward Island. *Can. J. Chem.* **67**, 481–490.

Yasumoto, T., Murata, M., Oshima, Y., Matsumoto, G. K., and Clardy, J. (1984). Diarrhetic shellfish poisoning. In E. P. Ragelis, ed., *Seafood Toxins*. ACS Symposium Series 262. American Chemical Society. Washington, D.C., pp. 207–214.

Yentsch, C. M. (1984). Paralytic shellfish poisoning: An emerging perspective. In E. P. Ragelis, ed., *Seafood Toxins*. ACS Symposium Series 262. American Chemical Society, Washington, D. C., pp. 9–23.

Yogasundram, K., and Shane, S. M. (1986). The viability of *Campylobacter jejuni* in refrigerated chicken drumsticks. *Vet. Res. Commun.* **10**, 479–486.

Yotsu, M., Yamazaki, T., Meguro, Y., Endo, A., Murata, M., Naoki, H., and Yasumoto, T. (1987). Production of tetrodotoxin and its derivatives by *Pseudomonas* sp. isolated from the skin of a pufferfish. *Toxicon* **25**, 225–228.

Zeissler, J., and L. Rassfeld-Sternberg (1949). Enteritis necroticans due to *Clostridium welchii* type F. *Br. Med. J.* **1**, 267.

THE ROLE OF THE FOOD AND AGRICULTURE ORGANIZATION OF THE UNITED NATIONS (FAO) IN FOOD CONTROL AND FOOD CONTAMINATION MONITORING PROGRAMS

John R. Lupien

Food Quality and Standards Service, Food Policy and Nutrition Division, FAO–United Nations, Rome, Italy

1. **Introduction**
   1.1. Introduction to Terms Used
       1.1.1. Food Quality Control
       1.1.2. Food Control
       1.1.3. Food Contamination Monitoring
   1.2. Function of FAO in Providing Technical Assistance to FAO Member Countries
2. **Principal FAO Activities in Food Control and Food Contamination Monitoring Programs**
   2.1. General Policies and Strategies
   2.2. Main FAO Activities in the Area of Food Control and Food Contamination Monitoring
       2.2.1. Food Control Infrastructures—Review
       2.2.2. Food Control Strategy Workshop
       2.2.3. Strengthening Food Control Programs at the National Level
       2.2.4. Training
       2.2.5. Urbanization—Review of Problems of Food Control

---

1.   INTRODUCTION

Problems of food contamination, adulteration, improper labeling, processing, packaging and handling can cause serious health and economic problems to all consumers, and damage the overall internal and export–import economies of all countries. The Food and Agriculture Organization of the United Nations provides advice and technical assistance to its member countries in the field of food control and food contamination monitoring programs. This appendix offers a brief description of the type of activities FAO has been carrying out in cooperation with FAO member countries in this field.

1.1.   Introduction to Terms Used

1.1.1.   Food Quality Control

Food quality control includes all steps necessary to protect the quality and safety of foods in the chain from agricultural and fishery production and harvesting, through processing and storage, to the marketing and preparation of food for consumption. Strictly speaking, this term refers to voluntary efforts made by the food industry or trade to ensure the quality and safety of the food produced or marketed by them.

1.1.2.   Food Control

Food control is regulatory activity based on the implementation of legislation, regulations, and standards that include compliance measures to ensure food is safe and offered for sale in an honest and fair manner to consumers. Food control also includes monitoring and surveillance program, for food hazards (e.g., microbial pathogens, chemical residues, mycotoxins, and radionuclides) that can provide vital information on the safety of national food supplies and serve as an indicator of environmental pollution problems. Food control is a mandatory activity enforced by national or local authorities to provide consumer protection and ensure that all foods domestically produced or imported and marketed conform to national requirements of quality and safety.

1.1.3. Food Contamination Monitoring

Monitoring consists of systems of repeated observation, measurement, and evaluation of substances or agents whose presence in, or on food, is considered to be undesirable, with the exception of substances normally produced naturally or by animals or plants themselves and intentionally added food additives. These systems are an important part of a food control program.

1.2. Function of FAO in Providing Technical Assistance to FAO Member Countries

The Food and Agriculture Organization of the United Nations was established in 1945 by nations determined to promote the common welfare by raising the levels of nutrition and standards of living of their people, securing improvements in the efficiency of production and distribution of all food and agricultural products, and bettering the conditions of rural populations.

The functions of the Organization include promoting and, where appropriate, recommending specific actions directed at conservation of national resources, improvement of methods of agricultural production, and for the processing, marketing, and distribution of food and agricultural products, as well as the improvement of education and administration relating to nutrition, food, and agriculture. It is also the function of FAO to provide such technical assistance as governments may request.

Within this overall mandate of FAO with regard to production, distribution, and marketing of food, and for nutrition improvement, the necessity of providing a safe, wholesome, and nutritious food supply, and the protection of this supply from losses, waste, deterioration, contamination, and adulteration is implicit to consumer protection against health hazards and economic fraud. FAO has therefore developed policies, advice, and programs to assist governments to orient their countries' policies and programs so as to include nutrition improvement and food protection objectives. *Food control* and *food contamination monitoring programs* are an integral part of this effort.

In FAO, the responsibility for the activities of ensuring quality and safety of food supplies and consumer protection is largely concentrated in the Economic and Social Department, and more specifically in that Department's Food Policy and Nutrition Division (ESN). However, work devoted to the prevention of food losses in postharvest handling and storage is also carried out within the

Agriculture and Fisheries Departments of FAO. As expected, there is close liaison and cooperation within FAO in these various activities.

The Food Quality and Standards Service of the Food Policy and Nutrition Division, where FAO's activities in the field of food control and food contamination monitoring are concentrated, is divided into two groups: the Food Standards Group, which provides the secretariat for the Codex Alimentarius Commission of the Joint FAO/World Health Organization (WHO) Food Standards Programme, and the Food Quality and Consumer Protection Group, which provides scientific and technical support to the Codex Alimentarius Commission and helps developing countries in the development of coherent national strategies for food control and food contamination monitoring programs, and the strengthening of the national infrastructure and capabilities.

For these activities also FAO maintains close liaison with other involved international agencies, particularly WHO, the International Atomic Energy Agency (IAEA), the General Agreement on Tariffs and Trade (GATT), the United Nations Development Programme (UNDP), the World Bank and regional development banks, the International Trade Centre, the United Nations Industrial Development Organization (UNIDO), and other regional specialized bodies (OECD, EEC, etc.).

2. PRINCIPAL FAO ACTIVITIES IN FOOD CONTROL AND FOOD CONTAMINATION MONITORING PROGRAMS

2.1. General Policies and Strategies

Developing a comprehensive national food control infrastructure is an extremely complex task that requires the participation, cooperation, and support of a wide range of institutions. Establishing a satisfactory food control system takes many years, and even then the system must continue to evolve and adapt as new technologies and socioeconomic realities emerge. Also, the components of a food control infrastructure come from several sectors (see Figure 1), and their integration is, sometimes, a difficult issue. FAO has developed an extensive series of food control policy and procedure guides and manuals (Exhibit 1) to assist Member Governments and food producers, processors, and consumers in reviewing and strengthening food control, quality, and safety activities. In view of this, the policymakers responsible for national food control programs have to make difficult decisions about where to place emphasis in allocating limited funding and manpower to have the most effective food control program. For that reason, they not only require a thorough knowledge of the national needs and resources, but also information and expertise coming from other sources, including international organizations, are helpful for them. The provision of some guidance based not only on basic knowledge but perhaps, equally as important, on practical approaches is much appreciated.

The general policy of FAO in the field of food control–food contamination is

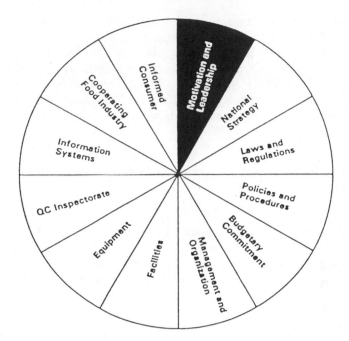

Figure 1. Sectors of a food control infrastructure.

to collaborate with its member countries in the identification of needs, the evaluation of the resources available in the countries, and the estimation of any needs for outside guidance and assistance. To the extent that its facilities and other obligations permit, FAO provides experts on request to survey conditions and to make specific recommendations concerning what a nation should do to modernize its laws and regulations, to improve its inspection and administrative services, and to strengthen laboratory technical facilities. Also it is possible to participate in developing national action programs for the improvement of food handling practices at all levels of the food chain and even within small village and household level food systems. Special attention is provided to the establishment of efficient systems for monitoring of food contaminants, both microbiological and chemical, including the contamination by radionuclides in products for national consumption and those involved in import–export trade and problems related to food control in connection with the rapid and extensive urbanization of some countries.

Once the countries establish their own priority, FAO also will collaborate with them in the definition and implementation of a specific program to strengthen the food control–food contamination monitoring situation.

The role of FAO at that moment is to supply the guidance for the establishment of a plan of action and the execution of the activities to provide experts for the technical backstopping for implementing them, to train the national groups to carry out the actions, to define, in collaboration with the national authorities, the criteria for evaluating the actions and to participate with the government in the evaluation of, then making the respective recommendations for their follow-up or improvement. The transfer of technology and "know-how" is the main aim of FAO's activity in order to encourage the self-sufficiency of the national officers. Also, the interchange of experience on an intercountry basis is strongly favored by FAO. Several regional and even interregional actions are encouraged by FAO and other international organizations.

2.2.   Main FAO activities in the Area of Food Control and Food Contamination Monitoring

2.2.1.   Food Control Infrastructures—Review

The strengthening of the food control infrastructure can be highly beneficial to government, industry, and, of course, consumers. All actions oriented toward improving the local situation are important; however, for an effective definition of priorities it is necessary that the Ministries of Health, Agriculture, Commerce, Industry, and so on have thorough discussions among themselves and with lawmakers and funding authorities. This also implies a thorough evaluation of the existing resources and trends and FAO is willing to participate in this activity if requested. To date, reviews of national food control infrastructure entailing recommendations for improvement have been carried out in the majority of the developing member countries of FAO. Some examples are Barbados, Bolivia, Colombia, Costa Rica, Guyana, Jamaica, Paraguay, Peru, Saint Lucia, Trinidad and Tobago, Uruguay, and Venezuela in the Latin America and Caribbean Region. FAO has also provided technical assistance for the review of the national food infrastructure in the Asia and Pacific Region in Fiji, India, Indonesia, Malaysia, Maldives, Mauritius, Nepal, Pakistan, People's Republic of China, Seychelles, Thailand and Tonga. In Africa: Algeria, Angola, Benin, Botswana, Cameroon, Cape Verde, Congo, Egypt, Ethiopia, Gabon, Guinea Bissau, Kenya, Liberia, Libya, Madagascar, Malawi, Maldives, Mauritania, Morocco, Mozambique, Republic of Central Africa, Senegal, Tanzania, Zaire, Zambia, and Zimbabwe.

In the Middle and Near East countries, such as Bahrain, Jordan, Kuwait, Qatar, Saudi Arabia, Syria, the United Arab Emirates, Turkey, and the Yemen Arab Republic, technical assistance has been received from FAO for evaluation and strengthening of their national food control infrastructure, among other countries.

During 1988 similar activities will be carried out in Bhutan, Côte d'Ivoire, Ecuador, Laos, Malta, and the People's Democratic Republic of Yemen.

2.2.2.   Food Control Strategy Workshop

Workshops conducted at both national and subregional levels have been held to consider, at a policy level, problems of food quality and safety and food supplies and consumer protection. The workshop sought to determine strategies that would ensure development of better food quality control systems, voluntary or mandatory, that would cover the entire food production, processing, and distribution chain.

In countries such as Cameroon, Cape Verde, Costa Rica, India, Indonesia, Liberia, Malawi, Pakistan, and others, FAO has recently sponsored national Workshops on Food Control Strategies. At the regional level some of the most recent workshops held are a workshop for the Caribbean subregion jointly sponsored by FAO, the Pan American Health Organization (PAHO), and the Caribbean Community (CARICOM). Another was the FAO Asian Regional Workshop on Export/Import Food Control Programmes, held in Bali, Indonesia.

In the Bali workshop, an examination was made of factors requiring consideration in deciding whether the establishment of an Export Inspection System (EIS) was justified. These included establishment and maintenance costs, the protection of the national reputation as a reliable supplier of quality products, the detrimental effects of the activities of dishonest and unscrupulous exporters, the levels of rejections and complaints from foreign markets, and the availability of professional staff and laboratory facilities to service an Export Inspection System. Types of Export Inspection Systems were also examined and the essential elements of a statutory system, including an export inspection law and regulations, the inspection service required to implement the law, administrative procedures, and certification provided, were considered in some detail. The workshop also considered and discussed the essential elements of a food import control system. Attention was paid to the legislation applying to imports, the inspection system implementing the legal requirements, the training of inspectors, and the analytical and support infrastructure.

Similar workshops are planned for the African, Latin American, and Caribbean regions.

2.2.3. Strengthening Food Control Programs at the National Level

Various FAO technical cooperation programs provide advice on drafting or updating comprehensive food laws and regulations, assistance on improvement of food inspection and analysis programs and facilities, training of technical administrators, inspection, chemicals, microbiology, laboratory technicians engaged in food control activities, and promotion of food quality control in the food production and distribution chain. These activities are carried out as part of the FAO regular program activities or through the activity of FAO acting as executing agency for projects funded by UNDP, other international agencies, or government donors.

Some examples of these projects at the national level are those on-going or recently completed: Angola, Burkina Faso, Costa Rica, Indonesia, Madagascar, Morocco, Mozambique, People's Republic of China, Seychelles, Syria, Thailand, Tonga, Turkey, and Zimbabwe.

2.2.4. Training

The training of personnel is one of the main FAO activities and it is carried out through the implementation of projects mentioned in Section 2.2.3 or through the realization of specific training courses or workshops dealing with specific items of food control: management, legislation, inspection and certification programs, or laboratories. Curricula have been developed for the various courses carried out that can be adapted to the differing circumstances of each specific group of trainees, region, food groups on contaminants to be covered, and so on.

Recent examples of regional activities are the UNDP-funded, FAO-executed Regional Cooperation Project for Establishing a Food Control Training Network in Asia, which has developed the capability of three training centers located in India, Indonesia, and Malaysia to train people from the region in several aspects of food control. To date courses on Basic Food Inspection Techniques, Low Acid Canned Food Inspection, and Inspection of Foods for Export have been held and the following are being planned: Management of Food Control Programs and another on Basic Food Inspection Techniques. The success of this regional effort is based not only on the strengthening of the capabilities of those food control officers that have been trained but also, and more important, on the established possibility of interchanging knowledge and experiences between the countries, the interchange of specialists, and the establishment of regional criteria. All these are components of the Technical Cooperation among Developing Countries (TCDC) Program that is one of the activities supported by FAO.

FAO, in cooperation with the Arab Organization for Standardization and Metrology of the Arab League, organized a regional seminar on quality control in the food industry with the participation of selected food quality control specialists from the countries of the region. The seminar identified a number of crucial issues that need to be addressed to promote the food industry and trade in the region. Also, a subregional Workshop on Management and Organization of Food Control Laboratories was organized by FAO in the United Arab Emirates to introduce newly acquired techniques in this field and promote technical cooperation among the food control laboratories in the Gulf region.

These types of activities at regional or subregional levels are also important for the Latin American and Caribbean Region where two activities have been planned in this field. The first is a Regional Workshop on Basic Food Law and the second a Regional Workshop for Inspectors on Food Sampling Techniques. The first is oriented towards food law-makers and food control managers dealing with the updating of their respective food laws–regulations in these countries. During the workshop the participants should analyze the present situation of the food legislation in the area, the importance of food legislation on the implementation of national food control programs, the main problems and constraints faced by its implementation, and the possibilities to adapt and adopt the FAO/WHO Model of Food Law. Recommendations for activities to strengthen the design and implementation of realistic and efficient food legislation are expected.

The Regional Workshop on Food Sampling Techniques will update the participants background in the most common food sampling techniques used for regulatory purposes. Also, the situation of national food control systems will be analyzed emphasizing the role of correct sampling for them.

Another planned activity is an Asian Regional Workshop in Laboratory Management and the preparation of background documents on food control in the Near East.

2.2.5. Urbanization—Review of Problems of Food Control

Rapidly growing urban populations are placing new and greater demands on food transportation and distribution systems, which often lead to shipment of foods over longer distances and, therefore, to increased problems of food spoilage, decomposition, and both chemical and microbiological contamination. This important subject was discussed at the last sessions of the Codex Committee for Latin America and the Caribbean (Havana, Cuba, 1983, 1985, 1987) and also for the Asia Region (Indonesia: Yogyakarta, 1985; Bali, 1987), which have given it high priority and asked international organizations to continue to support workshops and seminars on this subject.

In 1986, FAO, jointly with WHO, organized an Expert Consultation on Food Protection for Urban Consumers, to study the problems faced by local authorities in ensuring food quality and safety to rapidly growing city populations. The consultation discussed matters regarding the lack of local infrastructure and the need to tackle this problem and its effect, for example, faulty communication and coordination between national and local authorities, poor services, the existence of nonformal sectors in food production, and inadequate consumer education and information. The consultation recommended that national and local efforts be continued so as to limit the problem and had pointed out that it would be useful to implement certain measures, such as searching for alternative financing to provide the population with adequate water supply and sanitary services, training personnel involved in food production and distribution systems at local levels, promoting official recognition of, and attention to, the problem of street-vending of food; strengthening communication between national and local authorities, furthering actions carried out by international agencies with regard to legislation and standards, implementing simple mechanisms for reporting frauds or hazardous food, and establishing consumer organizations that would be involved in the search for solutions and in consumer guidance activities.

2.2.6. Street Foods Review

FAO assistance projects are being conducted to review and ascertain the quality and safety of street foods in order to determine what action should be taken to improve the current situation. "Street foods" is the term given, for want of a better description, of the wide variety of raw, cooked, semiprocessed, hot or cold foods and beverages sold by itinerant peddlers or from open-air food stalls in many developing countries.

Reviews are being conducted in several countries mainly in the Latin American and Asia Region: Colombia, Guatemala, Honduras, India, Indonesia, and the Philippines are some examples. Collection and microbiological analysis of samples of different types of street foods in these country reviews have shown heavy contamination with faecal coliforms in many samples and pathogenic bacteria in some samples.

Two regional workshops have been sponsored recently by FAO in this field. The first was the Latin American Workshop on Street Foods coordinated jointly by FAO and PAHO and held in Lima, Peru in October 1985. The second was held in Yogyakarta, Indonesia in November 1986 for the Asia Region. Throughout these workshops, the street food phenomenon was discussed and the factors that affect its growth in most of the urban cities of these regions were identified. Simultaneously, specific recommendations to review priority areas of concern were made particularly those referring to:

Improvement of technologies, process of preparation, equipment, distribution, and serving facilities.

Increasing capabilities of food handlers, physical infrastructure, and training activities.

Strengthening of regulatory infrastructure with recommendations for possible fiscal options.

Consumer education.

As a result of the recommendations made in the workshops, many actions have been implemented in this field. For example, it is interesting to mention that in Colombia and Peru FAO has financed two other studies on street foods oriented toward identifying better practices for preparation, handling, and services of street foods in the cities of Bogota and Lima and to develop models for the training of food vendors in these improvement techniques. The Colombian study also included the models for training of food control inspectors and public health promoters on street food control as well as the development of informative aids for the street food consumer. The models developed would be used in other countries of the region.

Moreover, additional efforts have been made to evaluate the socioeconomical importance of the street food phenomenon and the need to recognize it as a vast industry that provides employment to large segments of the population who might otherwise be unemployed. A specific study for the Asia region is planned for 1988.

Due to the lack of control by authorities and the lack of knowledge by the vendors of safe food handling practices, other activities have been oriented toward establishing a Code of Hygiene Practices for street food vendors. A primary proposal coming from the Codex Alimentarius Committee for Latin America and the Caribbean would be a starting point in this effort.

2.2.7. Food Standardization

The technical assistance for the establishment of food standards at both national and international levels is another activity in the field of food control that has been undertaken by FAO, in most of the cases in coordination with other international organizations as WHO.

Food standards are helpful for the orderly marketing of foods and for effective application of food control laws. Without food standards the purchaser may have no assurance that a packaged food will be of the identity and quality he expects. Traders in distant markets cannot buy with confidence if there are not standards by which they can specify the kind and quality of food to be delivered. The guidance provided by food standards is helpful to inspectors and essential to food analysts. Without standards the analyst knows neither what assays should be made nor whether his results indicate that a food is violative or nonviolative. The magistrate or court must have standards by which to judge whether foods, in fact, violate the law. In the absence of standards, cases involving adulterated foods may have to be dismissed. Uniform standards, intelligently applied, promote trade to the eventual benefit of producers, processors, traders, and consumers. They can be a powerful tool in upgrading the quality of foods. So, FAO's technical assistance in this field includes the provision of guidance for the implementation of national food standardization groups, the establishment of national standards, associated codes of practice, or other regulations and the activities oriented to verify its application through the food control system.

At the international level, FAO has been participating with WHO in this important field through two main groups: the Codex Alimentarius Commission and the Joint FAO/WHO Expert Committees on Food Additives (JECFA) and on Pesticide Residue (JMPR).

Following recommendations of the Eleventh Session of the FAO Conference, the Twenty-Ninth Session of the WHO Executive Board and a Joint FAO/WHO Conference on Food Standards held in 1962, the Codex Alimentarius Commission was established to implement the Joint FAO/WHO Food Standards Programme. The purpose of the program is to protect the health of consumers and to ensure fair practices in the food trade, to promote coordination of all food standards work undertaken by international governmental and nongovernmental organizations, to determine priorities and to initiate and guide the preparation of draft standards through, and with the aid of appropriate organizations, to finalize standards and, after acceptance by governments, publish them in Codex Alimentarius either as regional or worldwide standards.

At present, 133 member countries of FAO and WHO are also members of the Codex Alimentarius Commission, and membership is open to all Member Nations and associate members of FAO and WHO who are interested in international food standards.

The Commission has adopted its own Rules of Procedure and has established working procedures for the implementation of the Joint FAO/WHO Food

Standards Programme, which includes a Procedure for the Elaboration of Worldwide or Regional Standards General Principles for the Codex Alimentarius, and a Format for Commodity Standards. Exhibit 2 shows the Commodity Committee established until now and Exhibit 3 shows the list of standards and recommendations of the Codex Alimentarius Commission.

In the case of standardization of food additives and pesticide residue limits the Codex Alimentarius Commission implements the reports of the Joint FAO/WHO Expert Committee on Food Additives (JECFA) and Pesticide Residues (JMPR), groups of independent, internationally recognized experts, through the Codex Committee on Food Additives and the Codex Committee on Pesticide Residues.

The main objective of JECFA and JMPR are to provide independent scientific advice to the Codex Committees on Food Additives and Contaminants and on Pesticide Residues. JECFA and JMPR formulate general principles governing the use of additives and pesticides and evaluate the safety for the consumer of food additives, contaminants, and pesticide residues. The reports of JECFA and JMPR meetings are published in the WHO Technical Report Series, the toxicological evaluations of the substances are published in the WHO Food Additives Series, while FAO publishes the specifications on the purity and identity of the evaluated food additives in the FAO Food and Nutrition Paper Series.

2.2.8. Food Contamination Monitoring and Control

Food contamination monitoring and food contamination control are specific areas that have been receiving special support from FAO as part of its overall effort in food control.

FAO's technical assistance has been provided at global, regional, subregional, and national levels. Consultancies, training, supplies of laboratory materials and standards, and organization and sponsoring of workshops, courses, conferences, and so on have been the main tools used. The funds used have been obtained from FAO's regular program budget or from specific funds provided by UNEP, WHO, or other organizations or some government donors.

Below is a brief description of the main activities carried out in this field.

2.2.8.1.   National Level. Consultancies, both integrated in the framework of a food contamination monitoring project or as preliminary assistance. Consultants have been visiting countries at a government's request to assess existing facilities for food contamination monitoring and/or control of food contaminants. Also, they have determined priority areas with regard to personnel training in the areas of the management of this type of program, designs of national sampling plans, field inspection, laboratory analysis of both microbial and chemical contaminants, establishment of permissible maximum limits, and, in some cases, needs of training on specific measures to prevent and control food contamination.

Some examples of national projects implemented in this area are those carried out in the following areas:

Argentina: Assistance to the National Food Contamination Monitoring Programme

Cuba: Assistance for the Development of Inmuno-Assay for Mycotoxins

Chile: Study of Pesticide Residues in Priority Agricultural Products

Dominica: Control of Environmental Contamination in Food and Water

Guatemala: Aflatoxin Determination in Guatemalan Food Prepared from Corn

Peru: Preliminary Evaluation of the Chemical Contamination of Some Food Products in Peru

Uruguay: Establishment of a National Monitoring Programme of Pesticide Residues in Vegetable Foods

ASIA AND THE PACIFIC REGION

China: Strengthening the Food Contamination Control

Thailand: Strengthening Facilities for Food Analysis and Pilot Study for Food Additives and Contaminant Monitoring and Intake

Vanuatu: Aflatoxin Control for the Copra Industry

AFRICA

Malawi: Strengthening of Mycotoxin Monitoring Facilities

Swaziland: Fungal Contamination Control of Food and Human Health

Tanzania: Improvement of Mycotoxin Control in Tanzania

Zambia: Pilot Food Survey for Aflatoxin Contamination

2.2.8.2.   Regional and Subregional Level. At *regional* and *subregional levels* some of the most important activities carried out recently by FAO as an executing organization are as follows:

Latin America and the Caribbean Region

Training Course on Food Contamination Control in Latin America

Latin American Course of Pesticide Residues Analysis in Foods

Central American Training Course in Food Contaminants (Mycotoxin Analysis)

Africa

Control of Aflatoxins and Groundnuts ·

The FAO has assisted the African Groundnut Council through this project to train technicians and inspectors and upgrade the technical facilities of the member countries: Gambia, Mali, Niger, Nigeria, Senegal, and Sudan to control aflatoxins.

2.2.8.3.   Global Level

PESTICIDE RESIDUES, HEAVY METALS, MYCOTOXINS. At the *global level*, the training of personnel in proper laboratory methodology for analysis of food contaminants (pesticides, heavy metals, and mycotoxins) has been made through FAO/UNEP courses on International Training for Control of Environmental Contaminants in Food, held in Mysore, India (four courses) and two training courses on Food Contamination with Special Reference to Mycotoxins, held in the USSR. As follow-up to these courses, food contamination monitoring activities have been implemented in countries such as Brazil, Cuba, Guatemala, Mexico, and Tanzania.

The activities carried out by FAO at the global level also include the review of activities undertaken by international agencies and countries regarding the mycotoxin situation in order to define new policies and strategies to attempt to resolve this problem. In this regard, a Joint FAO/WHO/UNEP Second International Conference on Mycotoxins was held in Bangkok, Thailand from 28 September to 2 October 1987 to review the mycotoxin situation following the first conference held in Kenya in 1977. The main objectives of the conference were to reassess the nature and extent of the mycotoxins problem, its impact on human health, and its implication for international trade and to provide governments in developing countries with the following:

Recommendations for improved international harmonization of sampling analyses, monitoring, assessment, and control procedures with respect to mycotoxins.

A set of practical guidelines, strategies, and recommendations for monitoring, prevention, and control of mycotoxin contamination of food and animal feeds at the local level.

The conference recommended a number of priority activities to be undertaken by governments regarding prevention and control of food contamination with mycotoxins and urged international organizations to provide as adequate and substantial support to the programs as possible. FAO had planned to support training activities and projects at national and regional levels as conference follow-up but additional funds are required in order to be able to implement to some degree the conference recommendations.

As part of the global effort in food contamination monitoring FAO has been providing standards of mycotoxins and pesticides and other reference materials to many countries on a regular basis, and plans to continue doing so in order to support the studies and monitoring programs established.

RADIONUCLIDES. Following the April 1986 Chernobyl nuclear reactor accident, the FAO technical assistance on food contamination was expanded to the field of radionuclide contamination of foods. In response to requests from several member states to advise on actions that would need to be taken in respect of radionuclide contamination of foods, particularly those moving in international trade, in 1986 FAO organized an Expert Consultation that proposed the establishment of Interim International Radionuclide Action Levels for Foods Moving in International Trade (IRALFs). Further collaboration between FAO, WHO, and IAEA has eventually produced agreed-upon levels for unavoidable radionuclide contamination in foods (see Exhibit 4).

In addition, training, equipment, and supplies will be provided by FAO where possible on request of the governments, to assist food control officials in member countries to determine if food products were contaminated and, if so, at what level. Also, FAO is planning to sponsor regional workshops on Analysis of Food for Radionuclide Contamination, in regions or subregions where this problem is critical.

2.2.8.4. FAO/WHO Food Contamination Monitoring Programme. Simultaneously to the technical assistance that FAO provides to its member countries in the area of food contamination, the Organization also works with UNEP and WHO in this field.

One of the global activities started with UNEP support in 1976 was the Joint FAO/WHO Food Contamination Monitoring Programme. This program is one part of the overall UNEP Global Environmental Monitoring System (GEMS). The main objectives of the Joint FAO/WHO Food Contamination Monitoring Programme are as follows:

1. To collect data on levels of certain chemicals in individual foods and in the diet and to evaluate these data, review trends, and produce and disseminate summaries, thus encouraging appropriate food control and resource management measures.
2. To obtain estimates of the intake via food of chemical contaminants with a view to combining these data with those on exposure from other sources (air, water) thus enabling the evaluation of the risk to human health from multimedia exposure to chemicals.
3. To cooperate with governments of countries wishing to initiate or strengthen food contamination monitoring programmes.
4. To provide the Joint FAO/WHO Codex Alimentarius Commission with information on the levels of contaminants in food to support and accelerate its work on international standards for contaminants in foods.

NATIONALCOLLABORATING CENTERS AND PARTICIPATING INSTITUTIONS. In order to implement the program FAO and WHO decided to enlist the assistance of national food contamination monitoring programs in countries with strong ongoing monitoring activities or countries with plans to strengthen national food contamination monitoring capabilities. This network of FAO/WHO/UNEP Collaborating Centers and Participating Institutions has grown to 37 agencies and institutions in 34 countries and includes ministries of agriculture, health, national research, and national institutes of hygiene, nutrition, and agroindustrial development indicating the wide interest in food contamination and its potential adverse impact on agriculture, environment, food trade, and health. (See Exhibit 5 for current list of participating agencies.)

OPERATION OF THE PROGRAM

1. Data Collection and Evaluation. A principal goal of the program is to collect information on levels of certain contaminants in foods to evaluate trends in contamination and enable dissemination of summaries of reported data. In joining the network, collaborating centers and participating institutions agree to provide the program with data on a range of contaminants in foods such as organochlorine and organophosphorus pesticides, polychlorinated biphenyls, lead, cadmium, tin, mercury, and aflatoxins in different foods where residues are most likely to be found (Exhibit 6). Due to the size of national programs, available funds for monitoring, and other competing priorities, data submissions from various participating institutions have varied widely. Not all institutions have submitted data on all contaminants in all foods, and data submitted are based on numbers of samples ranging from 4–5 to 2000. This obviously can have an effect on the overall results.

Methods of collecting, processing, and reporting monitoring data on selected contaminants in individual foods and diet have been developed utilizing a computer storage and retrieval system located at WHO, Geneva. Data summaries by country, contaminants, and food groups have been issued and assessments of the data collected so far, in terms of time trends and where appropriate with respect to established guidelines or norms, have been carried out.

2. Guidelines for the Study of Dietary Intakes of Chemical Contaminants. "Guidelines for the Study of Dietary Intakes of Chemical Contaminants" have been prepared to assist countries in initiating studies of the dietary intake of contaminants by providing detailed procedures and methods by which such studies may be conducted. The Guidelines have been prepared under the Joint FAO/WHO Food Contamination Monitoring Programme, in collaboration with the Joint FAO/WHO Food Standards Programme and the relevant committees of the Codex Alimentarius Commission. The Guidelines describe several practical approaches for determining food consumption data of a population or of individuals. Based on such food consumption data, estimates of daily intake of a contaminant may be obtained using three different approaches.

Beginning in 1980 and as part of the data collection and evaluation activities mentioned previously, collection of dietary intakes data has been carried out in a systematic fashion. Data on the dietary intakes of certain contaminants have been received from 11 institutions and have been computer processed, summarized, and evaluated. The data received cover the period 1971–1983 and include information on the intakes of a series of organochlorine and organophosphorus pesticides, PCBs, cadmium, lead, and aflatoxins. Data collected have been compared to the acceptable daily intake (ADI) or provisional tolerable weekly intake (PTWI) of the contaminants in question.

3. Analytical Quality Assurance. Data submitted to the Joint FAO/WHO Food Contamination Monitoring Programme come from a wide variety of laboratories with different levels of experience in contaminant monitoring and analysis, as well as differing problems involving equipment, reagents, and technical support. To promote the quality and comparability of data submitted, interlaboratory studies have been initiated on

a. organochlorine compounds,

b. lead and cadmium, and

c. aflatoxins.

The National Food Administration, Sweden, the Ministry of Agriculture, Fisheries and Food, United Kingdom, and the International Agency for Research on Cancer, Lyon are involved in the cordination of these studies.

Results to date of the AQA studies have shown the need for greater emphasis on laboratory quality control procedures. Variations in results can be ascribed to equipment and reagent problems and also help identify areas where additional training may be needed.

4. Information Exchange. The objectives of the program information exchange activities are as follows:

a. To inform collaborating centers, participating institutions, scientists, and national health and agriculture staff about the activities of the program and the possibilities it offers for collaboration both in the assessment of chemical contamination of food and its control.

b. To provide rapid dissemination of useful information on global trends and levels of contaminants in food, exposure assessment of environmental pollutants, and methods for their determination and control. Some activities in this area include the following:

Visits by FAO and WHO staff and consultants to some 20 developing countries within the last 3 years, to explain the goals of the program, and enlist these countries participation.

Direct information exchange between participating institutions.

Distribution of documents issued under the Joint UNEP/FAO/WHO Food Contamination Monitoring Programme. These are listed in Exhibit 7.

Distribution of FAO guidelines and manuals covering different aspects of food control and food safety, for example, developing food control systems, food inspection, export inspection, food control laboratory, food analysis—chemical and microbiological, prevention of mycotoxins contamination, and surveillance of mycotoxins.

Distribution of Environmental Health Criteria documents published by WHO's International Programme on Chemical Safety, dealing with subjects of relevance to food such as mercury, polychlorinated biphenyls and terphenyls, lead, nitrates, nitrites, and nitrosamines, DDT and its derivatives, mycotoxins, tin and organotin compounds, manganese, arsenic, principles and methods.for evaluating the toxicity of chemicals, 2, 4-D, chlordane, heptachlor, paraquat and diquat, endosulfan, tecnazene, mirex, chlordecone, camphechlor, chromium, principles for the safety assessment of food additives and contaminants in food, carbamate pesticides, organophosphorus insecticides, and selenium.

Distribution of relevant publications of the International Agency for Research on Cancer in the series Environmental Carcinogens-Selected Methods of Analysis. The volumes on mycotoxins, N-nitroso compounds, and metals have been distributed to all institutions participating in GEMS/Food.

The WHO computerized mailing list now includes all 37 Collaborating Centers and Participating Institutions. Materials such as the Bulletin of WHO and FAO and WHO publications dealing with food safety are sent to these institutions.

This aspect of the program has been well received and further efforts will be made to increase the outreach of information to institutions and individuals interested in food contamination monitoring.

5.   Technical Cooperation. Examples of technical cooperation activities with developing countries include provision of chemicals and minor laboratory equipment, short-term training in contaminants analyses by assigning a consultant for a 2- to 3-week period to the institution requesting this service or, alternatively, by sending scientists from one institution to a Collaborating Center having the necessary expertise and facilities.

In summary, components of the program show that further work is needed to develop uniform levels of competence in contaminant analysis, as well as for uniform or comparable monitoring, surveillance, and sampling activities. The benefits of participation in this program and the other food contamination efforts made by FAO and other international organizations can be great, including improved reassurance to consumers about potential health risks from foods, and reduction of levels of contamination to assist in removing nontariff barriers to trade in foods.

2.2.9. Publications

FAO has published several guidelines and manuals (some of them jointly with WHO, UNEP, and other donor countries such as Sweden) covering different aspects of food control, and food safety, for example, developing a food control system, food inspection, export inspection, food analysis—chemical and microbiological, prevention of mycotoxin contamination, and surveillance of mycotoxins. The list of publications available is appended as Exhibit 1.

2.3.   Present and Future Situation

The activities that have been described show to a large extent the technical assistance that FAO has been providing to its member countries based on the country's needs and resources, the provision of funds allocated for project implementation, and the economic and technical background of the country's economy, including its food and agricultural resources, the state of health and nutrition of the population, and the status of existing food control and food contamination monitoring services.

The success of the technical assistance provided is the result of the concurrence of the expertise provided by FAO and the technical and financial ability of the countries to continue the activities once the initial phase is completed and outside aid is discontinued.

National involvement is essential and the benefits derived are directly related to the degree of involvement of the recipient government. Countries must give food control a high priority in order to obtain and initiate assistance in strengthening this activity.

The need for efficient food control systems to facilitate trade and protect consumers is still valid and much work needs to be done to cover the needs of the Member Government; thus assistance to developing countries requires increased and continuing support. Funds may also come from other international agencies, such as UNDP, from national "funds in trust" or from other sources. Additionally, strong support should be given to reinforce the Technical Cooperation Programme in the developing countries in order to strengthen the interchange of national experts and expertise, establish regional criteria, and reduce the cost for food control activities at national and regional levels.

3.   CONCLUSIONS

The FAO technical assistance program on food control and food contamination includes advisory services for project implementation; national food control or food contamination workshops; the drafting of up-to-date food laws and regulations including the design of national standards, the strengthening of the administration and inspectorate services, and the implementation of good manufacturing practices; the development of certification programs for food exportation; training courses in food analytical control and food contaminants analysis, monitoring, and control; the provision of laboratory equipment, publications in food analysis, chemical and microbiological, and in mycotoxin surveillance and prevention; supply of standards and reference materials; and so on.

This technical assistance is provided to FAO member countries on their request, directly or through projects funded by other international organizations or national funds. The success of this technical assistance is based on the concurrence of the expertise provided by FAO and the technical and financial ability of countries to continue activities once the initial phase is completed and outside aid is discontinued.

To the extent that its facilities and other obligations permit it, FAO has, and will continue to, provide this assistance under the framework of its overall mandate with regard to production, distribution, and marketing of a safe, wholesome, and nutritious food supply, in order to protect the consumers from health hazards and economic fraud and promote food trade.

EXHIBITS

Exhibit 1. FAO Food and Nutrition Papers

FAO TECHNICAL PAPERS
FOOD AND NUTRITION PAPERS:
1. Review of food consumption surveys, 1977
    Vol. 1 — Europe, North America, Oceania, 1977 (E')
    Vol. 2 — Far East, Near East, Africa, Latin America, 1979 (E')
2. Report of the joint FAO/WHO/UNEP conference on mycotoxins, 1977 (E' F' S')
3. Report of the joint FAO/WHO expert consultation on the rôle of dietary fats and oils in human nutrition, 1977 (E' F' S')
4. JECFA specifications for identity and purity of thickening agents, anticaking agents, antimicrobials, antioxidants and emulsifiers, 1978 (E')
5. Guide to JECFA specifications, 1978 (E' F')
5 Rev. Guide to JECFA specifications, 1983 (E' F')
6. The feeding of workers in developing countries, 1978 (E' S')
7. JECFA specifications for identity and purity of food colours, enzyme preparations and other food additives, 1978 (E' F')
8. Women in food production, food handling and nutrition, 1978 (E' F' S')
9. Arsenic and tin in foods: reviews of commonly used methods of analysis, 1979 (E')
10. Prevention of mycotoxins, 1979 (E' F' S')
11. The economic value of breast-feeding, 1979 (E')
12. JECFA specifications for identity and purity of food colours, flavouring agents and other food additives, 1979 (E' F')
13. Perspective on mycotoxins, 1979 (E' S')
14. Manuals of food quality control
    1 — The food control laboratory, 1979 (Ar' E') (Revised version 1986, E')
    2 — Additives, contaminants, techniques, 1979 (E' F')
    3 — Commodities, 1979 (E')
    4 — Microbiological analysis, 1979 (E' F' S')
    5 — Food inspection, 1981 (Ar' E') (Revised edition 1984, E')
    6 — Food for export, 1979 (E')
    7 — Food analysis: general techniques, additives, contaminants and composition, 1986 (E')
    8 — Food analysis: quality, adulteration and test of identity, 1986 (E')
    9 — Introduction to food sampling, 1988 (E')
15. Carbohydrates in human nutrition, 1980 (E' F' S')
16. Analysis of food and nutrition survey data for developing countries, 1980 (E' F' S')
17. JECFA specifications for identity and purity of sweetening agents, emulsifying agents, flavouring agents and other food additives, 1980 (E' F')
18. Bibliography of food consumption surveys, 1981 (E')
18 Rev. 1 — Bibliography of food consumption surveys, 1981 (E')
18 Rev. 2 — Bibliography of food consumption surveys, 1987 (E')
19. JECFA specifications for identity and purity of carrier solvents, emulsifiers and stabilizers, enzyme preparations, flavouring agents, food colours, sweetening agents and other food additives, 1981 (E' F')
20. Legumes in human nutrition, 1982 (E' F' S')
21. Mycotoxin surveillance — a guideline, 1982 (E')
22. Guidelines for agricultural training curricula in Africa, 1982 (E' F')
23. Management of group feeding programmes, 1982 (E' F' S')
24. Evaluation of nutrition interventions, 1982 (E')
25. JECFA specifications for identity and purity of buffering agents, salts, emulsifiers, thickening agents, stabilizers, flavouring agents, food colours, sweetening agents and miscellaneous food additives, 1982 (E' F')
26. Food composition tables for the Near East, 1983 (E')
27. Review of food consumption surveys — 1981, 1983 (E')

28. JECFA specifications for identity and purity of buffering agents, 1983 (E' F')
29. Post-harvest losses in quality of foodgrains, 1983 (E' F)
30. FAO/WHO food additives data system, 1984 (E')
30 Rev. FAO/WHO food additives data system, 1985 (E')
31/1. JECFA specifications for identity and purity of food colours, 1984 (E' F')
31/2. JECFA specifications for identity and purity of food additives, 1984 (E' F')
32. Residues of veterinary drugs in foods, 1985 (E/F/S')
33. Nutritional implications of food aid: an annotated bibliography, 1985 (E')
34. Specifications for identity and purity of certain food additives, 1986 (E' F''')
35. Review of food consumption surveys, 1985 (E')
36. Guidelines for can manufacturers and food canners, 1986 (E')
37. JECFA specifications for identity and purity of certain food additives, 1986 (E' F')
38. JECFA specifications for identity and purity of certain food additives, 1988 (E')
39. Quality control in fruit and vegetable processing, 1988 (E' F')
40. Directory of food and nutrition institutions in the Near East, 1987 (E')
41. Residues of some veterinary drugs in animals and foods, 1988 (E')
42. Traditional food plants, 1988 (E')
42/1. Edible plants of Uganda — The value of wild and cultivated plants as food, 1989 (E')
43. Guidelines for agricultural training curricula in Arab countries, 1988 (Ar')
44. Review of food consumption surveys — 1988, 1988 (E')
45. Exposure of infants and children to lead, 1989 (E')
46. Street foods, 1989 (E/F/S')

Availability: June 1989

Ar — Arabic
C — Chinese   ' Available
E — English   '' Out of print
F — French   ''' In preparation
S — Spanish

The FAO Technical Papers can be purchased locally through the authorized FAO Sales Agents or directly from Distribution and Sales Section, FAO, Via delle Terme di Caracalla, 00100 Rome, Italy.

Exhibit 2. Information on Codex Committees for Commodities and General Subjects

Commodity Committees of the Codex Alimentarius Commission

Reference Initials	Name	Host Country
CCCPC	Cocoa products and chocolate	Switzerland
CCS	Sugars	United Kingdom
CCPFV	Processed fruit and vegetables	United States
CCFO	Fats and oils	United Kingdom
CCEI	Edible ices	Sweden
CCSB	Soups and Broths	Switzerland
CCFSDU	Foods for special dietary uses	Federal Republic of Germany
CCFFP	Fish and fishery products	Norway
CCVP	Vegetable proteins	Canada
CCMH	Meat hygiene	New Zealand
CCM	Meat	Federal Republic of Germany
CCPMPP	Processed meat and poultry products	Denmark
CCCPL	Cereals, pulses, and legumes	United States
CCMIN	Natural mineral waters	Switzerland

Exhibit 3.   Codex Alimentarius Commission

FOOD AND AGRICULTURE
ORGANIZATION
OF THE UNITED NATIONS

WORLD HEALTH
ORGANIZATION

JOINT OFFICE: Via delle Terme di Caracalla 00100 ROME: Tel. 57971 Telex: 610181 FAO I Cables Foodagri

CX/GEN 85/1 (Rev. 1)   March 1988

LIST OF
FINAL CODEX TEXTS

SECTION I – CODEX STANDARDS

Subject	Reference	Publication
GENERAL INTRODUCTION		CAC/VOL I – Ed. 1
PROCESSED FRUITS AND VEGETABLES		CAC/VOL II – Ed. 1
Canned Tomatoes	CODEX STAN. 13–1981	"
Canned Peaches	CODEX STAN. 14–1981	"
Canned Grapefruit	CODEX STAN. 15–1981	"
Canned Green Beans and Wax Beans	CODEX STAN. 16–1981	"
Canned Applesauce	CODEX STAN. 17–1981	"
Canned Sweet Corn	CODEX STAN. 18–1981	"
General Standard for Edible Fungi and Fungus Products	CODEX STAN. 38–1981	"
Dried Edible Fungi	CODEX STAN. 39–1981	"
Fresh Fungus "Chanterelle"	CODEX STAN. 40–1981	"
Canned Pineapple	CODEX STAN. 42–1981	"
Canned Mushrooms	CODEX STAN. 55–1981	"
Canned Asparagus	CODEX STAN. 56–1981	"
Processed Tomato Concentrates	CODEX STAN. 57–1981	"
Canned Green Peas	CODEX STAN. 58–1981	"
Canned Plums	CODEX STAN. 59–1981	"
Canned Raspberries	CODEX STAN. 60–1981	"
Canned Pears	CODEX STAN. 61–1981	"
Canned Strawberries	CODEX STAN. 62–1981	"
Table Olives	CODEX STAN. 66–1981	"
Raisins	CODEX STAN. 67–1981	"
Canned Mandarin Oranges	CODEX STAN. 68–1981	"
Canned Fruit Cocktail	CODEX STAN. 78–1981	"
Jams (Fruit Preserves) and Jellies	CODEX STAN. 79–1981	"
Citrus Marmalade	CODEX STAN. 80–1981	"
Canned Mature Processed Peas	CODEX STAN. 81–1981	"
Canned Tropical Fruit Salad	CODEX STAN. 99–1981	"
Pickled Cucumbers	CODEX STAN. 115–1981	"
Canned Carrots	CODEX STAN. 116–1981	"
Canned Apricots	CODEX STAN. 129–1981	"
Dried Apricots	CODEX STAN. 130–1981	"
Unshelled Pistachio Nuts	CODEX STAN. 131–1981	"
SUGARS (Including Honey)		CAC/VOL III–Ed. 1
White sugar	CODEX STAN. 4–1981	"
Powdered Sugar (Icing Sugar)	CODEX STAN. 5–1981	"
Soft Sugars	CODEX STAN. 6–1981	"
Dextrose Anhydrous	CODEX STAN. 7–1981	"
Dextrose Monohydrate	CODEX STAN. 8–1981	"

w/S7102

Subject	Reference	Publication
Glucose Syrup	CODEX STAN. 9–1981	"
Dried Glucose Syrup	CODEX STAN. 10–1981	"
Lactose	CODEX STAN. 11–1981	"
Honey (European Regional Standard)	CODEX STAN. 12–1981	"
Powdered Dextrose (Icing Dextrose)	CODEX STAN. 54–1981	"
Fructose	CODEX STAN. 102–1981	"
Amendment to Explanatory Notes of Codex Standards for Sugars		Supplement 1 to CAC/VOL III–Ed. 1

PROCESSED MEAT AND POULTRY PRODUCTS AND SOUPS AND BROTHS   CAC/VOL IV–Ed. 1

Canned Corned Beef	CODEX STAN. 88–1981	"
Luncheon Meat	CODEX STAN. 89–1981	"
Cooked Cured Ham	CODEX STAN. 96–1981	"
Cooked Cured Pork Shoulder	CODEX STAN. 97–1981	"
Cooked Cured Chopped Meat	CODEX STAN. 98–1981	"
Bouillons and Consommés	CODEX STAN. 117–1981	"

FISH AND FISHERY PRODUCTS   CAC/VOL V–Ed. 1

Canned Pacific Salmon	CODEX STAN. 3–1981	"
Quick-Frozen Gutted Pacific Salmon	CODEX STAN. 36–1981	"
Canned Shrimps or Prawns	CODEX STAN. 37–1981	"
Quick-Frozen Fillets of Cod and Haddock	CODEX STAN. 50–1981	"
Quick-Frozen Fillets of Ocean Perch	CODEX STAN. 51–1981	"
Canned Tuna and Bonito in Water or Oil	CODEX STAN. 70–1981	"
Canned Crab Meat	CODEX STAN. 90–1981	"
Quick-Frozen Fillets of Flat Fish	CODEX STAN. 91–1981	"
Quick-Frozen Shrimps or Prawns	CODEX STAN. 92–1981	"
Quick-Frozen Fillets of Hake	CODEX STAN. 93–1981	"
Canned Sardines and Sardine-Type Products	CODEX STAN. 94–1981	"
Quick-Frozen Lobsters	CODEX STAN. 95–1981	"
Canned Mackerel and Jack Mackerel	CODEX STAN. 119–1981	"
Inclusion of Further Species in the Standard for Canned Sardines and Sardine-Type Products	CODEX STAN. 94–1981	Supplement 1 to CAC/VOL V–Ed. 1

LABELLING   CAC/VOL VI–Ed. 2

General Standard for the Labelling of Prepackaged Foods (Revised Text)	CODEX STAN. 1–1985	"
General Standard for the Labelling of Food Additives when Sold as such	CODEX STAN. 107–1981	"

Codex Guidelines on Labelling:
General Guidelines on Claims	CAC/GL 1–1979	"
Guidelines on Nutrition Labelling	CAC/GL 2–1985	"
Guidelines for Date-Marking of Prepackaged Foods for the Use of Codex Committees	–	"

COCOA PRODUCTS AND CHOCOLATE   CAC/VOL VII–Ed. 1

Cocoa Butters	CODEX STAN. 86–1981	"
Chocolate	CODEX STAN. 87–1981	"

Subject	Reference	Publication
Cocoa Powders (Cocoa) and Dry Cocoa-Sugar Mixtures	CODEX STAN. 105-1981 & Corr. (July 1986 - English only)	"
Cocoa (Cacao) Nib, Cocoa (Cacao) Mass, Cocoa Press Cake and Cocoa Dust (Cocoa Fines), for use in the manufacturing of Cocoa and Chocolate Products	CODEX STAN. 141-1983	Supplement 1 to CAC/VOL VII-Ed. 1
Composite and Filled Chocolate	CODEX STAN. 142-1983	"

QUICK FROZEN FRUITS AND VEGETABLES   CAC/VOL VIII-Ed. 1

Quick Frozen Peas	CODEX STAN. 41-1981	"
Quick Frozen Strawberries	CODEX STAN. 52-1981	"
Quick Frozen Raspberries	CODEX STAN. 69-1981	"
Quick Frozen Peaches	CODEX STAN. 75-1981	"
Quick Frozen Bilberries	CODEX STAN. 76-1981	"
Quick Frozen Spinach	CODEX STAN. 77-1981	"
Quick Frozen Blueberries	CODEX STAN. 103-1981	"
Quick Frozen Leek	CODEX STAN. 104-1981	"
Quick Frozen Broccoli	CODEX STAN. 110-1981	"
Quick Frozen Cauliflower	CODEX STAN. 111-1981	"
Quick Frozen Brussels Sprouts	CODEX STAN. 112-1981	"
Quick Frozen Green and Wax Beans	CODEX STAN. 113-1981	"
Quick Frozen French Fried Potatoes	CODEX STAN. 114-1981	"
Quick Frozen Whole Kernel Corn	CODEX STAN. 132-1981	"
Quick Frozen Corn-on-the-Cob	CODEX STAN. 133-1981	"
Quick Frozen Carrots	CODEX STAN. 140-1983	Supplement 1 to CAC/VOL VIII-Ed. 1

FOODS FOR SPECIAL DIETARY USES   CAC/VOL IX-Ed. 1

Foods with Low-Sodium Content (including Salt Substitutes)	CODEX STAN. 53-1981	"
Infant Formula	CODEX STAN. 72-1981	"
Canned Baby Foods	CODEX STAN. 73-1981	"
Processed Cereal-based Foods for Infants and Children	CODEX STAN. 74-1981	"
Gluten-Free Foods	CODEX STAN. 118-1981	"
Advisory Lists of Mineral Salts and Vitamin Compounds for Use in Foods for Infants and Children	—	"
Amendments to Codex Standards 53, 72, 73, 74 and 118 and to the Advisory Lists of Mineral Salts and Vitamin Compounds	—	Supplement 1 to CAC/VOL IX-Ed. 1
Codex General Staandard for the Labelling of and Claims for Prepackaged Foods for Special Dietary Uses	CODEX STAN. 146-1985	Supplement 2 to CAC/VOL IX - Ed.1
Amendments to Codex Standards 72, 73 and 74	—	"

FRUIT JUICES   CAC/VOL X-Ed. 1

Apricot, Peach and Pear Nectars	CODEX STAN. 44-1981	"
Orange Juice	CODEX STAN. 45-1981	"
Grapefruit Juice	CODEX STAN. 46-1981	"
Lemon Juice	CODEX STAN. 47-1981	"

Subject	Reference	Publication
Apple Juice	CODEX STAN. 48–1981	"
Tomato Juice	CODEX STAN. 49–1981	"
Concentrated Apple Juice	CODEX STAN. 63–1981	"
Concentrated Orange Juice	CODEX STAN. 64–1981	"
Grape Juice	CODEX STAN. 82–1981	"
Concentrated Grape Juice	CODEX STAN. 83–1981	"
Sweetened Concentrated Labrusca Type Grape Juice	CODEX STAN. 84–1981	"
Pineapple Juice	CODEX STAN. 85–1981	"
Non–Pulpy Blackcurrant Nectar	CODEX STAN. 101–1981	"
Blackcurrant Juice preserved exclusively by physical means	CODEX STAN. 120–1981	"
Concentrated Blackcurrant Juice preserved exclusively by physical means	CODEX STAN. 121–1981	"
Pulpy Nectars of Certain Small Fruits preserved exclusiverly by physical means	CODEX STAN. 122–1981	"
Nectars of Certain citrus Fruits	CODEX STAN. 134–1981	"
Concentrated Pineapple Juice preserved exclusively by physical means	CODEX STAN. 138–1983	Supplement 1 to CAC/VOL X–Ed. 1
Concentrated Pineapple Juice with Preservatives, for Manufacturing	CODEX STAN. 139–1983	"
Amendments to Codex Standards 44–49, 63–64, 82–85, 101 120–122 and 134	–	"
Guava Nectar Preserved exclusively by Physical Means	CODEX STAN. 148–1985	Supplement 2 to CAC/VOL X – Ed. 1
Liquid Pulpy Mango Products Preserved exclusively by Physical Means	CODEX STAN. 149–1985	"
Amendments to Codex Standards 44, 101, 122 and 134	–	"

FATS AND OILS

CAC/VOL XI–Ed. 1

General Standard for Fats and Oils not covered by individual standards	CODEX STAN. 19–1981	"
Edible Soya Bean Oil	CODEX STAN. 20–1981	"
Edible Arachis Oil	CODEX STAN. 21–1981	"
Edible Cottonseed Oil	CODEX STAN. 22–1981	"
Edible Sunflowerseed Oil	CODEX STAN. 23–1981	"
Edible Rapeseed Oil	CODEX STAN. 24–1981	"
Edible Maize Oil	CODEX STAN. 25–1981	"
Edible Sesameseed Oil	CODEX STAN. 26–1981	"
Edible Safflowerseed Oil	CODEX STAN. 27–1981	"
Lard	CODEX STAN. 28–1981	"
Rendered Pork Fat	CODEX STAN. 29–1981	"
Premier Jus	CODEX STAN. 30–1981	"
Edible Tallow	CODEX STAN. 31–1981	"
Margarine	CODEX STAN. 32–1981	"
Olive Oil	CODEX STAN. 33–1981	"
Mustardseed Oil	CODEX STAN. 34–1981	"
Edible Low Erucic Acid Rapeseed Oil	CODEX STAN. 123–1981	"
Edible Coconut Oil	CODEX STAN. 124–1981	"
Edible Palm Oil	CODEX STAN. 125–1981	"
Edible Palm Kernel Oil	CODEX STAN. 126–1981	"
Edible Grapeseed Oil	CODEX STAN. 127–1981	"
Edible Babassu Oil	CODEX STAN. 128–1981	"
Minarine	CODEX STAN. 135–1981	"

Subject	Reference	Publication
Amendments to Codex Standards 19–34, 123–128 and 135	–	Supplement 1 to CAC/VOL XI–Ed. 1

MISCELLANEOUS PRODUCTS   CAC/VOL XII–Ed. 1

Natural Mineral Waters (European Regional Standard)	CODEX STAN. 108–1981	"
Edible Ices	CODEX STAN. 137–198	
Food Grade Salt	CODEX STAN. 150–1985	Supplement 1 to CAC/VOL XII – Ed.1
Gari (African Regional Standard)	CODEX STAN. 151–1985	"
Amendments to Codex Standard 108	–	"

PESTICIDE RESIDUES

Codex Maximum Limits for Pesticide Residues adopted by the Codex Alimentarius Commission up to the end of the 16th Session (July 1985)	CAC/VOL XIII–Ed. 2	CAC/VOL XIII–Ed. 2

Guide to Codex Maximum Limits for
Pesticide Residues:

– Part 1 – General Notes and Guidelines	CAC/PR 1–1984	CAC/PR 1–1984
– Part 2 – Maximum Limits for Pesticide Residues	CAC/PR 2–1986 3rd PRELIM. ISSUE ENGLISH ONLY	CAC/PR 2–1986 3rd PRELIM. ISSUE ENGLISH ONLY
– Part 3 – Guideline levels for Pesticide Residues	CAC/PR 3–1986 2nd PRELIM. ISSUE ENGLISH ONLY	CAC/PR 3–1986 2nd PRELIM. ISSUE ENGLISH ONLY
– Part 4 – Classification of Foods and Animal Feed	CAC/PR 4–1986 ENGLISH ONLY	CAC/PR 4–1986 ENGLISH ONLY
– Part 5 – Recommended Method of Sampling for Determination of Pesticide Residues	CAC/PR 5–1984	CAC/PR 5–1984
– Part 6 – Portion of Commodities to which Codex Maximum Residue Limits Apply and which is Analysed	CAC/PR 6–1984	CAC/PR 6–1984
– Part 7 – Codex Guidelines on Good Practice in Pesticide Residues Analysis	CAC/PR 7–1984	CAC/PR 7–1984
– Part 8 – Recommendations for Methods of Analysis of Pesticide Residues	CAC/PR 8–1986 THIRD EDITION	CAC/PR 8–1986 THIRD EDITION
– Part 9 – Recommended National Regulatory Practices to Facilitate Acceptance and Use of Codex Maximum Limits for Pesticide Residues in Foods	CAC/PR 9–1985	CAC/PR 9–1985

FOOD ADDITIVES

Food Additives (comprises the following five parts):	CAC/VOL XIV–Ed. 1	CAC/VOL XIV–Ed. 1

– Part I   – Definitions
– Part II – General Principles for
             the Use of Food Additives

Subject	Reference	Publication
– Part III – Principle Relating to the Carry–Over of Food Additives into Food		
– Part IV – Guidelines for the Establishment of Food Additive Provisions in Commodity Standards		
– Part V – Food Additives Permitted for Use in Codex Standards [N.B. This is not an exclusive list of food additives]		

IRRADIATED FOODS

General Standard for Irradiated Foods	CODEX STAN 106–1983	CAC/VOL XV–Ed. 1

MILK AND MILK PRODUCTS STANDARDS
(See Section IV)   CAC/VOL XVI–Ed. 1   CAC/VOL XVI–Ed. 1

CONTAMINANTS

Contaminants – contains maximum levels for contaminants permitted in Codex standards adopted by the Codex Alimentarius Commission [N.B. This is not an exclusive list of levels for contaminants permitted in foddstuffs]	CAC/VOL XVII–Ed. 1	CAC/VOL XVII–Ed. 1

CEREALS, PULSES, LEGUMES AND DERIVED PRODUCTS

Wheat Flour	CODEX STAN 152–1985	CAC/VOL XVIII–Ed.1
Maize (Corn)	CODEX STAN 153–1985	"
Whole Maize (Corn) Meal	CODEX STAN 154–1985	"
Degermed Maize (Corn) Meal and Maize (Corn) Grits	CODEX STAN 155–1985	"

SECTION II – RECOMMENDED INTERNATIONAL CODES OF HYGIENIC AND/OR TECHNOLOGICAL PRACTICE

Subject	Reference	Publication
General Principles of Food Hygiene	CAC/RCP 1–1969, Rev. 1 (1979)	CAC/VOL A–Ed. 1
Code of Hygienic Practice for Canned Fruit and Vegetable Products	CAC/RCP 2–1969	CAC/RCP 2–1969 (Vol.D)
Code of Hygienic Practice for Dried Fruits	CAC/RCP 3–1969	CAC/RCP 3–1969 (Vol.D)
Code of Hygienic Practice for Desiccated Coconut	CAC/RCP 4–1971	CAC/RCP 4–1971 (Vol.D)
Code of Hygienic Practice for Dehydrated Fruits and Vegetables including Edible Fungi	CAC/RCP 5–1971	CAC/RCP 5–1971 (Vol.D)
Code of Hygienic Practice for Tree Nuts	CAC/RCP 6–1972	CAC/RCP 6–1972 (Vol.D)
International System for the Description of Carcases of Bovine and Porcine Species and International Description of Cutting Methods of Commercial Units of Beef, Veal, Lamb and Mutton, and Pork, moving in International Trade	CAC/RCP 7–1974	Out-of-Print

Subject	Reference	Publication
Code of Practice for the Processing and Handling of Quick Frozen Foods	CAC/RCP 8–1976	CAC/RCP 8–1976 (Vol.E)
– Method for Checking Product Temperature	– Annex I–1978	" "
– Code of Practice for the Handling of Quick Frozen Foods during Transport	– Annex II–1983	" "
Code of Practice for Fresh Fish	CAC/RCP 9–1976	CAC/RCP 9–1976 (Vol.B)
Code or Practice for Canned Fish	CAC/RCP 10–1976	CAC/RCP 10–1976 (Vol.B)
Code of Hygienic Practice tor Fresh Meat	CAC/RCP 11–1976	CAC/RCP 11–1976 (Vol.C)
Code of Ante–Mortem and Post–Mortem Inspection of Slaughter Animals	CAC/RCP 12–1976	CAC/RCP 12–1976 (Vol.C)
Code of Hygienic Practice for Processed Meat and Poultry Products	CAC/RCP 13–1976 Rev. 1 (1985)	CAC/RCP 13–1976 Rev. 1 (1985) (Vol. C)
Code of Hygienic Practice for Poultry Processing	CAC/RCP 14–1976	CAC/RCP 14–1976 (Vol.C)
Code of Hygienic Practice for Egg Products	CAC/RCP 15–1976	CAC/RCP 15–1976 (Vol.F)
– Microbiological Specifications for Pasteurized Egg Products	– Annex II–1978	" "
Code of Practice for Frozen Fish	CAC/RCP 16–1978	CAC/RCP 16–1978 (Vol.B)
Code of Practice for Shrimps or Prawns	CAC/RCP 17–1978	CAC/RCP 17–1978 (Vol.B)
Code of Hygienic Practice for Molluscan Shellfish	CAC/RCP 18–1978	CAC/RCP 18–1978 (Vol.B)
Code of Practice for the Operation of Radiation Facilities used for the Treatment of Foods	CAC/RCP 19–1979	CAC/VOL XV – Ed. 1
Code of Ethics for International Trade in Food	CAC/RCP 20–1979, Rev. 1 (1985)	CAC/RCP 20–1979, Rev. 1 (1985) (Vol. J)
Code of Hygienic Practice for Foods for Infants and Children (including Microbiological Specifications and Methods for Microbiological Analysis)	CAC/RCP 21–1979	CAC/VOL IX – Ed. 1
Corrigendum to Code of Hygienic Practice for Foods for Infants and Children (Microbiological Specifications)	CAC/RCP 21–1979	Supplement 1 to CAC/VOL IX – Ed. 1 (English and Spanish only)
Code of Hygienic Practice for Groundnuts (Peanuts)	CAC/RCP 22–1979	CAC/RCP 22–1979 (Vol.D)
Code of Hygienic Practice for Low–Acid and Acidified Low–Acid Canned Foods	CAC/RCP 23–1979	CAC/RCP 23–1979 (Vol.G)
Code of Practice for Lobsters	CAC/RCP 24–1979	CAC/RCP 24–1979 (Vol.B)
Code of Practice for Smoked Fish	CAC/RCP 25–1979	CAC/RCP 25–1979 (Vol.B)
Code of Practice for Salted Fish	CAC/RCP 26–1979	CAC/RCP 26–1979 (Vol.B)
Code of Practice for Minced Fish prepared by Mechanical Separation	CAC/RCP 27–1983	CAC/RCP 27–1983 (Vol.B)
Code of Practice for Crabs	CAC/RCP 28–1983	CAC/RCP 28–1983 (Vol.B)
Code of Hygienic Practice for Game	CAC/RCP 29–1983	CAC/RCP 29–1983 (Vol.C)
Code of Hygienic Practice for the Processing of Frog Legs	CAC/RCP 30–1983	CAC/RCP 30–1983 (Vol.C)

Subject	Reference	Publication
Code of Hygienic Practice for Dried Milk	CAC/RCP 31-1983	CAC/RCP 31-1983 (Vol.H)
Code of Practice for the Production, Storage and Composition of Mechanically Separated Meat and Poultry Meat intended for further Processing	CAC/RCP 32-1983	CAC/RCP 32-1983 (Vol.C)
Recommended International Code of Practice for Ante-Mortem and Post-Mortem Judgement of Slaughter Animals and Meat	CAC/RCP 34-1985 and Corrigendum	CAC/RCP 34-1985 (VOL. C) and Corrigendum

SECTION III- CODEX METHODS OF ANALYSIS

Subject	Reference	Publication
Determination of Total Solids Content (Oven-filter aid method)	CAC/RM 1-1969	CAC/VOL III-Ed. 1
Determination of Loss on Drying at 120°C for 16 Hours (USP method)	CAC/RM 2-1969	"
Determination of Loss on Drying at 105°C for 3 Hours (ICUMSA method)	CAC/RM 3-1969	"
Determination of Sulphur Dioxide (Monier-Williams method)	CAC/RM 4-1969	"
Determination of Sulphur Dioxide (Carruthers, Heaney & Oldfield method)	CAC/RM 5-1969	"
Determination of Colour	CAC/RM 6-1969	"
Determination of Polarization (ICUMSA method)	CAC/RM 7-1969	"
Determination of Conductivity Ash	CAC/RM 8-1969	"
Determination of Relative Density at t/20oC (BSI method)	CAC/RM 9-1969	CAC/VOL XI-Ed. 1
Determination of Allyl Isothio-cyanate Content (Indian Standards Institute method)	CAC/RM 10-1969	"
Arachis Oil Test (Evers) (BSI method)	CAC/RM 11-1969	"
Sesame Oil Test (Baudoin) (BSI method)	CAC/RM 12-1969	"
Determination of Soap Content (BSI method)	CAC/RM 13-1969	"
Determination of Iron Content (BSI method)	CAC/RM 14-1969	"
Estimation of Milk Fat Content	CAC/RM 15-1969	"
Determination of Fat Content	CAC/RM 16-1969	"
Determination of Water Content by Loss of Mass on Drying	CAC/RM 17-1969	"
Determination of Vitamin E (Tocopherols) Content	CAC/RM 18-1969	"
Determination of Sodium Chloride Content	CAC/RM 19-1969	"
Determination of Bellier Index	CAC/RM 20-1970	CAC/VOL XI (See CODEX STAN.33-1981)
Semi-siccative Oils Test	CAC/RM 21-1970	"
Olive-residue Oil Test	CAC/RM 22-1970	"
Cottonseed Oil Test	CAC/RM 23-1970	"
Teaseed Oil Test	CAC/RM 24-1970	"
Sesameseed Oil Test	CAC/RM 25-1970	"
Determination of Specific Extinction in Ultra Violet 1% (E 1cm)	CAC/RM 26-1970	CAC/VOL XI (See CODEX STAN.33-1981)
Soap Test	CAC/RM 27-1970	"

Subject	Reference	Publication
Determination of Drained Weight (Method A for Liquid Packing Medium)	CAC/RM 28-1970	CAC/VOL V (See CODEX STAN 37-1981)
Determination of Net Contents	CAC/RM 29-1970	"
Determination of Size	CAC/RM 30-1970	"
Determination of Water Capacity of the Container	CAC/RM 31-1970	"
Standards Procedure for Thawing of Quick Frozen Fruits and Vegetables	CAC/RM 32-1970	CAC/VOL VIII-Ed. 1
Standard Procedure for Cooking of Quick Frozen Fruits and Vegetables	CAC/RM 33-1970	"
Weight of Quick Frozen Fruits and Vegetables	CAC/RM 34-1970	"
Determination of the Alcohol-Insoluble solids Content (Quick Frozen Peas)	CAC/RM 35-1970	"
Determination of Drained Weight - Method I (AOAC method)	CAC/RM 36-1970	CAC/VOL II-Ed. 1
Determination of Drained Weight - Method II	CAC/RM 37-1970	"
Determination of Calcium in Canned Vegetables (AOAC)	CAC/RM 38-1970	"
Tough String Test	CAC/RM 39-1970	CAC/VOL II-Ed. 1 and CAC/VOL VIII-Ed. 1
Thawing and Cooking Procedure - Quick Frozen Fish	CAC/RM 40-1971	CAC/VOL V-Ed. 1
Determination of Net Contents of Products covered by Glaze - Quick Frozen Fish	CAC/RM 41-1971	"
Sampling Plans for Prepackaged Foods (1969) (AQL 6.5)	CAC/RM 42-1971	CAC/VOL VIII-Ed. 1
Determination of Total Soluble Solids Content of Frozen Fruits	CAC/RM 43-1971	"
Determination of Washed Drained Weight	CAC/RM 44-1972	CAC/VOL II-Ed. 1
Determination of Proper Fill in lieu of Drained Weight	CAC/RM 45-1972	"
Determination of Water Capacity of Containers	CAC/RM 46-1972	"
Determination of Alcohol Insoluble Solids (AOAC Method)	CAC/RM 47-1972	"
Method for Distinuishing Type of Peas	CAC/RM 48-1972	"
Determination of Mineral Impurities (Sand)	CAC/RM 49-1972	CAC/VOL II Ed. 1
Determination of Moisture in Raisins - AOAC Electrical Conductance Method	CAC/RM 50-1974	"
Determination of Mineral Impurities (Sand Test) in Raisins	CAC/RM 51-1974	"
Determination of Mineral Oil in Raisins	CAC/RM 52-1974	"
Determination of Sorbitol in Raisins and other Foods	CAC/RM 53-1974	"
Determination of Mineral Impurities in quick Frozen Fruits and Vegetables	CAC/RM 54-1974	CAC/VOL VIII-Ed. 1
Determination of Fats for all Infant Foods - Method 1	CAC/RM 55-1976	CAC/VOL IX-Ed. 1

SECTION IV - MILK AND MILK PRODUCT STANDARDS

Subject	Reference	Publication
PART I - Code of Principles concerning Milk and Milk Products		CAC/VOL XVI-Ed.1

Subject	Reference	Publication
PART II – Standards for Milk Products:		"
Butter and Whey Butter	A-1	"
(i) Butteroil and (ii) Anhydrous Butteroil and Anhydrous Milkfat	A-2	"
Evaporated Milk and Evaporated Skimmed Milk	A-3	"
Sweetened Condensed Milk and Skimmed Sweetened Condensed Milk	A-4	"
Whole Milk Powder, Partly Skimmed Milk Powder and Skimmed Milk Powder	A-5	"
General Standard for Cheese	A-6	"
General Standard for Whey Cheeses	A-7	"
General Standard for Named Variety Process(ed) Cheese and Spreadable Process(ed) Cheese	A-8(a)	"
General Standard for Process(ed) Cheese and Spreadable Process(ed) Cheese	A-8(b)	"
General Standard for Process(ed) Cheese Preparations	A-8(c)	"
Cream for Direct Consumption	A-9	"
Cream Powder, Half Cream Powder and High Fat Milk Powder	A-10	"
Yoghurt (Yogurt) and Sweetened Yoghurt (Sweetened Yogurt)	A-11(a)	"
Flavoured Yoghurt (Yogurt) and Products Heat-treated after Fermentation	A-11(b)	"
Edible Acid Casein	A-12	"
Edible Caseinates	A-13	"

PART III – International Cheese Standards:

Subject	Reference	Publication
Cheddar	C-1	CAC/VOL XVI-Ed.1
Danablu	C-2	" "
Danbo	C-3	" "
Edam	C-4	" "
Gouda	C-5	" "
Havarti	C-6	" "
Samsoe	C-7	" "
Cheshire	C-8	" "
Emmentaler	C-9	" "
Gruyere	C-10	" "
Tilsiter	C-11	" "
Limburger	C-12	" "
Saint-Paulin	C-13	" "
Svecia	C-14	" "
Provolone	C-15	" "
Cottage Cheese incl. Creamed Cottage Cheese	C-16	" "
Butterkase	C-17	" "
Coulommiers	C-18	" "
Gudbrandsdalsost (whey cheese)	C-19	" "
Harzer Kase	C-20	" "
Herrgardsost	C-21	" "
Hushallsost	C-22	" "
Norvegia	C-23	" "
Maribo	C-24	" "
Fynbo	C-25	" "
Esrom	C-26	" "
Romadur	C-27	" "
Amsterdam	C-28	" "
Leidse	C-29	" "
Friese	C-30	" "
Cream Cheese	C-31	" "
Certain Blue Veined Cheeses	C-32	" "
Camembert	C-33	" "
Brie	C-34	" "
Extra Hard Grating Cheese	C-35	" "

Methods of Analysis and Sampling for Milk and Milk Products:

Subject	Reference	Publication
Sampling Methods for Milk and Milk Products – General Instructions	B-1	CAC/M 1-1973

Subject	Reference	Publication
– Sampling of Milk and Liquid Milk Products (except Evaporated and Sweetened Condensed Milk)		
– Sampling of Condensed Milk and Evaporated Milk		
– Sampling of Dried Milk and Dried Milk Products		
– Sampling of Butter		
– Sampling of cheese		
Determination of the Fat Content of Dried Milk	B–2	"
Determination of the Fat Content of Cheese and Processed Cheese Products	B–3	"
Determination of the Acid Value of Fat from Butter	B–4	"
Determination of the Refractive Index of Fat from Butter	B–5	"
Determination of the Fat Content of Milk	B–6	"
Determination of the Fat Content of Evaporated Milks and of Sweetened Condensed Milks	B–7	"
Determination of the Salt (Sodium Chloride) Content of Butter	B–8	"
Determination of the Fat Content of Whey Cheese	B–10	"
Determination of the Dry Matter Content in Whey Cheese	B–11	"
Determination of the Phosphorus Content of Cheese and Processed Cheese Products	B–12	"
Determination of the Citric Acid Content of Cheese and Processed Cheese Products	B–13	"
Polarimetric Determination of the Sucrose Content of Sweetened Condensed Milk	B–14	"
Determination of the Fat Content of Cream	B–15	"
Milk Fat, Detection of Vegetable Fat by the Phytosteryl Test	B–16	CX 5/70–19th S.: – App. X
Milk Fat, Detection of Vegetable Fat by Gas-liquid Chromatography of Sterols	B–17	– App. XI
Cheese, Determination of Chloride Content	B–18	– App. XII
Cheese, Determination of Nitrate and Nitrite Contents	B–19	– App. IX–I
Anhydrous Milk Fat, Determination of the Peroxide Value	B–20	– App. IX–J
Butter – Water, Solids–non–fat and Fat on the same test portion	B–21	– App. IX–K
Caseins and Caseinates – Determination of Water Content	B–22	– App. IX–B
Rennet caseins and caseinates – Determination of Ash	B–23	– App. IX–C
Caseins – Determination of "fixed ash"	B–24	– App. IX–D
Caseins and caseinates – Determination of protein content	B–25	– App. IX–E
Caseins – Determination of free acidity	B–26	– App. IX–F
Milk and Milk Products – Determination of Lactose in the presence of other reducing substances	B–27	– App. IX–G
Dried milk – Determination of titratable acidity	B–28	– App. IX–H

Exhibit 4. Proposed FAO/WHO Levels for Radionuclide Contamination of Food in International Trade

I. *Purpose*

The aim of this document is to provide to the Codex Alimentarius Commission joint FAO/WHO recommendations to control foods in international trade that have been accidentally contaminated with radionuclides. The goal is to provide a system that can be uniformly and simply applied by government authorities and yet one that achieves a level of public health protection to the individual that is more than adequate in the event of a nuclear accident.

II. *Background*

Following the April 1986 Chernobyl, USSR nuclear reactor accident, large amounts of radionuclides were released into the atmosphere and carried by weather patterns prevailing at that time for many thousands of kilometers through Europe and the Northern Hemisphere. At the time of the Chernobyl accident there was a definite lack of comprehensive international guidance on radionuclide contamination and authorities responsible for agriculture, environment, health and trade were unable to take uniform action to control radionuclide contaminated food and feed. Differences between countries on acceptable levels of contamination of food led to confusion and disruption of trade.

Compared with background radiation from natural and man-made sources that existed before the Chernobyl accident, exposure to X-rays for medical purposes and other types of radiation exposure, radiation protection experts pointed out that exposure to Chernobyl-related radionuclide contamination would add only a small increment to pre-Chernobyl levels of exposure. Due to the known carcinogenic and mutagenic effects of radiation and varying estimates of increased rates of cancer from Chernobyl-related contamination, many consumers were not reassured by these statements.

For about four to six weeks after the Chernobyl accident confusion existed about whether or not to let children play outside, whether or not to plough under leafy green vegetables exposed to heavy fallout and whether or not interdiction of local and international shipments of foods and other agricultural products was warranted. Most countries that were directly affected by radioactive fallout from Chernobyl took significantly different and usually less restrictive approaches to control the levels of radionuclide contamination in food than those countries that were not directly affected.

Following the widespread confusion and concern that existed after the Chernobyl accident, FAO, WHO and IAEA took action to provide additional guidance to member countries on appropriate responses to nuclear accidents. Other bodies such as the Organization for Economic Cooperation and Development (OECD) and the Commission of European Communities (CEC) also took action to provide guidance to their member countries. The International Commission on Radiation Protection (ICRP) also undertook to review its previous guidance on nuclear accident responses.

Shortly after the Chernobyl accident, the Director-General of FAO called on the FAO Secretariat, working in close collaboration with WHO and IAEA, to develop limits for radionuclide contamination for foods in trade which could be accepted by the FAO/WHO Codex Alimentarius Commission and utilized by FAO and WHO member countries to assure orderly trade in foods in the event of accidental contamination with radionuclides. The FAO Secretariat commenced this work through preparation of papers examining various aspects of the problem, which were reviewed by the December 1986 FAO Expert Consultation on Recommended Limits for Radionuclide Contamination of Foods. This Consultation included food control, radiation protection, and safety experts from several countries. The recommendations of the FAO Expert Consultation were transmitted by the FAO Director-General in January 1987 to all FAO member countries, all United Nations agencies and to all other known interested parties so that the FAO recommendations could be used as interim guidance in controlling foods in international commerce until all consultations and final recommendations were available from FAO, WHO and IAEA.

The FAO Expert Consultation Report and recommendations were introduced into the Codex Alimentarius Commission approval and recommendation process by requesting the Codex Committee on Food Additives and Contaminants (CCFA) to consider the FAO report in its March 1987 meeting, prior to the June-July 1987 Session of the Codex Alimentarius Commission (CAC). The CCFA reviewed and generally endorsed the FAO Expert Consultation report, commended FAO on its rapid action, and requested FAO and WHO to convene a Codex Working Group prior to or during the June-July CAC Session so that Codex member countries could include appropriate expertise in their delegations to consider the FAO Report in depth before any action by the CAC. A Working Group was scheduled as requested by CCFA to meet during the CAC session but was subsequently cancelled at the request of WHO which suggested postponing the CAC review until after WHO had completed its work on developing guideline values. The June-July 17th Session of the CAC took note of the CCFA recommendations, commended FAO for providing the only available international recommendations for radionuclide contamination in foods in trade and urged speedy completion of the WHO work so that a joint FAO/WHO approach could be reviewed for approval by the CAC Executive Committee in its July 1988 session.

The FAO December 1986 Expert Consultation utilized food control principles to uniformly allocate the total amount of radioactivity from a dose of 5 millisieverts (5 mSv) over 100% of the food consumed. The FAO Expert Group assumed that all foods would be contaminated and utilized the most sensitive population group and body tissue in making its recommendations. On this basis, the group recommended interim international radioactivity action levels in foods which were considerably lower than those recommended by other groups. The FAO interim values were not significantly different from some national levels and those adopted by the Commission of European Communities (CEC) soon after the Chernobyl accident.

In assessments of acceptable contamination levels made by WHO, IAEA, OECD and the European Community Article 31 group in 1986-1987, approaches tended to concentrate on radiation protection and safety principles rather than food control and food law procedures. The dose level of 5 mSv was accepted by most groups as a basis for calculation. However, differences of assumptions about the percentage of food supply that might be contaminated and about which dose conversion factor should be used usually resulted in higher contamination levels than the FAO interim levels.

During late 1986 and 1987 WHO engaged several consultants and held a preliminary meeting in April 1987 to prepare the WHO recommended health-related approach to radionuclide contamination in foods. In September 1987, WHO held an expert consultation in Geneva and also invited participation of FAO, OECD, IAEA, ICRP and the Commission of the European Communities (CEC). The WHO Expert Consultation provided a methodology and guideline values which could be used by national authorities as a basis for setting their own levels. The reference level of dose was accepted as 5 mSv and food consumption was normalized to a hypothetical intake of 550 kg/y. The potentially contaminating radionuclides were divided into two main classes, the actinides such as Plutonium 239 and all others such as Caesium 137. Only food groups that were consumed in quantities greater than 20 kg/y were used in the calculation of the guideline values, and special values for infants were developed. Additivity of radionuclides contaminating one or more food groups was accommodated. These values, while assisting member states to develop their own levels, were considered too complex and unsuitable for application of international trade in food.

In January 1988, the WHO Executive Board urged the Director-General to continue to cooperate with FAO in developing uniform recommendations on maximum levels regarding radionuclides in food moving in international trade for consideration and adoption by the Codex Alimentarius Commission.

The principles applied to the control of contamination of foods moving in international trade are similar to those used in national food control legislation. These have been successfully applied by the Codex Alimentarius Commission in making recommendations about environmental contaminants such as lead, cadmium and mercury in food, and are the basis for current work on the establishment of guideline levels for aflatoxins. These food protection principles are based on the utilization of safety factors which assure the consumer of wide margins of safety beyond the basic levels derived from known health and toxicology research data. At the same time they provide national food control authorities with simple and uniform levels which can be applied to all foods moving in trade, whatever their origin, and whatever their destination in the distribution chain after clearance by control officials.

In most countries, national food law prohibits sale or shipment of food contaminated with poisonous or deleterious substances. However it is recognized that certain low levels of contaminants are unavoidably present in food and maximum levels for their occurrence have to be set to protect the safety of food supplies to all consumers. In arriving at a contaminant level, toxicological data on test animals are reviewed, and a series of conservative assumptions and safety factors are applied in setting the contamination level to be used for regulatory food control purposes. If a no-effect level has been demonstrated in controlled animal feeding tests, that level is the departure point for applying conservative assumptions and safety factors to arrive at a much lower contamination level for foods for human consumption. For contaminants such as radionuclides or mycotoxins where a no-effect level cannot be established, additional considerations are applied in setting contaminant levels which acknowledge the impossibility of avoiding all inadvertent contamination of foods with these substances.

The FAO Expert Consultation in December 1986 recommended interim limits for radionuclides in food. At that time, these were regarded as interim levels which would probably need revision at a later date as a result of the experience gained from the Chernobyl accident. It is recognized that both the FAO and WHO guideline values require specific knowledge of the profile of contamination and are not necessarily applicable to the control of future unknown accidental contamination through existing food control legislation.

It is therefore necessary to develop values that can be readily applied to future accidents under existing food control legislation.

III. *Derivation of values*

On examination, the approaches of WHO and FAO, and indeed of other organizations, are basically similar. They all assume a reference level of dose (usually 5 mSv), a total average food consumption rate, a dose per unit intake factor for various radionuclides and a pattern of food consumption, and calculate the levels by the following formula:

$$\text{Level} = \frac{\text{RLD}}{m \times d}$$

Where RLD = Reference Level of Dose (Sv)
m = mass of food consumed (kg)
d = dose per unit intake factor (Sv/Bq)

Controlling radionuclide contamination of foods moving in international trade requires simple, uniform and easily applied values. On this basis, it is clear that for regulatory purposes a simpler approach based on the measurement of total gamma activity is appropriate. This approach is one that can be uniformly applied by government authorities and yet one that achieves a level of public health protection to individuals that is considered more than adequate in the event of a nuclear accident.

In making these joint FAO/WHO recommendations the following assumptions have been made in calculating the recommended levels:

1. 5 mSv is the reference level of dose
2. 550 kg of food is consumed in a year, all of which is contaminated
3. Dose per unit intake factors for the radionuclides of concern (^{131}I, ^{137}Cs, ^{134}Cs, ^{90}Sr and actinides) can be conveniently divided into two classes, the actinides with a dose per unit intake of 10^{-6} Sv/Bq and all others with 10^{-8} Sv/Bq.

Applying these assumptions to the above formula, the level for "other" radionuclides will be:

$$\frac{5 \times 10^{-3}}{550 \times 10^{-8}} = 909 \, \text{Bq/kg}$$

which can then be rounded to 1000 Bq/kg. For the actinides this value would be 10 Bq/kg, as the dose per unit intake factor is 100 times larger.

It is recognized that the infant poses a special problem as a result of its sensitivity. It is felt that 1000 Bq/l does not adequately protect the infant when drinking milk contaminated with ^{90}Sr since the dose per unit intake factor is of the order of 10^{-7} Sv/Bq. For this reason, a level of 100 Bq/kg is proposed for ^{90}Sr in milk and infant food. ^{131}I in milk is judged to be of less importance even though the dose per unit intake factor is higher for infants. If the value of 1000 Bq/l is applied, the infant would need to consume approximately 18 litres of milk contaminated to this level to exceed the reference level of dose. At 0.75 l/day, this represents about 24 days of ingestion of milk contaminated to this level. This is not considered to be a likely eventuality and in view of its short effective half life no special values are suggested for ^{131}I in milk and infant food.

The proposed levels are tabulated below:

TABLE 1.   Proposed FAO/WHO levels for Radionuclides in Food moving in international trade

Actinides	10 Bq/kg
^{90}Sr in milk and infant food	100 Bq/kg
^{90}Sr in other food	1000 Bq/kg
All other radionuclides	1000 Bq/kg total gamma activity

NOTES: Dried or concentrated products should be considered on the basis of the food prepared for consumption. The level given should therefore be multiplied by the same factor used for dilution or reconstitution.

The levels are based on very conservative assumptions and are intended to be used as values below which no food control restrictions need to be applied. Measured values above these levels are not necessarily of public health concern but should alert the competent food control authorities for the need to assess the potential health detriment.

These levels are also intended to be applied to accident situations and not to naturally occurring radionuclides in foods.

The levels suggested are designed to apply particularly to foods in international trade, but both FAO and WHO have called attention in their expert meeting reports to special consideration which might apply to certain classes of food which are consumed in small quantities, such as spices. Some of these foods grown in areas affected by the Chernobyl accident fall-out contained high levels of radionuclides following the accident. Because they represent a very small percentage of total diets and hence would be very small additions to the total

Avda. Marathon 1000—Casilla 48
Santiago de Chile

China*   Dr. Chen Junshi
                          Institute of Nutrition & Food Hygiene
                          Chinese Academy of Preventive Medicine
                          29 Nan Wei Road
                          Beijing

Costa Rica   Ing. Jesús Gómez
                          Instituto Costarricense de Investigación
                              y Enseñanza en Nutrición y Salud—INCIENSA
                          Apartado 4
                          Tres Rios

Denmark*   Dr. N. Borre
                          Ministry of the Environment
                          National Food Institute
                          19 Morkhoj Bygade
                          DK-2860 Soborg

Egypt*   Dr. A. Mohieldin Zaki
                          Central Public Health Laboratories*
                          Ministry of Health
                          19 Sheikh Rihan Street
                          Cairo

                          Dr. (Mrs.) Salwa Dogheim
                          Central Agricultural Pesticide Laboratory
                          Ministry of Agriculture
                          Dokki, Cairo

                          Dr. (Mrs.) Khayria Naguib
                          Nutrition & Food Technology Division
                          National Research Centre
                          53 Mousadek Street
                          Dokki, Cairo

                          Dr. Ahmed Shaker
                          Centre for Occupational & Environmental Health
                          Imbabah, Cairo

Finland   Dr. Jorma Kumpulainen
                          Head, Central Laboratory
                          Agricultural Research Centre
                          SF-31600 Jokioinen

Germany,   Dr. P. Weigert
Federal Republic of* Centre for Surveillance and Health Evaluation
                         of Environmental Chemicals (ZEBS)
                      Bundesgeshundheitsamt
                      Postfach 33 0013
                      D-1000 Berlin 33 (West)

Guatemala*   Ing. (Mrs.) Marit de Campos
                      Unified Food and Drug Control Laboratory (LUCAM)
                      c/o INCAP
                      P.O. Box 1188
                      Guatemala City

Hungary*   Dr. (Mrs.) J. Sohar
                      Department of Toxicological Chemistry
                      National Institute of Food Hygiene and Nutrition
                      Gyali ut 3/a
                      H-1097 Budapest

India   Dr. B. S. Narasinga Rao
                      National Institute of Nutrition
                      Hyderabad — 500 007

Ireland*   Dr. T. O'Toole
                      Department of Agriculture
                      Agriculture House
                      Dublin 2

Japan*   Dr. Yukio Saito
                      Division of Foods
                      National Institute of Hygienic Sciences
                      1–18, Kamiyoga, 1-Chome, Setagaya-ku
                      Tokyo 158

Jordan   Dr. Omar Jabay
                      Organic Technology Section
                      Industrial Chemistry Department
                      Royal Scientific Society
                      P.O. Box 925819
                      Amman

Kenya*   Dr. J. N. Kaviti
                      National Public Health Laboratory Service
                      Ministry of Health
                      P.O. Box 20750
                      Nairobi

Netherlands*   Dr. P. A. Greve
National Institute of Public Health and
   Environmental Hygiene
P.O. Box 1
3720 BA, Bilthoven

New Zealand*   Mr. J. Fraser
Division of Public Health
Department of Health
P.O. Box 5013
Wellington

Pakistan   Chief
Nutrition Division
National Institute of Health
Islamabad

Peru   Dr. T. J. Aliaga Osorio
Instituto Nacional de Desarrollo
   Agroindustrial, INDDA
Av. La Universidad 595
La Molina Apartado 14–0294
Lima 14

Poland*   Professor M. Nikonorow
Department of Food Research
National Institute of Hygiene
24 Chocimska Street
00-791 Warsaw

Qatar*   Dr. A. R. Kotb
Regional Center
Food Contamination Monitoring
Ministry of Public Health
P.O. Box 42
Doha

Republic of Korea   Dr. D. Baik
Department of Food
National Institute of Health
Ministry of Health and Social Affairs
P.O. Box Sodaimun-Ku
5 Nokbun-Dong, Eunpyung-Ku
Seoul 122

Sudan

Dr. A. H. Ibrahim
National Chemical Laboratories
Ministry of Health
P.O. Box 287
Khartoum

Sweden*

Dr. S. A. Slorach
Food Research Department
National Food Administration
Box 622
S-751 26 Uppsala

Thailand

Dr. Ulit Leeyavanija
Department of Medical Sciences
Ministry of Public Health
Yodse, Bangkok 10100

Tunisia

Dr. Zouhair Kallal
Institut National de Nutrition et Technologie
Tunis

United Kingdom*

Dr. M. E. Knowles
Food Science Division
Ministry of Agriculture, Fisheries and Food
Great Westminster House
Horseferry Road
London, SW1P 2AE

United States of*
America

Mr. J. A. Burke
Division of Contaminants Chemistry
Center for Food Safety and Applied Nutrition

Department of Health & Human Services
Food and Drug Administration
200 "C" Street, S.W.
Washington, D.C., 20204

Uruguay

Dr. Eugenio Perdomo
Centro de Investigaciones Veterinarias
"Miguel C. Rubino"
Ministerio de Agricultura y Pesca
Brigadier Gral. Juan A. Lavalleja
Km 29
Pando

USSR*

Dr. V. A. Tutelyan
Institute of Nutrition
Academy of Medical Sciences
Ustinsky proezd, 2/14
Moscow 109240

Exhibit 6. 1984–1985 Monitoring Data Collected from the Collaborating Centers and Participating Institutions

sum aldrin and dieldrin, DDT-complex, total endosulfan, endrin, HCB, total HCH isomers, lindane (gamma-HCH), sum heptachlor and heptachlor epoxide, and PCBs.	cereals, whole fluid milk, whole dried milk (specify animal if known), butter, edible fats and oils (animal, vegetable origin and type), fish (canned or fresh), and eggs.
	total diet
On the above organochlorine compounds and on any other organochlorine compound which has been found to be present in this food	human milk
lead	canned: fruit, fruit juices including concentrates, infants food and juices, vegetables, milk, fish and meat (indicate whether cans are lead-soldered or otherwise)
	cereals, flours, legumes and pulses, fresh fruit, meat, fish, potatoes and other vegetables of major dietary importance, molluscs, crustaceans, kidney and spices
	total diet
cadmium	molluscs, crustaceans, cereal grains, cereal flours, potatoes, and other vegetables of major dietary importance and kidney
	total diet

tin (total)	food and beverages (beer, soft drinks, juices) in tin plate cans (indicate whether lacquered or not)
	total diet
mercury (total)	fish and fish products (excluding shellfish)
	total diet
total Aflatoxins or Aflatoxins B_1	groundnuts, tree nuts; legumes and pulses, maize and other grains of major dietary importance whether food or animal feed; spices and herbs
	total diet
Aflatoxin M_1	milk and milk products, eggs
Diazinon, fenitrothion, malathion, parathion, methyl parathion	cereal grains, vegetables and fruit of major dietary importance
	total diet

Exhibit 7.   Documents* Issued under GEMS/Food

1. Expert Consultation on the Joint FAO/WHO Food Contamination Monitoring Programme—Identification of Contaminants to be Monitored and Recommendations on Sampling Plans and Methodology. Rome 7–11 October, 1974 (out of print).
2. Expert Consultation on the Joint FAO/WHO Food Contamination Monitoring Programme—Development of System for Processing, Appraisal and Storage of Data, Geneva, 17–21 March 1975 (out of print).
3. Report of Meeting to Plan Phase II of the Joint FAO/WHO Food and Animal Feed Contamination Monitoring Programme. Geneva, 13–16 December 1976 (out of print).
4. Report on the Consultation on the Joint FAO/WHO Food and Animal Feed Contamination Monitoring Programme—Phase II, Geneva, 14–18 June 1977 (out of print).

*Available free of cost except where otherwise indicated) from the World Health Organization, Food Safety Unit, Division of Environmental Health, 1211 Geneva 27, Switzerland.

5. Report of the First Session of the Technical Advisory Committee on the Joint FAO/WHO Food and Animal Feed Contamination Monitoring Programme, Rome 13–17 March 1978. (out of print).

6. Summary Report of Data Received from Collaborating Centres for Food and Animal Feed Contamination Monitoring Programme. Phase II. Geneva 1979 (out of print).

7. Report of a Consultation on Analytical Quality Assurance—Joint FAO/WHO Food and Animal Feed Contamination Monitoring Programme—Geneva, 27–29 February 1980 (out of print).

8. Report of the Second Session of the Technical Advisory Committee. Geneva, 27 April–1 May 1981 (out of print).

9. Guidelines for Establishing or Strengthening National Food Contamination Monitoring Programme (English-French-Spanish) Geneva, 1979.

10. Joint FAO/WHO Food and Animal Feed Contamination Monitoring Programme—Analytical Quality Assurance of Monitoring Data, Geneva, 1981.

11. Joint FAO/WHO Food and Animal Feed Contamination Monitoring Programme. Summary of Data Received from Collaborating Centre—1977 to 1980 Part A—Countries; Part B—Contaminants, Geneva, 1981.

12. UNEP/FAO/WHO Summary and Assessment of Data Received from the FAO/WHO Collaborating Centres for Food Contamination Monitoring (English, Spanish) National Food Administration, Uppsala, 1982.

13. Report of a Joint FAO/WHO Meeting on Guidelines for the Study of Dietary Intakes of Chemical Contaminants. Rome, 16–21 December 1982.

14. Analytical Quality Assurance—II, Geneva, 1983.

15. Chapman, D. G. Information on Pesticides to Assist in Establishing Priorities for Food Contamination Monitoring. December, 1983.

16. Summary of Guidelines for the Study of Dietary Intakes of Chemical Contaminants, Geneva, 1984.

17. Report of the Third Session of the Technical Advisory Committee, Rome, 28 November–2 December 1983, Geneva, 1984.

18. Lindsay, D. G. Human Exposure Assessment to Pollutants in Food. EFP/HEAL/84.6 Geneva, 1984.

19. PCBs Residues in Food. A review of the 1980–83 data. Paper prepared for the 17th Session of Codex Committee on Pesticide Residues. March, 1985.

20. Mercury in Fish and Fishery Products. Paper prepared for the 19th Session of the Codex Committee on Food Additives, March, 1987.

21. Guidelines for the Study of Dietary Intakes of Chemical Contaminants. WHO Offset Publication No. 87, WHO Geneva (1985) (English, French, Spanish). Price Sw. Fr. 11.-

22. Galal Gorchev, H. and Jelinek, C. F. A review of the dietary intakes of chemical Contaminants. Bulletin of the World Health Organization, Vol. 63, No. 5, pp. 945–962 (1985).

23. Analytical Quality Assurance—III, Geneva, 1985.

24. Report of the Fourth Session of the Technical Advisory Committee, Geneva (1986).
25. Summary of 1980–83 Monitoring Data, Geneva, 1986.
26. Chemical Contaminants in Foods: 1980–1983, Geneva, 1986.
27. Dietary Intakes of Cadmium and Lead. Paper prepared for the 19th session of the Codex Committee on Food Additives, March 1987.
28. Dietary Intakes of Pesticide Residues and PCBs. Paper prepared for the 19th session of the Codex Committee on Pesticide Residues, April 1987.
29. Guidelines for Predicting Dietary Intake of Pesticide Residues, Geneva, 1988 (limited distribution).

INDEX

767